Building Adaptation

Second edition

James Douglas
Heriot-Watt University, Edinburgh, UK

ELSEVIER
AMSTERDAM • BOSTON • HEIDELBERG • LONDON • NEW YORK • OXFORD
PARIS • SAN DIEGO • SAN FRANCISCO • SINGAPORE • SYDNEY • TOKYO
Butterworth-Heinemann is an imprint of Elsevier

Butterworth-Heinemann is an imprint of Elsevier
Linacre House, Jordan Hill, Oxford OX2 8DP, UK
30 Corporate Drive, Suite 400, Burlington, MA 01803, USA

First edition 2002
Second edition 2006

British Library Cataloguing in Publication Data
A catalogue record for this book is available from the British Library

Library of Congress Cataloging in Publication Data
A catalog record for this book is available from the Library of Congress

ISBN-13: 978-0-7506-6667-1
ISBN-10: 0-7506-6667-6

For information on all Butterworth-Heinemann publications visit our website at books.elsevier.com

Typeset by Charon Tec Ltd, Chennai, India
www.charontec.com

Printed and bound in Great Britain

06 07 08 09 10   10 9 8 7 6 5 4 3 2 1

Building Adaptation

For Gloria, Errol and Estelle

Contents

List of figures

List of tables

Preface to 2nd edition

In the Preface to the first edition I expressed a hope that the book would elicit feedback. Thankfully I have received some constructive comments from a number of practitioners and students. Generally they have been generous in their comments about the book and have given positive responses to help improve its contents. Some of the modifications in this edition therefore are the result of such feedback. The reactions to the book and its sales confirm that such a text is needed.

Naturally, a primary aim of this edition is to make the necessary corrections in the first, especially in light of feedback and changes in legislation. It also includes new and revised material of recent developments in construction that impinge on building adaptations. Chapters 8 and 9 have been renamed to include the term 'refurbishment' rather than 'modernization', as the former is more widely used than the latter nowadays to describe improvements to buildings.

Even in the space of a few years a number of developments have occurred that have implications for the adaptation of buildings. Sustainability, of course, continues to gain increasing importance and this is reflected in many advances within the construction industry. The need for maximizing the use of environmentally friendly materials and processes, for example, is now well recognized. In particular the increasing importance of energy efficiency to sustainable refurbishment is such that it justifies a new separate chapter.

The prominence given to new construction rather than adaptation or refurbishment in built environment courses, though, continues unabated. For example, according to the www.prospects.ac.uk website (accessed 7/7/05) there are over 20 M.Sc. courses in the UK in construction management and related subjects such as construction real estate. These courses tend to focus almost exclusively on new build projects. On the other hand, there are currently only two postgraduate programmes in Britain dealing specifically with adaptation-related studies – viz, 'M.Sc. in Building Rehabilitation' at the University of Greenwich and 'M.Sc. in Built Resources Studies' at Oxford Brooks University. Even then the overwhelming emphasis of most of these courses is on the management rather than the technology of such work.

The continuing prominence given to new-build compared to adaptation schemes is of course easy to comprehend. It is more glamorous and tends to be larger in scale and cost than a typical adaptation project of the same building category. Major refurbishment schemes each exceeding £50 m to the Hilton Hotel and Royal Albert Hall in London after the year 2000, however, were notable exceptions to this rule.

The importance of new construction, of course, should not be overlooked. After all, today's 'high tech' is tomorrow's 'low tech'. The adaptability of modern buildings is just as important as that of older construction. New buildings are more flexible and loose fit compared to their traditional counterparts. These days it is important that the provision for future adaptation is taken into account in new build schemes. This is addressed in the revised Chapter 1.

Moreover, nearly half of all construction output in the UK is related to maintenance and adaptation work. There is therefore an increasing need for courses and textbooks dealing with this part of the industry's activity. This work aims to contribute to the growing need for information on this sector.

Preface to 2nd edition

[illegible]

Preface to 1st edition

This book was written to fill a perceived gap in the market for a text that deals with the fundamentals of building adaptation. It is not, in other words, just another tract on the rehabilitation and re-use of existing property. There is currently no single volume available that is specifically designed to provide a general introduction to the conversion, extension and refurbishment of property for students of building surveying and other cognate undergraduate programmes in the built environment.

There are, of course, many previously published books dealing with specific aspects of building adaptation. For example, the texts by Eley and Worthington (1984) and Highfield (1987) give an excellent grounding in the adaptive reuse of old industrial and commercial buildings. Moreover, the recently published two-volume work by Latham (2000) and the single volume by Highfield (2000) present a useful collection of case study examples illustrating different change of use and refurbishment projects respectively. However, there is no comprehensive introduction to the subject that covers the main physical interventions, excluding maintenance, to a property.

Adaptation is here interpreted in its widest sense as covering all the works to a building beyond maintenance. It therefore includes alterations, extensions, improvements as well as conversions and renovations. The glossary near the end of this book provides a list of definitions of these and other related terms used to describe major works to existing buildings. It also includes a list of common maintenance terms. This is followed by a Bibliography, which gives a list of the main textbooks and other useful sources of information on adaptation and related activities.

The range of adaptation options is enormous – from partial to full change of use, minor or major refurbishment, or small- to large-scale extensions. Thus this work cannot possibly cover all contingencies or situations. What it attempts to do, however, is give a comprehensive overview of adaptation. The texts listed in the Bibliography cover many of the topics in more detail.

This work also presents material relating to adaptation works not readily available within one construction textbook. For example, it includes jointing techniques for framed and unframed extensions. In addition, modernization measures such as input ventilation to overcome condensation in dwellings and over-roofing to improve the weather tightness of buildings are described in this volume. The aim of this work in other words is to complement books on building construction.

Only a few case study examples are presented in this work. There are numerous textbooks on the market, some of which are referred to above, that provide an excellent variety of actual change-of-use schemes. Rather, the focus of this book is on the key technical, legal and financial principles of building adaptation.

The importance of adaptation can be seen in the extent to which it contributes to the British construction industry's output. According to the BMI (1997b), maintenance and rehabilitation (one of several terms used to describe adaptation) accounts for nearly half of its workload. Indeed, overall maintenance and refurbishment accounts for 43 per cent of the total output of the European construction industry (ERG, 1999).

Despite the financial and functional significance of adaptation, the curriculum of most British undergraduate and postgraduate degree programmes in built environment departments concentrates on new construction. The coverage given to maintenance and adaptation, in other words, seems sparse in comparison. Even when they are included there is a tendency to focus on the management issues of these subjects than their technological aspects.

To a certain extent this overemphasis on new build is understandable. New construction is more prominent and ultimately more attractive. Dealing with an existing building, on the other hand, is often perceived as an unglamorous, onerous and unrewarding unless it is a prestigious historic property.

Another problem is that the prominence of building technology as a core subject of many construction programmes in higher education is being slowly but inexorably eroded. In recent times there has regrettably been a tendency to reduce the extent of this subject in undergraduate courses. There are a number of reasons for this state of affairs. As the importance of management subjects becomes increasingly recognized in construction, there is a great temptation to make room for these new topics in such courses at the expense of building technology. This inevitably tends to result in a dilution of the technical content of construction-related courses.

Some argue, anyway, that technology is advancing so rapidly that building professionals cannot know all its facets or keep up-to-date with every new development. Moreover, they may assert that there is no need for them to know the fine detail of the technical aspects. In contrast, others take the view that building technology is basically a craft-based subject and as such can be easily learned by experience.

What is often overlooked is that the building technology of today will need to be adjusted, modified or refined at some future date. This occurs through the process of adaptation. It is imperative that building professionals dealing with this increasingly common form of intervention are familiar with most if not all of the past as well as the latest construction techniques.

This book then takes a technological perspective of the adaptation of buildings. Still, it cannot ignore the management-related aspects associated with this form of construction activity. After all, adaptation is about managing change in the context of buildings. But construction professionals also need to have a good grasp of the technical aspects of building to enable them to design and manage any adaptation scheme in an effective manner.

Indeed, as greater recognition is given to environmental issues, especially global warming and pollution, the role of adaptation in helping this will become more important. Maximizing the re-use of old buildings is a key criterion of sustainability. In the developed world there is an increasing need to adapt obsolete or redundant buildings to continued or modified or new uses.

Building adaptation is taken from a British perspective in this book. However, other developed countries such as the USA are increasingly opting for adaptation rather than demolition and redevelopment. The principles outlined herein, therefore, should be applicable elsewhere so long as the climatic, constructional, and legislative differences are taken into account.

Disclaimer

Every attempt has been made to check that any of the advice given in this book is correct and up-to-date. However, neither the author nor the publishers shall be held liable for any loss, damage or other negative consequence, no matter how arising, occasioned by the implementation of such advice. Moreover, despite these efforts to ensure accuracy, they cannot accept responsibility for incorrect or incomplete information, changes in standards, changes in product ingredients or availability, undiscovered hazards in materials and components, or any adverse health or other effects.

Acknowledgements

As before, wherever possible every attempt has been made to acknowledge the numerous sources used in this book. The author would like to apologize in advance if there were any case where this has not been achieved. Due acknowledgement will be made in any subsequent edition.

I would like to reiterate my appreciation to the publishers Butterworth-Heinemann for their confidence in publishing this second edition. In particular, my thanks go to the commissioning editor Alex Hollingsworth for his support throughout the compilation of this edition, and to Lanh Te for her friendly and helpful assistance in finalising the draft manuscript. My thanks, too, go to Charlotte Dawes for overseeing the editing and proofing stages, as well as to the Project Management staff at Charon Tec for implementing efficiently the editing and correcting of the final proofs of this edition.

Finally, my gratitude must go to my family, friends and colleagues for their ongoing encouragement in completing this work.

James Douglas,
Edinburgh,
January 2006

1

Introduction

Overview

This chapter outlines the principles of building adaptation. It describes the nature and scale of as well as the reasons for the alteration, conversion, extension and refurbishment of properties. Adaptation is compared with maintenance, the other main branch of building performance management. The advantages and disadvantages of adapting buildings are examined. In this chapter the issues of obsolescence, redundancy, and vacant buildings are also addressed.

What is adaptation?

Definitions

'Adaptation' is derived from the Latin *ad* (to) and *aptare* (fit). In the context of this book it is taken to include any work to a building over and above maintenance to change its capacity, function or performance (i.e. any intervention to adjust, reuse or upgrade a building to suit new conditions or requirements). As regards existing buildings adaptation has traditionally come to have a narrower meaning that suggests mainly some form of change of use. The term has also been commonly used to describe improvement work such as adaptations to buildings for use by disabled or elderly people. In relation to this book, though, and as suggested above, it has a much broader connotation.

There are many other different terms that are used to describe interventions to a building that go beyond maintenance. Words such as 'refurbishment' or 'rehabilitation' and 'renovation' or 'restoration' are occasionally taken as being synonymous with one another, even by some in the construction industry (see Glossary). For example, Markus (1979) noted that 'in the world of building the terms "rehabilitation", "conversion", "remodelling", "restoration", "reinstatement" and so forth are unhappily confused'.

'Refurbishment', though, has gained widespread use in the UK as the most popular term to describe a wide range of adaptation work. The article 'Keeping up appearances' in the *Building* magazine supplement *Building Homes*, 25 November 1994, for instance, demonstrates this point. It described major alterations and improvements to 'a typical rundown Victorian terrace house in the throes of a conventional refurbishment'. The excellent series of regular features entitled 'Refurbishment' in the *Architects Journal* during the second half of the 1990s seems to reinforce this view. The articles in that series dealt with conversions, extensions and upgrades of significant buildings, as well as 'refurbishment' as defined in this work.

In one of the Energy Efficiency Office's Best Practice Programme publications (GIR 32, 1995), four types of 'refurbishment' were identified: major repair, acquisition and rehabilitation, conversion, and re-improvement. Including extensions, these comprise most of the adaptation work featured in this book.

Another contemporary example illustrates this broad use of the word 'refurbishment'. The writer recently encountered in the west side of Edinburgh a developer's signboard on the front wall of a redundant listed printing works that was being converted into flats. The sign read 'Coming Soon … A Refurbishment of this Historic Warehouse to Provide 1, 2 and 3 Bed Apartments'.

Occasionally some 'building adaptation' terms are used together. Certain construction companies, for example, advertise their services as 'specializing in renovating and refurbishing old homes'. Other contractors use the expression 'extensions and renovations' to indicate the range of works they undertake.

However, there is a technical as well as a semantic difference between these and other related general terms in building adaptation. 'Refurbishment' comes from the words 're', to do again, and 'furbish', to polish or rub up. Thus to refurbish something is to give it a facelift or a refit to enhance its appearance and function. In the context of a building it primarily involves extensive maintenance and repairs as well as improvements to bring it up to a modern standard. At a basic level refurbishment implies that the work involved is mainly superficial or cosmetic. It usually refers to upgrading the aesthetic and functional performance of the building. At the other extreme, though, refurbishment might include a lateral extension to the main part of the property (see Chapter 5) as well as major improvements to its fabric and services (see Chapter 9). The latter type of refurbishment is sometimes colloquially called 'giving a building a revamp or makeover' or 'revamping a building'.

Rehabilitation, on the other hand, because of its obvious relevance to the word 'habitation', is usually restricted to housing schemes. Like refurbishment, rehabilitation may include an element of modernization as well as some extension work. Still, unlike refurbishment it may comprise major structural alterations to the existing building as well (see Chapter 7).

Words such as 'recycling', 'remodelling' or 'renewal' are also sometimes used to describe major adaptations. 'Remodelling', for example, is commonly employed in the USA as an all-encompassing expression for these works. Such terms, whilst descriptive in their own way, only serve to blur the distinctions between the various interventions that can be done to existing buildings.

Moreover, in relation to building conservation, general terms such as refurbishment, rehabilitation, renovation and restoration lack precise technical meaning (BS 7913, 1998). As suggested above, 'restoration', is normally restricted to major adaptation work to dilapidated, derelict or ruinous residential or public buildings. In the writer's experience its use is very rare in the context of commercial properties. In addition, technically 'renovation' can occur to residential as well as commercial buildings but usually infers to less substantial works than 'restoration'.

Despite the absence of any universally agreed definition, however, 'building adaptation' is used in this book as an all-embracing term. In the author's view it is the one that best describes the full range of works to a property over and above maintenance.

Significance of building adaptation and maintenance

The importance of refurbishment/adaptation and maintenance in the UK can be gauged by its contribution to the output of the construction industry. Table 1.1 shows that this sector accounts for almost half of the UK industry's output. This is because of the extent and age of its existing stocks of buildings, the vast majority of which were built in the 20th century. Dilapidation, deficiencies in performance, sustainability of buildings are just some of the drivers that have stimulated and maintained the growth in building refurbishment and maintenance.

Table 1.1 Value of the building sector in the UK (Goodier and Gibb, 2004)

Sector	*Value (£bn)*	*%*
New build (excluding civil engineering)	53.3	54
Construction refurbishment and repair	45.0	46
Total UK construction	98.3	100

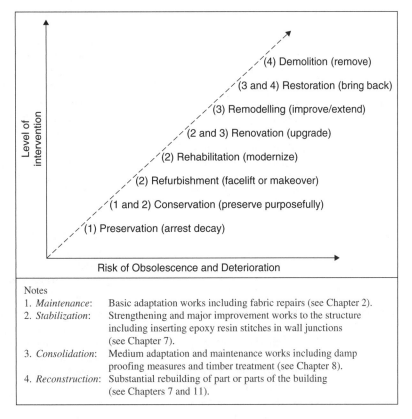

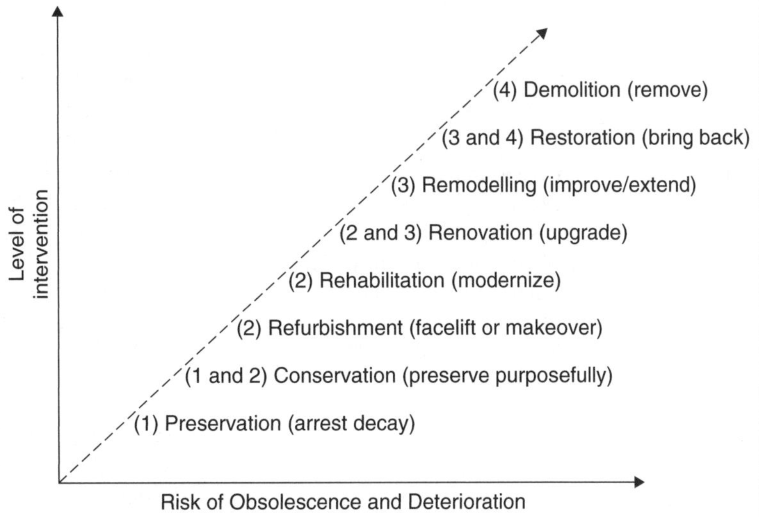

Notes
1. *Maintenance*: Basic adaptation works including fabric repairs (see Chapter 2).
2. *Stabilization*: Strengthening and major improvement works to the structure including inserting epoxy resin stitches in wall junctions (see Chapter 7).
3. *Consolidation*: Medium adaptation and maintenance works including damp proofing measures and timber treatment (see Chapter 8).
4. *Reconstruction*: Substantial rebuilding of part or parts of the building (see Chapters 7 and 11).

Figure 1.1 The range of interventions

Range of adaptation options

As suggested in the above section the scope of adaptation works is wide and depends on the extent and purpose of the change proposed to the building. Figure 1.1 shows that they can range from basic preservation works at one end of the spectrum to almost complete reconstruction at the other (see also Figures 1.8 and 3.2, respectively for the three main branches of adaptation and for increasing intervention). In between these two extremes, in approximate ascending order are interventions such as refurbishment, rehabilitation, remodelling, renovation, retrofitting, and restoration. The Glossary at the end of the book gives brief definitions of these and other related terms.

The differences between the various terms for adaptation options relate to both the extent and the nature of the change and intervention they describe. Table 1.2 highlights the scale of adaptation options.

Table 1.2 Scale of adaptation options and degree of change

Scale	*Degree of change*	*Type*	*Example*
Small	Low-key	Minor improvement of surfaces. Upgrading of fittings. Minor extension.	New floor coverings, re-roofing, painting/re-painting or rendering/re-rendering external walls. Replacement of doors, windows, and kitchen/toilet fitments. Porch, conservatory or small rear/side extension, and loft conversion. These may involve some minor structural work, such as forming new openings.
Medium	Substantial	Conversion scheme. Major upgrading of surfaces and elements. Major retrofitting of services. Enlargement of capacity. Structural alterations. Major change of use of an old building.	Change of use from office to flats or vice versa. Overcladding of walls and recovering of roofs with improved thermal qualities, and over-roofing flat roof, recladding of walls. New air-conditioning system, addition of lift/s and service cores. Major lateral or vertical extension. Removal/insertion of walls and floors. Conversion and renovation works to a derelict or wrecked property.
Large	Drastic	Extensive remodelling works. Reconstruction of new building behind existing main external walls. Extensive spatial and structural alterations to enlarge/reduce the building's capacity or change its use.	Restoration of a ruinous multi-storey building. Facade retention scheme. Major extension to as well as internal and external modification of existing building.

Levels of commercial adaptation

As seen in the above section the scope of adaptation works is wide and depends on the extent and purpose of the change proposed to the building. More specifically Table 1.3 shows the various levels of refurbishment for commercial premises, which are particularly influenced by market and lease considerations.

Timing and cost of typical adaptation projects

The duration and cost of an adaptation project depends, as with new-build work, on a variety of factors. The size, quality, complexity and location of the work will all influence the time it takes to complete the scheme as well as determine the level of expenditure required.

Table 1.3 Typical levels of commercial refurbishment (Based on Martin and Gold, 1999)

Type	*Cost £/m²* *(2005 prices)*	*Approximate time to carry out (months)*	*Approximate payback period (years)*	*Description*
Minor/ Cosmetic	180–410	1–3	2–5	This will involve redecorating, improving signage and lighting, replacing floor coverings, exterior painting and repair, minor changes to the fittings.
Services	200–400	3–6	5–15	Complete replacement of heating, ventilation and air-conditioning plant. Associated pipework, ducting, terminal units, controls and insulation may be replaced or upgraded as necessary. Typically takes place at 25-year intervals (control systems more frequently).
Structural	150–400	2–6	5–15	Addition of new lift shaft, escalators or risers, necessitating structural alterations.
Major	500–700	2–12	5–15	This will involve major changes to the services and interior fittings but without any significant structural alterations. May include addition of raised floor, improvements to core areas and entrance halls, new lighting, internal shading. Typically takes place at 25-year intervals and in conjunction with lease renewal.
Complete	800–1500	6–18	10–30	This will involve significant structural alterations, such as enlargement of the floor areas or partial demolition to create an atrium or stripping of the building back to the concrete (or steel) frame. New cladding may be fitted together with the installation of new services and full fitting out. Timing of a complete refurbishment is variable but likely to take place in conjunction with lease renewal.
New build	800–1500	18–24	10–30	Construction of a new building, excluding demolition of an existing building and loss of rent.

Another factor that influences when and how adaptation is undertaken is the phasing of the work. This is addressed in a little more detail later in the chapter.

The lists in Tables 1.4–1.6, respectively give an approximate indication of the time-scale and costs involved for typical range of small/medium/large-size adaptation schemes. They are based on typical projects but the lists are neither exhaustive nor precise and so should be used with care. (See also the list in Table 11.4 comparing the procurement options of various sizes of adaptation work.)

Table 1.4 Estimated time-scales and costs of typical small-scale adaptation schemes

Category of building	*Type of adaptation scheme*	*Approximate timescale*	*Typical cost range (@ 2006 Prices)*
Residential	• New or replacement central heating system in two-bedroom flat.	2–3 days	£3000–£4000
	• Typical domestic kitchen or bathroom refit.	2–3 days	£4000–£7000
	• Window replacement programme to detached dwelling (say eight windows and two doors).	2–3 days	£6000–£10 000
	• New or replacement central heating system in semi-detached two-storey house.	3–5 days	£6000–£10 000
	• Domestic loft conversion (one dormer and one bedroom + toilet).	4–6 weeks	£20 000–£30 000
	• Domestic loft conversion (two dormers and two bedroom + toilet).	6–8 weeks	£30 000–£40 000
	• Lateral single-storey domestic extension with mono-pitched roof (c. 30 m^2).	4–8 weeks	£25 000–£35 000
Non-residential	• Minor shop refit and refurbishment.	1 week	£40 000+
	• Small restaurant and bar refurbishment.	1–4 weeks	£60 000+
	• Painting of external walls of an office block or other multi-storey building.	2–4 weeks	£60 000+
	• Energy improvements to interior of office block.	2–4 months	£100 000+

Adaptability criteria

Adaptability is obviously a key attribute of adaptation. It can be defined as the capacity of a building to absorb minor and major change (Grammenos and Russell, 1997). The five criteria of adaptability are:

- Convertibility: Allowing for changes in use (economically, legally, technically).
- Dismantlability: Capable of being demolished safely, efficiently and speedily – in part or in whole.
- Disaggregatability: Materials and components from any dismantled building should be as reusable or reprocessable (i.e. recyclable) as possible.
- Expandability: Allowing for increases in volume or capacity (the latter can be achieved by inserting an additional floor in a building, which does not increase its volume).
- Flexibility: Enabling minor if not major shifts in space planning – to reconfigure the layout and make it more efficient.

Each of these five factors will be dealt with in the remaining chapters of this book.

Table 1.5 Estimated time-scales and costs of typical medium-scale adaptation schemes

Category of building	*Type of adaptation scheme*	*Approximate timescale*	*Typical cost range (@ 2006 Prices)*
Residential	• Basic restoration of a derelict single storey detached cottage.	2–4 months	£50 000–£90 000
	• Conversion of large detached house into four flats (or re-conversion of four-flat house into one dwelling).	2–3 months	£50 000–£100 000
	• Lateral two-storey domestic extension with pitched roof (c. 60 m^2).	3–4 months	£50 000–£70 000
	• Over-roofing scheme to school or office block.	3–4 months	£100 000–£150 000
	• Renovation and major repair to a two/three storey detached/ semi-detached Victorian villa.	3–6 months	£50 000–£150 000
	• Adaptive reuse of redundant mill into flats (say 10 No.).	3–6 months	£200 000–£250 000
	• Domestic single-storey basement extension (two rooms).	4–6 months	£80 000
Non-residential	• Over-roofing scheme to school or office block.	2–3 months	£100 000+
	• Extensive restaurant and bar refurbishment.	3–6 months	£100 000+
	• Adaptive reuse of redundant office into hotel (c. 30 rooms).	4–7 months	£100 000+
	• Lateral single-storey extension to an office block (c. 100 m^2).	4–7 months	£200 000+
	• Two-storey basement extension.	6–9 months	£200 000+
	• One/two-bay extension to superstore facility.	4–6 months	£300 000+

General objectives of adaptation

Background

Over time, the requirements of a property user may alter and can only be satisfied by interventions that go beyond maintenance (see section below on Performance Management). In other words, adaptation is needed so that the user continues to make beneficial use of the property over the long term (i.e. 5+ years). It can be done to achieve some or a combination of the following objectives:

Code compliance

Work to bring the property up to contemporary building standards, especially in terms of:

- facilities for disabled access;
- fire safety;
- sound insulation;
- structural stability;
- thermal efficiency.

Table 1.6 Estimated time-scales and costs of typical large-scale adaptation schemes

Category of building	*Type of adaptation scheme*	*Approximate timescale*	*Typical cost range (@ 2006 Prices)*
Residential	• Major overcladding or recladding scheme to large block of flats.	3–6 months	£1 500 000+
	• Extensive modernization works to a housing estate (say 100 houses).	4–6 months	£2 000 000+
	• Restoration of a derelict pre-18th century tower house or large multi-storey detached mansion.	6–12 months	£3 000 000+
Non-residential	• Overcladding and other major refurbishment of a large office block.	2–4 months	£1 000 000+
	• Recladding and other major refurbishment of a multi-storey office block.	3–5 months	£1 500 000+
	• Extensive refurbishment of multi-storey public building.	6 months+	£2 000 000+
	• Extensive refurbishment of a multi-storey retail/office complex.	6 months+	£3 000 000+
	• Major restoration of a multi-storey public building.	12 months+	£10 000 000+
	• Major refurbishment of a multi-storey hotel (e.g. London-Hilton in Paddington) or large shopping centre.	12 months+	£20 000 000+

Environmental enhancement

- Installation of new or upgraded services to improve internal comfort conditions and energy efficiency.
- Contribute to better sustainability (see Chapter 10).
- Higher performance standards for the indoor climate.
- Painting or repainting external facades of buildings to improve their appearance.
- External environmental improvements, such as new hard and soft landscaping, as part of an urban regeneration scheme.

Spatial modifications

- Adjusting the size of units (e.g. lowering the ceiling level).
- Vertical or horizontal division of a large building into smaller units (e.g. to compartmentalize it to form more, but smaller offices or rooms).
- Making units self-contained (e.g. as part of a conversion of a single occupancy into a multiple-occupancy property).
- Combining spaces (e.g. to form a larger space).
- Providing additional space (e.g. study).
- Expanding existing space (e.g. to form larger dining room).
- Providing common areas and circulation.
- Increasing accommodation (e.g. additional bedroom).
- Improving accommodation (e.g. additional bathroom or ensuite toilet).
- Providing space for special or new activities (e.g. work/computer room).
- Alterations for the elderly and disabled.
- Reconfiguration of internal planning for convenience.
- Changing the function of spaces (e.g. from lounge to kitchen or vice versa).

Structure and fabric upgrading

- Overcladding or recladding and over-roofing (see Chapter 9) to improve weather resistance, as well as aesthetic, acoustic and thermal performance of the building envelope.
- General repairs and improvements to damp proofing and timber preservation.
- Major remedial works to rectify flood and storm damage.
- Inserting new columns and/or beams for strengthening or improved load-bearing capacity (see Chapter 7).
- Repairing defective or substandard structural elements (e.g. underpinning foundations).

Building changeability

Problems and opportunities

Adaptation is about managing and controlling change in the context of the functional and physical attributes of existing buildings. It is based on the premise that buildings are not static in a use or condition sense over their service life. Even within the same use classification, the level of activity or intensity of occupancy is unlikely to remain constant over the building's whole existence. For example, the use of a detached dwelling occupied by a family of five will change even within one generation. As the children grow up and move out either to find work or go to college, one or more parts of the dwelling tends to become under-used or used for other activities.

Change, like death, is one of life's few obvious certainties. The nature and rate of change since the last quarter of the 20th century, however, has been unprecedented. All the indications are that this is likely to continue if not accelerate for the foreseeable future.

Much of this change has of course been facilitated by technological factors. Developments in automated building techniques and prefabricated components, as well as the use of synthetic and composite materials are among the most significant of these influences on the construction industry. All of these have impacted on its supply side.

Innovations in information technology, too, are having an effect on the use of buildings, despite the conservative nature of the construction industry. Modern microprocessors are much smaller than their original versions. They therefore take up less space but may require more sophisticated cabling.

Global issues concerning energy conservation and reduction in pollution to combat climate change, and the loss of finite resources have also played a part in both the demand for, and supply of, property. Sustainability is the primary policy response of governments both in the West and other parts of the developed world to these problems (see Chapter 10).

Economic growth and urbanization, though, are the key drivers of change in modern, developed countries. At the macro level all of these dynamic factors can trigger the need for urban renewal programmes. Building adaptation forms a significant part of urban renewal schemes, particularly in historic areas and housing estates. At the micro level these influences can prompt property owners to update and improve their building assets.

Building adaptation, therefore, is essentially about responding to changes in demand for property. It is for this reason that it is more prevalent in industrialized countries. As the stock of property ages and building use varies over time, adaptation has become more common. Any building that performs poorly in terms of energy efficiency, comfort conditions or environmental impact is a potential candidate for adaptation (Energy Research Group, 1999).

In the developing world there is a greater incentive to redevelop rather than adapt buildings because of the paucity or poor quality of the existing stock. Even in some developed countries, though, there has been a tendency to demolish many existing buildings and redevelop the site where the value of the latter is greater than the former. This more drastic response is not uncommon because such buildings are considered to have exceeded their economic life. The economic life of a building can only be best extended by adaptation rather than just maintenance (see Figure 1.2).

Nevertheless, even in countries such as the USA where redevelopment (i.e. demolition and rebuild) was, and to a certain extent still is, a common option, pressures from historic preservation bodies (the American equivalent of British conservation groups) have resulted in some old or landmark buildings being retained (Brand, 1994). For example, the General Services Administration, one of the USA's Federal Agencies, operates a Historic Building Preservation Program to manage the country's historic federal buildings.

Building change occurs through a variety of influences that can be categorized as either external or internal factors. In other words, these influences may respectively be considered as a result of exogenous and endogenous changes.

Exogenous changes primarily result from external factors such as the general economic climate or market influences. These often trigger or compound building obsolescence or redundancy (see below). For example, the decline of traditional industries such as cotton and steel resulted in many purpose-built properties becoming surplus to requirements in the second half of the 20th century.

Endogenous changes, on the other hand, emanate from factors that relate directly to the building itself. These changes are normally user generated. For example, lack of maintenance eventually results in the condition of the fabric becoming dilapidated (see below under Deterioration and Obsolescence). The building and its services and related installations may also be modified to meet the needs or expectations of the occupiers more closely. Some buildings can be subjected to various uses during their service life. Flats may be converted into offices and then reconverted back to flats over a period of time.

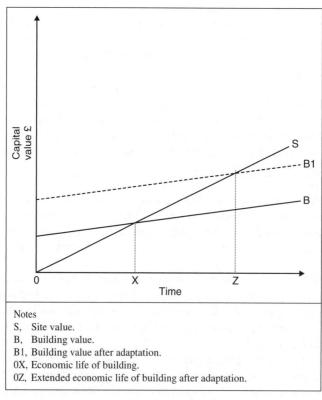

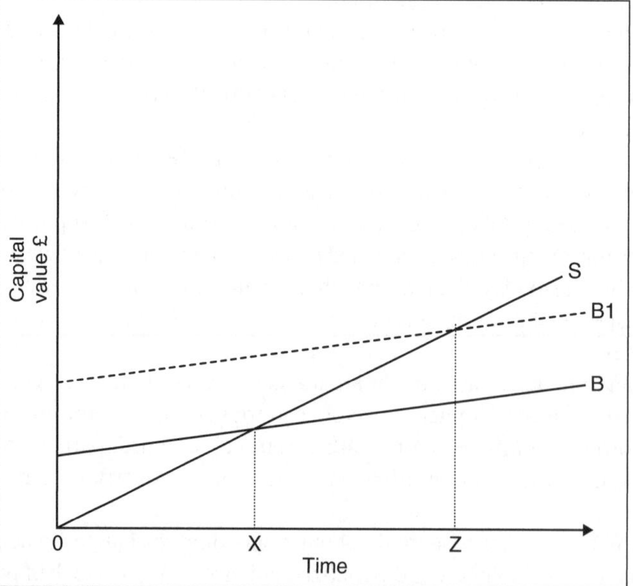

Notes
S, Site value.
B, Building value.
B1, Building value after adaptation.
0X, Economic life of building.
0Z, Extended economic life of building after adaptation.

Figure 1.2 Economic life of a building

Spatially, buildings change, too, though externally this is usually restricted by site constraints and planning controls. The size and capacity of buildings may need to be altered to cope with increases or decreases in demand for property space. In rarer cases this may even result in a reduction in the building's volume involving, for example, its partial or selective demolition. This could be done to reduce repair liability for surplus substandard floor areas or to make way for other activities such as more car parking on the site.

Nevertheless, change should also be seen in a positive light. It can provide the driver for making improvements in the built environment generally or in properties specifically. The scope and nature of change in relation to buildings is addressed in this book.

Degree of building change

As shown in Figure 1.8 building adaptation takes three principal forms: changes in function (e.g. conversion); changes in size (e.g. extension/partial demolition), and changes in performance (e.g. refurbishment). Respectively these are dealt with in Chapters 3 and 4, 5 and 6, and 7–10.

The degree of building change and the rate of response are, in the short run, dependent upon demand. In the long run, supply considerations, such as the availability, distribution, suitability and quality of the property stock, will also influence building changeability.

Building changeability and its measurement has only in recent years been considered in a systematic way (Henket, 1992). It consists essentially of two factors – changes in building use and changes in building condition. The rate of changeability of three common building uses is illustrated in Figure 1.3.

Traditionally, buildings have in the main been designed and described at one point in time (i.e. synchronically). Condition assessments, dilapidations surveys, structural inspections and other types of building surveys as defined by the Construction Industry Council (CIC, 1997) are examples of synchronic appraisals. They are in other words snapshot views of a building in time. As a result they only tell a partial story of its service life.

Duffy (1993) has argued that a more effective perspective is to consider buildings in terms of change over time (i.e. diachronically). This requires a more holistic approach to buildings to set them in their historical, temporal and physical contexts. Building adaptation is the process by which they are made to respond to these influences. Brand (1994) has shown that buildings have seven shearing layers of change (see Figure 1.4).

The cycle of maintenance, repair and adaptation to a building depends primarily on several factors:

- the purpose or function of the building (e.g. farm buildings are generally of a lower construction standard than dwellings);
- the quality of the building (e.g. its condition and architectural importance);

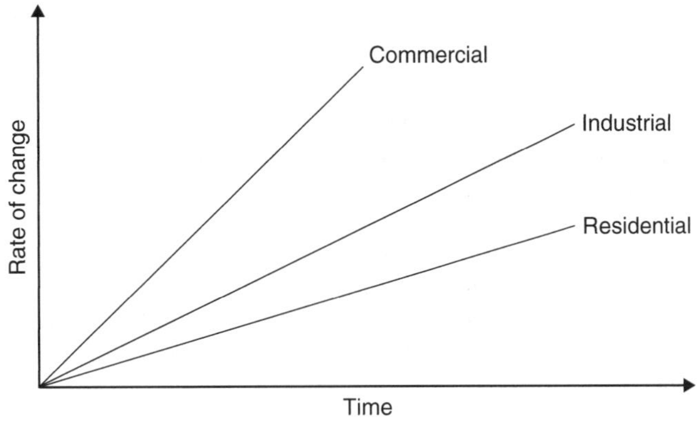

Figure 1.3 Varying rates of changeability of different buildings

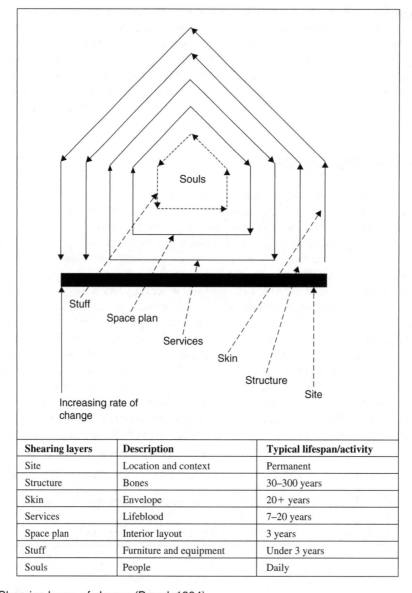

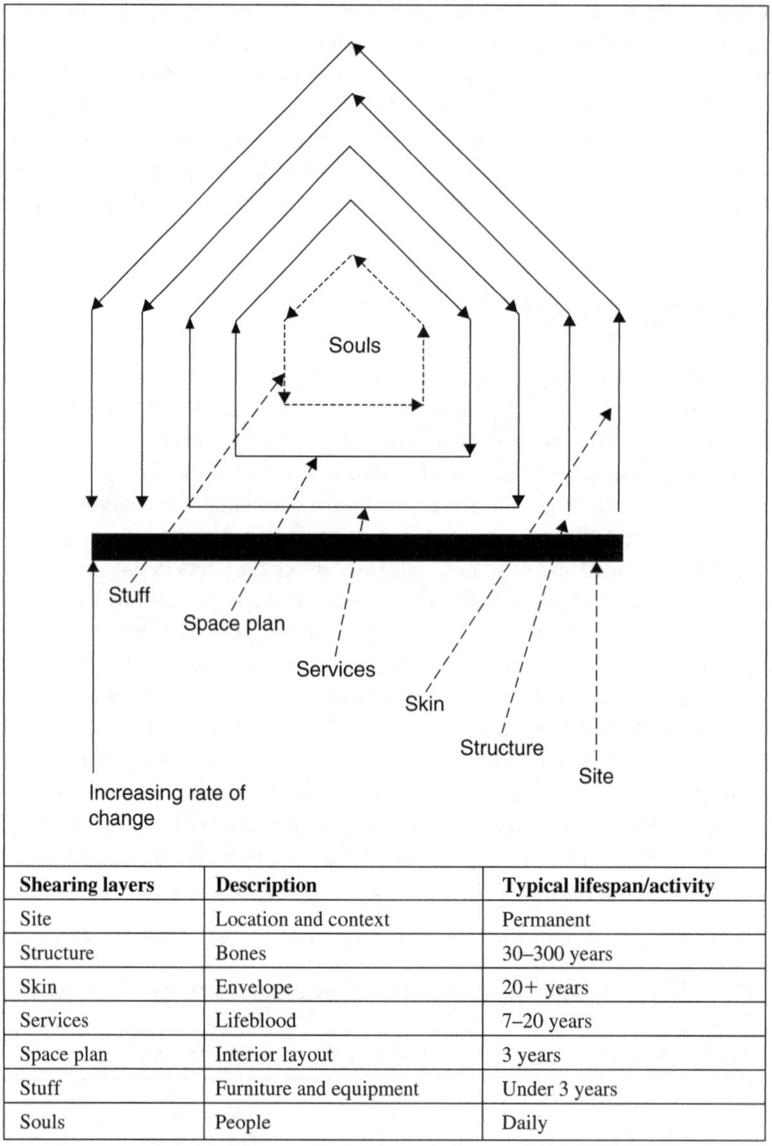

Shearing layers	Description	Typical lifespan/activity
Site	Location and context	Permanent
Structure	Bones	30–300 years
Skin	Envelope	20+ years
Services	Lifeblood	7–20 years
Space plan	Interior layout	3 years
Stuff	Furniture and equipment	Under 3 years
Souls	People	Daily

Figure 1.4 Shearing layers of change (Brand, 1994)

- the use, misuse and abuse the building is subjected to;
- statutory requirements, especially those relating to health and safety;
- the requirements and expectations of the users/owners.

The quality of a building is heavily influenced first of all by its initial costs, as demonstrated by the see-saw hypothesis (see Figure 1.5). This contends that a building having a low initial cost will other things being equal have a high level of maintenance cost. Generally speaking the reverse is also true. There are of course exceptions to this rule. Very large facilities with a high capital cost, for example, are usually relatively expensive to maintain.

As indicated by Duffy (1993), the capital cost of a building only accounts for about one-third of its total cost. Maintenance along with adaptation takes up the other two-thirds.

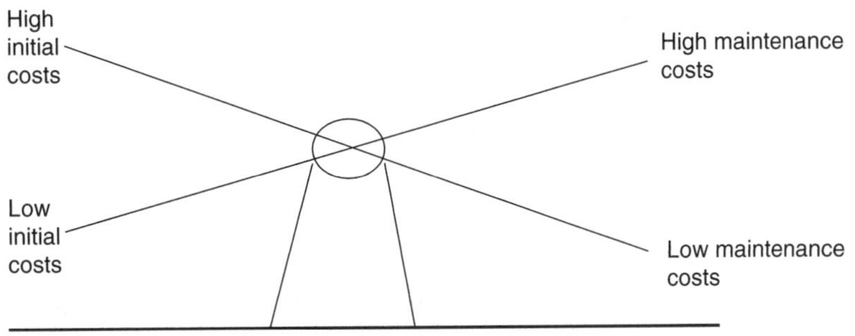

Figure 1.5 The maintenance seesaw hypothesis

The cycle of adaptation can be demonstrated in a simple way by comparing two contrasting building uses: residential and non-residential. Flats, apartments, and other dwellings tend to have a relatively long life expectancy (60+ years) with few major interventions (i.e. adaptations) over their service life.

Commercial, industrial other non-residential buildings on the other hand usually have a much shorter design life – in some cases as little as 20 years (e.g. McDonalds fast food outlets). This probably explains why these buildings are adapted more frequently than dwellings (as indicated in Figure 1.3).

The use of commercial property, of course, is primarily determined by economic factors – inflation, the level of economic growth, business cycles, tax levels, interest rates, and changes in customer requirements. All of these factors affect the demand for non-residential buildings. Moreover, such properties are normally subjected to a much higher level of wear and tear than housing. Competition can also drive companies to modify or modernize their property stock. This is most visible in the grocery-supermarket sector, where companies such as 'ASDA' and 'TESCO' are altering or refurbishing their existing stores on a medium-term basis (i.e. about every 5 years).

Two recent examples relating to these supermarket stores illustrate the point. One TESCO store was constructed in the edge of Edinburgh in the late 1990s. It was built to a standard five-bay size design. Within two years of the store's opening, the building was extended by about two-and-a-half bays. One possible reason for this is that planning permission may have been relatively 'easier' using this type of phasing. However, there would be no economies of scale in adopting such an approach within a few years of the first development. Phasing the construction of the building like this can effectively double the administrative costs (e.g. design fees as well as planning and building control charges). It will also involve work in and adjacent to an occupied part of the property – to connect the extension to the main building.

Most buildings, therefore, undergo a variety of cycles during their service life. Changes in the intensity or type of use will impact on how well a building performs. Interventions such as maintenance can stave off deterioration but by itself it cannot ensure continuing use of property. It is only through adaptation that the long-term utility of a building can be prolonged. This makes for a more sustainable built environment.

Reasons for adapting buildings

Background

Demolition was often the fate of redundant or obsolete buildings in British urban areas up to the late 1970s. The post-World War II recovery triggered a slow but inevitable rise in general prosperity and the boom in property up to that period. This in turn prompted rapid urban growth and redevelopment in both private and public sectors. Moreover, as car ownership increased there was a need to re-plan existing urban layouts to accommodate an expanding road system.

During the last half of the 20th century the adaptation of buildings has gained increasing recognition as a viable and increasingly acceptable alternative to new build. This was the result of a number of influences. The wholesale demolition and replacement of much of the defective but repairable housing stock in many urban areas, especially during the 1960s and 1970s, went too far in some cases (Latham, 2000a). It ripped the heart out of some communities and replaced much of the housing with poor quality or inadequate designs such as deck access blocks and multi-storey towers. Moreover, some residential construction methods such as the 'Skarne' system-building, which was used for multi-storey blocks of flats, have Scandinavian origins. Although these systems used good quality components and apparently worked well in their indigenous environment, they were often inappropriate for the cold and damp nature of the British climate. This was a classic example of the 'foreign solutions to domestic problems' approach. In other words, designs that work well in one environment may not do so in another, especially if it has a different macroclimate.

Apart from economic and legal developments, however, there are other reasons for adapting buildings as opposed to redevelopment. The main influences are discussed as follows:

Available grants

In some cases, particularly with housing and certain historic buildings, grants may be available to help with the cost of adaptation or improvement work. The extent and availability of such funding is restricted (see Chapters 2 and 11).

Timing

Remodelling work usually is quicker than new build. For example, the conversion of a detached mansion into multiple flats can take anything from 2 to 4 months, depending on the extent of works involved. The construction of a new dwelling of the same size with the same number of flats would normally take about double that time even if conventional techniques were used.

Deterioration

Adaptation work may be required to arrest deterioration owing to a number of influences (see section on Obsolescence and Deterioration). A sensible adaptation scheme should extend the economic life of a building, as shown in Figure 1.2. It shifts the building value line up because potentially a building is worth more after adaptation. This is in turn increases its service life.

Performance

The need to enhance the acoustic, thermal, durability or structural performance of a building is often a reason for carrying out adaptation work. Excessive energy consumption by a building may often prompt its refurbishment. This could include renewing the heating system as well as upgrading the thermal efficiency of the fabric.

Change of use

When a building is empty for a long period, its previous use may not be in demand. The building in other words is redundant. An adaptive reuse may be required to ensure the building's continued beneficial occupancy (see Chapters 3 and 4).

Legal restraints

An owner of disused building who is not allowed to demolish it because of planning constraints could leave the property alone and let it dilapidate naturally (subject to possible statutory notices by the local authority if the building is deemed to be in a dangerous or poor condition). Alternatively, the owner may adapt the building either in its existing use or to another use.

Conservation

The cultural as well as technical reasons can frequently influence the decision to adapt a building rather than redevelop the site. The architectural or historic importance of a building may be sufficient reason why it should be saved. These are addressed in Chapter 3.

Sustainability

Reusing or upgrading old buildings is a more environmentally friendly than redevelopment. The latter involves demolition as well as new-build activities, both of which expend more energy and waste than adaptation. The measures required to achieve sustainable adaptation are discussed in more detail in Chapter 10.

Advantages

Economic

It is usually much cheaper to adapt an existing building than it is to demolish it and redevelop the site (see discussion on cost implications of adaptations in Chapter 2). The reasons why adaptation normally costs less than new build are not hard to find. The adaptation process is generally much quicker than that of a new development. The existing infrastructure is already provided (i.e. the foundations, basic services and superstructure). Consequently, the contract period is usually shorter and borrowing costs for adaptation are much cheaper than for new build. It is for this reason that when interest rates are high the proportion of adaptation projects as opposed to new-build schemes tends to increase.

Adaptation obviates the need for wholesale demolition of the building. Demolition itself is an expensive, wasteful, messy, disruptive and dangerous activity (see Chapter 7). Hazardous substances and contaminated ground may be disturbed by such work, and these can be expensive to remedy or treat. The form of construction, the presence of deleterious materials (see Chapter 2) and the proximity of adjoining buildings will directly influence the ease and cost of demolition. This process also incurs more embodied energy in the overall development process.

Technical

The existing structure and fabric of the building can be fully utilized to provide an enclosure already in place. Little provision has to be made for protecting the works or storage of materials; the existing property itself usually provides an adequate enclosure. Thus the building only needs to be modified to cater for the proposed adaptation works.

The owner will also be able to make use of most if not all of the fittings and services in the building. This clearly can yield saving on the cost of purchasing and installing new components.

Spatial

The permitted 'plot ratio' for a proposed building may be less than that provided by the shell of the existing structure. In such a case an owner will maximize gains in floor space by retaining the property. In many instances large internal spaces can be reduced by subdivision without compromising the essential architectural quality of the whole property (Scottish Civic Trust, 1981).

The owner of a building can make full use of the existing site. In contrast with a new development the frontage may have to be set back to a different building line to satisfy the local authority's roads proposals.

Environmental

An enhanced appearance can be achieved for an adapted building. If the work has been designed and undertaken sensitively and carefully, the building should look better than before. This inevitably has a positive knock-on effect on surrounding properties. Moreover, an adapted building ought to be more energy efficient than previously, particularly when sustainability is a key policy criterion.

Because of their high thermal capacity and slow thermal response, some traditional buildings are relatively good at conserving energy. Older buildings tend to have thick solid walls, small windows and natural lighting and ventilation, which leads to economy in energy consumption (Scottish Civic Trust, 1981). However, this will depend on the U-values of the fabric of the building under consideration.

As we have seen adaptation is an important sustainability criterion. This is because it reduces both energy consumption and the generation of waste. It minimizes the need for using up fresh material resources and energy required in producing and transporting them. In other words, the embodied energy and transport energy consumption is much lower than with a similar size new-build scheme. Moreover, because it avoids demolition adaptation minimizes pollution and waste.

The adaptive reuse of derelict and redundant buildings also helps to relieve pressure on the development of greenfield sites. This encourages better use of existing scarce resources.

Social

Retaining the character of a streetscape is best achieved through adapting its buildings. Old buildings offer psychological reassurance because of their distinguishing characteristics (Scottish Civic Trust, 1981). It is with housing, however, that the social advantages of adaptation are more obvious. Within this category there are architectural, cultural and historic benefits of adapting buildings. It is for these reasons that building conservation is becoming increasingly important in the developed world.

It is of course not easy to measure the social benefits of adaptation. However, sensitive adaptation schemes such as the reuse of disused industrial buildings (see Chapter 4) can bring back life to run-down urban areas. The successful adaptation (whether refurbishment or adaptive reuse) of a redundant property can offer hope to a community devastated by the loss of traditional industries. Converting empty warehouses into, say, student residencies can be a successful way of rejuvenating life in an inner urban area. The centrality and convenience of such accommodation makes such schemes popular with students.

A well thought out adaptive reuse of a building of architectural or historic importance can bring considerable lasting prestige to its owner (Scottish Civic Trust, 1981). One example of this is the renovated Georgian terrace in Charlotte Square, Edinburgh, which is the headquarters of the National Trust for Scotland. This also involved a re-conversion to residential use of part of this terraced property.

Adapting a building can also result in health benefits for its occupants. It can help tackle dampness and poor indoor air quality, both of which can have a major impact on building-related illnesses (see health benefits of refurbishment in Chapter 8).

Disadvantages

Background

It is an implicit premise of this work that adapting buildings is generally a desirable objective. However, not all adaptation schemes are either necessary or worthwhile. In some cases, they can prove unsuccessful because the changes to the building are either inappropriate or of poor quality. This book, therefore, does not promote the notion of wholesale or universal building adaptation. Some old, dilapidated or redundant buildings are best demolished to make way for more suitable and attractive properties.

Accordingly, despite the various advantages of adapting an existing building, such an intervention often entails a number of drawbacks. The main disadvantages of adaptation are summarized as follows:

Functional

There is no guarantee that an adapted building will match the performance of a new purpose built facility. Restrictions as regards layout and heights may necessitate compromises and prevent full satisfaction of the users' needs. Adapting a building may prevent the site from being used more efficiently. Moreover, because the constraints of form, scale and aspect of the building have already been determined, refurbishment design is considered more problematic than new build.

Technical

The extended life of an adapted building is still only about half of that for a new facility. There is no guarantee that the adaptation works will overcome all the deficiencies in performance. Indeed, all existing buildings contain some latent defects that may prove difficult and expensive to resolve.

Economic

The maintenance costs of an old building, even one that has been refurbished, are usually still higher than those for new build. The rental income that can be derived from an existing building may not be as high as that obtained by a modern facility that fully meets the needs of today's building user. Moreover, the energy costs are likely to be higher as it is hard to match the insulation standards of new build. Some materials required for use in adaptation work to match existing are expensive and hard to come by.

Conservation requirements for the adaptation of old buildings may drive up construction costs. This may lead to the adaptation work costing just as much as a redevelopment scheme for the property.

Environmental

Not all adapted buildings result in an improved internal or external environment. The appearance or energy efficiency of the refurbished building may not be much better. The use may also not be compatible with surrounding properties in terms of density or nature.

Legal

Full code compliance with the building regulations may be difficult to achieve in some older properties. Spatial and constructional constraints with some of these buildings, for example, can inhibit the attainment of the required means of escape and level of fire resistance.

Planning restraints may limit the degree to which a property can be adapted. This is likely to have an impact on the viability of the proposal.

Performance management

The performance concept

The performance concept has been used in some areas of the construction industry since the early 1970s. It is based on a more systematic way of determining and achieving desired results by focusing on ends rather than means (CIB, 1993a).

Adaptation is one of the two primary elements of building performance management (see Figure 1.6). The other element is maintenance (which includes repair). Maintenance is the act of keeping a building in a pre-determined state. 'Repair' involves making good to restore to original condition. It may also mean replacement of failed components. Maintenance and repair may be extended to include minor works of beneficial improvements or upgrading that bring facilities up to an acceptable standard but falling short of adaptation.

For example, maintenance and repair, on a like-for-like basis, may not be adequate to satisfy user's current or future requirements. Thus such works may include some element of 'beneficial improvement' (e.g. replacement of windows on a like-for-like basis because they are beyond economic repair).

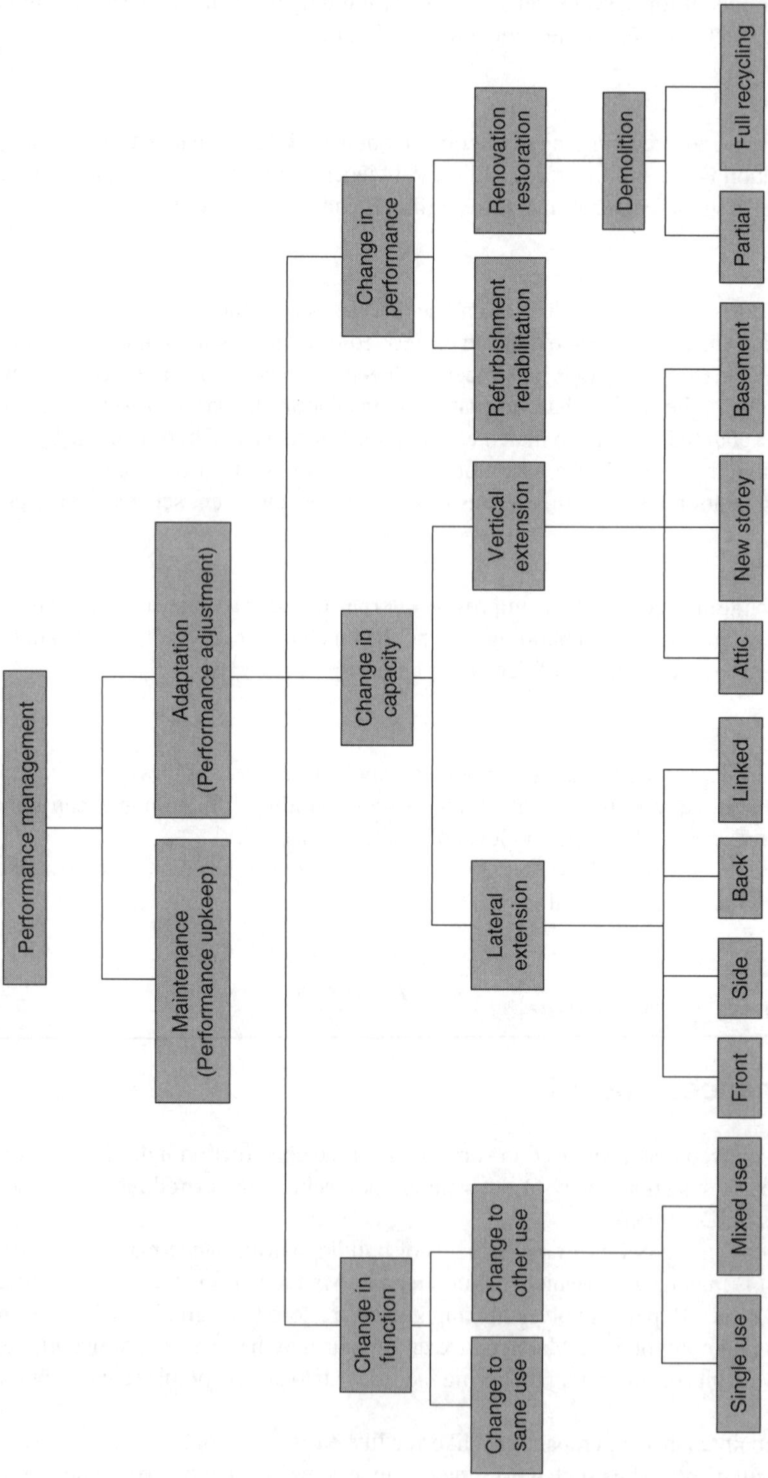

Figure 1.6 The two elements of performance management (adapted from Henket, 1992)

Adaptation on the other hand deals with 'substantive improvement'. Replacing timber windows with components having a higher performance in terms of appearance, thermal efficiency, air-tightness, sound attenuation and durability, such as double/triple glazed hardwood or resin-bound pultruded fibreglass, is an example of this type of improvement.

Adaptation is interpreted here as performance adjustment; maintenance, as performance upkeep. The influences of these two major forms of intervention are illustrated in Figure 1.7. But they are not the same for all building types. Figure 1.8, for example, contrasts the maintenance and adaptation cycles of residential and commercial buildings, which have different characteristics in terms of both use and form.

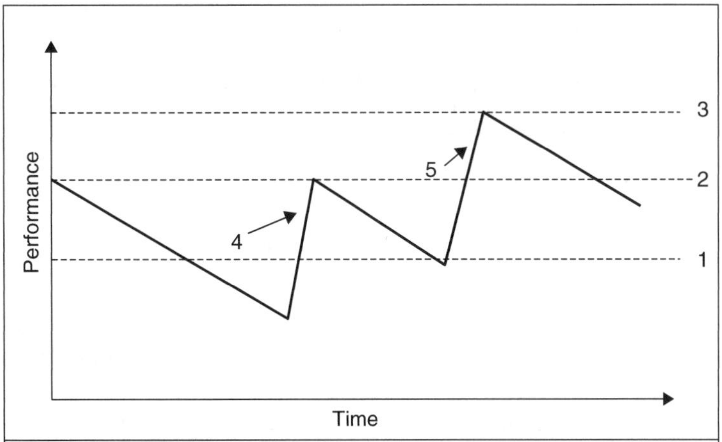

Notes
1. Minimum acceptable standard.
2. Original condition.
3. Optimal or maximum standard.
4. Maintenance.
5. Adaptation.

Figure 1.7 Influence of maintenance and adaptation on declining performance

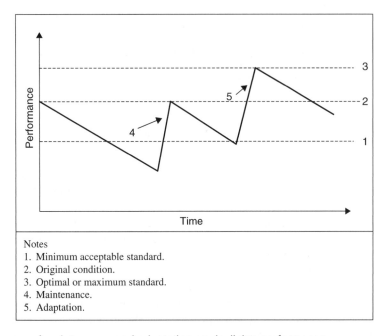

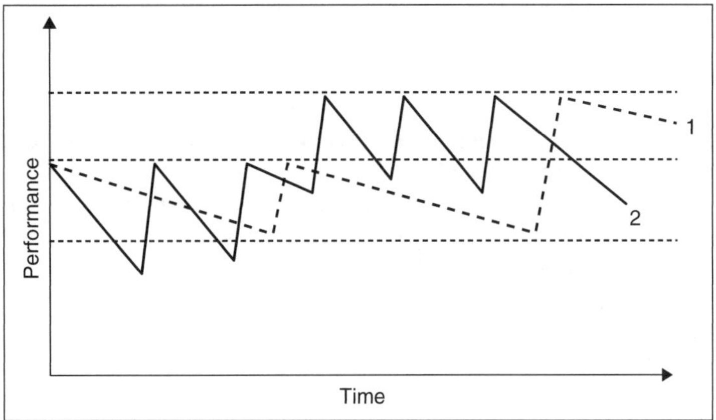

Notes
1. Residential.
2. Commercial.

Figure 1.8 Typical maintenance and adaptation cycles of different buildings

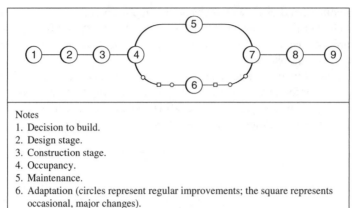

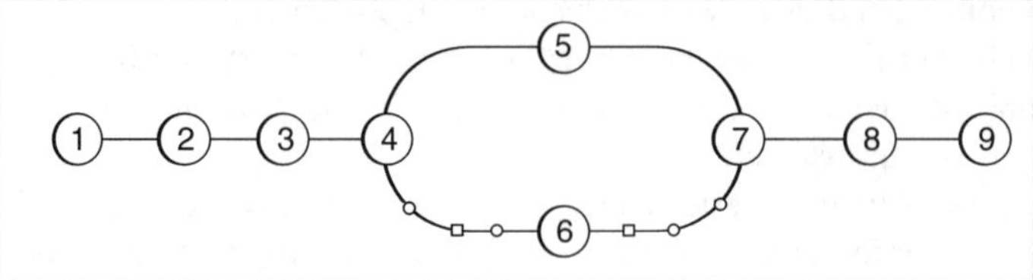

Notes
1. Decision to build.
2. Design stage.
3. Construction stage.
4. Occupancy.
5. Maintenance.
6. Adaptation (circles represent regular improvements; the square represents occasional, major changes).
7. Irreversible building obsolescence sets in.
8. Building fully obsolete.
9. Dispose/demolish.

Figure 1.9 A linear model of the whole life cycle of a building

The differences between maintenance and adaptation are further illustrated in Figure 1.9, which shows a linear model of the whole service life of a building. In contrast the typical life cycle of a building is outlined in Table 1.7.

The original use of the building may be sustained for decades by combinations of maintenance and adaptation. Many church buildings, for example, have lasted for hundreds of years. This is because of the robustness of their masonry construction coupled with some basic levels of maintenance having been consistently undertaken during the occupancy of those buildings.

For other types of buildings, however, the original use may not be sustainable for more than a few decades. This may be because of obsolescence or redundancy (see below). For example, the opening of a busy urban inner relief road can easily alienate a parade of good quality shops. Thus, in many cases it is functional change rather than age that causes obsolescence.

Adaptation and maintenance, therefore, are the two primary, though contrasting, interventions to buildings. The essential differences between these two aspects of performance management is summarized in Table 1.8.

Types of maintenance

There are essentially two main types of maintenance: proactive (i.e. planned) and reactive (i.e. unplanned). The main branches of these two are illustrated in Figure 1.10.

Planned preventative maintenance of course is more expensive to set up than unplanned maintenance. The latter, however, will result in more long-term running costs for a building maintained in this way (Chanter and Swallow, 1996; Wordsworth, 2000).

Even with planned preventative maintenance the overall performance of a building will inevitably decline. The acceptability of declining performance over time, however, is not only influenced by the level of maintenance the building receives. It also depends on the standards set by the building user/owner or through legislation/regulations, particularly regarding means of escape, access for disabled, health and safety at work, etc. Figure 1.11, for example, shows three different users of similar building types. Each user is likely to have different requirements and expectations, which invariably increase over time – more so with commercial buildings, which have to respond more and frequently to market influences. In contrast the condition or overall performance of the building is likely to decline over time, for the reasons given earlier (see Figures 1.7 and 1.8).

Table 1.7 Typical life cycle of a building (based on Eley and Worthington, 1984)

Stage	*Process*	*Results*
1. Birth	New activity or new process is housed by building shell. Degree of flexibility provided in design may vary.	Building user is accommodated.
2. Expansion	Uses expand to meet new requirements. New services introduced. Internal layout adapted.	Strain placed on fabric. Possible extensions. Change in function or spatial performance.
3. Maturity	Uses continue to fit/suit building. Or current needs exceed capacity.	Periodic maintenance and minor adjustments. New space taken elsewhere. Or more extensions and re-planning.
4. Redundancy	Change in: • Sources of power • Societies' cultural values • Market needs • Technology • Catchment areas	No maintenance: building in limbo. Building is becoming obsolete or potentially obsolete. Application for permission to adapt or redevelop. Building becomes empty – partially or fully. Attempt/s at letting. Vandalism: broken windows; fire or other damage. Risk of squatting. Consider mothballing. Decision to demolish (whole or part).
5. Rebirth/Demolition	Motivation to reuse. Variety of acceptable uses are matched with the building.	Building refurbished/restored (whole or part) cycle continues. Building made more sustainable. Make way for new building.

Table 1.8 The primary differences between adaptation and maintenance

Intervention	*Nature*	*Purpose*
Adaptation	Usually medium to large-scale work (e.g. change of use) to part or the whole of a building, done intermittently, over long periods. Involves substantive adjustments, alterations and improvements.	To improve or modify a building's performance. To satisfy changing user or legal requirements. To change or modify a building's use. To enhance a building's value or rental incore To enhance a building's image. To expand or reduce a building's capacity or volume.
Maintenance	Generally small-scale work to a building (e.g. cleaning out roof gutters or repainting external woodwork), done regularly or almost continuously, over both short and long periods. Usually repairs, but may involve some nominal or beneficial improvements.	To preserve the physical condition of the building. To sustain the value (rental and/or capital) of the building. To safeguard the investment potential of a building. To provide a satisfactory internal environment. To comply with statutory requirements. To protect a building's image/status.

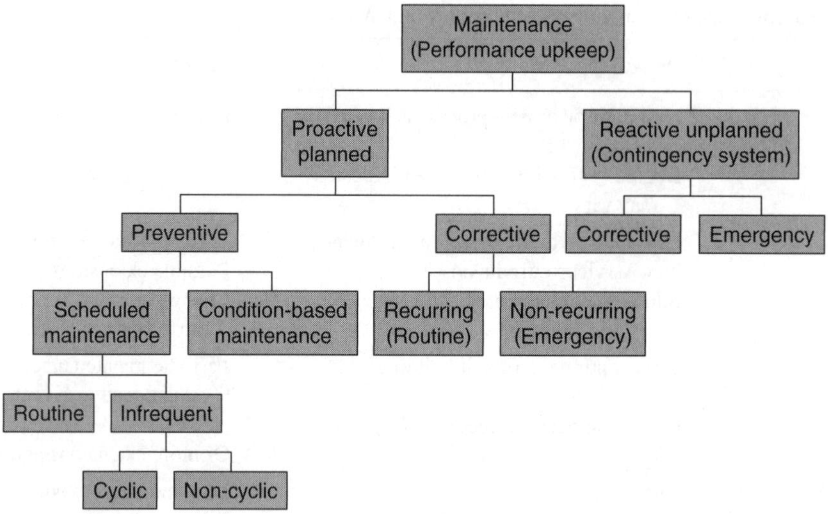

Figure 1.10 Types of maintenance

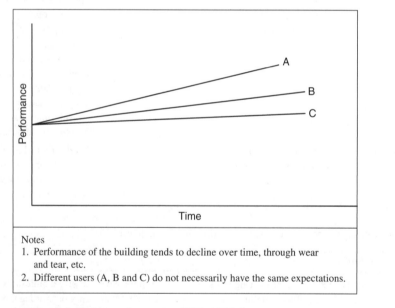

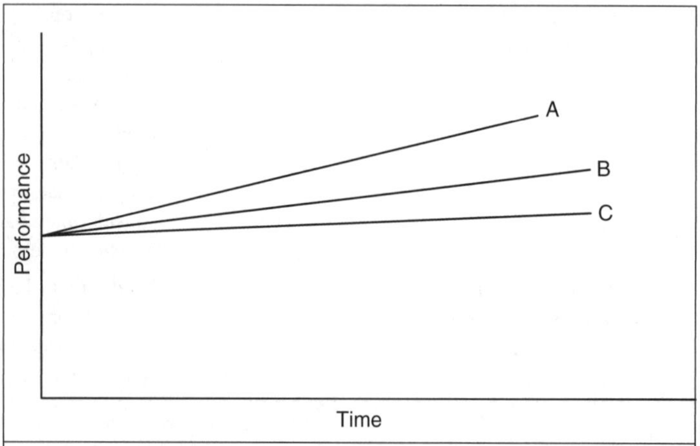

Notes
1. Performance of the building tends to decline over time, through wear and tear, etc.
2. Different users (A, B and C) do not necessarily have the same expectations.

Figure 1.11 Acceptability of declining performance (CIB, 1993a)

Market conditions affecting adaptation

Changes in residential demand

The government's Social Survey Division (SSD, 1998) monitors on a regular basis trends in UK household patterns. Since the 1970s there has been a significant change in the types of households in which people live. According to the SSD in the 1970s and 1980s, the average household size declined steadily from 2.91 in 1971 to 2.51 in 1989. Also, in the early 1990s, the average household size continued to decline but at decreasing rate.

Furthermore, over the last two decades of the 20th century there has been considerable change in the types of households in which people live. In the UK, for example, overall:

- One-person households increased from 17 per cent in 1971 to 28 per cent in 1995. In 1996, 27 per cent of households consisted of one person living alone, of which 59 per cent were aged 60 or over.
- The proportion of households consisting of married couples has decreased.

It is not only demographic influences that have resulted in changes in the demand for residential properties. Economic factors, too, play an important role in this regard. Increasing prosperity and availability of finance to fund these has encouraged homeowners to adapt their dwellings.

Many householders nowadays often want parts of their existing accommodation enlarged – such as kitchens; others require more ancillary accommodation – for example, a study, play room or garage. Working from home has also become more common. Spare rooms are being converted into mini-offices or house extensions are being built to accommodate such requirements.

The expectations of housing users have also increased, and as a result occupiers are demanding more from their buildings. They want additional facilities such as an en suite toilet with shower cubicle, as well as better thermal efficiency and greater comfort, and more attractive facades.

An increasing elderly population will also have significant implications in the demand for residential property. This will increase the use of facilities such as 'granny flats' within existing dwellings (see Chapter 3). There also likely to be a greater need for more residential and nursing homes, many of which will emerge from the conversion and extension of large detached dwellings.

Changes in non-residential demand

Rapid change in organizational structure and working practices has been a common occurrence in many larger businesses during the past 20 years. The type and overall size of building space that they require has changed as a result.

Whilst the nine-to-five shift is still common in many offices, the introduction of flexi-time and part-time working in the 1970s has had an impact on the use of building space. This has changed the churn rate of offices.

The rapid advancements of information technology have had a significant impact on the demand for office space. Computers, printers and other office equipment are smaller and lighter and thus take up less space.

Flexibility of space is a key requirement of new business organizations to accommodate new working practices. This is essential to meet the changing space needs of these burgeoning ventures.

Higher environmental standards demanded of modern working conditions have prompted the refurbishment of many office buildings. This offers an opportunity to create new facilities or refurbish existing ones. Such buildings, too, are often seen as reflecting the corporate image.

Responding to market needs means that retail outlets such as banks and supermarkets have to adapt every 5 years or less. The rate of commercial adaptation, in other words, may be faster than for other types of property as they are more sensitive to market forces.

Consequences of changes in demand

- In some sectors such as offices in expensive city centres demand has fallen. This has caused an over-supply of offices that has inevitably led to high levels of under-utilization of buildings. Partially occupied buildings are not uncommon in some urban and suburban areas.

- Long-term vacancy of building stocks. In the residential sector this has resulted in some 1 million houses in Britain being left empty (see below).
- Increasing obsolescence of old buildings that do not meet contemporary standards. Many system-built housing blocks have had to be demolished after only 20 or 30 years because of severe condensation and other dampness problems.
- Redundancy (see below) of many old industrial buildings in unsuitable locations.

It is of course extremely difficult to forecast with great accuracy every form of future demand for building space during times of rapid change. It is not so much the form that future demand will take that is uncertain but rather its extent. This problem is more acute for non-residential buildings, because economic conditions and market requirements predicate demand for such properties.

The future demand for housing on the other hand is more predictable because demographic changes are easier to calculate than long-term economic conditions. According to the government sponsored Urban Task Force (1999), it is estimated that 4 m more dwellings in the UK are needed by 2020 to cope with increases in housing demand.

The following is a suggested list of the main changes in demand affecting property that are likely to occur over the first half of the 21st century. Beyond that it is very difficult to predict in quantitative or qualitative terms with any degree of certainty the changes that may take place in the property market.

- Increased leisure facilities (e.g. sports/fitness centres; multiplex cinemas; ten-pin bowling; restaurants).
- More working from home (e.g. teleworking, with office facilities in the house).
- Greater use of information technology for home shopping, entertainment, etc.
- Greater diversity or ranges of retail outlets, such as coffee/wine bars; internet cafes.
- More single-person accommodation – either as bedsits or good quality single bedroom flats.
- More accommodation for the growing elderly population – especially single-storey accommodation.
- More responsive internal environments to cope with variable working conditions.
- Greater reliance on intelligent buildings – with sensors and 'smart' materials to respond more effectively to changing climatic conditions or users requirements.

Changes in residential supply

The overall state of repair of the much of the British housing stock is outlined in the English, Scottish and Welsh House Condition Surveys. These major housing appraisals are carried out on a quinquennial basis. The ageing stock of residential buildings is still deteriorating or unable to meet the needs of modern use despite some degree of maintenance. A survey carried out by the Association of Metropolitan Authorities in the 1980s estimated that the maintenance backlog of system-built housing was over £5 billion (AMA, 1985). This has demonstrated if not necessitated the need for modernization schemes to dwellings (see Chapter 8).

With the demise of the Parker Morris housing standards by the early 1980s, house sizes inevitably decreased. The slump in public sector house building during the last two decades of the 20th century also had an adverse impact on the housing supply.

As a result of these influences there is in many areas of the UK such as in London and the Southeast a severe scarcity of adequate housing. Unfortunately, too many vacant or dilapidated properties, which are the ripest for adaptation, are in the wrong areas. Ongoing increases in demand for accommodation are more likely to remain therefore in the south of Britain.

A report by the Urban Task Force (1999), led by Lord Rogers of Riverside, has advised that over £500 m must be spent on providing new housing as part of a major urban regeneration programme. According to that report adapting existing empty buildings could create 1.5 million new homes. The use of brownfield sites rather than greenfield sites is another of the report's key recommendations.

Changes in non-residential supply

- Many old buildings are not suitable for modern facilities, particularly information technology.
- The maintenance backlog for non-residential buildings in the UK runs into billions of pounds. For example, according to a recent study university laboratories in Britain 'are falling apart as a result of a £3 billion maintenance backlog' (JM Consulting, 2002).
- Improved environmental standards are required to meet the demands of modern working conditions.
- Buildings are often seen as reflecting the corporate image.
- Intelligent buildings are becoming more common.
- Sustainability is becoming more important as an issue or policy matter in corporate as well as public-sector schemes (see Chapter 10).

Consequences of changes in supply

- Greater pressure to adopt sustainable and lean construction techniques.
- Building conservation as a philosophy is likely to become more prevalent.
- More use of industrialized buildings and modular systems is likely.
- Increases in the use of composite and 'smart' materials (e.g. electrochromic glass).

Building obsolescence and redundancy

Background

The service life of every building is unique. This is because no two properties have exactly the same use pattern or exposure conditions. Sooner or later, however, obsolescence or redundancy threatens every building. Obsolescence is the process of an asset going out of use. It has been termed 'the fourth dimension in building' (Iselin and Lemer, 1993). This is because it determines the timing of adaptation or demolition of a property. Space occupies the first three dimensions (length, breadth and depth). Time occupies the fourth, which in the context of building use is crucial as well.

In a general sense, therefore, obsolescence is a measure of an object's usefulness over time. It indicates the tendency of assets and operations to become out-of-date, outmoded, or old-fashioned. It is the transition towards the state of being obsolete, or useless. In short, obsolete means antiquated disused, or discarded. An item that is broken, worn out or otherwise dysfunctional, however, is not necessarily obsolete. But these conditions may underscore the obsolescence of a building (Iselin and Lemer, 1993).

According to Nutt et al. (1976):

> … Building obsolescence refers to the degree of usefulness of a building relative to the conditions prevailing in the population of similar building stock as a whole. In other words, supply considerations as well as demand factors become important in building obsolescence …
>
> The degree of usefulness of a building will vary with time. Actions can be taken to increase the usefulness of buildings and hence reduce their relative obsolescence. Obsolescence should be viewed, therefore, as a function of human decision rather than a consequence of 'natural' forces. If no action is taken then an obsolete state is threatened. The overall likelihood of obsolescence will increase as the service provided by buildings decreases, either relative to requirements as demand changes, or absolutely as the building stock deteriorates but demand remains constant. So that any factor that tends, over time, to reduce the ability or effectiveness of a building to meet the demands of its occupants, relative to other buildings in its class, will contribute towards the obsolescence of that building.'

Redundancy, on the other hand, basically means 'surplus to requirements'. It is predominantly demand determined. A redundant building, therefore, is a good indicator of over-supply of a particular type of property.

According to ISO 15686-1:2000 refurbishment and upgrading are the major strategies to counter obsolescence. Thus the primary actions undertaken to sustain or increase the usefulness of a building are adaptation and to a lesser extent maintenance. Adaptations such as alterations, extensions and refurbishment can provide a degree of long-term security for buildings. By themselves, though, these interventions cannot guarantee its continuing beneficial use. It is demand that ultimately determines the level of occupancy of a property.

Thus, it has been argued that obsolescence of buildings is not synonymous with redundancy (Mills, 1994). For example, a building could be in a good condition but because of a slump in demand for that type of property, it would be classed as redundant but not necessarily obsolete for its intended purpose. Obsolescence, though, is often the trigger for redundancy of buildings.

The difference between redundancy and obsolescence therefore is subtle (see Table 1.9). But as indicated above redundancy is often a *consequence* of obsolescence. The causes of obsolescence and redundancy may be slightly different but the effects are the same: a building is under-utilized or becomes totally unused. In the context of adapting existing properties, however, the term 'redundant building' rather than 'obsolete building' is more usually used nowadays to describe a facility requiring or being ripe for major adaptation.

The distinction between obsolescence and redundancy can be further illustrated using an example. Take a 'traditional' but very uncommon form of wall construction in Britain such as wattle and daub. This is a relatively primitive method of building, which consists of mud plaster on a weave of reed or cane between timber framing. In a modern context it is classed as obsolete but not necessary redundant. It is obsolete because it no longer can meet current performance requirements such as thermal efficiency, structural capacity, weather-tightness and durability. A building comprising this form of construction, however, may still be required for conservation purposes or even for use as a display in a museum. In other words, it may not be completely redundant.

Table 1.9 Contrast between building obsolescence and redundancy

Factor	*Nature*	*Examples*	*Effects*
Obsolescence	Antiquated. Useless. Old-fashioned. Out of date. Outmoded. No longer fit for purpose. Mainly supply-determined. Essentially qualitative. Not easily controlled. Process is random. Cannot be forecast.	Buildings for special or unique uses – with short spans, low ceilings/roofs, small openings, and unusual or awkward configuration, making difficult to use or reuse in their present condition (e.g. telephone exchanges – especially the multi-storey concrete-framed versions built in 1970s).	Dilapidation. Partial disuse, eventually leading to redundancy and complete vacancy. Value decline that is not caused directly by use or passage of time (Ashworth, 1999).
Redundancy	Superfluous or surplus to requirements. Mainly demand-determined. Essentially quantitative.	Buildings no longer needed or are excess to requirements (e.g. corn mills, barns).	Complete vacancy. Prone to neglect and dilapidation. Vulnerable to vandalism and squatting.

Deterioration and obsolescence

The deterioration hypothesis

Defects in buildings have important implications for the adaptation of property. The systematic investigation and treatment of these is the discipline known as Building Pathology (CIB, 1993; Watt, 1999). This also involves the study and prediction of building life and durability. It is also concerned with issues such as changeability, maintenance cycles, service life, deterioration mechanisms, and failure rates.

The deterioration hypothesis posits that the condition of a building tends to worsen with age if left unattended. In one sense it is an example of the Second Law of Thermodynamics, which states that all processes manifest a tendency toward decay and disintegration, with a net increase in what is called the entropy, or state of randomness or disorder, of the system. Thus in the absence of any separate, organizing force, there is a tendency for things to drift in the direction of greater disorder, or greater entropy.

The Second Law of Thermodynamics is applicable to closed (rather than open) energy systems, which tend to fall into disorder (see Figure 1.12). (The First Law of Thermodynamics posits that matter can neither be created nor destroyed, only converted.)

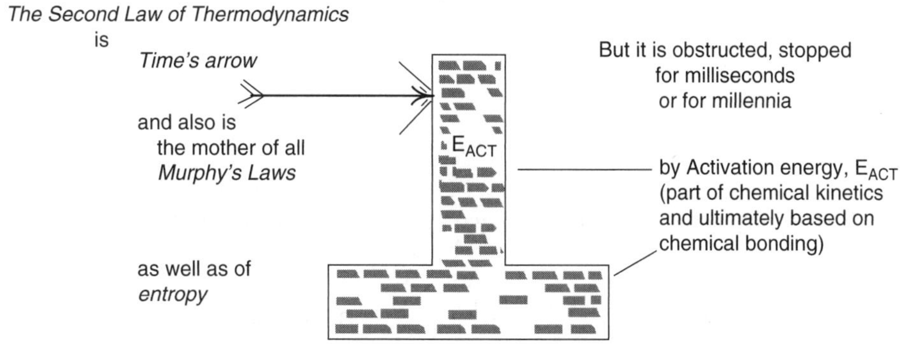

Figure 1.12 The effect of the Second Law of Thermodynamics (Source: F. L. Lambert, http://www.secondlaw.com/) (accessed June 2005)

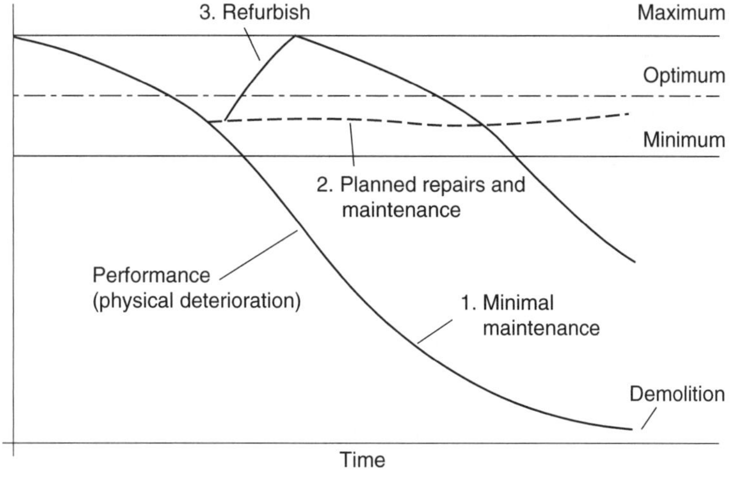

Figure 1.13 The impact of maintenance and adaptation on deterioration (Ashworth, 1999)

In the context of buildings the activation energy described in Figure 1.13 is maintenance and adaptation. These are the two primary interventions that are aimed at combating deterioration and depreciation as well as obsolescence (see Figures 1.14 and 1.15).

Deterioration usually goes hand in hand with obsolescence. However, the latter is less predictable and more difficult to control than the former, even through adaptation. Adaptation of buildings will always be needed to combat deterioration as well as obsolescence. Deterioration is inevitable as an ageing process. It is mainly a function of time and use, but can be controlled to a certain degree through maintenance and adaptation (Ashworth, 1999).

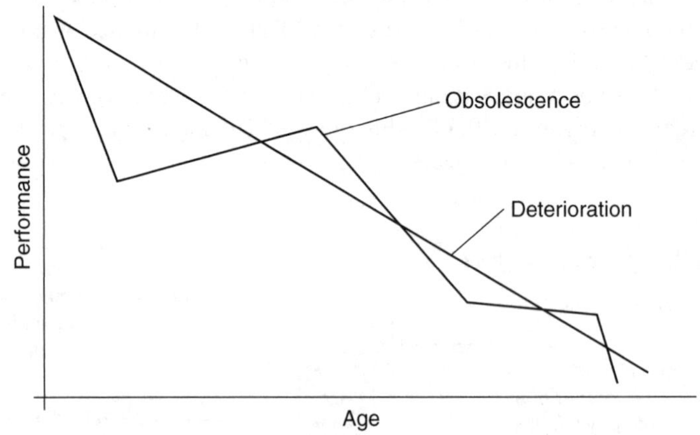

Figure 1.14 Relationship between deterioration and obsolescence (Ashworth, 1999)

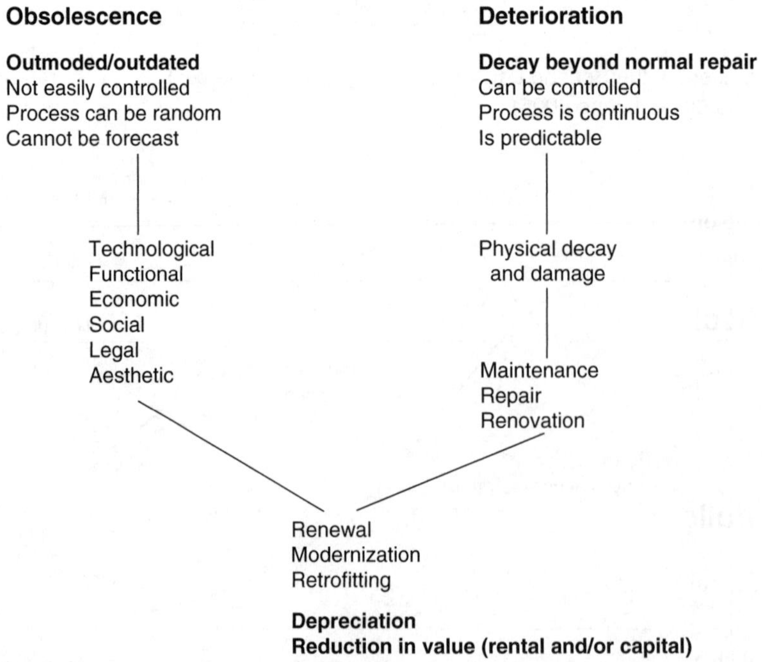

Figure 1.15 Obsolescence, deterioration and depreciation (adapted from Ashworth, 1999)

Ongoing rapid economic, demographic and technological changes are likely to increase the risk of obsolescence of many buildings. Designing flexible buildings is required to combat the early onset of this problem in new build as well as adaptation schemes.

The hypothesis that the physical condition of a building deteriorates with age (Nutt et al., 1976) is still applicable therefore. In relation to some building materials, of course, there are some specific exceptions to this rule. Good quality concrete, for example, tends to harden over time – but this is usually offset by carbonation, which causes corrosion of the steel reinforcement (see Chapter 7). Copper gains a green oxide coating known as verdigris, which acts as an attractive, protective patina. Carbon dioxide performs a similar function on sheet lead by forming a grey patina on its outer surface.

Deleterious physical change, however, is more common than other forms of change to materials and components. It can occur as a result of the influences of the elements, causing the appearance and the fabric of a building to degrade. Typically, transformations in the appearance of a building take the form of soiling of the external envelope. The accumulation of soot and grime on the facade of old buildings is an extreme example of this influence. Wear and tear as well as user abuse and vandalism also transform the features of a building – usually for the worse.

Deterioration of a building's structure and fabric is generally associated with one of three principal causes: dampness, bio-decay and movement (Addleson and Rice, 1992). All buildings to a greater or lesser degree are exposed to these three influences. Within each of these three causes, however, are numerous agencies or sources, such as condensation (i.e. dampness), dry rot (i.e. bio-decay) and subsidence (i.e. movement). As pointed out by Addleson (1992), the causes are few but the agencies (or sources) are many. This can often make the correct diagnosis difficult to achieve.

Primary deterioration influences

Thus, from the day it is completed, the performance of a building's original structure and fabric will be eroded or undermined by two primary influences:

1. *Climatic agencies (environmental influences)*: These particularly affect the exterior of the building, but ultimately impinge on its interior. For example, roof leaks can cause considerable damage and inconvenience within a building. Rainwater can easily ruin stock and equipment as well as damage furniture and finishings. It can also compromise the quality of the building's indoor environment.
2. *User activities (functional influences)*: These primarily affect the interior of the building, but can impinge on its exterior. Lack of maintenance, for example, will inevitably have an adverse affect on a property's external condition. This is most noticeable when vegetation grows profusely in uncleaned roof gutters. The gutters may clog after a few months and cause them to overflow, resulting in water and organic staining on the face of the building. Moreover, this increases the risk of fungal attack in vulnerable timbers embedded in the masonry nearby (Palfreyman and Urquhart, 2002).

Types of obsolescence

Determining when a building is obsolete is not always easy. It depends on the perspective taken in establishing the measure of a building's usefulness. There are essentially six main perspectives or types of obsolescence and these are compared in Table 1.10.

Causes of building obsolescence

In summary, then, buildings become obsolete through a combination of some if not all of the following influences:

- Technological change making existing building unsuitable for modern production processes. Advances in micro-processors and other forms of miniaturization of information technology, for example,

Table 1.10 Main types of building obsolescence and their influences

Type of obsolescence	*Criteria*	*Factors*
Economic (including Financial and Site)	Cost-effectiveness. Rate of return. Depreciation.	Rental income levels. Capital value versus redevelopment value. Oversupply or drop in demand.
Functional (including Locational)	Fulfilment of purpose. Degree of use. Technological adequacy.	Decreased utility. Inadequacy. Incapacity. Errors and omissions in the building's layout or form. Technical advances.
Physical (including Environmental)	Structural stability. Weather-tightness. Overall performance.	Structural failure. Physical deterioration. Dilapidation. Urban blight.
Social (including Cultural)	Satisfaction of human needs. Cultural requirements.	Demographic trends and shifts. Changes in taste and style. Changes in expectancy levels.
Legal (including Control)	Compliance with statutory requirements.	Changes in legislation or regulations. Changes in planning policies. Existing adverse legislation. Nuisances and hazards – dangerous buildings.
Aesthetic (including Architectural)	Style of architecture is no longer fashionable.	Office building designs of the 1960s.

have hastened the obsolescence of many multi-storey telephone exchanges built in the 1960s and 1970s.

- Deterioration of the building's structure and fabric. This reduces its ability to satisfy both user needs and performance requirements. It also adversely affects the appearance and thus the status of the building.
- Changes in the industrial base of the economy, with the manufacturing sector shrinking whilst the service sector is expanding, resulting in modified working practices.
- Available useable space inadequate in terms of layout and/or size.
- Non-viability of upgrading the building to comply with building and fire regulations because of the extent of remedial as well as improvement works required. The presence of extensive deleterious materials such as asbestos, for example, would be an influencing factor in this regard (see Chapter 8).
- Inflexible building, which is incapable of easy adaptation to the same or other use because of its restrictive morphology or structural configuration (see Chapter 2).

Causes of building redundancy

Some buildings are more prone to becoming incapable of further useful life than others. Old or listed buildings, for example, are especially susceptible to redundancy. This is because they are often tight-fit or inflexible buildings that do not lend themselves easily to reuse.

Buildings can become redundant for a variety of reasons (Scottish Civic Trust, 1981). The historic over-supply of certain types of buildings is probably the key factor. Other influences such as the movement of population from large urban areas and decline of the existing manufacturing base can accelerate redundancy of both commercial and residential buildings.

There are several contemporary changes that are leading to the stock of certain types of buildings becoming redundant and thus available for conversion. For example, changes such as:

- Tele-banking making many traditional banking halls surplus to requirements.
- Reduction in the size of the armed forces in the post-Cold War era has led to the disposal of underused barracks and other Ministry of Defence properties.
- Out of town shopping centres leaving empty retail premises in the centre of urban areas.
- Microprocessing of switchgear and terminals making many large multi-storey telephone exchanges built in the 1960s and 1970s surplus to requirements.
- Falling school rolls leading to surplus educational buildings in some areas. In some popular residential areas, of course, increasing school rolls may result in an expansion of an existing school complex (i.e. an extension).
- Decline of traditional manufacturing industry leading to factories and their associated buildings becoming empty.

Responses to building obsolescence and redundancy

Minimizing obsolescence and redundancy

The consequences of current and future obsolescence and redundancy need to be borne in mind in any adaptation proposal. These can be divided into three main groups: economic, technical and functional. Economic obsolescence occurs because maintenance has become unreasonably costly or disruptive, and when acceptable (cheaper) alternatives to maintenance are available (BS 7543, 1992). Depreciation of a built asset's capital/rental value is the primary economic consequence. Technical obsolescence implies that the performance of the building is deficient or otherwise lacking, leading to dilapidation and, if left unattended, dereliction. Functionally, a building usually becomes underused because of obsolescence. Complete vacancy, however, is the most noticeable effect of building redundancy.

Obsolescence and redundancy sometimes require different responses. Refurbishment and upgrading are the major strategies to counter obsolescence (BS 7543, 1992). Thus proactive adaptations such as alterations and modernization schemes can delay if not postpone obsolescence indefinitely. Redundancy, on the other hand, usually requires an adaptive reuse of the building, which tends to involve reactive as well as proactive work. This is because it is normally very difficult to anticipate precisely when and why a building is becoming redundant.

In any event, obsolescence and redundancy are either correctable or non-correctable, as indicated below.

Correctable obsolescence and redundancy

This is primarily achieved through adapting the property and remedying its physical problems such as:

- poor acoustic, fire and/or thermal performance;
- inadequate structural or spatial capacity;
- construction defects (e.g. dampness, fungal attack, etc.);
- inadequate or relatively inflexible layout;
- poor amenities or facilities;
- inadequate services (e.g. insufficient capacity).

Non-correctable obsolescence and redundancy

This type of obsolescence and redundancy is more likely to encourage redevelopment of the property. However, there may be legal restraints preventing its demolition. This situation occurs mainly as a result of economic and environmental influences such as:

- poor location (e.g. remote or inaccessible);
- inadequate building morphology (e.g. very inflexible layout or restrictive floor-ceiling heights);
- inadequate site (e.g. awkward or tight site access, irregular topography or contaminated land);
- unsatisfactory microclimate (e.g. severe exposure or pollution).

Obsolescence or redundancy may provide the impetus to consider either the adaptive reuse of the building or redevelopment of the site.

Future use of the building

Extending the useful life of the building and reducing component maintenance and replacements is a key criterion of adaptation. It also contributes to achieving sustainable development and preservation of scarce resources (BS 7543, 1992).

Vacant buildings

Reasons why buildings become vacant

Buildings or parts of buildings become vacant as a result of varied and complex influences. As indicated in the previous section, obsolescence and redundancy are the main triggers for this outcome. The main reasons for building vacancy, however, can be summarized as follows:

- Occupants moving to larger or better premises (e.g. because of obsolescence in the existing building and/or competition from other more attractive or more suitable properties).
- Cessation of activities or use because of economic influences (e.g. decline in business resulting in closure of property and redundancies – both in human and building terms).
- Seasonal activities resulting in only occasional or short-term use of the building (e.g. because of downsizing of business because of a slump in the market).
- Lack of new tenants or owners of empty buildings (e.g. because of better alternatives or a slump in demand for particular properties owing to market influences).
- Ignorance on the part of some vacant property owners: not being aware of the existence of some units in a large stock, particularly where property records are not kept; they may not appreciate the business case for bringing their vacant premises back into use; or the owner may not be concerned that the property is vacant.

Building vacancy can be divided into two main forms: temporary vacancy (0–5 years) and long-term vacancy (over 5 years). Temporary vacancy can occur as a result of short-term influences, such as decanting of housing stock during rehabilitation. It may also arise due to delays in obtaining a new tenant for a leased building or in obtaining a suitable purchaser for a vacant property.

Long-term vacancy is more likely to be caused by obsolescence or redundancy. It is therefore more difficult to resolve. Depreciation and dereliction are the inevitable consequences of long-term vacancy. This is because empty properties are more susceptible to the adverse effects of vandalism, squatting and theft. In addition, owing to little or no maintenance empty buildings tend to deteriorate at a much faster rate than occupied buildings. The lack of heating and cleaning allows increases in the levels of humidity and dust inside an empty building, both of which can lead to accelerated deterioration of its fabric and structure.

According to the ODPM (2003a) 'the longer a property is empty, the greater the likelihood that its physical condition will deteriorate, making it more difficult to bring back into occupation at a later date and increasing the likelihood that it will impact on neighbouring properties'.

In many cases of course the most obvious candidates for adaptation are vacant buildings. Vacancy is an obvious indicator of potential or actual redundancy. Not all buildings that are empty, however, are in this position because of neglect or persistent under-utilization. The vacancy of many properties is actually transitional, because they are in the process of being demolished, sold or adapted.

Vacancy of residential buildings

For the general public the effects of building vacancy are naturally more acute in relation to housing. The latest figures on vacant dwellings in Britain are as follows:

- England and Wales: 750 000 vacant dwellings, 3.4 per cent of the total stock (ODPM, 2003).
- Scotland: 109 000 vacant dwellings, 5 per cent of the total stock (Scottish Homes, 1997).

Part of the above findings was derived from the 1996 English and Scottish house condition surveys. These are stratified sample surveys of the housing stock carried out usually every 5 years. The figures indicate that about a third of the entire vacant stock are not being marketed and probably indicate long-term empty dwellings.

No cogent explanation has been advanced to the writer's knowledge as to why the proportion of vacant dwellings is higher in Scotland than in England. It may, for example, be a combination of poor quality stock in less-attractive areas. Another possible reason is that housing demand is generally lower in Scotland than elsewhere in the UK. Edinburgh, though, is one exception to this. It is one of the most expensive locations for property in Britain.

Vacancy of non-residential buildings

There is no comparable mass condition survey programme for non-residential properties. This is most likely because of the diversity of size and type of individual institutional, industrial, commercial, and agricultural buildings makes it difficult to both determine and present a total stock figure for these properties.

According to the Office of the Deputy Prime Minister (ODPM, 2003b):

> There were about 1.9 million business addresses in England and Wales in 1994, each on average occupying about 650 m^2 of building floor area. Total floor area is increasing at about 2 per cent per year. Building turnover (demolition and replacement) is somewhat higher than for dwellings, so the potential for energy performance improvement year on year for the stock as a whole is somewhat better. In 2000, the emissions of carbon from these buildings amounted to around 21 million tonnes, which is about 20 per cent of the national total. Offices, shops, hotels, industrial buildings and schools are the largest contributors.

The vacancy and under-utilization of many commercial buildings is indicative of an oversupply problem. Property analysts have shown that reckless over-investment in property was a major factor in several macro-economic crises. For example, several major financial crises such as the Savings and Loans scandal in the USA in the early 1980s; the UK property crash in 1992; and, the global economic crisis of the mid-1990s precipitated by the monetary problems in Asia. According to some economists all of these financial slumps share a common cause: they spring from the momentum of markets outstripping economic realities. In other words, these economic crises all have their roots in prices over-reaching values.

As a result of these financial crises of the 1990s, funding for property investments is now on a different basis. Investors are more cautious. Widespread, bold speculative developments in property are less common. The assumption that property will always be a good investment because of demand is no longer tenable. Instead, property development is usually restricted now to schemes involving either pre-letting or bespoke developments.

Upper-floor vacancy

A common feature of many town centres is the widespread disuse or under-use of upper-floor space above shops and other ground level commercial properties in some inner city streets (DOE, 1997). This problem can manifest itself in inner urban areas even where the demand for retail and office space is high. The result is that frequently large sections of three-or four-storey terraced blocks have few or no occupiers in the upper floors. Unlike other forms of partial under-utilization, upper-floor vacancy is often visible has a detrimental effects on the quality of a streetscape. In such cases windows are left unclean, with little or no curtains or, worse still, boarded up.

The under-use of upper floors is problematic not only in economic terms because of the reduction in rental income. It also adversely affects the quality of the streetscape, especially in historic town centres. Boarded up or unlit windows above ground-floor premises detract from the appearance of the property. In addition, roof leaks or other sources of dampness may go undetected leading to deterioration of timbers, etc.

Moreover, empty parts of properties are susceptible to vandalism and ingress by squatters or vermin. They may also provide an easier access route for thieves to the occupied premises below.

Upper-floor vacancy can arise for a variety of reasons, the most important of which are as follows:

Over-stringent regulations

The under-use of upper-floor space often relates to the difficulty and cost of satisfying Building and Fire Regulations and Environmental Health requirements (Scottish Civic Trust, 1981). These technical exigencies can inhibit the use of otherwise scarce commercial space. This is because, inevitably, mixed uses are often involved in such circumstances. Thus strict requirements for fire resistance of separating floors and walls, as well as adequate means of escape, make it technically very difficult to adapt vacant upper-floor properties.

Poor physical amenity

The accessibility of the upper floors and rear areas is often restricted in high street situations. This makes it difficult to provide separate access for alternative users. The lack of parking and delivery space at the rear is another restriction that inhibits the demand for upper-floor space in town centres.

Multiple-ownership

This can create its own set of problems such as sharing of responsibilities for cleaning and repairs. Tracing the owner of empty premises above street level units is sometimes difficult.

Commercial pressures

The desire of street level traders to maximize their display frontage can encourage them to remove or minimize separate front access to the upper floors. Many property owners are unwilling to become involved in the management of premises in multi-use, given the problems of dealing with several tenants in the same block.

Rent Act disincentives

The security of tenure provisions in the 1977 Rent Act made some landlords reluctant to rent out their upper-floor premises. This situation has of course eased with the introduction in the 1980s of Assured Tenancies and other short tenancy agreements.

Reducing the number of empty properties

Generally

Empty buildings are a wasted resource. They do not yield any rental income or property taxes. They are more prone to vandalism, occupancy by squatters, premature deterioration and intruder damage than occupied buildings. They take up valuable urban space, which could be redeveloped for more appropriate uses. In housing they are often a symptom of poverty and social depravation. In non-residential buildings they are indicative of blight and economic recession.

In some cases the only effective remedy is to demolish such buildings. This is particularly the case with housing. Long-term neglect, anti-social behaviour, and deprivation affect dwellings as well as the inhabitants in run-down housing estates. Even the most prestigious of rehabilitation schemes to such estates will not work unless there are adequate social infrastructures in place as well – such as a vibrant community centre, adequate training and education programmes and leisure facilities.

The criteria for retaining or removing vacant buildings are likely to comprise:

- the building's state of dilapidation;
- its overall performance;
- the architectural or historic importance of the property;
- the likely future demand for the property;
- the impact on surrounding properties (access as well as aesthetic considerations);
- the overall benefit to the community.

The Urban Task Force (1999) highlighted the environmental and economic burdens imposed by vacant properties. They stated that every local authority should be required to produce and maintain a comprehensive empty property strategy for their area. Identifying the extent of the problem in relation to all property market sectors (such as commercial buildings, private housing, etc.) would be the first step in addressing this problem.

Their main recommendation in this regard is to:

> Give local authorities a statutory duty to maintain an empty property strategy that sets clear targets for reducing levels of vacant stock. There should be firm commitments to take action against owners who refuse to sell their properties or restore them to beneficial use.

Upper-floor vacancy

There is no quick and easy solution to the problem of upper-floor vacancy. The ongoing nature of this dilemma is a testimony to its intractability in many inner urban areas. A suitable remedy requires a combination of some if not all of the following:

- Local authority input: As indicated above every local authority should have a statutory duty to produce and maintain a comprehensive empty property strategy for their area. A coherent policy for identifying vulnerable properties should be adopted to encourage owners to utilize their stock of empty buildings.
- Financial incentives: Grant aid can be made available for the proposed conversion of upper-floor space. These financial incentives can help to minimize the costs imposed by the requirements of the Building and Fire Regulations.

- Flexible planning policies: Local planning authorities can enhance the reuse of these empty storeys by encouraging mixed uses.
- Use and condition surveys: Before any conversion or other adaptation scheme is proposed a survey of the property to establish its state of repair and reuse potential is needed.

Options for property owners

Generally

Adaptation is only one of several options open to owners of a building. This is particularly the case where the property is empty or under-utilized, or when a long-term strategy for the property is being formulated to correct obsolescence or redundancy. Before the decision to adapt a building is made, therefore, clients and their professional advisers will consider one or more of the options listed in Table 1.11, each of which has its benefits and drawbacks.

An economic appraisal, as outlined in Chapter 2, can be used to determine which one of several options ought to be selected.

Provisions for future adaptations

To maximize a building's sustainability it is important to allow for future adaptations in its original design. This can be achieved by considering the following:

- Location and orientation of the original building: Lateral extensions will be heavily determined by the position of boundaries in relation to the existing building.
- Space around the building: Access (e.g. footpath and carriageway) and car parking space for the new or upgraded use.
- Selection and availability of products: Matching the new with the old to achieve an architecturally appropriate design.
- Foundation and basement design: Robust substructure capable of taking additional loadings involved when adding a vertical or lateral extension.
- Location and capacity of services: Suitability of existing drainage and supply pipelines to cope with additional or modified demand.
- Means of access and egress in respect of the site and the building: Ease of entry and exit for contractors when undertaking major works to the building to minimize disruption and inconvenience.
- Size and layout of the structural grid: This will impact on the design and construction of the extension.
- Legal restraints – planning, easements, covenants: May restrict extent and form of external changes – particularly if the building is listed or in a conservation area.

Mothballing

Mothballing is the process of deliberately vacating and closing a building, and protecting it until such time as the property can be brought back into beneficial use. As indicated in Table 1.12 it may be worthwhile to mothball a redundant building. The main purpose behind the somewhat drastic action of closing up a building temporarily is to protect it from the elements as well as secure it against pests, squatters and vandals.

Table 1.11 Options for property owners (adapted from Nutt, c. 1993)

Option	*Benefits*	*Drawbacks*
1. Do nothing but still occupy	No additional outlay in the short term. No immediate adverse consequences. Other options still open.	Defers problems to a later date, when they may become even greater. Maintenance backlog likely to increase at a faster rate. Existing deficiencies not addressed.
2. Maintain in current use	Preserves the asset. Sustains its use – in the short term. Ensures its ongoing service life.	Continues to require financial support for maintenance etc. Still liable for property taxes.
3. Mothball	Reduces property tax liabilities. Minimizes running costs such as cleaning, heating and lighting.	Property in put into temporary disuse – ideally less than a year. Costly to keep safe and secure. Vulnerable to vandalism and squatting. Dust and dirt accumulation. Higher humidity levels internally – resulting in greater incidence of dampness in the building. No rental income generated. Thus, this option may not be viable.
4. Redevelop	Provides a new building tailored to meet the needs of the user.	Takes time – resulting in delay in the availability of the facility. In the short term, this is the most expensive option.
5. Dispose	Realizes asset/site value. Reduces management and operating cost burdens.	Loss of a prestigious or potentially useful asset. Time-consuming – property is a relatively illiquid asset. Advertizing and professional fees to pay.
6. Market	Finds a suitable new tenant/owner. May ensure ongoing beneficial use of the property.	Requires funding for advertizing; may require investment. May require time to realize a suitable buyer.
7. Adapt	Enhances the physical and economic characteristics of the building. Retards deterioration. Postpones obsolescence – perhaps indefinitely. Reduces the likelihood of redundancy. Sustains the building's long term beneficial use.	Disruptive and may be expensive. Extended life is unlikely to be as great as new building. Upgraded performance cannot wholly match that of a new building.

There may be several reasons why mothballing is the only sensible short-medium term option. The building may not be immediately required. Time to arrange or plan its adaptation or future may be needed. Raising the necessary finance, including any grants, for the building's conservation, rehabilitation or restoration may take months or years.

Mothballing therefore is usually a last resort option when wishing to retain a redundant or obsolete building. It should not be considered as a long-term measure – ideally, 1 or 2 years at most.

Table 1.12 Categories of design life for buildings (Adapted from BS 7543, 1992)

Category	*Description*	*Building life category*	*Examples*
1	Temporary	Agreed period up to 10 years	Exhibition stands. Non-permanent site huts and temporary exhibition buildings.
2	Short life	Minimum period 10 years	Office internal refurbishment. Refurbishments to restaurants and retail units (involving mainly internal work). Temporary classrooms; buildings for short life industrial processes and retailing.
3	Medium life	Minimum period 30 years	Housing refurbishment (mainly external with some internal); most industrial buildings (again, primarily externally).
4	Normal life	Minimum period 60 years	High quality refurbishment of public buildings; new housing; new health and educational buildings.
5	Long life	Minimum period 120 years	Civic and other high quality buildings.

Notes
- Specific periods may be determined for particular buildings in any of categories 2–5, provided they do not exceed the period suggested for the next category below on the table; for example, many retail and warehouse buildings are designed to have a service life of 20 years.
- Buildings may include replaceable and maintainable components (see Table 3 in BS 7543, 1992).

Based on Park (1993) the nine stages involved in the process of mothballing a building are as follows:

- Documentation
  - Document the architectural and historical significance of the building. Undertake a record survey of the property in line with standard conservation practice.
  - Prepare a condition assessment of the building. Undertake a condition survey to determine the extent of repairs required.
- Stabilization
  - Structurally stabilize the building, based on professional condition assessment. Provide appropriate temporary/permanent support to vulnerable or defective structural elements.
  - Exterminate or control pests, including termites and rodents.
  - Protect the exterior from moisture penetration. Ensure any roof leaks are fixed and any vulnerable openings are closed off.
- Mothballing
  - Secure the building and its component features to reduce vandalism or break-ins. Install appropriately secured boarding over window and door openings. Painted mild steel or heavy gauge polythene panels reinforced with mesh may be considered more appropriate than plywood boarding over the fenestration.
  - Provide adequate ventilation to the interior. Keep some inaccessible windows slightly ajar provided the gaps closed with a suitable mesh to prevent ingress to the premises by birds.
  - Secure or modify utilities and mechanical systems. This should include turning off all supplies and taking electricity/water/gas metre readings. Any water fed systems should be drained down to avoid the risk of burst pipes from frost action in the unheated building.
  - Develop and implement a maintenance and monitoring plan for protection. Regular inspections of the premises to check for break-ins and other damage caused by vandals, storms, pests, etc.

If handled carelessly or insensitively, however, mothballing a building can give the owner a nasty shock upon reopening the property. Mothballing can easily have resulted in catastrophic damage to vulnerable components in the building's infrastructure. One of the inevitable primary problems with mothballed buildings is not having a constant presence of people for security, fire protection, weather-tightness, and the prevention of dust and mould contamination.

The decision to adapt a building

General requirements

Professional advisors such as architects and surveyors can make a significant contribution to the decision-making process in an adaptation scheme. It is important, however, that they are aware of the needs of the client at all times. For most users and owners of buildings, the following five key requirements are usually critical for a sustainable property:

1. the building should have a long life (i.e. be **durable** – to resist wear and tear);
2. it should have a loose fit (i.e. be **adaptable** – to accommodate future changes);
3. it should have low-energy consumption (i.e. be **thermally efficient** – low running costs);
4. it should be wind and watertight (i.e. be **weatherproof**);
5. it should provide a secure and healthy indoor environment (i.e. be **comfortable**).

Durability

For residential buildings an extended life span of at least 30 years following adaptation is to be expected. For non-residential buildings, where response to market influences is significant, the rate of adaptation or changeability will usually be much higher. This tends to lead to lower extended life span expectations.

Adaptability

Ideally the building should be capable of accommodating future change to permit modification or change to a different use. As the demand for building space varies properties will need to be able to respond to such changes to avoid becoming redundant.

Energy efficiency

Since the oil crisis in the early 1970s the Western world has been waking up to the importance of conserving non-renewable fuels. One of the main ways to do this is to improve the insulation and heating system within a building. This, along with other sustainability measures, is addressed more fully in Chapter 10.

Weather-tightness

A critical success factor of any newly erected or adapted building is that it should not leak. All too often, though, new or refurbished buildings let in rainwater or suffer burst service pipes. Not only does such leakage cause water damage to the fabric, finishings and services, it also can disrupt the use of the building.

Leaks in the fabric are usually attributed to badly installed flashings, cavity trays or damp proof courses (Endean, 1995). Chimneys and other projections, abutments, openings are the parts of a building most prone to water infiltration. Poor detailing of these and other vulnerable locations can increase the risk of moisture penetration, through rainwater or snow.

Burst water pipes can occur as a result of badly fitted joints, defective valves or frost damage to the pipework. The volume of water discharged through this type of leak can be considerable, particularly if it is undetected for a period or time.

Comfortable

Freedom from problems such as Sick Building Syndrome (SBS) is becoming increasingly important in some buildings, particularly commercial ones. There are also problems associated with glare, lighting levels, off-gassing of solvents, as well as indoor air quality (see Chapter 8). Adopting a sustainable construction policy in the adaptation design brief can help to prevent such problems (see below).

Influences on adaptation versus redevelopment

Background

There are many factors that can influence whether or not a property should be adapted or redeveloped. Again, of course, the needs and intentions of the property owner or user will be paramount. However, other influences will come into play as well, and these can be summarized as follows:

Alternatives

The quality of alternative options will have a bearing on the decision to adapt or redevelop the property. Demand factors as well as supply side factors will also influence the decision-making process.

Location of existing building

The location of the building is one of the main issues affecting its renewal. City centre sites with easy access are the most convenient in this regard. However, access restrictions or awkward entrances may inhibit a building's adaptation. Moreover, urban renewal considerations could encourage redevelopment instead. Even if an existing building is ripe for refurbishment, conversion or extension, its location may encourage a new development on the site.

Value of the existing site

The building may be situated in a site that is highly attractive. As a result the site may have a value exceeding that of its building. In other words, the building has reached the end of its economic life.

Building condition and morphology

The renewal of a building will be dependent upon its overall condition and suitability for major alterations. The existing layout and whether this is suitable to meet the required needs is also significant (Ashworth, 1997).

Operational factors

The degree of disruption that the proposed adaptation will entail can be an influencing factor. This will have implications for the anticipated contract time as well as the adaptation cost.

Occupiers' requirements

The degree to which the existing building can meet a variety of objectives for any prospective occupier, which include the provision for modern use and comfort, will be important (Ashworth, 1997).

Adaptation potential of building

The adapted building should be able to offer new efficiencies in terms of performance and technologies. These should be manifested in a more comfortable, attractive and accessible building.

Constructability

Constructability refers to the ease of construction attained through the optimum use of construction knowledge and expertise in planning, engineering, procurement and field operation to achieve overall project objectives (Pentagon Renovation Program, 2005).

Political influences

Government policy can have considerable influence over the use and reuse of property. This is mainly articulated via the various planning acts. However, recent legislation such as the Disability Discrimination Act 1995 is having a profound effect on the adaptation of buildings (see Chapters 8 and 9).

Statutory controls

Although building owners may prefer a new property they may be precluded from redeveloping their site because of planning controls. If the building is Listed or is in a Conservation Area approval for any demolition and redevelopment will have to be obtained from the local planning authority. In the event of a refusal an appeal to the Secretary of State for the Environment can be considered. All that this does of course is increase the time and money required to initiate development proposals. Even if permitted the planners may want to reduce the intensity of the new use of the site to a lower level than the original (e.g. less floor area or fewer parking spaces).

Cost and time considerations

The higher the rate of interest the more incentive there is to rehabilitate rather than redevelop. Generally speaking rehabilitation costs should be significantly lower than for new build. As a rule of thumb, if the rehabilitation costs are more than two-thirds of the redevelopment costs, then it is usually considered to be more viable to opt for the latter (see Chapter 2).

Accommodation

The standard of accommodation required by the client will be important. This relates to the available space and the quality of internal environment it can offer. In social housing quality can be measured by the popularity of dwellings in an estate. Rental levels can crudely measure quality of commercial premises.

Design life

The expected useful life of the adapted building is an important consideration. In the context of existing buildings, for example, adaptation should enable the building stock to last anything between 10 and 60 years depending on its category. The various categories of design life for buildings and components or assemblies, respectively were summarized in Tables 1.4 and 1.5.

Design process

The design process in adaptation is fundamentally different from that in new-build work. In the latter the process is one of synthesis rather than analysis (Gregg and Crosbie, 2001). According to CIRIA (1997) in new-build work '... The designer starts with a clean sheet and progressively builds up his design ab initio... In refurbishment work (i.e. adaptation) ... the task is more akin to detective work, with the designers endeavouring progressively to gather information on the existing asset and to develop the design of the refurbishment work as to be comparable with it.'

Adaptation design brief

Any medium to large size adaptation project should have a design brief to articulate the client's intentions for the project. Such brief sets out the basic design requirements for a project. The designer in charge (architect/surveyor/engineer) normally drafts it after full consultation with the client.

The project's design brief comprises the first part of the feasibility study. It should contain statements addressing each of the following matters:

- the purpose of the project (i.e. adapted building), its scope and content, and any necessary background information;
- a social brief indicating how and by whom the adapted building is to be used;
- the desired activities and functions, and the relationship between them;
- the constraints operating over the project (i.e. cost, planning, timing);
- special requirements (e.g. sustainability measures, height or other dimensional constraints); and in an Appendix, any Nominated Subcontractors;
- accommodation schedule, room data schedule and schedule of all surface finishes;
- any other information relating to requirements for the adapted building.

Phasing of adaptation

Background

It is not always possible to undertake the adaptation of a whole building all in one go. Budgetary restrictions as well as access problems or operational constraints may prevent the complete refurbishment of a property within a single contract. This is especially the case with very large facilities such as hospitals, hotels, schools and shopping centres.

Phasing of work, therefore, is a common feature of many medium- to large-scale adaptation projects. It is more concerned with the strategic pacing of a project and the overlapping between different activities in a block of activities (APM, 2000). In the context of the refurbishment of a commercial or industrial building phasing may allow part of it to remain occupied. This is especially beneficial in the case of leased property as such a response can ensure that it continues to generate rent during its adaptation. Similarly facilities such as hotels and shops, which rely on custom or sales for business, can still earn some income whilst they are being refurbished. For example, the refurbishment of a large hotel or factory is best achieved on a phased basis to minimize 'downtime' and subsequent loss of revenue. Careful planning and co-ordination of the work is required, however, to avoid disturbances as well as minimize safety and security problems.

In public facilities, too, phasing the work is often the only feasible way of undertaking adaptation schemes to minimize disruption. The use of the vacation periods is the optimum time for any adaptation work to educational complexes such as schools, colleges and universities. Campus sites particularly are best adapted by phasing (see late in this section).

Therefore phasing is influenced by the varying acceptability of declining performance and on the use of the building as dictated by the requirements of the occupier or owner. It can occur in both space and time, and in combination. The differences between these types of phasing are addressed as follows:

Temporal phasing

Generally

The frequency of adaptation depends on the needs of the occupier as well as the service life of the facility. Tables 1.12 and 1.13 list some typical timings of various forms of adaptation in relation to their design life.

The temporal phasing of adaptation works is either short term or long term.

Short-term phasing

Individual, small-scale adaptation schemes such as a shop refurbishment or a minor domestic extension take a relatively short period of time to complete. The former can be as short as a few days. The latter can take several weeks or months – depending on its size and complexity.

Such work is intermittent over the building's service life. Moreover, as with an extension some of these adaptations might be undertaken whilst the property is occupied. In many cases, though, short term phasing may result in the compete vacation of the premises undergoing adaptation – such as shop, bar or restaurant 'refits'. Larger adaptation projects such as a change of use, of course, can normally only be carried out when the building is entirely unoccupied.

See Chapter 11 for an indication of the estimated time-scales of typical adaptation schemes.

Long-term phasing

Extensive, more substantial works to an existing building, on the other hand, may last many years because of the extent of operations proposed. Budgeting and access restrictions on the degree of disturbance and the availability of capital could also extend the adaptation programme beyond the short term (i.e. say, over a year). In such cases the building or complex is usually still occupied to some degree. Examples of this would be the phasing of major extensions and refurbishment schemes to large building facilities such as shopping centres, schools, hospital complex or to an extensive estate of properties comprising many dwellings or warehouse units. In some cases such work can take up to 10 years or more to complete.

Spatial phasing

Vertical phasing

Multi-storey buildings can be refurbished on a floor-by-floor basis. This is more easily achieved where a single user occupies the whole building. One or more floors can be decanted at a time to reduce the

Table 1.13 Categories of design life for components or assemblies (BS 7543, 1992)

Category	*Description*	*Life*	*Typical examples*
1	Replaceable	Shorter life than building life and replacement can be envisaged at design stage.	Most floor finishes and service installation components.
2	Maintainable	Will last, with periodic treatment, for the life of the building.	Most external claddings, doors and windows.
3	Lifelong	Will last for the life of the building.	Foundations and main structural elements.

'downtime'. 'Downtime' is the period when a facility is not in use and therefore not earning rent or contributing to production.

Phasing the work on a floor-by-floor basis can also minimize the need to close off or vacate the entire building whilst the adaptation work is proceeding. Noise, demolition, dust and access problems between floors, however, can pose major risks with this type of phasing (see Chapter 11).

Vertical phasing can be done on a top-down or bottom up approach. The desired option selected is the one that keeps disruption and disturbance as well as costs to a minimum. If a top down approach is taken, separate vertical access should be provided to avoid excessive mingling of operatives and occupiers. Partial disconnection of the electrical, heating and plumbing services, however, may be impractical with this type of phasing.

Horizontal phasing

It may be prudent to undertake adaptation work to only part of a large low-rise building. Properties such as hospitals and schools covering a large area and comprising several wings may be too large and awkward to adapt in one go. Each wing of the property could be converted or refurbished individually as finance and convenience allows. Any proposed extensions can also be constructed on the same basis.

For large individual buildings horizontal phasing can be done on an east–west or north–south approach. Such a building can effectively be split into two or more parts at the most convenient positions. One half or part would be decanted to another section of the building or to another property, whilst the remaining half or part would still be in use.

Figure 1.16 shows the east–west phasing of the major refurbishment of a large public building in Edinburgh carried out between 2001 and 2004. In the event the building management team in charge of the project decided on a hybrid or incremental phasing scheme for this facility. This involved tackling the most pressing works, which were in the basement, first followed by an east and west phasing, all over a 3-year period.

Incremental phasing

Incremental refurbishment schemes comprise a combination of spatial and temporal phasing. Hospitals, educational properties and other large, complex or multiple-block facilities are best suited to this type of phasing. For example, the phasing of the refurbishment of school buildings can be organized to suit school use and holidays. The Easter and summer school breaks are particularly suitable for accommodating this type of work but it requires very careful design and tight construction management to minimize any disruption to the school activities.

Financial advantages can also be obtained by using an incremental refurbishment approach. It can, for example, tie in with the client's preferred cash flow rate. Monies allocated for the maintenance budget can contribute towards the refurbishment costs.

The billion dollar 'renovation' of the Pentagon building is probably the greatest example of this type of phasing in recent years in terms of scale, cost and time taken. Built in the early 1940s, the original building had an initial service life of at least 50 years. The US Department of Defense's Pentagon Renovation and Construction Program Office is organizing the current renovation project, which was originally scheduled to take 15 years and extend the building's service life by another 50 years (Pentagon Renovation Program, 2005).

The renovation project, the total cost of which will exceed a 1.2 billion dollars, is split into a number of phases related to the complex's fives wedges. The work to wedge 1 began in 1998 and involved structural demolition and the abatement of hazardous materials (in particular asbestos). Completed in March 2001, it was followed by the installation of new utilities and the build-out of tenant areas.

Option 1 – East–West phasing

	Advantages	Disadvantages
Services	• Heating can be dealt with vertically, i.e. complete removal of wet system throughout all floors at one time. • Electrical systems split into E and W risers giving a logical sequence of work. • Air conditioning can be installed easily in E–W spilt. • Works well in three phases.	
Staff	• Minimal number remaining at first floor could be decanted permanently. • Relatively small number to decant in first phase of ground floor. • Several functions could be decanted only once to an upgraded permanent location.	
Archives	• On completion of phase 1 works a large number of storage rooms come on stream in upgraded state requiring no further work – giving considerable advantages for both permanent storage and further decanting. • On completion of basement work a number of rooms become available for decanting for first phase.	• Need to move some archives at first floor to allow decanting of phase 1 staff.
General		

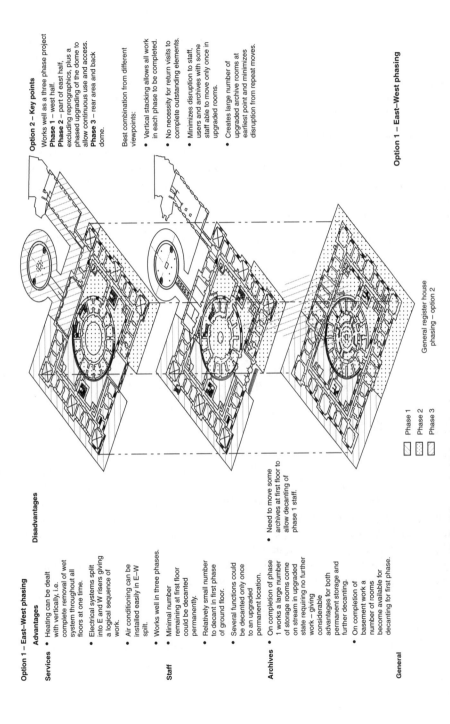

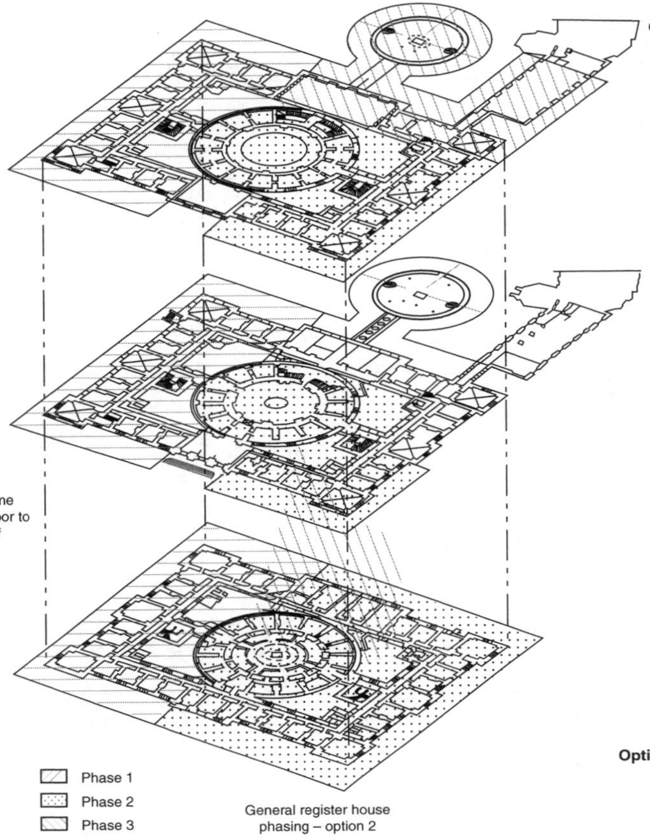

Option 2 – Key points

Works well as a three phase project
Phase 1 – west half.
Phase 2 – part of east half, excluding reprographics, plus a phased upgrading of the dome to allow continuous use and access.
Phase 3 – rear area and back dome.

Best combination from different viewpoints:

- Vertical stacking allows all work in each phase to be completed.
- No necessity for return visits to complete outstanding elements.
- Minimizes disruption to staff, users and archives with some staff able to move only once in upgraded rooms.
- Creates large number of upgraded archive rooms at earliest point and minimizes disruption from repeat moves.

Option 1 – East–West phasing

Figure 1.16 Phasing of a major public building refurbishment scheme

According to the Pentagon Renovation Program (2005):

> The renovation work involves the demolition and removal of all partitions, ceilings, floor finishes, mechanical, electrical, plumbing, fire protection, and communications systems. The basic structural system, as well as the stairwells and their enclosing walls, will remain. All electrical, mechanical, and plumbing systems will be replaced and a modernized telecommunication back-bone infrastructure will be installed. Utility connections will be made through the new Center Courtyard Utilities Tunnel without affecting the rest of the building…
>
> The project brings all remaining un-renovated areas of the building into compliance with modern building, life safety, ADA and fire codes. Work includes removal of all hazardous materials, replacement of all building systems, addition of new elevators and escalators to improve vertical circulation, and installation of new security and telecommunications systems. Renovated spaces will be modern, efficient, and flexible.

The other four wedges (2–5) are also being given a complete slab-to-slab reconstruction of the space. Replacement of all services (electrical, mechanical and plumbing systems) will take place in accordance with 'the new design and a modernized telecommunication back-bone infrastructure will be installed' (Pentagon Renovation Project, 2005). Prompted by the 9/11 attacks, the latest phase of the project has been underway since latter half of 2001. It is currently on an accelerated schedule for completion in December 2010, 4 years sooner than originally planned.

> The time savings on the project are being achieved through the use of a method of construction called SIPS (short interval production schedule). According to the Pentagon Renovation Program (2005):
>
> SIPS techniques have been traditionally used to construct buildings that require a great deal of repetitive activities, such as high rise offices buildings, apartments and hotels. It brings an assembly line approach to construction. With over 4.5 million square feet of office space remaining to undergo renovation, the Pentagon was a viable candidate for this technique. Each wedge will be divided areas of approximately 10,000 square feet. Each individual trade is allowed five days to complete their particular task before moving on to the next 10,000 square foot area. The sprinkler mains will be installed first, followed by additional plumbing work, the HVAC installation, electrical work, etc. The first area will be complete in 26 weeks with the remaining areas following every week thereafter. The SIPS schedule was tested in the USP Lab to determine the most efficient order for each segment of construction to take place. SIPS began in the A and B-rings on the fifth floor of Wedge 2 on June 24, 2002.

Summary

The pressure for more adaptation of existing buildings will probably increase in the developed part of the world owing to the growing need for more efficient and sustainable construction. Innovations in information technology, new working practices and stricter environmental controls are all causing a significant transformation in the type and extent of demand for property generally. Existing properties also have to respond to this shift in demand, from being tight fit, purpose-built facilities to flexible, 'greener' buildings.

New build projects are likely to become more expensive and less attractive in many cases. This is because of the expense of remediating brownfield and greyfield sites as well as the general lack of available urban land suitable for building in developed countries. Property owners therefore will have more incentive to adapt their buildings rather than demolish them and redevelop the site. Statutory influences such as listing and designation of Conservation Areas along with the increasing prominence given to building conservation will also encourage more adaptation than new-build projects.

Adapting buildings is an important component of any sustainability strategy. Along with adequate maintenance it is essential for ensuring the long-term prosperity of our built assets. Moreover, adaptation entails less energy and waste than new build, and can offer social benefits by retaining familiar landmarks and giving them a new lease of life.

The growth in adaptation schemes in recent years is having a major impact on the total output of the construction industry. It is now reckoned that along with maintenance it will soon match that of new build in many developed countries.

Altering, refurbishing or rehabilitating a building is just as, if not in some cases more, complicated than constructing new ones. There tend to be more constraints operating over an existing property than with constructing a new building. For example, the spatial restrictions imposed on an existing building in a congested site cannot be easily overcome. In some instances adaptation work involving changes to the exterior is not possible without substantial alteration or demolition of some or large parts of the structure. Moreover, building investigators cannot always fully determine the condition of a building prior to adaptation. This means that there is a greater level of uncertainty and hence risk with adaptation work than with new build (see Chapter 11).

Ongoing rapid economic, demographic and technological change is likely to sustain the adaptation of buildings. Such change will also increase the risk of obsolescence of many buildings. Obsolescence is much more difficult to control than deterioration, even through adaptation. Designing flexible buildings is required to combat the early onset of this problem in new build as well as adaptation schemes (Ashworth, 1999).

Adaptation, as with any part of the construction, is also affected by political changes. Following the General Election in 2001 there was a major shake-up of several central government departments that deal with construction and property. For example, the former DETR's functions, such as fire and building regulations, housing and planning, were taken over by the newly formed Office of the Deputy Prime Minister (ODPM). The Department for Trade and Industry (DTI) now deals with general construction industry matters. Environmental and farming issues, though, are handled by the Department for the Environment, Food and Rural Affairs (DEFRA). The Department for Culture, Media and Sport (DCMS) is responsible for architectural and conservation matters.

The evolution of buildings therefore occurs though adaptation. It is a process that enables the service lives of buildings to be preserved if not extended. The various strategies for achieving this are addressed in the following Chapters.

2

Feasibility

Overview

This chapter focuses on means of ascertaining the feasibility of adapting buildings. It considers the basis for making decisions in relation to the adaptation of both residential and non-residential properties. The aim is to provide a simple evaluation mechanism to assess the feasibility of undertaking such work.

Background

Feasibility criteria

Feasibility is a critical preliminary consideration in any construction project, whether involving new build or adaptation work. In this context it is concerned with the following three primary factors:

1. Viability (economic feasibility).
2. Practicality (physical feasibility).
3. Utility (functional feasibility).

These considerations are not of course of equal weighting. Depending on the needs of the client or the circumstances of the particular project, one factor will take precedent over the others. Ultimately, however, viability is usually the deciding factor, followed by practicality and then utility.

Viability

Viability is usually the most important and influential of all the assessment criteria because ultimately any development decision is based on financial considerations. Indeed, the main reason for the adaptation of non-residential property is to maximize income or asset value (Martin and Gold, 1999). In contrast, functional and personal considerations as well as financial matters play a role in the adaptation of residential property. According to Martin and Gold (1999) the main issues that should be considered to maximize income or asset value of a property undergoing adaptation are:

- aesthetics;
- need to upgrade services;
- change of use;
- change in regulations;
- to attract new tenants/retain existing tenants;
- to increase/improve the net lettable area.

A scheme may be feasible in other senses of the word, but it may not be viable. For example, it may be technically and legally possible to extend or convert a building. The cost of undertaking the work, however, may be so prohibitive that it does not render the proposed scheme viable. This may be due to either the poor condition of the building, necessitating extremely expensive remedial works, or the extent of the adaptation works required to comply with statutory requirements is too much. Historic buildings are highly susceptible to these problems and this along with their high maintenance costs helps to explain why many of them become dilapidated. Thus, the impact of adaptation costs on scheme viability depends on the property and the needs of the client.

Essentially any assessment of an adaptation scheme's viability compares the potential value with the projected costs. So long as the anticipated worth of the building is more than the former the proposed scheme can prima face be considered as financially feasible.

Economies of both scale and sale can be achieved, for example, by adapting a single occupancy mansion into several high-quality flats. The total price of four flats in the block is likely to exceed the value of the building adapted for a single occupancy.

It is important at this juncture, though, to distinguish between cost and value (RICS, 1995b). Cost is the amount of money it takes to manufacture an artefact or asset, and includes any basic profit. Value is the worth of something in monetary terms, which is influenced by market forces. In the short-run, value is determined predominantly by demand. Supply influences on value are usually manifested more in the long-term.

A simple example involving the refurbishment of a building illustrates this point. For example, a proposed modernization scheme costing, say, £10 00 000 for an office block may not necessarily increase the property's value by that amount. More especially, this outcome applies to residential adaptations. Homeowners often have to be reminded that spending tens of thousands of pounds extending or improving their dwelling does not always increase its value by the same quantity. It is the market that ultimately determines the value of property.

Some adaptations, however, are more likely to achieve added value to the building than other interventions. Installing double-glazing in a property will in itself not necessarily increase its open market value by the cost. The payback period for this type of adaptation may be over 10 years or more. Converting the loft of a detached/semi-detached dwelling is a more effective way of increasing its value in the short term (see Chapter 6).

A good starting point in assessing the significance of the estimated costs of adaptation is to compare them with those for redevelopment. The latter, of course, in the context of an existing building, inevitably involves demolition as well as the erection of the new building. Demolition costs can add considerably to the total redevelopment cost of a project if the building being removed is of an unusual or substantial construction. For example, if it comprises pre-stressed concrete members or is of heavily reinforced in-situ concrete construction, the building will be time-consuming and costly to demolish. The extent of salvageable and hazardous materials will also impact on demolition costs. If extensive salvageable materials (such as lead, copper, etc.) are present in the building being taken down, this could potentially reduce the demolition costs. On the other hand, the presence of extensive hazardous materials (such as asbestos – see below) would probably add considerably to the cost and duration of demolition. They will in other words affect the feasibility of adaptation.

The main quantitative methods of comparing adaptation costs with new-build costs are:

- adaptation costs less than the redevelopment costs;
- adaptation costs as a percentage of the redevelopment costs;
- total adaptation costs derived from using a formula (such as Needleman's – see below) less than the redevelopment costs; and,
- discounted cash flow analysis to determine which option has the optimal net present value.

Increasing order of accuracy

One of the critical attractions of adaptation work is that it normally costs less than new build. This is because as the infra-, sub- and super-structures of a building are already in place, major elemental cost savings are achieved. Another simple comparative decision tool could be used, this time focusing solely on costs. This works on the basis that so long as the adaptation costs do not exceed redevelopment costs the former should be adopted.

The straight comparison of adaptation and redevelopment costs is, of course, a crude and potentially flawed way of determining the most viable option. Even the most successful adaptation can never completely match the durability or other performance standards of a new building. Some clients therefore may object to having to pay almost the same for a building that will not be as effective as one bought new. In financial terms, too, the rental value of the adapted building is likely to be below that of new build. This is because a new building can always offer better facilities than a refurbished one.

Nevertheless, if the cost of the proposed adaptation works approach or exceed those for new build, there may still be reasons for choosing the former rather than the latter solution. The building in question may be Listed or in a Conservation Area. Any redevelopment of such a property requires planning consent. If the building is significant in historical or architectural terms, then permission to demolish and rebuild is likely to be granted only in exceptional cases. Also, the client may have a strong attachment to the building, and would be reluctant to dispose of it even it meant obtaining a modern property that meets the needs of the occupier better.

A more realistic approach is to consider adaptation costs as a percentage of new-build costs. Building Cost Information Service (BCIS), a trading division of RICS Business Services Ltd, continues to publish special reports on this matter (e.g. BMI, 2005). They show a fairly consistent relationship of around two-thirds on a comparison of rehabilitation (i.e. adaptation) costs with average new-build costs expressed on the same basis. The findings of that report were based on the analysis of contracts for over 2550 building projects.

Table 2.1 illustrates the comparison for some common types of buildings. According to the BMI, the only categories of building that show a significantly different relationship are community centres, where

Table 2.1 Adaptation and new-build cost comparisons (BMI, 2005)

Building type	*Adaptation mean cost (£/m²)*	*New build mean cost (£/m²)*	*Adaptation of new build (%)*
Factories	406	545	74
Offices	711	1155	62
Banks/building societies	814	1486	55
General hospitals	1007	1371	73
Public houses	835	1163	72
Community centres	852	999	85
Churches	943	1206	78
Primary schools	622	1011	62
Estate housing	389	670	58
Flats	660	779	85

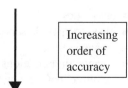

the average adaptation costs are as high as 85 per cent of the equivalent figure for average new-build schemes, and banks and building societies where the rehabilitation costs are 55 per cent of new build.

Naturally, some caution should be adopted when using these cost comparisons. Every construction project is different. Some adaptation schemes involve major structural alterations and extensions, which tend to increase the average cost £/m^2 of the project. Other adaptations involve relatively minor upgrading and refurbishment. For example, the conversion of bank premises to a café or restaurant does not entail substantial works, so basic construction costs can be kept to minimum.

Nevertheless, the values shown in Table 2.1 can prove extremely useful for strategic planning and early cost-advice purposes. They provide a sound datum from which to assess the financial viability of a proposed adaptation scheme.

What, then, should be the limit of the proportion of adaptation costs to those of new build? A useful starting point is obviously the rule-of-thumb indicated by the BMI (2005), which indicates that on average rehabilitation costs are around two-thirds of those of new build. At the upper end of the range, 75 per cent could be deemed a reasonable limit.

A variation on the percentage approach, however, is to compare the equivalent annual value of adaptation with that for redevelopment. With this method the two-thirds to three-quarters range can be set aside but still used for comparative purposes. The example shown below illustrates this approach:

Let: £x = adaptation cost; £y = redevelopment cost; 10 per cent = interest rate. Estimated remaining life of adapted building is 20 years, and the estimated life of redeveloped property is 60 years.

$$\text{Equivalent Annual Value (EAV) of adaptation} = \frac{x}{\text{YP}} = \frac{x}{1 - \text{PV/i}}$$

$$\text{EAV of redevelopment} = \frac{y}{1 - \text{PV/i}} = \frac{y}{\text{YP}}$$

For x and y to be equal:

$$\frac{x}{8.5} = \frac{y}{9.97}$$

$$x = \frac{8.5y}{9.97} = 0.85y$$

Therefore, using this approach, it would be uneconomic to carry out adaptation in this instance if its cost was more than 85 per cent of redevelopment costs. This approach, though, may still be considered too crude.

Needleman (1969) proposed a more analytical method of assessing the viability of adaptation over redevelopment. It is based on a formula that shows the maximum cost of modernization (i.e. adaptation) compared with the cost of rebuilding (i.e. renewal). His method, however, only takes economic factors into account. With this approach adaptation is deemed cheaper if the total costs are less than the redevelopment costs as derived from the following formula:

$$b > m + b(1 + \text{i}) - h\frac{r}{\text{I}}[1 - (1 + i) - \text{H}]$$

where: b = cost of rebuilding; m = cost of maintenance; i = rate of interest; h = useful life of adapted property; r = difference in annual repair and rent costs.

Practicality

The building should be capable of adaptation without major or long-term disruption to either its use or its structure and fabric. Clearly some buildings are easier to adapt than others are. Bespoke buildings such as specialist factories, and industrialized system buildings, and properties of unusual or uncommon structural configuration, are generally the most difficult to adapt. Thus the form of construction will be an important consideration (see section below on constructional characteristics of buildings).

Adapting a tight-fit building is likely to involve some partial demolition and reconstruction. In buildings of architectural or historic importance, this may be unacceptable if not severely curtailed. The advice and guidance of the appropriate conservation agency (e.g. English Heritage or Historic Scotland) should be obtained before undertaking such work.

The practicality of undertaking an adaptation scheme is influenced by the extent of access required for the work. Congested urban sites, for example, offer little background for contractors and their plant. This will necessitate careful planning of deliveries if there is no room to store materials and components 'on site'.

In residential adaptation schemes such as rehabilitation programmes habitability is a key criterion. Appendix A shows a checklist devised by the Building Research Establishment, BRE (GBG 9, 1991) for assessment of habitability.

Utility

In many buildings the existing spatial provision is more than adequate. Any successful adaptation project, however, must fulfil the spatial and environmental needs of the building's occupants. In spatial terms, the aim is to minimize wastage of underused space on the one hand, without overuse on the other.

As indicated in Chapter 1 provision for possible future change must also be allowed for in the scheme nowadays. Many new buildings have been designed to provide possible future change. This can be achieved for example by over-designing the foundations of a building to take additional storeys later on or providing space on the site for lateral expansion of the property at some future date.

The distribution of the housing stock by age gives some indication of the adaptability and hence utility of dwellings in England and Scotland. Nearly two-thirds were built before 1965. Table 2.2 shows this distribution.

Best value

Rationale

The report by Egan (1998) on 'Re-thinking Construction' prompted a drive for a fundamental change in the way construction projects are delivered. It affects everyone involved in the design and construction of adaptation as well as new-build projects. Reducing capital costs, defects and accidents, and improving predictability, productivity and contractor's profitability are its main aims.

Table 2.2 Summary of English and Scottish housing stock by age (DETR, 1998; Scottish Homes, 1997)

Dwelling age	*'000 dwellings England*	*Percentage*	*'000 dwellings Scotland*	*Percentage*
Pre-1919	4782	23.4	456	21.0
1919–1944	3900	19.1	318	15.0
1945–1964	8155	20.9	590	28.0
Post-1964	7433	36.5	759	36.0
Total	20,371	100	2123	100
Converted flats	798	3.9	39	2

This has triggered a virtual revolution in the procurement of projects in the public sector. It has led to the introduction of a number of initiatives, some of which are listed in Table 2.3.

It remains to be seen to what extent these requirements and techniques can be applied to adaptation projects. Still, a more coherent framework for this area is now emerging in the industry. This should provide incentives and guidance for those involved in the adaptation of buildings especially when using modern techniques and innovative materials.

Table 2.3 Examples of some initiatives in construction

Initiative	*Examples*
Benchmarking	Identifying and emulating best practice (see Chapter 11).
Best practice programmes (BPPs)	A move away from compulsory competitive tendering (CCT), which is often seen as inefficient and divisive, towards a more partnership approach with the participants in the construction process. BPPs involve some if not all of the other initiatives listed in this table and more.
Environmentally sound technologies (ESTs)	Solar panel heating, low water-use sanitary units, and other technologies, which help to achieve more sustainable construction.
Fast track construction	Speedier construction times using factory-made modular elements (e.g. toilet pods) and components (e.g. door sets), maximizing off-site fabrication and overlapping the design and construction phases.
Industrialization	Standardization and prefabrication can enhance quality and productivity. Modular units may be employed in some adaptation schemes (see Chapter 9).
Innovation	Adopting the latest proven techniques to enhance efficiency of the process (e.g. partnering) or product (e.g. positive input ventilation). The movement for innovation (M4i), formed after The Egan Report (1998), is pioneering radical changes of attitude in the design and construction industry (see Chapter 11 on M4i's six environmental performance indicators).
Lean construction	This is a relatively new approach to construction that has its roots in production management. It aims to reduce variability and waste, and increase productivity on construction sites by utilizing supply chain management (see below) and fast track construction.
Public/private partnerships (PPP)	A form of public and private sector partnering, which inculcates mutual objectives; team spirit; problem resolution; risk allocation. See section on PPP/PFI below.
Smart materials	Materials that respond to changes in environmental conditions (e.g. photovoltaics, electrochromic glass – see Chapter 8).
Supply-chain management	A way of working in a structured, organized and collaborative manner by all participants in a supply chain (e.g. client, consultants, contractor, subcontractors, manufacturers, etc.) to reduce waste and delays.

Value for money

It has become increasingly recognized that the lowest cost option is not necessarily the best or even cheapest way to procure buildings. Duffy (1993) showed for example that a building's initial cost accounts only for about a third of its total cost over its whole life. The other two-thirds are used on maintenance and adaptation.

In today's commercial world with its conflicting issues of scarce resources and increasing expectations there is greater pressure to obtain value for money (VFM) in construction projects. This applies to adaptation schemes as well as new build. Best VFM is defined simply as 'the optimal combination of quality (fitness for purpose) and lowest whole life costs to meet the customer's requirements' (Central Unit on Procurement (CUP), 1996).

According to the British government's Central Unit on Procurement (CUP, 1996), achieving this goal in construction needs a framework to ensure projects provide VFM, inter alia, by:

- defining the project carefully to meet the user needs;
- integrating value and risk management techniques within normal project management operations (see Chapter 11);
- adapting a change control procedure to deal with unexpected events;
- taking account of whole-life costing;
- avoiding waste and conflict through team working and partnering;
- appointing consultants and contractors on the basis of VFM rather on lowest initial price.

The best value for money level is determined by the combination of durability, optimal quality and lowest whole life costs as indicated in Figure 2.1.

Decision criteria

Determining the appropriate option for the adaptation of a building is one of the key stages in the decision-making process. According to Nutt (c. 1993) there are four key decision criteria to bear in mind

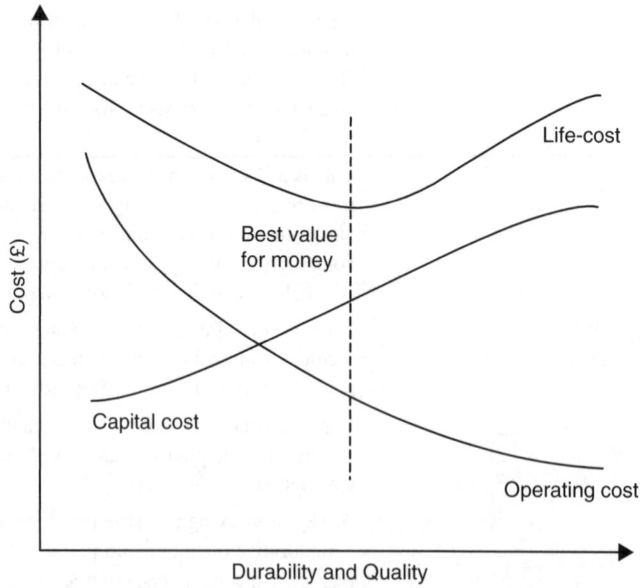

Figure 2.1 Value for money (Best and de Valance, 1999)

when assessing the feasibility of an adaptation proposal:

1. The relative cost of options: comparing the costs against some benchmark – such as BMI adaptation costs versus new-build costs for various types of buildings.
2. The relative value of options: Using discounted cash-flow analysis to determine the best value to the owner.
3. The probable risk of options: evaluating the business risks and the technical risks associated with each scheme.
4. The relative robustness of options: How well each scheme responds to, or can cope with, rises in interest rates or other changes in market conditions such as rental levels or demand for certain types of buildings.

In adapting a building, therefore, some if not all of the following questions have to be addressed:

What is the likely total cost of the adaptation?

The type of building and the extent of the proposed adaptation work will inevitably directly influence the construction costs. As Table 2.1 shows, the adaptation of different building categories involves a range of different costs. Adaptation projects consisting of the conversion of old buildings, particularly those involving conservation work, are usually more expensive than conventional change of use schemes. 'Hidden' costs such as administrative charges, professional fees, and repairs of hidden defects must also be considered.

What is the current use value (worth to the owner) and financial value (value in exchange or price) of the building?

The use value of a building is best obtained by some form of discounted cash flow (DCF) technique (RICS, 1995b). Such a calculation will determine, by using streams of discounted income and expenditure flows, the net present use value to the owner.

The value in exchange or price of property can be derived from a comparison of recent, similar property transactions in the area. For properties such as petrol stations and public houses, of course, the contractor's method or the profits method may be more appropriate.

If the Use Value exceeds the Financial Value, it would be more profitable to retain the building (and possibly adapt it) rather than dispose of it. Conversely, if the open market value of the property were greater than its worth to the owner, there is clearly a strong case for disposing of the property. Still, economic viability is not the only criterion. The social reuse of properties is often just as important as their profitable reuse.

Are there any planning restrictions affecting the building?

Change of use requirements will clearly influence the extent and nature of the reuse of buildings. Any doubts about Use-class changes can undermine the confidence in the proposal because it increases uncertainty regarding planning approval for the property. Density controls, height and width limits, and other restrictions will also have to be borne in mind for the adaptation scheme. These constraints can easily restrict the size and location of any extension.

How can the set of change options be identified?

The building's location, previous use and morphology will need to be assessed to determine the range of options for its alteration, conversion or extension. The professional adviser needs to establish the full range of options and present them to the client. This should include disposal.

Should alternative uses be examined?

It is possible that another use is more profitable than the original proposal. This will depend of course on the needs of the client. If the client is intending to be the user of the building, the type of business or other activity the client is involved in will determine the extent of alternative uses.

What is the building's potential for physical adaptation?

Can it be extended vertically or horizontally? Is it a tight-fit building? The pre-adaptation survey will not only establish the building's condition; it should also determine its flexibility and suitability for reuse.

What evaluation system is to be used to assess the options?

A simple rating system, using weightings if necessary, could be used as a preliminary method to identify the optimum option. Table 2.4 illustrates how this could be achieved when comparing three different adaptation schemes for the same building.

Ultimately, however, an economic option appraisal would need to be used to determine which one is the optimal solution. This can be done using a discounted-cash-flow evaluation to obtain the net present

Table 2.4 Adaptation analysis matrix (adapted after CUP, 1996)

Adaptation option	*Objectives/criteria*					*Total weighted score*	*Rank*
	Fire and statutory requirements (1) 0.3	*Extended service life (2) 0.1*	*Capital or Rental value (3) 0.3*	*Sustainability (4) 0.2*	*Functional suitability (5) 0.1*		
Option A							
Option B							
Option C							

Notes

Objectives/criteria include:

1. Accessibility for disabled users.
2. Condition as well as durability enhancement.
3. Increases in the rental value or property yield.
4. Energy efficiency, low-pollution.
5. Spatial efficiency as well as comfort conditions.

Generally: The number and types of Criteria and Options will vary from building to building, as these are ultimately dependent upon the needs of the owner or occupier. The rankings are also determined by the requirements of the client. They provide indicative rather than absolute guidance.

Options: A range of three or more options could be selected depending on the type of building and the needs of the client. For example, the following options could be used in the evaluation:

A = Refurbish to same use.
B = Convert to similar use.
C = Convert to another use.

Ratings: 1 = Poor.
2 = Fair.
3 = Average.
4 = Good.
5 = Excellent.

Method: Each option is scored against each criterion and the score multiplied by the respective weighting. Scores are tallied horizontally. Options are then ranked in terms of their total weighted score.

value (NPV) of each of the options being considered. If return on investment using an assumed or required rate of return is the key criterion the one having the highest NPV can be deemed the most viable. This is effectively the same process as a development appraisal.

Online tool as a decision aid for refurbishment/redevelopment of offices

Background

In 2002 the BRE launched a project funded by the Department for Trade and Industry (DTI) to develop a decision-making tool relating to commercial refurbishment. The project considers the relative sustainability of refurbishing and redeveloping office premises.

The result of the project was the development of the Office Scorer online tool. It is available for free on the Internet at www.officescorer.info/.

'Office Scorer' is a tool that is aimed at professionals involved in the strategic decision to refurbish or redevelop an office building. According to the Office Score web site referred to the above key questions relating to this tool are as follows:

What does it do?

The tool compares major or complete refurbishment with complete redevelopment, and redevelopment within an existing facade. It enables users to systematically compare and test the environmental and economic impact of different building design concepts for offices, and to identify sources of further relevant guidance.

Who should use it?

Planners, surveyors, architects and services engineers, building owners, property managers or occupiers can use the tool.

When should it be used?

The tool should be used early on in the decision-making process, when the options of refurbishing, redeveloping on the same site or redeveloping elsewhere are being considered. It can also be used to aid decision-making as the design of a building progresses.

How does the tool work?

BRE has modelled a number of building types over a 60-year life and has evaluated the environmental and economic impact of a range of factors. The results of this analysis are used within this software to quantify the likely environmental and economic impact of other refurbishment/redevelopment projects.

The tool is intended to provide indicative figures for the purposes of comparison, and only those parameters that have a significant impact have been included.

Some parameters, for example the impact of water use and the inclusion of lifts, have been disregarded because their impact is not significant, and is likely to be the same whether the building is being refurbished or redeveloped. To measure environmental impact, the tool uses BRE's ecopoints.

> BRE's Ecopoints are a single score which measure environmental impact. The average UK citizen would have the equivalent to 100 ecopoints impact a year, and the lower the ecopoints score, the lower the impact.

Economic impact has been measured using capital cost and whole life costs over 25–60 years. Whole life costs are discounted to prevent value over time, but reflect capital costs, and maintenance operation and disposal costs are the life of the building.

Constructional characteristics of buildings

Generally

The constructional characteristics of a building will have a major impact on the feasibility and potential of its adaptation. In particular it is important to appreciate the essential differences between traditional and modern construction to determine their suitability for adaptation.

Generally and simply speaking 'traditional buildings' are predominantly of load-bearing masonry construction. Timber-framed Period buildings are the primary exception to this rule. However, the fabric of most old buildings, regardless of their structural form, tends to be thick, heavy, porous and permeable. As a result their thermal mass is normally high.

Nevertheless, information on old or obsolete construction methods is not always readily available. It is often best obtained by consulting one of the available reprinted/facsimile or second-hand books such as Gilbert and Flint (1992), Gourley (c. 1910), and Nicholson (1823) and others as listed in the Bibliography.

The fabric of modern building construction, on the other hand, is normally thin, light, non-porous and impermeable (see Ross, 2005). Contemporary air-conditioned office buildings, for example, do not rely on much, if any, natural ventilation and offer little if any inherent thermal mass. These features affect the 'breathability' of their fabric.

'Breathability'

It is important to understand that all buildings need to breathe (Hughes, 1986). If this important principle is ignored there is a danger of entrapped moisture causing dampness-related problems in their fabric. Old buildings in particular (i.e. solid load-bearing masonry structures) achieve breathability naturally through their fabric, but in an uncontrolled manner, to avoid this problem. Any 'hard' or impervious external coating, covering, cladding or pointing therefore interferes with this process.

Even cement mortar in masonry joints can prove troublesome in this regard. The relatively hard, impervious cement mortar forces evaporation of moisture in a wall through the brick or stone. Crystallization of soluble salts during the evaporation process may lead to surface damage to stone and brick walls. This is because these crystals exert pressure on the internal structure of the brick or stone.

Therefore any adaptation work to an existing building involving major repairs or changes to the external fabric should take cognisance of this principle. Any repointing to the stone or brick fabric of an old building being adapted or repaired should be done using a mortar compatible with the existing masonry. Lime-based mortars are usually the best for use in repointing the masonry of buildings predating about 1920 (Holmes and Wingate, 1997).

New buildings, on the other hand, achieve 'breathability' in a more controlled or artificial manner – through background ventilation, forced ventilation, air conditioning, or a combination of these methods. The application of new external enclosures such as overcladding or cement rendering is thus unlikely to prove as troublesome on a modern building as it would on an old one, provided of course that the substrate is not already damp or deteriorated.

Traditional forms of residential construction

Traditional construction can be defined as those techniques that employ tried and tested methods using predominantly skilled labour on site (see AMA, 1983, for a more detailed description of traditional and non-traditional constructions). The materials and components are generally on a small scale, and are delivered to the site ready for installation. Some processing may of course still be required to these materials on site prior to installation.

Simpson (1992) offered a simple though limited taxonomy of traditional forms of low-rise construction – Types A and B. This helps to explain the dichotomy between old and new traditional forms of domestic construction.

Type A is medieval in origin, but was still common after the 16th century. Essentially the main elements are exposed in this type of construction. In other words, nothing is hidden. There is, in addition, no attempt to inhibit the movement of water vapour within the building. The natural balance between the internal and external environments was thus achieved by relying on natural draughts. Moreover, as was pointed out by Hughes (1986), the 'breathability' of such buildings is critical to their effectiveness in dealing with excessive levels of moisture that can permeate the fabric. Simpson (1992) noted that Type A construction required a great deal of natural ventilation to keep it dry – hence the reason for large open-hearth fireplaces and unsealed openings in the external fabric in buildings of this type.

According to Simpson (1992), Type B construction emerged sometime between the 1750s and the 1850s. Significant improvements in the quality of buildings arose during that period. However, the precise impulses for those advances are not clear. Simpson suggests that there could be three influences: changes in ideas of function and planning; changes of fashion or style; and changes in construction techniques. He does argue, though, that it is impossible to say which the main motivating factor was. No doubt, of course, economic and demographic influences prompted by the Industrial Revolution played an important part as well. Further research into the historical and technical factors involved is required to answer this point more fully.

If Types A and B are classified as traditional construction, Types C and D, two new categories proposed in this book, can be considered as relatively modern.

Type C is more sophisticated than its predecessors, and includes cavity wall construction. It reflects the improvements in the standard and performance of buildings that emerged in the late-19th century, prompted by the Public Health Acts of 1875 and 1890.

Type D, on the other hand, describes a more modern constructional style. It incorporates insulated-framed/panelled construction – even in ground floors. Overall, Type D places greater emphasis on thermal efficiency and weather-tightness as well as quality of finish. The aim is to provide a higher degree of comfort and achieve a more cost-effective design.

A comparison of these four categories is shown in Figure 2.2.

This taxonomy is of course fairly crude and hence may not be applicable in every case. However, it does offer a useful guide as to the development and pattern of basic constructional forms, which need to be borne in mind when proposing to adapt a building.

Modern forms of construction

Adaptation is, of course, not restricted to buildings of traditional construction. Modern properties, too, may require adjustment even a few years after they are built. This occurs especially with supermarket buildings, which often require regular adaptation to keep up with changing retail requirements. As indicated in Chapter 1 a standard TESCO supermarket building in the west of Edinburgh comprising six 8–9 m wide bays, for example, was extended in 2002, 2 years after completion of its construction.

Rented offices, too, often require refurbishment. However, this is usually timed to occur around the lease renewal period to make them more attractive to potential tenants.

Determining and understanding the form of construction of any building being adapted is essential if a successful scheme is desired. In most cases, with modern buildings this should be easy because the 'as built' drawings will be available either with the owner or in the local authority archives.

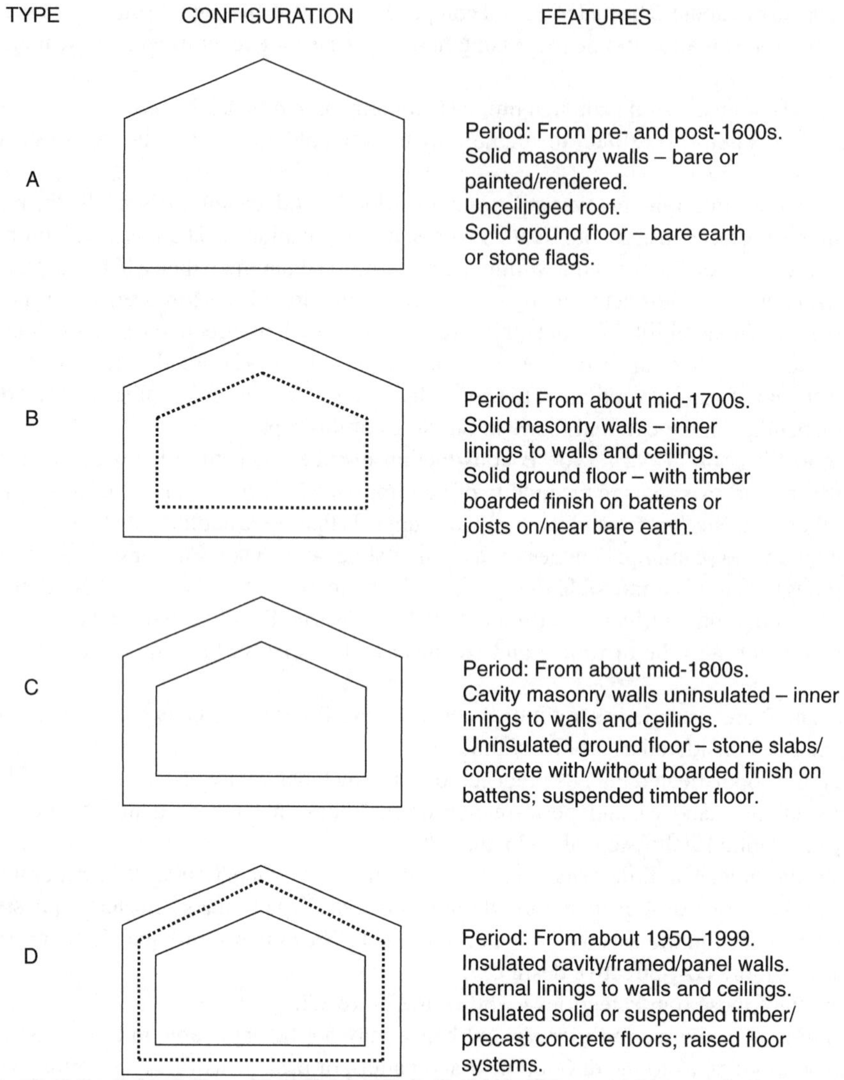
TYPE
CONFIGURATION
FEATURES
A
Period: From pre- and post-1600s.
Solid masonry walls – bare or painted/rendered.
Unceilinged roof.
Solid ground floor – bare earth or stone flags.
B
Period: From about mid-1700s.
Solid masonry walls – inner linings to walls and ceilings.
Solid ground floor – with timber boarded finish on battens or joists on/near bare earth.
C
Period: From about mid-1800s.
Cavity masonry walls uninsulated – inner linings to walls and ceilings.
Uninsulated ground floor – stone slabs/ concrete with/without boarded finish on battens; suspended timber floor.
D
Period: From about 1950–1999.
Insulated cavity/framed/panel walls.
Internal linings to walls and ceilings.
Insulated solid or suspended timber/ precast concrete floors; raised floor systems.

Figure 2.2 Basic taxonomy of forms of residential constructional forms (Douglas, 1997)

The modern forms of construction for residential and commercial buildings can be classified into the main categories shown in Table 2.5.

Commercial buildings

Building types

Before looking at the constructional characteristics of commercial buildings it would be useful to classify their main forms of construction. They can be divided into two broad groups: unframed and framed.

- Unframed — Solid or cavity wall construction (usually up to a maximum of six storeys) or column and slab construction (usually up to a maximum of about 12 storeys).
- Framed — Steel or reinforced concrete – skeletal form for multi-storey blocks.

Table 2.5 Typical forms of modern construction

Type of construction	*Original era*	*Comments*
Bunched/bundled tube	1970s	A form of steel or concrete-framed construction suitable for very high-rise buildings (Orton, 1992).
CLASP system	1950s	A form of system building for low-rise non-residential properties such as schools (see Chapter 9).
Column and plate	1950s	Once popular technique for multi-storey blocks of flats and offices.
Modular panel	1960s	Wall and floor panels prefabricated as storey high units and connected together on site. Large panel system was once a very common method of non-traditional construction for high-rise flats.
Modular pod	1980s	Prefabricated room-sized modules (often fully fitted out) bolted together on site to form the required plan shape and number of storeys. The pods are then clad externally with a single skin of brickwork or cladding panels to give the building the desired finish. They can also be used in 'renovation' projects (Lawson, 2001b) – see Chapter 9.
'No-fines' concrete	1950s	A popular form of non-traditional housing of solid wall construction, usually used for low-rise blocks but was used as the infill wall material in some high-rise flats (see Chapter 8).
Portal frame	1940s	Suitable for low-rise factory and warehouse properties, and for top storey extensions to flat-roofed commercial buildings (see Chapters 7 and 9).
Slip form wall and slab	1980s	A form of industrialized construction used for commercial buildings such as hotels.
Skeletal frame	1900s	'Stick-built' construction where members are connected on site (Lawson, 2001a). Very common conventional method for low-rise and multi-storey commercial and industrial buildings. It can be used as the superstructure for extensions.
Tunnel form reinforced concrete construction	1980s	This is a method of cellular construction in which a structural tunnel is created by pouring concrete into steel formwork to make the floor and walls.

The constructional characteristics of commercial buildings are varied. Generally, however, they are of either concrete or steel frame construction clad with stone or lightweight cladding. Table 2.6 summarizes the main differences.

Other adaptation issues

Site usage and density

A building or group of properties that covers more than about 60 per cent of a site does not allow much room for on-site access, lateral expansion or increased daylighting. The original intensity of use of the site might also have an impact on the permissible density of the building once it is adapted.

Table 2.6 Characteristics of different commercial building types (Martin and Gold, 1999)

Period	*Advantages*	*Disadvantages*
1900-preWW2	May have high ceiling. Designed for natural ventilation. Narrow floor plate aids this. Add solar shading and spot cooling in high heat gain areas. Heating, power and communications routed around perimeter.	May be listed (resulting in delays due to negotiation with English Heritage/Historic Scotland). Structural partitions – inflexible space, poor circulation, poor insulation. Low floor loading (may not accommodate current loads without strengthening).
Late 1950s/1960s	Open plan layout. Designed for natural ventilation. Narrow floor plate aids this. Add solar shading and spot cooling in high heat gain areas. Heating, power and communications routed around perimeter.	Low floor–floor height. Large glazed areas leading to high heat gain. Low floor loading (may not accommodate current loads without strengthening). Relatively lightweight partitioning. Poor insulation and high air infiltration rates. Single glazing. Addition of raised floor complicated around lifts and stairwells. Routing ducts and pipework may be difficult.
1970s	Larger floor–floor height than 1960s (for services). Deep plan, a disadvantage due to need for air conditioning and artificial lighting. Open plan layout.	Lightweight construction. Possible high air infiltration through facade, poor insulation. High heat gain through fully glazed facades.
1980s	Larger floor–floor height. Raised floor for services. Open and deep plan layout. High electrical power capacity.	Lightweight construction. Over-specified ventilation and air conditioning – difficult to control.

Layout and configuration

The shape and height of the building will have a bearing on its ability to accommodate modifications to its interior and exterior. Rectangular or square buildings are much easier to facilitate an extension or other external alteration. Internally, too, these layouts can be more conveniently reconfigured. Circular, curved or irregular shaped buildings, on the other hand, are more awkward and thus more expensive to adapt.

Building's adaptation potential

This relates to the alternative use and layout of the existing building. The property's location, condition, construction, morphology, and legal restraints will all influence the degree to which it can be adapted.

Purpose of adaptation

Once the decision to adapt a building has been made, the extent of the work will begin to be established. The adaptation could include a combination of works such as a conversion and extension to the building. For example, in Edinburgh a detached two-storey Victorian house used as a bed-and-breakfast facility was recently adapted into offices. The work consisted of a complete gutting out of the building, with only the main walls left. A new timber roof on a steel frame within the original shell was formed as well as a large side extension near the front.

If the future occupier is unknown, the proposed use of the adapted building will need to be considered very carefully. This will obviously entail a more speculative type of development. In urban areas conversions to residential use offer less financial risk because of the relative popularity of these schemes.

Economic factors

Impact on rental and capital values

Adapting a building is more likely to have a positive rather than negative impact on both its rental and capital values. A refurbished commercial property is more appealing to prospective tenants and therefore it can command a better level of rent. Similarly, the capital value of a refurbished property because of this improvement in its attractiveness is more likely to secure a prospective buyer. However, as indicated earlier there is no guarantee that the increase in value will match the outlay associated with the adaptation work.

Overt costs

The usual items of expenditure associated with any construction project, whether new build or adaptation, are reasonably straightforward to estimate. As already indicated, there are outline costs per square metre for the rehabilitation of the main categories of buildings. When the conversion involves either an alternative or mixed use, care must be taken when using those figures. Additional work to cover fire precautions and access provisions is required in mixed-use schemes.

Hidden costs and contingencies

One of the most awkward aspects of adaptation schemes, regardless of their size, is the likelihood of hidden costs arising during the contract. In particular, conversions tend to involve primarily internal works, because there is a higher risk of discovering unforeseen or hidden defects as the work proceeds. Problems such as deleterious materials, dry rot, dampness and defective services are often not apparent until the adaptation works are well under way.

Special needs

Increasingly the needs of users such as the elderly, the infirm as well as the disabled must be taken into account when adapting a building. The Disability Discrimination Act (DDA) 1995, which came fully into force in 2004, requires that all buildings serving the public should be accessible to all. This important piece of legislation (which has been amended by the DDA 2005 by extending access duties of service providers) is dealt with in more detail in Chapter 9.

Special needs, however, may require facilities that go beyond the requirements of the DDA. These, for example, could include adaptations such as:

- Stair or platform lift for infirm or elderly users (see Chapter 9).

- Special lifting hooks inserted into the ceiling of a bedroom or toilet to support lifting gear that facilitates easy use of bed or bath. This may include special plastic or glass fibre bath with watertight side opening door for easy access.
- Voice-activated controls of curtains or blinds as well as lighting and heating systems for severely disabled users.
- Intercom facilities for the elderly or infirm in their refurbished flats.

Pest control

Preamble

The urban environment is synonymous with large numbers of people. It represents potentially large sources of human spillage or waste, which in turn encourages the interest and the colonization of many pests such as birds, rats, mice and cockroaches.

Even occupied buildings of course can suffer from vermin infestation. Many old, vacant buildings ripe for adaptation, however, are more likely to be infested with a variety of pests. Not only is it detrimental to health, the presence of pests is also potentially damaging to the building itself and might inhibit if not prevent its adaptation. The opportunity to eradicate this problem should be undertaken when if not before the building is being adapted, otherwise the pest infestation could undo all the good work achieved by the proposed works.

However, it should be borne in mind that creatures such as bats are a protected species under the Wildlife and Countryside Act 1981. It is illegal under Section 9 of the Act for anyone without a licence to disturb or kill a bat. If bats are encountered in a building being adapted the Nature Conservancy Council must be consulted before work begins.

One of the most troublesome bird pests to urban buildings is the pigeon. Pigeon colonization of buildings in inner cities has now become a major problem and will no doubt continue to be so whilst little is done to prevent it. The feral or wild pigeon, for example, is parochial by nature and puts the priority of feeding first. Thus, if through discovery a constant source of food is identified these birds will remain within the locality of that food source for as long as it lasts.

Physical and economic effects of pest colonization

The excrement from pests such as pigeons can cause physical damage and decay of buildings, statues, clothing, etc. Birds can also cause damage to trees, plants and foliage, as well as fouling and spoiling of food or commodities stored or in transit from warehouses, shops, etc. Rats can cause damage to a building by chewing through cabling.

Pest infestation may be difficult and potentially expensive to eradicate. It can also have a detrimental effect on the capital or rental values of the property and makes it less attractive for adaptation.

Health hazards from pest infestation

There are a variety of different diseases that can be borne by pigeons, starlings and other urban birds, as well as rats and mice. Many of these diseases, such as salmonellosis and thrush, are infectious to humans. There are many ectoparasites that can be carried by pigeons and the many mites that can be found thriving on or around their areas of roosting. Inconspicuous corpses of dead birds are breeding grounds for flies and the inevitable contamination they create.

Bird droppings on and around buildings are hazardous in several ways. Pigeon excrement, which in obscure places can be as much as several centimetres deep, represents a danger to life and limb when displaced over walkways, pavements, fire escapes and public places by, for example, increasing the risk of slipping. Not only is this unsightly it can only cause deterioration of the stone. When dry this excrement can be blown as dust into the building via open windows or vents. This can lead to the occupants of such a building becoming ill after breathing in such dust because of the pathogens it contains.

Vermin such as rats and mice are often carriers of deadly bacteria and viruses. Fleas on rats can carry pneumonic plague, which is a highly contagious, often lethal, disease. As with house dust mites, minute vermin droppings can become airborne if disturbed and any humans in the vicinity may then inhale these particles. In rare cases vermin droppings could contain infectious pathogens such as the Seoul virus.

Like many birds, rats can act as carriers for parasites such as fleas and lice as well as infectious diseases. Rats, for example, can also transmit via contaminated water Weils disease, which is a severe, sometimes fatal, form of leptospirosis with fever, jaundice and muscle pain.

Control of pest infestation in buildings

The first measure is to identify the sources of infestation and roosting sites. For example, roof spaces, eaves and verges are vulnerable to birds; basements, cellars, manholes and subfloor voids are particularly prone to rat and mice infestation. The cause of any infestation such as debris and food waste should then be eliminated. See guidance in BRE Digest 415 (1996) on these issues.

Any vulnerable openings/areas in and around the building should be covered with a suitable wire mesh. A sound maintenance system with a strict cleaning programme is required to prevent any re-infestation.

In acute and ongoing cases of bird infestation on a building's exterior it may be prudent to install discrete polyester or metallic netting or wires on ledges of vulnerable parts of its façade. Positions such as ridge tiles, chimneys, copings, ledges and rails especially may require these measures.

Stainless steel anti-bird spikes may prove helpful in deterring birds from alighting on a small ledge but may trap waste paper and thus lose their effectiveness as well as make the building look untidy. Similarly, the application of a bird repellent paste at least 3-mm thick over vulnerable ledges may deter infestation but it can compromise the masonry's appearance.

Severe infestation of the interior of an empty property may necessitate fumigation. The building would need to be sealed as much as possible to minimize leakage and completely vacated. Smoke bombs containing an insecticide would be required in cases of extensive woodworm infestation. Chlorine dioxide gas can be used to sterilize a building's interior that has been contaminated with biological weapon such as Anthrax. This gas, though, is highly toxic and must only be undertaken by an experienced and licensed pest/biohazard control contractor.

An approved specialist contractor who is a member of the British Pest Control Association should carry out any eradication and repellent work. However, the effects on the appearance and maintenance of the building as well as any health implications for the occupants need to be considered before embarking on any of these measures.

Assessing the building

Analysing current provision

A thorough understanding of the building is necessary if its adaptation is to be successful. This will include establishing its:

- form of construction (traditional/non-traditional/composite?);
- structural morphology (load-bearing masonry, framed or modular?);
- physical condition (good/fair/poor/bad?);

- architectural or historic significance (Listed or unusual design/materials?);
- adaptation potential (flexibility/suitability for change of use?).

Strategy for adapting a building

Context

The context of the building and its surroundings will be established by considering the results in the investigations described above. This understanding should extend to changes that have taken place over time, the present levels of performance and the remaining service life of the building's main elements, components and services infrastructure.

Generally speaking, establishing the context of a building is vital if a full understanding of its condition and potential is required. This will necessitate an analysis of the property's history and any major changes that may have affected its aesthetics, structure and planning.

The stage of the property's life cycle must be ascertained. This will indicate the remaining service life of the building's main elements and components. It will also shed light on the past use and performance of the building.

Long-term needs

A strategy is a long-term plan so that the long-term needs of the building are properly identified. This will include two key factors:

1. Maintenance planning – determining in advance what remedial and other works are required to keep the building in good repair.
2. Future requirements for space, facilities and environmental conditions.

These aftercare issues are dealt with in more detail in Chapter 11.

Getting to know the building

Preamble

Before any adaptation work is undertaken it is vital that the building being converted or rehabilitated be fully investigated. According to Swallow (1999) this is required 'to achieve a thorough understanding of the building, including its construction, structural arrangement, physical condition, historical significance and suitability for a change of use'.

A survey of the building to be adapted will be required primarily to establish the building's condition and potential. If no record or as built drawings of the building exist, a measured survey will need to be undertaken. Alternatively, or for comparative purposes, the drawings from the measured survey could be used as a check against the original design drawings.

A condition survey will establish the building's state of repair. It will identify any major deficiencies in the structure and fabric. It should also highlight potential problems with the structural capacity and the capacity of the services.

In summary the main stages of the process of getting to know the building are as follows:

- Inception
- General feasibility
- Desk top survey

- Reconnaissance
- Condition survey
- Structural appraisal
- Diagnostic survey
- Evaluation of options
- Proposals.

(Reconnaissance and Condition survey: General physical inspection; Structural appraisal and Diagnostic survey: Specific physical inspection)

Inception

The decision by a client to adapt a building is normally taken after all other options have been carefully considered. Adaptation may only be possible because the adapted building will be able to fulfil the specific needs of the user. A rigorous evaluation of the client's requirements needs to be taken at this stage to see if supply (the building) matches demand (the client's requirements).

General feasibility

This stage considers the economic, technical and legal implications of the scheme, as explained earlier. Other influences, such as the client's requirements, are also likely to impinge on the decision-making process. It is important to establish the feasibility of the proposed adaptation in principle before committing time and resources to undertaking the stages indicated below.

Desk top survey

An examination of all documentation and other information related to the building is critical. This is best done before the site survey is undertaken. It prepares the surveyor more fully for the site survey.

Oral as well as textual information about the building can be revealing. Informal discussions with the current and previous users of the building as well as with local Building Control and Planning Officials may shed additional light on important historical or statutory aspects of the property. Consultations with adjoining owners may be prudent to ascertain if there are likely to be any possible concerns about or objections to the proposed adaptation.

Typical sources of information are:

- Title deeds or other property documents (e.g. to indicate any restrictive covenants).
- As built plans and drawings to show the original construction.
- Topographical maps of the area (e.g. to indicate its date of construction in the absence of other data.
- Geological maps of the area (e.g. Coal Board plans or British Geological Survey plans to give information on the presence of previous or existing mineworkings and typical ground conditions).
- The various Royal Commissions on Ancient and Historic Buildings – for photographic and other records of the property.
- Previous Building Regulation Consents (called Building Warrants in Scotland) or Planning Applications – to indicate any past major changes or interventions.
- Other local authority records – statutory notices or previously objected schemes by neighbours.
- Ecclesiastic records – particularly parish church documents relating to local inhabitants and their residencies.

Reconnaissance survey

Establishing the physical and spatial contexts of the building is critical. A number of issues may need to be considered at this stage:

- Is it detached/semi-detached? The extent to which a property is physically connected to another building will constrain the degree of external change.
- Is it part of a terraced block? If so, is it mid- or end-terraced?

- Is it a flat? If so, does it have only one level or is it a maisonette?
- What is its spatial relationship to adjacent/adjoining properties? Prior to undertaking any adaptation scheme it may be necessary to compile a Schedule of Condition of surrounding properties. This provides a valuable record of the state of repair of nearby buildings that may be physically affected by the adaptation work.
- Are there any major obstacles to a lateral extension?
- Are there any major obstacles to a basement extension (e.g. major underground services)?
- Do any of the surrounding properties display common faults that would suggest subsidence or other problems with the topography?
- Are there any access problems that may disrupt the adaptation work or prevent the building from being fully used?
- Location and capacity of services on site.

Condition survey

This comprises three stages: external inspection, internal inspection, and review of exterior and surrounding site. The external appraisal focuses on the main elements of the building, and records any significant defects (see Noy and Douglas, 2005 for sample condition survey schedules).

The internal inspection, on a room-by-room basis, records problems affecting the use and condition of the building and confirms to what extent if any defects found externally are affecting the building internally. In the absence of the original drawings, the on-site stage may involve a measured or dimensional survey of the property. The main condition survey may prompt the need for an appraisal of the building's structural capacity.

Structural appraisal

Some local authorities may insist on a structural appraisal of a building prior to its adaptation. This is especially likely in the case of listed buildings. A suitably qualified person such as a Chartered Building Surveyor (for structures not exceeding three storeys) or Structural Engineer (for more complex structures) with experience in appraising old buildings should undertake this appraisal. The appraisal will take the form of a technical report on the condition, strength and remedial measures (with supporting calculations) necessary to bring that part of the property up to the requirements of the building regulations (Coates, 2001).

A structural appraisal normally comprises an inspection of the following loadbearing elements:

- Substructure: Suspect foundations? For example, farm buildings and old structures may not have been built to an adequate standard. In such cases Building Control will demand exposure of the foundations at selective points using trial pits (Coates, 2001). This work is required to ascertain the condition and depth of the foundation, which may require underpinning (see Chapter 7) because it is either too shallow (i.e. less than 600 mm deep) or is otherwise defective.
- Walls: Any signs of distress such as bowing (e.g. owing to lack of lateral restraint), leaning (e.g. owing to roof spread), or excessive cracking (e.g. owing to subsidence or settlement)?
- Floors: Any excessive deflection (i.e. over 1/360th of the span), sloping, or biodeterioration? Like the roof the floor structure may require timber preservation treatment and strengthening work as part of the proposed adaptation. Solid concrete or stone flagged ground floors may lack a damp proof membrane or suffer from settlement.
- Roof structure: Roof spread or fungal/insect attack evident? The condition of the rafters may suggest that timber preservation treatment or even strengthening is required. A noticeable deflection in the ridge line is often indicative of roof spread.

- Chimneystacks: Although strictly speaking a chimneystack is not designed to take any imposed loads, it can still pose structural problems. A bowed chimneystack, for example, is indicative of sulphate attack in the brickwork. A leaning chimneystack, on the other hand, is usually caused by removal of a redundant chimneybreast in the roof space as part of internal adaptations to create more space.

Diagnostic survey

Major problems highlighted in the previous inspection may require further, more detailed investigation to ascertain the root cause. A diagnostic survey in such instances, whilst potentially disruptive, can reduce the risk of uncovering hidden defects during the course of the adaptation work. For example, an investigation could be undertaken by a building surveyor to trace the full extent of decay when dry rot is found during the initial survey.

Non-destructive investigations may form part of a diagnostic or general survey (GB Geotechnics Ltd, 2001). They can involve techniques such as:

- Displacement measurements: This is a technique used by engineers to assess the extent and progression of cracking.
- Endoscopy: This is a popular inspection tool for surveyors to inspect hidden voids and cavities for problems such as bridging, wall-tie corrosion and fungal attack, using an optical borescope (Noy and Douglas, 2005).
- Impulse radar: This specialist technique can be used to determine extent of any voids areas such as flue routes, hidden or blocked-up spaces.
- Micro-drilling: This is a very effective tool to determine the presence of decay in structural timbers using a special low-voltage hand-held drill that gives the result in the form of a resistograph. Seibert Technology Ltd (http://www.sibtec.com/) supplies this instrument (Noy and Douglas, 2005).
- Moisture analysis: To determine fabric and air moisture levels using electrical resistance moisture meters and hygrometers, respectively; meters with large pad sensors about the size of a domestic iron can be used to determine the location of leaks in flat roofs.
- Potential wheel/half-cell survey: This specialist technique is used for assessing the extent of reinforcement in concrete.
- Polarization resistance: Used for assessing reinforced concrete.
- Radiography: Another specialist technique used by engineers to identify the extent of fissures and voids within concrete.
- Thermography: To determine areas of heat loss, voids, or leaks using an infra-red imaging camera; it can also be used to pinpoint sources of leaks in flat roofs.
- Ultrasonics: To test the soundness of concrete and structural timbers.

Much of the equipment used for these non-destructive tests is very expensive to purchase. Most of them can, however, be hired from an appropriate supplier.

Evaluation of options

The condition of the building together with its spatial configuration and constraints will determine the extent to which it can be adapted. However, other factors such as rental potential and functional suitability will also determine its adaptability. As indicated earlier (see Table 2.4) an adaptation analysis matrix can be used as a preliminary technique to evaluate alternative schemes. Increasingly, though, an economic option appraisal will be required. This usually involves some form of discounted cash flow analysis, which calculates income and expenditure over the expected service life of the adapted building.

Proposals

Once the adaptation strategy (i.e. alteration extension/conversion/refurbishment) has been selected, scheme design proposals can be drawn up and costed. When the client has made the decision to proceed, the necessary approvals for finance and statutory regulations' compliance have to be obtained.

Problems with adapting buildings

Spatial constraints

Inflexible or tight-fit buildings are difficult to alter or convert to another use without major intervention to their structure and fabric. Because of this alteration work to such buildings often entails considerable expense. These issues along with the adaptation problems associated with old buildings are dealt with in more detail in Chapter 3.

Joining new work to old

In connecting an extension to an existing building a number of factors need to be taken into account. Allowance for differential settlement and lateral movement has to be made. The junctions between the old and the new parts of the building need to be made weather-tight and structurally sound. The compatibility of the existing and new constructions will have a bearing on this matter. These problems are addressed more fully in Chapter 5.

Adjoining properties

Care must be taken to ensure that the adaptation work does not adversely affect the use or condition of any adjoining buildings. Collateral damage, nuisance such as noise and dust, breaches in security, and interference with access are some of the main problems that adaptation works can create for adjoining buildings.

Code compliance

Most if not all adaptation schemes will require some form of statutory approval. Planning permission will be required for extensions and conversions if the proposed development is not exempted under a General Permitted Development Order (GPDO) 1995. Even when the work is within the provisions of a GPDO, planning permission may still be required if the building is Listed or is within a Conservation Area. Building control approval will almost certainly be required for adaptation works for a non-exempted building, particularly where they involve structural alterations.

With the full implementation of the Building (Scotland) Act 2003, the Scottish Executive is implementing a change in the definition of building work North of the Border. This is a fundamental alteration of the building regulations, which for existing buildings previously only extended to replacement walls and boilers. The new proposal relates to a widening of the meaning of 'material alteration', which includes work on existing buildings as already is the case in England and Wales.

Examples of the sections of the building regulations that apply to an adaptation proposal are given in the Chapter 3. The key requirements in any event are safety (especially in relation to fire and structural stability) and adequate standards of construction.

Table 2.7 The spectrum of building condition and the adaptation response

Rating	Bad	Poor	Fair	Good
Category	Ruinous	Dilapidated	Wind and water-tight	Optimal
Description	Only some walls left. Little or no roof structure remaining. No windows and doors remaining intact. Building often classed as a 'wreck'.	Extensive defects to the structure and fabric. Nearing dereliction or redundancy. Fabric crumbling.	Showing signs of wear and tear. Neglected and approaching being run down.	Reasonable state of repair. No major defects. Satisfies most user requirements.
Adaptation	Restoration Conservation Conversion to other use.	Renovation Conversion to same or modified use.	Modernization Refurbishment Rehabilitation Alteration Extension	Minor improvements Some alterations Extension

Minor adaptation works previously undertaken without official consent usually have to receive a 'Letter of Comfort' from the local authority confirming that they do not object to the illicit work that was undertaken (see Chapter 11). More extensive adaptations such as extensions and conversions would probably require a full application for a retrospective building warrant.

The effect of building condition on the adaptation response

Not all buildings of course are exactly in the same condition; some are worse or better than others are. A variety of factors will have a bearing on a building's state of repair: the level and standard of maintenance; the extent of wear and tear; the degree of user abuse/misuse; vulnerability to vandalism; the intensity of exposure to aggressive agents; its age and quality of construction.

Before any adaptation work is carried out to a building a condition survey should be undertaken to establish the full extent of its defects. The spectrum of building condition showing the main types of general condition, is outlined along with the adaptation response in Table 2.7.

Buildings in use

Clients are understandably reluctant to vacate their buildings completely for adaptation if they are the current occupiers. In cases where the use of the property has partially discontinued, owing to a downturn business activity, changing the use of the vacant parts may not prove too difficult. Out-of-hours working may help reduce the inconvenience for commercial clients.

A building can be susceptible to damage and intrusion by thieves during the time it is lying empty and even during the actual adaptation work. In some cases it is prudent to employ the services of a full-time watch-person.

Carrying out work to a building when it is occupied can be both disruptive and difficult for all concerned. If not handled properly adaptation operations can cause major safety and security problems for both client and contractor. In other words, the risk associated with such work is usually much higher than with adaptation works to empty buildings and new-build schemes (see Chapter 11).

In some cases a decanting of the occupier is the most feasible solution. This is usually the case in major housing rehabilitation programmes. The tenants are temporarily moved to alternative accommodation nearby. This can be achieved using transit houses or caravans; furniture and fittings are taken into storage.

Noise, dust control and security are the main issues that would need to be addressed if such work had to be carried out whilst the building is in use. Safety considerations would also be important (see Chapter 9).

The presence of deleterious materials in the building whether in use or empty is important. It would inevitably have an impact on its adaptability. The cost of removing such materials could render any adaptation proposal non-viable (see below).

Deleterious materials

Definition

The term 'deleterious materials' refers to any product in a building that might be hazardous to health or which may affect its performance or structure. Many existing properties were built using such materials. It is important they are identified in an existing building, therefore, before it is adapted.

The presence of extensive deleterious materials within an existing building can have serious implications for its adaptation. It can delay or even postpone such work and may affect the market value of the property. This is because the cost and disturbance involved in resolving a problem of deleterious materials in a building can adversely affect the viability of adaptation.

Buildings, of course, are not homogenous products. They contain a very wide range of different materials. Not all of them however are inert and innocuous. When disturbed some release harmful dust or when exposed to catalysts such as moisture, can lead to problems for the property owner or occupier.

Common types of deleterious materials

There are several types of materials that can be classed as deleterious, the most troublesome of which is asbestos (see below). This and the other main hazardous materials found in buildings are summarized in Table 2.8.

Asbestos

Description

Asbestos is a natural, fibrous material containing minerals such as silica and magnesium oxide. This naturally produced material is non-combustible, mechanically strong and is very durable. It has excellent fire and heat resistance qualities. It is also resistant to most acids, alkalis and other corrosive chemical substances. For these reasons it was used extensively in British buildings, especially those constructed or refurbished between 1950 and 1980. If a building from that period has a steel frame or boilers with thermal insulation it is more than likely to contain asbestos (DOE, 1986). In-situ or pre-cast concrete-framed buildings, because of their inherent fire resistance, are unlikely to have the same extent of asbestos as compared to other forms of construction.

The primary problem with asbestos is its potential hazardous nature. Asbestos is deleterious to health when it is cut, drilled, broken or otherwise damaged. This causes asbestos dust to accumulate in the air. Breathing air containing asbestos fibres can result in lung-related diseases such as asbestosis and mesothelioma. According to the Health and Safety Executive, HSE (1999) these ailments, some of which are carcinogenic, cause more deaths than any other single work-related illness.

Table 2.8 Principal deleterious materials and their treatment in adaptation work

Material	*Uses*	*Risk*	*Adaptation response*
Asbestos	See below.	Risk resulting from inhalation of fibres.	See below.
Fibre glass	Insulation quilt in floors, walls and roofs.	Skin irritant. Possible health risk similar to asbestos, but this is not proven.	Seal with polythene sheeting or replace with polythene coated glass fibre insulation.
High alumina cement concrete	Used for structural pre-cast concrete members such as 'T' or 'I' roof/floor beams, from 1954 to 1974.	Prone to premature deterioration in high humidity conditions (i.e. above 70%).	Rarely retained even if structurally sound and humidity levels can be maintained below 70 per cent. Its presence is likely to undermine any adaptation proposal. Strengthening would probably be required, however, in cases where any HACC elements are retained.
Urea formaldehyde foam	Cavity insulation.	Noxious fumes, which cause discomfort; they may be carcinogenic – but, again, this is not proven.	Seal gaps with suitable mastic filler to prevent escape of fumes.
Galvanized mild steel	Wall ties. Restraint straps.	Failure by corrosion, particularly in damp or marine (i.e. salt-laden) climates.	Install stainless steel replacement wall ties and restraint straps.
Lead	Cold water pipework.	Risk of contamination of (especially 'soft') drinking water.	Replace with copper or polyethylene pipework.
	Lead paint.	Risk of inhalation of lead dust during maintenance of lead painted surfaces.	Carefully remove and replace with a safe alternative paint (e.g. water-based paint).
Woodwool slabs	Decking in flat roofs. Permanent formwork in floors.	Premature degradation in high humidity conditions.	Reduce humidity levels to well below 70 per cent in soffit areas and cure roof leaks.
Calcium silicate	Bricks.	Higher risk of shrinkage than with clay bricks.	Care in designing the movement joints (e.g. 9.5 m centres for vertical joints).
Mundic blocks or concrete	Walling material – blocks or concrete.	Very rare, except in Devon and Cornwall. Loss of integrity in damp conditions.	Adequate damp proofing – DPCs, DPMs and surface coatings.
Brick slips	Cladding to conceal floor edges or nibs in cavity walls.	Poor adhesion or inadequate movement joints causing slips to arch or delaminate.	Provision for movement joints and stainless steel fixings to allow mechanical slipping.

The Asbestos (Prohibition) Regulations 1985 and 1992 banned the importation, supply and use of asbestos in Britain. Nevertheless, there are many properties built or adapted before these dates that contain some asbestos (HSE, 1993). It should not be assumed, therefore, that a building earmarked for adaptation contains no asbestos.

The main types and uses of asbestos found in buildings along with an indication of their risk are summarized in Table 2.9. Chrysotile was the first type of asbestos to be used in fibre cement products in buildings in this country. Its use increased gradually between 1910 and 1940, and then more rapidly up to 1960 (HSE, 1999). The other two types were used from the mid-1940s to the late 1970s.

The procedures for undertaking an assessment of asbestos in existing buildings are adequately described by the Loss Prevention Council, LPC (1993) and HSE (1993).

Where asbestos is found in buildings

Some asbestos-containing materials (ACMs) are more vulnerable to damage and more likely to give off fibres than others (HSE, 1993). Those materials containing a high percentage of asbestos are generally more easily damaged.

Sprayed coatings, lagging and insulating board are more likely to contain blue or brown asbestos. Asbestos insulation and lagging can contain up to 85 per cent asbestos and is most likely to give off fibres. Work with asbestos insulating board can result in equally high fibre release if power tools are used. On the other hand, asbestos cement contains only 10–15 per cent asbestos. The asbestos is tightly bound into the cement and the material will only give off fibres if it is badly damaged or broken.

The HSE (1993) has indicated that asbestos is more likely to be found in the materials listed as follows, approximately in order of ease of fibre release (with the highest potential fibre release first):

- Sprayed asbestos and asbestos loose packing – generally used as fire breaks in ceiling voids.
- Moulded or preformed lagging – generally used in thermal insulation of pipes and boilers.
- Sprayed asbestos – generally used as fire protection in ducts.
- Firebreaks, panels, partitions, soffit boards, ceiling panels and around structural steel work.
- Insulating boards used for fire protection, thermal insulation, partitioning and ducts.
- Some ceiling tiles.
- Millboard, paper and paper products used for insulation of electrical equipment. Asbestos paper has also been used as a fire-proof facing on wood fibreboard.
- Asbestos cement products, which can be fully or semi-compressed into flat or corrugated sheets. Corrugated sheets are largely used as roofing and wall cladding. Other asbestos cement products include gutters, rainwater pipes and water tanks.
- Certain textured coatings.
- Bitumen roofing material.
- Vinyl or thermoplastic floor tiles.

According to the HSE (1993) 'the new duty, part of the Control of Asbestos At Work Regulations (CAWR) 2002 will require those with responsibilities for the repair and maintenance of non-domestic premises to find out if there are, or may be, ACM within them. It also requires them to record the location and condition of such materials, and then assess and manage any risk from them, including passing on information about their location and condition to anyone liable to disturb them'.

Actions

The HSE (1993) suspect that there could be as many as 500 000 non-domestic premises in the UK that still contain asbestos materials. If asbestos is discovered or suspected in a building that is about to be adapted,

Table 2.9 Types of ACMs found in buildings

Type	*Risk rating*	*Common uses in buildings*
Crocidolite (Blue)	High This is the most dangerous but least common type of asbestos used in existing buildings. It is highly resistant to corrosive chemicals. Its removal is prone to produce high dust levels because of the material's friability.	Lagging to steam and high temperature/pressure water pipes. Lagging to boilers. Sprayed coatings for fire protection to steelwork. Used in buildings between 1950 and 1969.
Amosite (Brown)	Medium-high This is the second most common type of asbestos found in buildings. It is almost as dangerous as crocidolite but is more hazardous than chrysotile. Removal of this material can produce high dust levels.	Sprayed coatings on steelwork for fire protection as well as heat and noise insulation. Ceiling tiles and insulation boards. Asbestos cement high-pressure water pipes. Thermal insulation material. Used in buildings from 1945 to about 1980.
Chrysotile (White)	Medium-high This is the most commonly found form of asbestos in buildings.	Sheeting in doors and around steelwork for fire protection. Asbestos cement products: profiled roof sheeting, wall cladding and weather-boarding, roof tiles and slates, rainwater goods and accessories, moulded products such as cisterns, tanks and sewer pipes, floor and promenade tiles, and semi/fully-compressed flat sheet and partition board. Textured coatings, plasters and paints. Filters and gaskets. Used in buildings since the early 1900s to the mid-1970s.

a number of actions have to be undertaken. In the first instance it is important to notify the HSE. The presence of asbestos should be noted on plans and indicated on an asbestos register as part of the building's maintenance file.

The means of dealing with asbestos in an existing building depend on the type of material, its overall condition, the degree of damage if any, location and accessibility. Safe handling of the material using licensed, specialist contractors is essential. The following three actions may be taken, the first two of which are designed to manage the problem:

1. Monitor: If the asbestos is intact and well protected against damage or disturbance (before, during and after adaptation) it may be prudent to leave the material in place. This would need to be followed by the introduction of a management system to monitor the situation. Air samples may need to be taken to ensure that airborne fibre levels do not exceed 0.01 fibres per millilitre (F/ml) or cubic centimetre.

2. Encapsulation: Leave the material in place, and seal, cover or enclose it during the adaptation works. A monitoring system within the maintenance regime should be instigated after the adaptation work is completed.
3. Removal: This is a last resort option because of the disruption and costs involved. Once the building is vacated, remove and dispose of the asbestos before any major adaptation works are undertaken. Again, air monitoring should be undertaken to ensure that airborne dust concentrations are less than 0.01 F/ml.

Thus, the highest safety standards must be observed during its treatment or removal. Only a licensed contractor through the Asbestos (Licensing) Regulations 1983 (as amended by the 1998 Regulations) should undertake any work relating to asbestos. Moreover, it must comply strictly with the CAWR 2002, which introduced a new duty to manage asbestos in premises. This duty to manage asbestos in buildings came into effect in May 2004.

According to the HSE (1993), the duty to manage requires those in control of premises to:

- take reasonable steps to determine the location and condition of materials likely to contain asbestos;
- presume materials contain asbestos unless there is strong evidence that they do not;
- make and keep an up to date record of the location and condition of the ACMs or presumed ACMs in the premises;
- assess the risk of the likelihood of anyone being exposed to fibres from these materials;
- prepare a plan setting out how the risks from the materials are to be managed;
- take the necessary steps to put the plan into action;
- review and monitor the plan periodically;
- provide information on the location and condition of the materials to anyone who is liable to work on or disturb them.

It is ironic that the near hysterical rush to clear asbestos from buildings may cause more human exposure to mineral fibres and not less. Safe handling of asbestos removal, therefore, requires a number of precautions. The affected area or rooms must be sealed off from the remaining parts of the building to contain any asbestos dust released by the removal. Other precautions include:

- the use of hand tools only;
- suitable warning signs on approaches to affected areas and dangerous substance labels on waste bags;
- damping-down of affected areas with fine water spray to reduce dust levels;
- use of relevant personal protective equipment (PPE) including suitable respiratory gear and dedicated clothing which must be specially treated or disposed of after work;
- disposal of asbestos in labelled bags to an approved hazardous waste site.

Preliminary considerations

Appointing a professional adviser

Many factors will influence the choice of professional adviser for an adaptation project. Traditionally clients would instinctively go to an architect to handle the project. However, other construction professionals such as building surveyors and those within multi-disciplinary practices are now competing with architects for such work.

Normally the appointment of a consultant should be done with reference to an appropriate set of conditions of engagement. Both the Royal Institution of Chartered Surveyors (RICS) and Royal Institute of British Architects (RIBA), for example, publish Standard Conditions of Engagement. They also publish fee scales for consultants dealing with construction projects (see Chapter 11 on typical fee levels).

Expertise and reputation as well as experience in handling the type of work proposed are vital requirements. Similarly, such criteria, of course, apply to the selection of the contractor as well.

Once the decision to adapt a building has been made, the construction consultant involved has to plan, programme and co-ordinate the work. This requires good technical knowledge as well as sound management skills.

All building operations are often complex and full of uncertainties. Adaptation work is no exception. In some ways it is even more problematic than new build in terms of accidents and hazards (McLennan et al., 1998) – hence the reason for the higher level of risk associated with this work (see Chapter 11).

Sources of finance

Private sector

Clients will have to consider which sources of finance they wish to employ for their adaptation project. Some may be fortunate to depend solely on internal sources. For example, they may be able to rely on reserve funds or selling off some assets to raise some capital. These, though, are often never sufficient to cover the whole cost of the project.

More usually, therefore, the client will have to rely on external sources of finance. Table 2.10 summarizes the three main types of finance.

Public sector

With ongoing constraints on government spending public bodies have had to consider other ways of raising finance for their capital spending programmes. European Grants as well as block grants from central government can finance specific capital projects, for example medium to large adaptation schemes as well as new-build projects. These schemes usually require high levels of expenditure not only to upgrade the buildings but also to address their maintenance backlog.

Table 2.10 Types of finance for adaptation work

Type of finance	*Characteristics*	*Sources*
Short term	Payback period usually under 1 year. This is the most expensive type of finance (sometimes called credit). Bank overdrafts or short-term loans are typical examples of this type of finance. It should only be used to fund very minor adaptation schemes.	Banks. Building societies. Finance companies.
	Grant – for improvement, conservation, repair work.	Grant awarding body such as the local authority, English heritage, etc.
Medium term	Payback period usually between 12 months and 5 years. A term loan, usually up to a period of 3 or 4 years.	Banks. Building societies. Finance companies.
Long term	Payback period usually over 5 years (i.e. 15 to 25 years). Mortgage. Lease and sale-back.	Banks. Building societies. Finance/Investment companies. Local/central government funding bodies. Pension funds.

Public private partnerships

Until recently an option such as Private Finance Initiative (PFI) was considered one of the best ways to fund major adaptation work. PFI involved a partnership arrangement between the public authority and private sector participants (e.g. contractors, funding bodies such as banks and pension funds, and large companies, etc.).

PFI was launched in 1992 by the then Conservative administration. The current government did not abandon this scheme on taking office but rather developed and refined it. The new term is Public Private Partnerships (PPP). The primary purpose of PPP, however, remains the same: to increase the flow of capital projects against a background of restraint on public expenditure (RICS, 1995a).

The key features of PPP are as follows:

- There are two parties to a PPP project:
  - Public sector purchaser,
  - Private sector operator.
- The PPP project brief is based on a definition of service requirements.
- The project is defined by function (output specification) rather than a technical solution (input specification).
- The aim is to encourage innovation in bidders' proposals in meeting a functional specification.
- PPP entails a fundamental change for government's relationship with the construction industry: no major sole public sector capital investment; and shift to revenue expenditure on services.

Some local authorities are now using partnership schemes to fund large refurbishment projects. For example, since 2000 many local authorities throughout the UK have been using this method to procure a range of multi-million new build and refurbishment programmes to their stock of school buildings. Some of these refurbishment schemes involve extensions as well.

Major problems with implementing some PPP school refurbishment schemes, however, have led to some local authorities to reconsider their procurement options. The Construction News (Thursday, 26 May 2005), for instance, reported that Dumfries and Galloway Council had scrapped its £110 million schools refurbishment programme. It was said that 'refurbishment is the difficult part of a PPP scheme'. Chapter 11 indicates why this is so by addressing the risks associated with refurbishment schemes.

It is possible that this form of partnering will continue as experience is gained and lessons learned by its users. Curbs on public sector expenditure will continue to encourage the use of this method of financing of large capital projects.

Grants

Sources

Financial assistance may be obtained for undertaking a feasibility study as well as to support the building work associated with an adaptation scheme. Not all adaptation schemes, of course, can attract financial aid. Certain housing improvement projects may be eligible for a grant but this depends on the nature and extent of the works.

There are varieties of routes to help secure funding, especially for old, redundant buildings. Some of the most common sources are as follows:

- Architectural Heritage Fund.
- Central government departments (such as DTI for grants to fit solar panels).
- Churches Conservation Trust (for conserving redundant Anglican churches).

- Countryside Stewardship Scheme.
- English Heritage/Historic Scotland.
- Environmental bodies.
- European Regional Development Fund.
- Heritage Lottery Fund.
- Historic Chapels Trust (for conserving redundant buildings of all other denominations and faiths).
- Housing associations.
- Investment finance authorities.
- Local authorities.
- Local conservation bodies.
- Property finance companies.
- Regional Development Agency (RDA).
- Rural Development Commission (RDC).

However, most of these sources can guarantee funding for the entire project. Short-term grants up to 25 per cent, for example, might be available from the RDA for working farms. Moreover, any financial assistance given usually comes with strict conditions. Such conditions might be related, for example in the case of the adaptive reuse of an historic building, to the materials or methods used. Only Grade I listed buildings are likely to receive funding from English Heritage.

Conditions

The award of a grant is not automatic in most cases. There may be restrictions as to the use and sale of grant-aided premises. The RDC, for example, imposes strict conditions on the award of a redundant buildings grant. If the premises are sold or taken out of productive use within 5 years of the grant being paid, the grant must be repaid in accordance with the scale shown in Table 2.11.

Grants will normally not be available for works such as routine maintenance, any internal repairs except those relating to structural stability, extensions, and refurbishment.

Common types of grants

The following is a list of the common types of grants available for a range of adaptation works:

- Building conservation grant: Repairs to listed buildings, particularly those of Grade I (A) or II* (B), may be available in certain circumstances. Local authorities or English Heritage/Historic Scotland are typical sources of such funding.
- Common repairs grant: Repairs to multiple-occupancy properties such as tenements and blocks of flats can be partly funded by a common repairs grant from the local authority.

Table 2.11 Typical payback conditions on the awarding of a refurbishment grant (RDC, 1997)

Period within which sale occurs	*Grant repayable (%)*
1–12 Months	100
13–24 Months	80
25–36 Months	60
37–48 Months	40
49–60 Months	20
Over 60 Months	Nil

- Lead replacement grant: A grant of between 75 and 90 per cent may be available from the local authority to replace lead pipework used for the cold water supply, including the replacement of any lead tank. Households with a lead in water content at or above 50 ug/l with a child under 12 years old or with a pregnant household member are often given a priority 1 rating by some councils.
- Home improvement grant: Local authorities have some reserve funds for allocating home improvement grants. These are intended to help homeowners install basics amenities as prescribed in the Housing Acts and Housing Grants, Construction and Regeneration Act 1996.
- Redundant buildings grants: The RDC gives financial assistance for developing existing rural properties in England. It 'provides grants for the refurbishment of redundant buildings to bring them back into productive business use or to enhance their current business use. Grants cannot be given if work on the project has already commenced'.
- Renovation grant: Under Part I of the Housing Grants, Construction and Regeneration Act 1996 homeowners in England can apply to their local authority for a house renovation grant or disabled facilities grant (DFG).
- Solar panel grant: The Department of Trade and Industry via the Energy Saving Trust (www.est.org.uk/myhome) has allocated funds to encourage homeowners to install solar roofing panels as an energy-efficiency measure. For example, the grant available for a bolt-on or non-integrated system to replace roof tiles is the lesser of £2500/kWp or 50 per cent of total eligible costs per kilowatt peak installed.
- Other energy-efficiency-related grants: Energy-efficiency improvements are available (if referred by a health professional) from some local authorities to households where poor health is thought to be exacerbated by poor housing conditions. Edinburgh City Council's 'Warm + Well' is one such scheme.

Grants of up to £300 are available from most of the power utilities suppliers (e.g. Southern Electric, Scottish HydroElectric, etc.) towards cavity wall insulation. Grants are limited to mainland UK homeowners. Flats and mid-terrace houses receive £75. Bungalows, end-terrace and semi-detached houses receive £150. Detached houses receive £300.

A grant and advice programme from the Government Renewables scheme assists non-profit community groups (e.g. registered charities, LAs, HAs) to develop renewable energy projects. Technical Assistance grants of up to £10 000 and capital grants of up to £100 000 are available.

Constraints

Spatial

The space available for the proposed adaptation will influence its form and size, although some work requiring only a change of use may not entail any space gain or loss. In any event, a conversion scheme is still likely to involve adjusting the internal layout of the building to accommodate the new or modified use. For example, changing a single-storey barn building into a two-storey dwelling will necessitate the creation of a staircase to facilitate access to the new upper floor. This will inevitably have an impact on the both ground and first floor layouts.

The surveyor, architect or other building professional acting as the lead consultant in an adaptation will probably compile some schematic sketches of the floor layouts. These can help in the initial planning of the adaptation and give the client an idea of the proposed changes.

Technical

The building's form of construction and condition will determine the extent of any improvement works or repairs that may need to be carried out at the same time as the adaptation work. A condition survey or diagnostic investigation prior to compiling the main scheme drawings will be helpful in identifying the extent of repairs required.

Financial

The extent as well as the source/s of finance available for the project must of course be established at the inception stage. The staging or phasing of payments will depend on the expected duration of the project. Any monetary constraints on running expenses when the refurbished building is in full operation should also be clarified. These can influence the degree of thermal insulation and other energy conservation measures.

Temporal

The client may have a specific date by which the adaptation work must be completed. In addition, time restrictions during the project need to be ascertained. For example, the client because of potential security or disturbance problems may not permit night-shift working.

Legal

Buildings are subject to more statutory controls than almost any other asset. In the context of adaptations there may be legal restrictions on:

- the degree of adaptation of Listed Buildings or properties within a Conservation Area regulated by the Planning (Listed Buildings and Conservation Areas) Act 1990, in England and Wales, and the Planning (Listed Buildings and Conservation Areas) (Scotland) Act 1997;
- position of the extension, because of adjoining owners rights, building lines, planning approval, easements, restrictive covenants;
- shape and size of alteration or extension (e.g. planning approval, fire authority requirements);
- spatial requirements dictated by the Offices, Shops and Railway Premises Act 1963, etc.;
- circulation requirements influenced by the Fire Precautions Act 1971;
- materials used, construction standards [e.g. Building Regulations 2000 (as amended)];
- planning restrictions on change of use (e.g. Planning Acts 1990 and 1997);
- design requirements (e.g. Planning Act 1997).

Relaxation of the regulations may be required if full compliance is not feasible in certain circumstances. Tight vertical space for example may prevent the minimum 2 m headroom being achieved over the full length of a staircase. In such cases, however, a formal application for relaxation in England and Wales must be made to the Secretary of State for the Environment, Transport and the Regions. In Scotland the Scottish Executive administer relaxation matters.

Clients property requirements

Clients may have a policy as regards certain aspects of their adapted building. For example, sustainability may be a distinct company objective. The client may have specific requirements as regards the types of materials and fuel system to be used in the adapted building.

The client may require a minimum number of car parking spaces within the boundary of the property. This of course is only practicable for detached or semi-detached buildings. Access to the premises may be moved to a more convenient location. Alternatively the existing access could be enlarged.

The extent of security to the property in the form of fences and gates may require modification. This could also include addition measures such as security lighting and intruder sensors, and alarms.

Physical site constraints

Many urban sites are very tight. They do not allow for easy access to or increase in the existing building. The position of the services on and adjacent to the site may also influence the location and extent of any lateral extension.

Environmental

In compliance with sustainability, it is important to identify the building's potential for environmental improvement. A number of measures could be undertaken to enhance a building's sustainability, but this will depend on the type and extent of adaptation.

According to the Energy Research Group et al. (1999), the following measures should be considered:

- Increasing daylighting through roof lighting – using more or larger skylights rather than dormers on roof slopes in sensitive locations.
- Curbing overheating through the use of internal/external louvers or blinds (or electrochromatic glass).
- Reducing heating demand through installation of draught lobbies and by adding insulation to external walls and roof.
- Improving envelope performance by installing better windows and doors: double-glazed units with low-emissivity glass.
- Maximize natural ventilation by opening sections of windows and rooflights.
- Controlling ventilation and casual infiltration (i.e. leakage in) and exfiltration (i.e. leakage out) – using trickle vents.
- Ensuring good performance of active systems through better control: time clocks, thermostats, building energy management systems, and more efficient fittings: lights, heat emitters.
- Maintaining good indoor air quality (see Chapter 8) by substituting natural for synthetic finishes, such as linoleum, and water-based paints.

These and other energy-efficiency measures are described more detail in Chapter 10.

VAT implications

The value-added tax (VAT) regulations affecting building work are not straightforward. Currently, both new house building and conversion of commercial properties for housing are zero rated as far as VAT is concerned. VAT can, however, be reclaimed for the cost of alteration work (but not repairs and ordinary maintenance) on listed buildings provided that the rules laid down by the Customs and Excise are complied with. Such rules state that the building must be for domestic purposes or for use by an approved charity and the alterations must have been granted listed building work prior to commencement of the scheme.

VAT refunds, though, are not available in certain cases. For example, they are not available to developers or to those converting the building into a hotel. Professional fees are also not exempt from VAT.

In contrast, the full VAT rate of 17.5 per cent is imposed on the refurbishment or conversion of existing dwellings. Even adaptations such as the maintenance, conservation or change of use of listed and unlisted buildings incur VAT at the full rate. This is effectively a subsidy on new construction on greenfield sites, whereas work to existing buildings is penalized by the imposition of VAT (Urban Task Force, 1999).

The British government is looking into this VAT anomaly as regards building work. As it stands this situation militates against sustainability by inadvertently discouraging the retention and improvement of scarce resources such as dwellings and vacant non-residential buildings. Rather, property owners need to be encouraged to modernize their properties instead of redeveloping them or letting them become dilapidated. It is for these reasons that the Urban Task Force (1999) made the following recommendation:

> Harmonise VAT rates at a zero rate in respect of new building, and conversions and refurbishments. If harmonisation can only be achieved at a 5% rate, then a significant part of the proceeds should be reinvested in urban regeneration.

The UK government in late 2000 proposed a 5 per cent VAT rate on church repairs to help preserve an important part of Britain's rich-built heritage. Conservation bodies will be pressing the government to extend this to all listed buildings. Many consider that the current system is inequitable because it penalizes those who maintain our historic-built environment and encourages alterations to it. However, such a change to the VAT rules may conflict with the tax requirements of the European Union.

For more details of the VAT requirements relating to building works see the leaflet 'VAT: Protected buildings' (708/1/90), available from VAT offices.

The adaptation process

Participants in the adaptation process

The roles of the six main participant groups in the adaptation process are illustrated in Table 2.12.

Designing and procuring the adaptation of an existing building is not fundamentally different from the process of acquiring a new building. The main stages of the adaptation process are summarized as follows:

Client's brief

Most clients know what they want, but this is not always what they need. The needs of the client in any building project are paramount. It may be however that in relation to an existing building the client has not fully though through their requirements. Is extra space needed, for example? Can it be found within the existing building envelope, or must an extension be provided? If an extension is required, what size and where? The client's brief will always be influenced by budgetary concerns. Individuals as opposed to corporate clients are likely to have limited funds or limited access to funds. Therefore, translating a client's requirements correctly is crucial.

Choice of options

The range or extent of adaptation technically and financially feasible will need to be established. Consideration needs to be given to the extent of any volume changes, alterations and upgrading works. The main options can be evaluated using a matrix as indicated in Table 2.5.

Table 2.12 Participants in the adaptation process (adapted from Nutt, c. 1993)

Participant	*Involvement*	*Examples*
Investor group	Those who arrange capital to fund adaptation projects and purchase buildings.	Banks, finance companies, insurance companies, pension funds.
Producer group	Those who design, specify, cost and execute adaptation projects.	Architects, builders, engineers and surveyors.
Marketing group	Those who find users for buildings and buildings for users.	Estate agents and surveyors.
Regulator group	Those who ensure compliance with the statutory requirements.	Building control, fire authority, health and safety executive.
User group	Those who occupy, manage and use the building.	Individual users, Facilities and maintenance managers.
Developer group	Those who undertake some or all of the investor, producer and marketing roles above.	Contractors, development companies.

Outline scheme design

This will delineate schematically the layout, plans, sections and elevations of the proposed scheme for the client's consideration. The use of a cardboard model can be helpful to illustrate to the client the effects of an extension.

Prepare production information

Once the client has agreed the outline scheme, it can be developed to prepare more detailed design drawings to accompany applications for planning and building control approval.

Consideration should also be given to the compilation and form of tendering, contractor selection, preparation of specification and bills of quantities. These are dealt with in more detail in Chapter 11.

Monitoring building operations

Competent supervision of the works in progress is required to ensure adequate quality control. On large refurbishment contracts, the supervising surveyor or architect may employ a clerk of works to monitor the quality of work.

Formulate aftercare strategy

The building surveyor should compile a maintenance manual for the completed building. The information contained within any health and safety file under the Construction, Design and Management Regulations (CDM). In addition, to ensure the ongoing performance of the building a planned preventative maintenance programme would need to be instigated. This is dealt with in more detail in Chapter 11.

Design guidelines for adapting buildings

Preamble

It is often recognized that many remodelling projects, particularly those involving rural properties can be problematic. For example, the 'conservation' of many historic farm or vernacular buildings, has not always been wholly successful. One of the chief deficiencies with such schemes is that the original character of the building is completely lost during alterations or conversion to other uses. This can have an adverse effect on both the building and its surroundings. Similarly, unsympathetic extensions such as box dormers can interrupt or have a deleterious effect on the roofscape of an urban street.

To achieve a sensitive and successful adaptation of existing buildings some appreciation of architecture and an understanding of its importance in the built environment are required. This requires taking into account design principles such as balance, symmetry and style. In any case, pre-application enquiries of the local planning authority may be prudent to ensure that the proposal will satisfy planning policies and guidelines.

It is acknowledged that not everyone would appreciate the imposition of design requirements. Some architects and other building designers may perceive that prescribed guidelines are a constraint on their design flair. Others, however, welcome any such assistance if it helps to achieve a more sympathetic and successful adaptation scheme.

Design guidelines, however, are not intended to restrict innovative or exciting designs. Rather, they are meant to help achieve some consistency in adaptations so that inappropriate or incongruent designs are minimized.

In large adaptation schemes architects are often used because of their aesthetic and spatial planning skills. As the principal building design specialist, an architect can obviously play an important role in achieving a successful scheme.

Altering and extending buildings

Contexts

Local councils often have guidelines for alterations and extensions to residential buildings in their area. Such adaptation work to non-residential properties will normally be treated on its own merits. In any event, the following design principles will usually apply.

The location, nature and age of the building will influence if not determine the extent and nature of adaptation proposals. Is the property in a Conservation Area or is it Listed? Any previous reports on the property should be consulted as these can shed light on its history.

Policy

What is the local authority's policy as regards adaptations? The professional adviser should make enquiries to determine if the local council produces any written guidelines or requirements for adaptation proposals in its area.

Matching demand and supply

It is vital to match the user and the building to achieve a successful adaptation. If there is a mismatch, this suggests either the wrong user or the wrong building.

Degree of adaptation

It is important to minimize drastic changes to the existing layout if possible – unless of course this is forming part of the work. The extent and disruptive impact the proposed works should be established. If they are excessive this is indicative of a mismatch between the users needs and what the building can readily provide. It is always better to work with the building than against it – this will be cheaper and less likely to impact adversely on neighbouring properties.

Repair costs

To keep overall repair costs to a minimum, use most dilapidated areas for concentrations of re-arrangement of the building. This could include removing dilapidated rear or side extensions rather than undertaking expensive repair if sufficient accommodation will be retained, and the work does not cause any structural or legal problems.

Spatial factors

Circulation is the key to successful internal planning because it affects the way in which the building is used and laid out. It will be determined by user needs and statutory requirements (especially means of escape). The main aim is to make the circulation compact and well organized to avoid overuse of precious lettable space, without compromising fire safety. It is also important to respect existing proportions and construction when sub-dividing (e.g. column grids, beam casings, moulded cornices, panelling, etc.).

Basic design criteria

There are twelve key design criteria that apply principally to domestic adaptation schemes (based on Williams, 1995). They can however relate to non-residential properties as well so long as allowances are taken for differences in scale, form and use. The criteria are as follows:

1. *Harmonizing*: Any external modifications to the existing building, such as a proposed extension, should tone-in with the surrounding properties. The design, for example, can either match or contrast with that of the existing building. In any event, the extension needs to be consistent and compatible with the existing building in terms of scale as well as style. The alteration or conversion should not necessitate the erection of unacceptable new buildings to accommodate existing uses or contents.
2. *Overlooking Neighbours*: Providing minimum distances between parallel windows at the front, side and back so that 'overlooking' is minimized. This is a particular problem with rear extensions to mid-terraced properties.
3. *Dominance*: Any external works should form an unobtrusive and subordinate addition to the original building. The massing of the extension needs to have regard for the original building as well as surrounding properties. The buildings should be capable of alteration or conversion without any or with only localized minor areas of rebuilding. Where complete or substantial rebuilding is required the application will be treated as one for new development.
4. *Overdevelopment*: Local authorities usually require a minimum free garden space in relation to housing developments. Similarly, they may stipulate minimum car parking and access requirements. Similar restrictions on high densities may be imposed on non-residential adaptation schemes. The proposed use should be accommodated within the confines of the existing buildings without the need for substantial extensions or other incongruous additions such as garage blocks. An extension may be acceptable where it allows for a more sympathetic conversion of the existing building but will not be permitted where its purpose is to facilitate a high-density conversion.
5. *Existing building lines*: Front extensions are normally not encouraged because of the potential to encroach the building line. A distance of at least 1 m is usually required between a house extension and the existing building line.
6. *Minimum boundary distance*: Some authorities set a minimum distance to the various boundaries. One metre is a typical boundary minimum for residential schemes. This dimension is likely to be much greater for commercial buildings to accommodate vehicular access.
7. *Flat roofs*: It is preferable from a design perspective for lateral extensions to domestic buildings to have pitched roofs. Flat roofs for extensions over one storey in height are usually not encouraged for residential schemes. There is no reason why a flat-roofed finish should not be allowed for a single-storey residential extension at the rear.
8. *Dormer extensions*: Any new dormer should not project above the ridgeline of the existing property. Box dormers are unsightly and should be restricted to rear elevations in sensitive locations. Ideally dormers should have a pitched roof (see Chapter 5). As an alternative either recessed dormers or skylights should be considered in sensitive locations.
9. *Junctions*: Unsightly bonding on front elevations can be avoided by 'setting-back' the walling of the extension at least 102 mm – the width of a brick. Aesthetically and structurally it is more prudent to make the junction explicit. Attempting to mask the junction by, for example, rendering over the joint between the old and new walling may not work.
10. *Garages*: These must normally not project in front of the foremost part of the dwelling and should be at least a car's length away from the entrance. The site boundary will also have an impact on the location of the garage.
11. *Intrusive suburban alterations*: Intrusive suburban alterations such as domestic style external joinery, porches, dormer windows, brick or stone chimney stacks, and external services (meter boxes, soil and

vent pipes, etc.) must be avoided. Windows and doors in agricultural buildings are distinctive and different to domestic styles. This must be reflected in the joinery details of the conversion scheme. Intrusive suburban alterations to the setting of the buildings such as fencing in crew yards, non-traditional boundaries, non-indigenous planting and excessive paving must be avoided. In countryside settings it will be important to include proposals for planting with indigenous species to help screen the development.

12. *Retaining original character*: Alterations to the existing fabric should be kept to a minimum. Existing openings should be used and any new openings minimized. In the case of a residential use, new openings should be restricted to the minimum required to make the building habitable. Whilst a rooflight can be a useful alternative to a new window, a proliferation of such openings should be avoided. If possible they should be restricted to concealed roof slopes or other unobtrusive positions and fitted flush with the roof opening. They should be restricted to the smallest size required to give adequate ventilation. One-twentieth of the floor area of the room concerned will be the usual guide in assessing the need for new openings.

The use of innovative or modern-style materials on the external fabric of the proposed remodelled building should be avoided. Wherever possible materials that retain the vernacular or traditional appeal of the property should be used. For example, clay pantiles would be more appropriate than, say, aluminium profile roof cladding for the new roof coverings of a rural building.

Adapting old buildings

Objectives

Adapting old buildings brings with it problems over and above those encountered in ordinary adaptation schemes. Such buildings have more than their fair share of structural and fabric problems. They usually contain materials and construction methods that are uncommon if not obsolete, such as lath and plaster or solid (un-insulated) walling.

Thus the primary aim of any adaptation scheme involving the conversion or renovation of an old building is to achieve a sustainable and beneficial reuse. Such a scheme, however, must be sensitive and appropriate to the building's status, condition and capabilities.

The location of an old building and its relationship to other properties must have a bearing on any such adaptation scheme. The property, for example, may be situated within a large estate. It is likely to be surrounded by open fields and landscaping as well as possibly other buildings. The building being adapted may only be of secondary importance in the estate. The use and style of the main complex and other areas of the estate must also be taken into account as these may restrict or enhance the usefulness and potential of the building being adapted.

Adaptation criteria for old buildings

Whatever solution is made for the adaptation of an old building, it should be based upon carefully considered and selected criteria. The criteria adopted, however, will be dictated by the context of the building as well as by exigencies of the property owner. The following list, for example, describes some of these influences:

- Philosophical considerations – policy articulated by client; conservation requirements of heritage funding bodies; local planning authority's policy on extensions and conversions.
- Present location and proposed location within the new scheme.
- Relationship of building to other buildings and functions on the estate.
- Buildings or space required by the new scheme/use.
- Degree of protection in respect of Listed buildings and properties in Conservation Areas.
- The building's historical and architectural significance.

- The building's condition and flexibility for change.
- Size and form of the existing building.
- Economics:
  - viability;
  - current asset value;
  - value after adaptation;
  - costs in use.
- Functional potential of adjacent facilities and services.
- Legislation relating to new use, for example:
  - Building Regulations; Planning controls
  - Offices, Shops and Railway Premises Act 1963
  - Fire Precautions Act 1971
  - Health and Safety at Work Act etc. 1974
  - Disability Discrimination Act 1995.

Many local authorities have strict design guidelines for remodelling work associated with properties in important or sensitive situations (Williams, 1995). It is important therefore to consult the council concerned for their design requirements relating to old buildings. This will minimize any complications as regards obtaining the necessary statutory approvals.

Frequently adaptations to old buildings often involve or form part of a conservation scheme. The key conservation principles of minimum intervention and reversibility should be followed in most cases. 'As much as is needed as little as required' is a good rule to adopt. The following principles are typical requirements in this regard and should be adopted when dealing with the adaptation of old buildings, especially those which are Listed or in Conservation Areas.

Pre-adaptation building audits

Preamble

This section deals with pre-adaptation building audits. Post-adaptation building audits are covered in Chapter 11.

An audit is a check of a process and its product – whether financial, managerial or technical. The audit that most people are familiar with is of course an accounts audit. This where an auditor is reporting to show a true and fair view of the state of affairs of a business at a given date and of the profit and loss results for the period ended on that date.

In the context of buildings an audit verifies certain important aspects of their performance (see Appendix B). Like building appraisal and architectural design it is inherently a problem finding and problem-solving process.

A successful adaptation scheme is only the initial goal for any sustainable and viable property. Checks or audits need to be undertaken at regular intervals depending on the exigencies involved. These will help ensure a more sustainable future for the building. The main audits that can be undertaken on a building during its service life are summarized as follows:

- Access audit — To determine the extent to which the building offers a barrier free environment (see below).
- Condition audit — To determine the building's current state of repair (using a schedule of condition, and possible diagnostic investigation, as indicated earlier).

- Crime/Security audit — To identify areas around and parts of buildings susceptible or vulnerable to burglary and vandalism (see below).
- Design/Technical audit — To identify potential problems in key details in the proposed new building or adapted existing building (see below).
- Energy audit — To identify the building's potential for environmental and thermal improvement (see Chapter 10).
- Fire safety audit — To assess the fire risk and safety levels associated with the Building (see below).
- Maintenance audit — A review of the financial, technical, legal and organizational aspects concerning the aftercare of the building (see Chapter 11).
- Premises audit — May include a maintenance audit, but will also consider space management (i.e. space audit – see below) and environmental issues in the assessment (see Chapters 8 and 9).
- Safety audit — To identify those aspects or parts of the property or site that are potentially hazardous to health and safety (see Chapters 8 and 11).
- Space audit — To determine the amount and type of space available for a proposed conversion scheme (see Chapter 3).

Access audit

Preamble

As part of the building assessment prior to the adaptation of a property, a number of factors may have to be taken into account. An access audit is one such check. John Miller, a Member of the Centre for Accessible Environments (CAE), has defined an access audit as:

> A walkover survey to assess a building for ease of access and use by disabled people who may be employed in or visitors to the building. The report of the audit will take the form of a written report giving a detailed assessment against agreed criteria, together with detailed recommendations for modifications/improvements.

This involves among other things accessibility and facility improvements such as those listed in Table 2.13. The main design features of these disability requirements are shown in Figure 9.19.

Clients requirements

The following points will usually need consideration in the brief of most medium- to large-scale adaptation schemes:

- Who are the disabled people – are there any restrictions, what range of disabilities need be covered?
- Are there separate routes or areas for visitors and staff – is the audit required to be the same in both areas?
- What criteria are to be used to assess the building? CAE publications, CIRIA's *Buildings for Use by All*, Capability Scotland guidelines, BS 5810, or some other standard?
- Do the recommendations need to be costed and/or prioritized?
- Is the means of escape in case of fire to be considered?

Assessment criteria

The criteria described in the document *Designing for Accessibility* will normally be adequate for most access audit purposes. These are summarized in Table 2.14.

Table 2.13 Typical domestic accessibility and facility improvements

Improvement	*Means*	*Adaptation implications*
Ramps	Ramps with 1:12 slope inside and outside the building at strategic locations such as entrances and changes in floor level.	Space required. External door steps will need removing.
Handrails	Additional handrails inside and outside the dwelling.	Fixing provisions to walls.
Internal circulation	Doors widened or altered for better access, especially adapted kitchen, and other alterations to ease access.	New lintels required?
Vertical access	Hoist to the bathroom or bedroom. Stair-lift.	Space for hoist/platform lift. Forming new lift well through floor.
Bathroom	New bath/shower, or relocated bath/shower, or specially designed or adapted toilet or relocated toilet.	Modify existing layout or position of services.
Electrical	Modifications to lighting and power systems, for example.	Raised sockets or lowered switches.
Emergency alarm	Connection to an emergency alarm/intercom system.	Position for ease of maintenance.
Ironmongery	Easy to use door and window handles for the elderly and infirm as well as people in wheelchairs.	Rounded handles at reachable positions.
Sanitary ware	Low-level wash hand basins and WCs, with handrails supports.	New, specially designed WCs.
Worktops	Easy clean/reach surfaces.	Laminated plastic worktops at a height of not more than 900 mm.
Warden	Accommodation for a warden.	Modify existing space to include warden facility. Provide extension.

Output

Feedback as part of a technical audit or an access survey report could highlight deficiencies in accessibility, with suggestions as to how best these could be overcome. Any recommendations should of course be included in the design proposals for the adaptation work. These will not only help obtain the necessary statutory approvals but will also go a long way to achieving a more user-friendly and hence successful building.

Crime audit

A crime audit may form part of a modernization strategy to a housing scheme or commercial building. It would first of all highlight vulnerable or insecure parts of the building that are prone to intruders or vandalism. Recessed doorways, internal corners or other blind spots around the building, roof parapets, skylights, windows overlooking a flat roof, are all susceptible to these problems.

Such an audit would then identify measures that could be undertaken to reduce the incidence of vandalism and enhance security. This could include installing improved or additional locks on doors and windows. It may also involve installing halogen exterior lighting in vulnerable locations. The measures described in Chapter 9 to enhance school security can be applied with appropriate modifications to other vulnerable buildings being adapted.

Table 2.14 Assessment criteria for access audit

Requirements	*Standards*	*Examples*
Audio availability	Deaf loop system	Linked to load speaker system. Gives verbal prompts at key locations in the building.
Door types and furniture	Weight Robustness Ease of operation	Neither too light nor too heavy. Can resist attempted forced entry. Ease of opening – especially for the infirm and elderly.
Openings	Width Position Fire/escape doorway/s?	Minimum 900 mm Conspicuous To a designated point of safety (e.g. next to car park).
Ramps	Width Gradient Surface Position Handrails Protection	Minimum 800 mm Maximum 1:12 Non-slip Front entrance and external fire escape doorways. Balustrade and along wall Canopy? Barriers? Kerb?
Visual signage	Braille Large print Lit signs	Touch pads next to doors or changes in direction. Conspicuous and easy to read signage.

Fire safety audit

Preamble

Fire risk assessment is becoming increasingly recognized as being an important element of property management. The purpose of fire safety in adaptation or new build is to protect:

- Life — Human and animal.
- Property — Type of buildings and their contents.
- Processes — Minimize disruption to or cessation of the work or activities being undertaken in the building.
- Market share — Ensure the ongoing use of the property to allow occupiers to maintain their position in the market that they are operating in.
- Environment — Avoid smoke pollution, water damage, and waste resulting from a fire.
- Heritage — Preserve historic and cultural assets, whose loss would be irreplaceable.

The provision and maintenance of sufficient levels of fire safety is needed to protect employees in all workplaces. The same applies to occupiers of or visitors to public buildings. A fire risk assessment is the main way in which this is achieved. It involves assessing the fire safety standards in a building before, during and after adaptation. The purpose of such an assessment before and after adaptation is to highlight the need for maintaining current provisions or identify any essential improvements. A fire safety assessment during adaptation works needs to consider the risk management issues outlined in Chapter 11.

Assessment criteria

A fire risk assessment, which for workplaces must be in writing where there are five or more employees, should cover a variety of issues. Even in domestic premises a fire risk assessment can prove worthwhile.

In relation to a building before or after adaptation this process needs to be undertaken in the following stages:

- *Stage 1 Identify fire hazards* Consider the presence of sources of fuel (e.g. substances such as paints, solvent-based products, which will burn) within the building, and potential sources of ignition, such as faulty or high-voltage electrical equipment. Identify work processes that may entail a fire risk.
- *Stage 2 Identify the type and location of people at risk in the event of fire* The type of work or other activity within a building may increase the risk of a fire occurring or in the occupiers not responding promptly to a fire outbreak. Some occupiers may not be able to respond without assistance (e.g. children, elderly and disabled people or people asleep or inebriated with alcohol or drugs), or may be unfamiliar with the building (e.g. first time visitor). The type and use of building will clearly determine the type and mix of occupiers.
- *Stage 3 Evaluate the risks* Determine the adequacy of the existing fire safety measures. Instigate measures to reduce if not eliminate fire hazards. Remove or minimize the use of flammable substances or combustible materials in the building. Good housekeeping will help to achieve this. Consider the consequences of a fire occurring and undertake any improvements needed. Compensating for fire hazards can be done by providing an appropriate system of active fire protection (see Figure 4.15) or by re-arranging the use or work activity within the building, so that everyone present can detect any fire.
- *Stage 4 Record findings and action taken* Prepare an emergency plan. Inform, instruct or train respectively visitors and employees in fire precautions. A proper balance between active and passive fire precautions should be achieved. However, this may be difficult to achieve, particularly in the context of historic buildings (see fire safety and historic buildings in Chapter 4).
- *Stage 5 Monitor and review* Keep assessment under review (at least annually). This may need revising if circumstances in the building change.

Precautions

These will be influenced by a variety of factors, such as:

- The type of building. The building's height, size and form of construction will directly affect its level of fire safety.
- Number and size of exits will influence the extent of active fire precautions to compensate for any deficiencies in passive fire precautions.
- The contents/materials used or stored in the building – to help determine the fire and smoke loads (these loads refer to the fuel available for fire and smoke).
- Number of employees/occupiers and visitors in the building at any one time. The ability and condition of the occupants will also be an important factor. Many of the occupiers in a hotel or hostel, for example, may be asleep when a fire in the building occurs. Early warning therefore is essential in such cases. The fire safety system should be based on the assumption that some of the people in the building may not be familiar with its fire escape provisions.
- The need for assisted emergency evacuation – especially if any disabled or infirm people use the building.
- Means of fire detection and alarm.
- The extent and type of fighting equipment and installations – testing procedures.
- Access and facilities for the fire brigade – areas suitable for fire tenders, fire hydrants.

Design audit

Preamble

In this context a design or technical audit is a detailed check of critical parts of the proposed adaptation work. As part of the final evaluation stage of the design process it is becoming increasingly important to undertake this type of audit. The uses of Value Management and Total Quality Assurance have encouraged the need for rigorous control measures at all levels of the design and construction processes. These checks can help achieve a more cost-effective and better quality design.

A technical audit essentially involves an assessment of the proposed contract documentation. The process consists of a critical analysis of construction detailing and an examination of the specification. The person carrying out such an exercise must:

(a) have a detailed knowledge of building construction and adaptation work;
(b) be efficient and thorough;
(c) be methodical;
(d) be impartial;
(e) maintain good relations with his/her fellow professionals involved in the design process.

Main objectives

- To minimize potential defects, especially in the design of details.
- To improve the overall quality of the building.
- To ensure best value for money from the project.

Main drawbacks

- Potential for inter-professional friction – some designers may resent having their drawings and details checked by someone else.
- The process can be time-consuming and may cause undue delays if there is a dispute over the quality of the detailing.
- It increases the design costs.

Typical problems

Those undertaking a technical audit need to pay particular attention to the following:

- Substandard or inappropriate specification clauses.
- Insufficient or incorrect use of spaces.
- Poor detailing resulting in:
  - inadequate weather resistance (particularly at projections through roofs, junctions and abutments);
  - corrosion of metals (e.g. bi-metallic corrosion between dissimilar metals);
  - cold bridging (especially at openings in walls and abutments);
  - premature staining of the facade (see Parnham, 1996);
  - difficulties in cleaning certain areas (e.g. behind radiators, high up locations).
- Uneconomic life cycle costs of certain elements, components and materials.

Authoritative standards and references

Building consultants can refer to a variety of sources for reference when undertaking a technical audit. Table 2.15 lists some of the main sources.

Procedure

- *Step 1* Prioritize elements of the design or parts of the building requiring attention. These could be identified on the basis of their prominence, importance or susceptibility to failure.
- *Step 2* Highlight vulnerable or key details of those prioritized parts of the building. For example, abutments, reveals of openings, penetrations through roofs are particularly prone to failure.

Table 2.15 Sources of standards for design audits

Source	*Publication*
British Board of Agrement (BBA)	Certificates (e.g. 91/2682: 'Furfix Wall Extension Systems'; 99/3657: 'Fibrocem External Wall Insulation Systems'; and, 00/3664: 'Ultraframe Conservatory Roof System').
Building Research Establishment (BRE)	Digests/Information Papers/Reports. Good Building Guides/Good Repair Guides Defect Action Sheets.
British Standards Institution (BSI)	Codes of Practice (e.g. BS 8211: 1998: *Energy Efficiency in Housing – Part 1: Code of practice for energy efficient refurbishment of housing*).
Building Regulations	Approved Documents (England and Wales). Technical Standards and Deemed-to-Satisfy Requirements (Scotland).
Construction Texts	Everyday Details (Handisyde, 1991) Architects' Working Details 1 to 7 (e.g. Dawson, 1999 etc.).
Housing and Property Mutual (HAPM)	Component Life Manual. Defects Avoidance Manual. Guide to Defects Avoidance.
International Building Research and Innovation Council (CIB)	Working Commissions (e.g. W086 Building Pathology; W094 Performance Concept in Building). Journal: *Building Research and Information.*
National House Building Council (NHBC)	Standards for Conversions and Renovations (NHBC, 2005). Technical Standards (a two-volume reference source outlining the construction requirements for low-rise new-build residential properties).

- *Step 3* Identify risks associated with the details selected. These could, for example include problems such as cold bridging, rain penetration, or incompatibility. Undertake a life cycle cost analysis of suitable alternative designs to determine the optimal economic choice.
- *Step 4* Assess the risks. This means that the impact on the building of a failure in the detail should be anticipated. Determining or estimating the prognosis would also form part of this step.
- *Step 5* Control the risk. For example, to minimize cold bridging install insulated thermal breaks. Design adequate flashings and skirtings at flat roof abutments to reduce the risk of rainwater penetration at these details.

Summary

The decision to adapt or to redevelop (i.e. demolish and rebuild) a property will therefore be mainly determined by the economics of these two options. The potential value of the adapted building will also have a bearing on the viability of any proposed adaptation scheme. Other influences, of course, such as legal restraints will come into play as well but initially if not ultimately financial considerations will be paramount. If the decision is taken to adapt the building, its economic life can be extended well into the foreseeable future.

Accordingly, in adapting a building one or more of the following fundamental requirements must be met.

1. A new or modified use for the structure must be clearly established. The building should be capable of accommodating an improved or new use that has social benefits as well as being profitable. Ideally, or as a general rule the adaptation costs should not exceed the costs of redevelopment.
2. Finance must be available for the changes. Even simple adaptation schemes do not come cheap as allowance still has to be made for the costs of investigation and approvals before any construction work is done. Hidden and unforeseen costs need to be taken into account at the scheme design stage.
3. There should be sufficient space or land available to carry out the alteration or extension (i.e. both internally and externally). This also should include the space for access and site operations.

Adding value to an adapted building can be achieved by making a judicious choice of project. A successful scheme can produce positive results such as better rental returns, higher capital value, improved environment for users and the neighbourhood. This may involve conversion, extension, modernization or a combination of these interventions.

The next chapter addresses the conversion of buildings. It focuses on the principles that should be adopted when proposing to change the use of a property.

3

Principles of converting buildings

Overview

This chapter discusses the nature and rationale of building conversions and outlines the principal influences, options, methodologies and procedures for conversion projects. The focus in this section is on general conversion issues. Chapter 4 deals with the adaptive reuse of specific building types such as barns, churches and warehouses.

Background

Triggers for building conversion

Building conversion is usually stimulated by the need to ensure that properties have a continuing use. It is essentially a response to accommodate changes in the type or style of occupancy demand for a property. The intention to keep it in beneficial use is critical in such cases. Bringing new life to an old building threatened with disuse and eventual demolition is the ultimate positive objective of conversion.

For many occupiers a change of use is often seen as the cheapest and most convenient option for dealing with redundant buildings. As has already been stated in Chapter 1 there may be several reasons why an owner retains an existing property rather than demolishing and redeveloping the site. Additionally, external restraints may discourage if not prohibit the latter option.

The conversion of a redundant building offers the best way of prolonging, if not securing, its useful life expectancy. It also helps to increase the supply of certain types of properties when new build cannot meet the demand, for example when planning restraints make it difficult for house builders to develop on green-belt land.

The early 1980s saw a sharp decline in the house-building programme of the British public sector. This was caused primarily by central government curbing local authority spending on capital construction projects. The private sector, because of land and skilled labour shortages, has not been able to fill the shortfall in the supply of new housing.

It is for these reasons that housing conversion programmes are so attractive and have the added bonus of helping to achieve a more sustainable environment. Moreover, change-of-use schemes generate less energy and waste than comparable new-build projects (Energy Research Group, 1999).

Buildings have always been adjusted or modified to a certain extent. Although the adaptation of buildings is not a new process, their conversion to other uses on a regular, larger scale is a more recent phenomenon. Cunnington (1988), for example, found that there was evidence of only occasional building conversions from the Roman era up to the late-Medieval period, whilst between the 16th and 19th centuries conversions and other forms of adaptation became more common. This increase in the number of conversions was prompted by the rapid growth in population coupled with the various transformations in urban life through agricultural, industrial and religious developments during that period. By the late 19th century, however, the popularity of conversions declined in favour of new build. The second half of the 20th century has seen an upsurge in conversion schemes, primarily because of economic reasons but also because of the lack of adequate land for building in inner urban areas. This trend is likely to continue for the foreseeable future in the developed world.

Much of the existing building stock built in days long gone is of an inferior quality. This has necessitated replacement rather than the reuse of obsolete or redundant buildings. Moreover, owing to the paucity of suitable buildings, the rapid increases in demand for residential and commercial properties could only be met by new-build schemes.

Nowadays, however, many existing buildings that are potentially or actually redundant are still of good quality (Gregg and Crosbie, 2001). It would be a loss of valuable built assets to destroy such structures, because demolition is itself a wasteful, hazardous, polluting, disruptive and costly process. Those aspects of demolition associated with adaptation are addressed in Chapter 7.

In more recent times, demand changes have had, and are having, significant influences on the rate and nature of building conversion in both residential and non-residential sectors. In the housing sector, the focus is often on subdividing single-occupancy residential property into multiple dwellings. In many urban areas of the UK, for example, change-of-use schemes involving loft extensions to provide one- and two-apartment flats in redundant industrial buildings is a major growth area. As shown previously these conversions are primarily because of the shrinking of family sizes and increasing number of single people living alone.

In the non-residential sector, too, change of use is a popular form of adaptation nowadays. The massive expansion in British higher education in the early 1990s with the elevation of over 30 former polytechnics to university status, for example, had major property implications for these institutions. It forced many of these new universities to achieve a more economic use of their buildings as well as to acquire existing ones and convert them into academic and related facilities.

The reuse of redundant maltings, mills, corn exchanges, hospitals, mental asylums and other similar properties into residencies is an attractive option these days for private developers in urban and suburban areas. Even properties constructed for a specific purpose have been acquired for adaptation to a similar use. For example, the former Royal Naval College on the banks of the Thames at Greenwich, London was taken over by Greenwich University in 1998. The required adaptation involved substantial alterations to the interior of this Grade I listed building. Thus, although, in these cases, the uses are related, the spatial requirements are not the same. Such differences, though, have just as much to do with changes in function as they do with any other influence.

Sometimes these changes in demand can result in 're-conversions'. (See the planning implications of such conversions later in this chapter.) In some city centres, residential properties that were converted to offices in the 1960s and 1970s being converted back into dwellings again. Edinburgh has several good examples of this type of re-conversion. In Charlotte Square, for example, one of the city's finest Georgian areas, a major terraced block built in the 1820s was restored to part-residential, part-office use, which reflects its original character.

The nature and pace of change in property markets, of course, make it difficult to forecast demand with great accuracy. According to some researchers there are primarily two opposing schools of thought on the effects of such change. The first, the traditional view, is primarily from property professionals. It considers that the market is fairly cyclical and thus stable in the long term. This view assumes that the future will be much the same as the past.

The second school, predominantly from the general business sector and support services, takes a more radical position. It considered that much of the change affecting businesses is irreversible and as a result, the future is uncertain. The effect of this change, according to the second view, will fundamentally alter the structure of demand for property in both the public and private sectors.

As indicated in Chapter 1, regular and persistent change is an ongoing characteristic of modern living and working. Building conversion primarily involves modifications to the interior of the building to accommodate such change. Externally, therefore, little if any indication of a change of use is normally evident in some cases. However, some conversions to properties such as farm buildings can entail obvious alteration of their original character (see Chapter 4). In addition, where reuse requires an increase in floor area, an extension may be necessary to accommodate additional capacity. Such a conspicuous change has the potential to impact adversely on the external appearance of the building. Careful thought is therefore needed to maintain visual harmony with the original style.

Conversions can be classified into three distinct groups: adaptation to the same use, conversion to an alternative use and mixed use. The first two of these options are briefly examined in turn below. Mixed-use conversions are briefly dealt with in Chapter 4.

Adaptation to the same use

Adaptation to the same use usually entails some form of modification to the internal layout. Converting an under-used large single-occupancy house into several flats is a typical example of this type of reuse. In rare cases it may involve changing from a multiple-occupancy mansion back into a single-occupancy dwelling. It may also consist of exploiting empty or unused space within the building, such as the roof space or basement area.

Roof space conversions are still a very popular option for providing occupiers with additional accommodation in a convenient and relatively cost-effective way. This type of vertical extension is examined in Chapter 6.

Same-use building conversions may generate substantial internal alterations. For example, dividing a building into smaller units requires the installation of additional separating walls and floors. These elements must have adequate levels of sound insulation and fire protection.

Change to another use

Under this type of conversion a redundant building can be changed to a different use. One typical example of this is the adaptive reuse of an old church or warehouse into flats (see next chapter). Alternatively, non-residential projects such as banks into cafés, churches into shops or offices, or offices into hotels are other popular forms of adaptive reuse in urban areas.

Conversions to other uses are more awkward than same-use conversions. The new use may, for example, involves spatial and functional requirements that are quite different from the original. This often requires structural alterations to accommodate the change of use.

Choice of uses

Some buildings are more appropriate or ready for adaptation than other properties. The suitability for change of use is influenced by the condition of the property, its morphology, the proximity to other properties, as well as financial considerations. In general, the categories of buildings that are frequently the subject of a conversion scheme are listed in Table 3.1. It should be emphasized, however, that a redundant

Table 3.1 Range of typical conversion schemes

Original category	*Existing use*	*Typical new use*
Agricultural	Barn 'Dovecote' or 'Doocot' Threshing mill	Single/multiple dwelling; arts/crafts centre; coffee/snack bar; souvenir shop; local museum/gallery/centre
	Cart shed	Parish/community hall; hotel/leisure centre; function room
	Stable Smithy	Ditto
Commercial	Bank Public house	Coffee bar; public house; wine bar; new shop/office; restaurant; flats
	Shop	Ditto
	Office	Ditto
	Pavilion	Ditto
	Hotel	Office
	Corn exchange	Performing arts centre; drama/television studio
	Office block	Hotel; residential
Ecclesiastical	Church	Dwelling/s; arts centre; film theatre; lecture theatre
	Function hall	Community centre; office
	Manse/parsonage house	Restaurant; storage; workshop/garage; multiple flats; nursing/residential home
Industrial	Whiskey bond Mill, warehouse Maltings/distillery	Multiple flats; mixed use – shops and offices on ground floor, flats or small businesses on upper floors
	Railway station	Ditto
	Factory	Performing/fine arts centre; studio theatre; sports centre; offices
	Warehouse	Gallery; office; residential
	Windmill	Dwelling; office
Institutional	School	Community centre; flats
	College	Hotel
	Hospital	Educational facility; flats
	Mental asylum	Sports complex Youth/detention centre Offices Luxury apartments
Residential	Tenement Townhouse	Improved housing, with modified layouts/facilities
	Mansion house	Multiple flats Offices
	Medieval castle/tower house	Restaurant and bar Holiday accommodation Large single dwelling, or multiple apartments

building has the potential to be converted to numerous uses. The choice is only limited by the vision of those involved in commissioning, designing and constructing the adaptation work.

An example that illustrates the range of possibilities available is in relation to a redundant pavilion in a large country estate. Such a structure would have been regularly used in Victorian times (1837–1901) as a function room or summerhouse. Over time as recreational habits and occupancy frequencies changed, the pavilion would fall into disuse. The challenge for any adaptor is to find a beneficial and viable use for such a building.

Typical choices for reuse of this property and other similar buildings are summarized below. The reuse option selected and the philosophical approach and intervention it would involve is dependent upon a number of factors. The status of the building, whether or not it is listed, will clearly be influential. Secondly, the wishes and intentions of the client will determine the character and extent of any alterations:

- Garden pavilion/summer house – restored for similar use as originally designed, serving refreshments or used at functions – both private and public.
- Local gallery/museum/arts centre – as part of the estate's community links.
- Private meeting room – for various organizations and businesses.
- Board room – for the estate's management or hired out to businesses.
- Library – for the site or to house an important private collection on display to the public.
- Lecture room/small conference suite – for both private and public use.
- Function room – for both private and public use.
- Main reception and telephone office – for the entire estate.
- Telecommunications centre – for the business associated with the estate.
- Security office – for accommodation, facilities and storage.
- Plant/services room – to serve one or several of the nearby buildings.
- Storage area for gardens on the estate.

Single uses are all very well but may not be economically or functionally viable. A mixed-use approach is more likely to succeed in large redundant buildings such as storage depots or warehouses, particularly those on riverfronts and canals. In such cases the following mixed-use arrangement could be considered:

- Ground floor
  - canal mooring facilities and workspace-offices;
  - bars/cafés;
  - hotel;
  - museum or arts workshop and gallery space;
  - general shops.
- Upper floors
  - specialist retail;
  - flats;
  - hotel;
  - restaurant and bar.

Not all changes of use restrict the alterations to the interior of the building. Some conversions may involve increasing the size of a building. A granny-flat conversion, for example, could include a lateral extension (see below).

Granny-flat conversions

'Granny flats' are becoming popular in some new and rehabilitated housing projects. They involve accommodating parents and grandparents in adjoining flats. This is being seen as a useful way to address the needs of the young and old in a family simultaneously. It tackles the twin problems of caring for children and the elderly in the one household. According to R. Spitz, a French architect (reported by Sage, 2000), common but neutral zone should be provided in the new layout to separate the two occupancies.

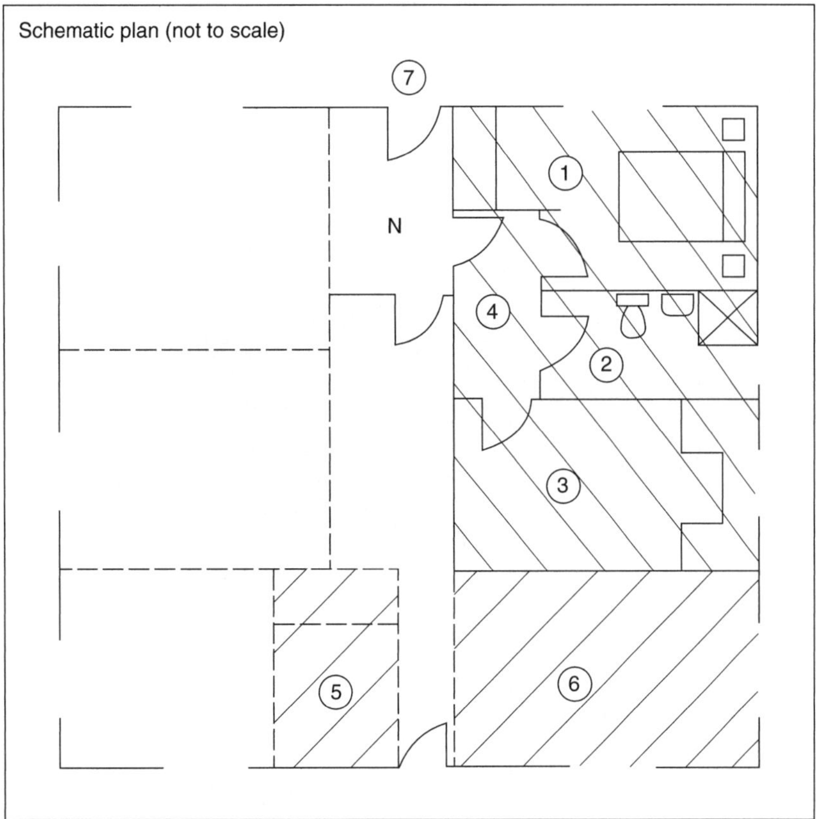

Notes

N = neutral territory, between the 'granny flat' and main dwelling.

Hatched areas represent new work – which involves adapting the existing ground floor bedroom and toilet to form the 'granny flat' accommodation as well as relocating original bathroom and kitchen.

1. Bedroom.
2. Toilet with shower cubicle (near position of original bathroom and kitchen).
3. Combined kitchen/lounge.
4. Hall.
5. Bathroom for main dwelling.
6. Kitchen for main dwelling.
7. Ramp and 'level' threshold to facilitate easy access.

Figure 3.1 Plan of a 'granny-flat' conversion within an existing detached bungalow

In the context of building adaptation, the formation of a granny flat can be achieved in one of two ways:

1. Adapt the existing space to contain the new accommodation (see Figure 3.1).
2. Provide a lateral extension to give a new ground-floor granny flat (see Figure 3.2).

The main part of the 'granny flat' comprises the bedroom, bathroom, and combined living room and kitchen. The area at the interface of this accommodation and the existing dwelling would form the neutral territory.

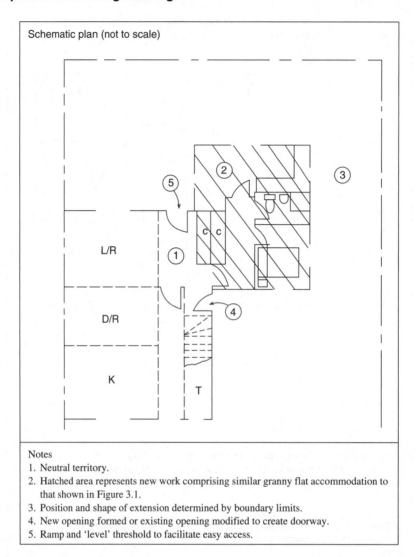

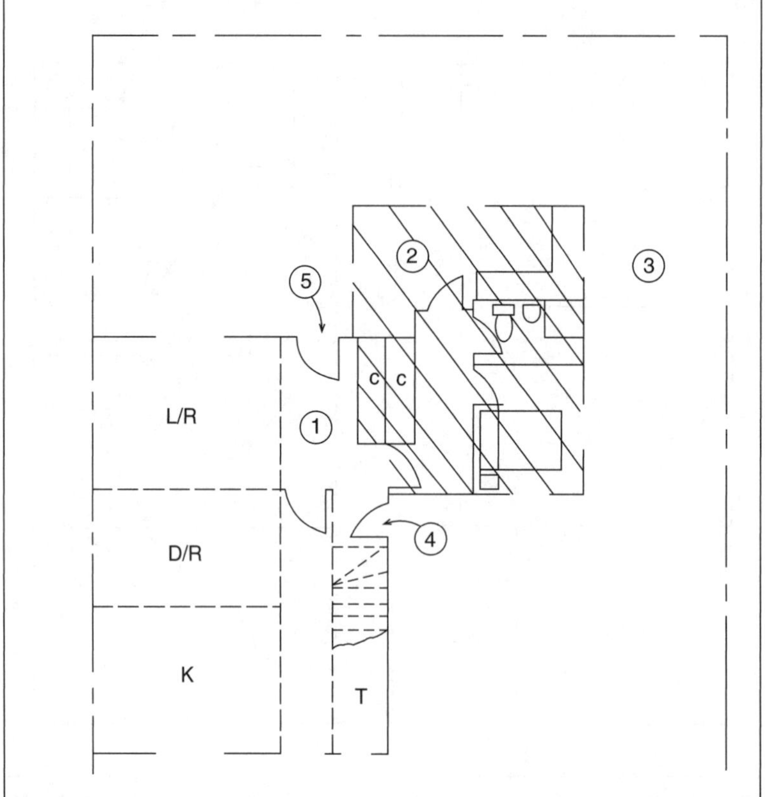

Notes
1. Neutral territory.
2. Hatched area represents new work comprising similar granny flat accommodation to that shown in Figure 3.1.
3. Position and shape of extension determined by boundary limits.
4. New opening formed or existing opening modified to create doorway.
5. Ramp and 'level' threshold to facilitate easy access.

Figure 3.2 Plan of a single-storey 'granny-flat' extension and partial conversion of a two-storey dwelling

This type of residential conversion is best restricted to the ground floor. It would only work in upper floors if a stair-lift facility (see Chapter 9) were provided for immobile or very elderly grandparents.

The building conversion process

Three essential requirements

The process of adaptation has already been addressed in general terms. The conversion process is no different but has certain characteristics that are peculiar to this category of adaptation.

Building conversion is basically another form of property development. Thus, as with any such venture there is an element of risk. The degree of risk will depend on a number of variables relevant to the proposed adaptation option selected. These variables include among other things:

- the location of the property;
- the condition of the building and the extent of works required to repair any defects;
- the length of time the conversion work will take;
- the general state of the economy, which impacts on interest rates;
- the level of demand for the use proposed at the time the property is marketed;
- the degree of legal and planning restraints operating over the property;
- the proposed cost of the change of use;
- the projected value of the converted building.

According to URBED (1987) there are three essential requirements and four stages of any successful conversion scheme: appropriate development approach, driving force and suitable building.

Appropriate development approach

Conventional approach: This involves either a conventional/institutional or entrepreneurial style. The first style already has a specific new use for the building. The market or client's requirements with such an approach are usually well known or can be easily determined. It requires skills in:

- spotting property market trends;
- negotiating (particularly with regard to acquisition of sites, planning permission and financial arrangements);
- understanding all the rules involved in adapting a building;
- co-ordinating for speed of completion of the project.

Entrepreneurial approach: The entrepreneurial approach, however, is more speculative. In this case the level of demand for the converted property is less well established, because the user might not be known. Because of increasing uncertainty in the business sector, this approach is being used less and less for commercial and industrial adaptations. Even with residential schemes, developers are unwilling to embark upon a major adaptation project unless there is an obvious market for the accommodation proposed. The main skills required for this approach are:

- perceiving viable end uses;
- creating enthusiasm and excitement among the participants;
- reacting to opportunities and being flexible;
- maintaining momentum over long periods (URBED, 1987).

Driving force

Ideally the scheme should have a project leader or cohesive team to initiate and co-ordinate the conversion. The driving force could be the client's or consultant's representative. A qualified construction professional with experience in adaptation work such as chartered surveyor, registered architect, chartered builder or engineer would be a suitable person for this role. In the case of public sector schemes, the driving force can emanate from a dynamic and resourceful group of officials or local councillors.

In the private sector, the driving force may be the developer or development company itself. Individual flair and entrepreneurship are important characteristics in this regard.

Suitable building

Ideally, the building should be suitable for the proposed change of use in locational, physical, spatial and legal terms. Many buildings become redundant because they are simply in the wrong place or have poor access. Changes in land use patterns, urban decay and new transportation routes can all help to make a building redundant.

A principal requirement for a successful conversion of any building, of course, is that it must be available for this form of adaptation. This means that there must be no legal or financial restriction affecting its reuse. Even buildings that are dilapidated can be converted even if it means substantial alterations to their structure and fabric.

Four stages of a conversion project

Stage 1: Incubation stage

The proposed new use must match the property. Building rehabilitation (i.e. adaptation) is about matching supply (the building) and demand (the uses). Not all supply and demand types match, of course. This is where the expertise, skills and perseverance of the building professionals involved come in.

A supportive local authority is essential for any proposed conversion scheme to be a success. This means that planning and building control requirements need to be flexible. In some cases, relaxation of parts of the regulations may be desired.

The preliminary appraisal of the scheme must establish what is both viable and exciting. This is especially the case with larger more substantial conversion schemes.

Stage 2: Negotiations stage

- Good property terms. These need to be established. This means making sure that any ownership issues are clarified.
- Space planning considerations. These must be taken into account at this stage to determine the extent of accommodation achievable in the building undergoing conversion (see below).
- Flexible financial package. This is to accommodate any foreseeable changes in conditions.
- Sound professional team. This team needs to have all the expertise required to handle the problems posed by the building being converted. This requires the input of architects, surveyors, engineers, and construction managers with the knowledge and skills to handle the problems posed by existing buildings.
- Competent contractor. It is advisable that only contractors with experience in adaptation work be invited to tender. Adaptations, as was highlighted earlier, have problems and constraints of their own that are just as if not more extensive than new build. Many large contractors are becoming heavily involved in major conversion and refurbishment schemes. Unless they are aware of, and sensitive to, the peculiar problems and constraints associated with adaptation work compared to new build, mistakes are likely to be common during the contract.

Stage 3: Construction stage

Efficient project management is essential. This requires good control over quality, time and costs. It needs to be flexible, when necessary, to take into account the needs of the occupier or owner (see Chapter 11).

Stage 4: Management stage

Communicating the scheme to all the parties involved – the client, contractor, consultants and approval bodies such as planning department and building control. This requires a committed management to organize and drive the scheme, as well as provide an adequate aftercare strategy for the converted building (again, see Chapter 11).

See Table 3.2 for a summary of these four stages.

Table 3.2 Stages in the conversion process (adapted from URBED, 1987)

Some inputs and tasks	*Key factors for success*
Incubation stage	
• Visualize the potential uses • Understand demand • Select the building • Obtain local authority support • Make survey and costings • Formulate the scheme • Obtain public support	• Uses to match the building (1) • Supportive local authority (2) • Viable and exciting scheme (3)
Negotiation stage	
• Negotiate building purchase • Obtain planning permission • Find appropriate professionals • Raise the finance • Maintain flexibility • Make detailed designs • Obtain building consent/warrant and quotations • Select reliable contractor	• Good property terms (4) • Flexible financial package (5) • Sound professional team (6) • Experienced contractor (7)
Construction stage	
• Manage the building work • Keep control of money • Keep control of time • Use a government sponsored workforce? • Deal with the unexpected	• Efficient project management (7) • Flexible when necessary (7)
Management stage	
• Market the scheme • Attract tenants • Organize continuing management • Create atmosphere of success • Plan further action	• Communicating the excitement (8) • Management committed to success (8)

Notes
1. Compatible and sustainable use.
2. Local planning authority backs the proposed scheme.
3. One that includes social as well as financial profitability.
4. No excessive easements or covenants.
5. Grants and other funding available.
6. Cohesive multidisciplinary team.
7. Production group familiar with the requirements and pitfalls of adaptation projects.
8. Dedicated and enthusiastic management team.

Space planning in conversions

Requirements

As indicated above, matching demand and supply is one of the key requirements in any conversion scheme (Markus et al., 1972). Space planning is about finding the most appropriate match between supply (the buildings and space available), and demand (the needs of the occupiers and the functions they and the space use).

An existing building that is scheduled for conversion has a set or limited amount of space to accommodate the new use. This is the supply side of the conversion equation. Any additional space required can be increased in two main ways:

- providing a lateral or vertical extension to the building;
- inserting mezzanine floors expand the floor space.

The demand side will be determined by the needs of the client. For example, in the conversion of an unused warehouse into residential accommodation, the number of dwellings will depend on each building's total internal floor area and whether the block is to consist of a mixture of single- or multi-apartment flats. Alternatively, the building could be designed to contain a range of only one- or two-apartment flats.

Planning controls, too, may impact on the extent and type of accommodation being offered. Local authority policy may determine that a combination of flat sizes should be provided.

Methodology

Firstly, a space audit of the property identified for conversion would need to be undertaken. A measurement or dimensional survey of the building forms an important part of such an audit. Any available plans and other drawings of the building could then be used to check the available space.

The space audit would establish the net useable area of the building and enable the designer to determine the amount of circulation/access and habitable spaces available. The client's brief may state the space standards that must be achieved if possible.

A simple procedure for ascertaining the number of apartments on each floor level is as follows:

- Step 1: Determine the net internal area of each floor level (i.e. the useable floor area in each level from the internal walls and excluding circulation space). Each flat will contain basic accommodation – bathroom/toilet, kitchen, living room and hall.
- Step 2: Establish the minimum floor space required for each flat. Firstly, this will depend on the apartment size (e.g. one, two or three bedroom). A mixture of one- or two-bedroom flats may be the client's preference. Secondly, the space standards set for each flat will influence their overall size. For example, a typical high-quality two-bedroom flat could have an overall floor area of about 56 m^2.
- Step 3: Determine how many groups of flats (or combinations of apartment sizes) can be accommodated on each floor. A redundant four-storey warehouse with a net internal area of each floor of 180 m^2, for example, could easily contain 4 two-bedroom flats or 3 three-bedroom flats. However, fitting in a fixed number of flats per floor level may not be practicable given the layout. A series of schematic layouts of each level of the converted building could be devised to set out and establish the optimal floor plan (Figure 3.3).

Converting historic buildings

Problems

Physical and operational constraints

As we have already noted in Chapter 2 adapting buildings is not straightforward. In particular older properties bring with them problems that are either absent or minimal in new construction. The architectural or

Schematic plans (not to scale)

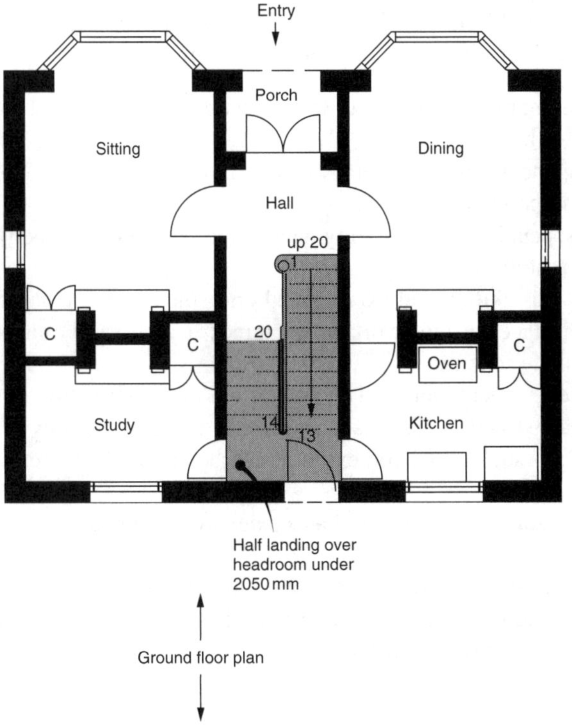

New stair in existing house

Case 3: Common stair

EXISTING large two-storey (detached) one-family house.

Floor to floor height	3400 mm
Number of rises	20
Each rise	170 mm
Each going	250 mm
Stair angle	34°
Stair width	950 mm

Spacious stair occupies a large floor area

House to be converted to 4 bedsitters (2 per floor)

The stair up to bedsitters 3 and 4 will be a 'common' stair

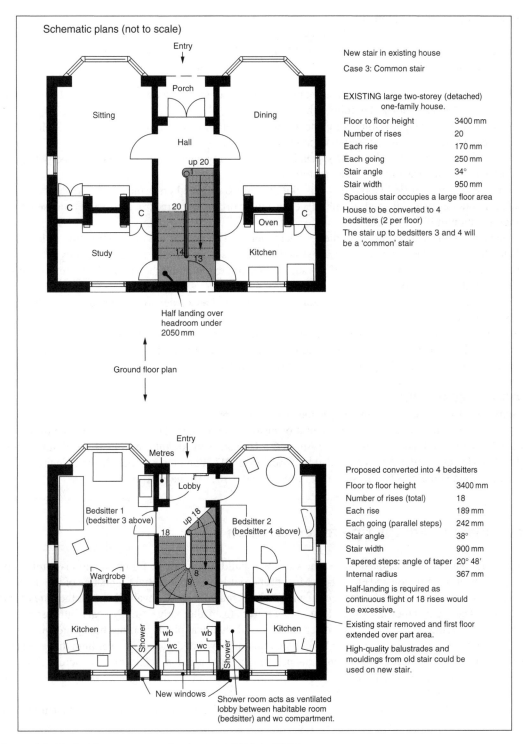

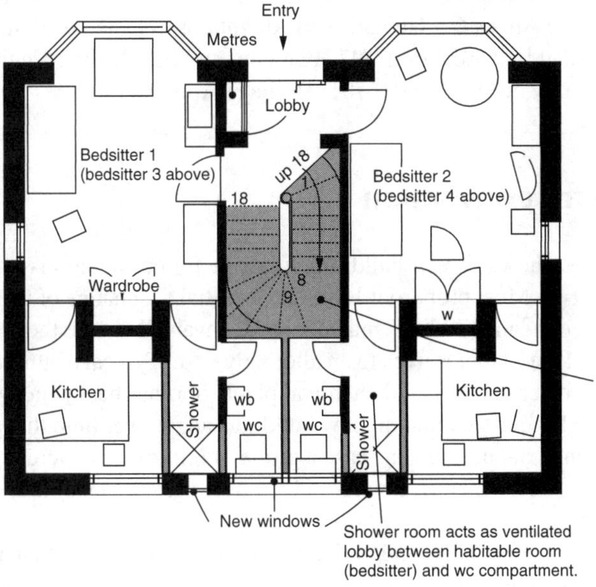

Proposed converted into 4 bedsitters

Floor to floor height	3400 mm
Number of rises (total)	18
Each rise	189 mm
Each going (parallel steps)	242 mm
Stair angle	38°
Stair width	900 mm
Tapered steps: angle of taper	20° 48′
Internal radius	367 mm

Half-landing is required as continuous flight of 18 rises would be excessive.

Existing stair removed and first floor extended over part area.

High-quality balustrades and mouldings from old stair could be used on new stair.

Figure 3.3 Typical house conversion layout (TRADA, 1978) historic sensitivity of the building is one important consideration. The likely presence of inherent defects or deleterious materials within an older property is another.

The following list (adapted from Gold and Martin, 1999) summarizes the main problems for designers and project managers in the conversion and refurbishment of historic buildings:

- The geometry of older buildings is commonly found to be unsympathetic to today's needs, although the use of wireless technology may relieve this.
- Difficulties in matching existing components and materials, which may no longer be available, or may only be made to metric as opposed to imperial dimensions.
- Replacements may have to be made specially, possibly using new materials, which are likely to have a different performance and may even cost more.
- The shortage of qualified labour, such as skilled and experienced craftsmen with the ability to work with traditional materials and the interface to tight work programmes, which can be caused by the incorporation of old craft processes in the scheme.
- Inadequate load capacity poses problems where there is a change of use or redistribution of loads.
- Uncertainty as to the actual as-built construction. This can give rise to unwanted discoveries such as concealed void areas, and hidden defects such as dry rot, excessive carbonation of concrete and corrosion of built-in steelwork.
- The building may be in such a poor condition that it necessitates extensive repairs as well as the refurbishment work.
- The proposed conversion scheme could damage or destroy the building's architectural or historic character if radical changes to its structure and fabric are being considered. Adherence to the conservation criteria listed below can help prevent these adverse effects.

Planning restraints

Planning Policy Guidance 15: Planning and the historic environment (1994) is an important document relating to adaptations to buildings of architectural or historic importance. The advice it contains in Sections 3.12–3.15: 'Alterations and Extensions and Annex C – Guidance on Alterations to Listed Buildings' should be heeded. However, PPG15 was amended by Circulars 14/97 'Planning and the Historic Environment – Notification and Directions by the Secretary of State' and 01/01 'Heritage Applications'.

Redundant buildings ripe for conversion

Despite the constraints mentioned above, the variety of building types ripe for conversion is limited only by one's lack of imagination and enterprise. Most properties have the potential for change of use. Granted, the practical or optimal reuse of some redundant buildings may not be achievable because they are of such a specialized nature that they do not easily lend themselves to an alternative use. Typical examples of these are bespoke buildings such as prisons, power stations, and chemical plants or other heavy industrial facilities. However, Brand (1994) showed in his classic book on how buildings adapt over time that even awkward structures can be converted to a beneficial use – in one case a group of reinforced concrete grain silos were converted into flats. The principle 'where there is a will (and the necessary finance) there is a way' applies to the conversion of old buildings.

In the view of URBED (1987), however, most of the industrial or commercial buildings that are ripe for conversion, in ascending order of viability, are:

- excess to requirements, with no obvious use as they presently stand;
- in poor locations, where businesses once flourished but no longer do; or alternatively,
- listed buildings in good areas.

Table 3.3 Three main grades of building conversion

Grade	*Characteristics*	*Potential new use*
High	Conservation philosophy adopted – defensible solution utilized; proposed use compatible with previous use; retention of the building's original character. Low density. Maximum social profitability to regenerate the local area.	Arts/crafts centre. Community centre – multi-function facility to maximize use. Museum. Residential/nursing home. Several luxury dwellings.
Medium	Some social benefit. Achieves some beneficial use of the property. A degree of compatibility if not similarity to previous use. Some improvement to the building's external appearance.	Community centre with some degree of mixed use. Restaurant and bar. Single dwelling. Sport centre.
Low	Unattractive or basic conversion and refurbishment of low-rise building or factory unit to industrial use. Short to medium service life. Minimal environmental benefit. Little improvement to the building's external appearance.	Garage. Warehouse. Workshop.

Finding new uses for redundant buildings, of course, requires ingenuity as well as determination. This is especially the case with properties in poor locations or where the demand for suitable reuses is already satisfied in the area. Although a building may be ripe for conversion, the quality of the new scheme may not always be consistent with the previous use. The practical reuse of buildings can result in one of three broad grades, as indicated in Table 3.3.

Naturally, the context of the building determines the grading of the conversion scheme. If the building is listed or is in a Conservation Area, only a high-grade conversion is likely to succeed and attract or generate grant funding.

Checklists for converting historic buildings

General points

Chapter 2 outlines the main procedure in assessing a building for adaptation. The process for doing this for historic buildings because of their architectural or historic sensitivity may require a less invasive and more sympathetic approach. The following points in the checklist devised by an American building conservation agency Heritage Preservation Services (see under National Park Service in Bibliography) summarizes the key points to consider:

- Check available documentation – via desktop investigation.
- Evaluate the historic character – the way it looks today (see next section).
- Assess the architectural integrity, including existing physical condition (also see below).
- Plan for the conversion work – reconfigure or modify the space to suit the new use.
- Check building regulations and other legal requirements.
- Check use of grants and review requirements and funding.
- Check available publications (e.g. technical guides published by English Heritage, SPAB, Historic Scotland).

SPAB Committee Recommendations

Many historic buildings have well-established and appropriate uses. Occasionally, though, some change is required to ensure a structure's future care, repair and protection. Creative adaptation can contribute positively to a building's history. Similarly, inappropriate re-use can fundamentally detract from its special interest and integrity.

The SPAB guidance therefore maintains that, where some change of use to a historic structure is considered essential, it is necessary to look carefully at the effect on all aspects of the building's character, fabric and setting. The following checklist is, therefore, intended to assist owners and their advisers in establishing what alternative use, if any, is most appropriate. The list should be read in conjunction with SPAB advisory documents, and other statutory and non-statutory conservation guidance:

- If the current use continues to be appropriate, can it be maintained? If not, why not?
- If the building is currently redundant, is there an immediate need for a change in use? Would minimal protective works, to allow 'mothballing', or a short-term low-key use, be a preferable alternative?
- Is the new use likely to secure a future for the building, and to be viable for a foreseeable period without major change or addition?
- Has the building been offered on the open market for its present use (if appropriate)?
- To identify what might have least impact on fabric and character, has the nature of the building been fully understood?
- An appreciation of the history and development of a place is essential in considering future use. Preparation of a Conservation Plan or Statement identifying the special interest of the building and its context may help in this.
- Have all potentially sympathetic uses been properly explored? It may be possible to argue for forms of re-use which have significant benefits for the special interest of the building and its settings, even if contrary to normal local plan policy. This is particularly so in the case of listed buildings and scheduled monuments.
- Can the building accommodate the requirements of the new use without seriously compromising the architectural character and/or historic fabric? Issues may include:
  - New openings: number, type, style and size.
  - Fire and safety: additional means of escape, protected stairs, upgrading of existing doors, partitions. Physical barriers to access. Thresholds, ramps, door furniture.
  - Subdivision of existing rooms/spaces: cornicing, panelling, plan form may be compromised.
  - Extension: scale, design, use of materials, abutment with the existing building.
  - Servicing: introduction of pipework, electrical cabling, altering internal environment.
  - The extent of rebuilding, if derelict or a ruin.
  - Floor loadings: strengthening of existing floor structures may be needed.
  - Sound insulation: increased insulation requirements to floors, walls and glazing.
  - Thermal performance: increased insulation requirements to floors, walls roofs and glazing
- Can the building accommodate the requirements of the new use without seriously compromising its setting? Issues may include:
  - Increased car parking: location, surfacing, entrance/exit, sight lines.
  - Division of open spaces: building groups, play areas/farmyards.
  - Separation from any historically linked curtilage buildings.
  - Fragmentation of the long-term management of the buildings.
  - Impact on standing or buried archaeological remains.
  - Effect on the broader character of the Conservation Area.

If circumstances change, can any alterations be reversed without damage to the building? Reversibility should not be an excuse for work of poor quality, and sometimes there may be an advantage in well-conceived and executed permanent alteration. However, the case for making change reversible should always be considered.

Identifying the visual character of a historic building

Before starting to convert or otherwise adapt a historic building, it is important to establish its visual character. The visually distinctive materials, features and spaces prior to work are much more likely to be preserved if they are identified prior to any adaptation scheme.

According to Heritage Preservation Services (2003) the visual impact of proposed changes to the exterior, interior and site can be assessed using a three-step approach. By applying the following method 'to evaluate a historic building from a distance, up close, and inside, you can begin to decide where alterations might reasonably take place – and which visual aspects you need to preserve':

APPRAISAL PROCESS: ASK ... LOOK ... IDENTIFY ... EVALUATE ... CONCLUDE

Step 1: Identify the building's overall visual aspects, by examining the exterior from afar to understand its distinctive features, and the building site, or landscape.

This first step involves looking at the building from a distance. Identifying the overall visual character of a building is nothing more than looking at its distinguishing physical aspects without focusing on its details. Such a general approach to looking at the building and site will provide a better understanding of its overall character without having to resort to an infinitely long checklist of its possible features and details. The major contributors to a building's overall visual character are:

- its shape – square, rectangular, L-shape, circular/semi-circular or a combination;
- roof and roof features, such as chimneys or cupolas;
- fenestration – openings for windows and doorways;
- the various projections and recesses on the building, such as porches that extend outward, or arcades that appear as voids;
- the exterior materials with their colour or patterning the trim and secondary features, such as decorative scrollwork;
- finally, the building's site (i.e. its immediate yard).

Some buildings will have one or more sides that are more important than the others because they are more highly visible. This does not mean that the rear of the building is of no value whatever, but it simply means that it is less important to the overall character.

Step 2: Identify the visual aspects of the exterior at close range by moving up very close to see its materials, craftsmanship and surface finishes.

The second step involves looking at the building at close range. This is where you will be able to see and appreciate the qualities and workmanship of exterior surfaces, that is:

- the building's specific materials;
- its craft details.

You have already learned about the character of a historic building from a distance. What distinguishes the close-up visual character is often the result of materials that differ sharply in their colour and texture. They often convey that sense of craftsmanship and age that distinguishes historic buildings from other buildings.

It is important to understand that many of these materials can be easily damaged or obscured by work that affects their surfaces. This can include painting previously unpainted masonry, rotary disc sanding of smooth wood siding to remove paint, abrasive cleaning of tooled stonework or repointing reddish mortar joints with grey Portland cement.

There is an almost infinite variety of surface materials, textures and finishes that are part of a historic building's character which are fragile and easily lost.

Step 3: Identify the interior visual aspects – spaces, features and finishes – by going into and through the building.

The third step involves looking at the interior. This needs to be done slowly in order to correctly identify its distinctive visual character. These are the visual aspects to be considered:

- Individual spaces and spaces that are related to each other.
- Interior features that are part of the building.
- Distinctive surface materials and finishes.
- Any exposed structural elements.

First, remember that the shape of a space may be an essential part of its character. In office buildings, this is generally the vestibules, lobbies or corridors. If the shape or plan is altered, the interior character is changed. With some buildings, the relationship between spaces creates a visual linkage, such as in a hotel – from the lobby, to the grand staircase, to the ballroom. Closing off the openings between those spaces would change the character dramatically.

Distinctive surface materials and finishes may be an aspect of the visual character, such as wooden parquet floors, pressed metal ceilings, wallpaper or grained doors. Exposed structural elements are less frequent (rooms with decorative ceiling beams or exposed posts, beams and trusses), but may be present in a church or train shed or factory.

Finally, so-called secondary spaces are not usually perceived as important to the visual character of the building. This is quite often where change can take place within a rehabilitation project.

Conservation criteria

General principles

Building conservation can be defined simply as 'preserving buildings purposefully'. It is sometimes taken as being synonymous with 'preservation'. Although closely related, the two terms imply different outcomes (see Glossary). Preservation in the UK primarily describes work associated with arresting or retarding deterioration, and is more commonly used in the context of structures such as monuments. However, it is used in the USA to mean conservation work.

The main reasons for conserving buildings are:

- Cultural: retaining a valued part of the built environment because of its architectural or historic significance.
- Educational: using the building as a learning resource.
- Heritage tourism: attracting visitors to an area.
- Historic variety: maintaining an urban area's character.
- Economic: conservation can create new jobs; it is more labour intensive than new build; and money spent on conservation schemes generally stays more local.
- Legal: complying with local and national planning policies and legislation.
- Technical: preserving the structure and fabric to minimize unnecessary repairs in future.

Changing the use or the renovation of old properties brings with it problems over and above those encountered in adaptation schemes to modern buildings. As indicated in BS 7913 (1999), an authoritative guide to the principles of the conservation of historic buildings, old properties need to be handled with care and sensitivity. This is to ensure that any proposed adaptation does not compromise their character or architectural quality.

In contrast, conservation goes beyond this by attempting to ensure that the building is put to beneficial use once the basic preservation works and required improvements have been completed. Coping with functional change, therefore, is a key attribute of conservation (Riyat, 1993).

A primary source of guidance on building conservation recognized internationally is the Burra Charter. It was published in the early 1960s and lays down seven key principles of conservation:

1. The place itself is important.
2. Understand the significance of the place.
3. Understand the fabric.
4. Significance should guide decisions.
5. Do as much as necessary, as little as possible.
6. Keep records.
7. Do everything in logical order:
   - Assess cultural significance.
   - Develop conservation policy and strategy.
   - Carry out the conservation strategy.

The Urban Redevelopment Authority and Preservation of Monuments Board of Singapore (URA and PMB, 1993) use the '3R Principle' for the conservation of their buildings of architectural or historic importance:

- Maximum *Retention* (to ensure that key features are not destroyed, damaged or hidden).
- Sensitive *Restoration* (to ensure that it blends in with the existing construction).
- Careful *Repair* (to avoid triggering further defects as well as arrest current failures).
  To which could be added a fourth R:
- Minimal *Reconstruction* (to avoid placing new work immediately next to old).

The key conservation criteria listed below should be applied wherever possible. Most importantly, any conversion scheme that contravenes these criteria is less likely to receive favourable treatment from the local planning authority or conservation funding body. Thus obtaining approvals such as planning permission, listed building consent or grants for such work is potentially made more difficult.

Meticulous recording

A full and detailed recording of the building should be prepared prior to it being subjected to any adaptation work. This will normally include full detailed drawings, a photographic and documentary record of the building being adapted. For example, the original building may have distinctive features such as roof slates laid in a fish-scale pattern. Ideally, these should be retained in any adaptation scheme if the roof coverings are being renewed.

Minimum intervention

The scale or form of the adaptation work must not be such that it destroys or adversely affects the character of the building. A degree of sensitivity, therefore, needs to be exercised when adapting old buildings. Overcladding an old stone-faced building with metal sheeting or recovering a pitched roof with tiles instead of slate, for example, would in this context be deemed excessive if not inappropriate. This is because such alterations would inevitably compromise the property's architectural merit. Internally, too, the original layout and arrangement of the building should be retained wherever possible.

Minimal loss of fabric

Many old buildings contain material of cultural value (e.g. original painted murals on walls and ceilings) or are unique in terms of their construction or architectural style. It is vital, therefore, that as little as possible of the heritage features of the building are damaged or lost as a result of a proposed adaptation. Once eliminated, they can be difficult, if not impossible, to reinstate or rectify.

Reversibility

This is based on the principle that nothing should be added to the building that later cannot be taken away. In addition, any alterations to a building should be capable of being removed and made good without too much collateral damage to the existing structure and fabric. Far too often, insensitive adaptation works have done irreversible harm to an old property. For example, slapping new openings in external walls or installing a large extension to the front or side of the building may be difficult to undo.

Compatibility of use

The new use for the adapted property must not be so different or inconsistent with the previous use as to radically affect the building's historical character or reputation. As indicated above, cultural as well as architectural considerations are likely to be important where the building is Listed or in a Conservation Area.

Explicitness of alteration

Any new work or other major alterations to an existing building are best made obvious rather than veiled. This, too, shows honesty in the adaptation work done to a building. The integrity of the building would be compromised by work that attempts to make artificial effects to the fabric.

Honesty and appropriateness of repair or restoration

All remedial works to historic buildings ought to be an honest attempt to rectify a defect or replace a missing section of their fabric or element or detail. Moreover, the repair must be appropriate to the nature of the construction. As shown in the previous chapter, repointing the external joints of solid masonry walling with a cement-based mortar inhibits the wall's ability to 'breathe' properly (Hughes, 1986). Similarly, applying a cement-based rendering system can trap moisture within the fabric of solid stonework. Both of these remedial works are likely to lead to a host of moisture-related problems in walls such as efflorescence, frost attack, fungal decay of built-in timbers and corrosion of metal fixings and reinforcement.

Sustainability

The current and intended use of the building should not adversely affect the chances of handing it down to future generations. The converted building, moreover, should not impose any additional energy or pollution problems on the local environment. How this can be achieved is addressed in Chapter 10.

Devise defensible solutions

The adaptation proposal selected must reflect the foregoing criteria. It must also be non-prejudicial to, if not consistent with, the building's evolution. Any conservation or new work should, therefore, respect its architectural, historical and cultural integrity, and comply with the principles outlined in Section 7 of BS 7913.

Ensuring the long-term success of the property would involve an appropriate aftercare strategy. To achieve this a properly managed planned maintenance system is vital. This would also entail compiling a maintenance manual to assist with the aftercare of the building (see Chapter 11).

As indicated in Chapter 1 the range of interventions to a building can be wide. With historic buildings, however, care must be taken to ensure that the level of intervention is appropriate and complies with the

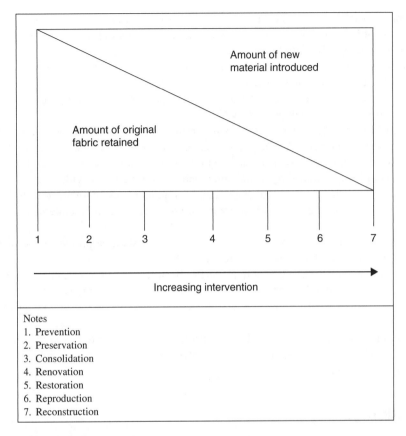

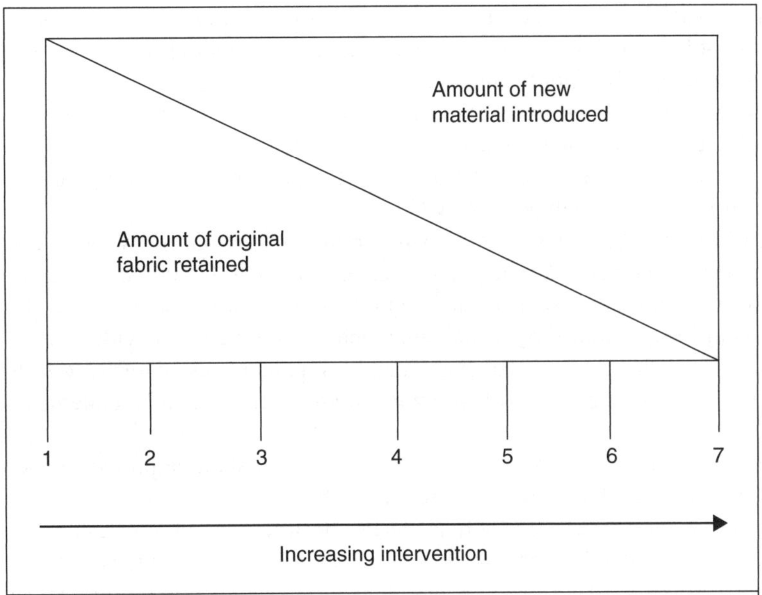

Notes
1. Prevention
2. Preservation
3. Consolidation
4. Renovation
5. Restoration
6. Reproduction
7. Reconstruction

Figure 3.4 Increasing levels of intervention

conservation criteria listed above. Figure 3.4 illustrates the effect of increasing intervention through the various measures indicated in Figure 1.1.

US rehabilitation standards

In the USA rehabilitation (i.e. conversions and refurbishments) of historic properties has to follow nationally applied standards, which are similar to the conservation criteria mentioned above. Such rehabilitation projects in the USA, therefore, must meet the 10 Standards (36 CFR Part 67) listed below, as interpreted by the National Park Service. Compliance with these Standards is necessary to qualify as 'certified rehabilitations' eligible for the 20 per cent rehabilitation tax credit.

The Standards are applied to projects in a reasonable manner, taking into consideration economic and technical feasibility:

1. A property shall be used for its historic purpose or be placed in a new use that requires minimal change to the defining characteristics of the building and its site and environment.
2. The historic character of a property shall be retained and preserved. The removal of historic materials or alteration of features and spaces that characterize a property shall be avoided.

3. Each property shall be recognized as a physical record of its time, place and use. Changes that create a false sense of historical development, such as adding conjectural features or architectural elements from other buildings, shall not be undertaken.
4. Most properties change over time; those changes that have acquired historic significance in their own right shall be retained and preserved.
5. Distinctive features, finishes and construction techniques or examples of craftsmanship that characterize a historic property shall be preserved.
6. Deteriorated historic features shall be repaired rather than replaced. Where the severity of deterioration requires replacement of a distinctive feature, the new feature shall match the old in design, colour, texture, and other visual qualities and, where possible, materials. Replacement of missing features shall be substantiated by documentary, physical or pictorial evidence.
7. Chemical or physical treatments, such as sandblasting, that cause damage to historic materials shall not be used. The surface cleaning of structures, if appropriate, shall be undertaken using the gentlest means possible.
8. Significant archaeological resources affected by a project shall be protected and preserved. If such resources must be disturbed, mitigation measures shall be undertaken.
9. New additions, exterior alterations or related new construction shall not destroy historic materials that characterize the property. The new work shall be differentiated from the old and shall be compatible with the massing, size, scale and architectural features to protect the historic integrity of the property and its environment.
10. New additions and adjacent or related new construction shall be undertaken in such a manner that if removed in the future, the essential form and integrity of the historic property and its environment would be unimpaired.

Listed buildings

Rationale for listing

Listing is the main means of identifying and protecting the UK's built heritage. It attempts to ensure that the architectural and historic interest of our important buildings is carefully considered before any alterations, either outside or inside, are agreed. Listing gives a building statutory protection against unauthorized demolition, alteration and extension. It is an integral part of the system for managing change to our environment through the planning process administered by local planning authorities and the Office of the Deputy Prime Minister.

According to English Heritage all properties built before 1700 that remain in anything like their original condition are listed. The majority of buildings dating from 1700 to 1840 are listed, though selection is necessary. Between 1840 and 1914 only buildings of definite quality and character are listed, and the selection is designed to include the principal works of the principal architects. Between 1914 and 1939, selected buildings of high quality are listed. A few outstanding buildings erected after 1939.

Listing is, of course, only the start of a process, rather than an end in itself. Its aim is to flag the significance of an asset so that its future management can enhance its contribution to local, regional and national life.

Responsibility for listing

Since 1 April 2005, English Heritage is responsible for the administration of the listing system in England. This involves introducing new notification and consultation procedures for owners and local authorities, as well as clearer documentation for list entries. According to English Heritage further changes will be made to the listing system throughout 2005–2006, including the introduction of new information packs for owners. The basic objective is to 'make the heritage protection system simpler, more transparent and easier for everyone to use'.

The Secretary of State for Culture, Media and Sport (referred to hereafter as the Secretary of State) is responsible for compiling the statutory list of buildings of special architectural or historic interest. English Heritage is responsible for providing expert advice on which buildings meet the criteria for listing, and for administering the process.

Categories of listing

Buildings are listed under the Planning (Listed Buildings and Conservation Areas) Act 1990 if they are deemed to be of architectural or historic significance. The relative gradings of listed buildings are as follows:

- **Grade I (Grade One):** Buildings treated as being of exceptional interest (= Category A in Scotland: Buildings of national or international importance). Only about 2 per cent of listed buildings thus far are in this category.
- **Grade II* (Grade Two Star):** These are particularly important buildings of more than special interest. (= Category B in Scotland: Buildings of regional importance). Around 4 per cent of listed buildings are in this grade.
- **Grade II (Grade Two):** Buildings of special interest, which warrant every effort being made to preserve them. (= Category C in Scotland: Buildings of local importance). Some 94 per cent of all listed buildings fall within this category.

Significance

Listing means that the approval of the appropriate national conservation agency such as English Heritage or Historic Scotland as well as the local planning authority is required before any works affecting the exterior or interior of a building are undertaken. Some financial advantages could, though, accrue from listing. As indicated earlier, VAT, for example, could be recovered on adaptation work to a listed building. Grants may be available from preservation trusts for work to a listed building. However, as indicated in Chapter 2, such funding will usually come with strict conditions as to repair methods, type and size of any extensions, restrictions on degree of alterations, required colour schemes, etc.

Building conversions primarily affect the interior of a building. In Scotland, for instance, some buildings are only listed internally. In such cases, this may seriously constrain any modifications that might be required to enable an effective change of use to proceed. Moreover, some buildings are listed only as far as their interior is concerned because of the quality or range of attractive, unusual or special architectural attributes. Features such as hardwood wall panelling, intricate plaster patterns and cornices on ceilings, fireplaces and their surrounds, and stairwells or ornate staircases may be protected under such listing. See Chapter 11 for details of the procedure required to obtain listed building consent.

Some external work, too, may form part of the building conversion. The formation of new openings or the blocking up of existing windows may be required to facilitate the new use. An external addition (see Chapters 5 and 6) sometimes forms part of the rehabilitated building.

The changes that may be restricted or prohibited by virtue of listing are summarized in Table 3.4.

Rehabilitating interiors in historic buildings

Requirements

When converting or rehabilitating old buildings it is important to avoid radically changing character-defining spaces. Such alterations could destroy the building's original features and may be difficult to rectify.

Table 3.4 Adaptations that may be restricted by listed building status

Change	*Examples*
External additions	Extra chimneys, large vent pipes – particularly if their height, position and style would be seen as obtrusive. Extension – lateral or vertical, especially if it is conspicuous or large. New dormers or skylights in the roof – more so at the front elevation. New window or door openings – again, especially at the front. Modern features such as canopies/awnings/shutters – which might detract from the building's original appearance. Modern equipment such as satellite dishes, aerials, masts, etc. Extra security measures such as barbed or razor wire.
Internal additions	Extra walls and ceilings to form more rooms – which could increase the building's density unduly or create excessive dead loads on the structure, as well as destroy the building's internal features. New suspended ceiling that covers a corniced or ornamental panelled soffit. New mezzanine floor – which will increase the useable area but reduce the large original space. New/repositioned staircase – which might destroy the original layout. Loft or basement extension – which would increase the property's density.
External removals	Removal of chimney or other roof feature such as final or pinnacle – which could compromise the original character of the building. Removal of pediment or other prominent façade feature – which again would have an adverse effect on the building's original character. Blocking up of window and door openings – could also undermine the original character of the building. Partial demolition (e.g. removal of early or more recent extension). Removal of external panelled hardwood doors or period style windows. Removal of cast/wrought iron ornamental features and railings.
Internal removals	Removal of fireplaces – especially Adam's or other ornate fireplaces. Removal of original chandeliers. Removal of ornate marble, mosaic or encaustic tile floorings. Removal of hardwood wall panelling (i.e. 'dado' or full height). Removal of timber joists and bressummer beams. Removal of cornices and ceiling patterned plasterwork. Removal of walls, or the formation of new or enlarged openings. Removal of panelled doors or their original ironmongery.

Recommended approaches for rehabilitating historic interiors

Jandl (PB 18, 1988, in National Park Service publications in Bibliography) provided a useful summary of the approaches that should be adopted when rehabilitating historic interiors:

- Retain and preserve floor plans and interior spaces that are important in defining the overall character of the building.
- Avoid subdividing spaces that are characteristic of a building type or style that are directly associated with specific persons or patterns of events.
- Avoid making new cuts in floors and ceilings where such cuts would change character-defining spaces and the historic configuration of such spaces.
- Avoid installing dropped ceilings below ornamental ceilings or in rooms where high ceilings form part of the building's character.

- Retain and preserve interior features and finishes that are important in defining the overall historic character of the building.
- Retain stairs in their historic configuration and to location.
- Retain and preserve visible features of early mechanical systems that are important in defining the overall historic character of the building, such as radiators, vents, fans, grilles, plumbing fixtures, switchplates and lights.
- Avoid 'furring out' perimeter walls for insulation purposes. Consider alternative means of improving thermal performance, such as installing insulation in attics and basements and adding storm windows.
- Avoid removing paint and plaster from traditionally finished surfaces, to expose masonry and wood – unless of course the coating is defective or trapping in moisture within the substrate.
- Avoid using destructive methods – propane and butane torches or sandblasting – to remove paint or other coatings from historic features.

Restoration

Preamble

Any type of adaptation, of course, involves an element of decision-making. This process, which involves the input of surveyors, architects and engineers, does not finish when the contractor starts work on site. It requires careful planning as well as constant monitoring and supervision to deal with problems and unforeseen events.

Restoration work differs from most other forms of adaptation in that it involves a combination of some, if not all, of the following:

- Significant change or modification of use.
- Major repairs to the structure and fabric.
- Substantial reconstruction work to replace missing or repair ruinous parts of the building.
- Possible extensive additions or removals.
- Constraints over the type of repairs and extent of any external and internal changes.
- Stabilization and consolidation work.

Stabilization

Old buildings undergoing restoration or other major adaptation work are often in a poor state of structural repair. Cracking, deterioration, distortion and instability are common structural problems in these properties. The following measures, some of which are addressed in more detail in Chapter 7, are often required to deal with such problems:

- Installing a temporary roof enclosure on independent scaffolding to keep water out during the adaptation work (see Figure 11.7).
- Diverting water away from the substructure by installing a field drain around the perimeter walls.
- Stitching and girdling of walls using stainless steel straps or resin rods to tie junctions, joints or other unstable sections of masonry (see Chapter 7).
- Underpinning inadequate, defective or eroded foundations (see Chapter 7).
- Removing plants such as ivy and other creepers that hold water too close to the walls.
- Securing a structure against intruding insects, animals and vandals.

Consolidation

In tandem with stabilization measures consolidation work is sometimes required to old or derelict buildings. This usually involves the following work:

- Repointing masonry: The joints of rubble walling in older buildings are often eroded. As a result they may need repointing. This should be done using a 'heritage' (i.e. lime-based) mortar.

- Grout consolidation: This work may have to be done using non-percussive drilling methods where the walling is unstable or vulnerable to vibration damage. Older buildings are especially prone to this. A viscous epoxy resin grout with a strength of around 10 N/mm^2 should be sufficient for this work.
- Filling disused chimneys: Most, if not all, flues in old buildings undergoing major restoration usually are redundant. If there is no intention of using them again they can be dealt with in one of two ways: blanked off and vented, and filled.
  - Blanked off and vented: The traditional way of dealing with redundant flues was to blank them off at top and bottom, but provide sufficient background ventilation using grilles at the fireplace position and grilled cowls at the roof end. Such voids, however, can act as a route for vermin and encourage interstitial condensation.
  - Filling: Grouting the flues with a pulverized fuel ash (PFA)/cement grout 15:1 mix is an alternative solution. This is done by drilling 30 mm diameter grout holes into existing joints central to the flue at 1 m centres. The grout is then injected at a low continuous pressure. It is carried out in 1 m lifts to prevent fluid overload.

As well as being more expensive, however, grouting flues also introduces more moisture into the fabric. This can increase planned drying out times (see section on drying out of buildings).

Tower houses

In rural and some suburban areas detached tower houses involve typical restoration. They can offer a challenging but successful opportunity for adaptation (see Latham, 2000b). Many of these buildings have fallen into a state of disrepair, so much so that some have become ruinous.

The restoration of a ruinous tower house can pose a formidable task for any adaptation consultant. Figure 3.5 shows the work required to restore such a property.

In many cases, emergency stabilization is necessary to ensure that a structure does not continue to deteriorate prior to a final treatment or to ensure the safety of current occupants, investigators or visitors. Although severe cases might call for structural remedies, in more common situations, preliminary stabilization would be undertaken on a maintenance level.

Such work could involve installing a temporary roof covering to keep water out; diverting water away from foundation walls; removing plants that hold water too close to the walls; or securing a structure against intruding insects, animals and vandals.

An old building may require temporary remedial work on exterior surfaces such as using reversible caulking or an impermanent, distinguishable mortar. Or if paint analysis is contemplated in the future, deteriorated paint can be protected without heavy scraping by applying a recognizable 'memory' layer over all the historic layers. Stabilization adds to the cost of any project, but human safety and the protection of historical evidence are well worth the extra money.

Drying out buildings

Causes of saturation

Old buildings in particular are susceptible to dampness to such an extent that parts of the fabric may be near saturation level. This and other sources of unwanted moisture must be removed before the finishings of any restoration or other adaptation scheme can proceed.

The main sources of this excess level of water are:

- Water from the drenching actions of the fire brigade following a major fire to the building.
- Rainwater percolating into the fabric of a derelict building, especially one that has had no roof for a long period of time.

Schematic cross-sections (not to scale)

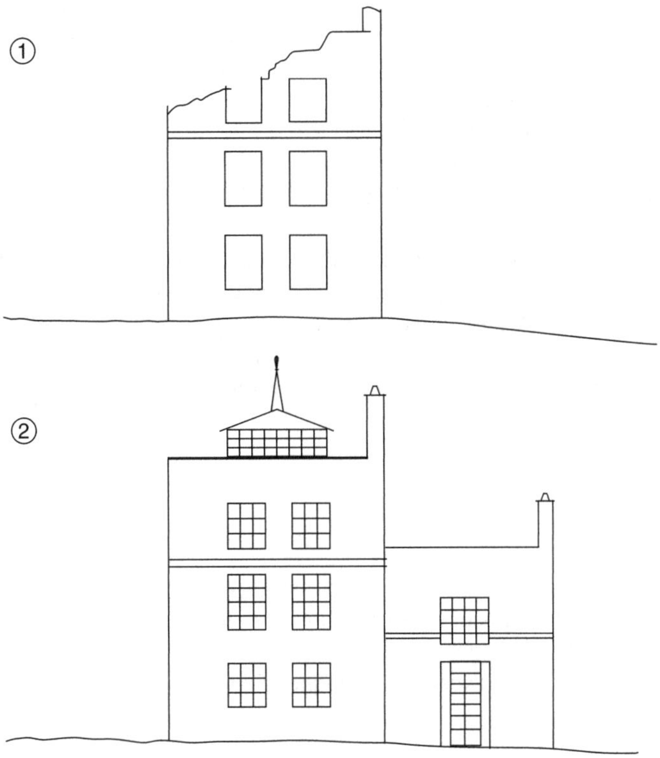

Notes

1. Existing ruinous tower house: 600–1000 mm thick random rubble masonry walls with some loose joints and no damp proof course (dpc); no windows or doors left; upper floors and roof structure missing; extensive overgrown vegetation and some debris at ground floor level inside building.
2. Renovated tower house: solid ground floor in lime concrete; upper floors of suspended timber construction with steel I beams to provide additional support; new flat roof complete with lantern-light style pitched roof extension. Breathable construction approach used (Hughe, 1986). No retrofit dpc installed immediately – instead, the basic remedial measures described after Table 8.6 should be implemented first. Walls rendered externally with a lime-based render or repointed with a lime-based mortar if masonry is in good condition. Walls panelled internally with 15 mm thick Fermacell dry lining board on 60 × 20 mm galvanized mild steel framing with mineral fibre insulation in between studs, with a 30 mm air gap between framing and face of masonry. New Georgian style good quality double glazed windows and doors. Possible extension that 'modulates the mass' of the tower house in a sympathetic manner.

Figure 3.5 Restoration of a ruinous tower house

- Construction moisture from processes such as new concrete screeds or floor slabs, grouting of walls and redundant flues, etc.
- Floodwater, as a result of the building being in a floodplain zone or from nearby overflowing river entering the building's basement and ground-floor areas.
- A burst water main flooding several levels of the building.

Rate of drying out

Elements such as floor screeds and slabs dry out from only one surface – the exposed face. Determining the rate of drying out of these elements, though, is reasonably straightforward. It is estimated, for example, that screeds dry out at a rate of about 1 mm a day (BS 6023). This means that a 50 mm thick screed would take about 2 months to dry out sufficient for floor coverings to be laid on it to minimize the risk of failure. A cold liquid-applied surface damp-proof membrane comprising a two-coat epoxy resin application can be used to suppress residual construction moisture in a concrete floor slab (see Chapter 9).

Estimating the rate of drying out of thick, solid masonry walls, however, is extremely complicated because of the number of variables involved. To the writer's knowledge there is currently no method for determining this precisely.

There are several reasons for this state of affairs. The actual composition of the solid wall construction in buildings cannot be determined with any degree of accuracy. For example, the size and extent of voids in walls, even after grouting, is not readily known. Moreover, the type of walling materials used, their density any size, and degree of saturation influence their rate of drying out. This makes it hard to identify the precise extent and levels of moisture within the core of thick walls (i.e. over 225 mm thick).

Indirect methods

These are the most common ways to dry out a building. They are basically of two methods recommended by the BRE (Digest 163, 1974):

- Dehumidifiers: Use industrial-grade (either desiccant or refrigerant types) and keep the windows closed. Some background heating is likely to be required to increase the rate of dry out. The containers need to be emptied regularly.
- Heaters: Industrial-grade hot air blowers, and keep the windows slightly open.

Direct methods

These are relatively rarely used because the extent of saturation is usually such that it can be controlled by indirect means. However, where bulk water is involved, more direct methods are needed to remove it quickly and efficiently. Again, there are basically two kinds:

- Pumping: Mechanical or electrical pumps can be used to remove bulk water from basements and other areas of a building affected by flooding. This method, though, will not remove moisture from either the air or the building fabric.
- De-irrigation: In some very rare cases where some parts of the fabric are saturated some form of de-irrigation might be needed. This could involve drilling 16 mm diameter holes at, say, 300 mm centres in the mortar joints to 'bleed' the excess moisture from the 'wet' masonry. Drill holes as deep as 400 mm would be needed in walls 1200 mm thick. The holes should be at least one-third of the wall's thickness. This should be deep enough to make contact with flue media or voids acting as a reservoir for any free water. The residual moisture can then be 'bled' from the masonry using a perforated copper tube not more than 15 mm in diameter inserted into the hole and linked to a hose, which in turn is connected to a mechanical suction pump. This is analogous to using well-points to extract moisture from very wet soil.

Façade retention schemes

Definitions

Façade retention involves leaving at least one elevation of an existing building to conceal a new, more modern interior (see Figure 3.6). It is a drastic solution because it entails major surgery to the property

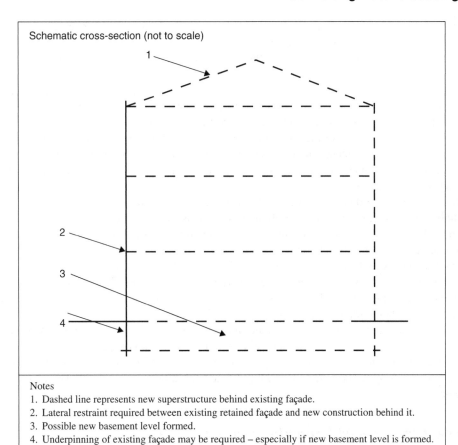

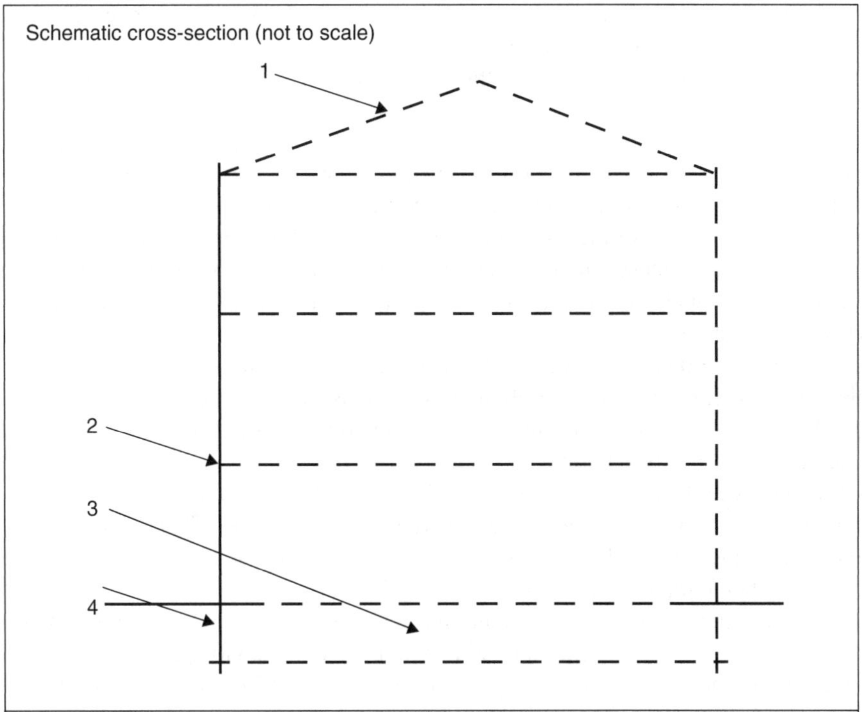

Notes

1. Dashed line represents new superstructure behind existing façade.
2. Lateral restraint required between existing retained façade and new construction behind it.
3. Possible new basement level formed.
4. Underpinning of existing façade may be required – especially if new basement level is formed.

Figure 3.6 Outline of façade retention

affected by such a scheme. The inside of the building is gutted out, leaving at most only the four main walls.

The façade of many older buildings in most urban areas is often listed or deemed important for architectural or historic reasons. Architecturally, such properties may have an ornate stonework external envelope or other unusual features. The materials used may in themselves have distinctive or rare qualities in terms of colour, texture or overall appearance. Historically, the façade may represent a distinctive landmark associated with a particular person, event or architectural style. Retaining it would at least help to preserve the streetscape.

In many cases, however, the internal configuration of old buildings is often inappropriate for the requirements of modern work and living. Maintenance costs on old buildings are normally much higher because of the need for skilled craftsmen and appropriate expensive materials for undertaking remedial works. Replacing the superstructure behind a façade with a matching development is in some cases an under-utilization of the site. This is because the storey heights are excessive, and the general state of repair of the building is such that simple alterations, refurbishment and/or remedial works are undesirable or uneconomic.

Façade retention can allow the new building behind to have a larger plan area and more storeys. It can be seen as a realistic compromise between the sometimes conflicting interests of the developer and conservation agencies involved. The finished project keeps the appearance of the original building by saving what is worth preserving. It also achieves a more efficient and economic use of the space behind by replacing what is not worth retaining. This can increase the viability of the project.

Nevertheless, it must be acknowledged that 'façadism' as it is pejoratively known in a few quarters, has its critics. Façade retention after all provides only a superficial solution to a building with architectural or historic qualities. Some architectural commentators, for example, are prone to criticising this option because it destroys the integrity of the building. Given that the façade is an integral part of the building, this is valid. The new building behind the façade can be thus branded a fake.

Essentially, 'façadism' is one of the issues of debate between modernists and traditionalists in building conservation. The former are those who take the view that buildings are primarily meant for people. They take a pragmatic or functional approach to the adaptation of buildings in that it can take any form so long as the property is being put to good use.

Traditionalists, on the other hand, see buildings in cultural as well as functional terms. They do not favour wholesale conversion or renovation of old buildings just for the sake of it. Respect for the past and adaptation works that are modest and sensitive and do not undermine the building's architectural merit are more important to them than conversion at any price.

A comparison between the two main options for façades is shown in Table 3.5.

Unfortunately, as with any controversy, extreme views on either side of this conservation debate can distort the issue. A balanced or moderate position can achieve a compromise, which retains the identity of the original building but at the same time allow its flexible reuse.

Depending on one's perspective, façade retention can either preserve or compromise the streetscape of an area. It can, for example, undermine the authenticity of the façade and townscape value by incorporating additional levels above the roof or across the original fenestration. Blinds or curtains can be used to prevent brightly lit, deep, open plan offices being highlighted on dull days or at night when all the building's lights are on.

Technical problems with façade retention

As with any construction project, each façade retention scheme is unique. The location of the property, the site history as well as the size and form of the existing building and retained façade will not be the same

Table 3.5 A comparison between façade retention and façade replacement schemes

Façade retention	*Façade replacement*
Usually restricted to old or damaged buildings of traditional load-bearing masonry construction.	Can be done to buildings of both modern and traditional construction.
Not feasible on thin-walled buildings (i.e. framed buildings with curtain walling).	Feasible as method of renewing the cladding of existing buildings.
Requires extensive lateral and some vertical support.	Requires vertical and some lateral support.
Preserves existing appearance of façade.	Allows for changes in appearance and performance of the façade.
Possible financial aid if Listed or in a Conservation Area, etc.	Unlikely to generate much, if any, grant aid.
May avoid or minimize planning restraints.	Requires compliance with the planning regulations.
Allows for better or even increased utilization of space behind the façade.	May provide minimal space change.
Existing façade may require strengthening (e.g. underpinning).	Allows façade to be structurally upgraded easily.
Overcomes problems of difficult internal remodelling work.	May necessitate internal alterations.

in every project. Still, to ensure that façade stability is maintained at all stages, the fundamental redevelopment issues remain:

- to determine the most appropriate form of temporary support for the existing façade;
- to facilitate the safe dismantling of the existing building behind the façade;
- the most appropriate construction of the new building behind the intact façade, including means of permanent support and provision for differential movement.

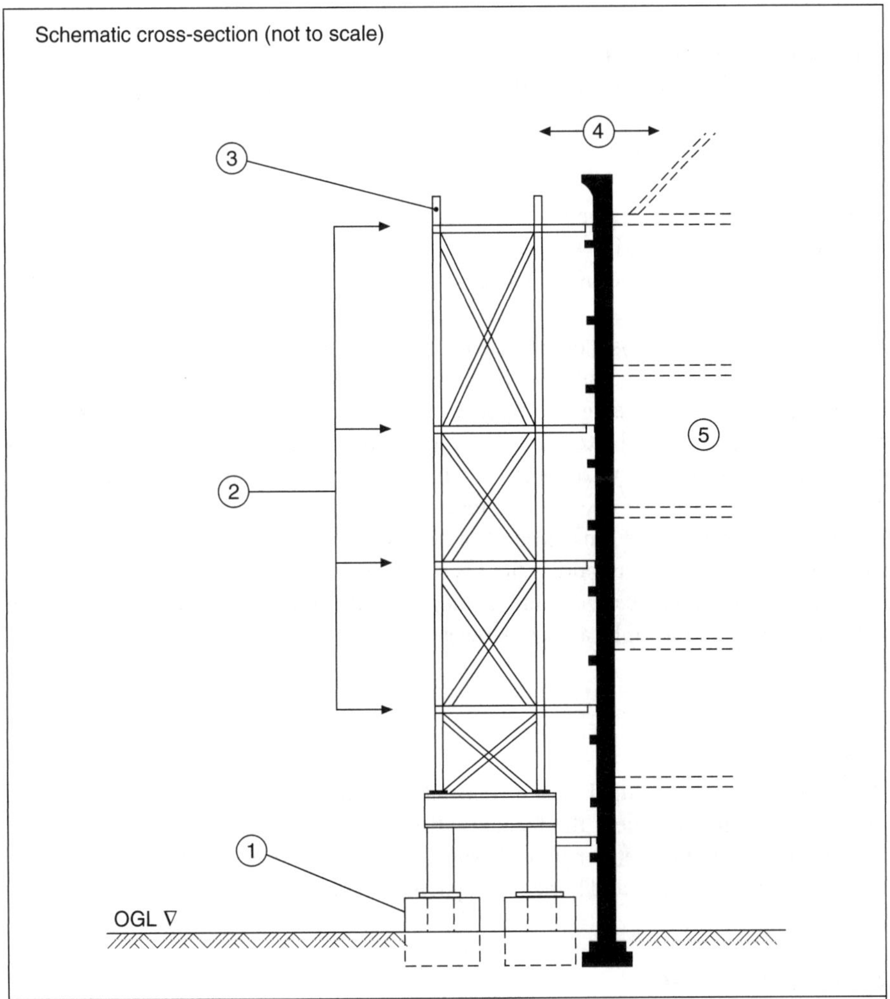

Notes
1. Concrete blocks on pavement or in-situ foundation (depending on soil conditions) for temporary bracing towers.
2. Lateral loads (due to wind) on building faÁade resisted by horizontal trusses spanning between braced towers.
3. External or internal façade retention framework.
4. Rotation caused by wind load, differential settlement or other factor.
5. Building behind demolished leaving façade connected to temporary frame for stability.

Figure 3.7 Temporary support for a façade retention scheme

During the demolition of the old and the erection of the new, the existing façade requires temporary support to prevent it collapsing (see HSE GS 51). The façade of any building, because of its slenderness, is highly unstable. Some vertical temporary support is required to prevent collapse of openings or balconies and other projecting features. In such cases strutting or dead shoring is required.

Lateral support is the most important form of temporary support required in façade retention schemes. The excessive slenderness of the retained wall makes it highly prone to rotation. Dead shores are required to provide vertical support, and raking or flying shores to prevent lateral movement. A structural engineer should design these.

Temporary support systems

According to Lunardi (1999) there are essentially two types of retention support system:

1. External 'exo-skeleton' system.
2. Internal prop system:
   - between other walls being retained,
   - tied back to temporary braced towers,
   - tied back to elements of the permanent structure constructed in advance of full demolition.

The 'exo-skeleton' system consists of framework of steel box/I-sections (see Figure 3.7). It obviously requires sufficient space around the building. This method is more suited to the buildings that are large in plan. Ideally, though the base of an external support system should be as wide as possible to for maximum stability without the need for much Kentledge. As a rule of thumb, therefore, 1:6 is the maximum the width-to-height ratio of the temporary support system. If it is more than this, say 1:7, the framework would be too slender. Moreover, narrower bases require more weight to stabilize the retention system. This in turn could create problems on weak-bearing ground or where there are voids such as basements or cellars below the pavement on which the support system is being erected.

An internal prop method, on the other hand, is more suited to buildings that are small in plan with two or more walls to be retained at right angles. However, this method takes up valuable internal working space (Lunardi, 1999).

Anchoring

Because of its slenderness the retained façade is unstable without permanent support. Even after the construction behind has been completed, the façade will still require tying in to the new work (Highfield, 2000). Stainless steel angle brackets or tie rods should be used at every floor level. Again, a structural engineer would determine the size and horizontal spacing of these fixings.

Differential settlement

The façade and the building behind it are essentially two different constructions, each with different loading arrangements. The façades of old buildings are usually of solid load-bearing masonry construction, with brick footings. However, the thickness of the walls will not be uniform throughout their height. Walls tend to reduce in thickness at each floor level to a minimum thickness of one brick at the roof eaves level.

The new building, on the other hand, is likely to be of framed construction – either concrete or, more commonly, steel. The loadings induced by the new building are likely to be higher than the façade to which it is connected. Any such 'connection' or linkage must allow for some settlement of the new building (Highfield, 2000). As much as 50 mm is not unrealistic in such circumstances for a building five storeys and under in height.

Substructure details

The foundation of the retained façade may require at the least remedial work. In more serious cases, underpinning may be required either to make good damaged sections of the existing footings or to take the foundations down to a lower level to accommodate a basement.

Façade replacement schemes

In some rare cases the façade of an existing building may be carefully taken down replaced with a different, but more usually facsimile, façade (see Figure 3.8). This is sometimes called recladding as opposed to overcladding (see Chapter 9). It is usually restricted to situations where there has been substantial damage to the façade of a building, for example as a result of a bomb or gas blast. Alternatively the work may be needed to replace a defective or otherwise deteriorated façade of a refurbished building. It may also be done where the client wants a completely new design for the external fabric.

Façade replacement schemes are relatively rare. They may be required where there has been substantial damage to the façade of a building, for example as a result of a bomb or gas blast. Alternatively, the work may be needed to replace a defective or otherwise poor-quality façade of a prestigious building.

Nonetheless, façade replacement schemes offer an opportunity to enhance the appearance and technical performances of buildings. They are often accompanied by over-roofing schemes (see Chapter 6).

Schematic cross-section (not to scale)

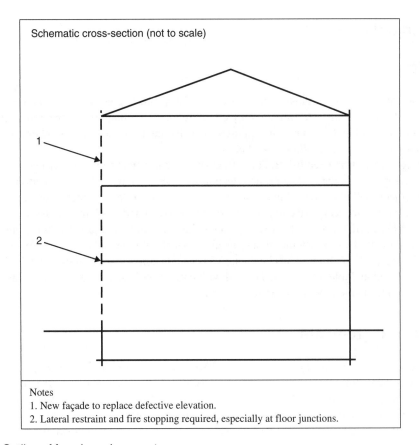

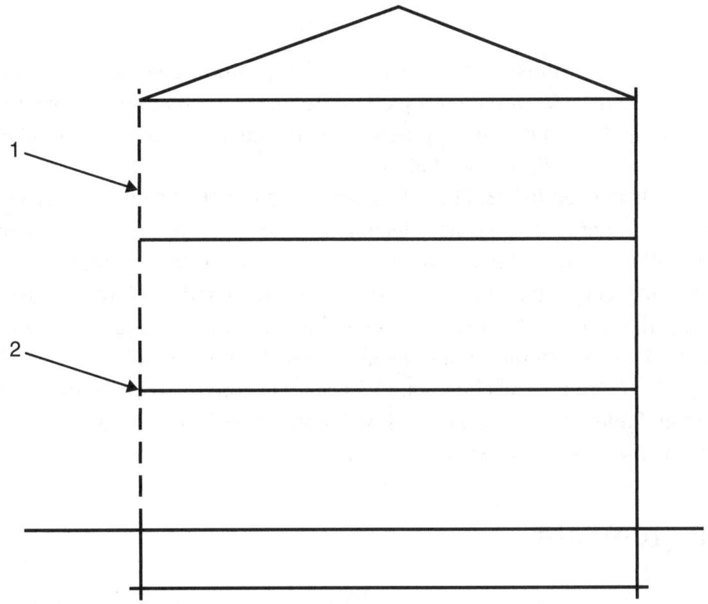

Notes
1. New façade to replace defective elevation.
2. Lateral restraint and fire stopping required, especially at floor junctions.

Figure 3.8 Outline of façade replacement

Façade relocation schemes

An even rarer occurrence than façade retention or replacement is a façade relocation scheme. In this case the façade of an old important building is carefully taken down stone by stone and reconstructed elsewhere or on another part of the site. Redevelopment takes place behind the relocated façade.

Exterior lighting

Installing exterior lighting can contribute to the visual success of the conversion or other adaptation of an old building. It is used in the evenings by positioning high-intensity lamps to highlight the whole building or its prominent features. The exterior lighting can also contribute to the enhancement of the building's security.

Sealed beam lamps with a built-in reflector should be adopted for floodlighting or spotlighting. Bulbs ranging from 500 to 1000 W rating are usually used for external use. Tungsten pressed armoured reflector lamps, tungsten halogen lamps or mercury fluorescent discharge lamps are common types for this purpose. The most suitable fittings are either of stainless steel or robust plastic and have toughened or polycarbonate glass covers with waterproof seals to minimize damage from moisture and vandalism.

In consuming energy exterior lighting, of course, does not directly contribute to sustainable construction. Its use, therefore, will probably be restricted to important properties, and even then only to certain periods throughout the evening.

Code compliance

Preamble

As already indicated, one of the most onerous aspects of any building conversion scheme is compliance with the statutory requirements. Conversion projects are required to obtain building control approval if not planning permission. A Use Class Order is required for a change from one use to another for certain premises, particularly retail units and offices (see below).

Both the English/Welsh and Scottish Building Regulations require that a material change of use must comply with the appropriate standards. The primary technical issues to be resolved are summarized below. The specific requirements will be addressed in a little more detail in the relevant chapters. For example, Chapter 4 deals with some of the practical code compliance problems encountered in adaptive reuse schemes.

It must be emphasized, however, that many of the building regulation requirements are highly complex. They do not lend themselves to easy interpretation and application. Every adaptation and new-build project is unique in this respect. Therefore, the checklist of basic code compliance requirements for building conversion, as presented in Table 3.6, should be used with care. It indicates the main provisions for England and Wales and contrasts them with those for Scotland.

Structural requirements

Part A and Section 1, respectively, of the English/Welsh and Scottish Building Regulations deal with structural requirements for adaptation work. Changes of use, particularly from residential to commercial, often involve increases in imposed loadings. For example, the standard imposed floor loading for dwellings is $1.5\,kN/m^2$, whereas for offices it can range from 3 to $6\,kN/m^2$ (BS 6399).

Inserting another floor or installing a new top storey will increase the dead load of the building. These alterations may necessitate strengthening of the building (see Chapter 7).

Table 3.6 Basic building conversion checklist for code compliance

Main provisions	*Parts/Sections*		*Typical adaptation requirements*
	E&W	*Scottish (1)*	
Structure	A	Section 1(2)	Structural calculations must be provided and checked for new or altered structural members (see Chapter 7).
• Structural stability			
Fire safety	B	Section 2	
• Means of escape	B1		See Table 4.7.
• Surface spread of flame	B2		Plasterboard or 'Fermacell' composite dry lining board is acceptable (see below).
• Fire resistance	B3 and B4		Largely concerns conversion of upper floor ceiling to a new floor construction, which must then achieve at least 30 minutes fire resistance. Two layers of 15 mm thick 'Gyproc Fireline' ceiling plasterboard, for example, gives 90 minutes fire resistance for separating floors.
• Access for fire service	B5		
Site preparation and resistance to moisture	C	Section 3	Adequate damp-proof membrane under upgraded existing solid ground floor.
Toxic substances	D	Section 6	Controls the use of urea formaldehyde cavity fill.
Sound	E	Section 5	Sound insulation required in separating walls and floors (see below and Chapter 9).
• Sound transmission			
Ventilation	F	Section 3	
• Rooms			Windows in habitable rooms must have an opening area proportional to the floor area. See Table 3.7.
• Roof space			See Chapter 9.
• Subfloor			1500 mm^2 per metre run of external wall.
• Vapour control			Incorporating a vapour control layer on the warm side of construction.
Hygiene	G	Sections 3 and 4	Location of WCs in relation to kitchens. Unvented hot water systems.
Drainage and waste disposal	H	Sections 3 and 4	Maximum permitted length for unvented branch connections from baths and wash hand basins. If exceeded branch pipes must be directly ventilated to the outside air.

(*Contd*)

Table 3.6 (*Contd*)

Main provisions	*Parts/Sections*		*Typical adaptation requirements*
	E&W	*Scottish(1)*	
Heat-producing appliances	J	Sections 3 and 4	Modifications to gas fires, cookers and boilers. Safety precautions and ventilation requirements.
Stairways, ramps and guards	K	Section 4	Dimensions and pitch of stair. Minimum 800 mm wide for domestic stair. Minimum 2000 mm headroom. Maximum pitch of 42° for standard domestic stairs; 38° for escape stairs. Maximum 16 risers in any straight flight.
Conservation of fuel and power	L	Section 6	Insulate around new accommodation only? U-value targets – see Chapter 8. New wiring.
Access for disabled people	M	Section 4	Applicable especially to buildings serving the public. Consideration must be given to dwellings as well.
Glazing – safety in relation to impact, opening and cleaning	N	Section 4	Safety glazing, provision for cleaning of windows.
Electrical safety	P	Section 4	Electrical installations, wiring, etc.

Notes
1. The Sections under Building Standards (Scotland) Regulations 2004, which replaced the old Parts A to T of the 1990 (as amended) regulations.
2. The Small Buildings Guide effectively covers the requirements of Section 1 for buildings not exceeding three storeys.

Downtakings and forming openings in existing walls have structural implications too. These involve changes in load paths, increases or decreases in loadings, strengthening of remaining load-bearing elements. Temporary and permanent support measures are also needed.

Temporary support measures may consist of propping of existing floors immediately before and during remedial or upgrading work. Walls offering lateral restraint may also require temporary bracing during the formation of openings in them.

Permanent support measures often involve the insertion of new beams to provide additional support under floors or buttresses to stiffen external walls. These issues are examined in Chapter 7.

Disabled access and escape provisions

Part M of the Building Regulations 2000 has requirements for disabled access and escape provision for certain categories of property such as shops, offices, factories and public buildings. As indicated in Table 3.6, the Scottish equivalent, Section 4 has tightened-up the requirements for disabled access.

In addition, the Disability Discrimination Act 1995 has strict requirements to achieve a barrier-free environment for all building users. These are addressed in more detail in Chapters 8 and 9.

Residential fire safety

Preamble

Fire safety is another key requirement that affects safety standards in buildings. Given its complexity the aim here is to outline the main requirements of domestic fire safety rather than cover all of the provisions contained in the building regulations. Again, it cannot be emphasized enough that these are only indicative.

New uses for old properties can mean more stringent requirements for the fire protection and sound insulation of the existing floors and walls. According to Haverstock (1998), for example, floors must have upgraded fire resistance in residential loft conversions, warehouse to office or hotel changes of use, and old house to institution conversions or change of use to flats. At least ½ hour fire resistance is required for most floors in domestic conversions, but a floor over basement may need to achieve a 2-hour rating in this regard. These and other forms of upgrading floors and walls are covered in Chapter 9.

Structural precautions

In achieving the required fire resistance to comply with Part B in England and Wales (Section 2 in Scotland), as a preliminary measure, it is necessary to check the following:

- Any member supporting a wall (e.g. a floor structure or balcony).
- Attic separating walls: check degree of seal with underside of roof finish.
- Any gable opening (e.g. roof window or oriel): check for effects.
- Gable wall: check its construction and condition.
- Chimney or flue: check the condition, location and its outlet position.

Thermal upgrading

One of the major requirements in any adaptation work is to enhance the energy efficiency of the building. This is usually best achieved by improving the level of insulation in the external walls, roof and ground floor of the building to reduce the overall U-value of the fabric (see Chapter 10).

Table 3.7 Requirements for soundproofing floors and walls

Sound transmission	*Element*	*Decibel rating*	*Remarks*
Airborne R_w (dB)	Walls	Minimum: not less than 53 dB	Existing 103 mm thick brick wall with plaster on each side is likely have a dB rating of about 35.
	Floors	Minimum: not less than 52 dB	Existing floors comprising with no soundproofing are likely to have a dB rating between 36 and 40.
Impact L_{wn} (dB)	Floors	Maximum: not more than 61 dB	Existing floors comprising with no soundproofing are likely to have a dB rating between 76 and 82.

Thermal upgrading often creates problems in conversions in historic properties. Internal insulation offers the only feasible way of improving the thermal performance of solid load-bearing masonry buildings (see Chapter 9).

Sound insulation

Building conversions comprising a change from single to multiple occupancy will require a high level of soundproofing to comply with the building regulations. Securing this in separating elements such as floors and walls can be difficult. Table 3.7 lists the main requirements in relation to both airborne and impact sound transmissions in these elements. How these can be achieved is shown in Chapter 9.

As with thermal efficiency, the regulations relating to sound insulation are being tightened-up. However, there is a conflict between achieving good levels of thermal and sound insulation. The former is best obtained by using lightweight materials or materials with low conductivity. This, however, does little to resist airborne sound transmission, which is one of the main sources of noise disturbance in dwellings.

The proposed changes will have implications for adaptations such as conversions as well as new-build work. According to Pearson (2000), they are essentially aimed at:

- restricting the use of some lightweight concrete blocks in wall construction.
- restricting the use of plasterboard on dabs on flanking walls (any wall in contact with a party wall) on any type of block.
- restricting the use of rigid foam plasterboard laminates on dabs to flanking walls for any block type.
- restricting the use of plasterboard on dabs to party walls.
- requiring two layers of 12.5 mm thick plasterboard to the walls and ceilings of all framed buildings to reduce sound transmission through walls.
- requiring façade insulation if eventual proposals include new requirements for attenuating sound.
- restricting the use of heavy-duty wall ties for larger cavities.
- restricting the use of beam and block floors for intermediate floors.

Ventilation

The correct balance between natural and mechanical ventilation in buildings is not easy to achieve. These days, there is increasing onus on designers and builders to minimize air leakage from buildings as a means of maximizing energy efficiency. The major drawbacks of this objective, however, are that eliminating background ventilation from a building can reduce the indoor air quality and increase the risk of interstitial condensation occurring in the external fabric. The latter can lead to moisture-related problems such as fungal

Table 3.8 Ventilation requirements for dwellings (Oliver, 1997)

Area	*Rapid ventilation (e.g. openable windows/doors)*	*Background ventilation mm^2 (1)*	*Extract rates: fans or passive stacks*
Habitable room	1/20th floor area	8000	(2)
Kitchen	Opening window (no minimum size)	4000	30 l/sec adjacent to hob 60 l/sec elsewhere or PSV (3)
Utility room (access via dwelling)	Opening window (no minimum size)	4000	30 l/sec or PSV
Bath/shower room	Opening window (no minimum size)	4000	15 l/sec or PSV
Water closet (separate from bathroom)	1/20th floor area or 6 l/sec mechanical extract	4000	–
Pitched roof	N/A, unless skylight is provided	10 mm continuous eaves ventilation if pitch is over 15° 25 mm continuous eaves ventilation if pitch is under 15°	–

Notes
1. Alternatively provide a minimum of 4000 mm^2 to all rooms in the above table with an overall average of 6000 mm^2 per room for the dwelling.
2. For internal rooms (i.e. those rooms not having any windows to the outside), provide either 10–15 min overrun to the mechanical extraction indicated in the above table, or PSV, or an open flued heating appliance may be acceptable. In all cases some form of air inlet is required.
3. PSV = Passive stack ventilation.

attack, mould growth, corrosion and efflorescence. Loss of thermal efficiency in any insulation materials because of their increase in conductivity when wet is another problem associated with this form of dampness.

It is often overlooked that all buildings need to breathe to some degree. Old buildings breathed naturally because of gaps in the construction and through the porosity of the fabric. Excessive draughtproofing can compromise this.

The basic ventilation requirements for dwellings are summarized in Table 3.8. Those for pitched roofs are listed in Chapter 9.

Services

Generally

Ascertaining the capacity, layout and condition of the existing services is one of the first considerations when converting or extending a building. Multiple-occupancy conversions will require separate supply and discharge services.

Figures 3.9 and 3.10 illustrate the typical plumbing layouts involved in basic conversion work.

Electrics

In relation to the electrical installation, the appropriate utilities authority will need to be consulted to assist in establishing the intake and, where necessary, the metering requirements. This will be further complicated

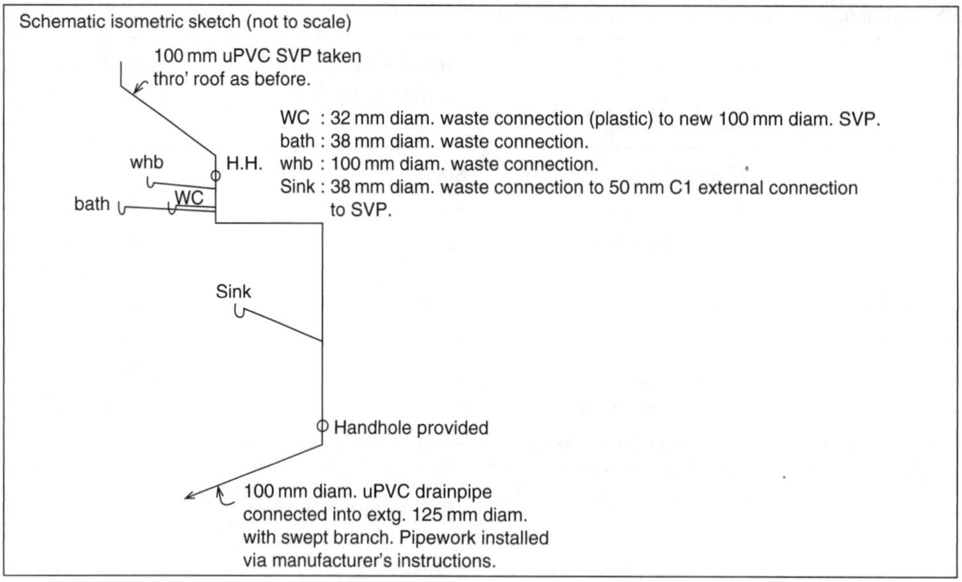

Figure 3.9 Layout of plumbing involved in small attic conversion

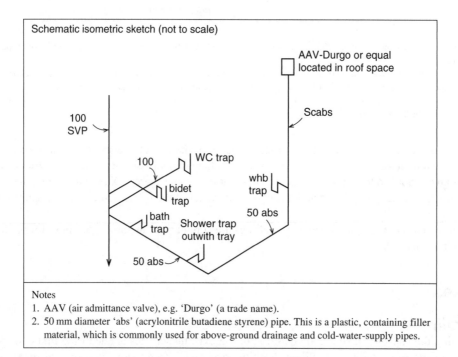

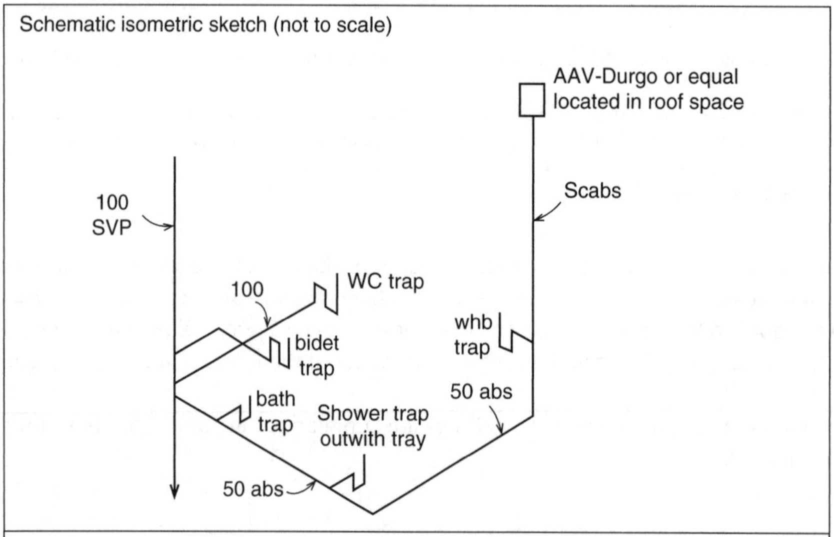

Notes
1. AAV (air admittance valve), e.g. 'Durgo' (a trade name).
2. 50 mm diameter 'abs' (acrylonitrile butadiene styrene) pipe. This is a plastic, containing filler material, which is commonly used for above-ground drainage and cold-water-supply pipes.

Figure 3.10 Layout of plumbing for new en suite bathroom in conversion scheme

if the capacity of the power supply may need upgrading from single phase to three phase if the change of use involves commercial or industrial activities.

Care must be taken to ensure that the correct wire and fuse ratings are used for all circuits and appliances. This is especially important to avoid overheating of wires and their connections. For example, a 10 kW shower unit should contain a minimum 40-amp fuse/contact-breaker and be fed using 10 mm gauge wiring.

Water supply

As regards water services, the cold water tank/s should be sufficiently high to obtain enough pressure to satisfy the intended use. This may need renewing to replace defective or undersized tanks. These days, polyethylene or resin-bonded glass-fibre tanks complete with vented lid are frequently used in place of lead or galvanized mild steel. Insulating the top and sides of the water tank is important to avoid it freezing in cold weather. Protecting all pipework with preformed polyethylene foam lagging is also necessary. Renewal of lead piping is dealt with in Chapter 8.

Heating system

It is likely that the opportunity will be taken to replace or upgrade the existing heating system. A wet-fed radiator system for each dwelling in a single-occupancy building converted into flats is one typical option for the heating system (see Chapter 8). Underfloor heating, however, is an effective alternative (as outlined in Chapter 5).

Drainage

The existing drainage system may not be adequate in terms of capacity or condition to cope with the new use. This may necessitate replacing deficient existing pipelines. In addition, multiple-use conversions will require additional drainage lines. These extra pipelines will require vertical and horizontal space.

Gas

The existing building may have a ready gas supply with redundant as well as 'live' piping made from mild steel. 'Block tin' was a common material for gas piping in pre-1990 dwellings. However, separate supplies to each flat will be required. This should provide the opportunity to renew any distribution gas pipework to and within the building.

Specialist services

Allowance may have to be made in the converted building for services associated with information technology. The rapid increases in the use of cable television, multiple-phone lines, security equipment and building management systems have prompted the need for buildings to accommodate these services.

Planning requirements

The GPDO mentioned in Chapter 2 defines certain categories of changes of use as not necessitating a planning application (i.e. those changes of use that are permitted development). Examples of changes that classed as permitted development are from:

- food and drink (Class A3) to shop (Class A1);
- food and drink (A3) to professional and financial services (A2);
- general industry (Class B2) to business (Class B1);
- car showroom to shop (A1).

However, changing from a shop (A1) to food and drink (A3) is not classed as permitted development. This is because the latter type of use would require facilities such as kitchen and preparation areas, which may not necessarily be present in the former. In such a case, planning permission as well as building control approval would be needed.

Even changes to the same use may require planning permission. Bringing a residential property back to its original use is an example of this. Therefore, converting two or more flats back to one house – a re-conversion – involves obtaining approval under the Town and Country Planning legislation.

The policy of the local authority involved and the supply of certain house types will influence the outcome of a planning application to convert or re-convert a property. If there is a scarcity of family dwellings in an area, then an application to re-convert a house back to this use will probably succeed. On the other hand, if there is glut of such properties or a shortage of smaller size flats, there is a strong chance that permission will not be granted.

As with the building regulations, in cases of doubt it is always best to consult the local authority on planning matters. Verbal feedback indicating that planning permission is not needed should be confirmed in writing. An application can then be made to the council's planning office for a 'Certificate of Lawful Proposed Use or Development'. Such a document can be used to prove that the local authority conceded that the proposed change of use did not require planning approval.

Inflexible buildings

Tight-fit characteristics

Traditionally buildings were designed and built with only one use in mind. Previously hardly any thought was given to the possible alternative use at a later stage in the building's life. Nowadays, though, the Construction, Design and Management (CDM) Regulations 1994 make it a requirement for designers to consider future use and demolition procedures for new buildings.

Buildings that cannot be altered easily to new use without massive alteration and structural change often beyond both economic and practical reason are termed 'inflexible' or 'tight fit'. This usually means that the structural configuration or morphology of the building does not easily allow alteration. This makes it difficult to undertake many modifications to the interior without the need for substantial support measures and some selective demolition.

Inflexibility is one of the main physical constraints affecting the effective reuse of any building. The feasibility of changing the use of a building is often determined by spatial constraints. 'Tight-fit' buildings are, by their very nature, difficult to reuse because of the extent of internal alteration required to make them better suited to other uses.

The client may wish to make the layout of the converted building radically different from the original. This is likely to have knock-on effects on the building's structure and fabric. The type and standard of finishes required by the user will also impinge on the degree of adaptation.

The factors that can make a building inflexible can be divided into external and internal:

External factors

- Restricted site, especially with a mid-terraced property, which constrains lateral and vertical extension or modification of the main building.
- The extent and location of available space about the building influences choices for expansion, off-street parking/unloading and erection of scaffolding. The façade of many buildings in urban areas often abuts the pavement, which may necessitate the installation of a gantry as well as scaffolding for the duration of the adaptation work.
- Restraints by local planning body or other authoritative parties on increases in, or modifications to, the proposed roofline.

Table 3.9 Main categories of inflexibility

Degree of inflexibility	*Examples*	*Characteristics*
Very inflexible (high)	Windmills; concrete grain silos Engine houses Chemical plants	Awkward or restricted plan shape (e.g. circular). Low or restricted floor to ceiling height. Engine bases comprising large concrete blocks, extensive pipework and other specialist services that are now redundant.
Fairly inflexible (medium)	Bespoke or unusual dwellings Churches and chapels Factories and warehouses Agricultural buildings Theatre and cinemas Whiskey bonds	High or awkward floor to ceiling height, which makes cleaning and heating the space difficult. Close column spacings (i.e. 3 m or less). Unsuitable capacity and layout of services for a new use.
Reasonably flexible (low)	Cotton mills Barns Large detached dwellings	Relatively easy to reconfigure the internal space. Externally, however, they may not be very flexible. Tight boundaries or roof extension restrictions may inhibit lateral and vertical extensions, respectively.

Internal factors

- Low floor-to-ceiling heights make it difficult to accommodate extensive services.
- High floor-to-ceiling heights and fenestration make the insertion of mezzanine floors difficult.
- Awkward floor plan shapes make it difficult to organize or articulate the space efficiently.
- Framed buildings with close column centres interfere with open-plan layouts.
- Deep-plan buildings where subdivision of space into smaller volumes denies natural light and ventilation.

Degrees of inflexibility

Because of their wide range of shapes and structural configurations, not all buildings have the same level of inflexibility. There are therefore degrees of inflexibility, and these are summarized in Table 3.9.

Building morphology and layout

The arrangement or configuration of the structure and fabric of a building has a major influence on its flexibility to adapt to other uses. Generally, building of load-bearing masonry construction are less adaptable in this regard than the modern framed building type.

The plan shape of a building can influence the degree to which it can be converted. The number of storeys is also significant. These in turn influence the:

- shape, size and juxtaposition of new rooms;
- number of floor levels required and available headroom;
- existing fire escape provisions;
- extent and articulation of circulation space;
- spatial efficiency of the building (see Chapter 9 for details of net-to-gross ratio).

Physical context

As indicated above the position of the building being converted will influence the degree to which it can be converted. Mid-terraced properties, by their very location, are constrained on three sides; and even the rear may not have much space for expansion. There is very little room for manoeuvre in relation to ingress and egress routes, for example.

Reducing or enlarging the useable space

As indicated earlier, old buildings often have high floor-to-ceiling heights. In many non-residential properties such as churches and function halls, there was only one storey, even though the roof was over 5 m above ground level. In economic, functional and environmental terms, this is a waste of space. The accommodation within such an interior is thus harder and more expensive to heat and keep clean. High ceilings are relatively inaccessible without tower scaffolding. This makes repainting and other maintenance work difficult and costly.

In such a case, therefore, it makes sense economically and spatially to insert additional floors in the building (see Figures 3.11 and 3.12). Undertaking this work, however, is easier said than done. There are structural and physical constraints that may inhibit the inclusion of more floors. For example, the walls may not be strong enough to support the increased loadings without additional support. A framework of steel columns and beams may be required in such circumstances. This is more likely where three or more additional floors are involved. Figure 3.13 illustrates schematically how this could be done.

The fenestration may be such that the proposed line of a new floor runs through the middle of a row of windows. This is very common in church conversions – see Chapter 4 for dealing with this problem (Figure 3.14).

Nevertheless, some conversion schemes can make full use of very high floor-to-ceiling heights. Certain bespoke buildings that become redundant such as old coal-fired power stations can offer spacious interiors that can be used for large art exhibitions or display shows. The conversion of the old Bankside Power Station in London into the Tate Modern Gallery is one of the most prominent recent examples of this type of adaptation.

Enlarging the space inside a building may be required when converting it from multiple occupancy to single use. This will normally entail removing walls and floors or inserting openings in these elements to articulate the previously separate spaces.

Askham and Blake (1999) described a three-storey property that was ripe for conversion. The building was a Victorian mid-terraced property of traditional solid load-bearing stonework with a slated half mansard roof, with 100 per cent site coverage. There was, therefore, no room to enlarge the building laterally. It was originally used as a service club and subsequently converted into offices, but these became vacant. The only possibility of increasing the useable floor space is to insert additional floors.

Figures 3.15a and 3.15b illustrate the possibility of converting a rural mill into high-quality apartments. In large schemes such as this, it may be financially prudent or necessary to go for a mixed form of tenures. One half or level could be offered for sale. The rest could be made available for rent.

The structural implications of these reductions and enlargements of space within buildings are dealt with in Chapter 7.

Different floor levels

Buildings on sloping sites can be difficult to convert if they depend on uniting adjacent properties horizontally. This is because of the likelihood of different floor levels across mutual or party walls. Moreover,

Plans (not to scale)

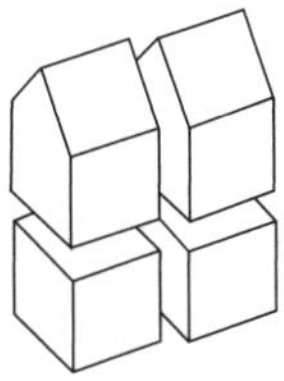

Kilmarnock and Loudoun District Council
Knockinlaw, Kilmarnock

1930/39 (traditional construction)
4-in-a-block

Four 4-apartment flats were converted to four 3-apartment terraced houses

In the original plan the entrances to both the lower and upper flats were at the side of the block.

One of these entrances was retained and a new one was created on the front façade to allow a private garden for each dwelling and to give more privacy and security.

Some non-load-bearing partitions were moved and existing walls upgraded to party wall standard in the same position on ground and first floors. The first-floor bathroom was retained but enlarged for one house and the bathroom for the other house was relocated on the first floor. More general storage space is required for the mid-terraced house but there is space to provide this. The results of the conversion are good terraced houses with private entrances and kitchen doors to back gardens.

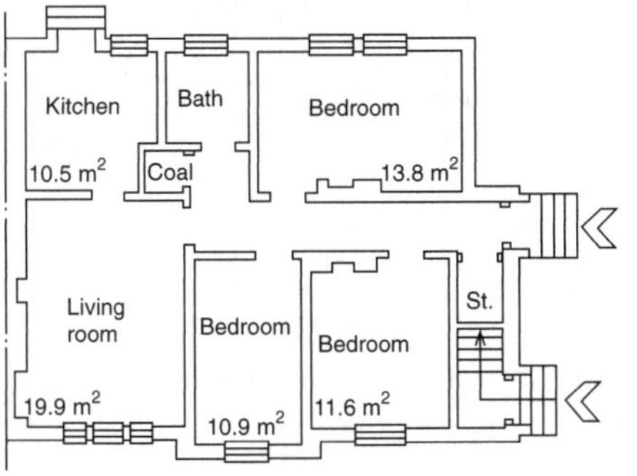

Original ground floor plan

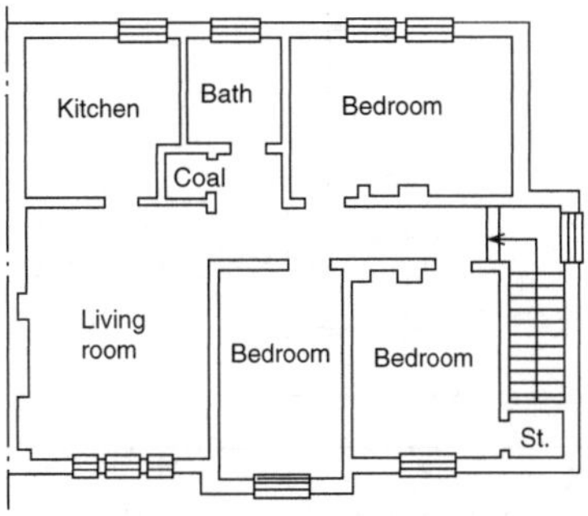

Original first floor plan

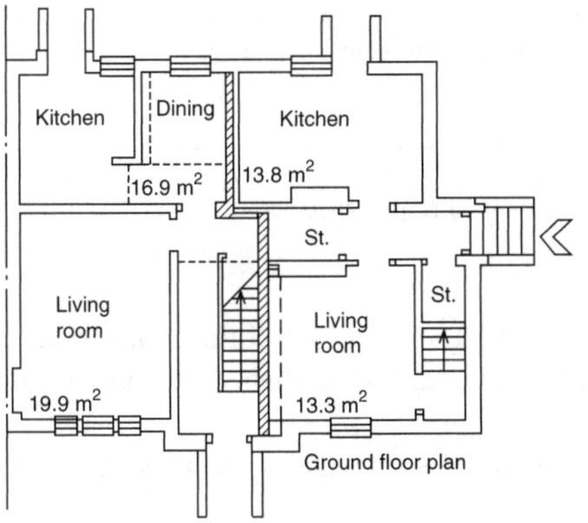

Converted ground floor plan

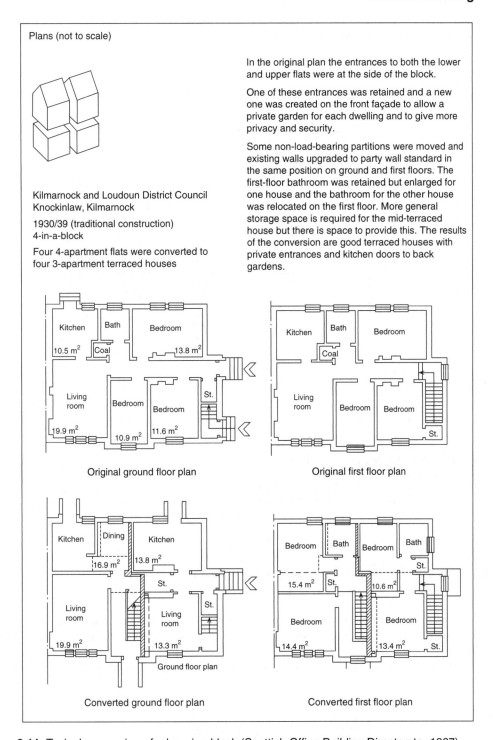

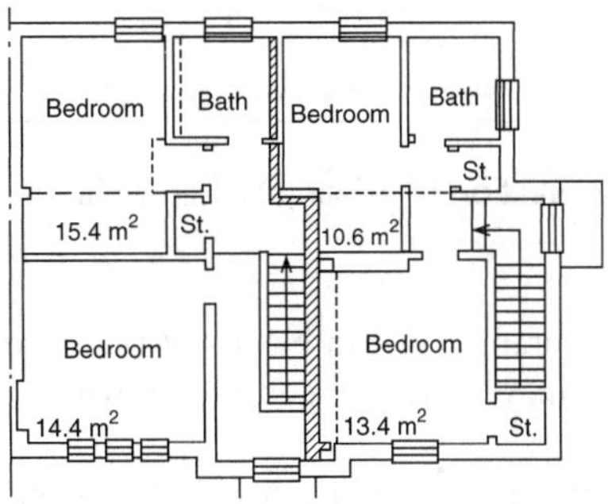

Converted first floor plan

Figure 3.11 Typical conversion of a housing block (Scottish Office Building Directorate, 1987)

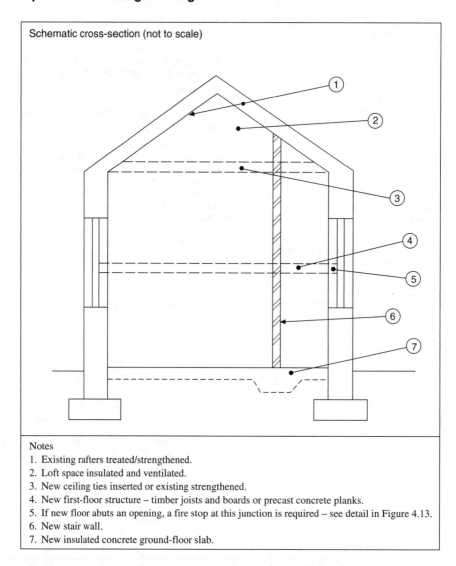

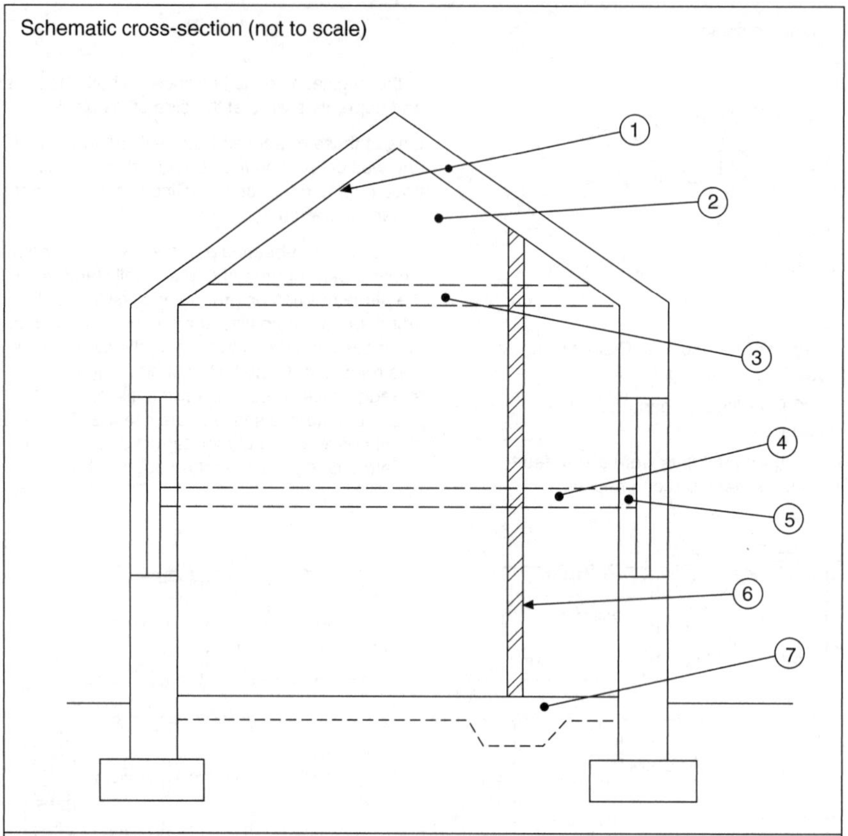

Notes

1. Existing rafters treated/strengthened.
2. Loft space insulated and ventilated.
3. New ceiling ties inserted or existing strengthened.
4. New first-floor structure – timber joists and boards or precast concrete planks.
5. If new floor abuts an opening, a fire stop at this junction is required – see detail in Figure 4.13.
6. New stair wall.
7. New insulated concrete ground-floor slab.

Figure 3.12 Section of existing building with new upper floors inserted

there is no guarantee that in blocks of flats, the adjacent floors will be at the same level. Differences in floor levels may only be confirmed when party walls are demolished (Scottish Office Building Directorate, 1987).

Adjustment of floor levels, therefore, may be necessary in some instances. This can be achieved in one of two main ways:

1. Increasing the lower floor level of one unit to that of the higher floor level.
2. Installing steps or a ramp to link the two different floor levels.

The first solution would only be feasible if the difference in level was small (i.e. <600 mm) and the area of floor being raised was small and not larger than the retained level. Installing steps or a ramp between the two different floor levels would take up space and require handrails or balustrades.

Schematic plan (not to scale)

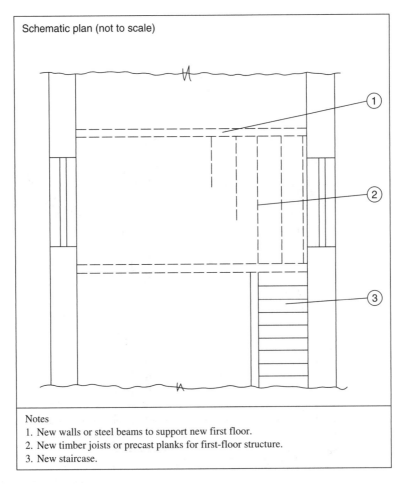

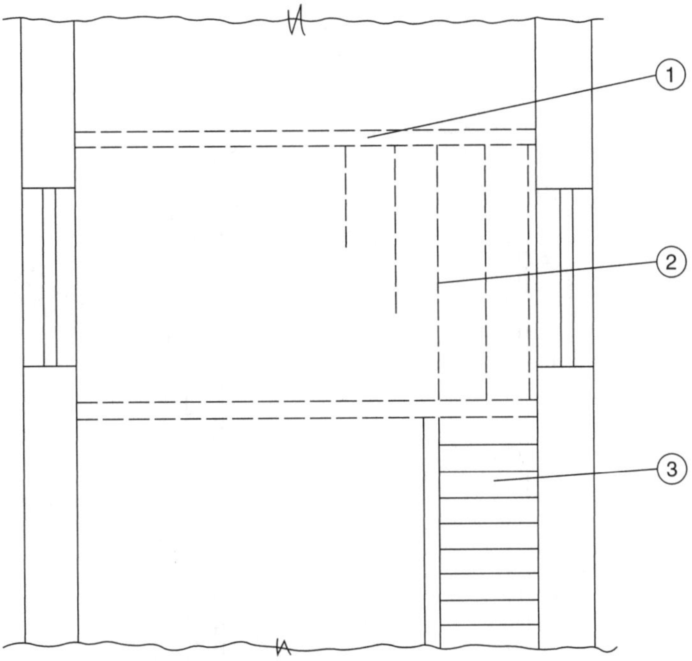

Notes
1. New walls or steel beams to support new first floor.
2. New timber joists or precast planks for first-floor structure.
3. New staircase.

Figure 3.13 Plan of building with new upper floor inserted

Summary

'New wine into old bottles' is an adage that applies to successful conversion schemes. A change of use is a natural part of the evolution of the life of many buildings. The dynamic nature of the occupancy of property means that any building is unlikely to have the same intensity or type of use throughout its service life. Conversion enables the owner or user to obtain maximum potential of the property. It provides the means whereby a building's use can be modified to suit changing user needs.

The conversion of a building may be partial or full. The former occurs when only a section of the building is directly affected. For example, the formation of a 'granny flat' on the ground floor of dwelling will require modification of the layout of the space it is taking over. Provision also has to be made of an interface between the new and existing uses. The rest of the property, though, can remain relatively untouched by the adaptation.

Full conversion of a building, on the other hand, brings with it other problems as well as opportunities. Extensive refitting of finishes and services is usually necessary. As a result, it is unlikely that much, if any, of the property could be kept in use during the change of use work.

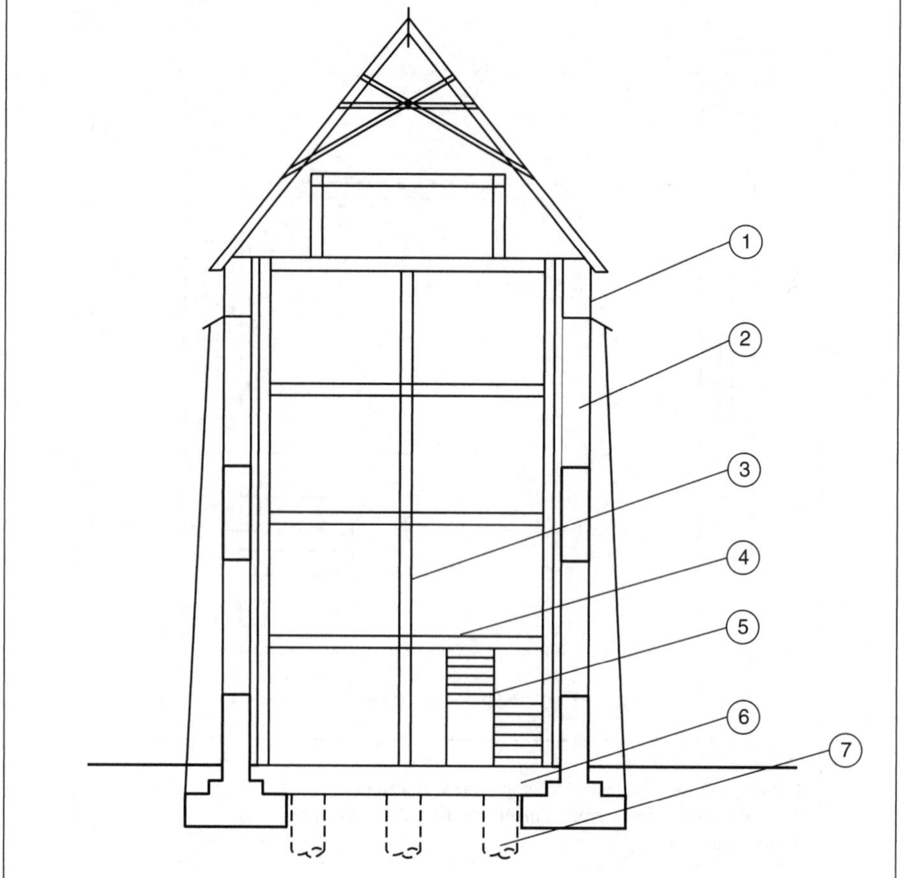

Notes

1. Existing church building of solid masonry construction with very high floor to ceiling space. (See also issues raised Figure 7.24.)
2. Existing fenestration retained. (Figure 4.13 shows how detail at floor/window junction is addressed.)
3. New structural steel framework connected to existing masonry but with slip bolt connections to allow for any differential settlement – similar to façade retention as described earlier.
4. New upper floor slabs of composite construction – profiled metal decking with reinforced concrete structural topping would be more suitable from a buildability point of view than precast floor planks.
5. In-situ reinforced concrete stairs – again, better than precast stairs for buildability in this case.
6. Thick raft reinforced concrete foundation with soft joint where it abuts the existing masonry footings. The raft is cantilevered to make it independent of main structure.
7. Replacement piles if deemed necessary by engineer – depending on loadings and ground conditions.

Figure 3.14 New structural steel framework to increase number of floors

Schematic sketch (not to scale)

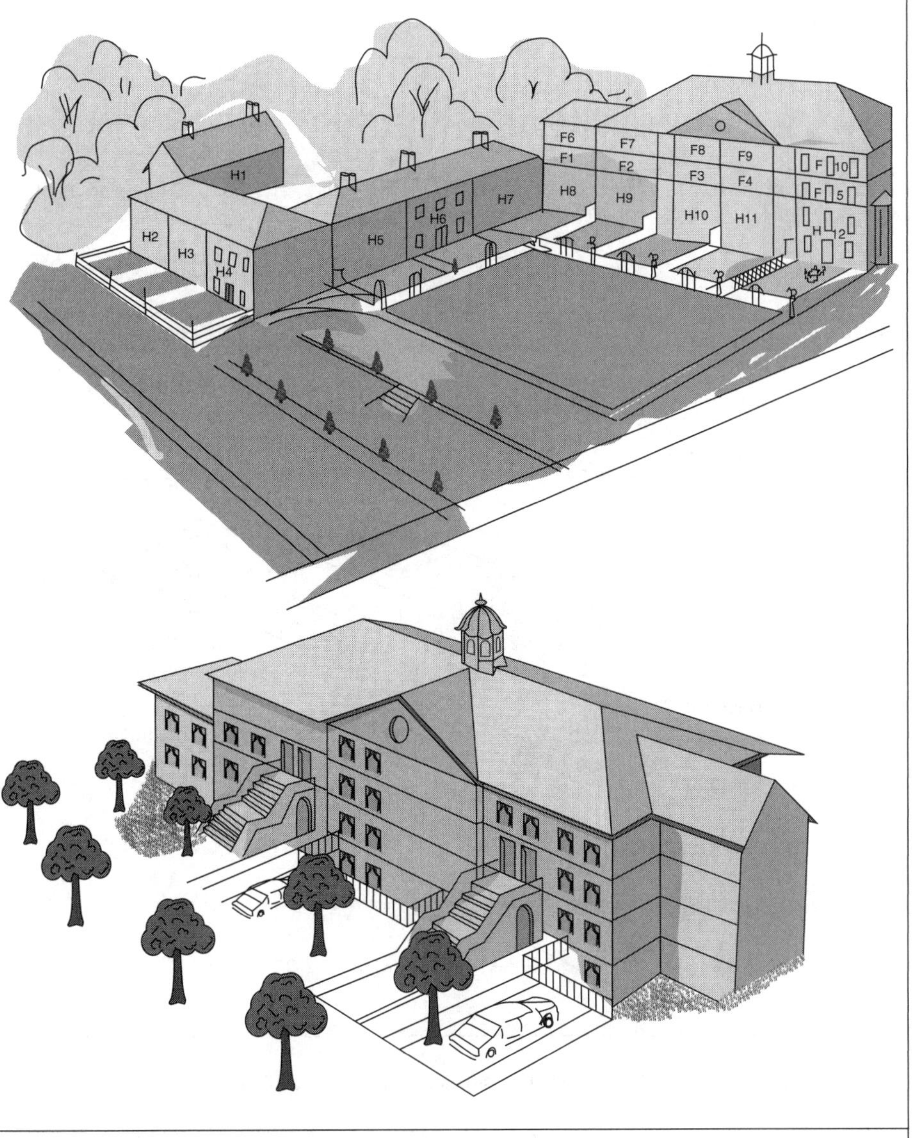

Notes
1. F1 to F10 = flats.
2. H1 to H12 = houses.
3. Artist's impression of completed development.

Figure 3.15a Sketches of proposed conversion of rural mill to apartments (small schemes) (Binney et al., 1990)

Schematic sketch (not to scale)

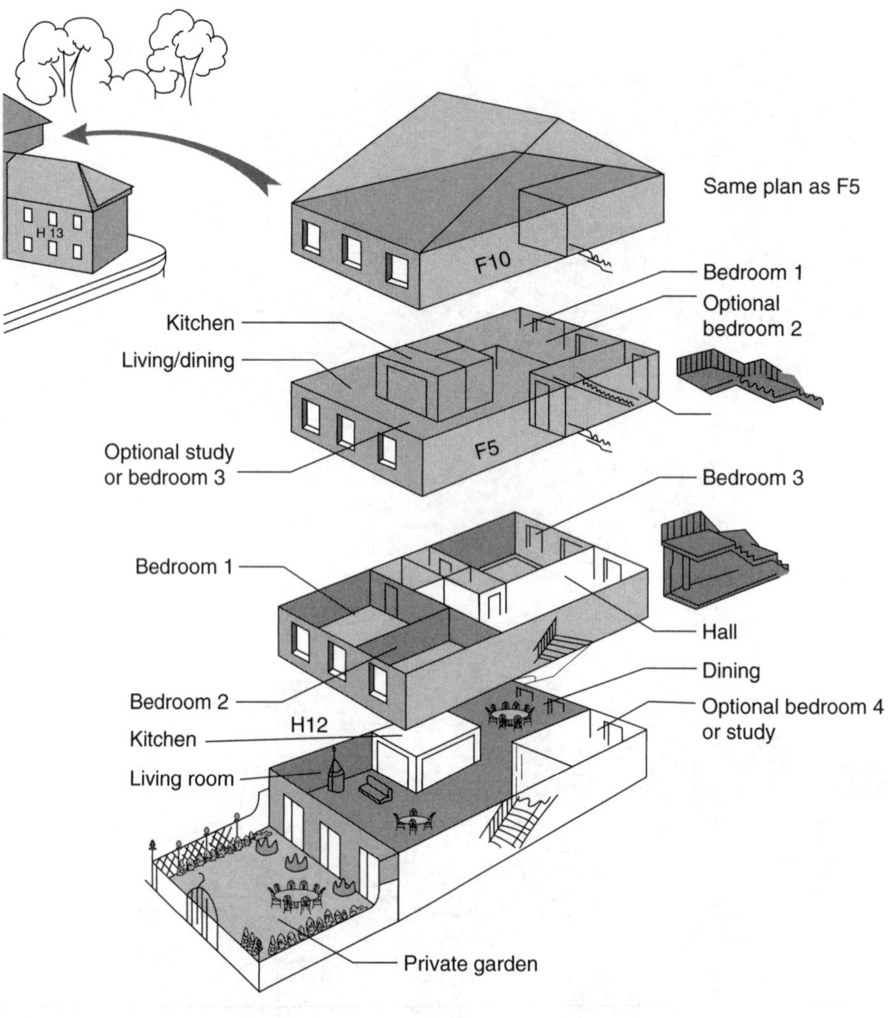

Notes

1. The mill and adjoining buildings could be restored and adapted as a mixture of houses, cottages and flats.
2. Every window has splendid views down the valley while lawns and terraces could be communally maintained, ensuring the mill has a beautiful setting.
3. Small private gardens, with adjacent car-parking spaces, could be provided in front of the building, giving residents privacy.

Figure 3.15b Sketches of proposed conversion of rural mill to apartments (large schemes) (Binney et al., 1990)

The best or most sustainable solution for saving an obsolete or redundant property is conversion to a more beneficial use. It provides an adaptation opportunity that can achieve a social as well as profitable reuse of old buildings.

The optimal conversion scheme involves minimal change to the external appearance of the building. Most of the significant alterations occur within the property. Changes in layout, finishes or services are likely with such work.

Conversions of older properties such as tenements and others built before 1950 are often prone to timber decay. For example, the presence of fungal attack within separating floors has often resulted in patch repairs or complete replacement of these elements. If such works are not done properly this can trigger a further outbreak at a later date. It can also lead to a marked reduction in the sound insulation performance of the floor (CIRIA, 1987). The methods for improving sound insulation in such cases are described in BRE IP 6/88 (see also Chapter 9).

New uses for old properties can mean more stringent requirements for the fire protection and sound insulation of the existing floors and walls. According to Haverstock (1998), for example, floors must have upgraded fire resistance in residential loft conversions, warehouse to office or hotel changes of use, and old house to institution conversions or change of use to flats. At least fi hour fire resistance is required for most floors in domestic conversions, but a floor over basement may need to have 2-hour protection. These and other forms of upgrading floors and walls are covered in Chapter 9.

4

Adaptive reuses

Overview

This section develops the principles outlined in the previous chapter. It deals in more detail with the adaptive reuse of buildings. It also explains the principal technical and other support factors for the feasibility and detailed design of conversion projects. As indicated earlier, bespoke and other tight-fit buildings are the most challenging for a change of use proposal. This chapter, then, will focus on the adaptive reuse of five typical categories of redundant properties: church, farm, industrial, office and public buildings.

Background

Rationale of adaptive reuse

In Chapter 3 the focus was on the general principles of converting buildings other purposes. The adaptive reuse of property is considered in more detail in this chapter. It is concerned with converting buildings into other, more effective and efficient uses. More effective here means that the adapted property serves the client's requirements better and gives the building an extended useful life.

For example, this may be because the building's appearance is not in keeping with the corporate image of the company. An adaptation scheme in such circumstances would focus on refurbishing the property internally as well externally. Effectiveness therefore relates to the degree to which the building satisfies the business or social needs of the occupiers.

More efficient, on the other hand, is related to the performance aspects of the building. It means that its spatial and technical characteristics are enhanced or are in keeping with the needs of the user. The layout of a building, for instance, may require reconfiguration to make it suit new or modified living or working practices.

Adaptive reuse therefore is about overcoming obsolescence and redundancy in buildings. It is also about ensuring the long-term future of buildings threatened by dilapidation, vacancy and eventual demolition.

Ideally, of course, it requires a relatively flexible, responsive and viable building. Potentially most types of building can be adapted to another use. It is not only the number of buildings available for conversion that limits the range of possibilities, but also their location and form of construction.

In Chapter 1 it was noted that the nature and rate of change is challenging everyone dealing with the provision of property. All of that change impacts on both its demand and supply. On the demand side, it has created a greater degree of uncertainty in property markets. As a result, anticipating the future needs of organizations and users is becoming increasingly more difficult.

Uncertainty, of course, is allied to risk (see Chapter 10). A greater level of uncertainty in any venture inevitably increases the risk. In the UK property market crashes of the last quarter of the 20th century, for example, many developers and other property investors lost millions of pounds because of speculative developments that failed to realize the anticipated yields.

Funding of property development these days is on a different, more cautious basis. Instead of speculative projects – whether new build or adaptation schemes – developers and builders are tending to focus on securing pre-letting arrangements or bespoke buildings for specific clients before major investment takes place. This is also applicable to commercial adaptation schemes.

Speculative development nowadays is just as likely in the housing sector. This pertains to residential adaptation schemes as well. Conversions to residential use are very popular in many inner urban areas. However, there is still a financial risk in undertaking such work.

As indicated in the previous chapter, building conversion is the primary adaptation response in anticipating and accommodating changes in the demand for and the use of existing buildings. Research undertaken in this field during the last quarter of the 20th century at London University's Bartlett School of Architecture, for example, confirmed that very significant changes have taken place in the property market's commercial sector (Nutt, c. 1993). Those changes occurred particularly on the demand side.

There are five primary reasons for demand side changes in business attitudes towards property. The first and probably the greatest driver of such changes in demand is, of course, economic prosperity or decline. The shorter cycles between economic recessions over the last 30 years has forced many companies to rationalize their activities and curb their costs drastically. Deregulation of the financial services sector has also had an impact on how people invest in and use property.

Property is a significant cost-centre, the second largest after salaries. However, it is a relatively illiquid asset that is expensive to both manage and maintain. In addition, there are more legal restraints operating over property than with any other type of investment. Any major investment in property such as a conversion, therefore, requires careful planning and managing.

Secondly, the impact of information technology on society is increasing at an accelerating rate. This has not only seen rapid improvements in the performance of personal computers in homes and offices. It has also meant that equipment is reducing in size thanks to the advances in micro-technology. This results in, for example, smaller and faster processors, less use of paper-based transactions and more e-mail and other electronic modes of communication.

Thirdly, increasing international competition, triggered by the dismantling of international trade barriers, better transportation and more efficient communication networks, has created a global market. This has encouraged companies to assess the effectiveness of their fixed assets such as property as well as the quality of their products more rigorously.

Fourthly, property users are becoming more sophisticated in their expectations and requirements for buildings. Occupiers are less tolerant these days to deficiencies such as leaks, uncomfortable indoor air quality and unresponsive heating systems.

Fifthly, and allied to the previous item, there is a growing recognition that there is a direct link between the quality of the workplace and the effect it has on the performance of its most costly resource – its employees. Problems such as sick building syndrome (SBS) and building-related illnesses (BRI) are becoming increasingly recognized in some office buildings. Chapter 8 contains a fuller examination of these issues in the context of adaptation generally.

Consequently, demand side influences have resulted in changes in corporate objectives and work practices. According to property researchers such changes have lead to new organizational structures, flexible employment arrangements, novel working practices, different office, retail and leisure requirements, and to changing demands for transport facilities. 'Teleworking' and 'hot-desking' are common features of many modern offices. These changes result from the recognition that space is the measure of building efficiency.

These businesses influences are affecting the level and type of demand for commercial property clients in particular and cannot be ignored. Internally, they have an impact on the following building-related issues:

- churn rate (i.e. the rate at which the office layout changes each year), which gives rise to increasing staff turnover; proliferation and diversity of technology in the workplace;
- greater emphasis on the quality of the workplace and its effect on productivity;
- higher need for better staff amenities at the workplace (e.g. staff lounge, fitness rooms);
- greater portability of equipment and facilities;
- increases in non-conventional modes of work (such as desk-sharing or 'hot-desking');
- more use of modern methods of communicating in the workplace using video conferencing;
- increases in the relative proportion of professional and managerial staff; and, above all
- greater focus on the needs of the customer: the user or consumer.

Externally, these changes are likely to suppress demand for conventional commercial office space. This in turn tends to prompt the need to refurbish and subdivide existing office blocks or convert them to new uses. Subdivision of large buildings into smaller units became popular in the 1980s, for example, when many redundant school and factory buildings were converted into multiple workshop units.

Changes in the demand for building types inevitably have supply side implications of course. For example, as household sizes decrease, the greater is the need for one- and two-apartment dwellings than for the traditional three to five apartments. Even with non-residential buildings, which are more likely to suffer from obsolescence and redundancy, there is an increasing need to adapt to alternative uses to respond to changes in the property market.

The impact of facilities management on adaptation

The emergence of facilities management in the last quarter of the 20th century has highlighted and attempted to address changes in business needs. Facilities management is a holistic, commercial-orientated discipline that administers, controls and organizes the non-core activities (such as maintenance, adaptation, cleaning, security, catering) to support the core business of an organization. It is an umbrella term that covers four key areas as shown in Table 4.1.

As can be seen from the above table, adaptation plays an important role in building operations and maintenance functions of facilities management. It also impinges on facility planning because the demand for and

Table 4.1 The main areas of facilities management (after Thomson, 1990)

Primary functions	*Real estate and building construction*	*Building operations and maintenance*	*Facility planning*	*General/office services*
Examples	New build Acquisition and disposal	Run and maintain plant Maintain building fabric Manage and undertake adaptation	Strategic space planning Space management Define performance measures	Provide and manage support services Office supplies Security and catering services
Levels	Strategic	Tactical	Strategic	Tactical
Characteristics	Hardware		Software	

use of space in buildings is not constant. Space planning and space management inputs are needed to anticipate such stringent demands.

Space planning is about determining the efficient and proper use of a building. Space is one of the primary measures of efficiency of commercial premises. This is discussed in more detail in Chapters 8–10.

Space management, on the other hand, is about arranging and handling the allocation and shifts in the use of space within a building. According to McGregor and Then (1999), the three basic requirements of space management are as follows:

1. Appropriate and agreed 'guidelines' (e.g. space standards).
2. Good design in the first place (e.g. to avoid or minimize the occurrence of underused or overused spaces).
3. Appropriate format and level of information management (e.g. to communicate requirements and decisions regarding space allocation).

Space management at any level is organization led in that the requirements of the user are paramount. However, to be effective it has to have a professional set-up and complete the loop of brief, analysis, implementation and feedback.

McGregor and Then (1999) state that a very important principle in space management is the interface with space users. Modifications are more likely to be accepted by involved if they feel that they have participated in the decision-taking process. Thus consulting the users can reveal practical information about their specific space needs, which could avoid expensive and disruptive mistakes. This can have a positive knock-on effect for the morale of the workforce because the participants feel that their opinions are being respected.

Adaptive reuse options

Generally

There are a number of possible combinations for changing a building's use nowadays. Generally, however, as shown in Table 4.2 below, the range of possibilities is limited by the change from same use to a different use. In relative terms, single use to mixed-use adaptation is more likely than the reverse type of conversion. This is usually because the configuration and size of building may not suit a single occupancy. For example, a large workshop with offices above would not easily suit conversion to a single dwelling or office. Smaller units of this type may, however, lend themselves to a single residential or commercial use.

Residential conversions

Conversions to residential use are the most popular type of change of use. This is mainly due to the continuing high levels of demand for good quality dwellings in familiar settings. Given the restrictions on building on greenbelt areas and lack of buildable urban land for housing, this type of conversion is likely to remain popular for many years.

The formation of new openings (called 'slapping' in parts of Britain such as Scotland) is a common feature of converting small dwellings into larger units. In many cases, walls or parts of walls have to be removed to accommodate the change in layout. New support systems such as beams are required to span these openings. Strengthening works may also be required to existing walls and floors (see Chapter 7).

Table 4.2 Typical combinations of changing a building's use

Existing use	*New use*		
	Residential	*Non-residential*	*Mixed*
Residential	Multiple flats in a tenement being formed into smaller size unit to increase rental or sale yield from properties.	Three-storey town houses in inner urban areas being converted offices.	High street tenement/town house buildings – ground floor converted to office or retail use and the upper floors as flats.
Non-residential	Redundant office block or mill converted into high-quality flats.	Disused educational building converted into workshop units for small businesses. Redundant telephone exchange or office block converted into a hotel.	Conversion of post-war clothing factory into sports facilities and workshop units for small businesses.
Mixed	Redundant workshop and office premises converted into flats.	A derelict row of retail units with flats above converted into offices.	Refurbishing a mixed-use development in the high street involving shops on the ground floor and offices on the upper floors.

The focus in this section is the conversion of existing dwellings to other residential use. The main examples of these as summarized by Scottish Office Building Directorate (1987) are as follows:

- Conversion of two or more dwellings into fewer larger dwellings. This is less common nowadays because of the shrinking family size.
- Conversion of two or more dwellings into a greater number of small dwellings. This is the most common type of housing conversion.
- Conversion of flats or maisonettes to terraced houses.
- Conversion by 'room transfer' to create a variety in the size of adjoining dwellings without changing the numbers of actual dwellings.
- Internal conversion by reducing the number of apartments and using the area of a room to upgrade the kitchen, bathroom and storage areas of dwellings not meeting current space standards.
- Enlargement of an existing dwelling, usually adding a kitchen or bathroom externally or utilizing space in the original roof void.

Housing conversions will always occur because of ongoing demographic and social changes, which alter the demand for dwellings. For example, as life expectancies increase, so will the proportion of elderly people living together or on their own. As more and more people want to live alone the need for single-person flats will increase.

According to the Scottish Office Building Directorate (1987) it is usually more complicated and expensive to convert from large to smaller dwellings than it is to form larger dwellings from small ones. This is because the subdivision of buildings involves the insertion of separating walls and floors, which can be difficult and disruptive. Also, when more dwellings are created within a restricted volume, a certain number of bed spaces are inevitably lost. These days, of course, with smaller household units being the norm, this is usually not a problem for the developer.

Figures 4.1 and 4.2 illustrate some of the options available for dwellings that can be converted to other residential use.

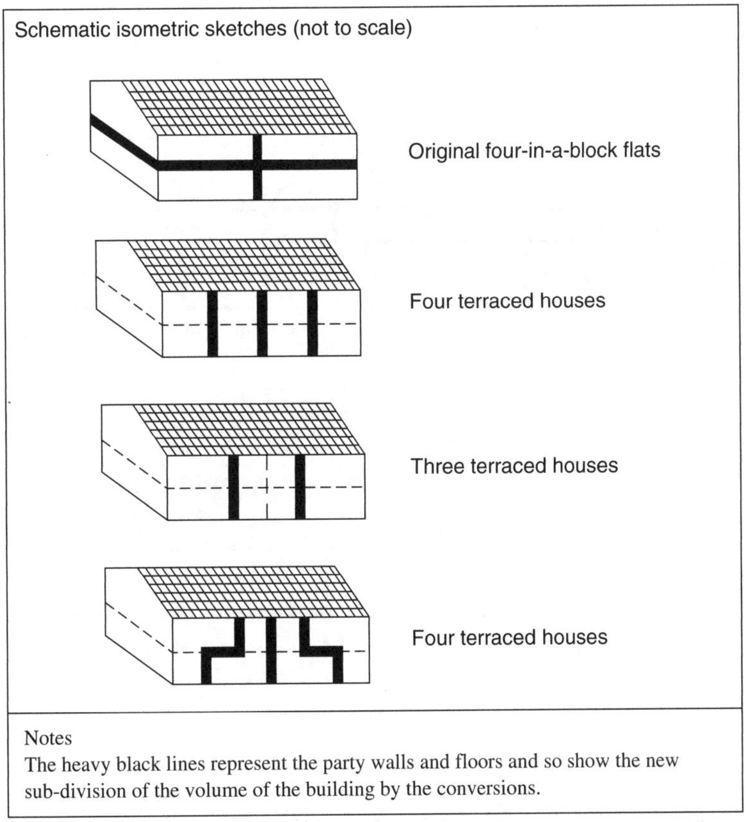

Figure 4.1 Some of the ways four-in-a-block flats can be converted (Scottish Office Building Directorate, 1987)

Conversion of housing to other residential use is not the only means of increasing the supply of dwellings. Many non-residential buildings afflicted by redundancy such as mills and barracks have the potential for conversions to dwellings. For example, Figure 4.3 shows existing and proposed ground-floor plans of a redundant army barracks block. One adaptive reuse option would be to convert it into high-quality flats – at least two on each floor, depending on the length of the building. Alternatively, the block could be converted into either a small hotel or office block (with or without an extension).

Non-residential conversions

In urban areas in the 1960s through to the 1980s the boom in office sector prompted the conversion of many residential buildings to commercial use. With changes in the demand for office space in city centres, however, this trend is being reversed in some cases. Re-conversion of offices back to residential use is becoming popular in many older inner urban areas. This often requires a great deal of reinstatement work to walls and floors to bring the property back to its original character.

Many commercial conversions these days involve a change of use within the same building category. For example, redundant bank premises in town centres are proving ideal premises for conversion to cafes and restaurants. Such changes of use require a relatively low level of adaptation work. As a result, disruptive operations such as structural alterations and changes to the external fabric are kept to a minimum.

Schematic plan (not to scale)

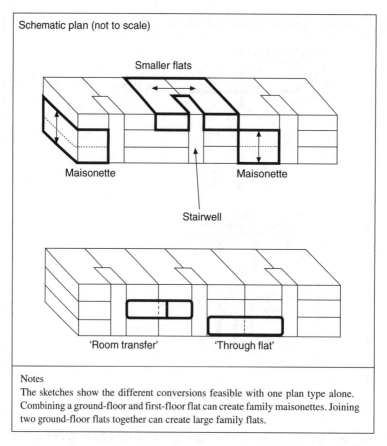

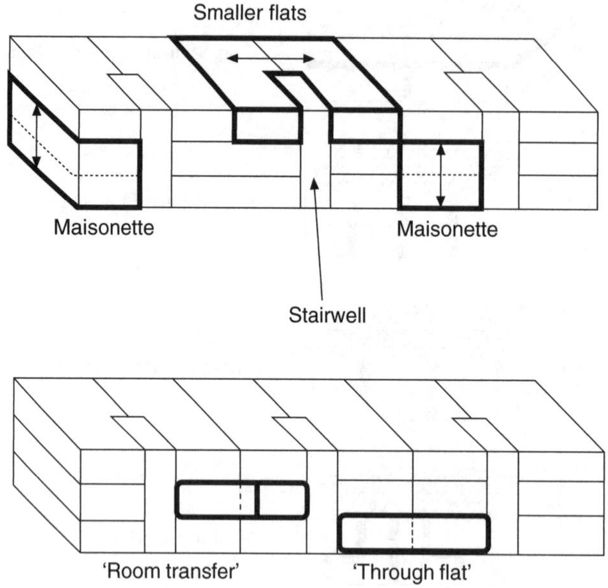

Notes
The sketches show the different conversions feasible with one plan type alone. Combining a ground-floor and first-floor flat can create family maisonettes. Joining two ground-floor flats together can create large family flats.

Figure 4.2 Some of the ways pairs of tenemental flats can be converted (SOBD, 1987)

The existing layout of the barracks block in Figure 4.3 would not need to be radically changed to convert it into a small hotel. The central corridor layout could be retained. The configuration of the accommodation, however, would need to be modified to include en suite bathrooms for each room. Normally these would be placed back to back to facilitate easier connection of the services and minimize noise problems between rooms.

Mixed-use conversions

This involves the formation of two or more different uses within the same building. Converting a mid-terraced office building into, say, retail use on the ground floor and flats on the upper floor is one form of mixed-use. These types of conversions, however, are the most complicated of all, because of the more rigorous technical requirements such as increased fire protection and sound insulation.

There are many potential benefits in going for a mixed-use in a building. It is an efficient way of maximizing the use of a property that might otherwise remain empty or partly occupied. Financially the risks are spread across different types of occupancy, and as a result several sources of funding can be tapped. According to Stratton (1999) in mixed-uses complementary functions such as residential, commercial and cultural can 'feed off each other, making a scheme more attractive to all users and giving it a long-term vitality'.

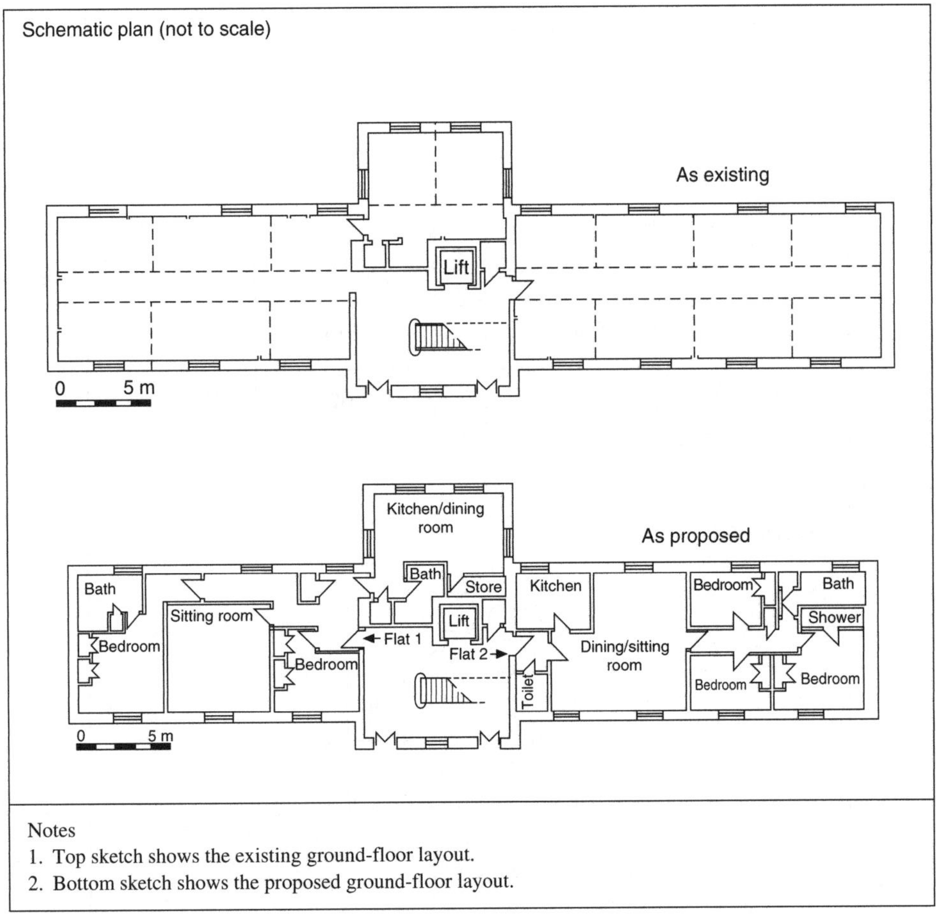

Figure 4.3 Plan of proposed residential conversion of a military barracks block (Anon, 1994)

Accordingly, one of the optimum ways of overcoming the financial risks associated with problems of same/different use conversions is to go for a mixed-use scheme. Such schemes, though, tend to involve the most complex range of problems. Because they involve more than one-occupancy class within a single building, more stringent planning requirements and building standards are likely to apply in each part. Moreover, the local authority is likely to evaluate critically the compatibility of the proposed uses.

Figures 4.4 and 4.5 illustrate the different use mix options available for a four-storey Victorian block in the main high street of a typical town.

Common adaptive reuse options

The most suitable adaptive reuse option will depend on the original use and intended use of the building. Table 4.3 compares housing conversion with commercial/industrial conversion.

As indicated previously, however, the type and form of adaptive reuse of a property is constrained as much by the imagination as well by economic and legal restraints. The photographs in Figures 4.6–4.9, for example, illustrate some imaginative reuses of redundant buildings.

The next five sections deal respectively with the adaptive reuse of churches, farm buildings, factories, offices and public buildings. These are the main types of property that can sustain an adaptive reuse.

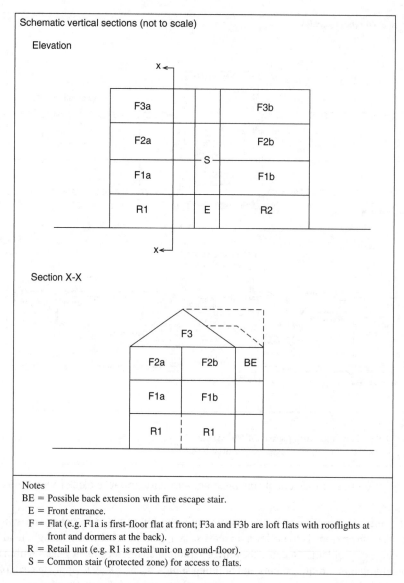

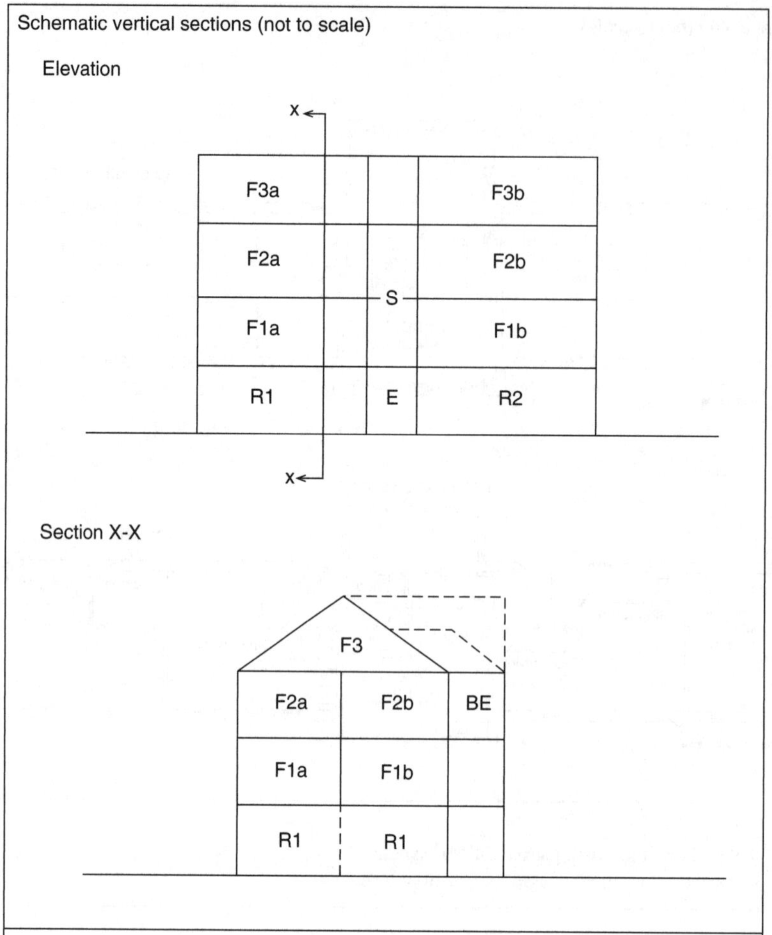

Notes
BE = Possible back extension with fire escape stair.
E = Front entrance.
F = Flat (e.g. F1a is first-floor flat at front; F3a and F3b are loft flats with rooflights at front and dormers at the back).
R = Retail unit (e.g. R1 is retail unit on ground-floor).
S = Common stair (protected zone) for access to flats.

Figure 4.4 Vertical section of urban block showing option 1, mixed-uses

Adaptive reuse of farm buildings

Problems with farm buildings

Modern methods and patterns of farming have resulted in the widespread under-utilization of old agricultural properties such as barns and steadings. These antiquated buildings are often not spacious enough to take newer and larger plant and equipment or are too tight to house the day-to-day use of the latest farming machinery. Thus, again, technology has been one of the main influences in prompting the obsolescence, and subsequent change in the use, of these buildings.

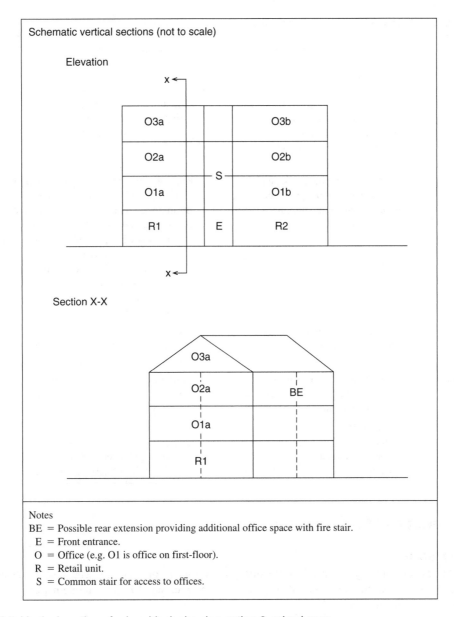

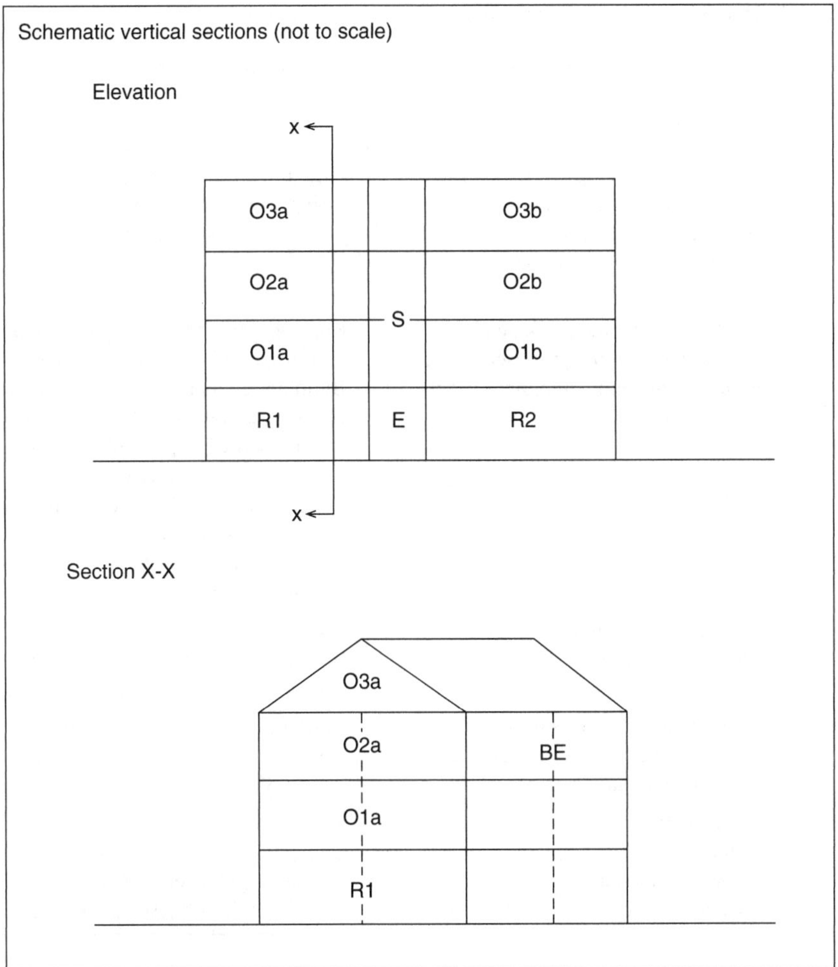

Notes
BE = Possible rear extension providing additional office space with fire stair.
E = Front entrance.
O = Office (e.g. O1 is office on first-floor).
R = Retail unit.
S = Common stair for access to offices.

Figure 4.5 Vertical section of urban block showing option 2, mixed-uses

Restricted spans and low headroom are common internal features of old farm buildings. Such buildings are usually small scale and of traditional construction. This means that it is difficult for such buildings to fulfil their original purpose without radical alteration. An alternative use may be the only feasible option for saving them. Once empty for a prolonged period, the risk of dilapidation leading to a ruinous condition is very high. This of course makes their beneficial reuse more difficult.

Buildings related to agricultural activities can also require change of use resulting from farm amalgamations. Rationalization of groups of nearby properties related to the farm often results in some buildings becoming surplus to requirements. These business influences can thus make blocks such as complete steadings or isolated properties redundant.

Table 4.3 Comparing housing conversion with commercial/industrial conversion

Factors	*Housing conversion (dwelling/s to dwelling/s or non-residential building to dwelling/s)*	*Commercial/industrial conversion (residential to non-residential or commercial to other commercial)*
Technical	Low imposed loadings = 1.5 kN/m^2. Separating floors and walls required between flats. Stringent thermal and sound insulation requirements.	High imposed loadings =>3 kN/m^2. Stringent fire protection and means of escape requirements.
Legal	Permitted development? Depends on the extent of the proposed work. Building regulation compliance required.	Change of use permit may be required if property is outwith Use Class Orders. Building regulation compliance required. Fire Precautions Act 1971 needs to be complied with, particularly where a fire certificate is required. Offices, Shops and Railways Premises Act 1963 is applicable for commercial schemes. The Factories Act 1961 is relevant to industrial conversions.
Economic	Several small flats more lucrative than one dwelling.	Dictated by office/retail/industrial market. Building may not be totally suitable for new commercial or industrial use.
Environmental	Re-conversion of offices back to dwellings becoming popular in some inner urban areas.	Additional external works may be required to minimize the impact of this type of conversion.
Functional	Separate entrances? Disabled access? Car parking, including disabled spaces.	Storage areas. Circulation space. Car parking/delivery points. Access for disabled/fire authority.

Characteristics of farm buildings

Configuration

Farm buildings tend mainly to be of single-storey or low-rise construction. The residential and non-residential units can be classified as either traditional (before, say 1945) or modern (say, after 1945) construction as outline in Table 4.4.

Inferior construction standards

Construction standards in farm buildings are generally much lower than other types of buildings. The fabric of sheds and barns, for example, will normally be of thin, single-skin construction. Even dwellings on agricultural property are deficient in terms of thermal efficiency (i.e. high fabric U-values), lack of damp proofing to walls and ground floors, and inadequate electrical supply for lighting and power, heating system and other services common in urban housing.

Poor levels of maintenance

Neglect or an overall lack of maintenance is a common feature of farm buildings. This is often compounded by the extensive abuse or wear and tear to their structure and fabric that is characteristic of agricultural

Figure 4.6 A former hosiery factory in Leicester, built as a six-storey worsted spinning mill c. 1844. Recently converted to office use as part of a City Challenge Scheme. To meet the requirements of the tenants, the Land Registry, an extension to the front, overlooking the Grand Union Canal, has been provided in a modern style. To the rear, glazed staircase/lift towers have been added

work. These factors, along with the poor quality of construction means that more often than not farm buildings require extensive initial remedial works as part of any adaptation scheme.

Contamination

Contamination of floors and walls where livestock or chemicals have been housed is a common occurrence in farm property. Faecal and other waste materials, fertilizers, milk and other by-products of agricultural production can cause damage to concrete floor slabs and brick/stone walling. Moreover, corrosion of metals from farm chemicals and damp conditions is common.

Lack of services

Buildings such as barns, steadings and storage sheds have little or no services infrastructure, such as proper lighting, heating, power, water supply, foul drainage, etc. High humidity and low temperature levels are, therefore, normal in these buildings – in some cases they are little different from those in the external environment.

Ill-considered alterations

Structural alterations to farm buildings are often ill considered because of the remoteness these properties and their exemption from some aspects of building and planning controls. Such work is often done piecemeal and undertaken by unskilled workers. Thus the quality of many agricultural barns and shed is very poor.

Adaptation potential

Most of the industrial types of farm buildings such as portal frame sheds do not have much potential for conversion to another use. In particular they are unlikely to prove suitable for housing conversion. Their main adaptation potential therefore is for use as a storage facility.

Figure 4.7 The former Anglican church of St John the Divine, Leicester 1853–1854 by G. G. Scott. After several years of being redundant the church has been converted to flats. Note how the building has been disfigured by the inclusion of roof lights and clumsy alterations to the windows

Nevertheless, despite their deficiencies many redundant residential farm buildings of load-bearing masonry construction often have excellent reuse potential for several reasons. Being out of town makes them desirable as country residencies. They have large internal spaces that allow easy adaptation. They also have generous external spaces, which can easy accommodate extensions. Many farm dwellings are single storey, which can make them highly suitable for sheltered housing schemes.

Historic farm buildings

Because of the problems mentioned above older farm buildings are more likely to suffer from fabric as well as structural deficiencies. Consequently, the walls and roofs of such properties are usually run down

Figure 4.8 A former print works in Leicester adapted to office use. The fenestration has been remodelled and replaces a very bland façade of dull industrial windows

if not derelict. Any agricultural adaptation scheme therefore will also have to include major repair and upgrading works such as damp proofing, timber treatment, upgrading the insulation of the fabric, remedial works to the rendered or pointed stone or brick walls.

Many old farm properties, however, such as barns and cart sheds are of some value. This is not only because of their architectural character, but also because of their historic significance. Some might be linked to an important place or event or to a famous person. Moreover, such buildings are often structurally stable and can offer excellent residential conversions opportunities. Redundant farm buildings in areas where there is a housing shortage are highly desirable because of their adaptation potential.

The adaptive reuse of these agricultural buildings is a means by which they can continue to contribute to both an area's landscape and the needs of the local community. This results in the adaptive reuse having

Figure 4.9 A former congregational chapel in Leicester dating from 1865 converted to a Jain temple in the 1980s. (Jainism is an ancient Indian religion which teaches love to all creatures and non-violence.) The interior has been remodelled and the exterior clad in marble and other stone sculpted in India

'added value'. The alternative is to allow redundant farm buildings to become derelict or demolish them if they are deemed dangerous and/or an eyesore. These dilapidated properties can attract children and even squatters and vandals. In many cases such buildings that have been left ruinous have perished because of this form of deliberate attrition.

It has to be acknowledged, however, that the conversion of a farm building inevitably involves a degree of loss whether of the property's fabric, character or relevance. According to English Heritage's statement (May 1993) residential conversion of historic farm properties can potentially:

- destroy much of the original fabric by making new openings and be replacing and removing structural elements such as timber frames, and even by demolishing and rebuilding whole stretches of wall;

Table 4.4 Basic constructional characteristics of farm buildings

Age	*Use*	*Characteristics*	*Adaptation potential*
Old	Residential	Solid load-bearing masonry Vernacular style architecture in isolated developments. Polite style architecture in later, 18th/19th century, large, landed estates.	Possible conversion to other residential use or country hotel and/or restaurant.
Modern	Residential	Load-bearing masonry cavity construction. Similar to ordinary dwellings but more utilitarian in style.	Suitable for conversion to other residential use.
Old	Non-residential	Solid load-bearing masonry. Roofs clad with corrugated steel or asbestos cement sheeting. Vernacular style of architecture.	Suitable for conversion to residential or other use such as crafts centre, country hotel and/or restaurant.
Modern	Non-residential	Framed construction in steel or pre-cast concrete. Clad with corrugated aluminium sheeting.	Possible conversion to storage facility.

- disrupt walls and roofs with new doors and windows, and break up rooflines with dormers and chimney-stacks;
- block interior spaces with inserted floors and partition walls and remove original fittings;
- create an enclosed plot, cluttered with amenities such as garages, fuel tanks and hedges, that disrupt the agricultural setting and the integrity of any farmstead group; archaeological deposits may be destroyed by new foundations, drain or even swimming pools.

The extent to which some or all of the above negative effects could be tolerated depends on the context of the property, the type of proposed adaptation, and, above all, the policy of the local planning authority. In any event, adherence to the basic principles of conservation is likely to yield the more successful conversion schemes involving old buildings. The key building conservation criteria were addressed in Chapter 3.

Figures 4.10 and 4.11, respectively, illustrate examples of insensitive and sensitive domestic conversions of farm buildings.

Adaptive reuse of church buildings

Characteristics of church buildings

Church buildings still form a distinctive part of the urban and rural landscape in many parts of Britain. With the trend towards a more secular society, however, this is becoming no longer the case in some areas.

The Victorian era was the heyday for the erection of church buildings in Britain. During that period of course the Industrial Revolution and the British Empire were at their zenith. The rapid rise in population in the 19th century encouraged the expansion of towns. This in turn stimulated the demand for religious buildings to serve the spiritual needs of the community.

In the second half of the 19th century many Anglican church buildings were built. Roman Catholic churches between 1850 and 1875 increased by some 450 (Powell and de la Hey, 1987).

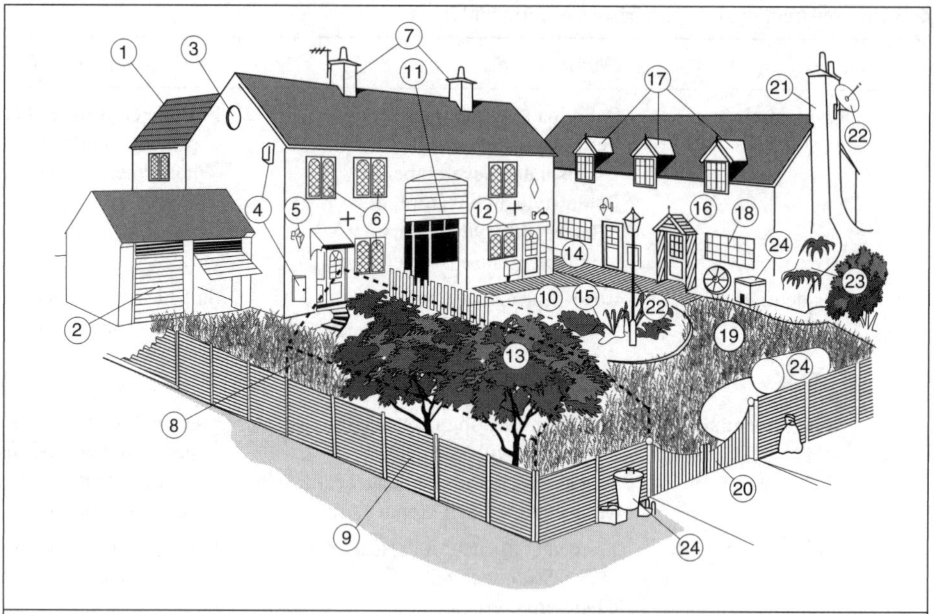

Notes

1. Inappropriate or large extensions.
2. Metal up and over doors.
3. Blocking in of original barn feature.
4. Domestic 'clutter' on elevation.
5. Over-ornate lighting.
6. Standard narrow framed windows and lead applied.
7. Chimney-stacks breaking ridge. Chimneys should be strictly avoided unless they already exist.
8. Demolition of one farmbuilding group.
9. Excessive use of larch-lap fencing.
10. Attempt to subdivide plots.
11. Horizontal 'panel' detail and mixed infill.
12. Ugly lintel detail.
13. Inappropriate 'ornamental' trees.
14. Standard door and side panel to fill gap.
15. 'Soft' planting.
16. Elevation broken by 'cottage' style porch.
17. Breaking eaves line with dormers should be avoided.
18. Window proportions alien to original format.
19. 'Sea' of tarmacadam (an eyesore in any rural development).
20. Unsuitable and useless gate design.
21. Tall outside stack ruins plain gable. Such stacks have no place on farm buildings.
22. Unsuitable and incongruous features, such as satellite dishes.
23. Unnecessary detail hiding original steps.
24. Domestic clutter on ground area.

Figure 4.10 Example of insensitive conversion of a farm building (Ryan, 1995)

The increases in membership of Christian churches during that period, however, were mainly in the Nonconformist denominations. This resulted in a massive rise in demand for ecclesiastic buildings. According to Powell and de la Hey (1987), for example, between 1860 and 1900 nearly 4000 Methodist chapels were opened.

There two broad types of old church buildings in terms of size and style: (i) simple and (ii) ornate. The former is usually associated with the Nonconformist or reformed traditions, which reacted against the existing elaborate building styles. Their churches were consequently simpler and more domestic in character and size. The latter, on the other hand, were usually on a built on a grander scale.

In general, though, old church buildings tend to have the following common characteristics:

- thick, solid stone (sometimes brick) walls built in lime/cement mortar, with regular masonry buttresses;
- tall, narrow fenestration, usually glazed with stained glass and leadwork;
- ground floors of stone flags laid directly onto the bare, mainly in pre-1850s buildings;

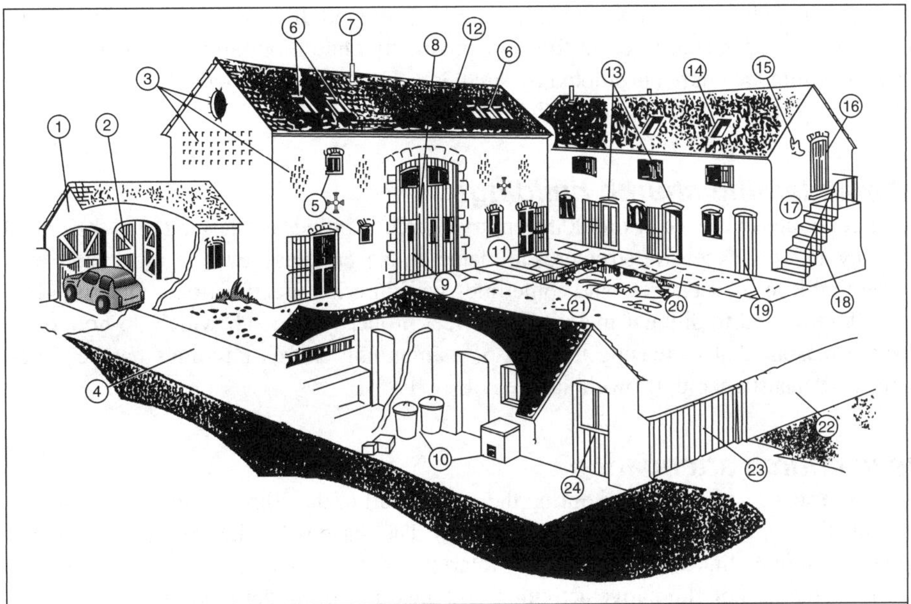

Notes

1. Good use of cart shed for garaging.
2. Sturdy ledged and braced doors.
3. Retention of elevation details (ventilation pattern).
4. Use of foldyard wall for containment.
5. Respect of original ‘haphazard’ openings.
6. Minimum use of rooflights (which must be flush and not raised) or glazed roof tiles, and flush patent industrial roof glazing.
7. Simple metal flues.
8. Threshing doorway retained with vertical emphasis.
9. Well-designed, simple, postern door.
10. Useful storage of domestic clutter retaining format of original buildings.
11. Successful infilling of unwanted openings, fully glazed openings retaining doors as shutter.
12. Unbroken ridge lines.
13. Two examples of door/shutter treatment.
14. Rooflights correct size.
15. Sensible outdoor lighting.
16. Retention of loft door.
17. Retention of original steps.
18. Sturdy wood or metal handrail detail of simple design.
19. Simple boarding-up of unused doorways.
20. Respect for original raised walkways.
21. ‘Hard’ surfacing material of texture.
22. Wall retained.
23. Gate design reflects original use.
24. Boarded height to retain sill line.

Figure 4.11 Example of sensitive conversion of a farm building (Ryan, 1995)

- suspended timber ground floors in churches built after the 1850s;
- steep duo-pitched roofs covered with slates;
- roof structure consisting of heavy timber members;
- high floor to ceiling dimensions (i.e. over 5 m).

Reasons for redundancy of old church buildings

Excess supply of buildings

Over-capacity is a problem that is not restricted to ecclesiastical buildings. Since about the middle of the 20th century buildings associated with traditional industries – such as coal-mining, textile and steel manufacturing – have been similarly afflicted by changing socio-economic factors. This has inevitably led to a surfeit in the number of such buildings.

The stock of church buildings is now well in excess of demand. It is not surprising therefore that many church buildings are redundant. Even in times of prosperity church buildings are underused because of fewer activities within the church prompted by smaller congregation numbers. As congregations dwindle old church buildings offer an attraction to other faiths (see below).

Cost of maintaining church buildings

Although the continued existence of pre-20th century church buildings is a testament to their durability and robustness, they are very expensive structures to maintain. Nevertheless, even in those days many church buildings were built to tight budgets and sometimes with inferior specifications. Such buildings require specialist craft skills and traditional materials to effect proper repairs, all which do not come cheap. It is for all these reasons that increasing the cost of maintaining church buildings has been described as 'burdensome and unsustainable' (Powell and de la Hey, 1987).

Decline in church attendance

Churchgoing as a social activity halved during the second half of the 20th century. The orthodox churches have been wrestling with this dilemma for many years. The reasons for the decline in church attendance are varied but seem to be linked to increasing secularization within society in the developed world, combined with the attractions of alternative activities, and pressures of modern living.

Changing demographic trends

Many churches have shrinking congregation through natural wastage. The majority of churchgoers in the orthodox traditions are elderly. The effect of this is to make it even more difficult to attract younger people into the church. Apart from the various evangelical churches, there are not enough younger members coming through the system to offset this loss. Added to this is the shortage of clergy in many of the established denominations.

Rise of modernist architecture

The popularity of modernist architecture in the 1960s and 1970s prompted a move away from church buildings of traditional load-bearing masonry construction to new framed and cladded buildings, using newer materials such as laminated timbers. This also reflected changes in liturgical practices, which moved away from traditional formats.

New churches

Nowadays the predominantly Evangelical/Charismatic/Pentecostal branches of the Christian religion can claim to be maintaining if not increasing their congregation numbers. These denominations, which are theologically very conservative, also tend to attract and hold a higher proportion of young people, and are more committed to spreading their faith. They also are inclined to use modern facilities or old buildings adapted to more modern use.

Options for redundant church buildings

Preservation as a monument

Where suitable use cannot be found for a redundant building, preserving it as a monument may be the only option to save from demolition. Buildings that lie empty or otherwise unused, however, tend to deteriorate at a faster rate than occupied ones because of lack of both cleaning and heating (ODPM, 2003a).

Cognate religious use

One of the best examples of a redundant church building being out to an alternative though related use is the Scottish National Bible Society headquarters in Edinburgh. The property was formerly the home of the Murrayfield Parish Church. In 1980, owing to a decline in parishioners, the building became redundant. It was 'converted' to its present semi-commercial use in the late 1980s.

Some church buildings have been bought by other faith traditions – Muslim Mosques, Hindu Temples and Sikh Gurdwaras. Many Nonconformists churches are relaxed about their redundant properties being converted for other religious uses. The Church of England, however, requires that the new religious use must be Christian-related.

Continuing use as a place of worship

Church buildings, through a combination of lack of maintenance and high repair costs, are expensive to maintain. However, in this more ecumenical age, different congregations or even different churches may find it more convenient to pool their already meagre resources by sharing a church building.

Community use

Many churches favour redundant church building being put to community or social use. For example, the building may be converted to a nursery or play-group facility, community centre, or exhibition/concert/meeting hall.

Commercial use

Redundant church buildings can be put to a varied range of commercial uses. At the lower-quality end such a building may be put to storage use as a last resort to save it from demolition. The danger with this type of adaptation is that there is a temptation to carry out insensitive alterations such as inserting a large loading bay in the external fabric of the building. Uses that require large open spaces such as restaurants are in many ways ideal candidates for church conversions.

Recreational use

Church buildings, because of their high ceilings offer excellent venues for certain indoor-sports activities. For example, some redundant churches are being used as mountaineering training facilities, complete with mock climbing faces.

Residential use

According to the Church Commissioners (1993) about 10 per cent of the Church of England's redundant buildings are converted to dwellings. Usually, single-occupancy church conversions offer potentially the least interventionist option. This type of adaptive reuse, however, is best suited to the smaller church buildings (see Figure 4.12). It is not very common for the larger church buildings because of the large sums of money required to convert and maintain them. Multiple-occupancy conversions, on the other hand, offer a more viable case for developers in the latter case. The adaptive reuse of large redundant churches into several flats is probably the most appropriate option in terms of viability and sustainability.

Mixed residential/church use

One innovative way of reusing part of the church building is to convert part of it into another compatible use such as a dwelling. The spire for example, could be converted into a flat or several flats, which are closed off from the main part of the building. This option could save an otherwise potentially redundant property.

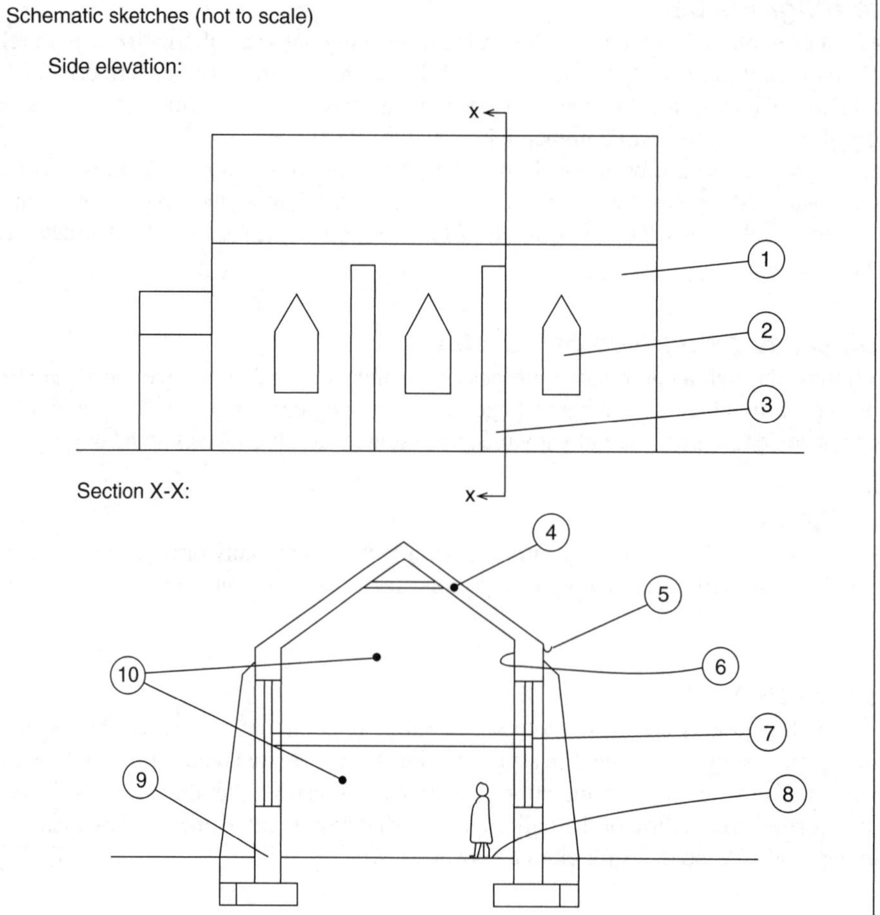

Notes

1. Solid walls may require pointing externally and thermal upgrading internally.
2. Existing fenestration retained – with possible secondary glazing.
3. Existing buttresses retained.
4. Roof coverings renewed to match existing but with new sarking boards and felt. New skylights fitted to provide daylighting to upper floor. Integrity of rafters checked and repaired if necessary.
5. Rainwater goods renewed in cast iron.
6. Internal wall finish left fair faced if condition and thickness of walls is sufficient to minimize heat loss.
7. New first-floor level abuts openings – see Figure 4.13.
8. Existing stone-flagged/earth floor replaced with 150-mm-thick solid concrete insulated floor slab on dpm.
9. Main walls checked for rising dampness – outside ground level should be lowered if necessary.
10. Layout of accommodation dictated by available space and client's requirements. See Figures 3.11a and b for an indication of the basic principles of planning of internal space.

Figure 4.12 Typical conversion of a small church building into a single dwelling

Industrial use

During the last quarter of the 20th century workshop or other light industrial use was common conversion option for many redundant church buildings. Nowadays church groups and local authorities see this type of use as less attractive.

Retaining the fenestration

As indicated earlier many old church buildings have tall, narrow windows, which were designed to maximize internal daylighting. Any sensitive conversion scheme whether to a church or other redundant building with these openings would want to retain them. However, if an additional floor is being proposed inside such a property, careful consideration needs to be given to the detailing at the windows to avoid disrupting the architectural integrity of the building.

Figure 4.13 shows one way in which a new floor structure that comes in the middle of a tall window opening can be accommodated.

Internal frameworks

In larger church buildings, with floor to underside of roof height in excess of 5 m, it may not be feasible to install several floors using new beams inserted into the existing walls. Instead, it is more feasible to install a skeletal framework of structural steel (see Figure 3.14). This is similar to the new framework required in a façade retention scheme. The main difference is that the existing roof in this instance is often still in place. This can make it difficult to achieve full access the interior for erecting the main structural frame.

The installation of an internal framework requires the following:

- Adequately sized fenestration to install framework elements. If the openings are not large enough to admit such elements, they could be inserted into the building through makeshift gaps in the roof.
- Pad foundation bases for the new columns. This will involve digging through the existing ground floor and installing pad bases about 1 m × 1 m square and 0.3 m deep, depending on the loading conditions.
- Fire protection to the steelwork.
- Tying-in of the existing walls to the new frame, using stainless steel anchors.
- New floor structures, slabs of pre-cast concrete are ideal as this provides inherent fire protection and good soundproofing provided there is space to insert them into the building; if not profiled galvanized steel decking topped with an in-situ concrete structural 'screed' would suffice.
- New stairs, again, best in pre-cast concrete if access allows; if not, in-situ concrete stairs would be needed.

Adaptive reuse of industrial buildings

Characteristics of pre-1900 industrial buildings

Prior to the end of the 18th century agriculture was Britain's main economic base. The Industrial Revolution changed all that. By the early 1800s Britain became one of the world's greatest industrialized economies. This caused a massive boom in the construction of factory and warehouse buildings throughout the country.

Mills were among the most common types of industrial buildings in 19th century Britain. They consisted primarily of textile mills, though others such as flour mills and steel mills were constructed as well.

Schematic vertical cross-section (not to scale)

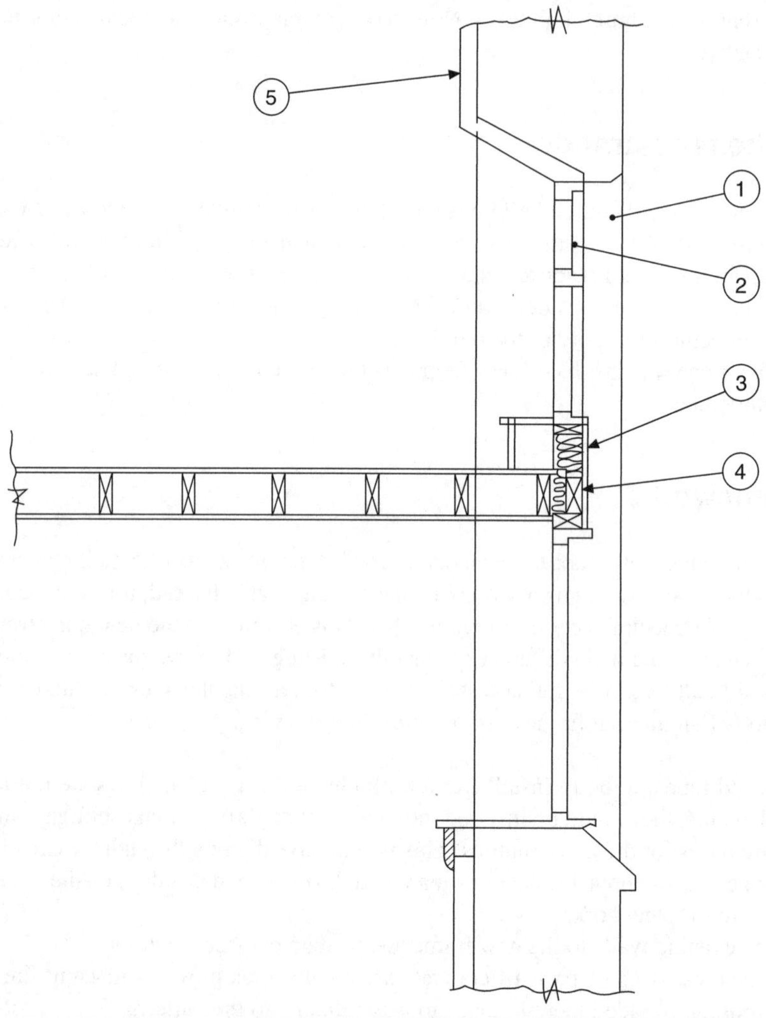

Notes

1. Existing openings retained but stained glass or other original windows removed.
2. New double-glazed hardwood windows cleanable from the inside fitted to these openings. First-floor windows have safety rail at 1.05 mm above floor level. The dead lights below this level to be glazed with safety glass. All windows to have horizontal/vertical DPCs and be pointed in silicone mastic sealant.
3. Floor structure masked behind lead cladding panels on building paper on 9 mm water and boil-proof (wbp) ply. Framework behind the lead cladding to consist of 100 × 40 timber studs infilled with 100 mm thick quilt insulation, with vapour control layer on warm side of construction.
4. Butting floor against window will not achieve an adequate fire stop.
5. Internal insulation using rigid polyurethane backed plasterboard on dabs.

Figure 4.13 Detail of new floor junction at existing window opening

Despite their differences in function or size, most 19th century mills and warehouses have several characteristics in common with one another:

- multi-storey (i.e. usually over three storeys), with long front and or side elevations of solid load-bearing masonry;
- external walls punctuated at regular intervals with large windows;
- wrought iron or cast iron members in regular bays forming an internal frame supporting the floors;
- built to last over 100 years;
- sited in urban areas adjacent to power source such as water or coal, because the area became urban as a result of the industry attracting people, housing, etc.;
- large spans and open spaces internally;
- structurally robust;
- adequate external space around the building, for access and car parking;
- occasionally, they contain a basement area suitable for car parking.

Characteristics of post-1900 industrial buildings

By the early 1900s mild steel replaced wrought iron in framed construction (Yeomans, 1997). This became the principal structural material for factory and warehouse buildings. Concrete in either pre-cast or in-situ form also was used but could not achieve the large spans as economically and efficiently as steel.

After World War II industrial buildings comprised mainly of skeletal steel frame construction with lightweight cladding for the walls. The roofs were either flat, saw-toothed (i.e. mono-pitched with glazed gable) or northlight pitched.

The common characteristics of industrial buildings built between the 1950s and 1980s are:

- They were often designed and constructed to minimal thermal standards.
- They sometimes comprise a group of connecting buildings, occasionally of differing forms of construction.
- Few industrial buildings are within a city centre. Most of them are situated in the suburban areas.
- They tend to have large roof areas, often either flat or shallow pitched roofs. These are covered either by roofing felt, corrugated asbestos or steel sheeting placed on lightly insulated substrates.
- Industrial buildings are likely to contain most forms of asbestos.
- The fabric of industrial buildings is generally lightweight and thin-skinned, which usually requires thermal upgrading.
- The main factory or warehouse areas are normally single storey with generous floor to ceiling heights.
- Partial demolition internally (e.g. of columns) may be required to accommodate new uses.
- Load-bearing capacity of the building may need to be increased to suit new use or accommodate additional, intermediate, floors.
- Industrial properties are frequently subjected to higher levels of user use and abuse than other types of buildings.
- Levels of contamination in floor slabs and soils are higher than other properties.
- The electrical power supply is single phase, which may not be suitable for some uses.
- Maintenance standards in industrial buildings are generally much lower than for other types of property.

Options for reusing redundant industrial buildings

The decline in the manufacturing sector coupled with the increase in the service economy has inevitably resulted in the redundancy of many industrial buildings. In contrast to commercial and residential conversions, however, the adaptive reuse of some industrial buildings has traditionally been less common.

Nevertheless, many traditional multi-storey mills and similar structures have been turned into offices, university buildings, as well as housing.

The scope for the adaptive reuse of more recent industrial properties, however, is not as great as other non-residential buildings. The size, condition and structural configuration of these are not conducive to an adaptive reuse for housing or offices. More usually newer factory and warehouse buildings are converted to an alternative industrial use.

The effects of economic and technological influences have left legacies of often empty and sometimes unusable industrial buildings. Redevelopment of course is an obvious option but this is constrained by financial and legal factors. Moreover government now policy tends to encourage development on brownfield sites.

As seen in Chapter 1, though, adaptation is a key aspect of sustainable construction. This is pervading all aspects of civic life. The various civic trusts and conservation groups in the preservationist lobby are promoting sustainability considerations linked to the built environment and transport.

Some of the common options for the adaptive reuse of industrial buildings are as follows.

Preservation as a monument

Because of their large size, most modern industrial buildings are not suited to this form of virtual mothballing. They would still require some maintenance and security to prevent them from falling prey to squatters, thieves and vandals. These are the main sources of building degradation instigated by humans. Human neglect of course can allow the forces of nature to attack an empty building. Wind and rain penetration is the likely consequence of neglect.

Industrial museum

As an alternative to mere preservation the conservation of redundant industrial buildings to use as museums can often prove a viable option in both economic and social terms. However, this option is primarily suited to older industrial buildings. For example, disused mining buildings and mills built in the 19th century make excellent living examples of old fashioned if not extinct working practices.

New industrial use

Post-1900 industrial buildings are more suited to industrial reuse than their pre-1900 counterparts. The former can be converted into smaller units for workshop or storage purposes. Alternatively they can be adapted to accommodate a similar use to the one previously. Some alterations involved in such a conversion are enlargement or reduction of entrance doorways to main factory or warehouse area, and strengthening parts of the structure to accommodate increased crane or vehicular loads.

Residential use

Many pre-1900 industrial buildings such as whiskey bonds and jute mills are ideal candidates for conversion to flats. Their large spacious areas offer good opportunities to provide generous accommodation. Although their structural capacity is more than adequate for residential use, the frames frequently require fire protection. Chapter 7 shows how this can be achieved.

Offices and hotels

Certain types of industrial buildings in inner urban areas often lend themselves to a change of use to offices. The degree to which such buildings can take separate occupiers, depends on the configuration of the building and the access provision or potential.

Sports centre

Some redundant factory buildings can be converted into sports facilities and other ancillary community uses. The capacious spaces often make some industrial buildings suitable for these uses. Part of such a complex could contain a community centre facility with meeting rooms complete with offices, kitchens and toilets, but might involve converting the office accommodation or adding an extension to the main factory area.

Art galleries

One of the common features of bespoke industrial buildings such as power stations is their large open spaces and volumes. These areas can provide ideal settings for displaying art exhibits. One of the most recent prominent examples of this type of adaptive reuse is the Tate Modern Gallery in London, which successfully exploits the voluminous spaces of the redundant Bankside Power Station to create a large international art museum.

Adaptive reuse of office buildings

Characteristics of office buildings

Up until the 1950s most traditional office buildings are generally of solid load-bearing masonry construction. They were either purpose built for commercial use or were converted from residential use.

By the second half of the 20th century framed buildings became more popular for offices. The construction standards of early modern office buildings (i.e. those built in the 1950s and 1960s), however, were usually relatively modest. They often have awkward physical characteristics such as close column spacings (i.e. <5 m) and low floor to ceiling heights (<3 m). Large toilet, lift lobby, stair and other non-working activity spaces as well as cellular offices are characteristic of early contemporary commercial premises. These physical constraints limit the ability of such buildings to meet today's demand.

Duffy (1993) pointed out that the specifications of the late 1980s British office buildings differ in many respects from those of the preceding decades. More modern offices tend to have:

- larger floorplates (up to 3000 m^2), more easily reconfigured to accommodate change in layout;
- more generous storey heights (up to 4 m, floor to floor);
- bigger, more accessible ducts (up to 2% of floor area);
- access floors (typically 150 mm clear);
- more cooling capacity (up to 30 W/m^2 for small power);
- more finely zoned air conditioning (25 m^2 is not unusual);
- more capacity for cellularization (up to 50% of useable area).

In general, though, modern office buildings comprise the any of the following three constructional forms:

1. Reinforced concrete frames with in-situ floor slabs and beams, with brick, stone or curtain wall cladding.
2. Reinforced concrete plate slab and columns with brick infill panels.
3. Steel frames with in-situ or pre-cast concrete floor slabs, and aluminium curtain walling.

Options for reusing redundant office buildings

Preamble

The adaptive reuse of office blocks is more feasible than most other non-residential buildings. However, the size, condition and structural configuration of these are not always conducive to an adaptive reuse for housing or factories. More usually they are converted to an alternative commercial use.

Other offices

As indicated previously, office work practices are now more dynamic. Flexibility of the workspace is critical. This is required to accommodate the differing and changing space demands of different occupiers. Thus a major refurbishment programme to a commercial building may be required before it is suitable for reuse as multiple- or single-use offices.

Hotel

Office blocks, because of their typical multi-storey form and regular floor layouts, lend themselves to reuse as hotels. Many important urban areas such as Edinburgh, for example, have increased their supply of hotel accommodation not by building more but by converting redundant office buildings.

Department store

Potentially, a multi-storey office building could be converted into a single-occupancy department store. However, the demand for this type of retail use is only likely to be high if at all in town centres.

Flats

Low-rise office blocks (i.e. those not exceeding about three storeys) can sometimes be converted to residential use. In such a case the building best suited is of traditional load-bearing construction.

If the building, regardless of its height, has a deep plan, however, accommodation in the central part will not be able to have windows for natural lighting or ventilation. Alternatively, in such circumstances the space could be partly taken up by an atrium. This in turn, though, would bring its own problems in terms of fire safety and security.

Adaptive reuse of public buildings

Characteristics of public buildings

The term 'public buildings' is used here to describe institutional properties such as hospitals, museums, postal headquarters, assembly halls and the like. It is a generic term for buildings that are used by the public for local or central government administration and/or assembly purposes.

Corn exchanges, for example, were a common feature in many large market towns in the 19th century. Like mills these were of substantial construction but also had their unique characteristics, such as:

- single storey, with relatively narrow frontage with an extensive depth;
- solid load-bearing masonry construction with cast iron or heavy timber trussed rafter roof;
- large spans and tall floor to ceiling heights.

Table 4.5 Adaptation potential of public buildings

Type	*Original use*	*Adaptation potential*
Low-rise end/mid-terraced or corner building	Offices Local dignitary's official residence	High-quality flats
Low-rise detached building	Local or national dignitary's official residence	New office use Residential/nursing home
High-rise end/mid-terraced or corner building	Offices	Flats Hotel
High-rise detached building	Archives Offices	New office use Hotel

Options for reusing redundant public buildings

Changes in mental health care have resulted in the closure of the Victorian mental asylums and other similar large facilities. They are ripe for conversion into luxury flats. Other related institutional uses, such as a nursing home, and hospital or college/university accommodation, can also prove viable alternatives.

Other public buildings that may become redundant because of occupants' moves to newer more appropriate facilities are council offices and post office sorting properties. These buildings may not easily lend themselves to an adaptive reuse without substantial alteration and upgrading. Again, over-capacity of certain facilities such as hotel of offices in the area may inhibit a potentially viable alternative adaptation to these uses. Even with the benefit of a façade retention scheme to the property, the market may not be able to sustain a beneficial use for the property.

Problems with the adaptive reuse of old buildings

Integrity of the building

Generally

In this context integrity is a concept that relates to what extent or degree the building retains its original characteristics, form, nature and use. There are four key aspects of building integrity: architectural, constructional, cultural and structural, each of which must be considered in any adaptation scheme, particularly one involving a change of use.

Architectural integrity

An adaptive reuse whilst involving little alteration to a building's exterior, could involve propose radical change to its interior. This may result in the building's architectural integrity being destroyed or compromised.

An ill considered or badly designed extension or external alteration to a building can easily compromise its original style and profile. The adaptation of old buildings in particular must be carried out in a way that not only preserves the structure and fabric of the building in a sensitive manner, but also respects the overall character of the property and its surroundings.

Constructional integrity

The proposed adaptation work must not reduce or adversely affect the performance the existing building. This is especially important in relation to weather-tightness and fire protection. For example, absent or poorly installed cavity trays in an abutment or connection interface between a building or extension can allow wind drive rain penetrate through the outer skin of the cavity wall and down into the room below where the outer skin above becomes an inner skin below.

Any adaptation scheme to an old building should try to incorporate the latest energy efficiency measures but care must be taken to ensure that there will be no conflict with, or devaluing of, its historic detail. In many cases good environmental practice can go hand in hand with building conservation.

Structural integrity

Adaptation schemes such as conversions to other uses can result in increased loadings and changes in load paths (Coates, 2001). These can lead to failure unless precautionary strengthening or underpinning measures are taken (see Chapter 7). For example, commercial use entails at least double the floor loading to that of a residential property. This can result in sagging floors because of overloading.

Cultural integrity

This relates to the use of the building. Ideally, any new use should be complementary to the previous one.

Building evolution

Preamble

When dealing with the adaptive reuse of a historic property it may be necessary to establish its stage of evolution if possible as part of the 'getting to know the building'. Determining the dating and sequencing all parts of the property so that sensitive repairs and alterations are undertaken is not easy. In this regard it is best to adopt a systematic approach. The aim is to establish the extent and timing of any various phases of development. Figure 4.14, for example, illustrates how one old building has been adapted over a period of 400 years.

Dating a building can be achieved using direct and indirect methods, as summarized below.

Direct dating methods

- Building marks (e.g. confirming foundation stone laying ceremony, date plaque).
- Ordnance Survey plans (e.g. a plan dated showing the presence of the property).
- Documentary evidence (e.g. title deeds, as built drawings).
- Carbon dating (e.g. used by Archaeologists investigating old buildings to determine the date of organic materials, to within ±50 years).

Indirect dating methods

- Architectural style (e.g. bay windows are a common feature of Victorian buildings).
- Form of construction (e.g. cavity wall did not become popular until the early 1920s).
- Dendrochronology (i.e. counting the growth rings of timber members to determine their age by reference to a database of tree types).
- Local place names (e.g. Robert Burns Street is unlikely to have existed before the poet was born in 1759, unless of course it was renamed at a later date or relates to another person of the same name).

Dilemmas

Philosophical approach to adopt?

Once the building is fully understood in all of its aspects, an appropriate philosophical approach should be adopted. This is normally articulated through the client's or funding body's conservation policy. It will depend on factors such as:

- Conservation criteria selected: See Chapter 3.
- Extent of change: As seen in Table 3.4, statutory designations such as listed building or conservation area status will severely restrict the degree of change allowed. In adapting such buildings, therefore, a decision has to be made as to the extent of any modifications to the original fabric.
- Repair/replace with same or different materials? Many old buildings are constructed of materials that are either no longer available or are available but not at an economic cost. In building conservation projects it may be a requirement of the client or funding agency (e.g. English Heritage or Historic Scotland) that only 'traditional' materials such as lime mortar be used in the adaptation work.

Ground-floor plan of property at Broad Street, Stamford, Lincolnshire

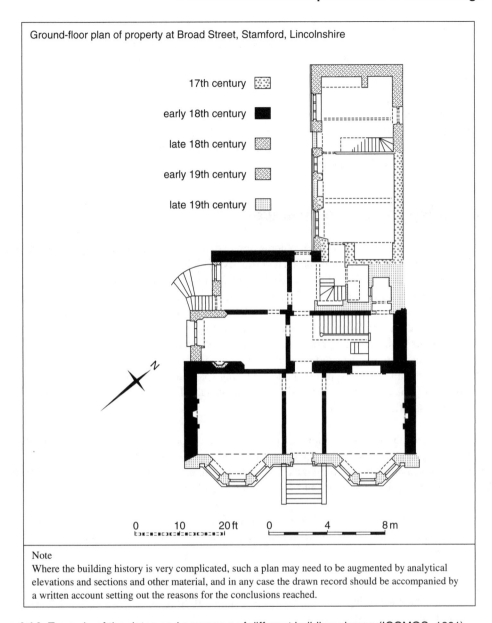

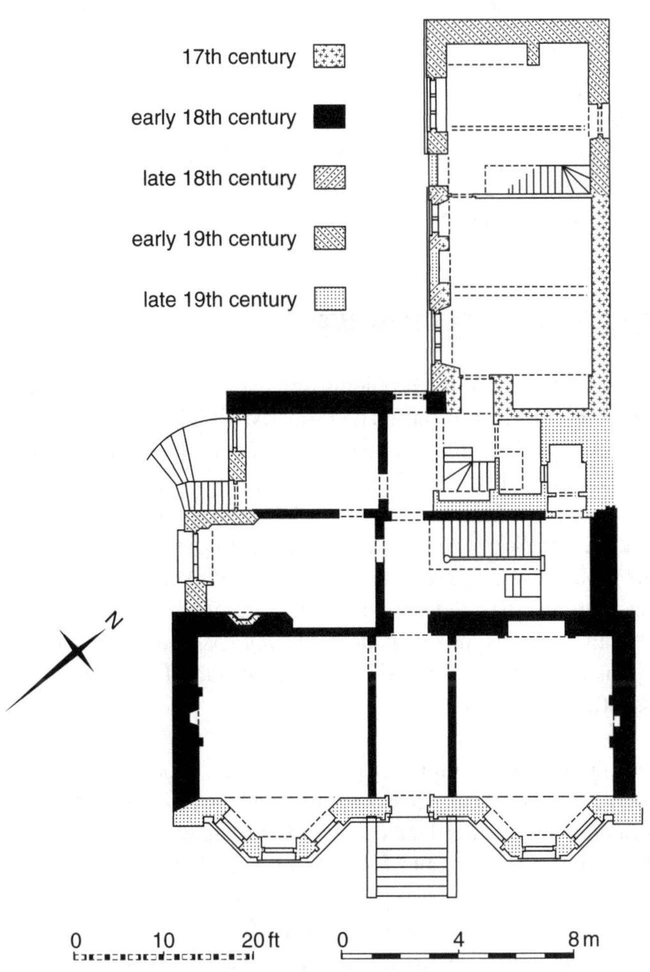

Note

Where the building history is very complicated, such a plan may need to be augmented by analytical elevations and sections and other material, and in any case the drawn record should be accompanied by a written account setting out the reasons for the conclusions reached.

Figure 4.14 Example of the dates and sequence of different building phases (ICOMOS, 1991)

- Substitute materials: The extent to which substitute materials can be used depends on the age, construction and condition of the building as well as on the repair philosophy prescribed by the client or conservation body involved. For example, the use of ordinary Portland cement for repointing of masonry walling may be proscribed.
- Facsimile of original acceptable? Many architectural purists balk at the use of finishes that attempt to copy the genuine article. Imitation stone cladding is a classic example of this kind of 'façadism'. It is an option that is often frowned upon as a way of achieving an ashlar or coursed rubble finish.

Performance requirements?

Old buildings cannot always match new buildings in terms of certain performance aspects. For example, thermal efficiency or sound insulation levels similar to those in new buildings may be difficult to achieve because of constructional constraints.

Hazardous materials?

Many old industrial buildings in particular contain hazardous materials such as asbestos, which could adversely affect the feasibility or timing of an adaptation scheme. These must be dealt with in an appropriate manner as is indicated in Chapter 2.

Building regulations compliance

The code compliance and space planning problems of an adaptive reuse can be highlighted in the proposed conversion of a large office building to flats. The property, which was built in the late 1930s, is shown in Figure 4.15. Its of Art Deco style and has Grade 2 listed status. The following building elements therefore require minimum intervention (i.e. the original appearance to remain) and retention of the original interior features of the entrance lobby, hall and stairwell. The fenestration and associated features are to be retained. In terms of the building's construction, the steel columns and core as indicated on the plans remain, but associated interior walls, not referred to above, will be removed.

The design process requires firm and explicit briefing criteria that determine the scheme design. In converting buildings, certain comparisons are made regarding space standards and back-up facilities (e.g. car parking provision, concierge/porter/domestic waste disposal). For luxury development the following assumptions and essentials are applied:

- Standards of occupation for the number of people: assume 2-bedroom flats equates 3 bedroom spaces (1 double, 1 single); 1-bedroom flats equates either 2 bed spaces (1 double) or 1 single bedroom.
- Developer's return on capital require maximizing the number of flats but minimizing compromises to upmarket space standards and fit-out. The letting agent's view would always be sought in these circumstances.
- The basement provides car parking, but the number of spaces (refer to the attached plan) maximizes 9 spaces (no double parking plus disabled parking, provision for waste bins, services intakes and central heating boiler).

The proposed plans should work within the assumptions referred to above. Details of influences of the existing building plan (circulation, fenestration, columns configuration) are described after the following main influences determining the planned solutions:

(i) Building Regulations Part B1 Means of Escape.

(ii) Third floor is more than 4.5 m above ground level, so alternative means of escape from each flat would normally be needed. Given that this is a tight site (the building is enclosed at two elevations, the rear and one of the gables), there is no scope to construct an external staircase. The only alternative would be to install another staircase in the building – at the opposite end to the existing. This would be an expensive and disruptive option that would take up valuable domestic space.

(iii) Accommodation is in excess of the standards applied in the Housing Act 1985 (ss. 352, 357), the now defunct Parker Morris Standards (mandatory for public schemes 1967–1981), and the current Housing Corporation floor area bands (1998). See also space standards in Adler (1999).

(iv) Travel distances in protected corridors not exceeding 7.5 m.

(v) Travel distances in flats not exceeding 9 m. Where this cannot be achieved within the existing layout, relaxation of the regulations might be considered. However, any granting of relaxation will come with conditions, usually additional fire precaution measures such as alarms or even sprinklers.

Figures 4.16a–c shows, respectively, the sketch plans detailing:

- Ground floor: 5 flats
  - 2 × 1-bed (1 bed space),
  - 3 × 1-bed (2 bed spaces).
- First floor: 5 flats
  - 1 × 2-bed (3 bed spaces),
  - 3 × 1-bed (2 bed spaces),
  - 1 × 1-bed (1 bed space).
- Third floor: 5 flats
  - 3 × 1-bed (2 bed spaces),
  - 2 × 1-bed (1 bed space).

All flats have alternative means of escape.

Normally general arrangement plans are required to 1:100 scale. Drawings should be clear, properly labelled and indicated main fixtures (sanitary provision, kitchen surfaces, basic room furniture layouts) including door swings, fire doors and protected corridors. Again, any downtakings should be shown in red.

The main relevant sections of legislation determining the construction standards and particularly space planning are contained in the Building Regulations 2000 (as amended), as follows:

- Part B1: Means of escape
- Part E: Resistance to the passage of sound
- Part F: Ventilation (especially to kitchens and toilets)
- Part G: Hygiene
- Part H: Drainage and waste disposal
- Part L: Conservation of fuel and power
- Part N: Glazing – safety on impact, opening and cleaning
- Part P: Electrical safety

North of the border the following requirements of the Building Standards (Scotland) Regulations 2004 would apply as follows:

- Section 2: Fire (means of escape, fire resistance)
- Section 3: Environment (hygiene, health, ventilation; especially in kitchens and toilets)
- Section 4: Safety (safety of glazing in use; electrical installations)
- Section 5: Noise (resistance to the passage of unwanted sound)
- Section 6: Energy (energy economy and heat retention)

Other legislation relevant to this proposed adaptive reuse may include:

- Construction (Design and Management) Regulations 1994
- Construction Products Regulations 1991 (the quality of materials)
- Construction (Safety, Health and Welfare) Regulations 1995
- Control of Asbestos at Work Regulations 2002

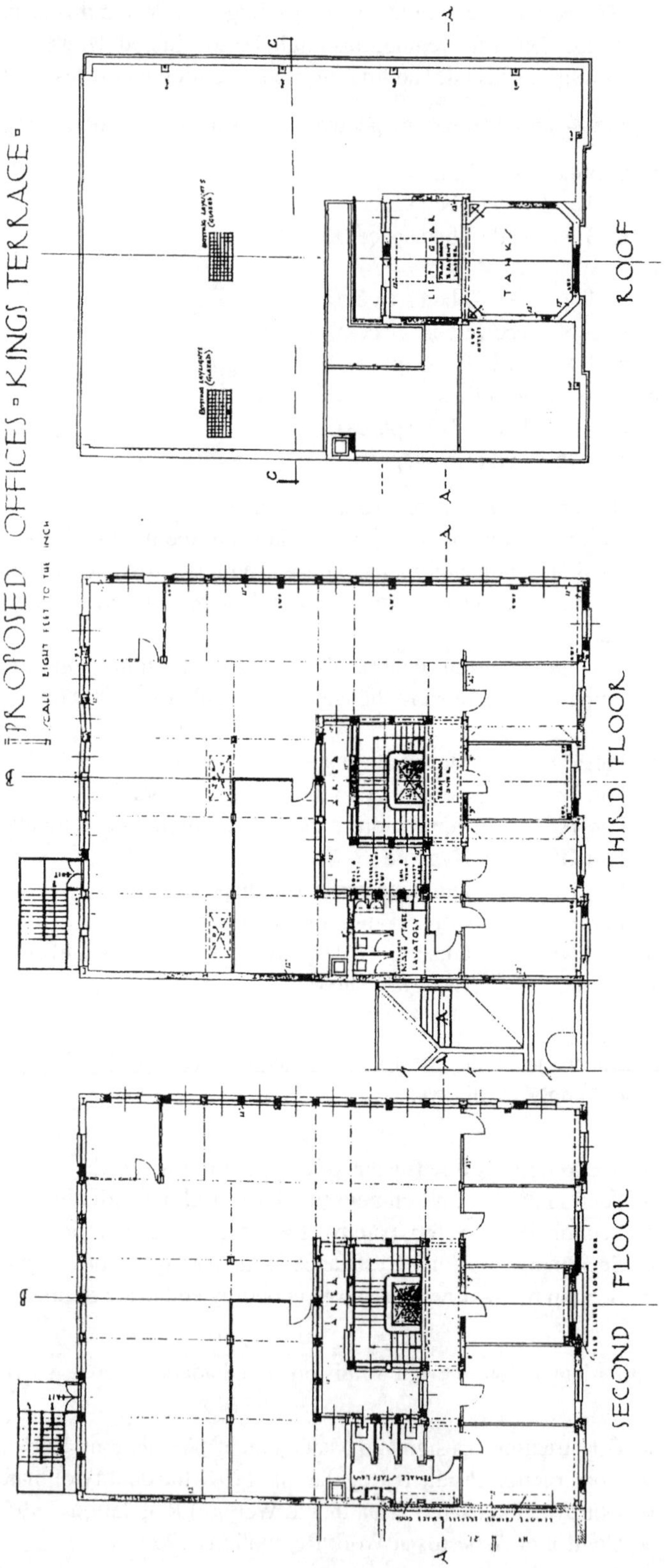
PROPOSED OFFICES - KINGS TERRACE
SCALE EIGHT FEET TO THE INCH
ROOF
LIFT GEAR
TANKS
THIRD FLOOR
MALE STAFF LAVATORY
SECOND FLOOR
FEMALE STAFF LAVS

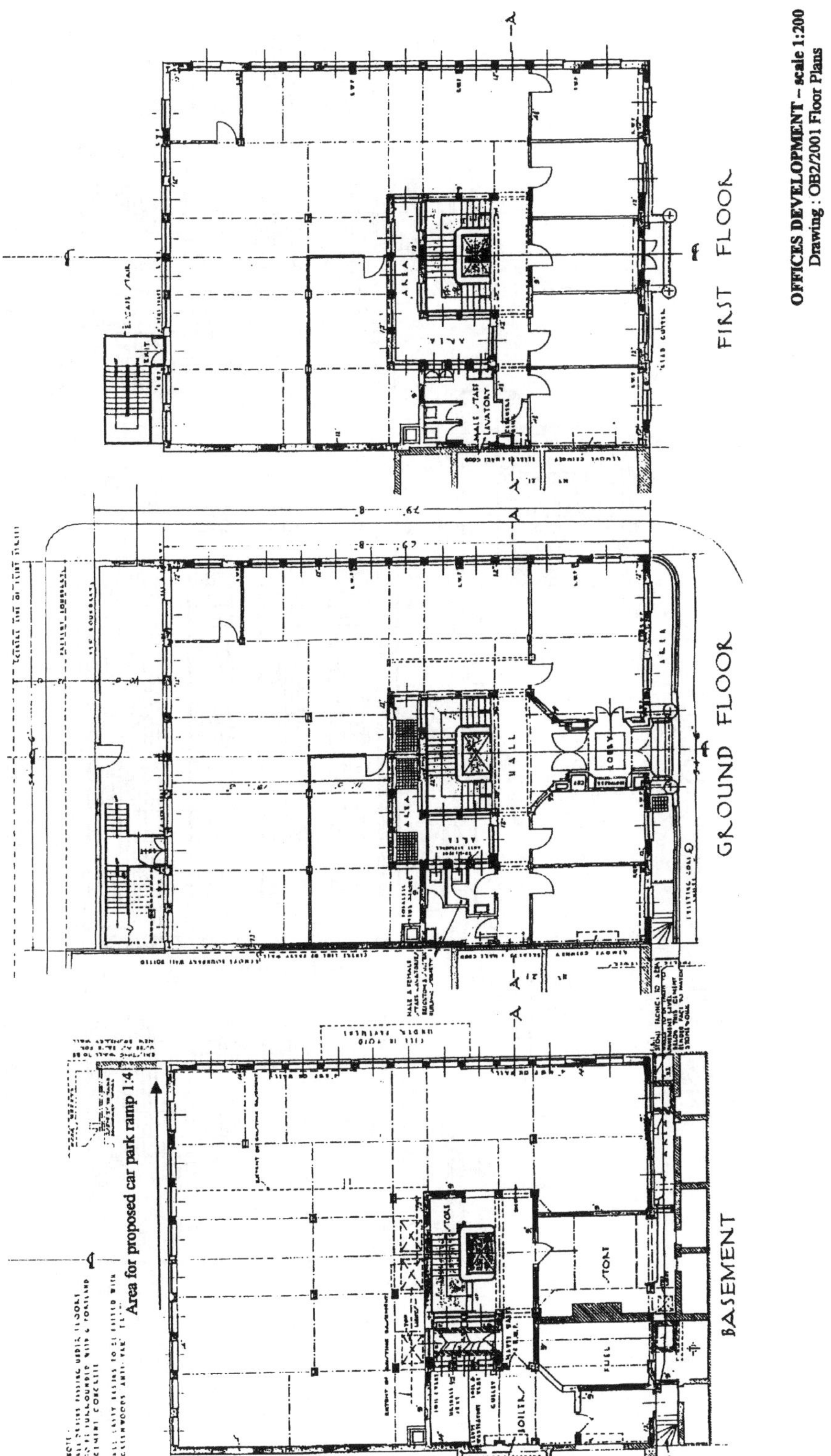

Figure 4.15 Existing office block ripe for an adaptive reuse (courtesy of CEM)

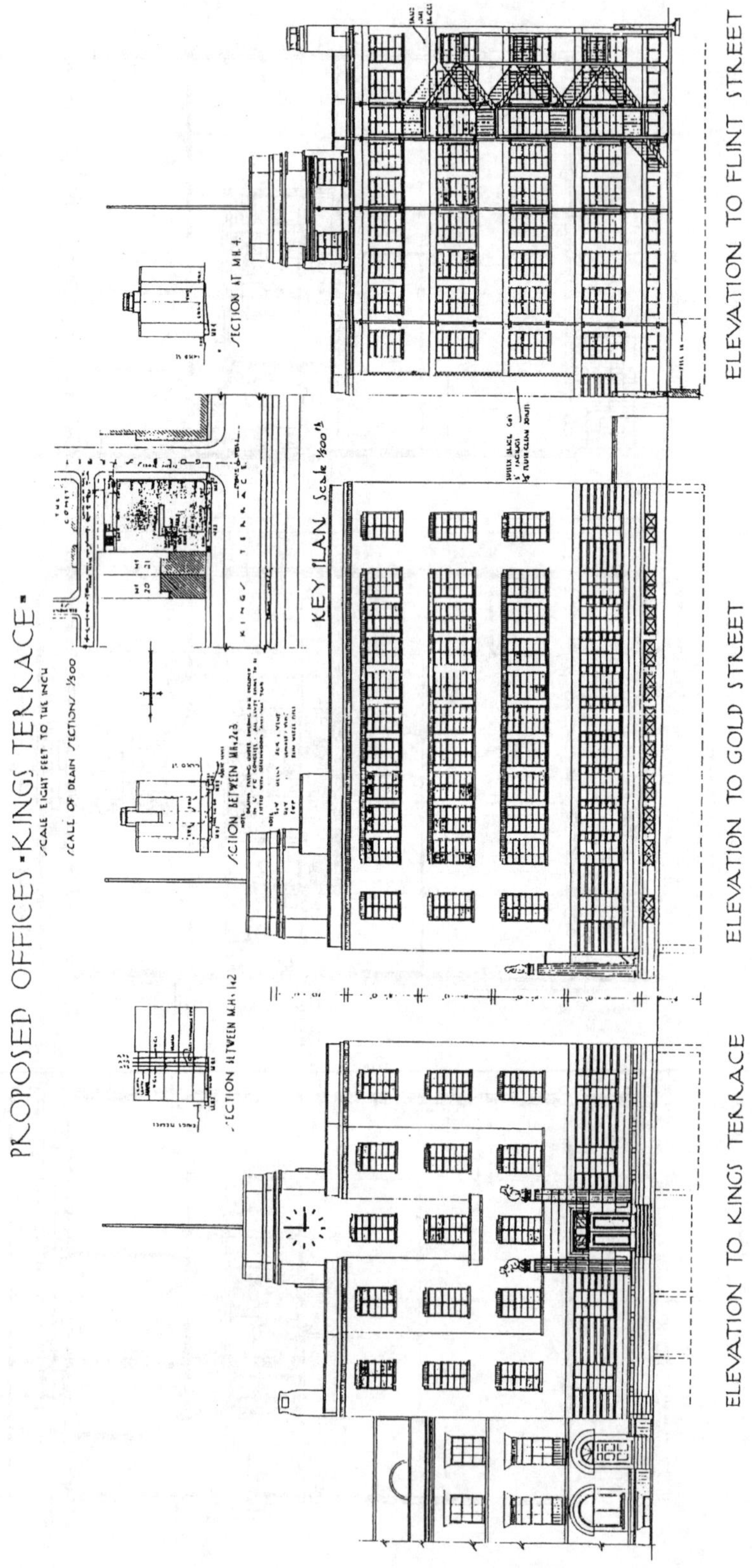

PROPOSED OFFICES - KINGS TERRACE -
SCALE EIGHT FEET TO THE INCH
SCALE OF DRAIN SECTIONS 1/300
KEY PLAN Scale 1/500 th
SECTION AT M.H.4
SECTION BETWEEN M.H.s 2 & 3
SECTION BETWEEN M.H.s 1 & 2
ELEVATION TO KINGS TERRACE
ELEVATION TO GOLD STREET
ELEVATION TO FLINT STREET

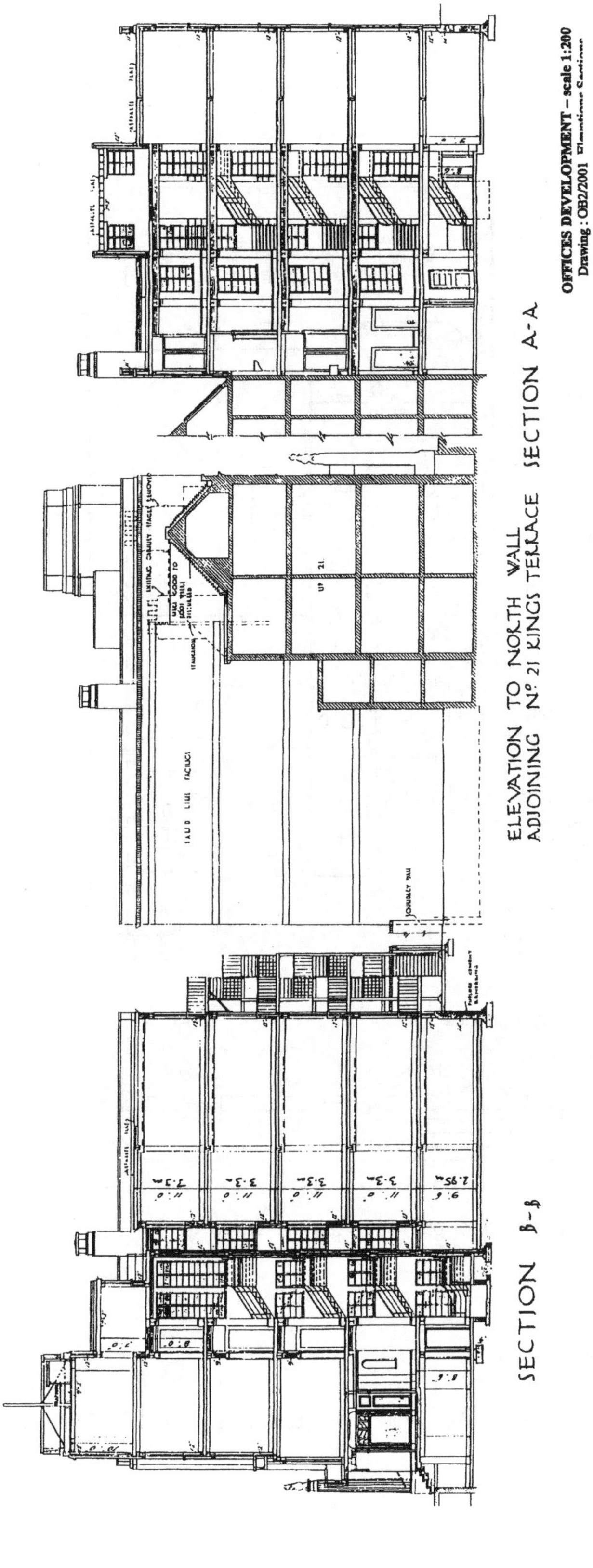

Figure 4.15 (Continued)

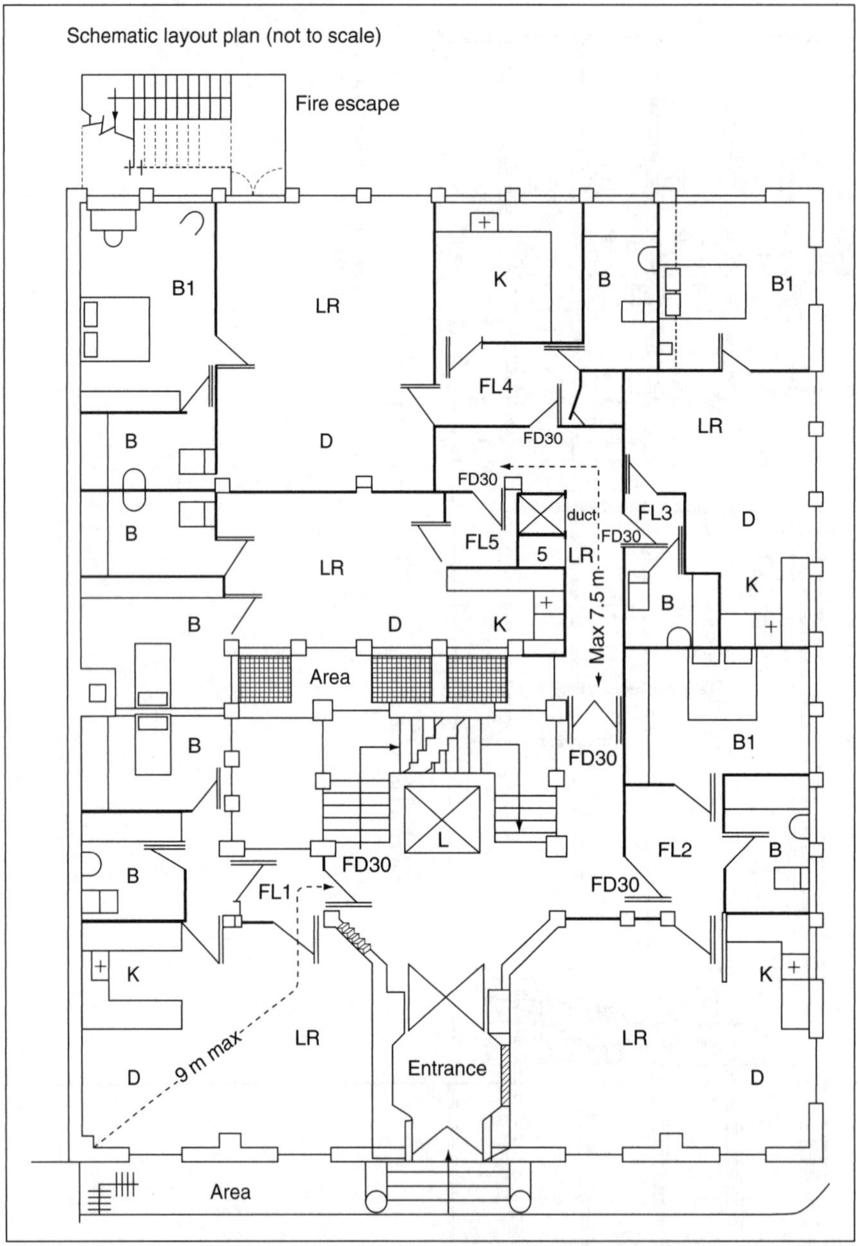

Figure 4.16a General arrangement drawing – ground-floor plan (courtesy of CEM)

- Disability Discrimination Act 1995
- Gas Safety (Installation and Use) Regulations 1994
- Housing Act 1955 (standards of accommodation)
- Planning (Listed Buildings and Conservation Areas) Act 1990 or the Planning (Listed Buildings and Conservation Areas) (Scotland) Act 1997
- Water Industry Act 1991

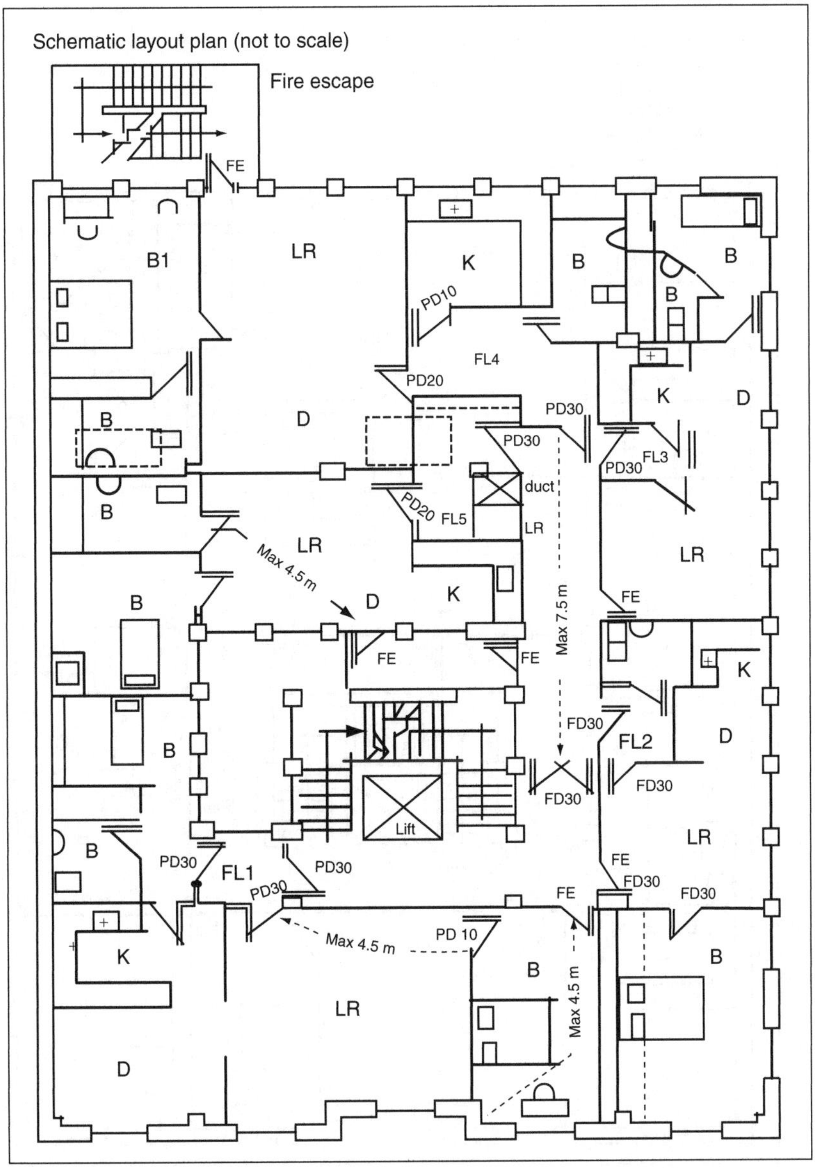

Figure 4.16b General arrangement drawing – first floor plan (courtesy of CEM)

The outline specification for the proposed adaptive reuse scheme should include builders' work fulfilling the following design and construction features:

- Part A: Under Part A1 'Loading' in the approved documents, the existing structure (designed for office floor loadings) will more than adequately cope with reduced domestic loads
- Part B: Protected lobbies to flats on third floor, travel distances within flats not exceeding 9 m and protected corridor 7.5 m. Flats located on the third floor are 4.5 m above ground level and require an alternative means of escape. Other issues include the fire resistance (FR) between levels (at least 1 h); 2 h between basement and ground floor

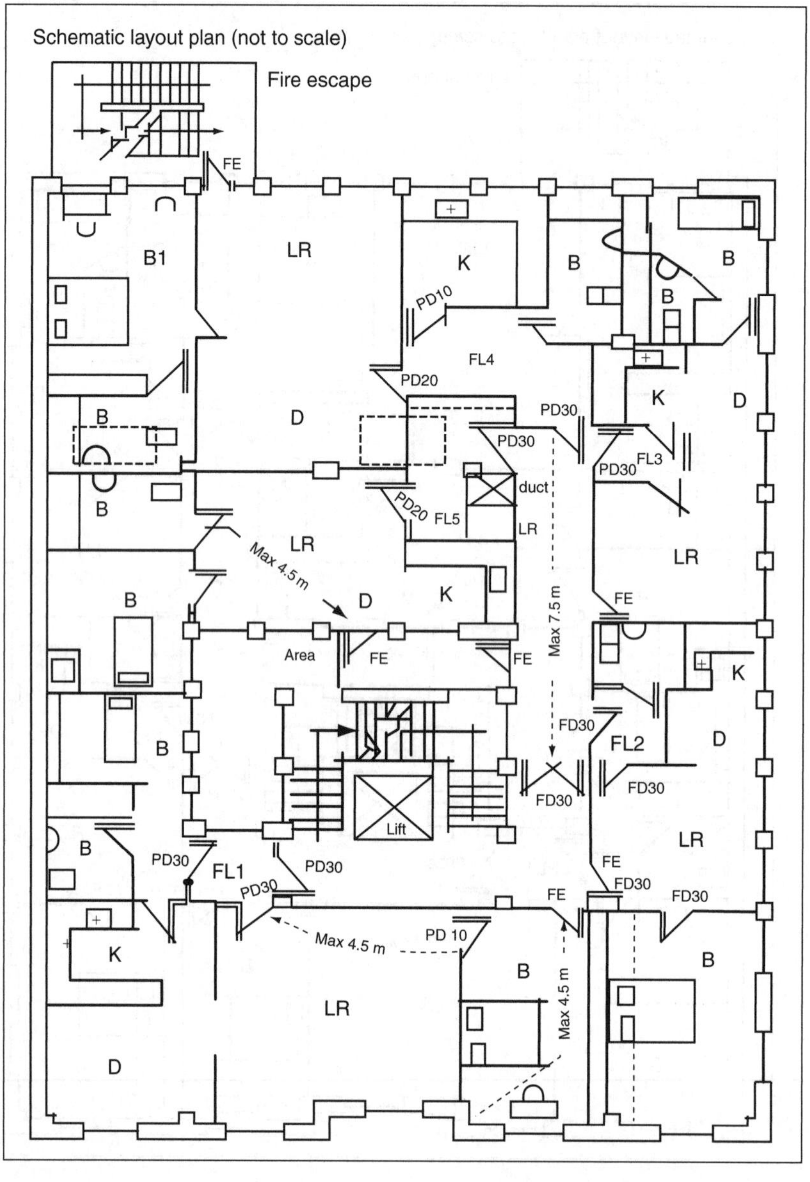

Figure 4.16c General arrangement drawing – third floor plan (courtesy of CEM)

- Part E: Upgrading floor specification for impact sound resistance (separating floor) and party wall (separating wall) construction between flats
- Part F: Ventilation to habitable rooms (mechanical to bathrooms and kitchens), available daylight
- Part G: Provision of sanitary, washing and hot water storage
- Part H: Basement refers; bins and provision of sink disposal units including drainage constructions to existing, horizontal service to suspended ceiling voids
- Part L: Replacement window provision (double glazing, thermally broken), insulated overcladding to walls and insulation to refurbished flat roofs
- Part N: Reversible or extended casement hinges for internal window cleaning

- Part P: Electrical system installed and tested by certified by a prescribed competent person (e.g. an approved contractor of the National Inspection Council for Electrical Installation Contracting)

Historic interiors

Finishings

Many old buildings have attractive and distinctive or unusual features and finishings that wherever possible should be retained in any sensitive adaptation scheme (Lander, 1982). Listed building status of course may protect these, but in other buildings without this form of protection valuable interiors may inadvertently be destroyed. Replacing these on a like-for-like basis may not be easy, but architectural salvage companies may be able to supply appropriate replacement items.

The following features are especially worthy of retention and refurbishment:

- Wall coverings: The original painted, papered or textiled coverings on plastered walls may still be in adequate condition to warrant preservation. The lath and plaster substrate may, however, require remedial work as well.
- Dado panelling: Hardwood or ornate softwood panelling up to height of about 1 m from floor level is a common feature in the drawing room, lounge or library of many prestigious buildings. These finishes may, however, require refurbishing – by a French Polisher as well as a joiner. In any case, if natural-finished timber panelling is being retained or installed in an adaptation scheme, it must be coated with an approved fire-retardant lacquer.
- Cornices and ceiling: Elaborate plaster cornices and ceiling patterns can add a great deal to the character of a room. These should be retained and made good using a suitable plaster of Paris or equivalent proprietary rapid-setting filler.
- Panelled doors: These can usually be refurbished by dipping them in a bath of mild caustic solution to remove any unwanted paint coatings. The woodwork can then be revarnished using an appropriate fire-retardant lacquer. New brass ironmongery should be fitted if necessary.
- Picture rails: Hardwood or ornate softwood picture rails may be worthy of retention, provided that they are unaffected by woodworm or fungal attack. Removal of these rails can cause damage to the plaster substrate, which may be difficult to patch repair properly.
- Floorboarding: Many pre-20th century buildings have their suspended floors covered in tongued and grooved boards made from hardwood or good quality softwood. It may be worthwhile to consider sanding and varnishing such floors, provided the joints are reasonably tight rather than open. Again, a suitable fire-retardant varnish or finish should be used.
- Parquet flooring: Some, more prestigious, dwellings may have hardwood parquet flooring as the finish in the main lounge and bedrooms. This type of timber finish may require sanding down and revarnishing. Tenting or general uplifting of the parquet blocks may be caused by inadequate allowance for movement at the perimeter or through failure of the adhesive caused by dampness. These deficiencies would of course have to be rectified to prevent any reoccurrence of the problem.
- Ornate fireplaces: Many Georgian and Victorian properties have ornate surrounds made from cast iron, hardwood and marble, or a combination of these materials. The Adam's Fireplace is a common mantlepiece with fluted support legs that was used for surrounding the open hearth in Scottish Georgian buildings. See Davey et al. (1995) for more information on the repair and refurbishment of ornate fireplaces.
- Encaustic and other ceramic tile flooring: This style of floor finish was common in many public buildings in the early 1800s. Patterned ceramic floors in reasonably good condition can be refurbished using a soapless detergent cleaning agent. The more heavily stained or tarnished floors would probably need to be refurbished using a proprietary deep cleaning agent such as 'HG Remover'.

Table 4.6 The nature of different spaces within historic buildings (Adapted from Stratton, 1999)

Nature of space	*Building type*	*Number of storeys*	*Attributes*	*Alternative use*
Small single space	Small church Meeting hall	Single storey	Character for specialized uses	Single or double dwelling Workshop
Large single space	Warehouse Corn exchange Large church	Single storey	Flexibility and easy movement	Sports complex Multiple flats
Small repeated spaces	Workshops Offices	Up to four storeys	Small units, poor access	Multiple offices Flats Small hotel
Large repeated spaces	Mill Warehouse/bond	Multi-storey	Flexibility, but difficult to subdivide	High-quality flats Hotel Offices

- Staircases: An ornate hardwood or stone staircases often forms the central feature of a building's interior. This is especially true if the staircase is broad and curved. Fungal attack and woodworm are the two main enemies of timber staircases. However, as with stone stairs, timber staircases can suffer from neglect or excessive wear and tear, resulting in worn treads and tarnished finishes. The insertion of heavy duty wearing strips in the treads, along with general renewal of the finishes, is likely to form part of the refurbishment of timber or stone staircases in an historic building.

Uses

Traditionally buildings were designed for only one use. In particular the nature of the interior of non-residential buildings constructed prior to the mid-1900s usually reflects their original function. This inevitably has an impact on their adaptability to another use, either in spatial or cultural terms.

Table 4.3 illustrates the main types of space and their appropriateness of different new uses.

Fire safety in adaptations

Development of a fire

A fire can only start if there is an adequate supply of fuel, an adequate supply of oxygen and a means of ignition. The fire develops by, first of all, the fuel igniting. As more oxygen becomes available more fuel is consumed and the temperature rises significantly, making more fuel available. In other words materials that do not ignite easily become ignited as the temperature rises.

Fires can be triggered deliberately or inadvertently. Arson accounts for a large proportion of fires in buildings (Kidd, 1995). A tight security regime as well as an appropriate fire safety engineering system is required to minimize the incidence of malicious fire-raising.

A further factor which influences fire development is the nature of the burning material. Some plastics, for example, produce droplets of molten material as the fire consumes them. These drop onto other fuel materials and can cause further fires to start.

Fires can occur during the course of an adaptation scheme. For example, reckless use of gas guns for torching on felt or soldering lead and boilers for bitumen and asphalt can cause fires when used in adaptation as well as new build schemes. Care therefore must be taken when specifying materials or components that may require the use of these hot-working processes. The fire risk of these is sufficient to prompt a ban or severe restrictions on their use in certain conservation projects (see Table 4.9 for details of typical fire risks associated with adaptation schemes).

After the building has been adapted the fire risk remains. The main sources of inadvertent ignition, in ranking order, in residential buildings are cooking appliances, electrical equipment, smoking and matches. In other building types electrical equipment, matches, smoking are the main inadvertent sources of ignition.

Fire safety engineering

Principles

In Chapter 2 the main fire safety criteria were identified. Fire safety is normally considered to cover the safety of people, the protection of property, both in the building concerned and in the surrounding area, and the processes involved. Fire safety engineering develops this theme by a means of preventing and controlling fires within buildings. It is based on a fire safety design philosophy that is holistic. Such an approach considers the building as a complex system, and fire safety as one of the many interrelated subsystems, which can be achieved through a variety of equivalent strategies.

In a fire smoke is the biggest killer – primarily from carbon monoxide poisoning. It is estimated that over 70% of all fatalities in a fire are smoke-related (Drysdale, 1994). Smoke and fumes are not only is hazardous to people, they are also deleterious to buildings and building products as well. For example, smoke causes extensive soiling and disfigurement to decorations and furnishings. As smoke often contains aggressive particles it can also erode materials such as polyvinyl chloride (PVC) sheathing of electrical wiring even if the heat does not melt them (see Chapter 1 about the fire risks associated with PVC).

In addition, smoke reduces visibility and contains irritants that make eyes and nose stream and induce coughing. It also causes oxygen vitiation and transmits heat.

There is no reason why fire safety engineering cannot be adopted in adaptation schemes. It is very helpful in dealing with the problems of fire in historic buildings because they involve potentially higher risks. In general, however, this approach would be mainly appropriate for the large adaptive reuse or refurbishment projects, especially those containing an atrium or other high-risk fire zone (see Figure 4.19).

The key precautions and management policies for an effective fire safety strategy are summarized in Figures 4.17 and 4.18, respectively. The objectives, tactics and components of a fire safety strategy are illustrated in Table 4.7.

Requirements

The two critical fire safety requirements in any adaptive reuse project are structural protection and means of escape. Approved Document B in the regulations for England and Wales and Section 2 in the Scottish regulations deal with these complex issues. The requirements for fire precautions depend on the category and size of building involved. Each new build or adaptation scheme, therefore, has to be considered on its own merits.

The Building Regulations of course are primarily concerned about protecting people – neighbours, members of the public and fire-fighters as well as the building users. In the context of commercial and industrial premises, however, the property insurers may demand higher standards than the Regulations when the structures themselves or the contents are highly valuable, such as works of art, precious records, expensive equipment, etc. Moreover, for these non-residential buildings there may be a requirement to follow the Loss Prevention Council's rules (Fire Protection Association, 1992).

Thus, new steel support beams or party floors in a residential conversion scheme may need to be given at least 1-hour protection against fire. This could be doubled or even quadrupled in the case of a commercial or industrial building conversion project.

Nowadays, it is usual to apply a fire safety engineering approach to any commercial new build or major adaptation scheme. This will help to determine the balance between the necessary active and passive fire precautions. However, this is not easy to achieve, particularly with historic buildings (see below).

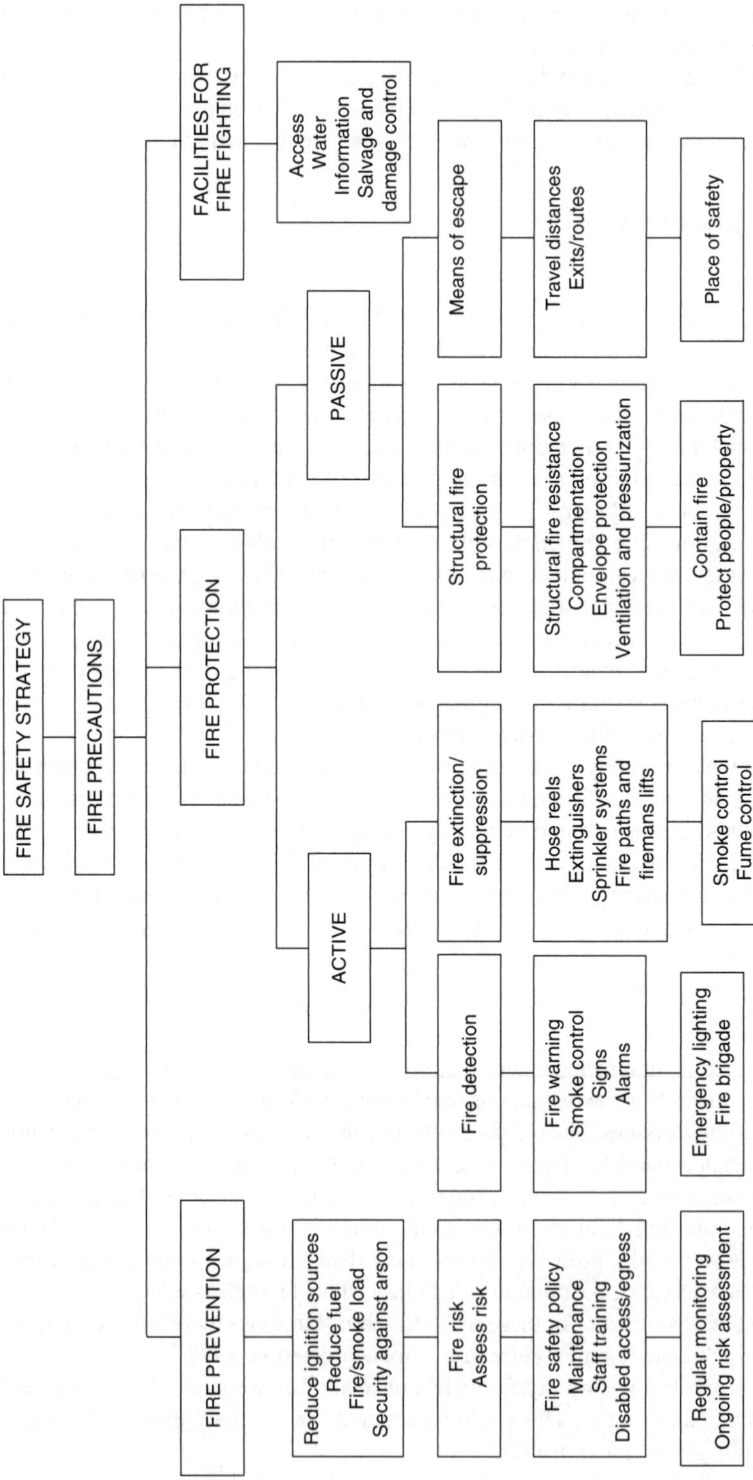

Figure 4.17 Key precautions for a fire safety strategy (adapted from Fire Protection Association, 1992)

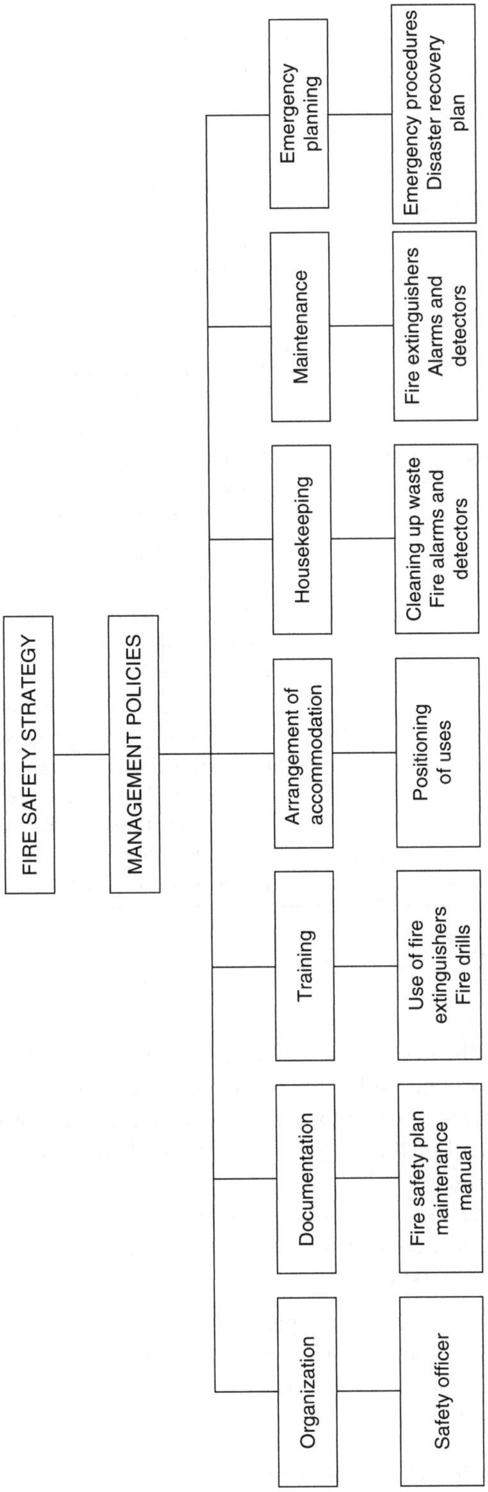

Figure 4.18 Key management policies for a fire safety strategy (adapted from Fire Protection Association, 1992)

Table 4.7 Fire safety hierarchy

Hierarchy	*Requirements*	*Means*
Aim	Fire safety	Coherent policy. Regular monitoring and review.
Objectives	Life safety considerations	Safeguarding escape routes. Protecting occupants in adjacent compartments.
	Preventing or minimizing conflagration	Limiting the extent of damage. Preventing fire spread. Facilitating fire fighting.
Tactics	Prevention	Identify hazards. Assess risks. Implement control measures. Monitor.
	Escape	Sufficient number and capacity. Adequately protected and lit.
	Containment	Prevent spread to other parts.
	Communication	Within building – to warn occupants. Outwith building – to notify fire service.
	Extinguishment	Curtail spread and effects of fire.
Components	Occupants	Type, age/ability range, etc.
	Building	Category, age, form of construction.
	Furniture	Combustibility of coverings and foam infill.
	Fittings	Types of fabric and insulation – low toxicity.
	Contents	Flammability. Toxicity. Fire/smoke load.
Risk	Ignition	Electrical fault; gas explosion; cigarette stub; hot-working processes.
	Fuel	Gas, petrol and other flammable substances – affects fire load and smoke load.
	Smoke	Thick, toxic fumes – affect vision as well as breathability.
	People	Able-bodied/infirm; awake/asleep/intoxicated.
	Building contents	Flammable/valuable?

Active fire precautions

Fire detection

This is becoming critical in virtually all categories of buildings. Even in domestic adaptation schemes such as extensions there is now a need to install hard-wired smoke detectors in the new parts of the property. Battery operated smoke detectors have proven to be unreliable in residential properties. This is because of the failure of many occupants to replace the batteries when required. Some tenants, for example, have even been known to deactivate the alarm.

The position of any fire warning devices in an adaptation scheme should be carefully considered. These should be located at ceiling level near combustible appliances or other sources of fire or smoke.

Detectors may be of the following types:

- Heat: normally for residential, commercial and public buildings;
- Smoke: normally for residential, commercial and public buildings;
- Infrared: normally for industrial buildings operating where radiation is emitted during a fire;
- Ultraviolet: normally for industrial buildings operating where radiation is emitted during a fire.

Fire suppression and extinction systems

The design of a fire suppression system is aimed at:

- reducing the amount of oxygen available to the fire;

- minimizing the amount of water needed to extinguish a fire;
- cooling the environment within which the fire takes place, thus reducing the available supply of heat and fuel;
- transforming the flammable materials into non-flammable materials by saturation.

Improvements in the fire extinction facilities may have to be considered when a building is being adapted. This will depend on the category and intensity of use, the condition and form of the building, as well as the fire risk.

The main problems of retrofitting fire suppression and extinction systems in an existing building are:

- minimizing collateral damage to the structure and fabric;
- obtaining available supply and adequate pressure;
- accommodating pipe runs and pumps without adversely affecting the interior of the building;
- disruption to the services and use of the building whilst the installation is being undertaken;
- it involves another item added to the maintenance requirements.

Sprinklers

The case for using sprinklers as a major active fire precaution in all types of buildings is widely acknowledged. Fire brigades, for example, are keen advocates of sprinklers – even in domestic premises. There has never been a reported fire death in the UK in a building fitted with a sprinkler system (Scottish Executive, 2000). The main advantages with this type of active fire precaution are:

- sprinklers raise the alarm, control the fire and may put it out;
- speed of operation is quick, they are activated within seconds;
- only the heads nearest the fire will operate, thus minimizing the use of water to combat the fire;
- few moving parts, maintenance once a year;
- accidental operation of sprinklers is a 1 in 16 million chance;
- sprinkler heads and pipework can be concealed from view (see Chapter 3);
- fire and water damage to property is minimized;
- hazards faced by firefighters such as flashover and backdraught, are reduced;
- local authority and insurance costs will be substantially lower.

Sprinklers perform a dual function of fire detection and suppression. The sensor on the sprinkler head (e.g. a quartzoid bulb) is an effective detector, which breaks when the ambient temperature exceeds about 68°C thus activating the system. The released water suppresses any fire within its range, which for each sprinkler is usually an area of about 9 m^2 (Stollard and Abrahams, 1995).

The use of sprinklers should also be given consideration in any residential scheme undergoing adaptation. High-risk dwellings such as sheltered housing, homes for people with special needs or social problems (e.g. pyromaniacs under supervision released into the community), may necessitate the installation of sprinklers in such buildings. The installation of sprinklers in housing costs about £2000 per dwelling (Scottish Executive, 2000).

Smoke control

In some conversions the local authority may require smoke vents at the top of stairs in an attic conversion or in the roof top floor of a refurbished building. These can comprise automatic vents installed in the flat or pitched roof. The vents are released on activation of the fire alarm. Proprietary smoke hatches such as Thermolatch® II can feature a curb-mounted fusible link housing that allows the latch to be quickly and easily reset from one end of the vent at the roof level.

In large facilities such as hospitals double fire doors in corridors have to be opened regularly. This can be an inconvenience for users and is likely to reduce the service life of the doors. Such a problem can be overcome by using hold-open devices on the doors. These are usually magnetically operated connections that are released during a fire. The doors, which should have seals around their perimeter, then close to perform their fire and smoke barrier function. Pressurization of escape routes may be required to prevent them from becoming clogged with smoke. Alternatively, especially in atria and top floor corridor areas automatic smoke vents as indicated above can be used.

The main danger with providing ventilation in a fire is that it might inadvertently fuel the fire if a back-draught effect is created. Smoke control in single use as well as in mixed-use buildings is complex. Thus, a specialist fire engineer should determine the position, size and type of smoke vents being considered in any adaptation scheme.

Passive fire precautions

Preamble

Passive measures use the form, structure and fabric of the building to protect the occupants and the property from the effects of fire. The aim is to prevent any spread of flames and smoke or further damage to the premises as a result of a fire outbreak. Containing the fire and facilitating quick and safe exit by the building users are paramount.

Fire safety is important for all categories of buildings. In any non-residential adaptation scheme, however, the statutory requirements relating to fire safety are likely to be extensive. This is because the intensity and complexity of use in commercial and industrial premises is usually greater than with dwellings.

Compartmentation

The means of dividing a building into a number of compartments or sub-compartments is called compartmentation. It comprises a series of fire- and smoke-tight areas that are designed to contain the spread of fire and gain time for the occupants to escape.

An adaptive reuse scheme involving mixed uses will usually necessitate a building's compartmentation. Obviously the smaller the compartment the easier and simpler the fire suppression system can be. In a small compartment a fire risk may be satisfactorily dealt using a single extinguisher. In a large compartment a full sprinkler system may be required.

The importance of a compartment is that it limits the amount of fuel available to a fire. It also to a large extent, limits the amount of oxygen available and makes it easier to prevent the fire reaching other sources of fuel because the compartment must have an appropriate level of fire resistant construction. A properly constructed compartment will prevent the fire moving either laterally or vertically. Clearly a fire is more likely to break through a construction vertically upwards than horizontally, but if it is contained it will attempt to break out anywhere it can including vertically downwards.

Means of escape

As far as means of escape is concerned the objective here is to allow people to evacuate a compartment without moving through the fire if that is possible within 2½ min. In a large compartment this means that several exits may be necessary. In a small compartment it means that the exit should be kept as remote as possible from the major fire hazards. In a little kitchen, for example, we may consider that the cooker is the greatest fire risk, in which case, the escape door should be remote from that appliance so that people do not have to pass close to a fire in order to escape.

The key requirements for means of escape are:

- The principles, get out safely and quickly (i.e. within 2½ min to a safe zone externally).
- Factors influencing design, building type/risk/occupants.

- Design features, continuous protection; services to facilitate escape, provision of alternative means of escape, travel distances (see below).

The key sections of Part B of the Building Regulations and their implications for adaptation schemes are summarized in Table 4.8.

Table 4.8 Passive fire precautions and their adaptation implications

Building Regulations Part B (England and Wales)	*Key element*	*Key requirements*	*Adaptation implications*
B1: Means of escape	Travel distances.	For residential and assembly: 9 m in one direction; 18 m in more than one. 4.5 m for apartments on third floor and above.	New escape routes or openings may be required. Additional signage and lighting may also be necessary.
	Number of escape routes.	At least one.	
B2: Internal fire spread (linings)	Inhibit spread of fire and rate of heat release through internal linings.	Ideally Class 0 should be the objective.	Overcladding of surfaces with plasterboard or Fermacell gypsum fibre composite.
B3: Internal fire spread (structure)	Maintain stability for reasonable period and restrict spread of fire or smoke within building or to other buildings or parts of buildings.	Compartmentation: maximum area of compartment depends on the number of storeys and category of building. Sub-compartmentation. Firestops. Openings. Glazing.	Upgrade existing walls and floors that become separating floors to 1 h fire resistance. Individual rooms, etc. – usually upgraded to ½ h fire resistance. Cavity barriers in over cladding – at every storey and at maximum 8 m centres horizontally. Seals around pipes and fire shutters in ducts. Doors should be self closing and match fire resistance of walls. Georgian wired glass.
B4: External fire spread	External envelope designed and constructed to resist spread of fire over surface or to another building.	Fire resistance Surface spread of flame.	Position and size of extension or new openings. Type of wall or roof overcladding.
B5: Fire service	Access and facilities for the fire brigade.	Design and construct to facilitate the work of the fire services in gaining access and placing appliances.	Adequate width of access road – not compromised by extension or other adaptation. New fire hydrant.

Table 4.9 Common causes of fires in historic buildings

Period	*Possible cause of fire*
During adaptation	Arson (including malicious ignition following break-in and burglary). Carelessly discarded cigarettes or matches igniting combustible material. Flammable material (e.g. unprotected curtain) too near a heat source (e.g. electric light fitting). Heat from oxy-acetylene torch being used to weld leadwork on roof. Heat from butane blow-lamp igniting combustible material nearby. Naked flame from blowtorch or other source igniting combustible material in the vicinity. Any accident that triggers an ignition (e.g. spark from electrical equipment or grinding work).
After adaptation	Any of the above, plus: Lightning strike. Sparks or heat from an electrical fault, such as a badly wired connection or overloaded power socket. Sparks or heat from a faulty heating system boiler. Flames or sparks or from open fires, woodstoves or cooking processes. Poor or inadequate maintenance allowing some of these deficiencies to occur.

Emergency exit hatches, known in the USA as roof scuttles, can be installed in floors and flat roof decks of commercial premises. They usually come with a counterbalanced spring mechanism of easy operation to comply with the CDM manual handling regulations.

Fire precautions in historic buildings

Preamble

Like adaptation work to historic buildings generally, the effects of fire precautions in these properties should be discrete but effective. The aim is to achieve an effective level of fire protection that does not adversely affect the character of the building. This, of course, is easier said than done. Installing sprinklers and alarm sensor cabling in old properties can be difficult, especially where there is an absence of free void spaces within the building.

Common causes of fires in historic buildings are listed in Table 4.9.

General protection measures

As well as a combination of suitable active and passive fire precautions, other measures should be considered when adapting an historic building. First of all, a coherent policy on fire safety should be articulated and incorporated within the building's maintenance manual. This should also result in the publication of a plan of action in the event of a fire. Adequate maintenance of fire safety procedures as well as equipment is vital if a basic level of fire protection is to be achieved. For example, 'hot work' (i.e. operations involving blowtorches and other high fire-risk work) should only be allowed if there is no other appropriate technique available. If this is not possible 'a hot work permit system should be implemented, and adherence to this scheme should be a written requirement within the work contract' (Historic Scotland, 1998, 1999a).

Secondly, it is important to achieve an appropriate balance between active and passive fire precautions in historic buildings. In the case of the latter, the following should be considered in any adaptation scheme:

- Improve the proper fitting of all doors (to reduce excessive gaps) and ensure that they have adequate self-closing mechanisms. This could also involve increasing the depth of doorstops to 25 mm using hardwood straps glued and screwed to the doorframe.

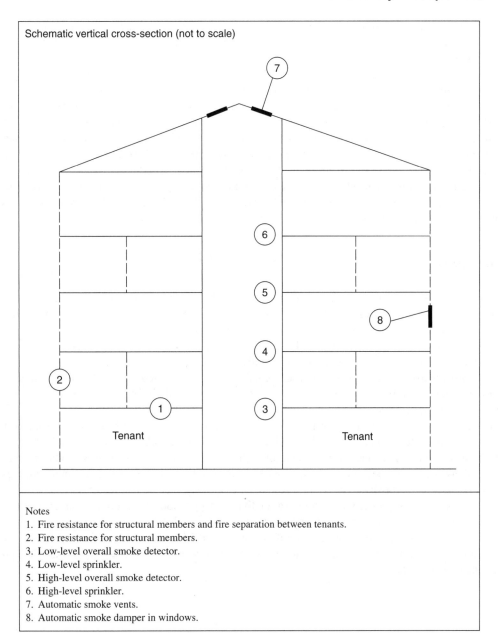

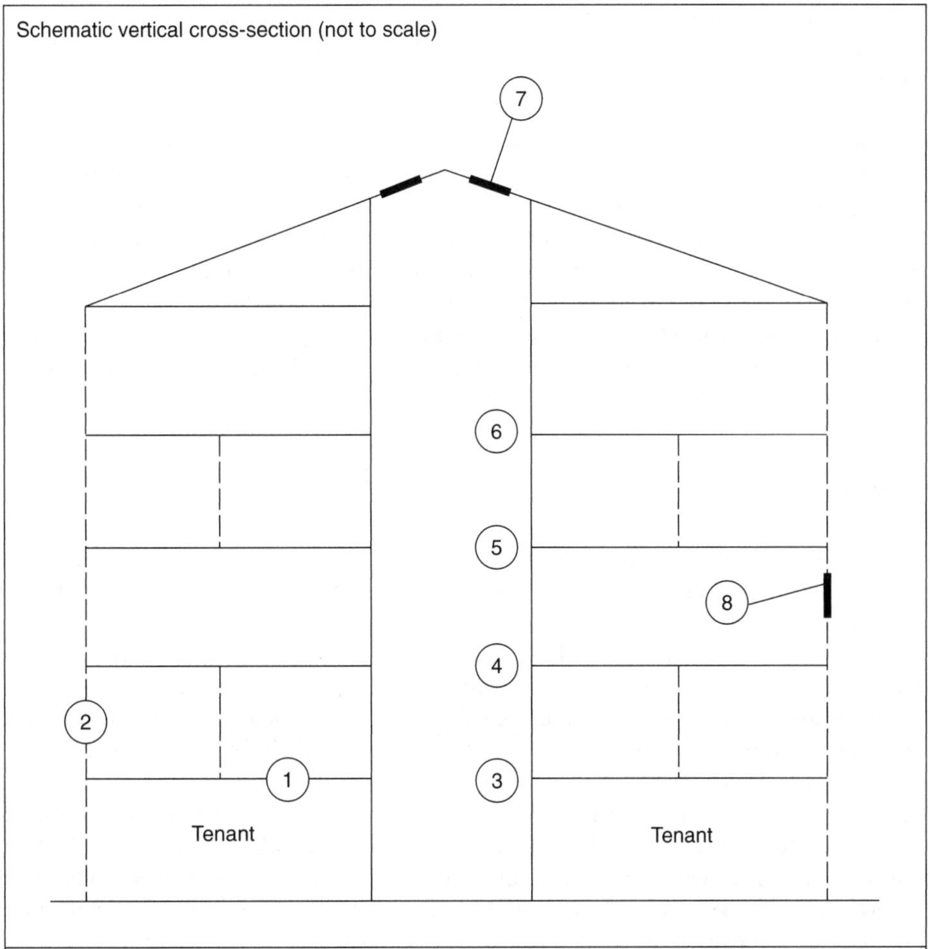

Notes
1. Fire resistance for structural members and fire separation between tenants.
2. Fire resistance for structural members.
3. Low-level overall smoke detector.
4. Low-level sprinkler.
5. High-level overall smoke detector.
6. High-level sprinkler.
7. Automatic smoke vents.
8. Automatic smoke damper in windows.

Figure 4.19 Typical fire precautions for an adaptive reuse of building with atrium

- Upgrade the fire resistance of any thin wall constructions (e.g. partitions).
- Minimize any structural discontinuity to avoid potential instability between elements during a fire.
- Identify possible hidden wall and floor voids (e.g. using impulse radar) and if necessary install fire barriers or if this is not possible insert early warning devices in these areas.
- Increase the level of compartmentation within the building. This could include installing fire barriers to separate large roof voids into smaller spaces.

Active fire precautions in historic buildings can involve the discrete use of sprinklers and detection equipment. Minimum intervention, one of the key conservation criteria, should always be the goal when considering such work.

There is a strong case to consider using sprinklers and smoke detectors in high-risk areas such as roof voids, important rooms or sections of the building containing flammable materials (Historic Scotland TAN14, 1998). Sprinklers can be either exposed or hidden, the latter type being preferable in most cases for historic buildings. However, the installation of hidden pipework systems may result in a great deal of disturbance to the fabric and structure of a building unless it has easily accessible voids in floors and ceilings. Where this is not possible, exposed pipework systems with sidewall sprinklers can be installed just under or above the cornice of rooms to reduce the visual impact of sprinkler heads.

The location of sprinkler heads and smoke detectors in a ceiling can be made discrete by installing them in ventilation grilles, ceiling rose or within ornate ceiling work. The design of sprinkler head in terms of colour and style should also be such that the surrounding finishes obscure it.

Summary

The adaptive reuse of buildings is a key strategy of sustainable construction. It provides an economic and socially advantageous way of giving otherwise disused buildings a new lease of life. It minimizes the need for wasteful and disruptive demolition, which is almost irreversible.

However, it is important not to overlook the many and varied legislative influences on the adaptive reuse of buildings. For example, under the Chronically Sick and Disabled Persons Act 1970, local authorities must provide assistance for adaptations designed to secure a disabled resident's 'greater safety, comfort or convenience'. Still, local authorities may be entitled to take their financial resources into account when deciding their priorities as regards what adaptations to complete.

An adaptive reuse scheme can offer added value in terms of the benefits it brings to the local community or environment. It can, in other words, maximize the social profitability of the existing building stock.

Obsolescence and redundancy of many building types is likely to continue owing to changes in technology and in demand. These major influences will fuel the need for adaptive reuse of sound buildings that would otherwise be demolished. Demolition is in such cases very much a last resort option.

The adaptive reuse of property is of course a relatively new phenomenon. It did not emerge as a viable alternative to the redevelopment of existing property until around the second half of the 20th century. Other interventions such as extensions, alterations and refurbishment work are frequently included in many adaptive reuse schemes nowadays. Our attention is now turned to these further forms of adaptation in the remaining chapters.

5

Lateral extensions

Overview

This chapter considers the factors that need to be addressed in the design and construction of lateral extensions to buildings. It outlines the requirements and precautions for enlarging a building's volume horizontally. Conservatories as well as conventional domestic and commercial extensions are examined in this part of the book.

Background

Rationale for enlarging buildings

Increasing the capacity of a building is one of the most visible forms of adaptation. The way this is normally done is of course by extending the building horizontally. An extension is here defined as any addition that is physically as well as functionally linked to an existing building. 'Extensions' that are separate from a building are not considered in this book as these are really new buildings and involve little if any alteration to the existing property.

Lateral extensions often entail an obvious adaptation to a building. They are very conspicuous compared to other types of adaptation such as conversions and even some vertical extensions. This is because the impact of a change of use is mainly to the interior. Thus, the design of an external addition to a building must be handled with care.

Naturally, the need to increase the spatial capacity of a building is the primary reason for extending a building horizontally. This may be to provide additional facilities not present in the building or to expand its existing accommodation. For example, a detached bungalow may lack a study or play room. The owner may not have enough free accommodation within the dwelling to, say, convert one of the bedrooms into a study or play area.

Another reason for enlarging a building laterally is that this may be the best way to re-configure the internal space. The existing layout, for example, may lack adequate circulation, fire escape or suitable access. Similarly, the current building plan might need rearranging to make it more suited to the changing work or living patterns and use densities prompted by the occupier.

Rising property values is one of the main drivers for domestic extensions. The potential cost of moving to a larger and thus usually more expensive property can often encourage house owners to extend their property rather than relocating elsewhere. Moving elsewhere is an expensive process because of the professional fees and removal costs incurred.

Moreover, moving house is generally considered to be one of life's most stressful activities. Avoiding such pressures by extending one's present dwelling can obviate the need to relocate to another home, and minimize the resultant financial and emotional costs.

All of these drivers are slightly offset by the temporary disruption of having one's home as a partial building site during the contract. Regardless how clean and tidy the contractor is, there is inevitably some inconvenience and mess involved.

There is no universally accepted definition as what in volumetric terms actually constitutes an 'extension' to a building. For the purposes of this book, though, a lateral extension will be taken as any external horizontal addition to a building exceeding, say, a $2\,m^2$ single-storey porch and not greater in size than the original building. As a rule of thumb, a lateral extension normally should not exceed 50 per cent of the existing building's volume. This of course is an entirely arbitrary limit but it applies to most domestic extension schemes. Extensions to non-residential properties, in contrast, are can be sometimes almost the same size of the original building. In the case of many hypermarkets or superstores it is not unusual to see them being extended by two or three bays, even within a few years of the original building's construction (see section on Connection required to framed buildings).

Indeed, it may be argued that a new building connected to an existing property that matches or exceeds the latter in size is strictly speaking not a true 'extension'. Moreover, some have argued that because lateral extensions predominantly consist of new-build work, they are not really adaptations (Friedman and Oppenheimer, 1997). Lateral extensions, however, regardless of their size, involve work that is not found in new build: connections between old and new constructions, forming openings in existing elements to provide access between the two units, and directly making good if not improving the use of the existing building.

Even if a new building is abutting an existing one, of course, this does not necessarily mean that it is a real extension. It may merely be a new property with another owner forming the continuation of a terrace or block. Thus, the basic criteria that may be used to determine whether or not such a development is an extension are as follows:

- The new attached building is ideally if not usually smaller in size than the original.
- Access between the two buildings is normally essential (either for fire escape purposes or general access to each part of the whole building). This is usually achieved by inserting openings in the abutting wall of the existing building. Alternatively, a link corridor at ground level or at an upper floor may be formed to connect them via an existing or new opening in the building being extended (see below).
- The ownership or occupancy of the two buildings is the same even if the uses are not.
- The design of the new adjoining building complements if not matches the original.
- There is some degree of physical attachment of the extension to the main building. The degree of linkage, however, depends on the size, nature and position of the extension. Certain requirements such as weather-tightness and allowance for differential settlement, however, are critical.

Decision to extend

The decision to extend a building will be influenced by requirements that help to make it more effective (i.e., meet the short–medium-term needs of the occupier) and sustainable (meet the long-term needs of the occupier and society as a whole). It can be achieved by considering the following issues:

- Advantages of extending over moving:
  - Savings in cost and time.
  - Less stress.
  - Familiarity with existing building may give preference to extending it.

- The disadvantages of extending:
  - Temporary accumulation of builder's plant and waste.
  - Temporary disruption to the use of the building.
  - Spatial and legal restraints may prevent full compliance with requirements.
  - Possible conflict with neighbours over boundaries or overshadowing of adjacent properties.
- The philosophical arguments:
  - Impact on the habitation or business: extension needed to increase production capacity of the company.
  - Impact on the value of the property.
  - Impact on the residents or workforce.
  - Impact on the building's architectural character: change in appearance for the better?
  - Impact on the structure: can additional loadings be accommodated by the existing walls and foundation?
  - Impact on the surrounding landscape and streetscape.

Practicalities of extending a building

In particular the following points should be considered:

- Economic considerations: Budget; adaptation costs; maintenance costs; impact on capital/rental value.
- Timing: Starting date; phasing of operations; duration of contract.
- Suitability of the existing construction and site: Soil conditions; access; ease of connecting to existing; required flexibility.
- Disturbance and interruption to habitation or processes: Extent of noise and dust generated; storage and accommodation requirements; waste generation and disposal.
- Compatibility of the design: Impact on aesthetics of existing building; acceptability of proposals to the planning authority.
- Compatibility of the construction: Consideration of the availability of the materials and components; form of construction of extension identical to original building.
- Code compliance: Extension work should continue to be required to meet the same standards as for new build. Where this is not possible relaxation of the regulations would have to be sought. If granted, however, any such relaxation will come with conditions.

Provisions for future extensions

An increasingly important requirement relating to new-build work is that at the design stage there should be provision for future extensions. This helps to make buildings more sustainable and can be achieved by considering:

- The location and orientation of the existing property. Future extensions will be facilitated if there is sufficient space around all or part of the original building.
- The flat-roof construction and detailing to facilitate another storey.
- Use of space around the building.
- Selection of materials and components.
- Foundation and basement design, substructure should be over-designed to take additional loads without the need for strengthening measures such as underpinning.
- Extent, condition, position and capacity of services.
- Means of access and egress in respect of the site and the building.
- Possible decanting of occupants at interface between old and new.

Forms of lateral extensions

Preamble

The type of extension required is dictated by functional and spatial factors as well as technical, financial and legal considerations. First and foremost are the needs and available budget of the client. A change in the size or demands of a family or business may provide the trigger for residential or commercial extensions respectively. In any event, the range of lateral extension is usually limited to the forms outlined below.

Front extensions

Extensions at the front of a building are normally restricted to small additions such as porches and vestibules. This is because front extensions are the most conspicuous and thus can radically alter if not destroy the character of the building. Planning authorities are therefore not keen to sanction large front extensions. Moreover, because the site boundary is usually very near the front of a building, an extension at this position can easily encroach on the plot's limits. Properties where the front elevation abuts directly onto the pavement are a prime example of this constraint. Still, in some cases, a front extension may form part of a larger side extension (see below).

End extensions

An end extension is a form of side extension. It can occur if a building is being elongated or where an extension is being installed at the gable of an end-terraced block. In such circumstances the available space for expansion is likely to be limited. However, extensions to single-storey-framed buildings such as large supermarkets or warehouses are often two or more bays in length because of the spare space. This may of course encroach on the existing car park, the layout of which would have to be modified to accommodate the extension.

Figures 5.1 and 5.2 show two possible types of end extension for a domestic property. There are of course other possible styles such as those indicated for rear extensions, but there is likely to be greater planning restrictions with side extensions because they are more conspicuous.

Side extensions

In residential property, an attached garage often forms the main type of side extension. In other instances, the side extension may be required to provide an enlarged or additional room. More usually, however, a two-storey extension comprising, say, a garage on the ground floor and a bedroom on the upper floor forms the main type of addition incorporating these functions. Figure 5.3 illustrates a typical example of a side extension.

A major design restraint is that many local planning authorities are keen to avoid the 'terrace effect' with side extensions (see Figure 5.4). If the space between individual dwellings or blocks of dwellings becomes overdeveloped, this can give the impression of a terrace. Side extensions in closely situated dwellings can destroy the individuality of the streetscape. This may also create access problems to the rear of the premises.

Schematic isometric sketch (not to scale)

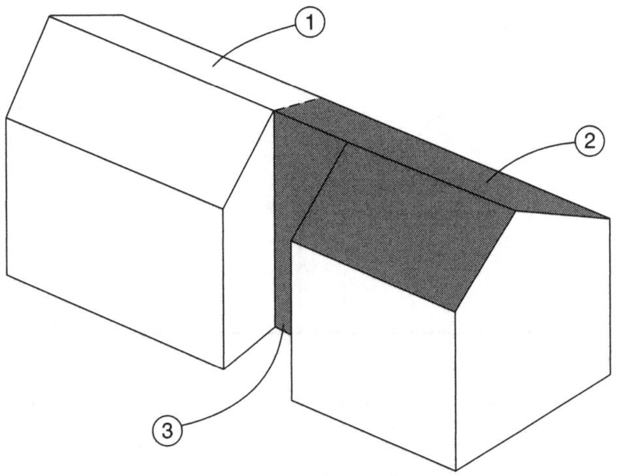

Notes
1. Existing gable-end building – low-rise domestic or high-rise commercial property.
2. New extension with pitched roof continuing existing roof profile.
3. Glazed- or panelled-diaphragm wall.

Figure 5.1 End extension with recess at junction at front elevation

Schematic isometric sketch (not to scale)

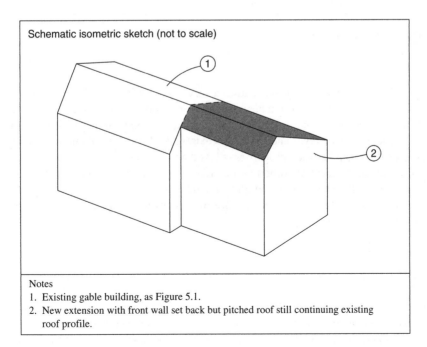

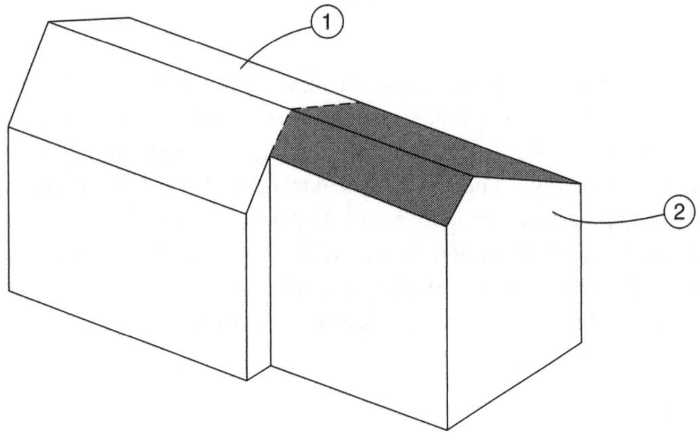

Notes
1. Existing gable building, as Figure 5.1.
2. New extension with front wall set back but pitched roof still continuing existing roof profile.

Figure 5.2 End extension set back at front elevation

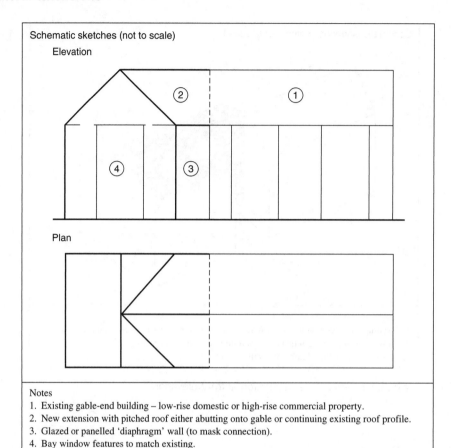

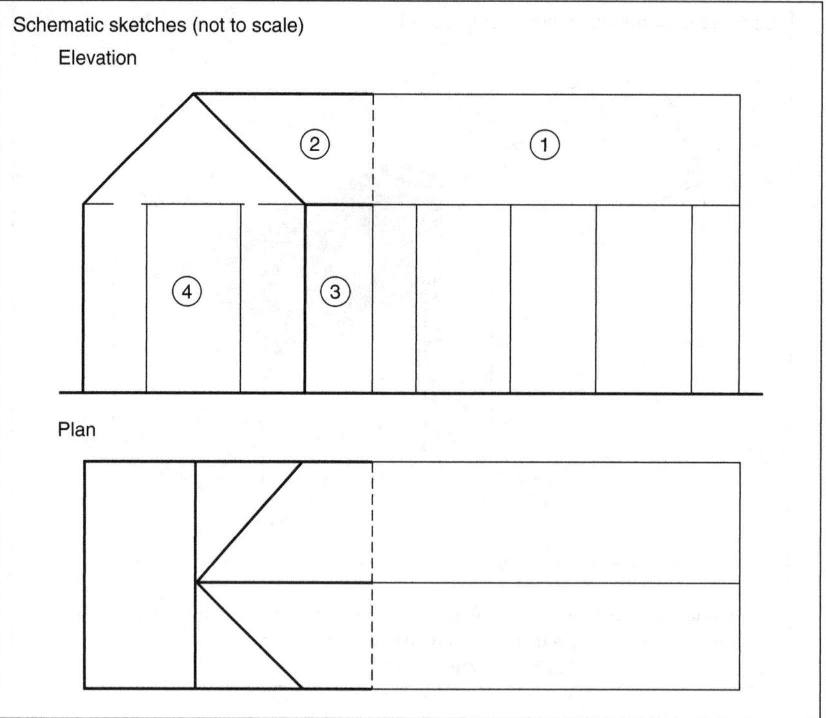

Notes
1. Existing gable-end building – low-rise domestic or high-rise commercial property.
2. New extension with pitched roof either abutting onto gable or continuing existing roof profile.
3. Glazed or panelled 'diaphragm' wall (to mask connection).
4. Bay window features to match existing.

Figure 5.3 Typical side extension

Wherever possible such an extension that causes a very close gap between it and the adjoining property should be avoided. In some cases it has not been unknown for a side or end extension to be installed within less than 100 mm of the next-door neighbour's house. Such a development may cause friction between the occupiers if not handled sensitively. In addition, it is extremely difficult to obtain access for maintenance and other purposes if the gap between properties is less than 600 mm.

Side extensions to non-residential properties are often on a larger scale than this type of enlargement in housing. The impact that they have on the building and its surroundings is therefore greater. For example, detached Victorian two- or three-storey houses are often converted into commercial use, which often necessitates a lateral extension.

Rear extensions

These extensions are very popular in mid-terraced residential dwellings because of the physical site constraints. Moreover, the planning constraints are usually less stringent at this part of the property. However, in mid-terraced properties neither the foundation nor the sidewalls should encroach on the adjoining neighbours' land. A reinforced raft foundation with downstand edge beam taken to the boundary can allow

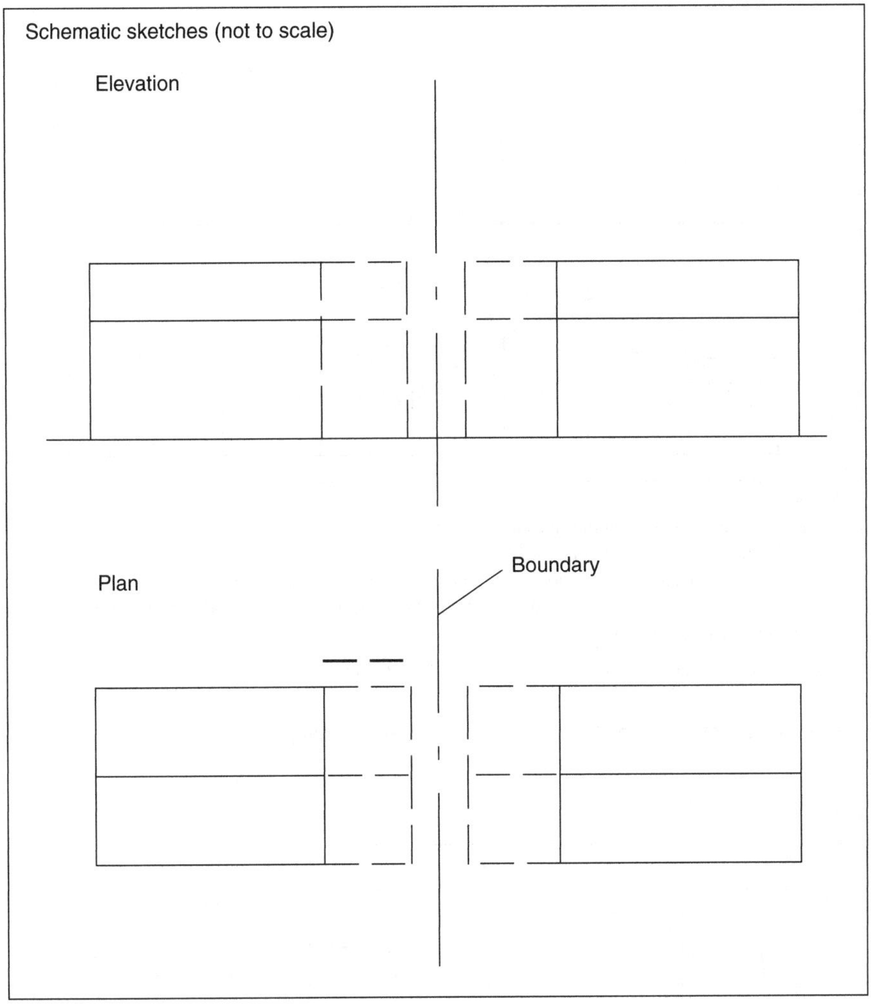

Figure 5.4 Example of the terrace effect with lateral extensions

the sidewall to be built to this position. Alternatively, the sidewalls can be set back at least 150 mm so that the scarcement of the extension's foundation does not cross over the boundary line (Williams, 1993).

Figures 5.5–5.7 show some examples of rear extensions. Some indication of the roof design of the various styles shown is given.

A conservatory is a very popular form of small extension usually restricted to the back or sometimes to the side of a dwelling (see below). Stephen (2002) gives examples of acceptable and unacceptable rear extension designs.

Corner extensions

As indicated above, front extensions sometimes form part of a larger side extension. This solution allows the front of the building to have a minor increase in size along with a larger addition at the side (see Figure 5.8).

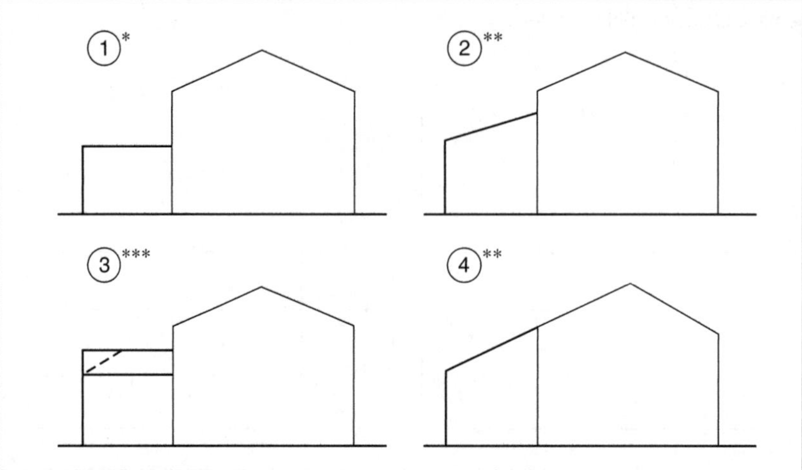

Notes

1. Single-storey extension with flat roof.
2. Single-storey extension with lean-to roof.
3. Single-storey extension with duo-pitched roof with gable or hipped end.
4. Single- or two-storey extension with lean-to roof (containing possible skylight or dormer) continued from main roof.

Indicative design ratings: * = Poor; ** = Fair; *** = Good.

Figure 5.5 Some schematic sections of single-storey rear extensions

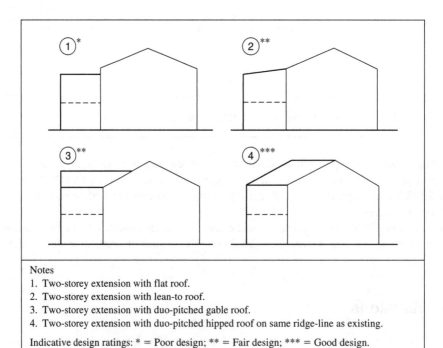

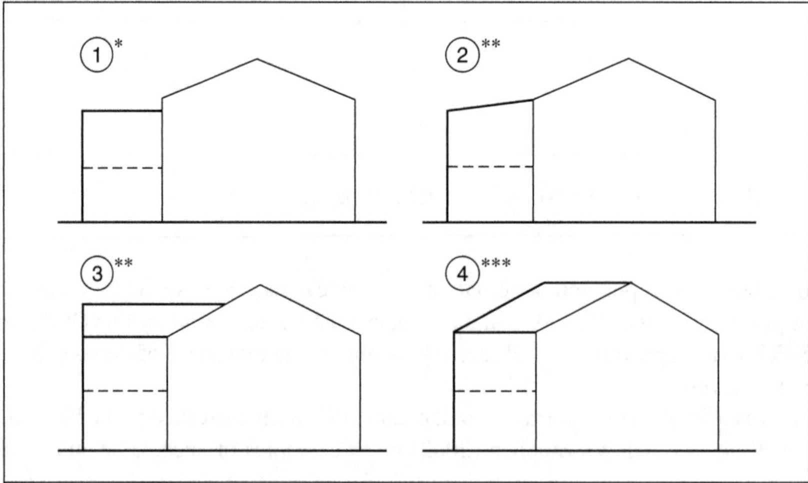

Notes

1. Two-storey extension with flat roof.
2. Two-storey extension with lean-to roof.
3. Two-storey extension with duo-pitched gable roof.
4. Two-storey extension with duo-pitched hipped roof on same ridge-line as existing.

Indicative design ratings: * = Poor design; ** = Fair design; *** = Good design.

Figure 5.6 Some schematic sections of two-storey rear extensions

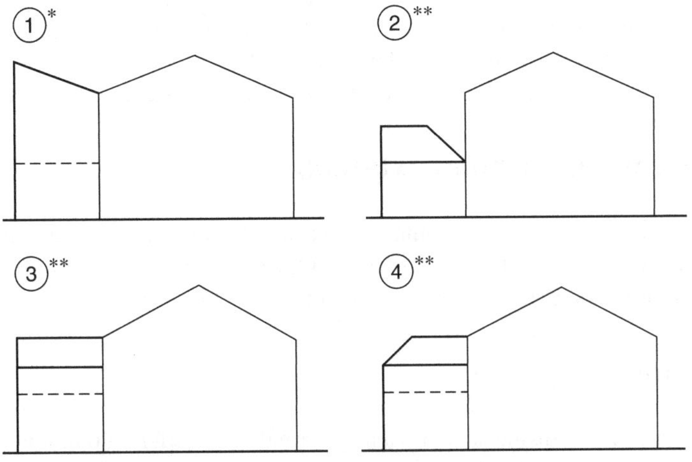

Notes
1. Two-storey extension with lean-to forming valley gutter with rear slope of main roof.
2. Duo-pitched roof with hipped end at abutment on single-storey extension.
3. Single- or two-storey extension with barrel vault or other curved roof.
4. Single- or two-storey extension with half mansard roof (i.e., roof with pitched sides and flat top).

Indicative design ratings: * = Poor; ** = Fair; *** = Good.

Figure 5.7 Some other rear extension designs

Schematic isometric sketch (not to scale)

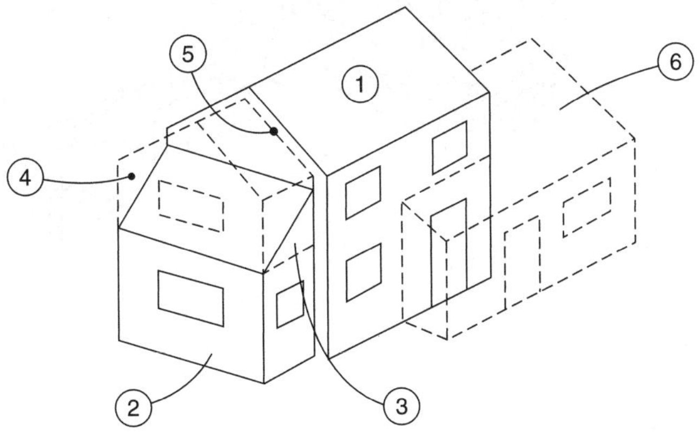

Notes
1. Original detached dwelling.
2. Existing small side extension.
3. Walling forming lean-to removed to facilitate construction of new storey.
4. New first floor level side extension.
5. Roofline of extension set back from main building.
6. New front and side extension with mono- or duo-pitched roof.

Figure 5.8 Combined side and up or front, and side extensions

Corner extensions are sometimes referred to as 'side returns'. The main distinguishing characteristic of such extensions of course is that they involve a difficult corner junction. The design and detailing of this point has to be handled with care to avoid problems such as lack of weather-tightness owing to the awkward abutment especially with pitched roofs (see section Impact of extension on existing construction).

Combined lateral and vertical extensions

In rare cases an extension may involve an enlargement of the property both horizontally and vertically. Figure 5.8 shows two types of combined extensions. Other examples of combined lateral and vertical residential extensions are illustrated in Chapter 6 (see Figures 6.17 and 6.18).

Linked extensions

Occasionally in commercial extension schemes a large or radically different building from the main property is sometimes proposed. In such cases, it may be more appropriate to separate the two properties with a link corridor or block (see Figure 5.9). The link section may only be at ground level but in larger

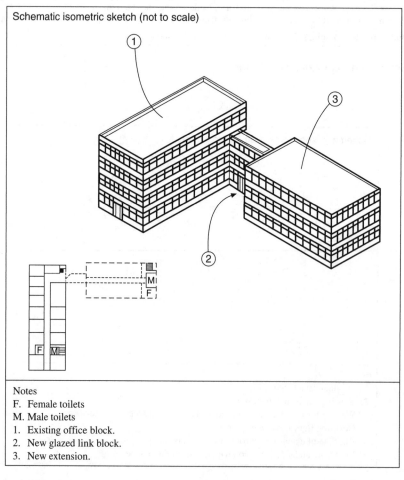

Schematic isometric sketch (not to scale)

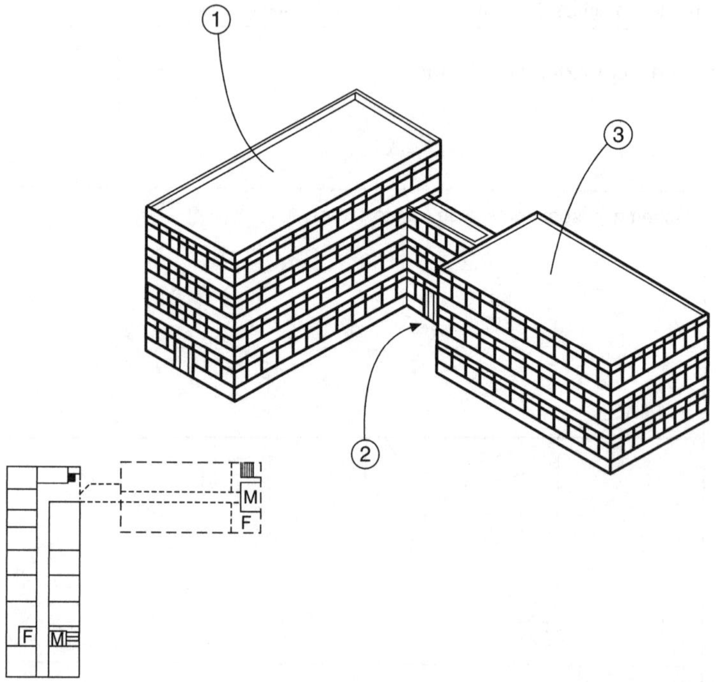

Notes
F. Female toilets
M. Male toilets
1. Existing office block.
2. New glazed link block.
3. New extension.

Figure 5.9 Linked extensions

commercial buildings it may in some cases be at first or a higher floor level if not the same height as the two properties.

Link corridors allow a degree of flexibility in both the use and design of the extension. For example, although an extension to the main building using a link corridor is usually for the same user, the connection between them can easily be severed so that the two parts can be used separately. This can be done either by closing off one end of the corridor or removing it in its entirety and making the good link openings to both the buildings.

Unusual or non-standard extensions

Occasionally a designer may want to create a more unusual style for an extension. This could relate to the location or plan shape of the extension as well as its height and overall profile. For example, Figure 5.10

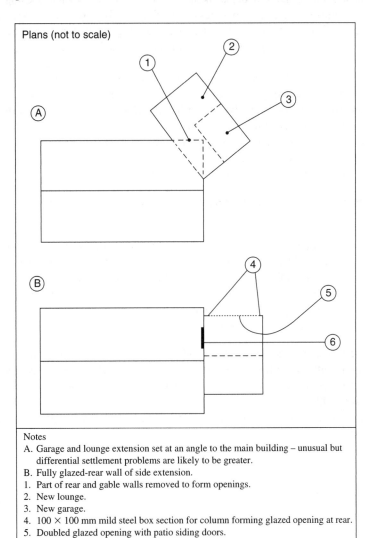

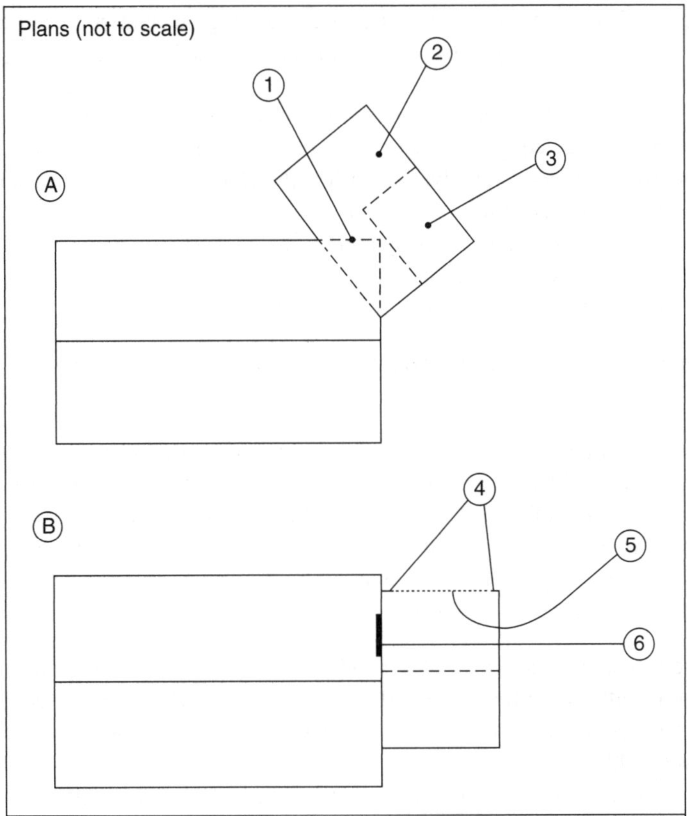

Notes
A. Garage and lounge extension set at an angle to the main building – unusual but differential settlement problems are likely to be greater.
B. Fully glazed-rear wall of side extension.
1. Part of rear and gable walls removed to form openings.
2. New lounge.
3. New garage.
4. 100 × 100 mm mild steel box section for column forming glazed opening at rear.
5. Doubled glazed opening with patio siding doors.
6. Existing openings in gable enlarged/repositioned.

Figure 5.10 Non-standard domestic extension plans

shows a lateral extension that is set at a 45° angle to the main building. Architecturally, this results in an unusual or innovative design. However, it brings with it problems of achieving adequate weather-tightness. It also requires proper allowance for differential settlement and at the skewed junction.

The other example shows a side extension with the rear wall constructed using a mild steel box section framework. This allows for maximum use of glazing for daylighting.

Statutory requirements

Town planning

Consulting the local authority's planning department is a useful first step when proposing to extend a building. Many planning authorities have strict policies controlling the size, shape and style of domestic extensions (Williams, 1993). Conservation Areas and Listed Buildings in particular are likely to have more stringent controls. In any event, the key aims of such policies are usually:

- To prevent the extension from blocking out the daylight to neighbouring properties.
- To protect other aspects of amenity of adjacent occupiers (e.g. access).
- To ensure that the extension does not adversely affect the character of the original dwelling.
- To retain the essential character of the streetscape.
- To prevent the creation of dangerous highway conditions.
- To prevent overdevelopment of urban plots.
- To safeguard the provision of a reasonable private garden space.

Generally, for 'permitted development' the extension must fall within the following parameters (according to Haverstock, 1998):

- Terraced houses: Not be more than 10 per cent or more than 50 m^3 of the volume of the original dwelling (whichever is the greater) with a maximum of 115 m^3. The maximum height of such an extension with a pitched roof is 4 m (if the roof is flat, 3 m) when within 2 m of a boundary.
- Other houses: Not be more than 15 per cent of the volume of the original dwelling, or more than 70 m^3 or 10 per cent (whichever is the greater) with a maximum of 115 m^3. Again, the maximum height of such an extension with a pitched roof is 4 m (if the roof is flat, 3 m) when within 2 m of a boundary.

As with a change of use, a 'Certificate of Lawful Proposed Use of Development' can be applied for from the local planning authority if there is any doubt about the suitability of a proposed extension. If the property is within a Conservation Area, the permitted development rules may be suspended by an article 4 Direction (Haverstock, 1998).

Appendix G shows some typical planning policy guidance that a local authority may publish on extensions and similar adaptations.

Building regulations

Generally

Extensions to an existing building are classed as 'new work'. This means that all of the main parts of the building regulations apply. Moreover, a proposed detached building of more than 10 m^3 will be treated as an enlargement of the dwelling if it is within 5 m of the original house (Haverstock, 1998).

However, certain types of extensions are granted complete exemption from building control. For example, the following ground level extensions of less than 30 m^2 floor area normally do not require building control approval:

- greenhouses;
- conservatories (not greater than 8 m^2 – see below);
- porches (again, not greater than 8 m^2);
- covered yards or ways or a carport open at least on two sides.

The main proviso with these 'exempted' extensions is that they must not contain a solid fuel (including coal, wood or peat), gas or liquid fuel, oil, paraffin heating appliance.

The general statutory requirements for small lateral extensions such as conservatories are summarized below. Table 3.2 provides a summary of the key building regulation requirements that can also be applied to lateral and vertical extensions. See Chapter 11 for more information on the statutory controls relating to larger extensions and other adaptation works.

Impact of extension on existing construction

An extension may require the modification of the original building's construction or services. For example, the section of existing external wall at the abutment with a domestic lateral extension more often than not contains window or door openings. In such cases, the existing room/s at this position will become an 'internal room' (i.e., room without windows), unless of course it is being enlarged by the extension. In any event, the adapted room area and openings will have to meet the requirements of the building regulations as regards ventilation and daylight. Mechanical extract ventilation will be required if the space in question is or becomes a kitchen or toilet. This involves having to extend the existing ventilation ducting through the extension to the outside air. A new duct would need to be formed in the external wall of the original building to accommodate the extended route of the ventilation duct.

Figure 5.14 illustrates some of the points to watch for when designing a lateral extension of domestic scale. Similar problems might occur with commercial and other large-scale lateral extensions. The major difference with commercial extensions is that a superstructure of framed construction with lightweight cladding is likely to be used.

Fire safety

The general principles of fire safety were dealt with in Chapter 4. Extensions, however, entail some specific requirements relating to fire safety. These include amongst other things the size and position of new openings in relation to adjoining properties, and adjustments to existing fire escape routes.

Ramp access

Single-storey extensions are unlikely to require any staircases. However, to facilitate disabled access or connect two ground floors at different levels (e.g., on a sloping site), ramps should be used wherever possible.

The new building regulation requirements for pedestrian ramps are as follows:

- ramps steeper than 1 in 12 are no longer permitted;
- 1 in 12 permitted for ramps up to 5 m long;
- 1 in 15 permitted for ramps up to 10 m long;
- 1 in 20 or shallower permitted without restriction;
- raised kerb required on exposed edge (except for dwellings).

Design and other influences on extensions

Preamble

Granted, on the whole the actual construction process of lateral extensions, regardless of their size, is almost the same as that for new build. There are, however, several major design and construction considerations that occur with both lateral and vertical extensions but not in new-build projects. These and other issues are dealt with as follows.

Client's requirements

Addressing the requirements of the client is of course the most important initial factor in any construction scheme for the professional involved. Establishing the precise reason for an extension therefore is critical. This is necessary to evaluate all the options. It may be that, contrary to the client's thinking, an extension is not necessary if the existing internal space can be altered to accommodate the required space change.

Building clients normally have a good idea what they want but are unclear about the feasibility or means of achieving their goals. This is where the building professional comes in to advice clients of the options open to them.

Implications

Extending a building has a number of consequences. The value of the property may increase but not necessarily by the amount of money being spent on the extension. Some lay people make the mistake of equating cost and value. The former is determined by production factors such as labour rates and prices of materials. The latter, in the short run, is determined by the market (i.e., by demand). Additionally, a building that is enlarged is likely to have a higher council or other property tax rate levied against it.

The work to extend a building is likely to be subject to value added tax (VAT). This will inevitably increase the cost of the project for the client.

Spatial constraints

Position

Apart from budgetary restrictions, the availability of space imposes the severest constraint on any lateral extension scheme. This is especially true in the case of mid-terraced properties. Clearly in such cases only a front or rear extension would be possible. As already shown, front extensions are often limited in size because boundary and planning restrictions are likely to be greater at the front than elsewhere around the property.

The distance between the building being extended and the boundary will clearly limit the plan size and shape of the extension. The impact of the extension on access, daylighting and privacy in relation to the surrounding or adjoining properties must be considered.

Internal factors

The internal layout, however, may influence the precise location of the extension, to accommodate linking of key zones within both structures. For example, if the kitchen is being enlarged, this will determine to a greater or lesser extent the position of the extension. Such an extension will be located along one or both of the two external walls forming part of the kitchen.

The position of circulation areas is also influential on the design of an extension. For example, new spaces in an extension will have to be accessed from the existing circulation space.

Extensions to detached buildings

Some of the technical issues that often accompany simple lateral extensions to isolated detached dwellings are illustrated in Figures 5.14 and 5.18. It has to be borne in mind, however, that every construction project is unique and may bring with it its own peculiar problems and difficulties.

Extensions to detached blocks in rows

Some of the design issues that can affect simple lateral extensions to detached dwellings in rows were dealt with in Figure 5.4. The 'terraced effect' is a critical consequence of end extensions to detached blocks of houses in rows. Because such an effect destroys the original streetscape, local authorities are anxious to avoid extensions in such circumstances.

Fenestration

Ideally the fenestration of the extension should harmonize with the existing arrangement in terms of size and position. The extent and arrangement of existing openings in the elevation onto which the extension is being placed can influence the position and design of an extension. The aim is to avoid blocking of light or ventilation to existing rooms. The existing openings if possible should be used to form link to new-build section.

In some cases, however, window and door openings may fall within the line of the intended position of the extension. There are essentially two ways of dealing with this problem:

(i) Block up partially or fully the openings that cross the junction between the roof/wall of the extension and the original building (see Figure 5.11).
(ii) Modify the extension where its roof/wall profile crosses any openings that are being retained (see Figure 5.12).

Aesthetics

Lateral extensions can be contrasting or matching in their appearance. It is important, however, to avoid unsightly or overbearing additions to a building. The extension, in other words, should be designed to harmonize with the main building in terms of size, shape, colour and texture.

Massing

The bulking effect that the extension has on adjoining or surrounding properties is called 'massing'. This is particularly important in the case of a lateral extension that is over half the volume of the original building. An excessively large extension can dwarf the adjoining properties, particularly at roof level. The term 'modulating the mass' describes a solution where the design of a new lateral or roof extension avoids imposing itself on the main building.

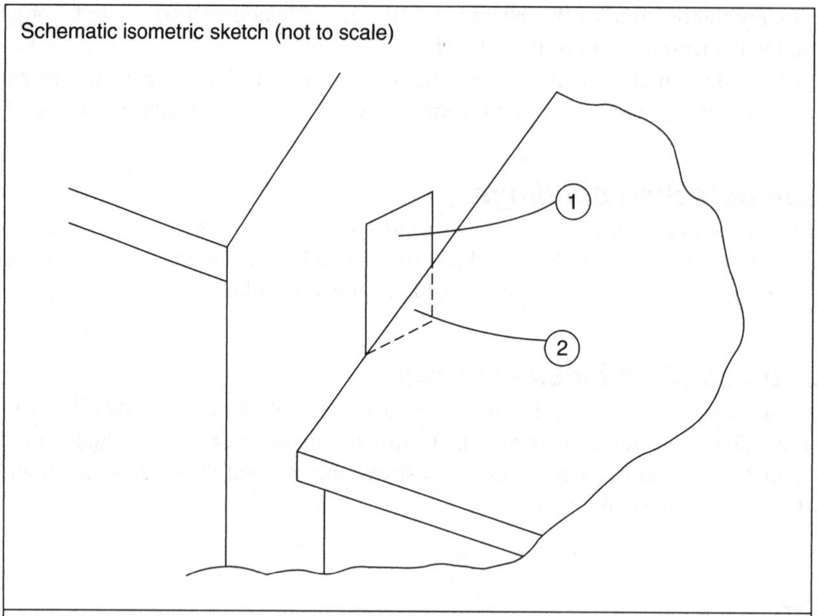

Notes
1. Existing opening blocked up or partially retained with sloping sill above roof of extension.
2. Roof of extension partially covers former opening.

Figure 5.11 Solution 'A' for fenestration at extension junction

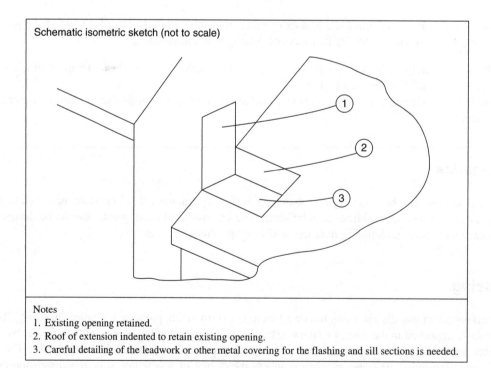

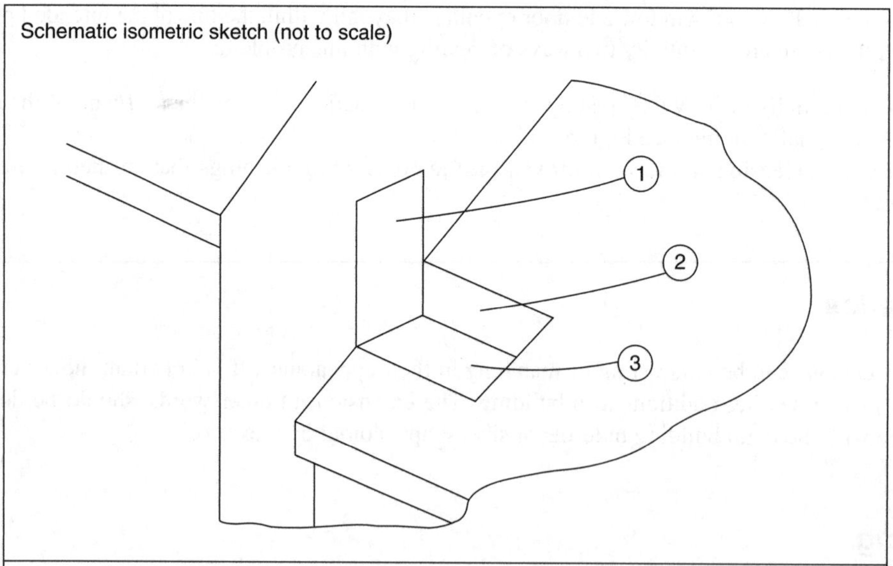

Notes
1. Existing opening retained.
2. Roof of extension indented to retain existing opening.
3. Careful detailing of the leadwork or other metal covering for the flashing and sill sections is needed.

Figure 5.12 Solution 'B' for fenestration at extension junction

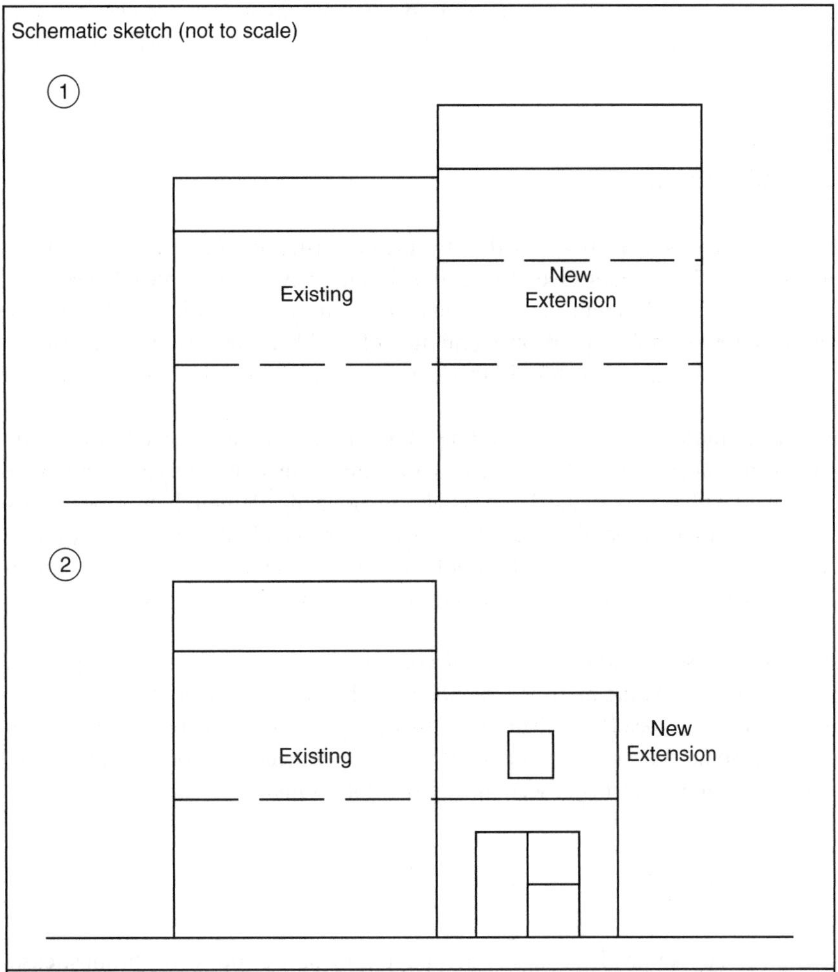

Notes
1. Existing building over dominated by new extension.
2. Extension takes secondary position – an effect known as 'modulating the mass'.

Figure 5.13 Examples of massing with lateral extensions

Modulating the mass is especially important where the extension is to an old property. Good and bad examples of massing are shown in Figure 5.13.

Roof styles

A flat roof is the cheapest way of finishing the top of a building. Although this type of roof gives a building an incomplete look, using parapets could mitigate it. In rainwater disposal terms it is not as efficient as a pitched roof. It may be justified on aesthetic grounds if the main building has a flat roof as well.

A flat-roofed extension, however, may be necessary to avoid blocking up light through existing windows or avoid cutting into existing decorative features. Moreover, complex plan shapes may be difficult to cover with a pitched roof.

A pitched roof, though, is the preferred option, for aesthetic as well as weathering performance reasons. The design of an extension's pitched roof will also be influenced, however, by the roof style of the main building.

Form of construction

Generally, the form of the original building dictates the construction of an extension. Ideally, the former should be compatible with the latter (see also section Joining new work to old). However, some modern extensions may comprise a different constructional form from the existing building. Even small-size domestic extensions, for example, can involve a mixture of load-bearing masonry and framed construction (see Figure 5.10). The advantage of this is that a more unusual or innovative modern design can be achieved.

As a general rule, the extension to a building of steel- or concrete-framed construction should be preferably of the same morphology. If not, there is a greater chance that difficulty in the connection and differential settlement will be experienced. It has to be acknowledged, though, that these days brick/block-clad timber-frame constructions are being used for extensions to buildings of solid masonry construction. Time will tell if there are any incompatibility problems with such an arrangement. It is highly unlikely, of course, that solid load-bearing masonry construction would be used as the extension to a timber-frame building.

There may, however, be potential problems of constructional and structural compatibility between a main building of solid load-bearing masonry extended by a building of modern timber-framed construction. The latter has a much lighter dead load than the former per unit area. This reduces the risk of differential settlement occurring. Some allowance for minor differential settlement, however, would need to be made (see points under connection requirements in section Construction forms).

Cost

A lateral extension of course involves substantial amount of new-build work. In most cases therefore it is reasonably easy to ascertain the cost of a domestic extension. As a rule of thumb, at prices in the year 2006, a single-storey residential lateral extension of traditional brick- and block-cavity construction not exceeding 30 m^2 would normally cost between £25 000 and £35 000, excluding VAT and the cost of furnishings. The cost of a two-storey extension of the same plan size is likely to come within twice that range.

The cost of an extension to a non-residential building is more difficult to identify immediately owing to the wide range of sizes and types available. Chapter 1 gives some indication of typical costs and timings of a range of adaptation schemes.

Juxtaposition

The shape and form of an extension naturally impinges architecturally as well as physically on the building it is attached to. Thus, the design of the extension must have regard to design of the existing property. Is the intention to match or contrast with the existing? As indicated above, the issue of massing may also be critical. Moreover, the impact that the extension will have on access to and around the property needs to be considered. Some of these considerations may also apply to a new-build scheme but usually they are less stringent.

There are essentially two ways of dealing with the juxtaposition of the extension. The selection will depend on the size of the extension and on the context of the existing building. One option is to make the addition conspicuous by some forming a distinct break in the construction at either roof or wall junctions or at both. Stepping the extension back or forward from the existing facade usually does this and makes the connection obvious.

The other way is to make the extension discrete by attempting to hide or mask the connection. 'Toothing' the brickwork for the extension into the existing wall may conceal the junction but this depends on the match between the old and new constructions.

Interface

The joint is often the weakest link in any structure or system, and this is no exception with lateral extensions. The junction between the old and the new involves not only a physical connection but also the formation of openings to link the access between the two buildings. This interface must be achieved without affecting the constructional and structural integrities of both the existing building and the extension. Proper detailing of the junction of the two buildings at the roof is important for the weather-tightness and stability of the adapted property. The incorporation of appropriate cavity trays and damp proof courses (dpcs) is critical in this regard (see below).

Occupancy

There is normally no major reason why the occupancy of an existing building being extended cannot continue to some degree, if not fully, during the adaptation works. However, achieving an uninterrupted use of the property that is in the process of being extended is often fraught with difficulties. The work can create considerable disturbance for the occupiers. Allowance, for example, must be made for any noise and dust nuisance control during these works. In addition, preventing the security of the building from being compromised during the adaptation works is an important objective in the project.

Accordingly, careful phasing of this work is essential to avoid disruption or inconvenience. For example, breaking through from new to old should be deferred as late as possible (see section on this later in this chapter).

Upgrading works

Naturally, upgrading and altering the existing building being extended often accompanies extension work. It provides the ideal opportunity to undertake additional measures such as improvements to the fabric and services. In such cases, economies of scale as well as access opportunities are available. New fire escapes or access routes, for example, might have to be provided in the existing building. This may be because the extension has affected the existing provision or the building does not meet with current requirements. Code compliance, of course, need not apply to the whole original building. It will only be mandatory for the new work and where it impinges on the existing construction.

Single-storey lateral extensions are usually much easier and hence cheaper to construct than vertical extensions. This is because the latter always require staircases and forming additional openings in the roof or separating floor to provide access between the old and new constructions. In addition, sometimes strengthening works to the existing structure are required particularly if there is a change of use involving increased floor loadings.

Two-storey extensions, however, are even more difficult because their structural and aesthetic implications are often greater. More substantial foundations are needed in such cases. The design of the roof will also require more care to harmonize with the existing building.

Structural and non-structural upgrading works respectively are dealt with in Chapters 7 and 9.

Sustainability factors

Ideally the extension should provide an opportunity to enhance a property's environmental performance. The client may have a policy for the extension that requires the maximum use of indigenous materials, low energy use services and high thermal efficiency of the fabric. See Chapter 10 for more details of sustainability measures.

Harland (1998) indicates the main requirements in this respect for residential adaptation schemes are as follows:

- Design for longevity, reusability and recyclability (e.g., by using recyclable metals, roof tiles and wood and lime mortar).
- Obviate the need for air conditioning if possible (e.g., by using passive cooling/ heating) (see Chapter 9).
- Use chlorofluorocarbons or hydro-chlorfluorocarbons (CFC/HCFC)-free materials and processes with low-volatile organic compounds (low-VOC) levels. Non-plastic materials as listed in Table 1.1 should also be considered.
- Utilize as many local materials as possible but avoid recycled materials with high transport energy costs. This not only reduces costs but also helps ensure that the extension is compatible with the existing construction.
- Make the fabric highly insulated to reduce heating costs. Chapter 8 describes the various methods of how this can be done.
- Retain any heritage features of the main building. Do not design an extension that will compromise the character of the original property, for example, a flat-roofed modern-style extension that covers distinctive features on the elevation of an historic stone-built property.

Common issues with domestic extensions

Some of the typical issues to consider with the design and construction of extensions to dwellings are listed below. Although primarily related to rear extensions, many of the points are also applicable to front and side extensions (Figure 5.14).

- Existing drainage pipelines that pass under proposed extensions have to be encased in concrete unless they are rerouted away from any construction. For example, linking any drainage lines from toilet or kitchen extensions to existing soil vent pipe may prove difficult to achieve without either modifying or rerouting the pipe layout.
- If an existing manhole comes within the proposed extension, check whether building control will allow this (Haverstock, 1998). At the very least the manhole cover should be airtight, with bolted-down and double-seal cover.
- If the floor of the extension is a solid concrete slab, ducted ventilation is required to any adjacent sub-floor voids (per BS 8102, 1990).
- If the extension involves building on a party wall a Party Wall Agreement is probably needed (Haverstock, 1998).

Schematic plan (not to scale)

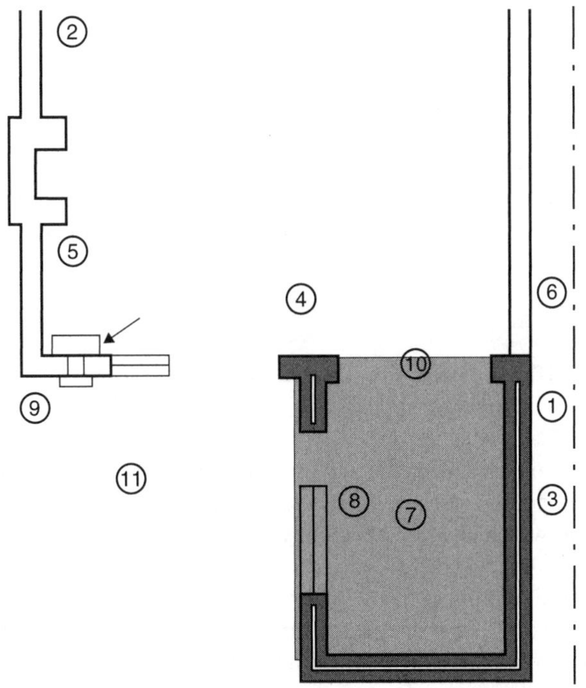

Notes

1. Distance from boundary and height of extension?
2. Does the extension involve work to a party wall?
3. Will most of the neighbours light be taken away?
4. Is the inner room ventilated?
5. Is there a place for a balanced flue?
6. How good is the condition of the existing masonry?
7. Check U-values of existing walls and windows.
8. Show on drawings: drains to be removed; new manholes inside the extension to be made airtight; position, height and type of any trees and other large vegetation; ducted ventilation to underfloor spaces; details of beams, piers and foundations; and dpc levels.
9. Are bearing pads provided for lintels?
10. Any existing mechanical extract ducts at the abutment will have to be extended and exited through the extension.
11. How do new foundations relate to existing walls and (if necessary) underpinning?

Figure 5.14 Typical issues related to rear extensions (Based on Haverstock, 1998)

- The new extension should not deprive the adjoining property for sufficient light for ordinary purposes. Check the local planning authority's guidance on daylighting, privacy and sunlight.
- If an existing room adjacent to an external wall becomes an 'inner room' by virtue of the extension it has to be ventilated. According to Haverstock (1998) the room areas and openings will have to meet the requirements of Building Regulations F1 either by having a permanent opening of 1/20 of the combined floor areas between them to be counted as one room, or by meeting the requirements for closable openings and ventilation openings.

- There should be a suitable wall area available to take any balanced flue appliance (Haverstock, 1998).
- Check to ensure that the existing brickwork particularly at the abutment position is in good condition. DPC levels must be maintained and shown. External walls to all habitable rooms must have a U-value of 0.3 W/m^{2}K.
- Double beams supporting the first-floor joists over openings need bearing pads cast in concrete and diaphragming as in BS 449 Part 1969 Clause 18 (Haverstock, 1998).
- Check building control opinion on requirements for foundations and their connections. According to Haverstock (1998) some authorities consider foundations should pick up existing footing levels even if they step down away from the buildings to normal levels for the area (see BRE GBG 53). Other building control departments might insist on the use of steel dowel bars to link the old and new foundations.

New exterior additions to historic buildings

Conservation concerns

The design of any lateral extension must be handled with care to avoid any negative impact on the existing building. This is even more important in the case of older buildings that are of architectural or historic interest. Any exterior addition will inevitably have conservation concerns, which must be addressed to achieve a successful outcome.

Lateral extensions to historic buildings, if not designed properly, can damage or destroy significant materials and change the character of such properties significantly. Such additions should only be considered if they are vital to the building's ongoing beneficial use and only after it has been determined that the new use cannot be achieved by altering non-significant, or secondary, interior spaces.

Key conservation requirements for new additions

According to Weeks and Grimmer (1995) there are three main requirements to bear in mind when planning a new exterior addition to an historic building:

1. Preserve significant historic materials and features:
   - Avoid constructing an addition on primary or other character-defining elevation to ensure preservation of significant materials and features.
   - Minimize loss of historic material comprising external walls and internal partitions of floor plans.
2. Preserve the historic character:
   - Make the size, scale, massing and proportions of the new addition compatible with the historic building to ensure that the historic form is not expanded or changed to an unacceptable degree.
   - Place the new addition on an inconspicuous side or rear elevation so that the new work does not result in a radical change to the form and character of the historic building.
   - Consider setting an infill addition or connector (i.e., link block) back from the historic building wall plane so that the form of the historic building(s) can be distinguished from the new work.
   - Set an additional storey well back from the roof edge to ensure that the historic building's profile and proportions are not radically changed.
3. Protect the historic significance – make a visual distinction between old and new:
   - Plan the new addition in a manner that provides some differentiation in material, colour and detailing so that the new work does not appear to be part of the historic building. The character of the historic resource should be identifiable after the addition is constructed.

Conservatories

Introduction

Conservatories are a common form of small-scale extension in residential properties nowadays. This is because they are cheaper and easier to erect than more substantial forms of lateral extensions. In domestic cases they are normally restricted to small single-storey structures having not less than three-quarters of the area of their roof and not less than one-half of the area of their external walls made of translucent material.

This type of lateral extension, however, is sometimes also installed in commercial properties such as public houses, guesthouses and hotels, but on a larger scale. In these cases they are used as lounges or as part the restaurant area.

According to ECD Partnership (1986) domestic conservatories have three possible roles:

An extra, pleasant living space.
A place for delicate plants.
An energy saving house addition, through:

- extra insulation for walls, windows and doors;
- draught lobby for door and windows;
- ventilation pre-heating solar effect.

In general, conservatories are light and flimsy in comparison to both the main building and other types of lateral extension. They are designed to make full use of glazing for maximum daylight. Their main function is for lounge or light recreational use, particularly during the spring and summer so that maximum advantage can be taken of the longer periods of daylighting.

Double-glazing systems using unplasticized polyvinyl chloride (uPVC) or hardwood framing in dwellings have become very popular during the last quarter of the 20th century. This has been accompanied by a growth in the number conservatory installations. The latter use a modular system for the superstructure. The foundation and upper base construction is built using conventional concrete and brick.

Statutory controls

Normally, planning permission is not required for a small-scale residential conservatory provided that it is located at least 1 m from any boundary and is not greater than 8 m^2 in area. Nevertheless, in cases of doubt it is best to contact local planning authority for any specific design guidelines that they may employ for conservatories. These could relate to the position and style of the conservatory, particularly in sensitive situations (e.g., conservation or special area).

Conservatories are usually exempt from building regulations if they are:

- attached to a domestic building (usually at the rear or side) and located at least 1 m from any boundary;
- separated from the rest of the dwelling by a door (usually a patio door or French window);
- glazed with safety glass;
- less than 8 m^2 (which used to be 30 m^2) in floor area;
- restricted to ground-floor level.

Technical factors

Constructional forms

Conservatories were originally viewed as merely glorified greenhouses. Granted, the construction of these two structures is similar. The former, however, requires a higher standard of construction and comfort conditions. As a result they need double or even triple glazing, whereas greenhouses only have single glazing. The superstructures of many conservatories, like greenhouses, are designed to be dismantled and re-erected elsewhere.

Schematic sketches (not to scale)

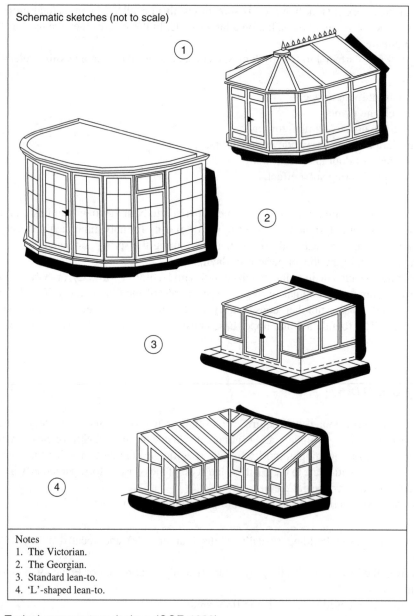

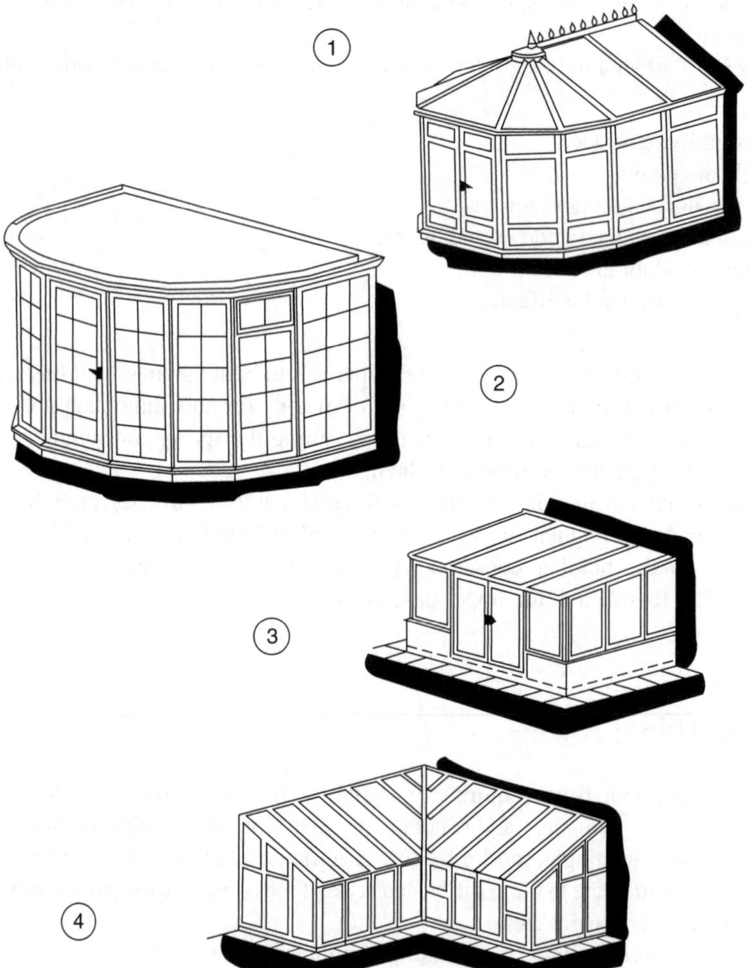

Notes
1. The Victorian.
2. The Georgian.
3. Standard lean-to.
4. 'L'-shaped lean-to.

Figure 5.15 Typical conservatory designs (GGF, 1999)

There are essentially two forms of conservatory superstructure: the full height glazed system and the partially glazed system. Off-the-shelf or pre-fabricated conservatories can reduce costs. They can have a variety of roof styles ranging from the standard hipped duo-pitched to the mono-pitched or even hexagonal dome type (see Figure 5.15).

As a general rule the connection requirements for conservatories are the same as those for extensions of conventional construction.

The primarily technical considerations for conservatories are summarized as follows:

- Allowance for minor differential settlement.
- Weather-tightness of joint between conservatory and main dwelling.
- Allowance for ventilation – both background and rapid.
- Protection against excessive heat loss and heat gain – using low-emissivity glazing (e.g., glass that has been given a metal coating that reduces transmission of solar energy) and roller or venetian blinds (which can be mechanically operated).
- Adequate security to the conservatory which does not compromise the main building's security.

The shape of the conservatory and the extent of glazing involved will affect heat gains and losses. For example, 'a conservatory with 11 m^2 of south facing single glazing can provide about 1000 kWh of useful energy per year (1440 kWh if double-glazed). However, conservatories cannot be justified in energy saving terms alone' (ECD Partnership, 1983).

A useful checklist for the conservatory design for dwellings and public houses is shown in Table 5.1. A conservatory in such properties is a good way of reducing heat losses. A conservatory provides a

Table 5.1 Checklist for conservatory design (based on BS 8211)

Item	Design points
Orientation	Select an orientation that achieves the best solar access. Domestic conservatories should face south ±45°. In public houses and restaurants, west to south is preferable.
Fabric	Materials within the conservatory should preferably have a high thermal mass and be absorbent to solar radiation, such as brick or blockwork, concrete paving or tiled floors. Thermal mass is needed inside, where sun falls.
Glazing	Supporting framework should incorporate thermal breaks to reduce the risk of condensation. Glazing should be double-glazed, if possible with low-emissivity glass or argon-filled cavities.
Ventilation	Locate the extension so that as much of the ventilation for the dwelling as possible enters through the conservatory, to take advantage of all solar gain. Provide temperature-operated ventilation at top and bottom to 10% solar-glazing area to prevent overheating. Provide controllable vents – via cords, winder handles or electronic controls. In addition to normal trickle ventilation, a conservatory should have high-level working windows or openable rooflights. In larger conservatories sweeper (i.e., destratification) fans can help to distribute warm air from roof level down to the occupied space in winter.
Shading	Provide light coloured or metalized blinds to minimize summer overheating and glare. Lightweight curtains, banners or plants may be used as an alternative to blinds.
Location	Position domestic conservatory adjacent to living room and to doors, where possible.

thermal buffer space between indoors and outdoors as well as an attractive customer area during busy periods.

As with window replacement work it is advisable to ensure that the contractor involved in undertaking this work is a member of the Glass and Glazing Federation (GGF). This means that in the event of the contractor going into liquidation, the federation can indemnify the client's losses.

Figure 5.15 shows some typical conservatory designs. There are of course other styles, but the one selected will be based on the grounds of cost and suitability for the property.

Under-floor heating

Small domestic conservatories normally may not require any direct heating source as they are primarily used in the warmer months. Large conservatories, on the other hand, need to have an efficient heating system to achieve an adequate level of thermal performance throughout the year. Thin panel radiators with micro-bore piping can offer an adequate heating system in such circumstances but these take up space.

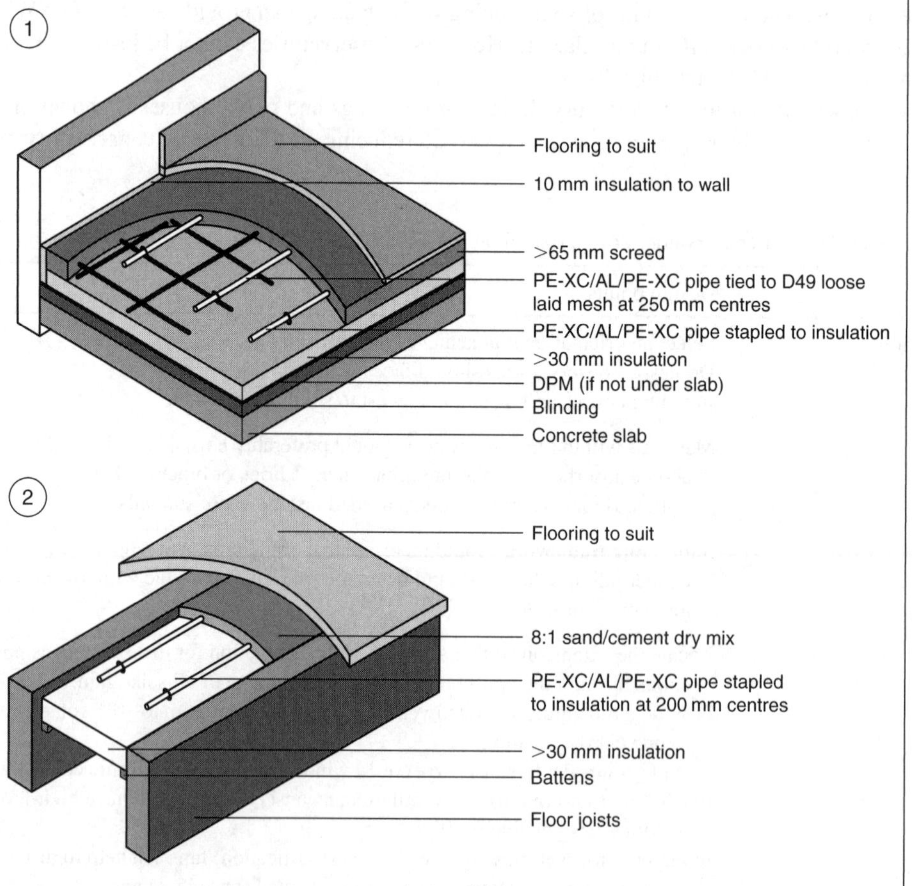

Notes
1. Insulation over existing slab.
2. Between joists in existing suspended timber floor.

Figure 5.16 Typical under-floor heating (www.continental-ufh.com)

A more suitable type of heating for a conservatory, therefore, is under-floor heating. It can also be considered in the ground floor of large extensions. Nowadays a low temperature hot water supply through a coil of polyethylene high performance pipes forms the most common type of system. A pumpable anhydrite screed (see Chapter 9) is well suited for this kind of under-floor heating application. This type of topping is laid much thinner (i.e., around 40 mm) than traditional screeds, with only 25 mm cover over pipes or wires being required (Figure 5.16).

The primary advantages of under-floor heating are as follows:

- space efficiency (no usable space is wasted to accommodate heating units, etc.);
- heat efficiency (more even heat distribution);
- low fire risk (no exposed combustible products or fittings);
- aesthetics benefits (the system is discrete – no pipes or radiators to spoil the interior's appearance);
- comfort factors (it gives the floor a warm, comfortable feel);
- health factors (it minimizes the risk of condensation and mould growth);
- energy efficiency (it directs heat to where it is needed most – the area at and just above the floor);
- low maintenance costs (embedded pipes need little if any attention);
- safety (no hot pipes or radiators are present to scald vulnerable users such as children and old people);
- security (it contains few visible parts to damage);
- durability (the system can last the life of the building).

Despite these benefits there was a drop in the popularity of under-floor heating in the UK between the 1960s and 1990s. The main reasons for this were primarily associated with the following problems:

- installation and system failures (poor workmanship difficult to resolve once the system is installed);
- inadequate control (slow response to changes in temperature requirements);
- poor accessibility (embedded pipes difficult to alter or repair without disturbing the floor and its coverings);
- high capital cost (over 10% more expensive than conventional radiator system).
- flooring comfort problems (e.g., 'hot foot' syndrome).

Joining new work to old

Preamble

Joining new work to old brings with it a number of risks and potential problems. There are essentially three key technical factors to bear in mind when designing and building lateral extensions: engineering factors, constructional factors and operational factors.

Engineering factors

Soil conditions

The soil conditions of the property being extended must also be taken into account. The ground under which the extended is being placed may be of a different type and quality from that of the main building. Therefore the soil conditions should be investigated if there is any dubiety about their capacity and stability.

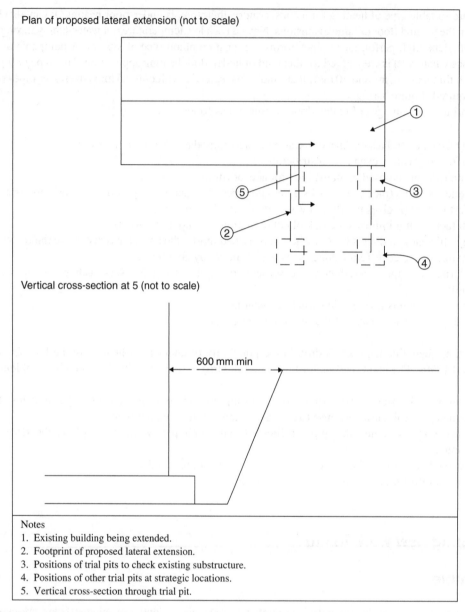

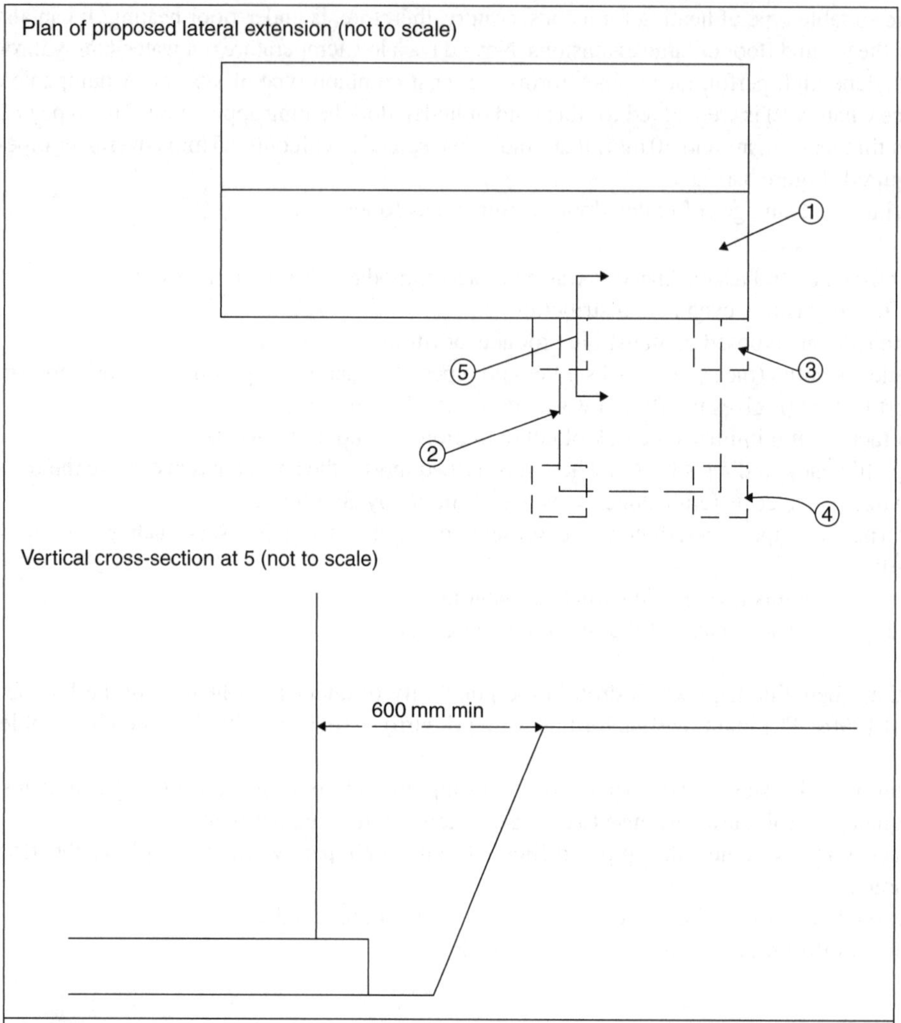

Notes
1. Existing building being extended.
2. Footprint of proposed lateral extension.
3. Positions of trial pits to check existing substructure.
4. Positions of other trial pits at strategic locations.
5. Vertical cross-section through trial pit.

Figure 5.17 Exploring ground conditions for lateral extension using trial pits

A basic investigation involves digging a trial pit about 1 m × 1 m in plan to the formation level at a selected position along the wall where the building is being enlarged (see Figure 5.17). This will allow the following checks about the ground conditions:

- Confirm the depth and condition of the original foundation. The existing footing checked for defects such as cracking, sulphate attack, etc.
- Existing substructure masonry checked for defects such as erosion, frost damage, missing/defective dpc, etc.

- Condition and type of subsoil checked. If a handful of soil remains in a ball-like shape after being compressed by the hand, the soil is likely to be clayey (i.e., cohesive) soil. The appropriate precautions for building on such ground therefore need to be taken. Clayey and other cohesive soils may pose particular problems because of their sulphate content and hydrophilic characteristics. They tend to swell when they absorb water. Conversely, any trees in the vicinity of the proposed extension can cause shrinkage of clayey soils (see Figure 5.18). Alternatively, trees cut down in that position can cause the soil to heave as a result of rehydration.
- If, however, the soil collapses in one's palm after being compressed with the fingers, then the soil is likely to be sandy/gravely (i.e., non-cohesive). This type of soil has better, more stable load-bearing properties.
- Depth of trial pit should not be greater than the bottom of the existing foundation otherwise there is a risk of undermining the substructure of the building.

In complex cases, however, a more detailed investigation involving taking bore holes and soil samples for analysis by a geotechnical engineer may be required. The expense and trouble incurred in carrying out this work would be much less than the cost and disruption involved in rectifying a subsequent subsidence problem.

The level of the water table should be established at the same time the soil conditions are being determined. This should highlight any potential risks of flooding during the excavation work and damp penetration. It will also have a bearing on the type of tanking required should a basement form part of the proposed adaptation work.

Where an extension is proposed over a site with a high water table certain precautions should be taken. A field drain should be installed around the new part of the building. The type of tanking required may involve a drained system to ensure no water as a result of hydrostatic pressure enters the basement.

Differential movement

It is very important to consider the structural effects of any extension on the foundations and superstructure of the original building (see Figure 5.19). Differential settlement occurs because a new building after completion is in the process of settling – which can take up to about 6 months or more – whilst the original building has already settled. Any joint therefore should allow a degree of vertical downward movement (see below).

Changes in the load paths of the original building will also be encountered if an opening in the wall connecting it with the extension is being made. This may over-stress narrow bands of walling – especially those less than 600 mm wide at corners or between openings.

Differential movement can also occur where dissimilar materials abut each other. For example, if cast in situ concrete slabs, pre-cast concrete beams or floor units and lintels bear on to dissimilar materials such as blockwork or brickwork, a separating layer should be formed at the bearing. This can be achieved using two sheets of some smooth, incompressible material, such as heavy gauge polythene.

Tilting or sinking

Disregarding poor or inadequate soil conditions when extending a building can give rise to tilting/sinking and differential settlement. Tilting of the new work is one of the most significant structural problems following the extension of a building. Lack of adequate foundation of the extension and proper tying-in of the new to the old are the two main reasons for this failure. Sinking is also caused by differential settlement. These types of structural problem are illustrated in Figure 5.19. See also Figure 6.10 in Chapter 6 for the effects that a roof extension can have on differential settlement.

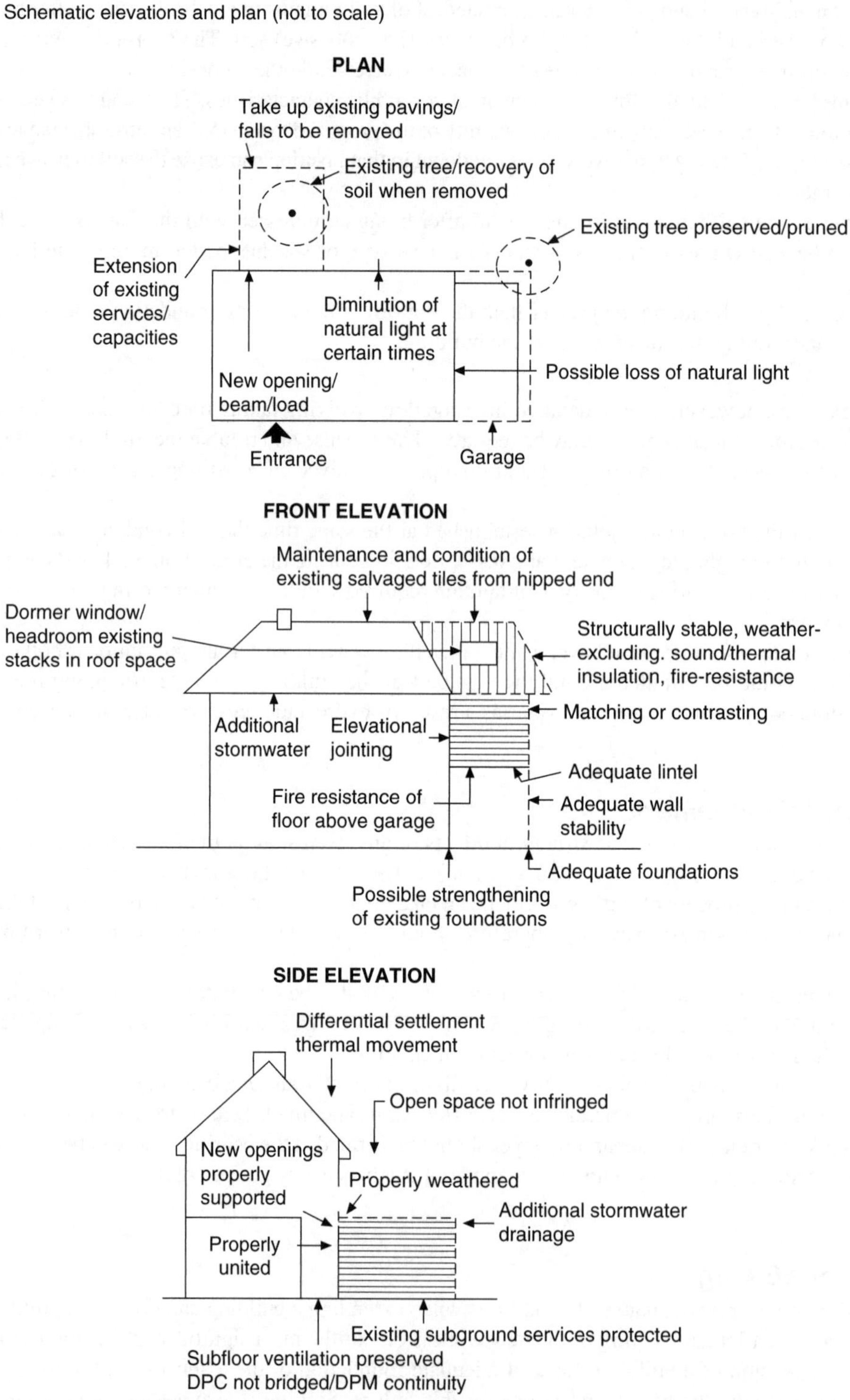
Schematic elevations and plan (not to scale)
PLAN
Take up existing pavings/
falls to be removed
Existing tree/recovery of
soil when removed
Existing tree preserved/pruned
Extension
of existing
services/
capacities
Diminution of
natural light at
certain times
Possible loss of natural light
New opening/
beam/load
Entrance
Garage
FRONT ELEVATION
Maintenance and condition of
existing salvaged tiles from hipped end
Dormer window/
headroom existing
stacks in roof space
Structurally stable, weather-
excluding. sound/thermal
insulation, fire-resistance
Matching or contrasting
Additional
stormwater
Elevational
jointing
Adequate lintel
Fire resistance of
floor above garage
Adequate wall
stability
Adequate foundations
Possible strengthening
of existing foundations
SIDE ELEVATION
Differential settlement
thermal movement
Open space not infringed
New openings
properly
supported
Properly weathered
Additional stormwater
drainage
Properly
united
Existing subground services protected
Subfloor ventilation preserved
DPC not bridged/DPM continuity

Figure 5.18 Other technical issues with small lateral extensions (adapted from CEM notes)

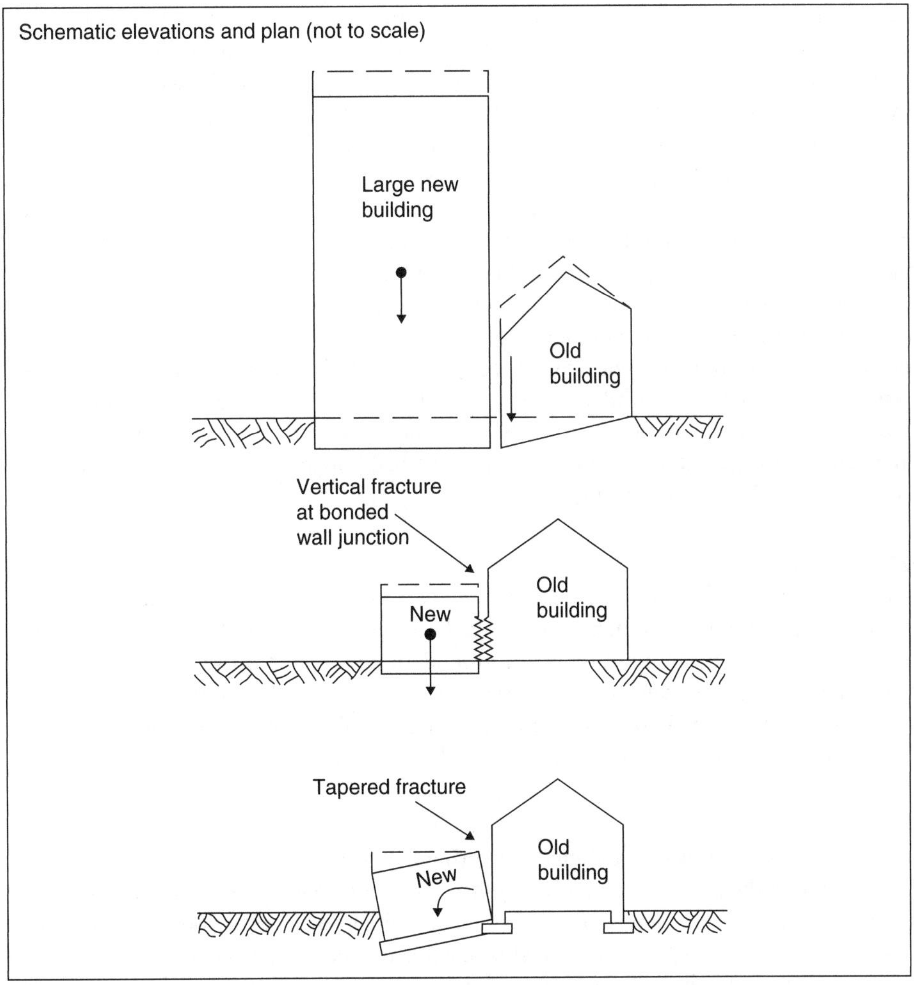

Figure 5.19 Examples of differential settlement between old and new construction (Richardson, 1996)

Foundation requirements

Normally, if the soil conditions across the site are consistent and stable the consultant (architect/surveyor) will specify a foundation for the extension that matches the existing building in terms of type, size and depth. A suitably qualified engineer, however, should design foundations for extensions in sites with variable or uncertain soil conditions. These may necessitate the extension having a different foundation from that of the original building. For example, if the only feasible position for the extension were over made-up ground or on soil with poor bearing capacity concrete piles or a raft foundation may be required.

Movement joints in walls

The usual degree of movement between a lateral extension and an existing building should be not less than about plus or minus 10 mm in both horizontal and vertical directions. In bigger adaptation schemes the structural design may require larger allowances.

For large extensions movement joints must maintain fire separation, thermal insulation and weather protection. The spacing of vertical movement joints depends on the type of walling. For clay brickwork the maximum spacing is 12 m; 9 m for calcium silicate (flint lime) brickwork; and 6 m for concrete block or in-situ walls. These spacings should be halved respectively in the case of parapet walls.

The common form of movement provision is by using straight movement joints. They can be formed by abutting the bricks or blocks against both sides of a strip of 10 mm thick rigid filler. This filler is about 10 mm less than the width of the wall so that the recess it forms at the front face can be filled with a polysulphide mastic. In plastered or rendered walls movement joints should be continued through the all surface finishes using proprietary stop beads at both edges of the joint.

Internal partitions will require movement joints too. These will have to be carefully detailed so that uncontrolled cracking and movement do not deface the internal finishes.

Continuity across a movement joint may be required for stability reasons. This can be achieved using slip ties set in every second or third bedding joint. These normally consist of stainless steel strips, $200 \times 25 \times 3$ mm, set parallel to the plan of the wall. One end of the tie should be sleeved or wrapped in a debonding material to allow it to move freely within the bedding joint. Proprietary types of slip-ties with one end sleeved are available for this purpose.

Wall junctions

Unless a movement joint is required, walls are usually bonded in at returns and junctions (e.g., separating walls). However, differential movement may occur between walls on different footings. For example, where the external wall skins are built off a strip foundation and the partition walls of a non-suspended ground-floor slab (i.e., independent of the external wall foundation), differential movement is likely. In such cases, the walls should be butted and tied together rather than bonded at their junction. This allows the partition walls to be constructed after the main walls.

Services

The impact on services is another important engineering factor. A drain line or water supply pipe that runs through the position of the proposed extension has to be either rerouted or encased in concrete. Any manhole being retained in this location has to be fitted with a bolted-down cover containing a double seal. The integrity of any discharge or supply pipes must be established before any covering over work to them is done to minimize leaks. Making good a defective service pipe after the work is undertaken is much more difficult and disruptive.

The extendibility of existing services is also important. Lack of capacity of the existing services may necessitate their upgrading or the installation of additional cables or pipes.

It may be necessary to allow for movement of the services between the extension and the existing building. For drainage pipes the usual method of providing movement joints is an O-ring connection. However, if excessive movement of more than plus or minus 10 mm is anticipated, a 'flexible' connection or double bend should be used.

The type of pipe material used for the services between a lateral extension and the original building should also be selected with care. Generally, relatively small-bore copper and iron pipes can accommodate some bending to take up the new position of the extension. Specific expansion joints are required for large cast iron and PVC pipes. Brittle pipe materials such as vitreous clay should not be allowed to pass from one construction to another.

Constructional factors

Preamble

The constructional aspects of joining new work to old are concerned with buildability, fitness for purpose and weather-tightness. Treatment to achieve these requirements has to be considered at foundation, walls, floors and roof. For example, the junction between the old and new has to be carefully designed to minimize rainwater penetration as well as tilting. The type of connection will depend on whether the extension is small or large (see below) as well as the form of walling involved.

Site preparation

In many cases the ground at the locus of a proposed lateral extension is usually a grassed area. This means that the site preparation is similar to that for a new-build scheme. In such instances removal of the topsoil and marking out of the foundations may proceed without too much disruption.

However, in some urban sites the area of the proposed extension may require substantial preparation. This is particularly the case where the area is covered with a hardstanding or redundant lean-to extension. For example, in the case of commercial properties such as supermarkets the new extension will often encroach on the existing car park. Any concrete or tarmacadam surfacing would have to be dug up in its entirety and removed before any substructure work proceeds.

It is not advisable to built directly off or through any hardstanding as it might contain inadequate damp proofing or encourage heave at a later date. The quality of the construction cannot be trusted to act as a suitable base for any slab or foundation. Similarly, any redundant constructions over the position of the proposed extension would have to be demolished (see Chapter 11).

Retaining walls

The erection of a lateral extension on a sloping site is more problematic than on a flat site. Landslip and groundwater movements are two of the main risks associated with developing on sloping sites.

A side or rear extension of a building on a slope will almost certainly entail the erection of a retaining wall. This would be required to keep the raised ground away from the new part of the building and allow for access around it. Figure 5.20 shows schematic cross-sections of a proposed extension on a sloping site.

Wall junction

It is usually quite difficult to achieve a seamless or invisible connection between the two buildings. This is because the facade of the existing building will show unmistakable signs of grime as well as wear and tear. In contrast, the extension will seem bright and clean by comparison.

For brickwork it will be important to match the new with the old for colour, texture, size and mortar joint finish. A 'risband' (i.e., straight butt joint) movement joint as described above should be formed in such instances. However, if a good match between the two walls is not possible, masking the joint with a change in construction or finish could be considered.

Setting the front face of the extension back or forward, say, at least 100 mm is a good way of masking the joint between old and new in subtle way. The return created in the walling can act as a distinctive feature of the adapted building.

An alternative way of masking the joint is to fit a cover strip over the joint. Externally, this could comprise an aluminium strip cover fixed to one side of the joint to allow for movement. Internally, a timber or plastic cover strip fixed on one side of the joint can be used.

The joint between the columns of a framed building and a framed extension is dealt with below.

Roof abutments

The connection arrangement between the flat roof of a building and an extension of the same construction is shown in Figure 5.21. The detail involving floors with downstand edge beams is basically the same, except for the parapet upstands.

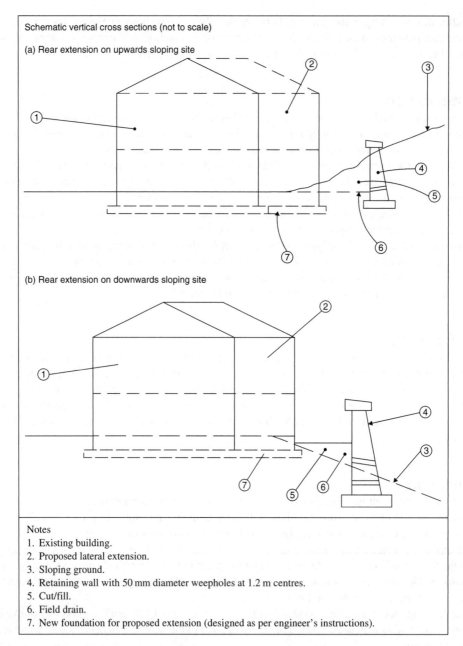

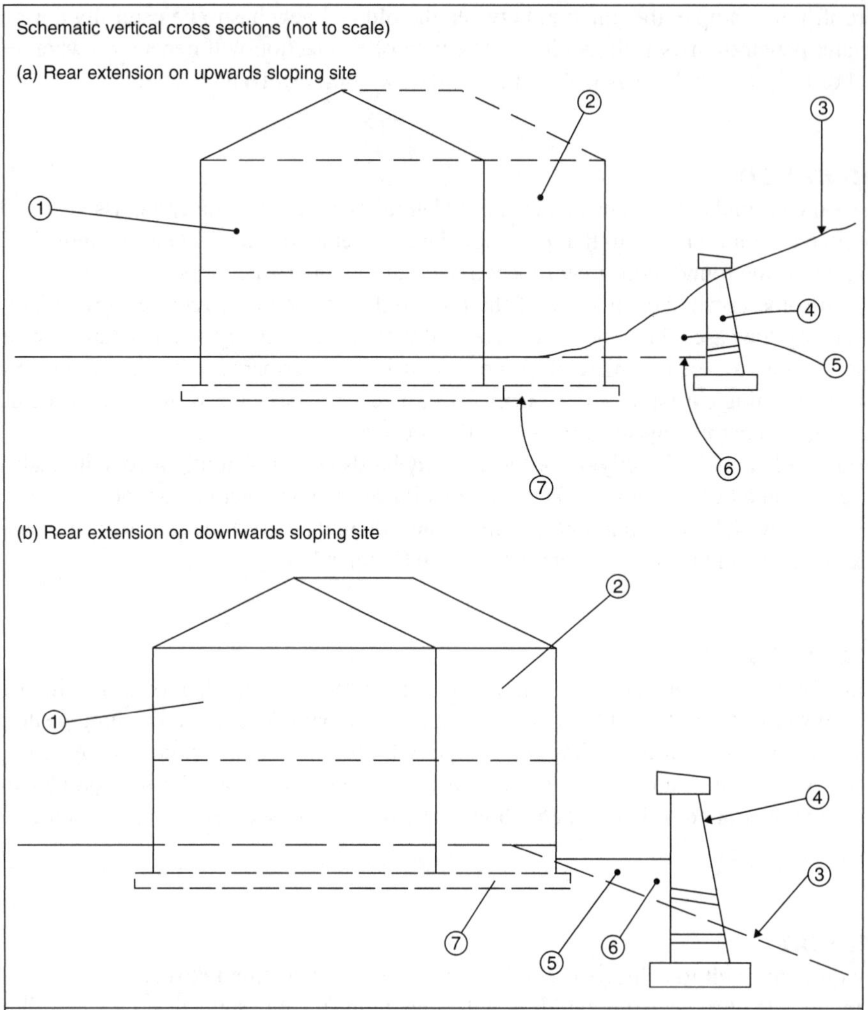

Notes
1. Existing building.
2. Proposed lateral extension.
3. Sloping ground.
4. Retaining wall with 50 mm diameter weepholes at 1.2 m centres.
5. Cut/fill.
6. Field drain.
7. New foundation for proposed extension (designed as per engineer's instructions).

Figure 5.20 Lateral extensions on a sloping site

Matching the pitched roof coverings of an extension to those of an existing building is often difficult to achieve. The slate or tile coverings of the new roof are inevitably cleaner (and probably sounder) than the existing. This problem is especially noticeable with end-terrace extensions. Where the terrace is long, for example, it would be uneconomic as well as impracticable to replace the entire roof finish. The consent of the various owners to contribute to such an improvement would be difficult to obtain.

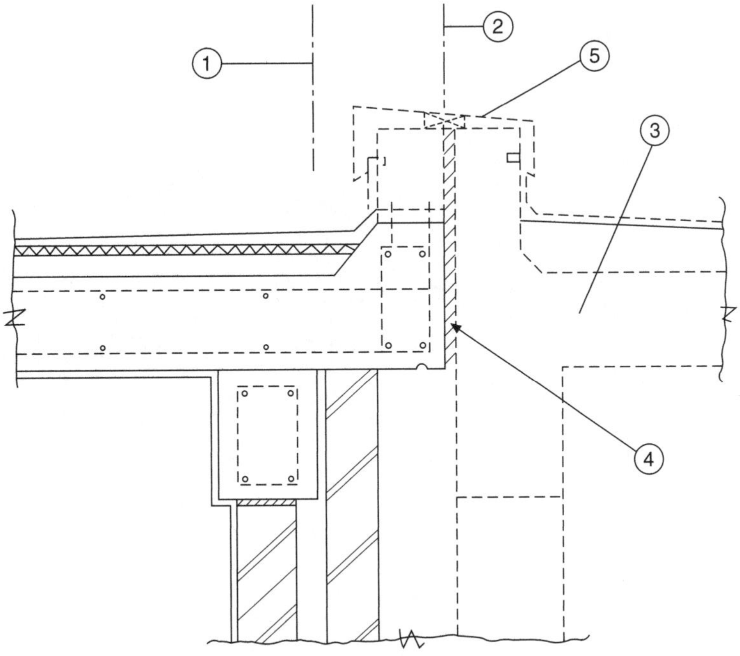

Notes

1. Option 1: Existing reinforced concrete flat roof covered with an external skin of wall cut back to this line. This option is disruptive as it involves removal and making good of a section of the slab at the verge. A new reinforced concrete kerb would have to be formed at this position to provide an abutment similar to that shown for Option 2. The existing roof may be upgraded to match that of extension. The butt joint between the old and new foundations could have dowel bars inserted in a similar manner to that shown in Figure 5.28.
2. Option 2: Existing edge detail retained but verge extended to form new kerb to match that of extension. New reinforced concrete kerb with dowel bars grouted into existing slab. This option obviates the need for cutting the concrete roof slab but necessitates a cantilevered foundation at the abutment with the doweled into the existing. The whole or part of the skins of the existing cavity wall along the abutment with the extension may be removed. All new work is shown by dashed lines.
3. New reinforced concrete flat roof of extension covered with asphalt or high-performance felt on tapered insulation. Reinforcement not shown for clarity.
4. Joint between the old and new roofs filled with polystyrene insulation board.
5. Construction joint covered with aluminium capping. 150 mm wide aluminium cover plates under butt joints of capping to provide watertight joint. Unfixed timber bearer under capping at centre to prevent sagging of aluminium.

Figure 5.21 Flat roof construction joint at verge

Consideration should still be given, though, to replacing the entire pitched roof coverings of small/medium-size, detached buildings with a new extension. This is so that a more homogeneous finish is achieved with the extension and original building. The overall effect can mask the new extension, as could be the case with Figures 5.22 and 5.23.

The junction of a flat-roofed extension onto a pitched roof is vulnerable to rainwater penetration. The detailing and construction at this intersection therefore has to be handled with care. A common error occurs

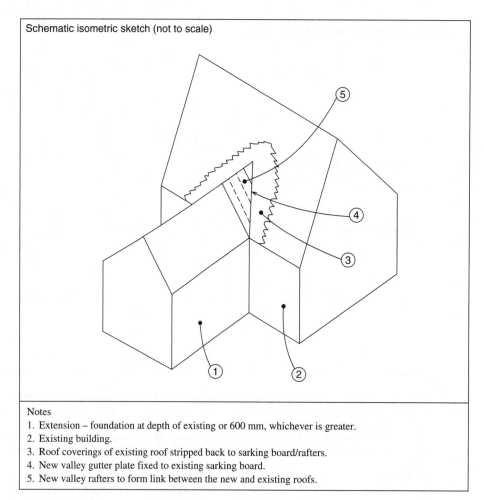

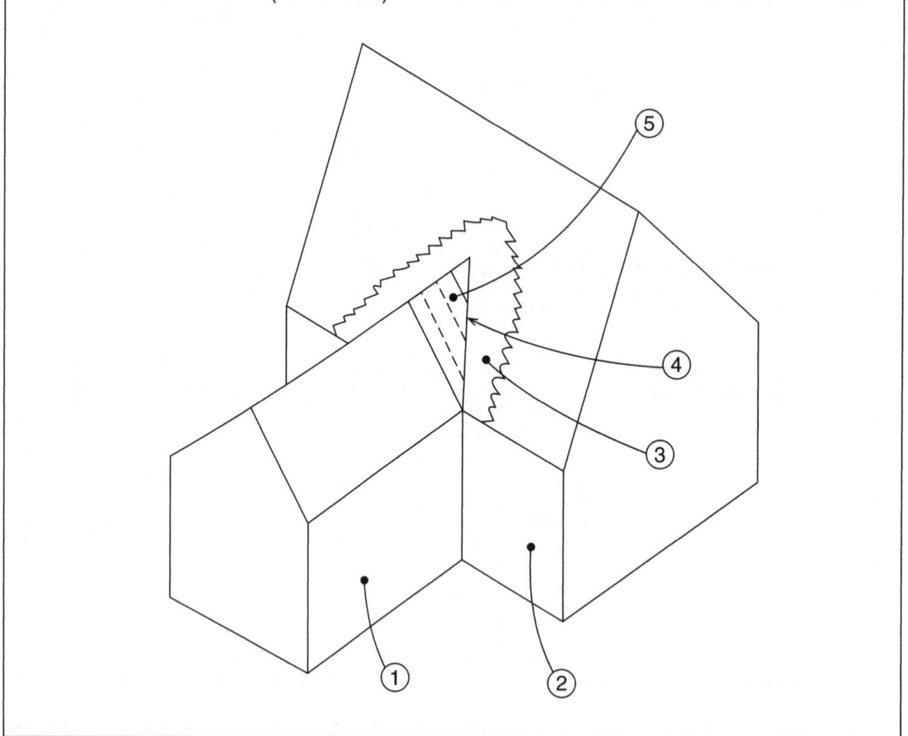

Notes
1. Extension – foundation at depth of existing or 600 mm, whichever is greater.
2. Existing building.
3. Roof coverings of existing roof stripped back to sarking board/rafters.
4. New valley gutter plate fixed to existing sarking board.
5. New valley rafters to form link between the new and existing roofs.

Figure 5.22 Lateral extension with pitched roof continued onto existing

when the sarking felt is tucked under rather than over the new flat-roof membrane. As a result, if there is a leak in the pitched-roof section above the flat roof, rainwater will inevitably drip down the sarking felt and enter the extension by dripping through the exposed edge of the membrane.

In the case of a flat-roofed extension to a bungalow consideration needs to be given to matching the ceiling height of the adjoining rooms. It may be difficult to achieve this where the existing eaves is at or near the same height as the existing ceiling line. Warm-deck flat-roof constructions, although preferable in most cases, are inevitably deeper than their cold-deck counterparts. This may necessitate a downstand beam above the opening connecting the old and new rooms because of differences in ceiling levels between these areas.

Floor abutments

In some instances, however, the abutment at the various floor levels may comprise a slab without a downstand edge beam. The connection requirements in this case are illustrated in Figure 5.24. In any event, allowance for movement in the floor slab should be made. Providing a 10 mm wide vertical joint filled with rot-proof board in the slab at abutments or construction joints can fulfil this function. The edge of the

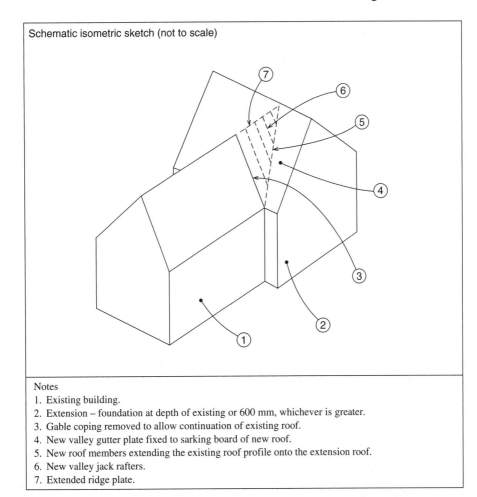

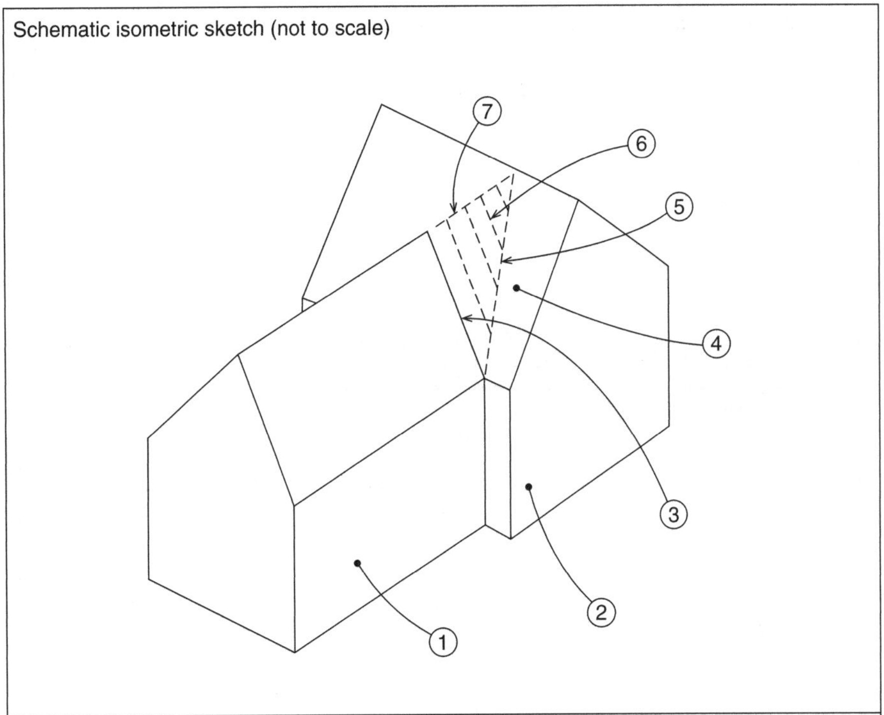

Notes
1. Existing building.
2. Extension – foundation at depth of existing or 600 mm, whichever is greater.
3. Gable coping removed to allow continuation of existing roof.
4. New valley gutter plate fixed to sarking board of new roof.
5. New roof members extending the existing roof profile onto the extension roof.
6. New valley jack rafters.
7. Extended ridge plate.

Figure 5.23 Existing building with pitched roof continued onto lateral extension

board should be recessed about 20 mm from the face of the slab to allow the joint to be finished with a polysulphide mastic filler.

Weather-tightness

The weakest part of an extension in terms of weather-tightness is the joint between it and the existing building. This is not a problem if the profile of the existing roof is being continued without any obvious break. However, if the rooflines are at different levels, an abutment has to be formed. As indicated earlier, abutments must be detailed carefully to avoid rainwater penetration problems. If the existing building is of cavity wall construction, ideally a cavity tray should be installed along the line of the abutment because the outer leaf at this position becomes an inner leaf (see Figure 5.25).

Admittedly, installing a cavity tray in an existing building is not easy. Sloping cavity trays at the abutment between a new pitched roof and an existing wall are even more difficult to install than horizontal ones involved in an extension with a flat roof.

In any event, the cavity tray should be installed in staggered bays 900 mm long to avoid causing any settlement in the brickwork outer skin above. Proprietary cavity trays are available for this purpose. They

Schematic vertical cross-section (note to scale)

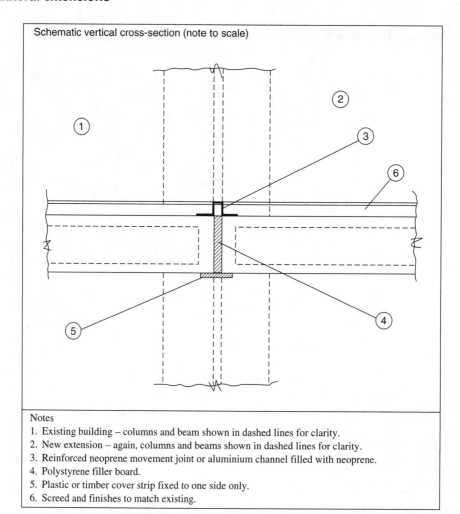

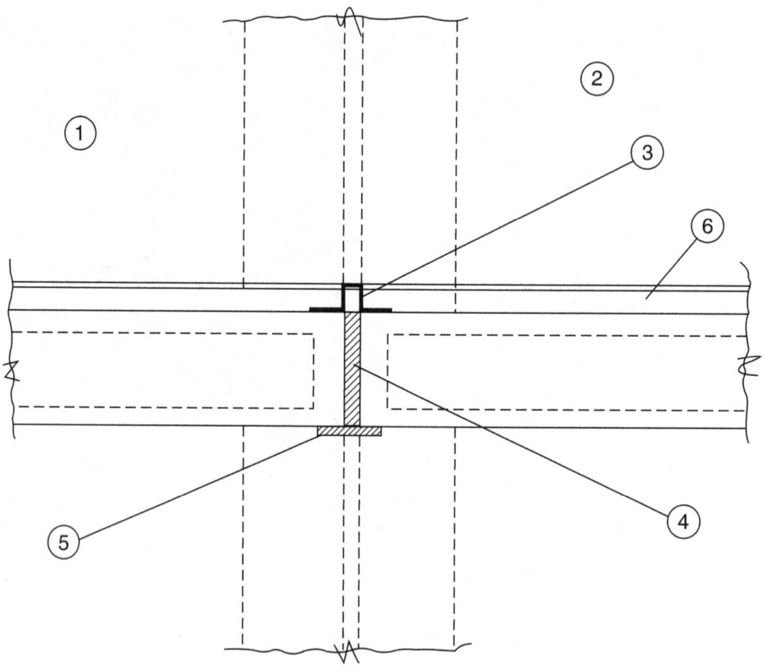

Notes
1. Existing building – columns and beam shown in dashed lines for clarity.
2. New extension – again, columns and beams shown in dashed lines for clarity.
3. Reinforced neoprene movement joint or aluminium channel filled with neoprene.
4. Polystyrene filler board.
5. Plastic or timber cover strip fixed to one side only.
6. Screed and finishes to match existing.

Figure 5.24 Upper-floor movement joint

should be tucked into inner leaf at nearest joint at least 100 mm above the level of the dpc at the outer leaf, and have a minimum 75 mm lap at the end joints.

Moisture bridging can also occur at the abutment of the extension and the external wall of the existing building. This may manifest itself as moisture staining on the internal wall surface and can easily be mistaken for rising dampness. It occurs if the existing rendering is not removed at this position and is instead plastered over. Such rendering during the winter months will be relatively damp and as a result acts as a wick for moisture to permeate through the plaster. This can result in damp staining and possibly mould growth. The presence of the latter indicates that the problem is unlikely to indicate true rising dampness (Oliver, 1997).

Dampness

Leaking drains or water supply pipes in the area of a lateral extension may result in dampness problems. The presence or chlorides but not nitrates in the water-contaminated mortar or masonry would confirm this. Even a small fracture in a waste pipe that originally was originally external but becomes located internally if located within a lateral extension can cause dampness to appear on the wall or floor at the abutment.

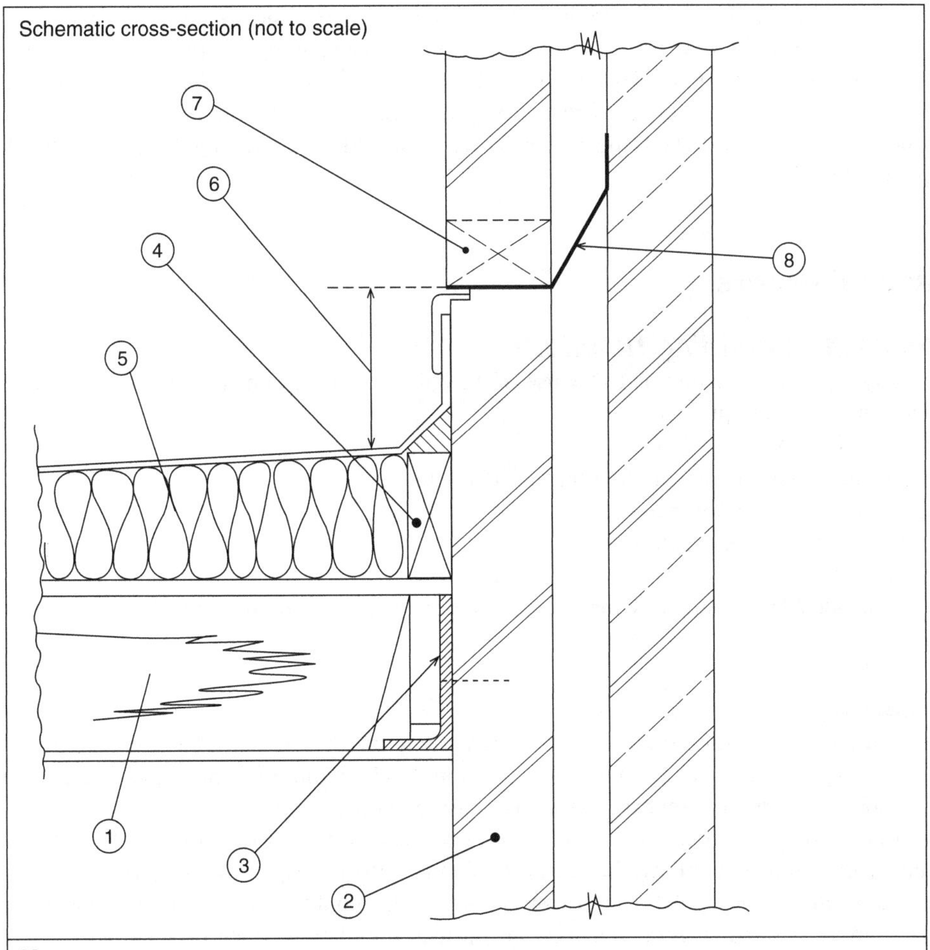

Notes

1. Flat roof of new extension.
2. Outer skin of gable wall of existing building becomes an inner leaf below flat roof.
3. Joists on hangers on 150 × 500 mm timber bearer or resting on mild steel angle plate (c. 125 × 75 mm) bolted to wall of existing building with 12 mm diameter expanding bolts at 900 mm centres.
4. Timber bearer at abutment to provide adequate fixing substrate for angle fillet.
5. Insulation board with vapour check on warm side of construction, tapered to achieve 1:40 fall.
6. Minimum 150 mm high skirting with flashing tucked under cavity tray DPC.
7. Proprietary plastic weep hole inserted in perpends every 900 mm, above cavity tray.
8. Proprietary cavity tray with upturned edge with self-adhesive surface that sticks to inner skin, to prevent penetrating dampness running down cavity from abutment.

Figure 5.25 Cavity tray at abutment

Openings

The enlargement of an existing opening or formation of a new opening is inevitable in most lateral extension schemes. With the former a window opening may be extended downwards to form a new doorway. Alternatively, an existing opening could be widened – but this will necessitate a new lintel. The operational and structural implications of this work are dealt with in the next section and Chapter 7 respectively.

The constructional implications of forming openings during an extension or other adaptation work, however, must also be taken into account. The method of forming an opening and the quality of workmanship involved in this work is important. This is especially the case with cavity walls. If such work incorrectly undertaken or inadequately controlled, there is a danger that debris such mortar lumps and broken brick can fall into the cavity causing it to be bridged. This in turn can lead to penetrating damp after the adaptation work has been completed.

Operational factors

Maintenance of building functions

There is likely to be some disruption to the use of the building during construction of the extension. This may impact on the need to provide:

- Storage space in the existing building for the contractor.
- Access for legitimate building users.
- Security against uninvited visitors.
- Freedom from nuisances such as dust and noise.
- Protection against hazardous activities such as welding, cutting, grinding, etc.

Disturbance

The effects of the extension on the use of the original building before and after the adaptation work is implemented must be taken into account. This is required to help minimize problems of disturbance due to dust and noise, breaches of security, and safety considerations.

A number of basic precautions can be taken to keep such disturbance to a minimum. Doorways near the disturbance work can be temporarily blocked off using polythene-covered screens to limit the spread of dust and debris. Heavy-duty polythene or canvas dustsheets should be laid over floor coverings in and around the work area to reduce scratch marks and waste accumulation on the floor surface.

Breaking through

The timing of the 'breaking through' from an existing to a new building is of especial importance in lateral extension work. Ideally, it should be undertaken late rather than early in the contract. In other words, before this is done the new work should be fully secure and weather-tight. The breakthrough work must therefore be done with due care to minimize inconvenience to the building and its occupants.

Forming a new opening in an existing wall must be done with due care to avoid danger to the operatives and minimize the risk of collateral damage to the surrounding construction. A 'Stihl' or other similar diamond-toothed circular saw about 300–400 mm diameter can be used to cut the profile of the opening in the existing masonry.

For cavity walls each leaf can be cut out individually and removed course by course from the top down. In the case of solid walls one side should be cut as deep as possible and carefully removed in brick-size pieces, again from the top down, using a hammer and chisel or pneumatic hammer. A similar cut should be formed in the remaining masonry on the other side, which can then be removed in the same manner as the first section.

Breaking through an existing floor to form a new stairwell may necessitate some temporary support under the section being removed. The main aim is to prevent sudden and uncontrolled collapse of the floor structure, which could cause serious injury as well as damage to the surrounding construction.

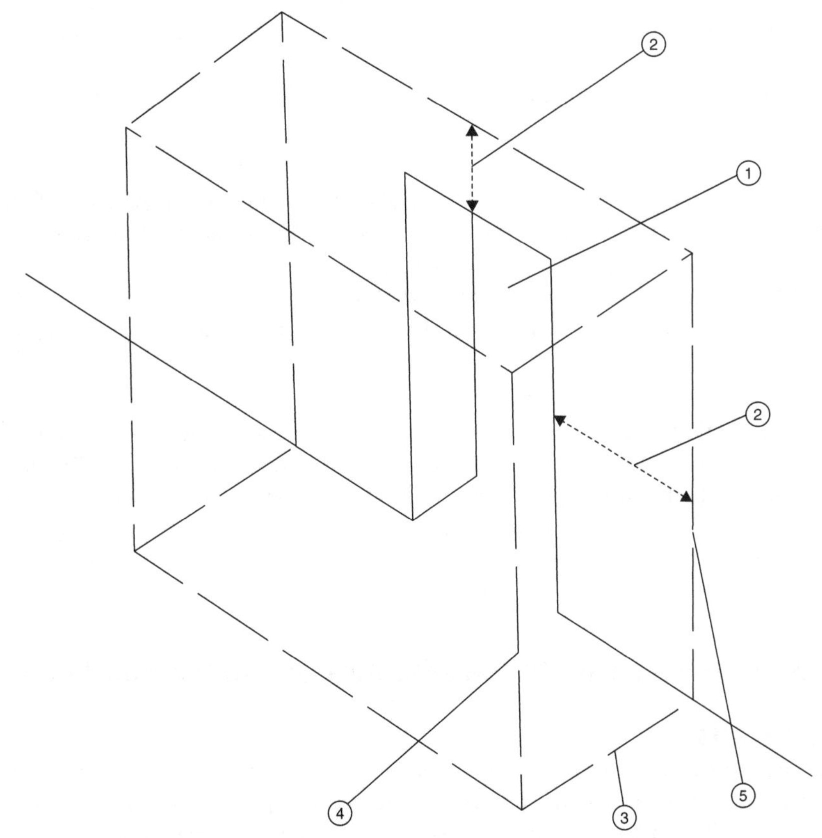

Notes
1. Proposed or adjusted opening.
2. 600 mm minimum.
3. 600 mm working space.
4. 50 × 50 mm softwood framing covered with heavy gauge translucent polythene sheeting reinforced with polyester mesh.
5. 'Duck tape' to seal joints in sheeting and gaps between enclosure and wall surface.

Figure 5.26 Temporary enclosure during breaking through work

To reduce dust and noise transmission it may be necessary to form a temporary enclosure on either side of the proposed opening. This is especially useful where the building is being occupied during the alteration work. Figure 5.26 illustrates a common method of forming an enclosure around the proposed formation of an opening in a wall. Once the breaking through work is complete, consideration should be given to the making-good the reveals in the wall and the junction between the new and existing floors.

Similarly, when removing the gable wall of a building next to an end extension some precautions are necessary. A temporary screen at least 1 m away from the gable and up to its full height will be required for security as well as dust and noise control.

Safety regulations

The CDM Regulations, which govern health and safety on building sites, are unlikely to apply to small domestic extensions. However, on larger-scale lateral extensions, such as those to commercial buildings, they will apply because the number of operatives on site is likely to exceed five.

Nevertheless, the following health and safety issues should be considered regardless of the size of the extension:

- Protection of building occupants and the public: Padded wrappings around props, scaffolding poles and other temporary works where the extension work is adjacent to public footpaths. These should be marked with red and white tape to make them conspicuous.
- Maintaining safe access: Clear and well-sign-posted routes.
- Fire safety and means of escape: Fire extinguishers and alarm stations at every level or at strategic positions.

A risk assessment should be carried out before breaking through and other hazardous work commences. In any event, it is important that operatives wear basic personal safety equipment (PPE) such as facemask and goggles, as well as high visibility vest, hard-hat, and safety boots whilst undertaking these operations (see Chapter 11). Extraction equipment may be necessary in cases where large amounts of dust are generated during the adaptation works.

Connection requirements to load-bearing masonry buildings

Small extensions

Low-rise extensions providing additional accommodation such as a bedroom or kitchen usually need to be connected up the entire length of wall. High-rise extensions comprising a small plan area such as a new staircase also are required to have continuous connection with the existing building to avoid any risk of detachment (see below). Figures 5.30 and 5.31 show the basic connection requirements.

Large extensions

In commercial and industrial properties extensions are often on a large scale, that is, over half the plan size of the existing building. In such cases, the connection requirements may be different for those for small extensions. Figures 5.32 and 5.33 illustrate some of those requirements.

Generally, there should be no need to remove the external scarcement of the main building to accommodate both small and large extensions. Removing this is disruptive and time-consuming. It would even increase the risk of vibration and impact damage to the remaining parts of the structure. A stepped foundation for the extension can be installed to overcome this problem (see Figure 5.38).

Extensions with basements

In some cases an existing building without a basement may have an extension involving the formation of a basement. This requires the incorporation of underpinning along the abutted wall of the existing building, which is dealt with in Chapter 6.

Foundations

The type of foundation for the extension is dependent on a number of factors. If the main building has simple strip foundations (Figure 5.27) it is unlikely that the extension will need to be any different. However,

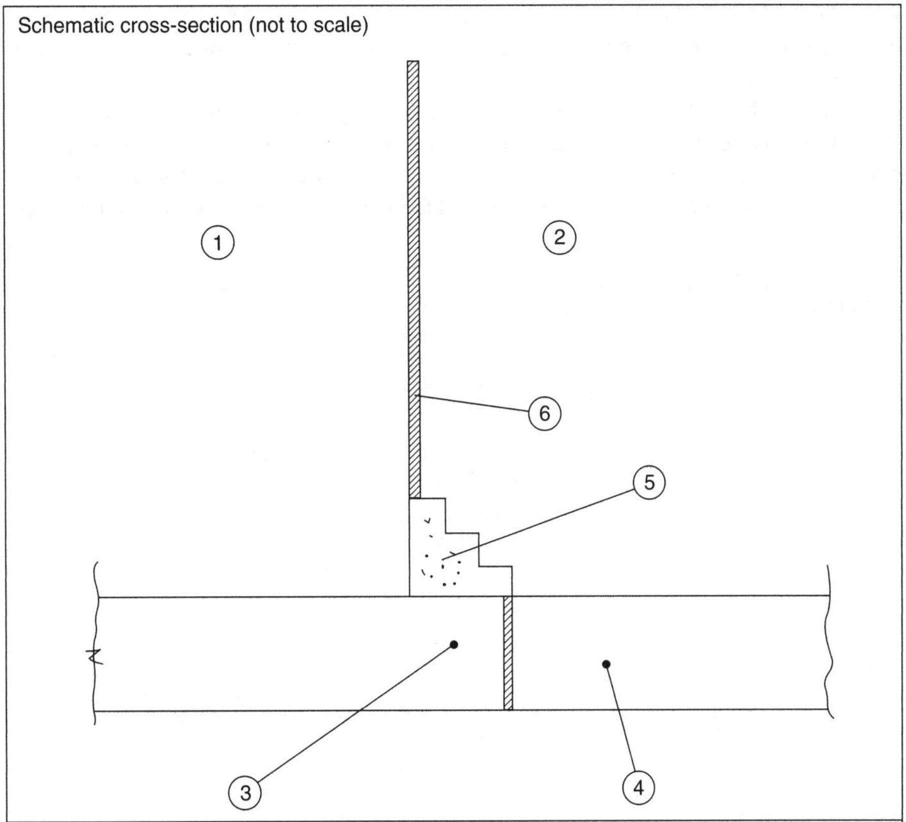

Notes
1. Existing building on soil of good bearing capacity.
2. New extension on soil of similar quality as existing, therefore, minimal risk of differential settlement.
3. Scarcement of existing foundation retained. There would be little advantage in removing the scarcement to allow easy abutment between the old and new constructions. Indeed, doing so would be time-consuming and may cause damage to the existing building.
4. New simple strip foundation for extension butt jointed against filler board to existing footing.
5. Weak mortar mix – e.g., cement:sand.
6. Rot-proof compressible filler board 20 mm thick to allow for differential movement in between 'Furfix' or other proprietary connection plate as shown in Figure 5.29.

Figure 5.27 Traditional method of connection at strip foundation level

poor ground conditions or increased loadings may necessitate a foundation design for the extension that is not exactly the same as that of the main building. Some short-bored piles, for example, may be required if the soil conditions are variable or the stratum for a safe bearing capacity is lower than the 600 mm minimum depth.

The extent of tying-in needs to be considered at both foundation and wall levels. Normally, the foundation of an extension is at least at the same level as the existing building – at least 600 mm. This is to prevent frost heave and soil erosion next to the footings.

As a general rule the foundation of the extension should be simply butt-jointed against the existing foundation. Allowance for differential movement is made either by stepping the brickwork and bedding it in a weak mortar mix, or by providing a 25 mm thick soft joint over the existing scarcement under the new brickwork (see Figure 5.19).

Some building control authorities may, however, insist that the two foundations should be tied-in using metal dowel bars. This might be because of the risk of excessive differential settlement owing to poor soil conditions within the district.

In such cases, with simple strip foundations, three 16–20 mm diameter mild steel dowel bars about 500–600 mm long placed in the middle and spaced at about 150 mm centres – depending on the loading (see Figure 5.28) – would usually be sufficient. These bars are inserted into pre-drilled holes filled with an epoxy grout in the existing concrete. The end embedded in the foundation is wrapped in a debonding tape

Schematic cross-section (not to scale)

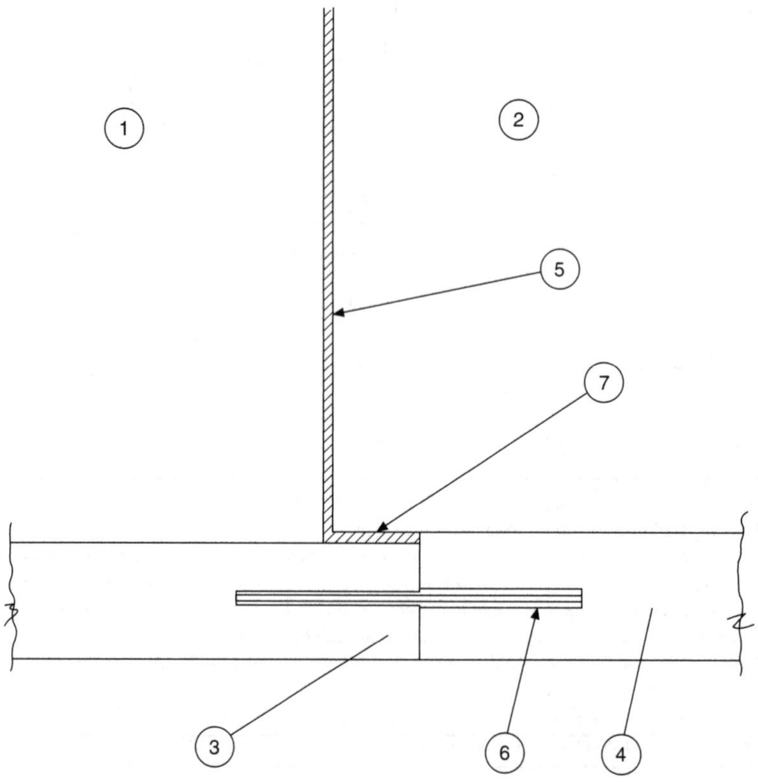

Notes

1. Existing building on soil of poor/medium bearing capacity.
2. New extension on soil of similar quality as existing building, therefore, slight task of differential settlement anticipated.
3. Again, scarcement of existing foundation retained.
4. New reinforced concrete raft or wide strip foundation for extension butted directly to existing footing.
5. Same 'Furfix' connection detail as in Figure 5.27.
6. 16 mm diameter mild steel dowel bars 600 mm long (drilled and grouted into 25 mm diameter holes – to allow for a little vertical movement – 300 mm into existing strip foundation). The remaining 300 mm length, which is embedded in the strip foundation of the extension, is coated with a debonding compound to allow for lateral movement. About 3 No. dowel bars at each connection position should be sufficient for most low-rise domestic buildings.
7. 20 mm thick rot-proof filler board to accommodate some differential settlement.

Figure 5.28 Alternative method of tying-in at foundation level (e.g., 'Denso' tape) to allow for contraction in the concrete and slight differential settlement. This could be used to tie-in two raft foundations, with the dowels spaced at about 200 mm centres. A similar method of tying-in ground beams to pad foundations of framed buildings can be used (see below, Figure 5.28).

Walls

- Wall profile plates: For low-rise extensions and large multi-storey extensions of traditional construction, the tying-in of walls is best achieved using profile plates bolted to the existing building. 'Furfix', 'Helix Spiro-Starters' and 'Catnic ProCon 250' are three typical proprietary names of these connection plates (see Figure 5.29). Expanding stainless steel fixings such as 100 mm long 'Rawlbolts' can be used to anchor the profile plates to brickwork and stonework. Some authorities, however, may insist on the use of resin-bonded stainless steel bolts when fixing the plates to no-fines concrete or blockwork walls.
- As indicated above, small-size multi-storey extensions are sometimes required to form a new means of escape staircase or provide a suitable protected enclosure to an existing exposed steelwork fire stair. In such cases, full tying-in up the complete height of the extension is essential to avoid it leaning away from the main building (see below). This may require more robust fixings such as lateral anchor bolts into the existing frame if present in addition to any wall profile plates. A structural specialist would need to determine the size, spacing and location of such connections.
- Toothing or block-bonding: In older buildings with small extensions toothed joints were used to connect new brick walls to old (Catt and Catt, 1981). Even if a suitable flexible mortar (e.g., one containing lime) is used, there is still a risk that the brick toothing will shear in the event of any differential settlement.

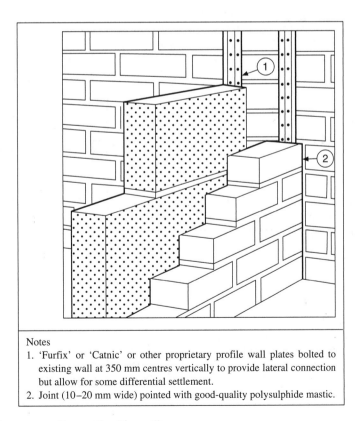

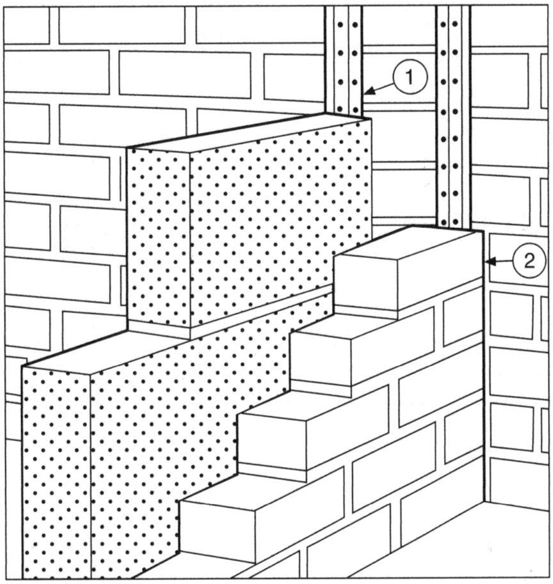

Notes
1. 'Furfix' or 'Catnic' or other proprietary profile wall plates bolted to existing wall at 350 mm centres vertically to provide lateral connection but allow for some differential settlement.
2. Joint (10–20 mm wide) pointed with good-quality polysulphide mastic.

Figure 5.29 Modern connection method for walls

- If both the existing building and its proposed extension have a brickwork finish, alignment and matching of the bricks will need consideration for such toothing to be successful. There could, for example, be a problem of mismatch in brick size and/or finish of new work and existing – particularly if the only available bricks are metric and the existing are imperial. A change of material or cladding may be required to mask the construction detail.
- Block-bonding, however, is better than brick-toothing. It can be used in stone- as well as brick-built walls. Another advantage of block-bonding over toothing is that the key obtained with the latter depends on the integrity of single bricks, which are liable to be damaged in the cutting process. However, adjoining existing quoins may have to be cut out in order to maintain bond 'closers' (Catt and Catt, 1981).
- Risband joint: In some crudely built extensions a risband (i.e., butt) joint would be used between the old and new constructions. However, this method is highly prone to failure because of the lack of any proper tying-in and its resultant inability to accommodate differential settlement. According to Catt and Catt (1981) structural stability may require the restraint of tops of walls by strapping joists to the existing structure even if the walls of an extension are not bonded in. However, for these reasons neither risband joints nor toothed- or block-bonding joints should be used nowadays.

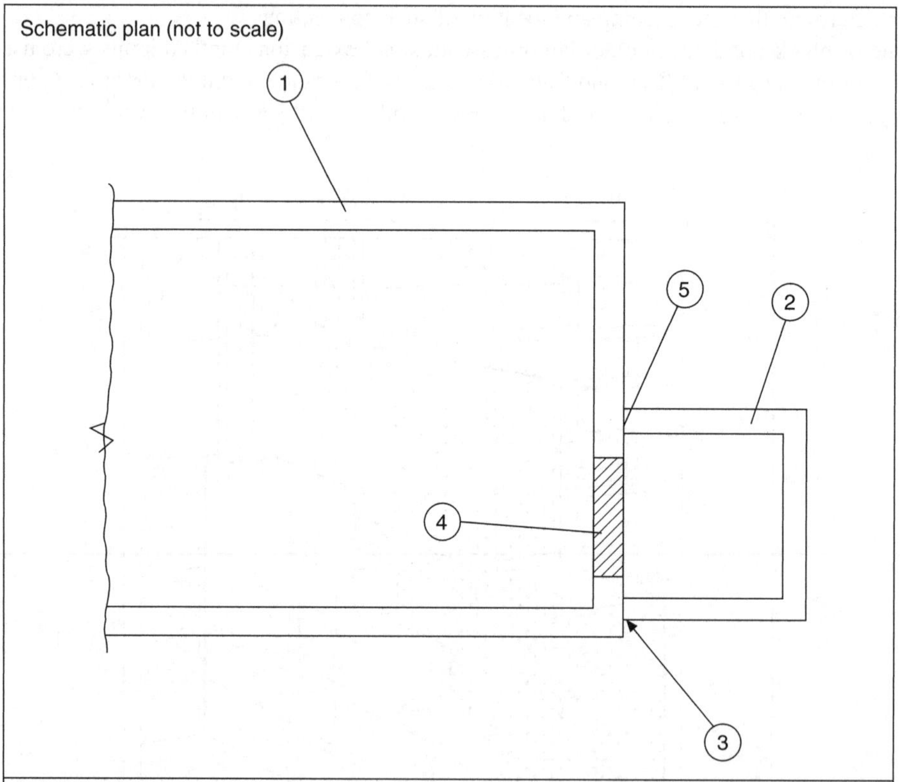

Notes

1. Existing building.
2. Small side/front/rear extension (i.e., less than full width or under 30 m^2).
3. New work usually set back from face of existing building to mask joint.
4. New opening formed or existing opening enlarged in external wall of original building. A lining to the reveal of the opening could be used to cover joint between old and new and mask any movement.
5. Walls tied using standard connection method (see Figure 5.29).

Figure 5.30 Connection requirements for small extensions

Form of construction

The type and degree of connection between an existing building and its extension depends on the form of construction of the two structures. As a general rule with lateral extensions the two constructions should be compatible. Thus, if the existing building is of load-bearing masonry, the extension ideally should be of a similar form. Still, extensions of modern timber-framed construction with an outer leaf of brickwork are often adopted as the form of construction for an extension to a building of traditional solid load-bearing masonry. On the other hand, it is unlikely these days that an extension of this traditional form of construction would be erected against a timber-framed building.

Schematic plans (not to scale)

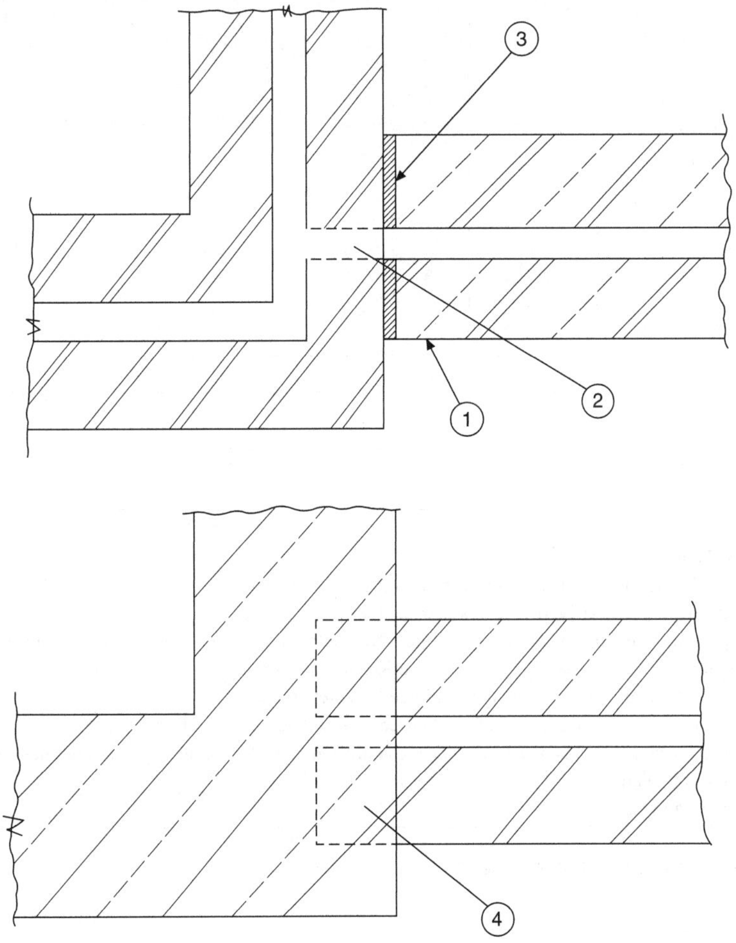

Notes

1. Cavity-wall extension set back from but abutted to cavity wall of existing building.
2. Cavity should be made continuous or have vertical dpc inserted to avoid moisture bridging laterally.
3. Extension connected to existing using 'Furfix' or other proprietary wall profile plate.
4. New cavity wall of extension may be toothed into existing solid masonry every two or three brick courses using lime or other weak mortar or better still connected using the proprietary wall profile plate described above.

Figure 5.31 Connection details at solid and cavity walls

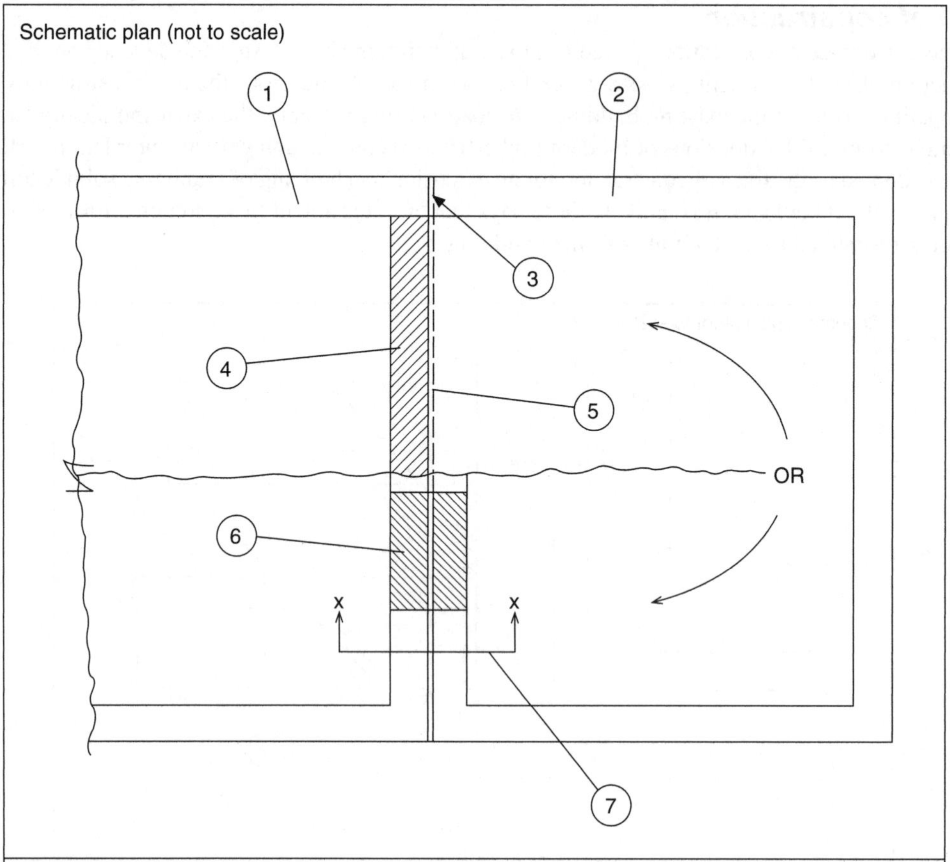

Notes
1. Existing building.
2. Large side/end/rear extension (i.e., at or near full width or over 30 m^2).
3. Wall connection method can be the same as for small extensions.
4. It may be possible to remove most of the existing wall abutting extension to maximize opening space but this would depend on the form of construction and any imposed loads involved. This may necessitate the installation of new steel or reinforced concrete lintels.
5. 20 mm wide movement joint formed in floor between old and new constructions.
6. New opening/s formed in walls at abutment between old and new. Again, these will require new steel or precast concrete lintels.
7. See detail in Figure 5.33.

Figure 5.32 Connection requirements for large extensions

Connection requirements to framed buildings

Timber-framed dwellings

When a timber-framed dwelling is being extended it is normal to design the extension having the similar form of construction – which is normally platform frame. Figure 5.34 shows how a new timber-framed extension can be attached to an existing building of the same construction.

Vertical cross-section at X–X (not to scale)

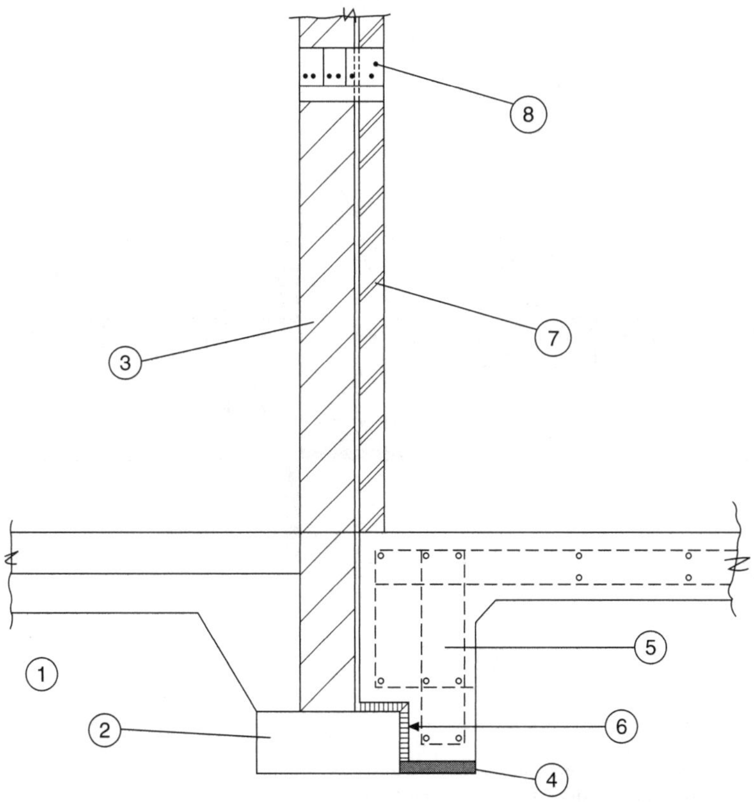

Notes

1. Building regulations compliance not shown.
2. Existing foundation – underpinned or made good if ground beneath it has eroded or is otherwise suspect.
3. Existing solid wall of main building.
4. Blinding layer.
5. New stepped reinforced concrete foundation of extension.
6. 25 mm minimum thick rot-proof filler board.
7. New wall of extension abutting existing building. A single skin may be all that is required to mask the existing wall if it is in a poor condition and the new wall is not taking any imposed loads. If, however, support is required for an upper floor structure or the roof of the extension, a more substantial wall construction such as cavity or one brick thick wall may have to be provided.
8. New precast concrete lintels on concrete pad stones.

Figure 5.33 Cross section of detail at independent connection

Steel-framed buildings

Extensions to framed buildings are normally of a similar structural configuration (see Figure 5.35). As indicated earlier, it is highly unlikely that an extension of load-bearing masonry would be attached to a framed building. However, framed buildings nowadays are often used for extensions to buildings of traditional construction – such as a timber-framed extension to a solid structural masonry dwelling.

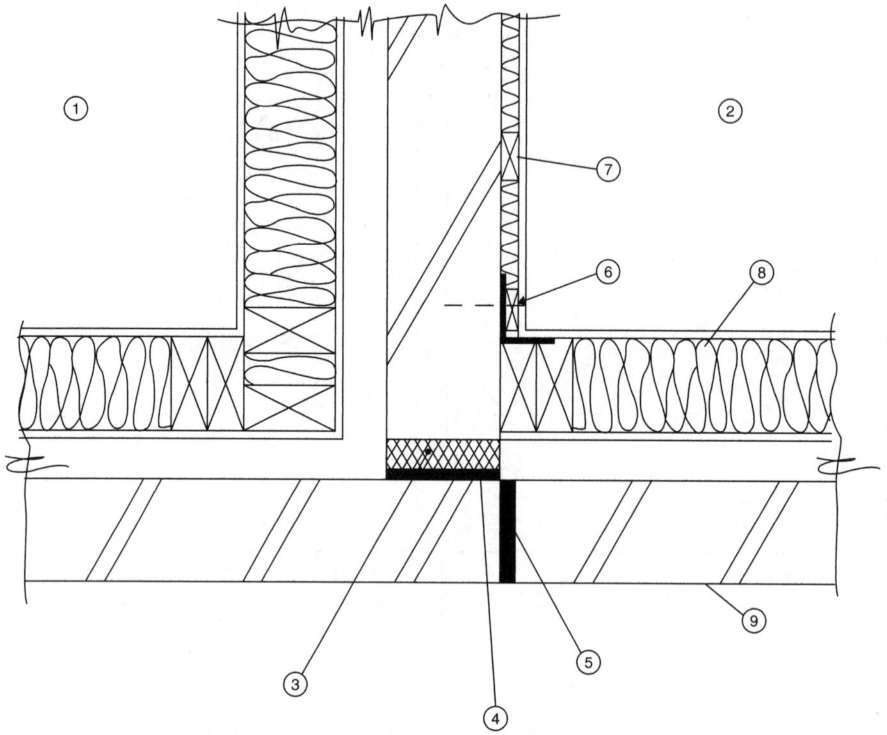

Notes
1. Existing building of timber-framed construction.
2. Lateral extension also of timber-framed construction.
3. Vertical gap cut in brickwork at junction with new timber-framed construction of extension using circular carbornundum/diamond tipped saw to inserting of polyurethane batt to act as a thermal break.
4. Vertical dpc to provide moisture break.
5. 'Furfix' or other wall tie profile plate bolted to existing outer leaf to provide connection for extension.
6. 75 × 75 × 75 mm stainless steef angle plate 2 mm thick with 50 mm slot in face against existing brickwork to allow for differential movement.
7. 35 × 25 mm softwood framing with 25 mm thick polyurethane insulation between battens and covered with gypsum dry-lining plasterwork.
8. New timber-frame inner leaf.
9. New facing brickwork outer leaf.

Figure 5.34 Junction of lateral extension to timber-framed dwelling

In most cases with framed buildings, the extension is of a similar if not identical construction. Thus, if the original building is of steel-frame construction, it is preferable to make the extension of the same morphology and materials. The proximity of an extension of framed construction to another framed building will depend on a number of factors such as the similarity of the frameworks, the loading imposed on the extension, load paths being taken, etc. These differences in construction are outlined in Figure 5.35.

If, however, the extension to a building of traditional load-bearing masonry is of framed construction, the structure of the latter will have to be built almost directly abutting the former. This is required to provide support for the extension and continuity of the fabric. Because of the scarcement of the existing building's foundation, the extension's foundation will need to be cantilevered.

Figures 5.35–5.37 illustrate some of the connection details for steel-framed buildings. The requirements for concrete-framed buildings at substructure level are similar to those for steel (see Figures 5.38 and 5.39).

Schematic elevations and plans (not to scale)

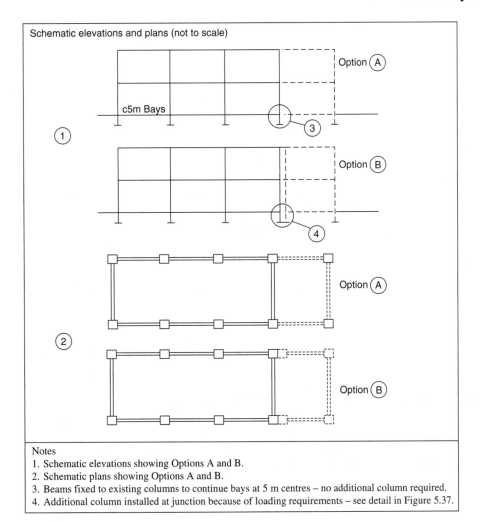

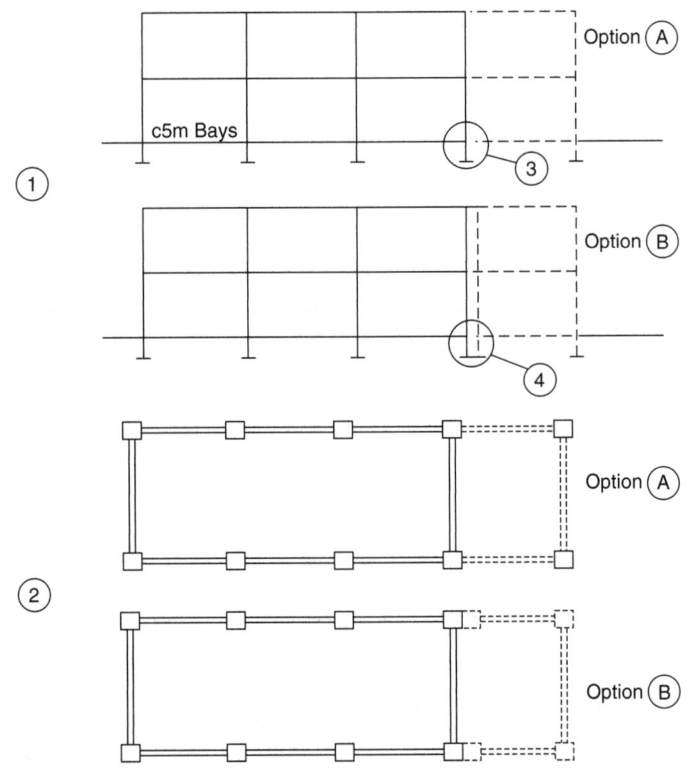

Notes
1. Schematic elevations showing Options A and B.
2. Schematic plans showing Options A and B.
3. Beams fixed to existing columns to continue bays at 5 m centres – no additional column required.
4. Additional column installed at junction because of loading requirements – see detail in Figure 5.37.

Figure 5.35 Options for connecting a framed extension

The extent of connection of the superstructure of a framed extension to the existing building of similar construction will depend upon the risk of differential settlement. Where the existing bays are continued the beams of the extension are bolted to the columns of the existing building (see Option A in Figure 5.35). Little if any connection is required above ground level where the superstructure of the extension is effectively an independent structure (Option B in Figure 5.35). A plan detail of the column joint in such circumstances is shown in Figure 5.40. However, where there is a risk of differential settlement, building control may insist that the pad foundations of the existing and extension frameworks at the abutment are tied using steel dowels in a similar manner to connection detail shown in Figure 5.29. The dowels will resist vertical settlement but allow some lateral movement.

Summary

Lateral extensions offer property owners the ideal opportunity to expand their facilities without the need to move to alternative premises. The work can, with only some disruption, be undertaken whilst the main building is still being occupied.

Schematic vertical cross-section (not to scale)

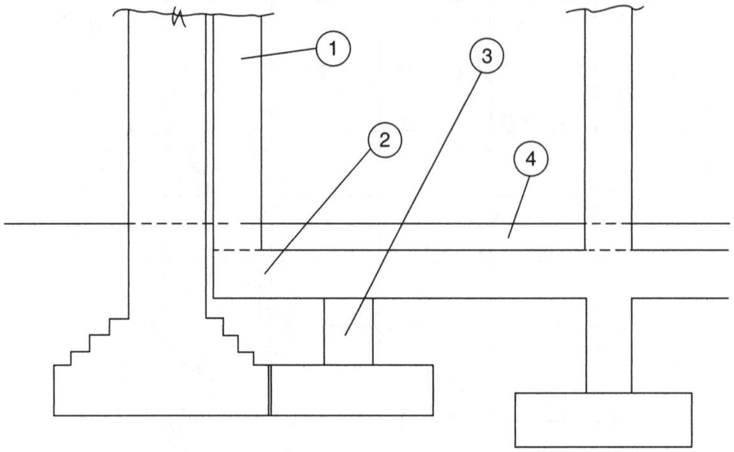

Notes
1. Column of new extension.
2. Cantilever beam.
3. Short column under cantilever beam linked to pad foundation.
4. New ground-floor slab of extension with 20 mm wide construction joint at junction with existing building.

Figure 5.36 Framed extension to load-bearing masonry building

Schematic vertical cross-section (not to scale)

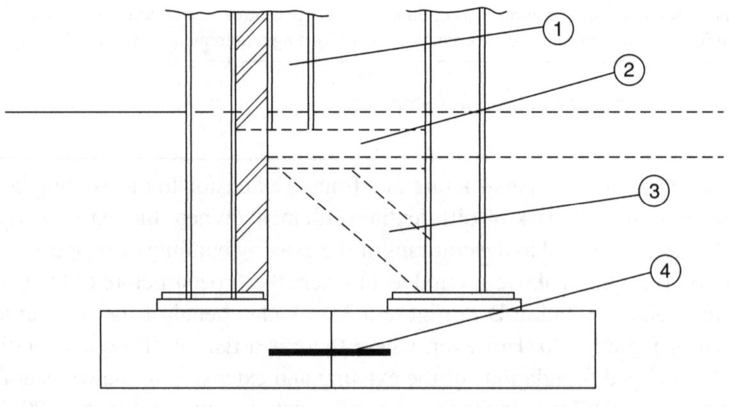

Notes
1. New steel stanchion – if considered necessary for structural reasons.
2. Cantilevered steel beam welded or bolted via angle sections to stanchion (if necessary).
3. Steel strut welded at connections, to stiffen cantilever beam.
4. Approximately 3 No. 16 mm diameter steel dowel bars at 350 mm centres drilled and plugged using epoxy mortar.

Figure 5.37 Framed extension to framed building

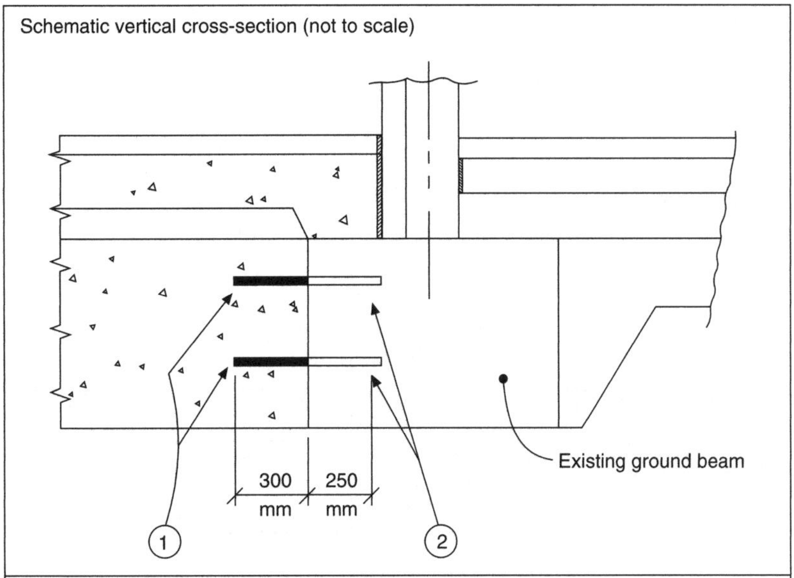

Notes
1. Dowel bars to be treated with debonding compound in new foundation.
2. 5 No. 16 mm diameter dowel bars 550 mm long drilled and plugged using epoxy mortar (3 at top row at 350 mm centres and, two at bottom row at 350 mm centres).

Figure 5.38 Typical ground beam connection to existing foundation in middle of bay

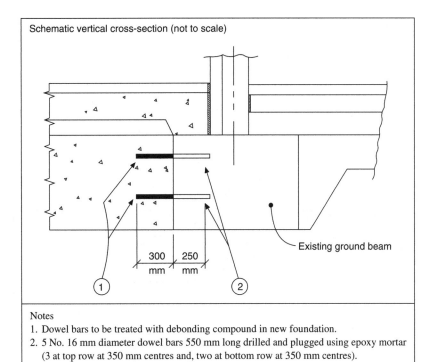

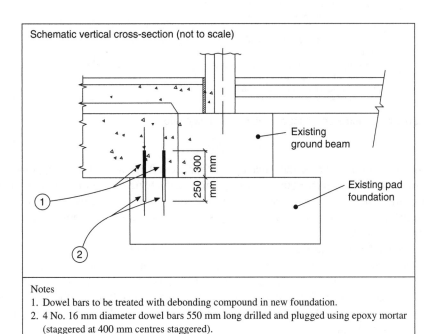

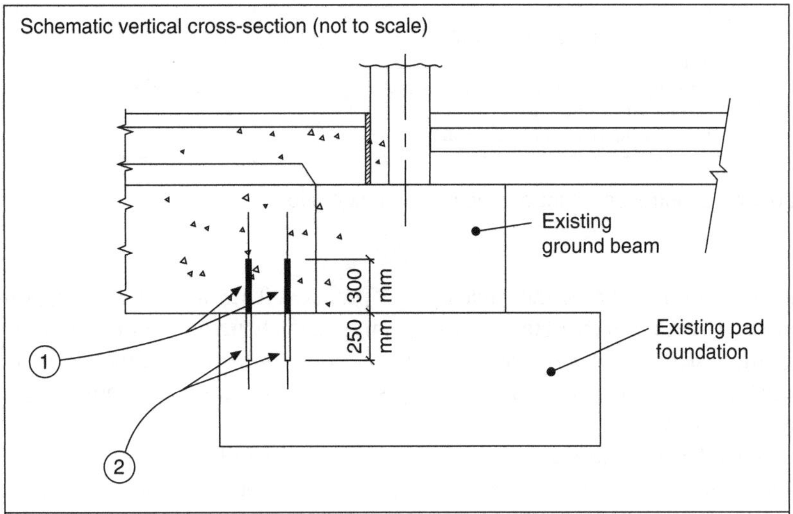

Notes
1. Dowel bars to be treated with debonding compound in new foundation.
2. 4 No. 16 mm diameter dowel bars 550 mm long drilled and plugged using epoxy mortar (staggered at 400 mm centres staggered).

Figure 5.39 Typical ground beam connection to existing construction at pad foundation

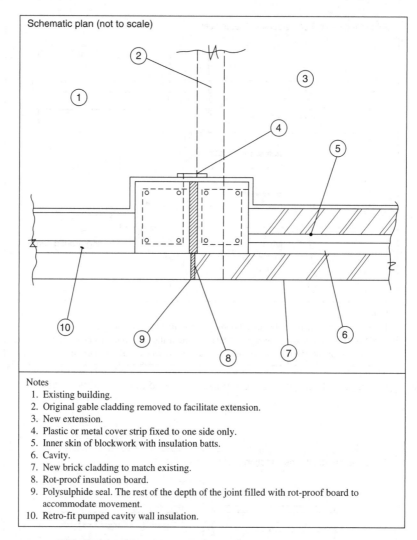

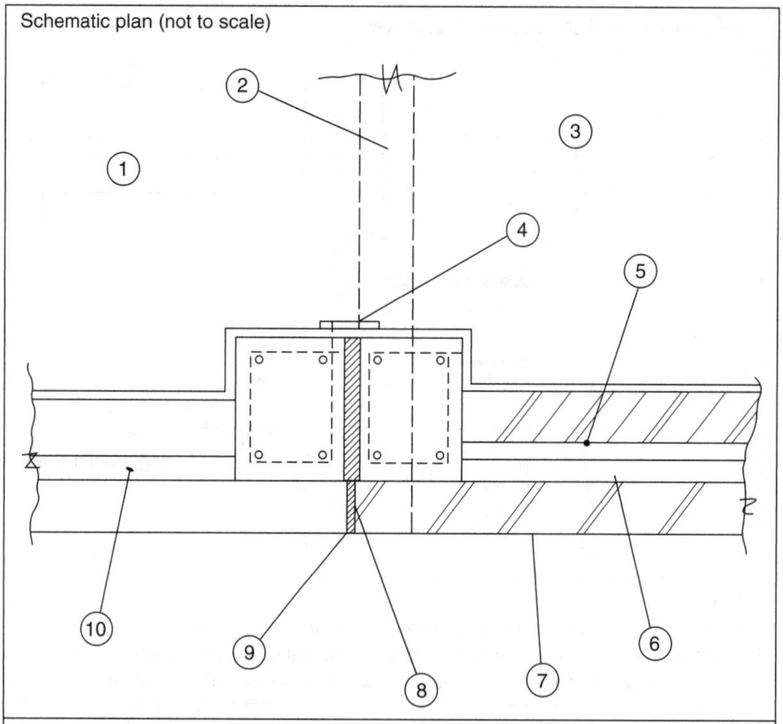

Notes

1. Existing building.
2. Original gable cladding removed to facilitate extension.
3. New extension.
4. Plastic or metal cover strip fixed to one side only.
5. Inner skin of blockwork with insulation batts.
6. Cavity.
7. New brick cladding to match existing.
8. Rot-proof insulation board.
9. Polysulphide seal. The rest of the depth of the joint filled with rot-proof board to accommodate movement.
10. Retro-fit pumped cavity wall insulation.

Figure 5.40 Column construction joint at junction with extension

Such extensions not only enlarge the capacity of a property. They also offer an opportunity to create better use of existing space within a building to accommodate changing working or living patterns. The increase in facilities such as 'granny flats' and study rooms is likely to generate more extensions to dwellings. Similarly business expansions and changes in work practices can prompt lateral extensions to commercial buildings as well.

Nevertheless, lateral extensions must be carefully designed and built if avoidance of major planning difficulties and constructional problems are to be avoided. It also involves considering impact of this type of extension on the building regulations relating to the existing part of the building affected by the extension. The interface between the old and new constructions, therefore, probably poses the biggest challenge to designers and contractors involved in such adaptation work. This is especially important in the case of lateral extensions to historic buildings.

Lateral reductions, although rare, inevitably have structural implications. These and vertical reductions are dealt with in Chapter 7.

6

Vertical extensions

Overview

This chapter covers the factors that need to be addressed in the design and construction of roof space adaptations and other forms of vertical extensions to buildings. Loft extensions especially are the most popular form of this type of building enlargement. The requirements and precautions for increasing a building's volume both upwards and downwards are outlined in this chapter.

Background

Vertical extensions can often offer the most discrete and economical way of increasing the capacity of an existing building. They are often formed either in the roof space or basement (see Figure 6.1). Such changes are of course less obtrusive than lateral extensions. In many cases the only obvious signs of, say, a loft conversion is the presence of roof windows or dormers. Basement extensions, being below ground, are even less noticeable externally unless a light-well at this level is formed.

Most vertical extensions are upward because there is nearly always free space to expand in that direction. In contrast, downward vertical extensions are limited because of the physical restrictions imposed by the substructure and soil conditions. On rare occasions the latter type of extension may still be feasible if a basement is required in a very tight site (see below).

The constructional and structural implications of vertical extensions, though, are often greater than for other forms of additions to buildings. This is because vertical extensions require staircases and possibly a lift for vertical access. Staircases or lift-shafts of necessity take up space, which must be accommodated in both the new storey and the one below it. They may also be required to have a minimum fire resistance of 1 h.

Vertical extensions also inevitably entail increases in the loading on the existing building. In such cases, strengthening work to structural elements such as walls, floors and foundation may be necessary (see next chapter). However, the substructure of most buildings is usually over-designed, some with a built-in allowance for additional loading, so that they can easily accommodate an extra storey without substantial strengthening works.

Moreover, because of their impact on the extra requirements for stairs and services in the building, vertical extensions are sometimes more complex than lateral additions. For example, linking any drainage lines from toilet or bathroom facilities in the loft space to the existing soil vent pipe may prove awkward. This will inevitably increase the project's costs.

Vertical extensions also are likely to require a higher level of fire and sound insulation between floors. This is especially so if they are to become separating or party floors (see Improvements described in Chapter 9) because of mixed or multiple use within the adapted building.

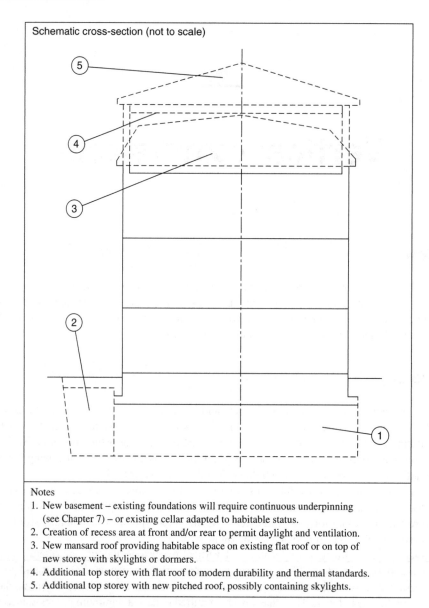

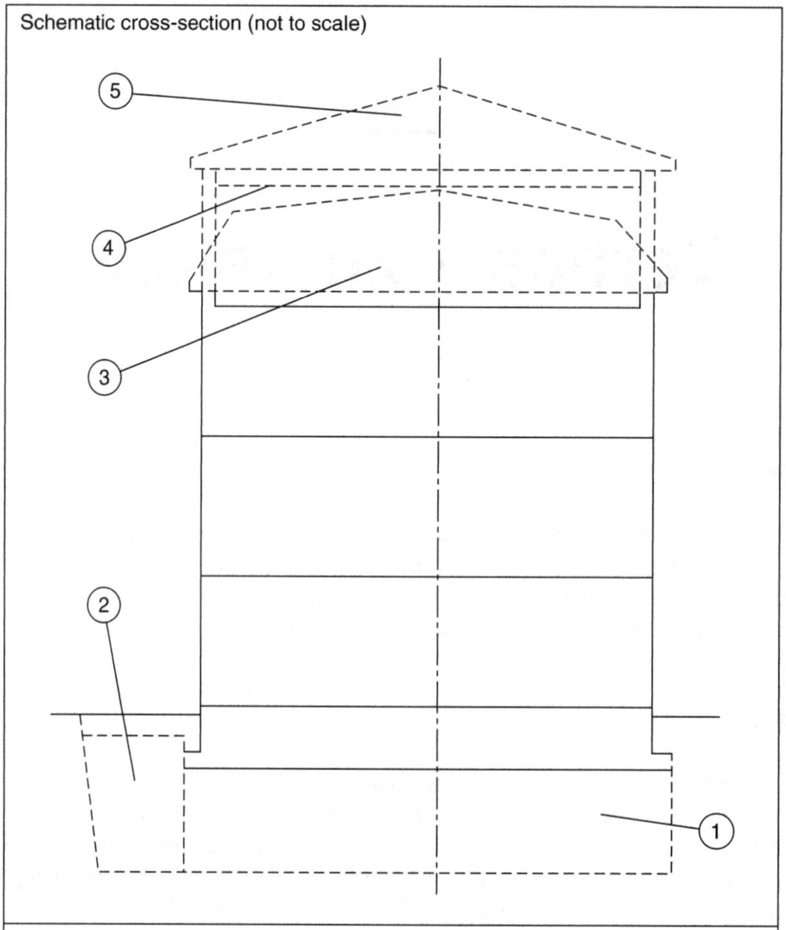

Notes
1. New basement – existing foundations will require continuous underpinning (see Chapter 7) – or existing cellar adapted to habitable status.
2. Creation of recess area at front and/or rear to permit daylight and ventilation.
3. New mansard roof providing habitable space on existing flat roof or on top of new storey with skylights or dormers.
4. Additional top storey with flat roof to modern durability and thermal standards.
5. Additional top storey with new pitched roof, possibly containing skylights.

Figure 6.1 Scope for extending a building vertically

Loft extensions

Preamble

Adapting the roof space is usually the simplest and thus cheapest way of adding another storey to a building. This is why 'attic or loft conversions' are by far the most common type of vertical extension. The reason for this is that the roof space of traditional buildings is the largest unused void for potential expansion of use *within* a property. It can provide attractive internal spaces, particularly for residential use. In avoiding any

Schematic cross-section (not to scale)

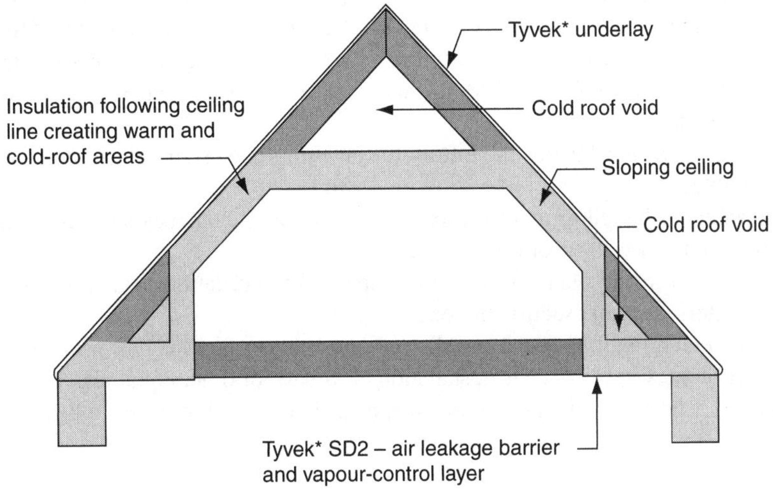

Figure 6.2 Modern 'room in the roof' style of loft conversion using sealed system

obvious enlargement of a building, a loft extension still achieves an expansion in the habitable space without the need for conspicuous external alterations. Thus, its unobtrusiveness is one of the key advantages of a loft extension.

Loft extensions are normally restricted to pitched roofs of traditional construction (see types below). They are not feasible in modern trussed rafter roofs. This is because of the spatial and structural constraints such constructions pose. Modern pitched roofs for economic reasons have a shallower pitch and contain lighter sections than their traditional counterparts. It is thus difficult to achieve the necessary headroom within the roof void of the former type of roofs (see next section).

The pitch of a modern roof can be as low as 17° if covered with interlocking tiles. The standard pitch, however, is nearer 30°. Moreover, the members in the trusses making up roof structures prefabricated these days are designed within strict tolerances. This is so that the size of timber required is minimized, which results in savings in materials costs as well as lighter roof dead loads. However, these sections cannot be easily cut or adjusted without seriously compromising their load-bearing performance.

Pitched-roof structures of traditional construction, in contrast, are more substantial in size and section. Their timbers (such as ties) are therefore easier to cut and remove or reposition and still retain their structural integrity. See Figure 6.3 for an indication of the spatial constraints involved in converting a roof space to habitable use.

If an upward vertical extension is required in a pitched-roof of modern construction, however, major alterations to this part of the building are required. The only feasible means of achieving this is to remove the existing roof structure and replace it with either a traditional purlin/close-couple roof or a modern trussed roof designed to contain habitable space between the structural members. In either case, the cost might be at least double that of a conventional loft conversion.

In any event, loft conversions entail the following:

- Installing decking on and sound insulation between the ceiling ties: These ties then become joists, which may require strengthening or even replacing (see next chapter). The decking can be either tongued and grooved (t&g) chipboard sheets or standard t&g softwood boards. Care should be taken when doing this work to avoid nailing through cables or pipes that are lying over the joists/ties.

- Providing natural light into the loft via dormer window/s or rooflights: The type and style of window used might be dictated by economic as well as planning considerations. A rooflight, for example, is much cheaper and less intrusive than a dormer with a similar size window. It is about half the cost of the latter.
- Improving existing stair access or providing a new stair: The original access to the roof space might be via a 'Ramsey' ladder (i.e. a folding hatch ladder) or very narrow (i.e. $<$600 mm) and steep (i.e. $>$45° pitch) stair.
- Reserving space in the level below the loft for the new/improved stairwell: A small loss of space on this floor is inevitable.
- Cutting an opening in the ceiling to form a stairwell: This work is both messy and disruptive. It also involves trimming the ceiling joists to accommodate the staircase.
- Inserting insulation in the plane of the roof slope, with ventilation to avoid condensation if placed between the rafters (see next section but one).
- Modifying and enlarging the capacity of the services to the loft: This will probably involve repositioning the hot- and cold-water tanks. The installation of 'Stuart' or other equal and approved hot- and cold-water pumps may be required where there is an insufficient head of water to the loft.

Options for loft extensions

Ideally a roof pitch of about 45° is required to achieve adequate headroom without the need to involve major adjustments to the roof such as installing a dormer. Figure 6.3 shows the desirable floor to ceiling dimensions for the roof space.

Many roof pitches however are well below the 45° optimum pitch for a loft conversion. As indicated earlier the usual pitch of a modern roof is around 30°. This is too low for forming habitable accommodation in a roof space unless a flat-roofed dormer forms part scheme (see Figure 6.4). Moreover, the structural configuration of modern roof trusses is such that they cannot easily accommodate a loft extension.

The pitch of the roof will also dictate the shape of the usable space within the loft. 'Coom'-ceilings (i.e. sloping ceilings from the walls) are inevitable in standard-size roof spaces. However, the extent of such ceilings is limited in that they should not start less than 1.2 m up from floor level.

Exposing the rafters to achieve a sloping ceiling (i.e. sometimes referred to as a 'skieling'). This can help achieve greater space and give added character to the roof space. However, this can only be achieved if the

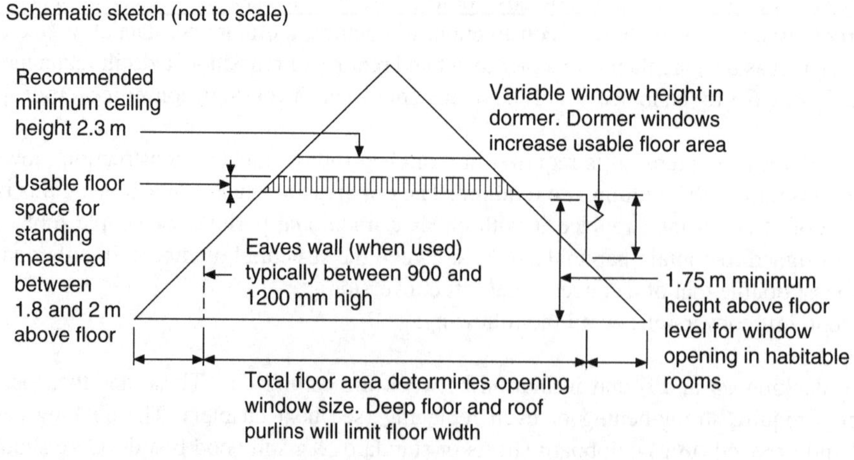

Figure 6.3 Desirable room dimensions for an attic (TRADA, 1993)

rafters are deep enough to accommodate the 50 mm minimum ventilation gap, insulation and plasterboard finish.

Adequate headroom is often a problem in loft areas. Recessed lighting in the ceiling should be used where the floor to ceiling height approaches the 2 m minimum dimension.

There are basically six options for achieving the optimum dimensions in a roof space. These are described in Table 6.1.

Ventilation and insulation requirements for loft extensions

Traditional room-in-the-roof design, using the ventilated roof approach

Forming accommodation in a previously empty loft space usually necessitates upgrading the thermal efficiency of the roof slope. When the insulation is placed between the rafters a 'warm roof' void is formed. This requires a minimum 50 mm continuous vent gap between the topside of the insulation and the underside of the sarking board (per BS 5250, 1995). This is analogous to the ventilation requirements for 'cold-deck' flats roofs – except, of course, that the ventilated space is on the slope rather than in the horizontal.

To ensure an adequate level of ventilation in a pitched roof where the insulation is placed between the rafters, tile or slate vents on the slope as well as at the eaves may be needed with this method. However, where careful attention to the roof finish is required on conservation grounds discrete vents should be used. These may still appear conspicuous on slate roofs because of their relatively 'flat'-slope profile. Alternatively a 'sealed pitched-roof' design can be used to obviate the need for such background ventilation (see Chapter 9).

Modern warm room-in-the-roof design, using the breathing roof approach

With the rising thermal standards imposed by the building regulations it may be increasingly difficult to achieve the maximum U-value without temporarily removing the roof coverings. For example, if the rafters are 125 mm deep, the maximum depth of insulation could only be 75 mm to maintain the 50 mm continuous vent gap to the underside of the sarking. This would probably be an inadequate depth for ordinary mineral fibre or polystyrene insulation to satisfy the U-value requirements of the building regulations (see Chapter 9).

The breathing or sealed approach uses a vapour-control layer on the warm side of construction plus a breather membrane (such as 'Tyvek®' – as described in Chapter 9). The breather membrane, which is placed over the insulation and sarking board, acts as a wind-waterproof layer and allows moisture to continuously escape from the roof. There is therefore no need to provide a 50 mm continuous vent gap above the insulation, which can thus be the full thickness of the rafters (see Figure 6.2). The cold roof voids at the eaves and ridge also do not need to be ventilated.

Stairs

Statutory requirements

Part K of the Building Regulations 2000 (as amended) (Section 4 in Scotland) covers the main standards relating to stairs serving a single dwelling are summarized in Table 6.2.

Style of stairs

Stairs have either open or closed risers. Non-enclosed stairs, which are permissible up to two-storeys, are normally designed with open risers – provided the gap is not more than 100 mm. Closed stairs, on the

Table 6.1 Options for accommodating an attic (Clarke Associates, 1989)

Option	*Implications/Comments*
1. Raise the pitch to accommodate the required dimensions 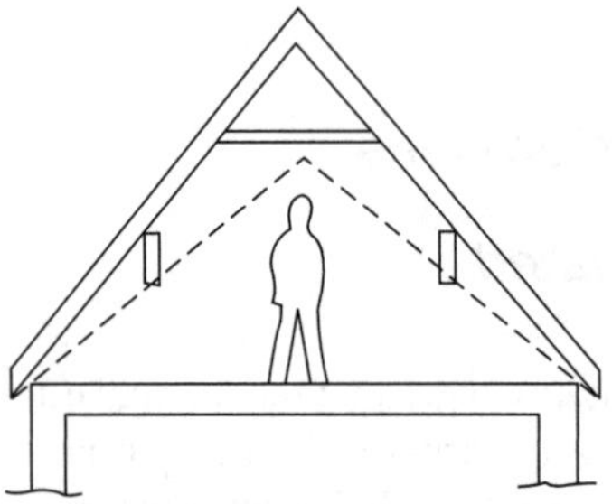	Existing chimneys will need extending. Junction with adjacent roofs requires treatment. Purlins required to support rafters. For large spans either laminated timber or steel UB sections would have to be used. The former are lighter and thus easier to install.
2. Raise wallhead 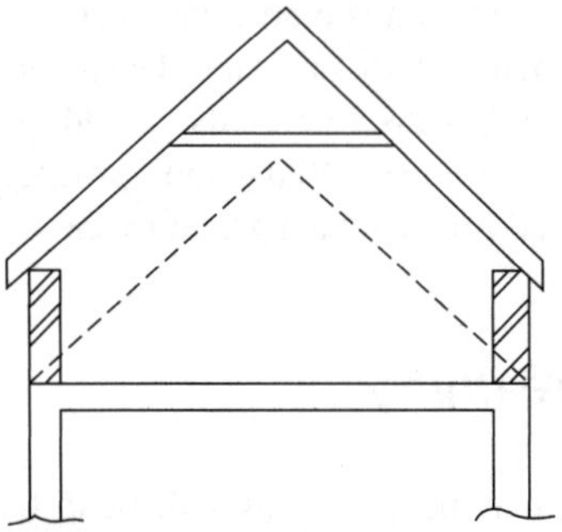	Coping will have to be removed and exposed wallhead made good. Difficulty of brick matching and problems of eaves alignment.
3. Flat-roof extension to rear of existing pitched roof 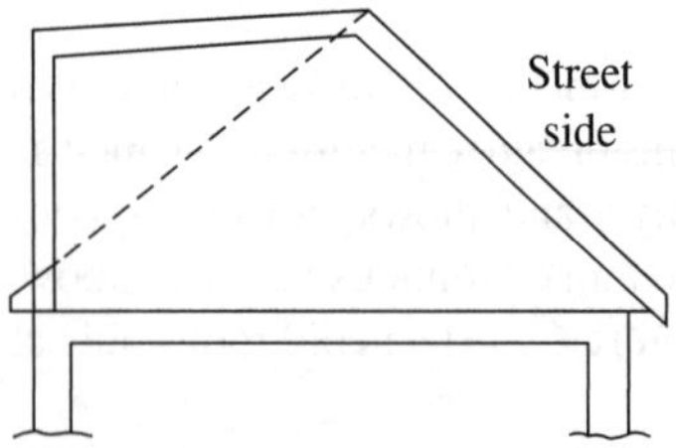	Major alterations required to existing roof structure. Result gives a continuous dormer effect. Very intrusive to the roofscape at the rear.
4. Rear mansard style roof extension 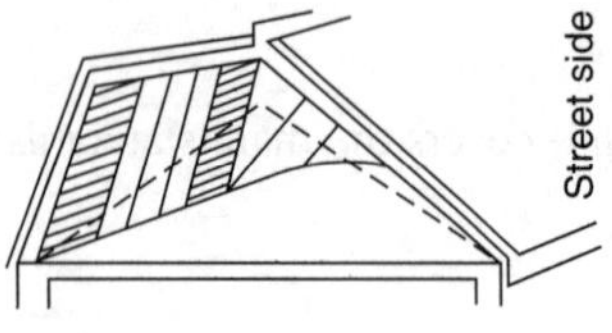	Again, major alterations required to existing roof structure. The result is less intrusive and better use of the roof space.

(*Contd*)

Table 6.1 (*Contd*)

Option	*Implications/Comments*
5. Half mansard 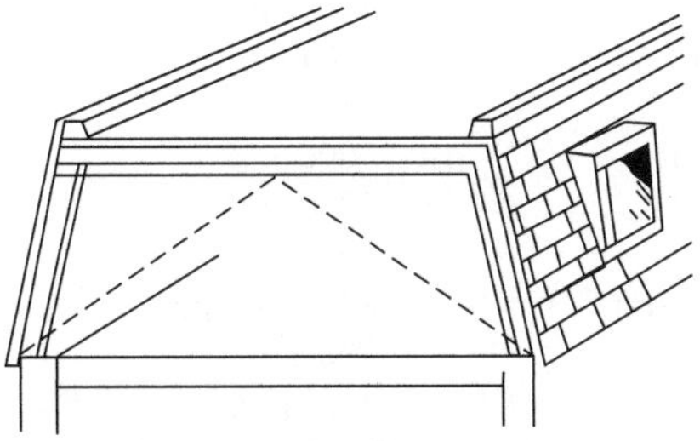	Possible long-term flat-roof drainage and maintenance problems.
6. Full mansard 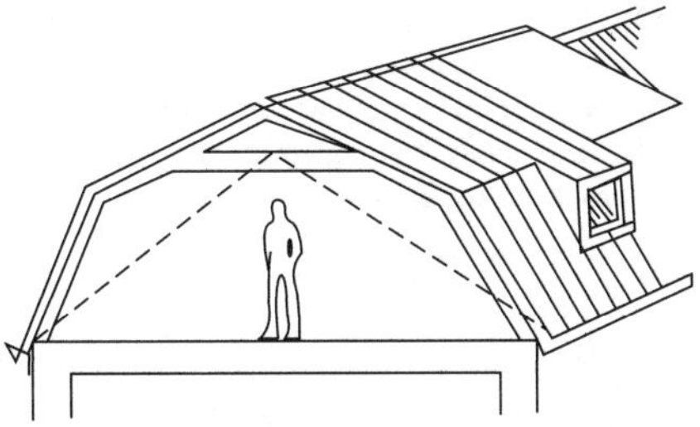	'Real' mansard is less intrusive, but would involve substantial roof work.

Schematic sketch (not to scale)

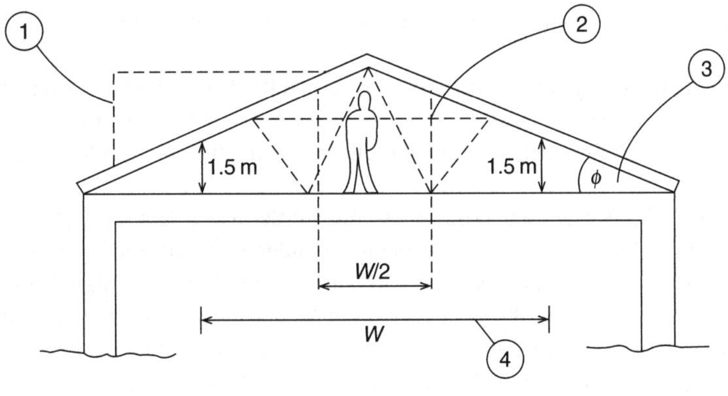

Notes

1. Possible new dormer if pitch is sufficient (i.e. >30°).
2. Possible existing ties and struts. These roof members would have to be repositioned or if they are removed the roof structure may require strengthening.
3. W = minimum usable width of room.
4. Where there is a shallow pitch <30°, only large span roofs would be able to accommodate usable space.

Figure 6.4 Typical pitched-roof slope

Table 6.2 Statutory requirements relating to stairs

	Type of Stair	
Key requirements	*Private stair*	*Any other stair*
Maximum rise (1)	220 mm	170 mm
Minimum going (2)	225 mm	250 mm
Maximum pitch	42° (3)	38°
Minimum headroom (4)	2000 mm	2000 mm
Minimum width (5)	800 mm	1000 mm
Maximum number of steps	16	16
Height of balustrade or handrail	840–1000 mm	840–1000 mm
Maximum gap between balusters and risers	100 mm	100 mm
Minimum length of landing	900 mm	1200 mm

Notes

1. Stairs, except in a dwelling, must have profiled risers to minimize tripping, and nosings distinguished through contrasting colours.
2. In the case of tapered treads, the minimum width of the narrow side is 50 mm.
3. The combination of maximum rise and minimum going will result in a pitch steeper than this.
4. In some buildings it may be difficult to achieve the necessary headroom because of restrictions such as the underside of a major support beam or roof structure. In such cases, relaxation might be granted to allow a headroom limit of 1.95 m. However, this will normally come with conditions attached. These may involve the installation of additional lighting and smoke alarms at the landings.
5. Between handrails. A private stair providing access only to one room.

other hand, can easily achieve an adequate level of fire protection but this makes the staircase dark and uninviting.

Buildability, cost, safety and available space are the primary determinants of a loft extension's stair design. The most popular styles of stairs for loft conversions, listed in order of safety and accessibility, are:

- Straight flight: This is relatively the easiest and most cost-effective to install. A baluster rather than a solid banister or partition could be formed on the balcony and outer side of the stair to open it up and give more light. Using toughened glass or transparent acrylic for the balustrade can create a light-well effect.
- Winder and straight flight: Space constraints may obviate the use of a simple straight flight. In such instances these restrictions can be overcome by using a winder at the top and/or bottom of the stair. The greater complexity of this type of stair makes it more expensive than the straight flight stair.
- Dogleg: This may be required where the shape of the room where the stair is being positioned prevents a straight flight. Alternatively, a dogleg stair may be considered appropriate where there is sufficient space to have a more elaborate staircase – usually in commercial premises or larger dwellings.
- Spiral: Because they do not need anchoring to a wall spiral staircases are frequently considered as space-savers. The minimum diameter of such a stair, however, will be about 1830 mm plus 50 mm clearance all round for finger clearance. As with winders in stairs, the minimum width of tread at its narrow side is 50 mm.

The weight of a spiral stair is carried on a single point through a central pole. In such a case, strengthening the floor to take the point load may be required. For a ground-supporting floor in a single-storey dwelling having a loft extension this could involve providing a $300 \times 300 \times 10$ mm (or larger) mild steel base plate bolted to the floor to spread the load. In the case of upper floors, which are more likely to be of suspended timber construction, strengthening can be achieved by installing an additional joist or set of noggings under the proposed newel position.

It is more economical to use a standard-design 'kit' for a spiral stair. Some spiral stair manufacturers can offer bespoke staircases but these are inevitably more expensive – ranging in cost from about £5000 to supply and install for an ordinary single-storey steel stair to tens of thousands of pounds for the more elaborate types serving three floors.

Materials for stairs

- Timber: Softwood is the most common material for domestic stairs either in new build or loft conversion schemes. The exposed surfaces of the stair are usually finished with a matt or gloss paint. Exposed treads and risers will require resilient finishes such as hardwearing plastic or metal strips. In more prestigious schemes stained or varnished hardwood such as mahogany or teak offers a more attractive, natural finish. However, hardwood is much more expensive than softwood and requires more screwed fixings.
- Metal: Mild steel is an attractive material for stairs, especially the spiral type, because of its lightness and strength. However, like most other metal stairs, it is cold to the touch and requires a painted finish for protection against corrosion and to enhance its appearance. New or reclaimed cast iron stairs are a popular choice for spiral stairs in older buildings. The capacity of the floor would need to be checked, though, to determine whether it could support such a relatively heavy material. Stainless steel offers a modern, stylish finish to a stair, whether straight or spiral. Slender but rigid sections can be used for handrails and balusters. However, as with other stairs it is best to avoid locating the rails in the horizontal position to prevent children using them as a makeshift ladder.

Fire safety in loft extensions

Single/two-storey dwellings

As well as structural and thermal requirements, fire safety is a key criterion in loft extensions. Overall, for attic conversions of only one storey with one or two rooms and a total area of 50 m^2 or less, the basic fire safety requirements are:

- at least one means of escape;
- modified half-hour fire resistance for loft floor (see Chapter 9);
- half-hour fire-resisting enclosure to loft stair (ditto);
- new doors on to the new staircase to have half-hour fire resistance;
- habitable room doors to be made self-closing and glazing to be made fire resisting (e.g. with Georgian wired glass). Rising butt hinges are acceptable as a means of making doors self-closing in dwellings but not normally elsewhere (Haverstock, 1998);
- mains smoke detector and fire alarm (to BS 5446 Part 1).

Dwellings over two storeys

For these properties there are no exceptions in the regulations. Accordingly, the relevant full fire resistance (i.e. 1 h) will be required for the floors. How this is achieved is addressed in Chapter 7.

In the case of four or more storeys Clause 4.4 of BS 5588: Part 1: 1990 should be followed (TRADA, 1978).

In addition, normally an alternative means of escape must be provided. Self-closing fire doors with at least half-hour fire resistance should be fitted to all new doorways. Mains-linked smoke detectors at strategic positions (e.g. top and bottom landings) should also be installed as part of the loft extension work.

External fire spread

This is usually not significant for loft extensions. However, the following circumstances may require attention:

- The size, extent and location of new openings, both in the wall and roof, may be restricted.
- If the gable wall of a property becomes the external wall of a previously unused loft, it may be necessary to upgrade the wall's fire resistance. This will depend, though, on the wall's proximity to the boundary.
- The quality of the roof covering forming part of an attic conversion should match that of the original.

Cost of domestic loft extensions

A loft extension is, in many respects, the most convenient way of enlarging a dwelling's accommodation. Depending on the size of the space and facilities required, the average domestic loft conversion comprising two bedrooms or one bedroom and bathroom costs anything between £25 000 and £50 000 (not including fixtures and fittings) at current (2006) prices. Typically, it takes about 6–8 weeks to complete this type of contract.

The main cost items with construction works for loft conversions are as follows:

- Daylighting: installing a dormer or skylight – the former being about twice the cost of the latter – and making good the roof coverings.
- Decorations: including making good existing decorations damaged during and as a result of the adaptation work.
- Finishings: wall, floor and ceiling finishings; if t&g timber 'vee' boarding is used it must be finished with a fire-retardant varnish.
- Floor deck: including insulating and possibly strengthening floor structure.
- Services: electricity for lighting, CO/fire detectors and power, and (if a toilet/bath is provided) water supplies and drainage.
- Staircase: including forming stairwell and making good surrounding areas.

More elaborate or substantial domestic loft conversions, however, can easily exceed £50 000. This may be as a result of alterations to the existing roof structure to accommodate the new use, or a more elaborate stair design, or increasing the capacity and quality of services. The cost of timing of these and other adaptation projects are outlined in Chapter 11.

Retrofit dormers

Preamble

Dormers are vertical windows projecting fully or partly from or into the slope of a roof and having their own roof. They can dramatically improve the character of the roof, which would otherwise be plain and unattractive. Ideally, though, the chosen design should reflect the style of any dormers in and around the building being adapted.

Another advantage of having dormers in a roof is that they can help increase the potential accommodation space within the loft. The floor area within a dormer helps to open-up the space by enlarging the loft area whilst still achieving the minimum headroom.

There is a wide range of styles of dormers that can be used. The type selected depends on a variety of factors, such as:

- the style of dormers in the area;
- the local authority policy on roof extensions;
- the pitch of the existing roof;
- the morphology of existing roof construction;
- the client's new accommodation requirements;
- the extent and quality of natural lighting required.

Designing a retrofit dormer

Types of dormer

The main types of dormers are described in Table 6.3. However, not all of them are appropriate for retrofit dormers (i.e. a term used to describe new dormers installed in an existing building to form part of a vertical extension). For example, installing a half dormer would entail cutting through the wallhead to form part of the new window opening. This is very rare in practice.

The design of a retrofit dormer therefore should reflect the type of building and take cognisance of similar constructions in surrounding properties. Box dormers and flat-roofed dormers should be avoided at front elevations and other conspicuous positions as their plain design can seriously compromise the appearance of the existing roof.

Positioning the dormer

Consideration should be given in the design to locating the dormer in a suitable position provided existing dormers or chimneys do not get in the way. According to Clarke Associates (1989), the best position to put the dormer, particularly if is on the front elevation, is at any of the following:

- centred over the lower windows (and ideally no wider);
- on the centreline of the building;
- centred on a projecting bay or feature.

These design principles are illustrated in Figure 6.5.

Installing a retrofit dormer

Adjacent to a chimney

The size of the roof and the extent of any chimneys it contains will influence the position of any retrofit dormer. If a chimney is in the vicinity of the proposed dormer, there are basically three ways of dealing with it:

1. Abut the chimney to the dormer.
2. Enclose the chimney in the body of the dormer.
3. Position the dormer at least 600 mm away from the chimney (although the writer has seen examples where this gap is only 300 mm).

Table 6.3 Styles of dormers (Clarke, 1989)

Type of dormer	*Comments*
Flat-roof dormer 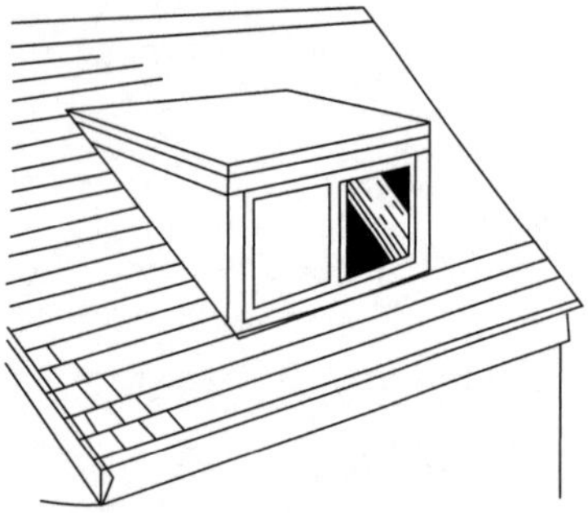	The most common but aesthetically an inferior type of dormer. Its popularity is primarily due to it being the easiest and thus cheapest dormer to construct.
Pitched-roof dormer (with or without hipped front) 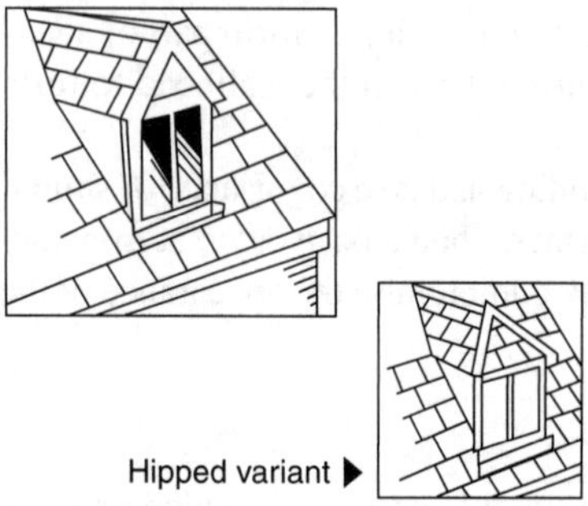	A more architecturally appropriate style as it matches better the profile of the main roof.
Dormer gable 	Similar to the preceding dormer, but this style is more common in Georgian and Victorian buildings.
Splayed dormer 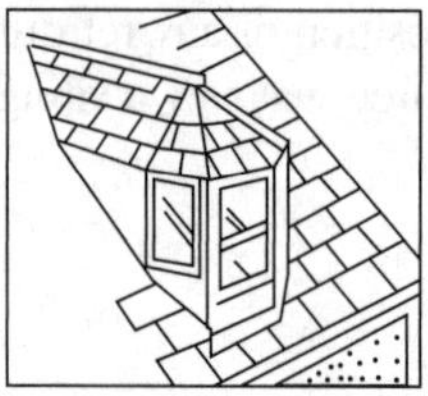	This style of dormer is found in many Victorian buildings with bay windows. It is common in certain parts of the country such as Scotland and North-east England.

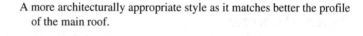

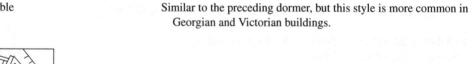

(*Contd*)

Table 6.3 (*Contd*)

Type of dormer	*Comments*
Eyebrow dormer 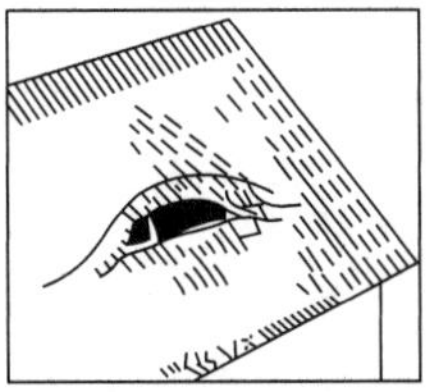	This type is more likely to be used for small dormers in thatched roof buildings. It is therefore predominantly found in rural areas.
'Catslide' dormer 	Another style of dormer predominantly found in rural areas. It is usually associated with pantiled roofs and dormers with painted or stained weatherboarded cheeks.
Barrel-vault dormer 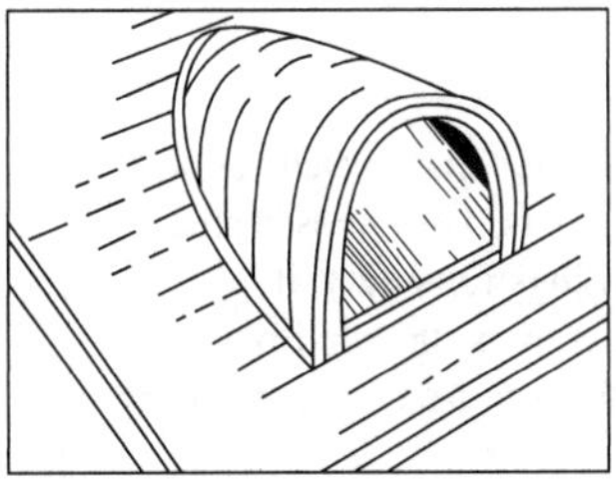	This type of dormer is basically a variation of the pitched-roof dormer. It is occasionally found on the roofs of late Victorian buildings in urban areas. The curved roof is usually covered with either sheet lead or bituminous felt.
Triangular dormer 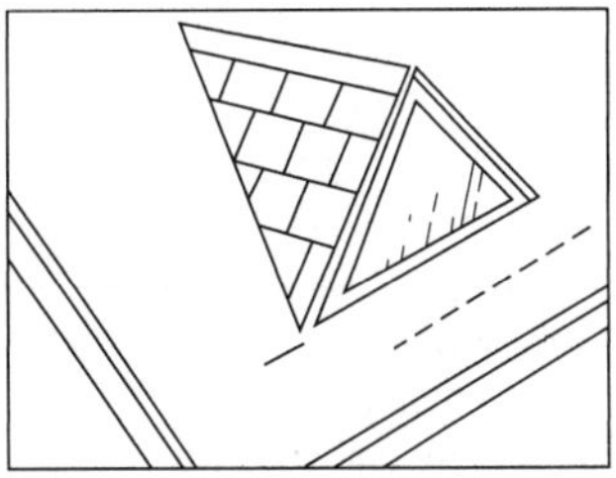	This style is usually used with smaller 'skylight'-type dormers. Its coverings match those of the main roof.
Mansard-roof dormer 	Usually restricted to rural areas but sometimes found in urban areas in dwelling having a wider roof span and lower pitch.

(*Contd*)

Table 6.3 (*Contd*)

Type of dormer	*Comments*
Gambrel-roof dormer 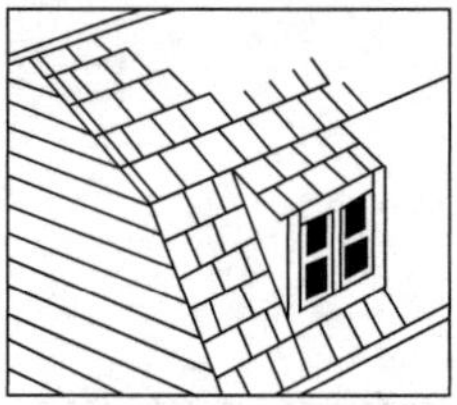	Widely used in weatherboarded cottages and houses in south-east England. Mainly rural areas.
Belvedere dormer 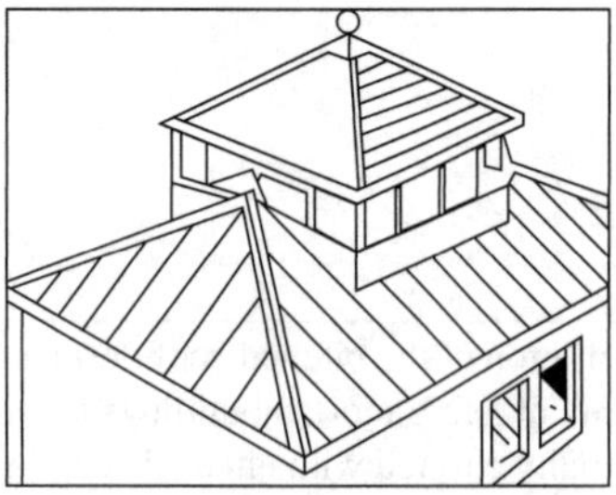	This can be used in roofs with short ridges to form a type of bell tower.
Recessed dormer 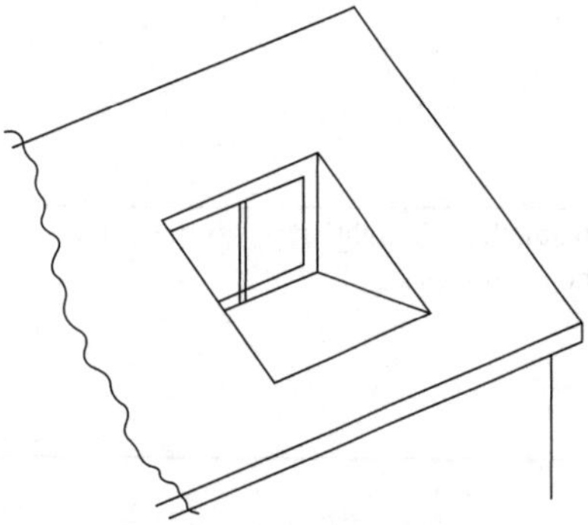	This type of 'dormer' is only found in larger span structures, as it cannot be accommodated on smaller roofs. The recessed area can also be used as a balcony, with railings for safety. It is ideal as the window opening and balcony on the top floor of penthouse flats. In smaller span roofs a semi-recessed dormer may be possible.
Box dormer 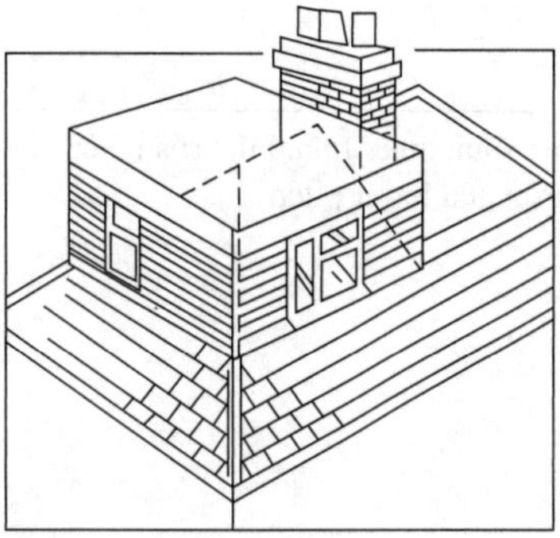	Architecturally, probably the worse type of dormer. Its blunt unattractive appearance can destroy the original profile of a roof. This type tends to be constructed in roofs that cannot offer adequate headroom without major disruption to their structural form.

(*Contd*)

Table 6.3 (*Contd*)

Type of dormer	*Comments*
Continuous dormer 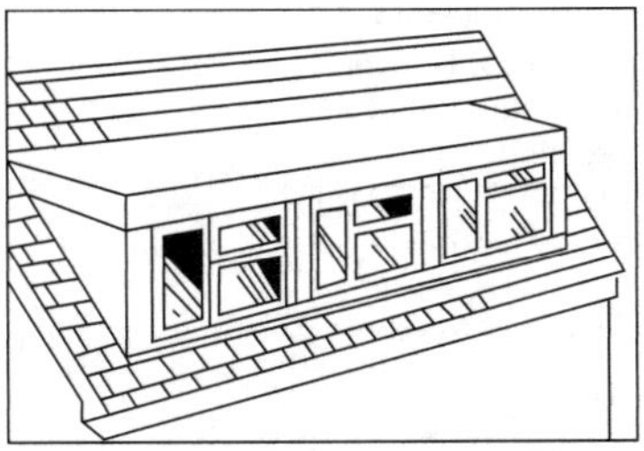	This occurs when two separated dormers in the one roof are joined together or where one dormer covers the length of the roof. It is sometimes encountered in existing chalet villas containing as-built dormers running the whole length of the roof.
Half dormer 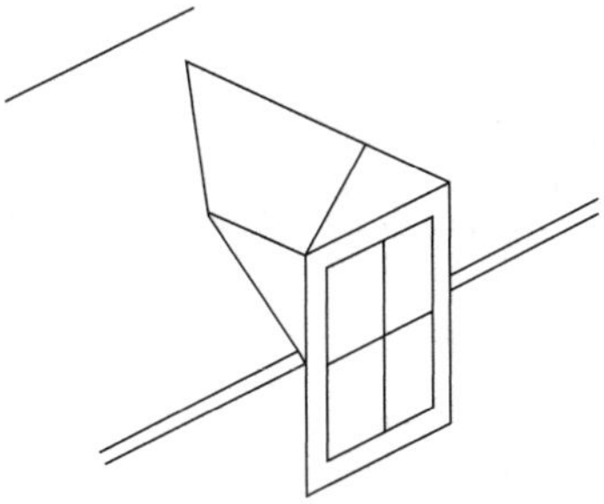	This type is common in some Victorian buildings. It is not really feasible as a retrofit option as it would mean cutting through the wallhead to form the window openings.

The first two options will require the construction around the chimney to have the required fire resistance. This can be achieved using a non-combustible material such as brick or block. However, detailing to ensure an adequate, weatherproof junction will be difficult.

Temporary protection

Once the position of the dormer has been established and approved the installation can then proceed. Allowance must be made for temporary protection during the opening-up work and the erection of the dormer carcassing. There is a high risk that rainwater penetration will occur in adverse weather conditions during this stage. Care needs to be taken to ensure that this temporary covering is adequately secured and has a generous overlap at the edges of the exposed area to minimize leaks. The client should be warned of this and advised to take sensible precautions such as removing vulnerable fittings below the position of the new dormer and having polythene sheeting under sufficient buckets ready to deal with any persistent wind-driven rain ingress.

Technical issues

It is normally less troublesome to install a dormer during the construction of the original pitched roof. Inserting a dormer in an existing roof entails a variety of disruptive works. Firstly, the roof coverings over the position of the dormer have to be removed carefully and laid aside for reuse. Secondly, a hole the size of the dormer's plan shape needs to be formed in the roof structure. This entails cutting through the sarking

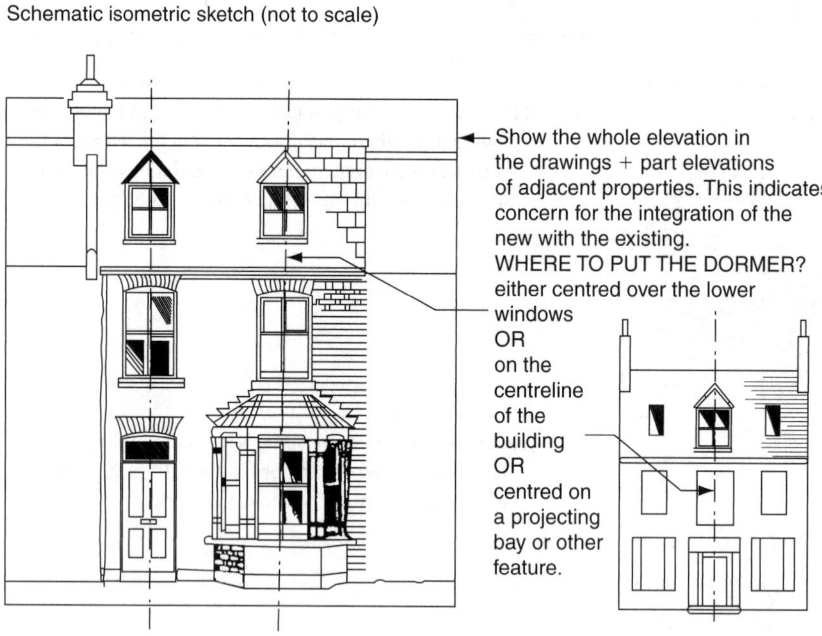
Schematic isometric sketch (not to scale)
Show the whole elevation in the drawings + part elevations of adjacent properties. This indicates concern for the integration of the new with the existing.
WHERE TO PUT THE DORMER?
either centred over the lower windows
OR
on the centreline of the building
OR
centred on a projecting bay or other feature.

Figure 6.5 Positioning a retrofit dormer (Clarke Associates, 1989)

boards and trimming the rafters once the covering is removed. The trimming rafters at the cheeks of the dormer need to be doubled-up to strengthen the roof structure at this position (see Chapter 7).

Erecting the carcassing of the dormer is the next stage of the process. There are essentially two ways in which this can be done. One is to erect the framework for the cheeks of the dormer from the loft-floor structure. The other is to build the framework of the dormer from the rafters of roof structure at either side of the opening.

Figure 6.5 illustrates the main technical issues involved when installing a retrofit dormer. The design of dormer selected will be made primarily on the grounds of cost. For this reason the flat-roof dormer is the most common choice. This is because its relative simplicity of shape makes it the cheapest option.

External finishes

The external finishes to dormers are also influenced by economic and aesthetic factors. The cheeks of many dormers, for example, are normally tiled or slated to match the existing roof coverings. In some cases a small dead-light window is incorporated into each cheek. In older properties the sides of dormers are clad with zinc or lead sheeting. Nowadays they are often clad with either timber weatherboarding with or an insulated panel system finished with a render to match the existing walling. In any case both the cheeks and roof of the dormer need to attain a U-value of at least 0.35 W/m^2. Whatever finished is used lead flashings should be installed at the junction of the cheeks and roof slope for weather-tightness.

The top of a flat-roofed dormer is normally covered with a single-ply polymeric membrane or a two-layer high-performance elastomeric felt. In some cases lead, copper or stainless steel sheeting is used.

Pitched-roof dormers can be slated or tiled to match the main roof. The extent and position of any rainwater goods on the dormer will depend on the direction of its roof slope. Dormer flat roofs that slope from their eaves to the junction with the main roof do not need gutters or downpipes – but this option does not encourage the disposal of rainwater away from the main roof. Moreover, the junction between the flat- and pitched-roof slopes must be carefully designed and constructed to prevent rainwater ingress.

A better design for the rainwater disposal is to make the roof slopes towards the eaves. In such a case, however, a half round gutter will be required to prevent rainwater cascading over the fascia.

Retrofit rooflights

Introduction

A more discrete way of providing natural light into an attic is to insert windows that follow the pitch of the roof. These are called rooflights, skylights or roof windows. As a general rule rooflights can suit most pitched roofs. They are by far the least obtrusive and most cost-effective way of achieving sunlight, daylight, ventilation and solar gain into a loft. In contrast, dormers are more expensive to install and more prominent in terms of their overall effect on the roof. Rooflights are better suited to roofs with pitches from 45° to 85°.

The building regulations (B1, in England and Wales and Section 2 in Scotland) require that windows within 6 m of the boundary must provide AA fire rating. If the windows are of the large double-glazed type, they will need laminated glass on the inside.

The building regulations for England and Wales and Scotland require that openings for escape or rescue purposes should have an unobstructed opening of at least 850 mm × 500 mm. Figure 6.6 illustrates the key requirements for fire safety in dormers and rooflights.

Roof windows are especially appropriate in the front elevations of properties in architecturally sensitive areas. Installing, say, a dormer in the roof at such a location as part of a loft extension would disrupt or spoil the profile of the roof (Figure 6.7).

Rooflights are now supplied under a variety of proprietary names, for example, Colt, Manor Joinery and Velux. The last named is probably the most popular market brand in Britain (see Figure 6.8).

Materials

Traditionally, rooflights were made from either good-quality softwood or mild steel. Nowadays, however, the frame and sash construction of roof windows consists of any of the following combinations:

- Timber only: High-quality softwood such as Scandinavian pine or hardwood such as mahogany, coated with a high-performance acrylic water-based exterior grade sealer.
- Timber/metal composite: High-quality laminated pine covered externally with a thin preformed sheet of dark grey aluminium; with natural stained wood finish internally.
- Plastic/timber composite: Polyurethane with a timber core.
- Plastic/metal composite: Polyurethane with stiffened steel profiles.

Plastic rooflights are often available but there are limitations on their use if they have a Class 3 or lower surface rating (e.g. each a maximum 5 m^2 in area and at least 3 m apart). Less commonly glass-reinforced plastic (GRP) is used.

The glazing can consist of a single pane or double or triple layers. The latter two types are preferable because of the acoustic and thermal insulation benefits that multi-paned-glazing units offer. Laminated or toughened glass may be considered for security or safety reasons. Low-emissivity glass can also be

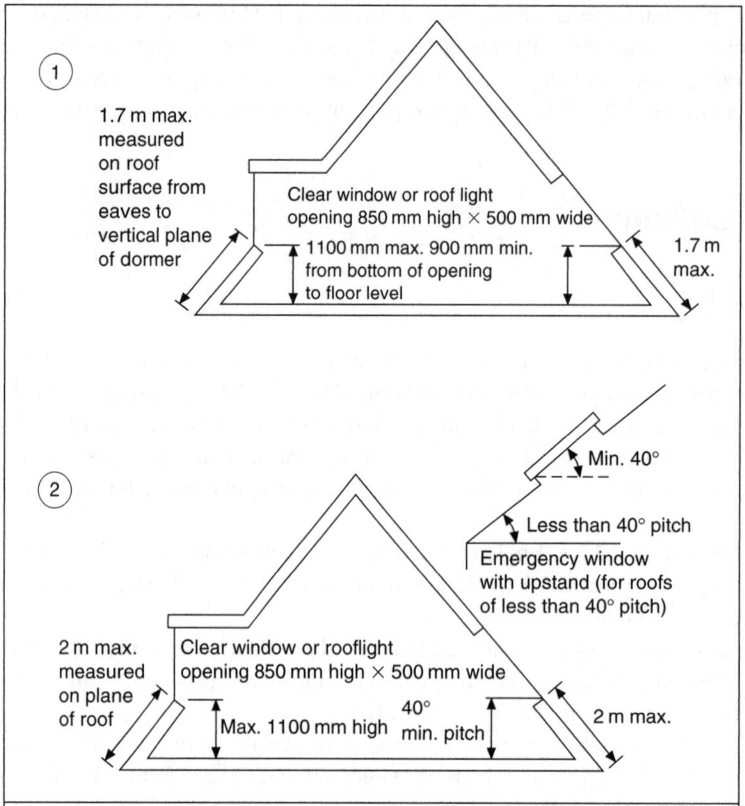

Notes
1. Requirements for England, Wales and Northern Ireland.
2. Requirements for Scotland.

Figure 6.6 Dormer or rooflight used as means of escape (TRADA, 1993)

included to provide a degree of solar protection. Alternatively, a solar reflective coating can be applied to ordinary glass.

The flashings and frame surrounds are normally of thin-section aluminium coated with a metallic paint. This material can be easily moulded to follow the profile of the roof tiles.

The hinges must be robust and durable. Anodized mild steel or, preferably, stainless steel should be used for these fittings. The hinges should allow the window to reverse for cleaning.

Accessories

A range of accessories is now available with rooflights, which can enhance their performance. These include features such as:

- Blinds: either roller or Venetian, to give shading or privacy, and a degree of sound insulation against pelting rain and hailstones.
- Window opening mechanism: ordinary lever handle or if the rooflight is high up, by rod or cord, or even mechanically operated.

Schematic isometric sketch (not to scale)

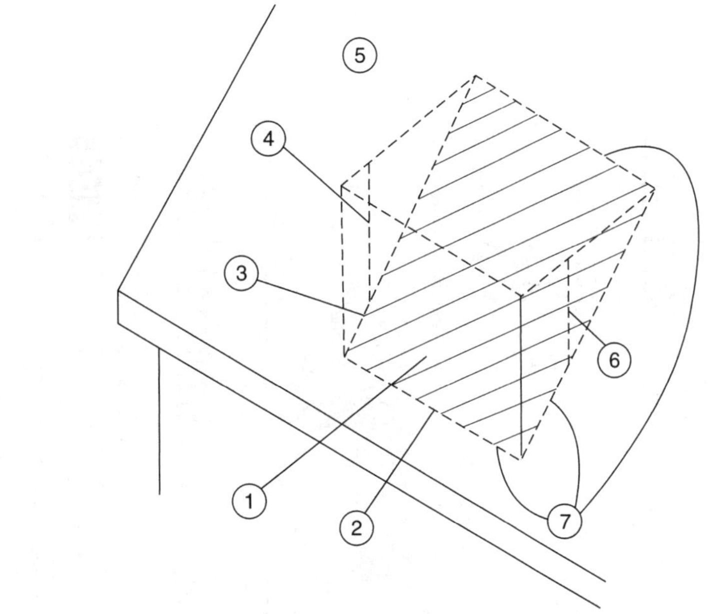

Notes

1. Existing roof covering and decking or sarking board (shown as hatched area) must be removed carefully. Tarpaulins or other temporary coverings should be used at the end of each working day during the opening-up work to minimize the possibility of rainwater penetration.
2. Trimmed rafters cut carefully and temporarily supported whilst 200 × 75 mm trimmer is installed at top and bottom of dormer (i.e. at 7).
3. Additional rafter installed on either side of trimming rafters to strengthen roof structure.
4. Main vertical timbers 100 × 50 mm at 300 mm centres, and doubled-up at corners of dormer. These struts can either be taken from the roof slope or the attic floor structure.
5. For small/standard-size dormers 150 × 50 mm roof joists – doubled-up at window opening – at 400 mm centres, should be sufficient. Warm deck-roof construction should be used for flat roof or duo-pitched roof. This will require the removal of more slates/tiles of existing roof covering to accommodate the increased depth of dormer roof structure.
6. Cheeks of dormer clad with sarking board, insulation, felt and finished with slates, timber weatherboarding or render on lath.
7. Proper lead flashings at the peripheral junctions of the dormer are critical for weather-tightness.

Figure 6.7 Installing a retrofit dormer

- Locking: can comprise key operated type, with ability to lock in the partially open position. Childproof safety catches are essential on most if not all roof windows.
- Vents: these should give at least 8000 mm^2 trickle (background) ventilation.

Problem areas

Preamble

As with any penetration through a roof the installation of a rooflight creates a vulnerable part of the building. As a result it is susceptible to the following problems if adequate precautions are not taken.

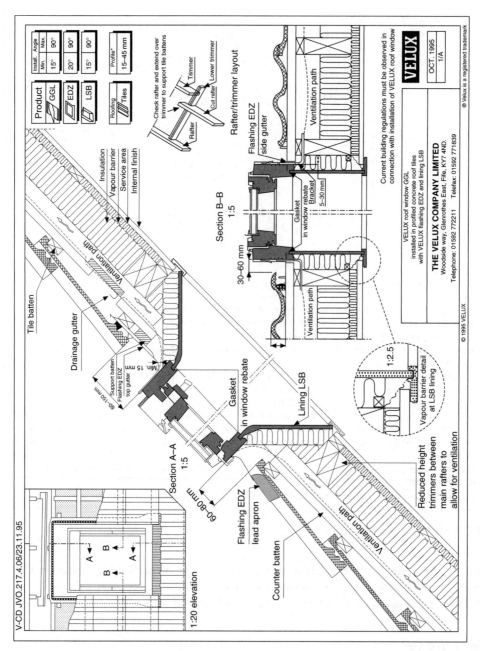

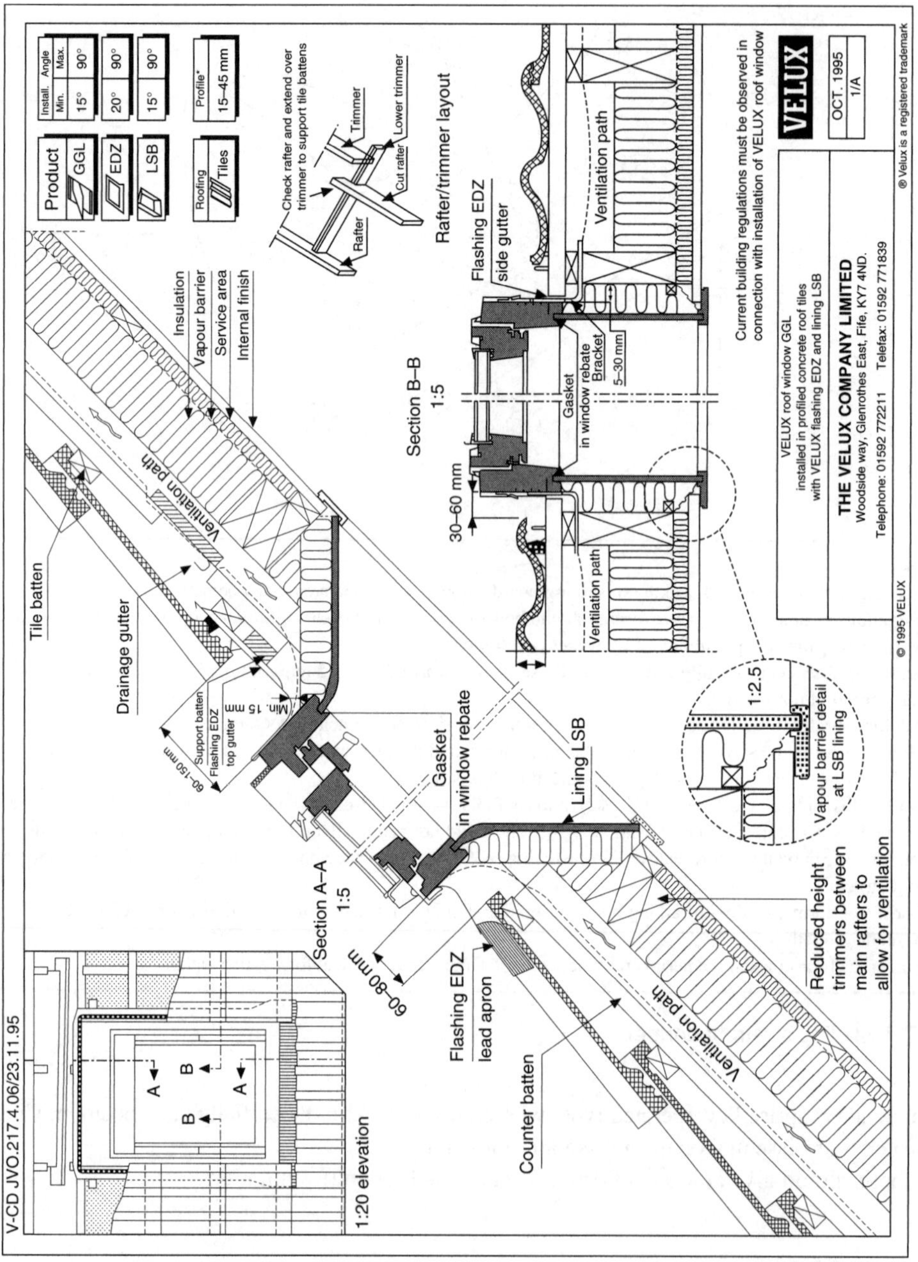

Figure 6.8 Rooflight details (Velux, 1995)

Leaks

These are obviously the most troublesome failure in a roof window. Adequate flashing properly fitted around the frame is critical to avoid this problem.

Condensation

This problem can be reduced on the inside surface of the glass by using double/triple glazing. It can be further assisted by increasing the airflow over the window by splaying the head and sill to improve warm airflow across the glass from a radiator under the window.

Heat gain/loss

Heat gain/loss from large rooflights can make the new attic area uncomfortable unless blinds are used for south-facing windows. In more severe cases outside awnings may be used, but these require regular maintenance and may be vulnerable to wind damage in exposed locations. Double or triple glazing in conjunction with blinds can keep heat losses to a minimum.

Cleaning

Using reversible sashes or windows that can be cleaned from the inside facilitates this. Safety, however, is the key requirement.

Noise

Noise owing to heavy rainfall on the glazing can be a nuisance for some occupiers. This can be dampened but not necessarily eliminated by using blinds and triple glazing.

Maintenance

This is more troublesome where there are more mechanical parts to the rooflight, such as remote operating devices.

Roof light-wells

Skylights (i.e. non-opening rooflights) can be fitted to pitched roofs to provide natural light to the loft space when it is used for storage. Light-wells, on the other hand, offer a cheap means of allowing more natural light into the habitable parts of a building via a skylight. They use no energy and require little if any maintenance. There are basically two methods of forming a light-well in the roof of a building: the traditional method and the modern method. The latter is described in more detail in Chapter 10.

The traditional method of doing this is to form a light-well which gives increased daylighting to the floor level below the roof space. This is particularly advantageous in deep plan areas on the top storey where natural light via windows is not available in the centre of the building.

Figure 6.9 illustrates schematically how a traditional light-wells is formed.

Roof extensions

Means

In tight sites the capacity to extend a building laterally can be severely limited. Mid-terraced properties for example can only allow an extension of a suitable size at the rear. Even so there may be little if any room

Schematic vertical cross-section (not to scale)

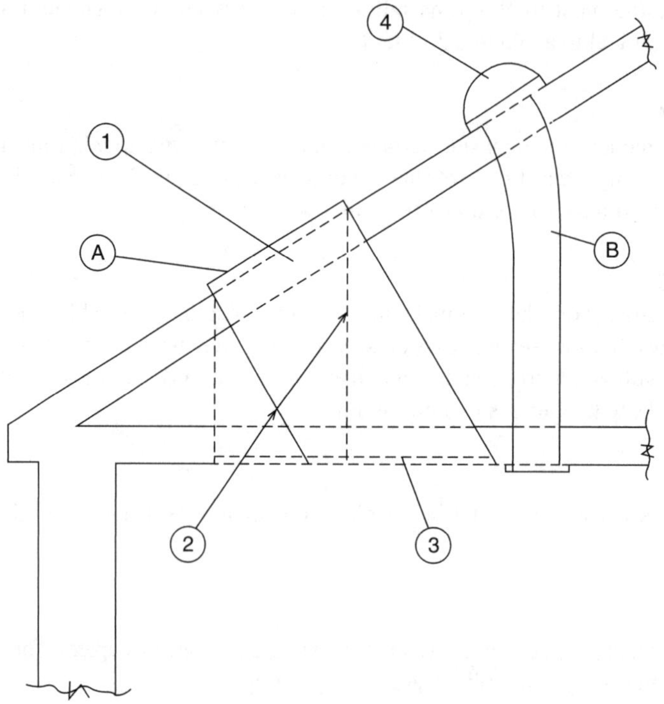

Notes

A. Traditional method of forming a light-well with skylight.
B. Modern method of forming a light-well using a reflective duct (e.g. Sun Pipe).
1. New opening formed in existing roof. A new rafter is added to the existing on either side of skylight to strengthen the roof structure if more than one rafter is trimmed.
2. Duct consisting of framing and plasterboard/plywood to form vertical or slanted light-well.
3. Security grille.
4. Hemispherical glass can come complete with flashing.

Figure 6.9 Methods of forming a light-well

to extend the property at the back. Planning controls as well as site constraints often preclude such extensions, especially in inner-urban areas. Moreover, a flat-roofed building in such circumstances cannot offer any loft conversion capabilities unless it is being over-roofed (see next section on Over-roofing). The substructure conditions or the proximity of tunnels or other encumbrances such as major services or utility lines may not make a downward vertical extension possible.

The only means of increasing the building's capacity in such circumstances is to add another one storey or more. This may include the provision of a pitched roof on the extension. However, as over-roofing a flat roof is a form of improvement that does not necessarily involve any increase in habitable space, it is included under the chapter dealing with refurbishment buildings.

In refurbishing framed commercial buildings, adding another storey is usually best achieved by using a steel structure. This approach can be adopted for steel and reinforced concrete-framed constructions. In the case of a reinforced concrete- or steel-framed building, for example, the new storey could comprise a steel portal frame supporting a curved roof. The portal frame could set back from the existing parapet to minimize disruption to this part of the building as well as provide an access path around the top storey.

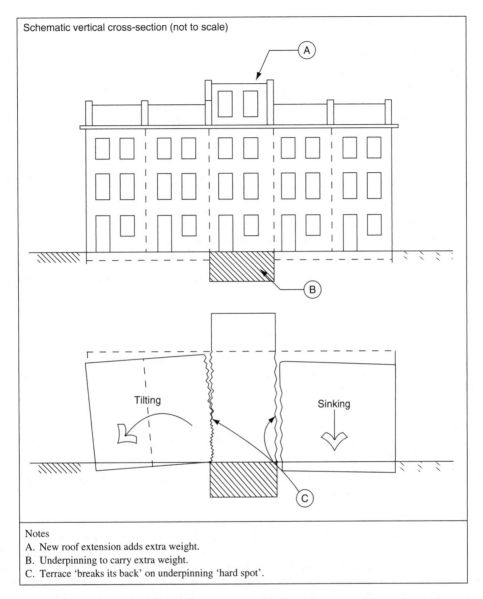

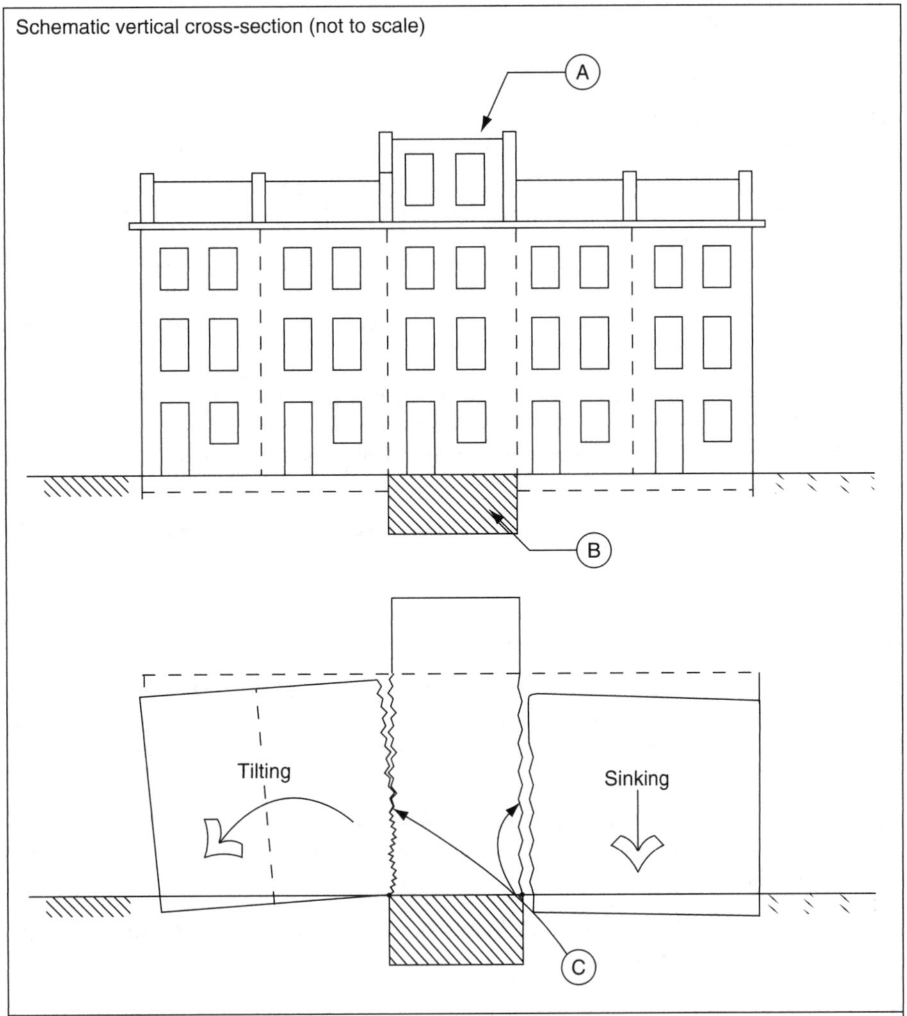

Notes
A. New roof extension adds extra weight.
B. Underpinning to carry extra weight.
C. Terrace 'breaks its back' on underpinning 'hard spot'.

Figure 6.10 Problems with partial underpinning to take new roof extension (Richardson, 2000)

The constructional requirements for adding a new pitched roof to a flat-roofed building are shown in Figure 6.10. The drawing illustrates how a steel portal frame can be used to form an extra storey with a new curved roof on a reinforced concrete-framed building. The new storey is set back from the existing parapet to provide a distinct break in the design as well as to give an access route for maintenance. Alternatively, a similar style of roof could form part of an over-cladding scheme to a building (see Chapter 9).

There are essentially three designs for a roof extension. In order of likely preference these are:

1. an additional storey incorporated within a new pitched roof (see next section on Over-roofing);
2. an additional storey (or storeys) with pitched roof;
3. an additional storey (or storeys) with flat roof.

Structural implications

A building that collapsed in Middlesex, England, during its adaptation in the mid-1990s highlighted the risks to health and safety associated with this type of work. The case in question was a three-storey commercial building that was halfway through a refurbishment contract in 1995. During the early afternoon of 1 August of that year a substantial part of the structure suddenly collapsed, killing four construction workers.

The following extract from the official report into the tragedy (HSE, 1999b) neatly summarizes the background to and causes of this tragedy:

> … The building was constructed in 1969/70, initially as a single-storey structure. It was extended upwards in 1970. In 1995 it was being refurbished by building contractors advised by a structural engineer and an architect.
>
> As refurbishment progressed, three men were engaged in removing panel walls between the key structural supporting columns. They discovered that a column at first-floor level was supported on a single lightweight concrete block. Within minutes of this discovery the building collapsed, killing the three men and the site agent who had been called to investigate.
>
> Subsequent investigation by inspectors of the Health and Safety Executive (HSE) and staff at the Building Research Establishment (BRE) revealed serious defects in the original construction of the vertical extension of the building. Initially, the building had a flat roof with a low parapet wall round the edge. When the building was extended top three storeys, the lightweight concrete blocks forming the bottom course of the parapet wall were left in position and used to support the load-bearing column at first-floor level.
>
> The collapse was caused by the failure of one or more of the lightweight concrete blocks supporting the brick columns at first-floor level. This led to consequent rapid and catastrophic collapse of two-thirds of the building. The extent of the collapse was much greater than might have been expected because of the lack of continuity(ties) between key structural elements.
>
> Examination of four brick columns still standing at first-floor level showed no externally visible signs of the lightweight concrete blocks in the columns. They had effectively been hidden by the facing brickwork, internal plaster, and the inclusion of infill brickwork …

Two recommendations (82a and 82b) were made in the above report. They are worth stating because of their relevance to many types of adaptation work:

> 82(a) Clients, planning supervisors, contractors and their advisers when renovating, refurbishing, extending or demolishing a building, particularly if it was built before the Building (5th Amendment) Regulations 1970 took effect, should address the possibility that it may not be robust, and that damage to a key structural element could lead to disproportionate collapse. Where this is the case the risk assessment should include an evaluation of the risks of such collapse. For instance, if heavy plant is to be used near to key structural elements it may be necessary to provide barriers to prevent contact with the building. Planning systems would be needed with crane operations and propping of the building could be appropriate.
>
> 82(b) The defects discovered in the brick columns in this building reflect either gross incompetence or total irresponsibility. Everyone in the construction industry needs to be vigilant against such blatant malpractice. DETR Building Regulations Division will issue advice shortly to local authorities and others on the inspection and appraisal of open plan precast concrete and masonry structures built before 1970.

Structural implications are, therefore, inevitable, regardless of which of the three options listed above is selected. To avoid problems such as overstressing, distortion or even disproportionate collapse, consideration must be give strengthening the main structural elements affected by the additional storey (see Chapter 7). It is possible, of course, that the existing building was designed and built to accommodate an additional storey. In such a case, the structure may require little or no strengthening, but this would also be dependent on the building's overall condition and on the extent of internal alterations associated with the adaptation work such as forming new openings in the old roof to accommodate the new access stair.

Thus, the issues that need to be taken into account when adding another storey to a flat- or pitched-roofed building are as follows:

- In the case of a building with a pitched roof, the removal of the old roof structure may be needed – to avoid risk of bearing new structure on defective substrate.
- In the case of a building with a flat roof, the removal of the entire existing parapet wall may be advisable – to avoid risk of bearing new structure on defective substrate.
- The joint or connection of any new load-bearing element to the existing substrate.
- Matching or contrasting wall finish of the new top storey.
- Detailing the junction between the new/modified existing wallhead and new extension – to ensure weather-tight construction.
- Sealing and masking the junction between the old and new parts of the building – to maintain structural integrity and resistance to rainwater penetration.
- Forming an opening in roof deck for well of new stair – to provide improved access to the roof space.
- Anchoring the new roof to the existing wallhead – to minimize the risk of wind uplift, particularly if the new roof has a wide overhanging eaves.
- Possible underpinning of the building along the length of the roof extension or inserting of new columns to support the additional loadings (see Figure 7.18) for strengthening the structure to take the increased loading. However, care must be taken to avoid creating a hard-spot in the substructure that could encourage differential settlement. Figure 6.10 illustrates the substructure problems associated with adding a roof extension to part of a building. As highlighted earlier, there is a risk of differential settlement occurring in such circumstances.

Over-roofing

Preamble

Over-roofing usually forms part of a modernization scheme rather than a vertical extension to a building. In the former there is usually no attempt to increase the property's usable space. In this sense the over-roofing is still an extension but only of the building's height, rather than its occupational capacity.

In the 1950s and 1960s with the third phase of non-traditional construction, flat roofs were the norm in many new housing developments. Flat roofs were very popular for the simple reason that they are much cheaper and simpler to install than their pitched counterparts.

As indicated in any elementary construction textbook, flat roofs have a number of other advantages over pitched roofs. They offer easy access for maintenance and can easily accommodate complex plan shapes.

In the past flat roofs have, of course, had a poor reputation in terms of weather-tightness and thermal efficiency. Early versions of such roofs usually contained only minimal insulation (often not more than 25 mm thick) under a traditional three-layer felt system. As a result they were prone to excessive heat loss, ponding and leaking, all of which resulted in their having high maintenance costs. Moreover, they were not very durable as their service life was usually between 15 and 20 years.

Modern flat-roof designs have done much to rectify these deficiencies using high-performance single-ply membranes on thicker insulation in a conventional 'warm-deck' or 'inverted warm-deck' system laid to better falls (i.e. 1:40). Such systems have a life expectancy of at least 30 years.

Another difficulty with re-roofing a flat roof, however, is the knock-on effect of increasing the thickness of the insulation. This involves the need to raise the entire parapet, which is a very expensive option and creates a potential mismatch with the finishes at the wallhead.

As a result of these problems with flat roofs many designers and clients consider over-roofing rather than re-roofing as an option for dealing with their flat-roofed buildings (see Hillier et al., 1998; see also Figure 6.11). Over-roofing frequently forms part of a comprehensive modernization scheme that includes over-cladding (see Lawson et al., 1998). This form of roof improvement is not restricted to flat roofs or single buildings. As shown in Figure 6.12 it can be used to integrate a number of small roofs or different shaped roofs into one composite roof. It can also be used to cover over an open courtyard. The design of the new over-roofing system can adopt a straight- or curved-pitch profile for the slope (see below).

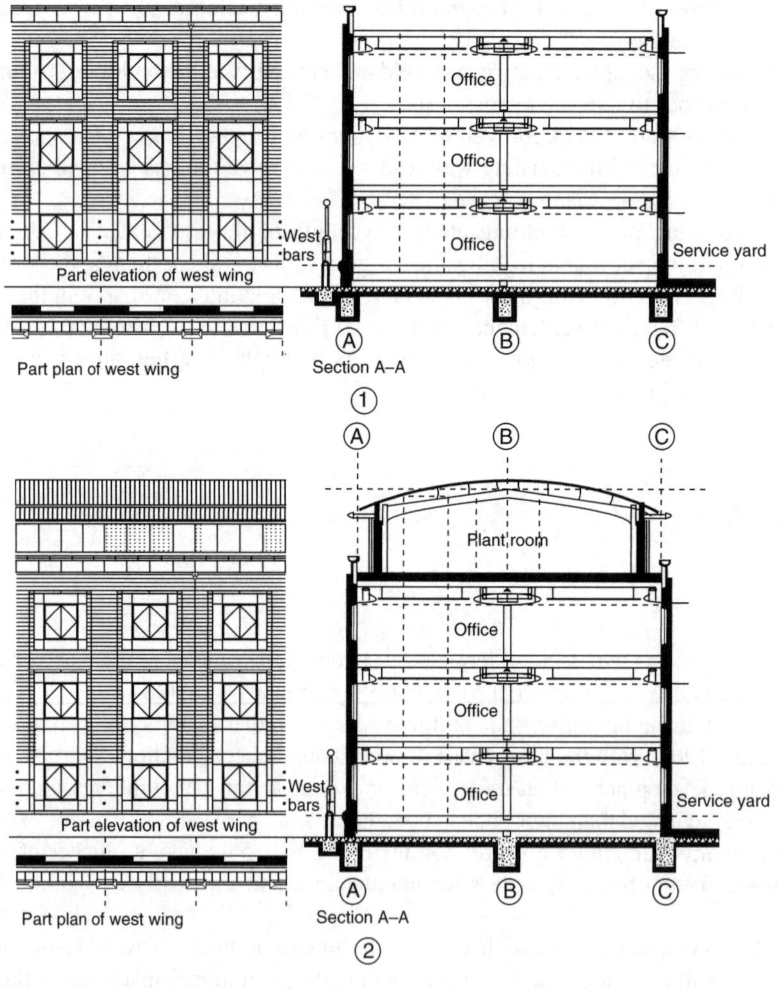

Notes
1. Existing building.
2. Refurbished building with new storey and barrel-vaulted roof.

Figure 6.11 A vertical extension involving a steel portal frame on a flat-roofed building (from drawings of refurbishment of Post Office building in Chesterfield)

Assessing over-roofing

Advantages of over-roofing

- Enhances as well as modifies the overall appearance of the building.
- Improves the efficiency of the roof's rainwater disposal and overall weather resistance.
- Loft area can be designed to provide space for additional accommodation, storage or services plant and equipment.
- Internal courtyards or spare external space can also be over-roofed to increase the floor space capacity of the building.
- Extends the service life of the building as well as improves its thermal performance.
- Improves the marketability of commercial or residential buildings for all of the above reasons.
- Re-roofing work can be undertaken with minimal disruption to existing building.
- Existing roof covering can remain intact, which avoids removal and disposal costs.

Disadvantages of over-roofing

- It is an expensive installation (see section below for information on costs).
- Access to high-rise blocks, which may entail a great deal of scaffolding, makes this an awkward as well as costly measure.
- Increases the roof height, which may be deemed inappropriate in an inner-urban area.
- Roof maintenance will in future be more difficult. Thus, 'access to the original roof and maintenance cradles must be provided in the over-roofing' (BRE report BR 185, 1991).
- New rainwater gutters and downpipes are required.
- The existing deck and/or structure may require strengthening (see Chapter 7).
- There is a risk of interstitial condensation occurring in roof spaces containing several voids with different environmental conditions – the space above the ceiling, the original flat roof void, and the space created by the new pitched roof.
- There is a risk of fire spread and flanking sound transmission within a long-roof space (see below).

Existing roof conditions

- A detailed roof survey should be undertaken to establish existing roof construction and connection details to the existing load-bearing elements. If any deterioration of the existing supports is suspected, a full inspection of the roof should be undertaken by a building surveyor/engineer.
- Calculations and feedback from a structural engineer are required to establish the layout of proposed structure and the holding down arrangements for the existing structures (Kalzip, 2003) (see www.kalzip.com).

Over-roofing schemes

Residential schemes

In housing, over-roofing can be divided into low-rise and high-rise schemes. Low-rise schemes are usually implemented on non-traditional two- and three-storey housing such as 'Orlit' pre-cast concrete-framed construction or 'Hawthorn Leslie' steel-frame construction (NHCA, 1999b).

Over-roofing of high-rise dwellings often forms part of a modernization scheme to large panel or other multi-storey blocks (see BRE report BR 185, 1991). The opportunity can be taken to radically enhance the appearance of these buildings – such as with barrel-vault roofs containing glazed- or multi-coloured-panelled gables.

In rarer cases, it may be considered appropriate to decapitate the redundant top floor of a housing block of three- or more storeys. An over-roofing scheme can form part of the building's modernization programme where its height has been reduced (see Figure 6.15).

Commercial schemes

Multi-storey office blocks can benefit from an over-roofing scheme. Curved- or mono-pitched-roof profiles are now common in this form of roof upgrading. Proprietary systems such as described below are suitable for commercial as well as residential buildings.

Schools, colleges and other institutional buildings can also benefit from over-roofing. In some cases the opportunity is taken to roof over existing inner courtyards as part of a refurbishment scheme to these facilities (see Figure 6.12). A glazed system for the covering of such an over-roofing scheme is usually used to maximize the available natural lighting in that area.

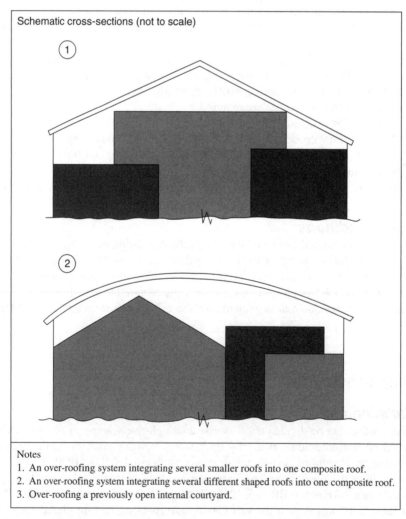

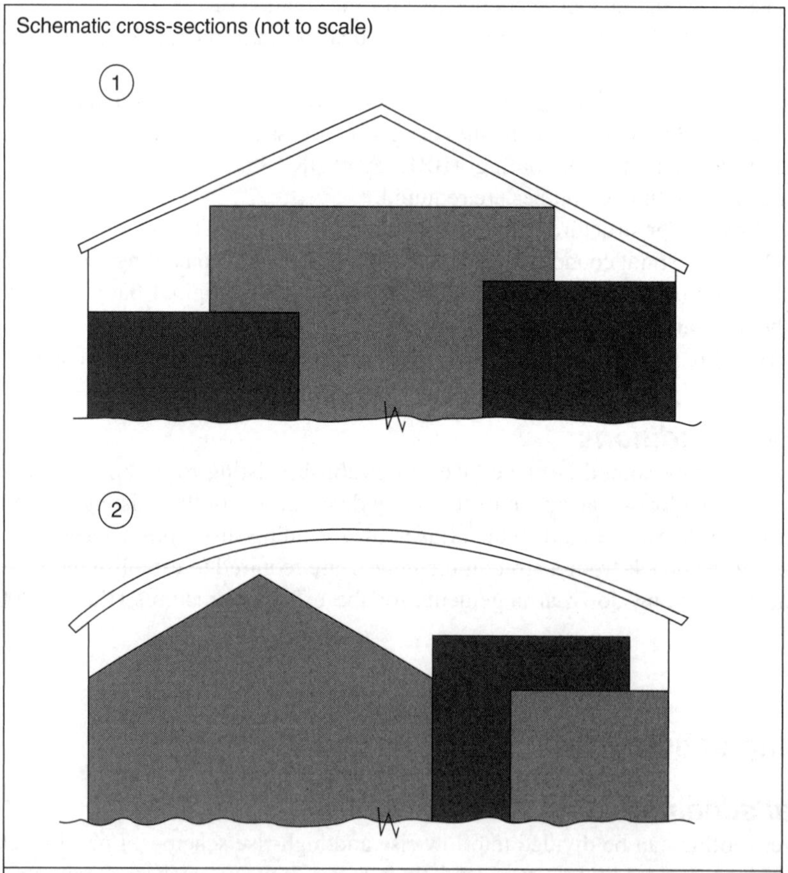

Notes
1. An over-roofing system integrating several smaller roofs into one composite roof.
2. An over-roofing system integrating several different shaped roofs into one composite roof.
3. Over-roofing a previously open internal courtyard.

Figure 6.12 Opportunities for over-roofing (adapted from Ashjack™, 2001)

Dangers of over-roofing

As indicated earlier, care needs to be taken when over-roofing a building to ensure that a complex internal environment in the roof space/s is not created. This occurred in a building that was over-roofed in the early 1990s (Carter and Skipper, 1992a,b).

The problem related to the over-roofing of a 60 m × 30 m single-storey air-conditioned, commercial-type property accommodating electronic equipment. The building was constructed in the late 1970s and is of steel-frame construction with 275 mm cavity walls and a profiled steel-deck flat roof covered with asphalt.

Figures 6.13 and 6.14 show the 'original' and 'existing' constructions, respectively. Note the presence of the air-conditioning plenum duct.

The remedies considered by the investigators, in chronological order, were as follows:

- A three-layer HT elastomeric roofing felt system to provide a waterproof membrane.
- A new single-skin aluminium over-roof (which was to include modification of roof drainage).
- Removal of the over-roof, supporting structure and three-layer felt system back to the original asphalt surface.

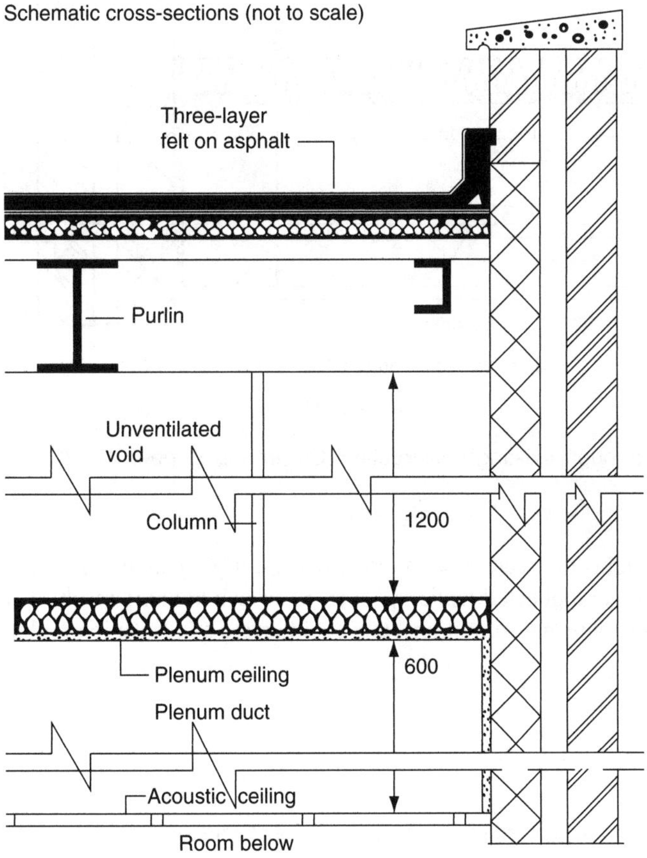

Figure 6.13 Section through 'original' construction (Carter and Skipper, 1992a,b)

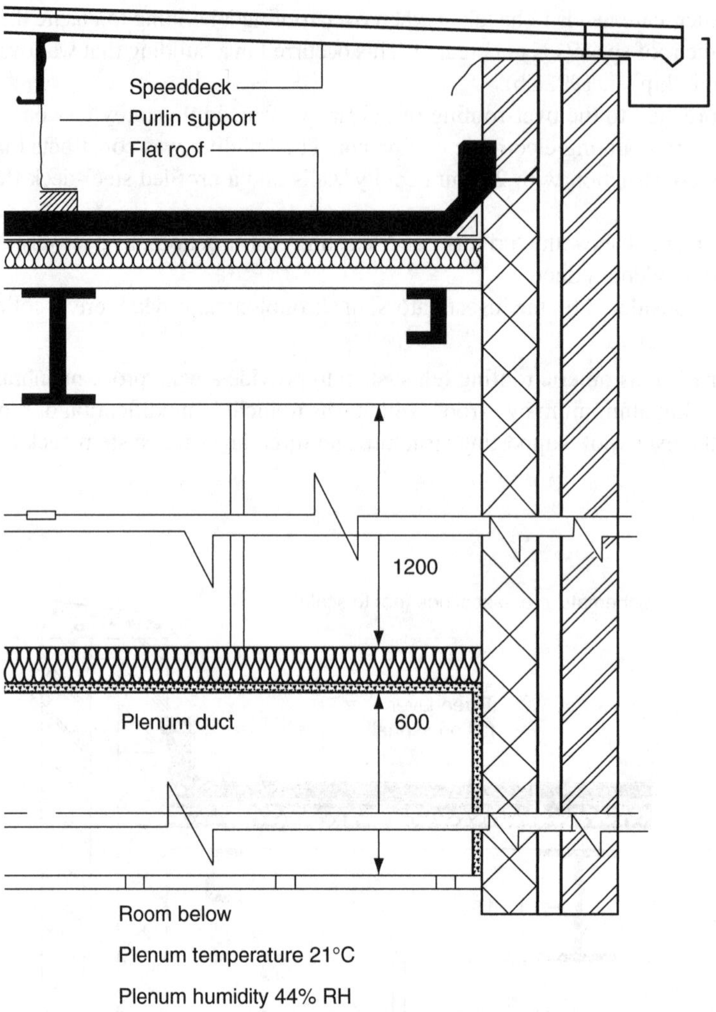

Figure 6.14 Section through 'existing' construction (Carter and Skipper, 1992)

Clearly this case highlights the need to be careful when specifying an over-roofing system. If the existing flat-roof structure has a complicated make-up, such as air-conditioning resulting in several separate void areas with different environmental conditions, this could create condensation problems in the roof space.

Design options

Roof style

In the 1950s and 1960s with the third phase of non-traditional construction, flat roofs were the norm in many new housing developments. They were very popular in those days for the reasons stated above. Flat roofs also allow for intricate plan shapes, which require more complex-pitched-roof profiles (see Figure 6.15).

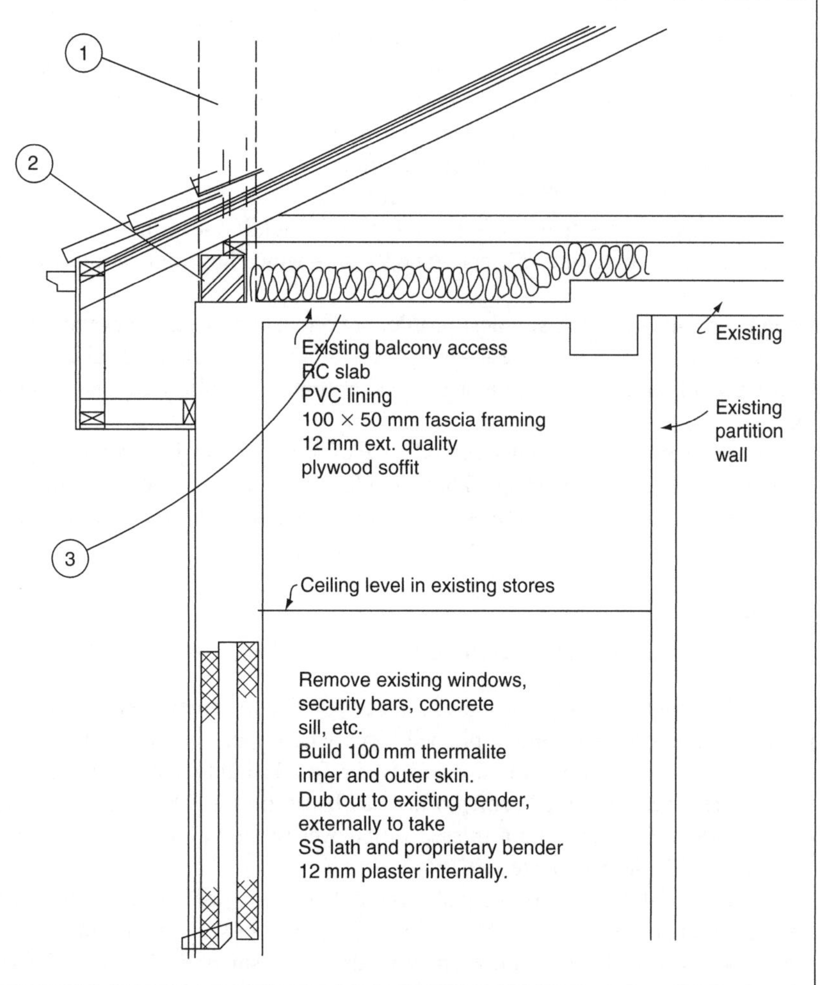

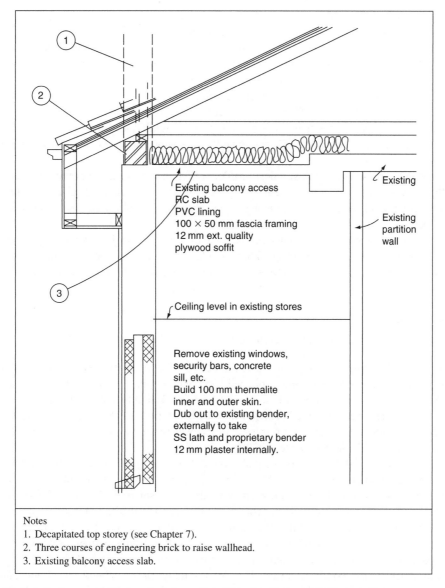

Notes
1. Decapitated top storey (see Chapter 7).
2. Three courses of engineering brick to raise wallhead.
3. Existing balcony access slab.

Figure 6.15 Detail at eaves of flat roof of decapitated block with new pitched roof

Pitched roofs, of course, have a number of other advantages over flat roofs:

- more efficient rainwater disposal and avoidance of ponding;
- more durable coverings, which have about twice the expected service life of traditional flat roof coverings (i.e. minimum 30 years guarantee usually available with proprietary systems);
- better finish to replace an out-dated style of roof.

The main designs that can be used to over-roof a flat or pitched roof are listed below:

- Duo-pitch roof: Gable ends finished with brickwork extended up to the verge; or new diaphragm gable finished with timber or plastic t&g weatherboarding on insulated framing/brickwork.

- Northlight pitch roof: Variation on the standard-duo-pitch theme but with two different roof slopes – the steeper of which can be finished with glazing or photovoltaic panels.
- Hipped roof: This is a more attractive option than the gabled version. It is more expensive because the inclusion of hips increases the complexity and hence the cost of the installation.
- Truncated hipped roof: This is may be more suitable for roofs where accommodation space is being created. The design, however, still results in a 'flat' rooftop.
- Mansard roof: A mansard roofs offer better scope for using the roof space for habitation. It is, however, a complex, and thus more expensive, roof structure to construct. If one of the surfaces slopes at an angle of 70° or more to the horizontal it is deemed to be a wall – which has implications for compliance with the building regulations.
- Mono-pitched roof: This gives the simplest profile, but only slightly improves the roof's appearance.
- Butterfly roof: This option can be used where there are legal restrictions raising the height of the roofline. It is especially useful in tight inner-urban sites that do not allow any increases in the height of the building.
- Curved roof: A barrel, ski-slope or other curved style of roof can transform a building's appearance by giving it a more modern finish and shape. For example, a parabolic curved mono-pitched roof covered with stucco embossed aluminium, stainless steel or other metallic profile cladding, is a popular choice for many modern commercial developments (see Figure 6.11).

Roof structure

There are basically four types of roof structure that can be used in over-roofing schemes:

1. *Trusses*: This is the simplest arrangement and can be provided to achieve spans of up to 40 m or more. It is used with relatively low-duo/mono-pitched roofs where there is no intention to occupy the loft space. A mansard version can, however, use a trussed rafter design to achieve a habitable space within the roof void. Barrel-vault or curved-roof profiles can be used as an alternative to straight mono/duo-pitch types. 'CoverStructure' provide a trussed rafter over-roofing system that comes with a British Board of Agreement (BBA) certificate (99/3645).
2. *Propped rafter*: This configuration is adequate where there is no need or desire to provide accommodation in the new roof space other than for water tanks and other services. Lightweight steel or timber sections can be used in a finely engineered structural design. 'Ashjack™ SC (Spot Cleat) Framing' (see Figure 6.16), 'SpeedDeck' (see their informative guide, published in 2000, on this option), 'System One Frames' and 'dibsadeck®' are just four of the many proprietary metal over-roofing systems available on the market. These systems usually use cold-formed galvanized steel sections for the rafters, purlins, posts and braces. Vertical channel section posts are welded to base plates, which are bolted to the structural deck of the existing flat roof. They are normally spaced at between 1.2 and 2 m centres to support the rafter frames, which in turn support purlins. The base angles are usually laid at right angles to the direction of the existing roof joists to ensure even distribution of the new roof load. These are secured through the existing roof covering with the appropriate fixings dependent on the existing roof construction. The structure is braced using channel sections bolted to the posts.
3. *Purlin*: This can be used in a duo-pitched roof with gable ends. The masonry at each end is built up to form the triangular gable peak and provide support for the purlins. The type of purlin used depends on the span between supports. Timber beams can be used for short spans (i.e. under 5 m). Steel 'I' beam or laminated timber purlins can be used for spans exceeding 5 m between the gable and any cross walls. New rafters as per Table A1 of the Building Regulations are then installed over the purlins.
4. *Portal frame*: Where maximum habitability of the roof space is required, the portal-frame option provides the best method to attain this objective. This is because a steel portal frame is highly efficient in structural and spatial terms. It can also be formed in mansard, curved or straight roof profiles. Figure 6.11 illustrates one example of this option.

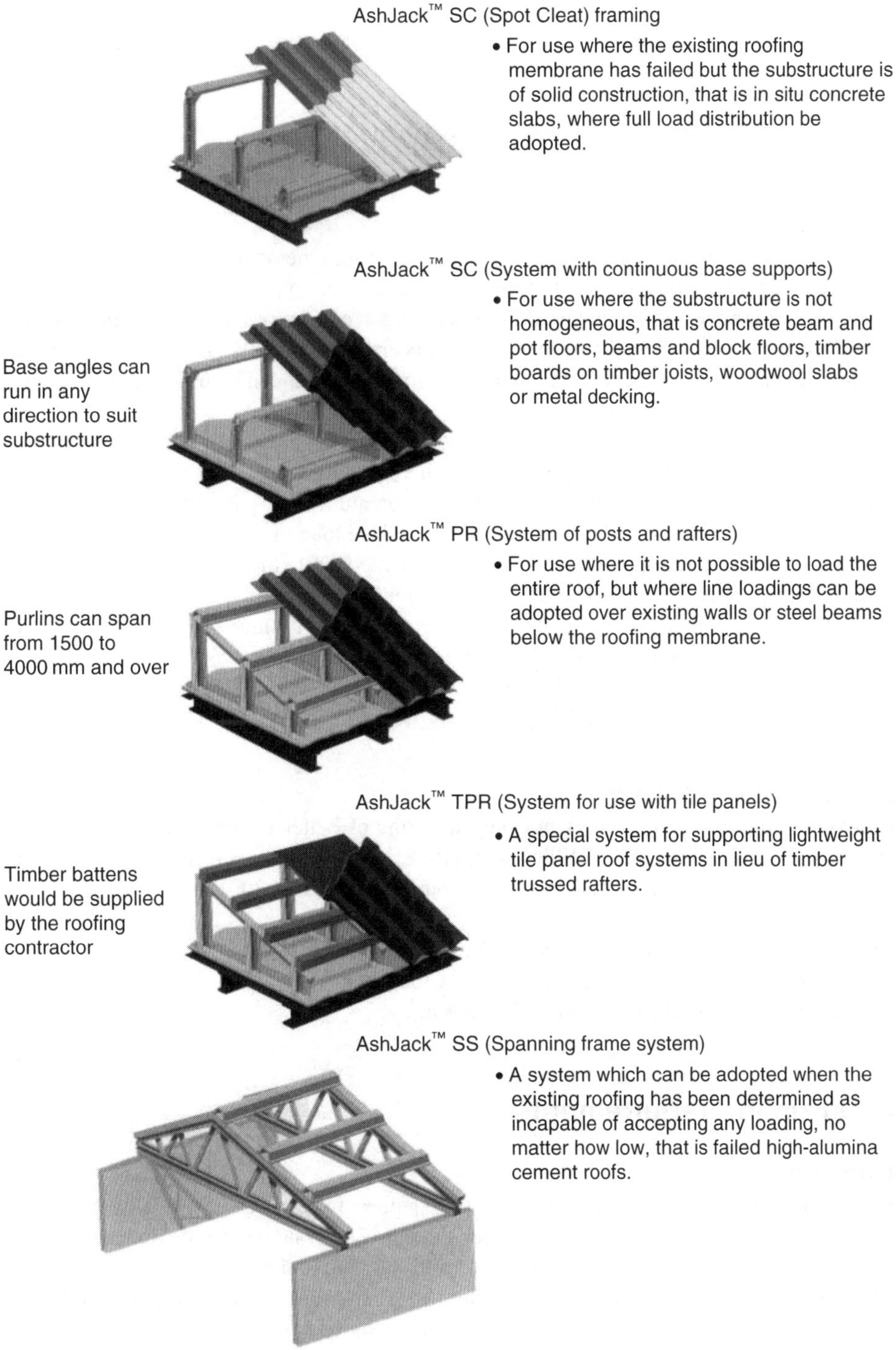

AshJack™ SC (Spot Cleat) framing

- For use where the existing roofing membrane has failed but the substructure is of solid construction, that is in situ concrete slabs, where full load distribution be adopted.

AshJack™ SC (System with continuous base supports)

- For use where the substructure is not homogeneous, that is concrete beam and pot floors, beams and block floors, timber boards on timber joists, woodwool slabs or metal decking.

Base angles can run in any direction to suit substructure

AshJack™ PR (System of posts and rafters)

- For use where it is not possible to load the entire roof, but where line loadings can be adopted over existing walls or steel beams below the roofing membrane.

Purlins can span from 1500 to 4000 mm and over

AshJack™ TPR (System for use with tile panels)

- A special system for supporting lightweight tile panel roof systems in lieu of timber trussed rafters.

Timber battens would be supplied by the roofing contractor

AshJack™ SS (Spanning frame system)

- A system which can be adopted when the existing roofing has been determined as incapable of accepting any loading, no matter how low, that is failed high-alumina cement roofs.

Figure 6.16 Over-roofing framing systems for different flat-roof substrates (courtesy of Ashjack™, 2001)

Roof finish

A variety of roof finishes in terms of colour and texture are available. The choice will be dictated by several factors:

- the context of the building (e.g. listed or modern?);
- the pitch of the roof;
- the client's requirements.

The choice of the type of roof covering for an over-roofing scheme will depend on cost as well as aesthetic influences. Certain finishes such as slate, for example, may be considered the most appropriate finish in older urban areas where the surrounding buildings have this type of roof covering. In any event, it should match the performance of the original roof in terms of appearance, durability and water-shedding efficiency.

The following is a summary of the main roof coverings that can be used on an over-roofing structure:

- Corrugated or profiled light aluminium/steel – which comes ready-crimped for barrel-vault or other curved-roof shapes. 'Kalzip' supply and install a standing seam version of this in a wide range of alloys and finishes such as aluminium, traditional zinc on aluminium ('AluPlusZinc') and stainless steel.
- Concrete tiles: good durability, but increases the dead load of the roof.
- Clay tiles: good durability and slightly lighter than concrete tiles, but expensive.
- Glazed, fully or partially, best for roofing over inner courtyard.
- Lightweight metallic/synthetic slates/tiles (see recovering options for pitched roofs in Chapter 9).
- Natural slates – attractive and durable, but expensive.
- Artificial mineral fibre/cement slates (e.g. 'Eternit').
- Titanium-cladding panels (see Chapter 8).

Costs of over-roofing

Over-roofing can result in savings not only in terms of better energy performance but also in reducing maintenance costs. According to Hillier et al. (1998), the following is a list of indicative costs for typical over-roofing projects for a normal rectangular building based on 2006 prices:

- Over-roofing without habitable space: £80 to £150 per m^2.
- Over-roofing with habitable space: £500 to £700 per m^2.
- Replacement flat-roofing system: c. £50 per m^2.

Construction requirements

Anchoring

Modern roof structures, because they consist of slender, finely engineered components, are much lighter than their traditional counterparts. In the latter, there is usually no need to tie down or anchor the roof structure because its dead load is often sufficient to prevent wind uplift.

In contrast, it is important that any modern roof structure is adequately secured to a suitable load-bearing element. In new-build work this is normally achieved using vertical galvanized mild steel restraint straps 900 mm long fastened to every second truss and fixed to the inner leaf of the wallhead brickwork. Alternatively, the trusses can be anchored using special fixing brackets bolted directly onto the wallhead. This latter method, however, is not as effective as the former.

Portal-frame roof structures are usually anchored directly to the flat-roof deck. The base plate at the foot of each stanchion is bolted to the concrete flat-roof slab or to the structural timbers of the flat roof.

Insulation

As much as 25 per cent of a building's heat loss occurs through the roof (Cairns, 1993). Thus, the opportunity to enhance this element's thermal efficiency should form part of any modernization programme. The type of roof structure selected will determine the position as well as the amount of insulation required. See Chapter 9 on methods of increasing the thermal efficiency of roofs. However, consideration must be given to providing background ventilation if a non-sealed roof is being specified.

Ventilation

The thermal characteristic of the new roof (i.e. whether it is a cold roof, a warm roof or a 'sealed' roof) will dictate the extent of ventilation required. Traditionally roofs have to be ventilated to expel warm moist air that rises through the house into the loft space. To minimize the risk of condensation forming in the roof space a gap equivalent to a 10 mm continuous opening in the eaves was needed to provide background ventilation. As indicated in the next chapter BS 5250 provides the basic guidance on this requirement. However, modern construction practice is tending to favour the 'sealed' pitched-roof method, which unventilated (see Chapter 9 for more details of this system).

Fire spread

The large space created within an over-roofing scheme is a potential hazard in terms of fire spread. To minimize this risk fire barriers are required within the roof space at regular intervals – usually every 8 m.

This can be achieved using a blanket of 75 mm thick mineral fibre reinforced with chicken wire. It is hung from the new rafters/underside of the sloped roof soffit/structural deck down to the ceiling level to provide a 'fire curtain'.

Sound transmission

As with fire spread, the creation of a new roof space may encourage flanking sound transmission. Noise, like fire, from one room can be transmitted up through the ceiling and along the roof space and down into adjoining rooms. Heavier construction than a fire blanket (such as timber studding filled with mineral fibre insulation and clad both sides with two layers of 9 mm thick plasterboard) will be needed, though, to achieve an adequate level of sound insulation.

Security

Forming a 1 m wide overhanging eaves can provide additional security for the over-roofed building. This design helps improve its anti-burglar/vandal features.

Existing chimneys

Any projections in the existing flat roof will require extending through the new pitched roof – unless, of course, these projections have become redundant. For example, chimney-stacks and vent pipes that will continue to be used require extending and sealing through the new roofline. Care in detaining and installing the flashings around these projections should be taken. Special preformed collars can be used for pipes and round flues that project through the roof.

Accessories

New gutters and downpipes unplasticized polyvinyl chloride (uPVC) or powder-painted metal can provide relatively maintenance-free fascias and soffits.

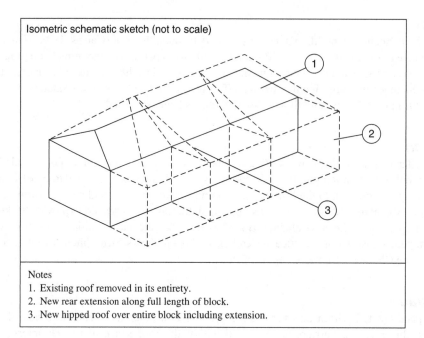

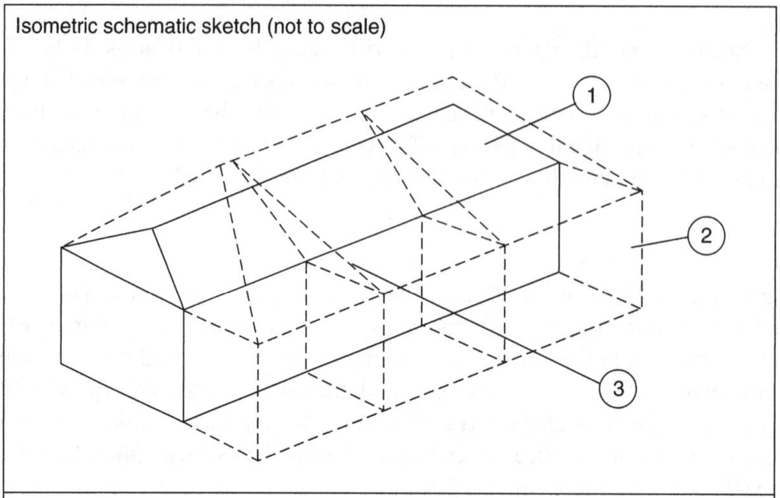

Notes
1. Existing roof removed in its entirety.
2. New rear extension along full length of block.
3. New hipped roof over entire block including extension.

Figure 6.17 Combined vertical and lateral extension to terraced block of three dwellings in a rural area

Combined vertical and lateral extensions

We have already seen that not all extensions need to fall within the boundary of one elevation. Some, such as corner extensions, may encroach on more than one side of a building. This may be required to accommodate two or more uses in the extension, such as an enlarged front dining room and a new side garage.

In rarer cases an extension may involve an enlargement of the property both horizontally and vertically. Two examples of combined extensions are illustrated in Figures 6.17 and 6.18.

The scheme shown in Figure 6.17 indicates an extension built onto an existing single-storey extension with a lean-to roof. This necessitates the forming of an opening in the end wall of the existing property to allow access to the new first-floor room (see next chapter on Structural Alterations). It also requires the sloping wallhead of the lean-to section of walling to be either stepped or levelled at first-floor level to accommodate the masonry of the additional storey. The wall construction of this extension should match that of the masonry below.

Another version of this is the extension onto a side garage or previous lateral extension to form a bedroom or other accommodation above it. In this case it is possible that the foundations and walling of the garage would need to be upgraded to sustain the additional loadings. It is not unusual for garage walls to consist of either a single skin of brickwork (i.e. 105 mm thick, with one-brick wide piers at 1.2 m centres) or pre-cast concrete panels (c. 50 mm thick). Neither of these constructions is inadequate to support another storey. Moreover, if the garage were being converted into living accommodation, the walls would need to be upgraded to satisfy Building Regulation requirements for thermal insulation, fire resistance, sound insulation, as well as structural integrity.

If the original garage structure is inadequate to cope with these it is prudent to demolish the structure and rebuild it to a proper standard. At the very least the roof of the side extension would have to be removed to prepare the wallhead for the additional storey.

Schematic front elevation sketch (not to scale)

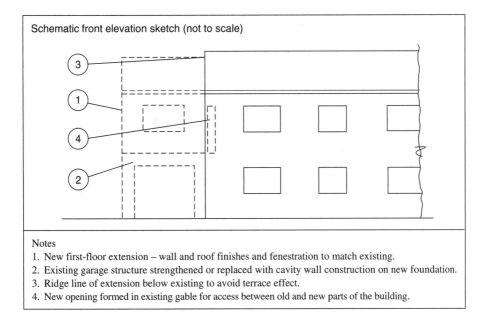

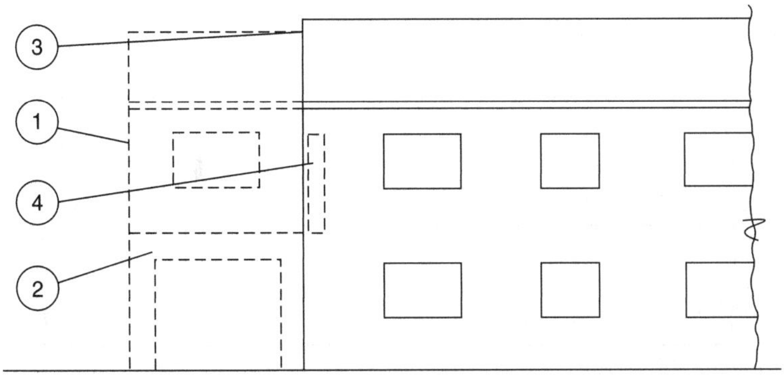

Notes
1. New first-floor extension – wall and roof finishes and fenestration to match existing.
2. Existing garage structure strengthened or replaced with cavity wall construction on new foundation.
3. Ridge line of extension below existing to avoid terrace effect.
4. New opening formed in existing gable for access between old and new parts of the building.

Figure 6.18 Combined lateral and vertical extension involving add-on to garage

Figure 6.18, on the other hand, shows a novel means of enlarging a block of terraced bungalows in a rural area. Owing to the scale of the work involved, however, it cannot be done whilst the occupants are in the property. In other words, they have to be decanted during this adaptation. It involves taking off the existing roof in its entirety. In the first instance the slate-roof coverings are carefully removed and laid aside for reuse. Up to about 50 per cent of the stripped slates can be recycled in the rehabilitated block. The next phase involves constructing a full-length rear extension to enlarge the kitchen accommodation of each dwelling. A new trussed rafter roof can thus be installed over the old and new parts of the block. The benefits of this scheme are that the accommodation is enlarged with a new roof structure that is little different from the original.

Basement extensions

Potentials

Downward vertical extensions are the least common of all enlargements to existing buildings. This is because of the costs involved in excavating and providing temporary support followed by tanking and permanent support in the basement area. Forming a new basement where one previously did not exist is the only feasible way of extending a building vertically downward. It can be an attractive option for increasing the floor area of a building in a tight site situation or where planning controls would not allow an extension either upwards or horizontally.

The conversion of cellars into valuable habitable space has become increasingly attractive in built-up areas where other forms of extension are neither feasible nor permissible. It is, however, the most complex type of extension to construct because it usually involves lowering the ground, underpinning to an engineer's specification, structural supports and waterproof tanking. As these operations constitute almost 90 per cent of the contract, forming a basement extension requires experience and care to avoid many of the risks that can beset adaptation work.

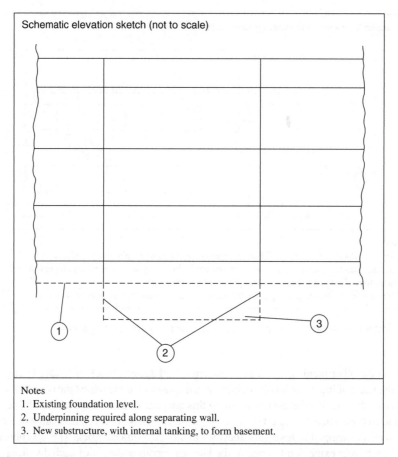

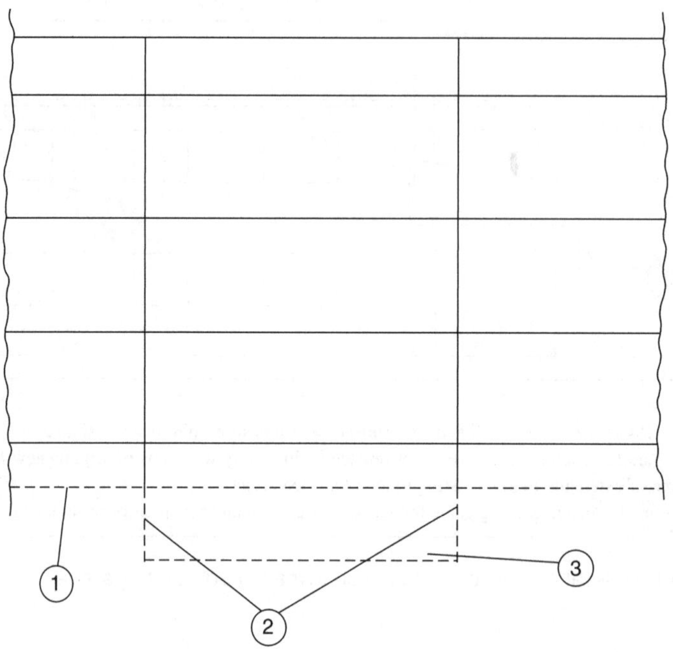

Notes
1. Existing foundation level.
2. Underpinning required along separating wall.
3. New substructure, with internal tanking, to form basement.

Figure 6.19 Basement extension and underpinning requirements to mid-terraced block

The most likely circumstance where a retrofit basement is required is as part of the renovation or redevelopment of a mid-terraced property. A restoration project may include a new basement as part of the work. In a redevelopment scheme, say following the major destruction of a mid-terraced property damaged by fire, the opportunity to form a basement may be taken (see Figure 6.19). Planning regulations may restrict any roof extension because of the effects of this on the streetscape.

In any event, provision for both temporary and permanent support must be taken in forming a new basement. The temporary support may involve raking shores or propping to restrain the existing wall abutting the extension. Permanent support is very likely to entail underpinning to comply with the requirements of the Party Wall Act 1996. In this context mass concrete underpinning would be the most appropriate solution.

Basement extensions will inevitably require some form of tanking to control, if not prevent, any groundwater ingress. The grade of tanking used will depend on the use to which the accommodation formed by the basement will be put. CIRIA (1997) prescribes four grades of basement tanking, as shown in Table 6.4.

Forming a basement in a two-storey building can increase its floor space by up to 50 per cent. For a three-storey building the increase is as much as a third. If a basement extension is combined with a loft extension, the space in a dwelling can be expanded by as much as 70 per cent.

Expanding the capacity of one's property has positive economic as well as functional knock-on effects. This is the type of adaptation that is most likely to increase the value of a property. A basement extension is

Table 6.4 Grades of basement tanking (BS 8102: 1990 and CIRIA, 1997)

Grade	*Description*	*Requirements*	*Construction*
1	Basic utility	Minor seepage may be acceptable.	Brickwork walls and rc floor slab with no tanking.
2	Standard	No visible damp patches permitted.	Engineering brickwork or ordinary-quality rc walls and rc slabs with basic tanking.
3	Habitable	No dampness acceptable (the most likely grade required for basement extensions).	Good-quality rc walls and slabs with good-quality tanking. Drained system may be needed in flood-prone areas or in ground with high water table level.
4	Special	Very high degree of moisture resistance and humidity control.	High-quality rc structure with full tanking and field drain in pebble backfill. Again, a drained tanking system, with sump and/or pump facility may be required in water-logged ground or where there is a risk of the basement flooding.

rc: reinforced concrete.

particularly advantageous in built-up urban areas where roof and lateral enlargements are not possible owing to site and planning restrictions.

New basements in existing buildings, however, have other benefits as well. Given that they accommodate increased density without reducing the space between buildings, basements are good in terms of land conservation.

Thermal efficiency is another advantage because of the thermal mass offered by the basement walls themselves. This results in the basement areas retaining warmth and using less heat than above-ground rooms.

Problems

Excavating the space for a basement, however, is both expensive and messy. The work and cost involved in cost of digging out a basement varies depending on the size of the basement as well as the soil conditions and groundwater encountered. The presence of any services within or below the soil being removed must also be taken into account. Removal of the soil usually generates a great deal of dust and debris, and requires regular disposal to an approved dump.

One important building regulations requirement for basements is that they must have alternative escape routes. These could be in the form of an escape window 850 × 500 mm within 1.6 m of the floor. The ironmongery and opening light should facilitate easy escape without compromising security.

The provision of natural lighting and ventilation to a basement may be difficult to achieve fully. As indicated in Figure 6.1 earlier a light-well at the front or back would be the best way to achieve this, provided of course that the planning authority has no objections and there is sufficient space available at these locations.

Another potential problem is that the basement floor could be below the sewer invert. If this were the case and foul water disposal was required, some form of pump system would be required to discharge the foul water up to the existing sewer level. There are cost implications for the maintenance as well as installation of

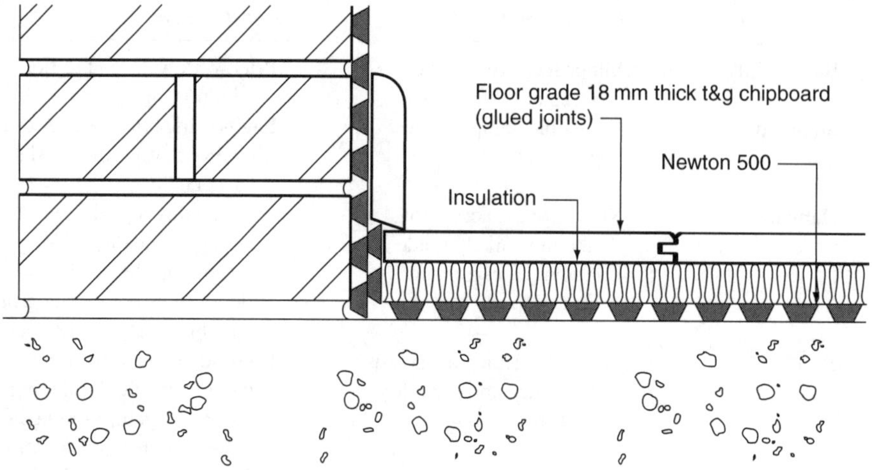

Figure 6.20 Membrane system for tanking of retrofit basement (John Newton & Co Ltd, 2001)

this facility. Moreover, the pump will take up space, which may require some form of enclosure to provide concealment from and soundproofing for any adjoining habitable rooms.

Structural waterproofing

Tanking of existing or retrofit basements to make them habitable or usable is likely to form a major part of a cellar or vault refurbishment scheme. Traditionally, applying a two-coat asphalt system to the prepared floor and walls of the basement would achieve this for building substructures in relatively dry-ground conditions. Modern methods involve the application of acrylic/elastomeric/polymeric thin-coat systems. The latter are more preferable if the available space in the basement is restricted.

Where the water table is high or flooding is likely, however, ordinary tanking systems are inadequate. Depending on the risk of water seepage one of the following three classes of structural waterproofing may have to be used:

- Class C: Membrane-only system (see Figure 6.20): provides a reasonable level of basement waterproofing – to Grade 2 or 3 standard (as per Table 6.4).
- Class B: Membrane system with drain (see Figure 6.21): suitable for providing a dry basement where some seepage is possible – to Grade 3 if not 4 standard.
- Class A: Membrane system with pumped sump (see Figure 6.22): required in sites where there is a high risk of flooding or persistent water seepage owing to hydrostatic pressure – to Grade 4 standard.

There is a wide range of proprietary basement waterproofing systems on the market covering these three classes. 'Trace Basement Systems™', 'Sovereign' and 'Newton System 500' are just some of the many companies who offer this specialist service for dealing with persistently damp basements. However, some of them, such as John Newton & Co Ltd, no longer undertake any below ground work without drainage.

Installation work

Forming a basement extension is another type of adaptation work that is best undertaken by a specialist contractor. Ideally, the contractor should be a member of the British Structural Waterproofing Association.

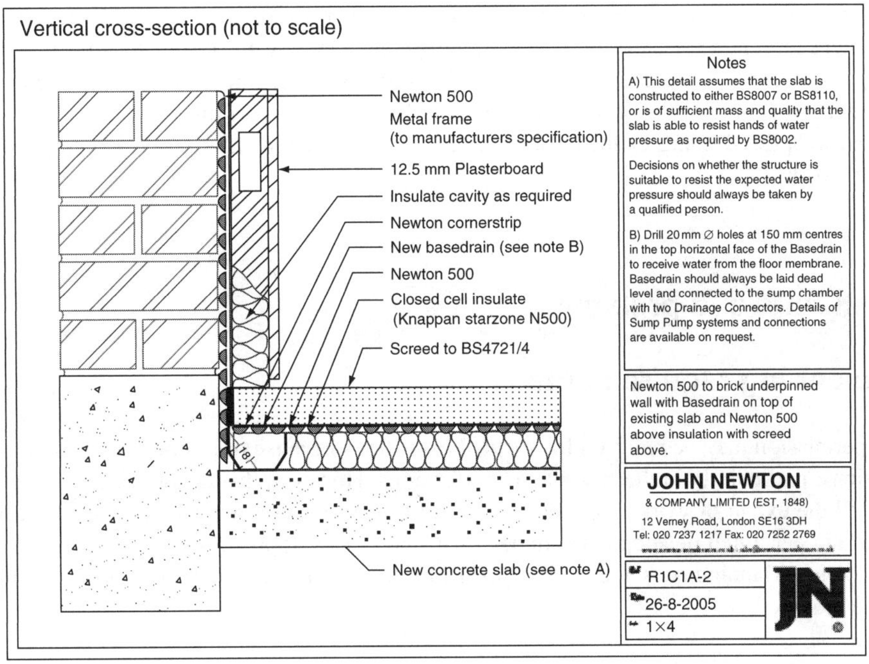

Figure 6.21 Membrane system for tanking with drainage channel in retrofit basement (Courtesy of John Newton & Co Ltd, 2005)

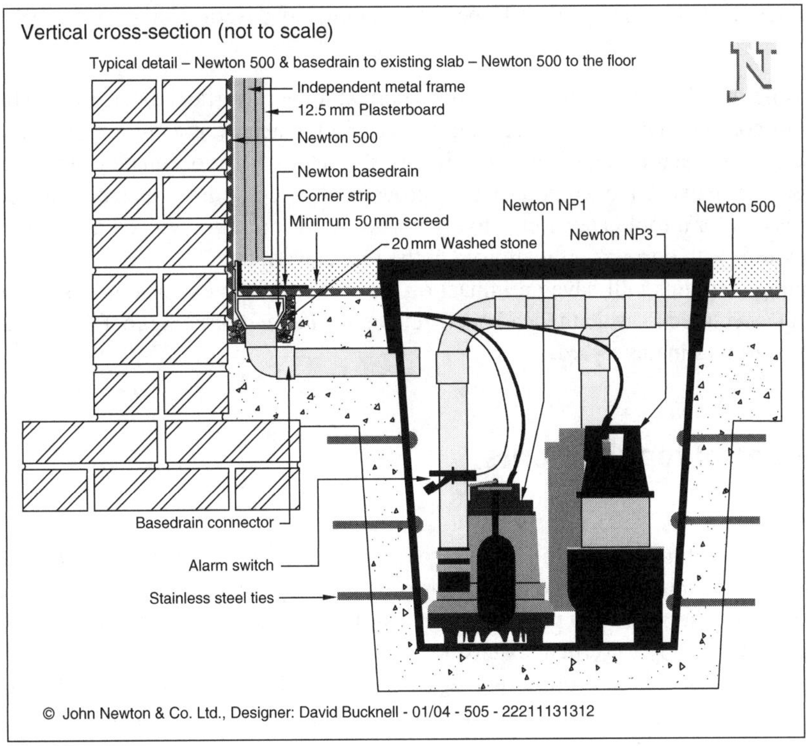

Figure 6.22 Membrane system with sump pump for retrofit basement in waterlogged site (Courtesy of John Newton & Co Ltd, 2005)

Owing to its high-risk nature basement extension and waterproofing work should not be subcontracted. Only experienced contractors holding a substantial Public Liability Insurance scheme (i.e. £1 m+) and employing competent operatives should be used. Such provisions according to one contractor are a) likely to be required for re-sale of the property and b) give the client peace of mind that a professional who has been trained and passed exams on the proper use of these basement installations and waterproofing systems is doing the work.

Constraints and problems

Town-planning requirements

Planning permission may be required for vertical extensions because of aesthetic considerations or density limitations, particularly in the case of roof extensions. However, this depends both on the category of building and the extent of work proposed. Simple loft extensions on residential properties, for example, are classed as 'permitted development' under the General Development Order 1995 as provided by the Town & Country Planning Acts. Generally, this means that such works on houses do not require planning permission provided that:

- they are for domestic use only;
- they are not proposed for a listed building or other special case;
- they are not higher than the existing domestic building;
- the enlargement is within the 50 m^3 or 10 per cent parameters described in Chapter 10.

It is not surprising that those planning restraints can affect upward vertical extensions. After all, the roof is often the most conspicuous part of a building. Increasing the height of a building will not only alter its profile but also change the roofline of the street. It may also intensify the occupancy of the building. In conservation areas such an extension would probably prove contentious. Under such circumstances refusal of planning permission for a roof extension is likely unless the work only involved a loft conversion entailing low-density use, with dormers only allowed at the rear elevation.

Thus, planning restraints will have an impact on the design of the extension. For example, consideration of massing is required to avoid the 'carbuncle' effect of a large, protruding roof extension. Box dormers are particularly notorious in this regard.

Building-control requirements

Generally

Building-control consent is required for any type of vertical extension. The building regulations are concerned with achieving a minimum standard of construction. They are also about the safety implications of proposed extensions, in terms of fire precautions and structural stability. Again, though, the requirements listed in Table 3.6 may apply to vertical extensions.

Even minor works such as inserting a rooflight or placing decking in a loft would require building-control approval. Figure 6.23 illustrates the statutory requirements for loft extensions.

The regulations relating to fire safety have largely been dealt with earlier. In summary, however, the some of the other key sections of the regulations affecting vertical extensions are dealt with below.

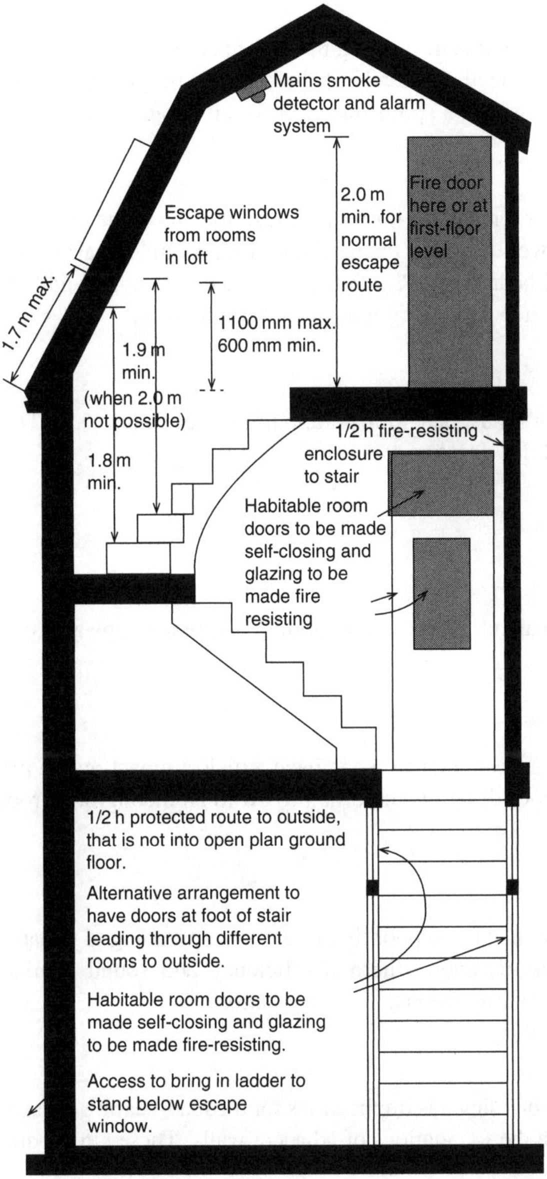

Figure 6.23 Typical statutory requirements for loft extensions (Haverstock, 1998)

Structure

Additional or modified loadings need to be taken into account. Changing the roof coverings from slate to concrete tiles, which are much heavier, could overstress the roof structure, particularly if it is of modern design with its slim sections. This can result in a sag in the rafters or even roof spread.

Insulation

The insulation should contain a vapour-control layer on the warm side of construction. The aim is to achieve a U-value of 0.25 W/m^{2}K. Ventilation, however, is essential where any unheated void areas are formed (see Chapter 9).

Staircase

The maximum pitch for the stair is 42°. Being the principal means of escape, the stair must be enclosed to provide at least half-hour protection. Ideally, it should be a straight flight stair to facilitate ease of egress. In more restricted areas, however, a spiral stair or stair with dogleg and/or winder may be permitted.

Space standards

There is no minimum headroom height for rooms prescribed now. Ideally, the headroom should be at least 2.3 m and certainly not lower than 2 m (the minimum headroom for stairs). However, the old regulation Q6 in Scotland allowed small bedrooms ($<14.9\,m^3$) to have the ceiling height less than 2.3 m provided that three-quarters of the plan area had a height of not less than 1.9 m.

Windows

The position of the windows must be within certain limits (see Figure 6.2). This is required for fire escape, general safety and cleanability reasons.

Technical

The technical factors that can influence how a building is extended upwards or downwards are summarized as follows.

Architectural

The effect of an upward vertical extension can have a major impact on the profile of a building, for good or ill. Again, compatibility of design is important to avoid an incongruous roof extension detracting from the rest of the building.

Constructional

The condition of the fabric may be so poor that it requires upgrading to enhance key performance features such as durability, weather-tightness, thermal efficiency and sound insulation. A roof extension can achieve this as well as provide additional accommodation.

Structural

Adding another storey to a building has implications for the load-bearing capacity of the building. Extending downwards can undermine the foundations of adjacent walls. These effects are discussed in more detail in the next chapter.

Spatial

Space is technically always available for an upward vertical extension. However, a restrictive covenant on the property prohibiting an increase in roof heights above a certain level can thwart an extension in this direction. Similarly, obstructions such as tunnels or services near the foundations of an existing building can inhibit any basement extension.

Environmental

An expansion of the property's capacity will intensify its use, which may result in an increase in demand for car parking. This in turn can affect the provision of soft as well as hard landscaping around the adapted building.

Economic

The professional adviser should of course inform his/her client about the financial implications of the adaptation proposal. These will normally include:

- the effect of extension on the capital value of the property;
- the capital cost of the proposed extension;
- the whole life costs of the building with the extension;
- professional and other fees associated with the work (see Chapter 11);
- the value added tax (VAT) requirements of the adaptation works;
- the cost and availability of funds for the scheme.

Summary

The easiest and most discrete way of increasing a building's capacity is usually done through a vertical extension. Traditional buildings in particular because of their heavy-roof construction easily lend themselves to a loft conversion, the most common type of vertical extension. Modern roof structures on the other hand are more problematic owing to the tightly designed nature of their structural members.

Vertical extensions, however, bring with them their own problems. The need for an adequate staircase/s to facilitate access to the new accommodation must be borne in mind. This reduces slightly the usable space on the floor immediately below the new level.

The siting and provision of services, too, can be problematic with vertical extensions. This is especially the case where toilet and/or bathroom facilities are required in the new level accommodation.

Basement extensions and conversions are becoming increasingly popular as a means of expanding a property's internal useable space. Given the complications that they entail, however, basement extensions are much more expensive than roof or top-floor extensions.

Modifications to load-bearing elements are usually required when extending a building vertically. Reducing the capacity of a building vertically has similar structural implications. These and other structural alterations are dealt with in the next Chapter.

7

Structural alterations

Overview

This chapter deals with the typical types of alterations that are often undertaken to a building in association with conversion and other adaptation schemes. It explains the principal technical and other support factors in such projects involving changes to the structural characteristics of a building.

Background

Significance of structural adaptations

Adaptation schemes often entail some modifications to the layout, configuration or morphology of buildings. In many cases these are of a structural nature or will have structural implications. In other words, they impact on the building's load-bearing parts. This means that they may entail increasing, decreasing or redirecting loads through the affected element/s. Such changes can result in cracking or distortion or more seriously partially/total collapse if they are not adequately accounted for in the design.

Removing existing structural elements or inserting new ones may form part of a major conversion project. Even a small extension can entail some structural alteration. One common example of this is inserting a new opening in an existing load-bearing wall to facilitate passage to and from the extension and the main building. The effect is not only to change the layout of the existing building but also to redirect load paths and even increase the intensity of the loading on the remaining elements. This is especially the case if the wall above the proposed opening is taking any imposed loads from the floors or roof above.

Precautions have to be taken when removing walls that provide lateral restraint. A spine wall, for example, may offer lateral restraint to the gable end of a building. Removing the former may destabilize the latter. This could also result in cracks and bulges in the gable and adjoining walls.

Partial removal of a chimney-stack up to the attic within a building is sometimes carried out carelessly if not illicitly. This is usually done on the pretext that because the solid fuel fireplace is no longer used removing the chimney-breast increases the available floor space. However, this can cause instability in the stack if little or no support for the upper sections of the remaining structure is provided. Brick corbelling, timber joists or gallows brackets do not offer a suitable support mechanism in such circumstances.

Inserting suitable size steel 'I' beams (UBs) under the remaining section of the stack is the most effective way of supporting the remaining part of the chimney. These new beams have to achieve adequate bearing on the nearest appropriate load-bearing walls.

Some adaptation schemes may involve the removal of a redundant chimney-stack above the roofline. In such cases any remaining parts of the flue should be ventilated and the surrounding masonry and roof coverings made good.

Significance of non-structural adaptations

Even non-structural alterations can have structural implications. For example, changing the coverings on a pitched roof can increase the dead load on the structure. This is particularly common when slate roof coverings are replaced with concrete interlocking tiles. Typically the latter has a dead load of approximately 1 kN/m^2, whereas the former has a self-weight of around 0.5 kN/m^2. Some old roofs that have been subjected to this alteration have sagged afterwards. This in turn leads to other problems such as disruption of the roof coverings and the encouragement of moss and other organic growths on the slates/tiles.

Another structural effect of a non-structural alteration occurs when the fire protection to new and existing elements is disturbed by modifications in the use or construction of the building. The integrity of these elements' fire resistance could be compromised or inadvertently reduced by such work.

Many ground-supported concrete floors prior to the 1960s may not have an effective damp-proof membrane (dpm). It would be prudent therefore in any major adaptation scheme to install either a surface dpm over the finished slab or a new damp-proofing layer under a finishing screed. The problem with the latter option is that it would take about 2 months for a 50 mm thick screed to dry out so that a floor covering can be laid without fear of failure. According to the BS 8204 a cement-based screed dries out at a rate of about 1 mm a day. It also states that flooring should not be laid until a reading using a hygrometer indicates that such a screed or concrete slab has a relative humidity of 75 per cent or less before a floor covering can be laid.

With timber floors, a major problem is excessive notching of joists to accommodate electrical and heating services. This, along with creep, can cause the joists to sag excessively if the notching is greater than one-eighth of their depth (see below).

Removing a chimney can have non-structural as well as structural implications. The existing pitched-roof coverings have to be extended over the gap filled by the stack. Matching the existing coverings in terms of shape and finish is not always easy. The remaining part of the chimney has to be provided with background ventilation to avoid interstitial condensation occurring in the redundant flue.

Replacing defective windows and doors with new unplasticized polyvinyl chloride (uPVC) double glazed units can also have structural implications for older properties. This is especially the case with bay windows, which are common in many Victorian buildings. Figure 7.1 illustrates the structural effects of such work. A method of strengthening the new mullion is shown in Figure 7.2.

In addition, removing masonry mullions from window openings to facilitate a larger glazed area should be avoided on structural as well as conservation grounds. It can compromise the original design of the fenestration and overload the stone lintel above. This can result in cracking and eventual sagging of the lintel. The window framework below the lintel will also be susceptible to distortion and may require strengthening – usually by using galvanized mild steel stiffener channels within the top-rail section.

Structural and non-structural alterations can therefore affect both the serviceability and stability of a building. Safety of its occupants and the public is of course the primary criterion. The best way of ensuring this is for the design of any structural alterations to undergo a rigorous check to confirm their adequacy. A thorough survey before design works start should be undertaken so that factual rather than assumed data are used. It is for these reasons that building control approval is required for any adaptation work involving structural alterations (see end of this chapter).

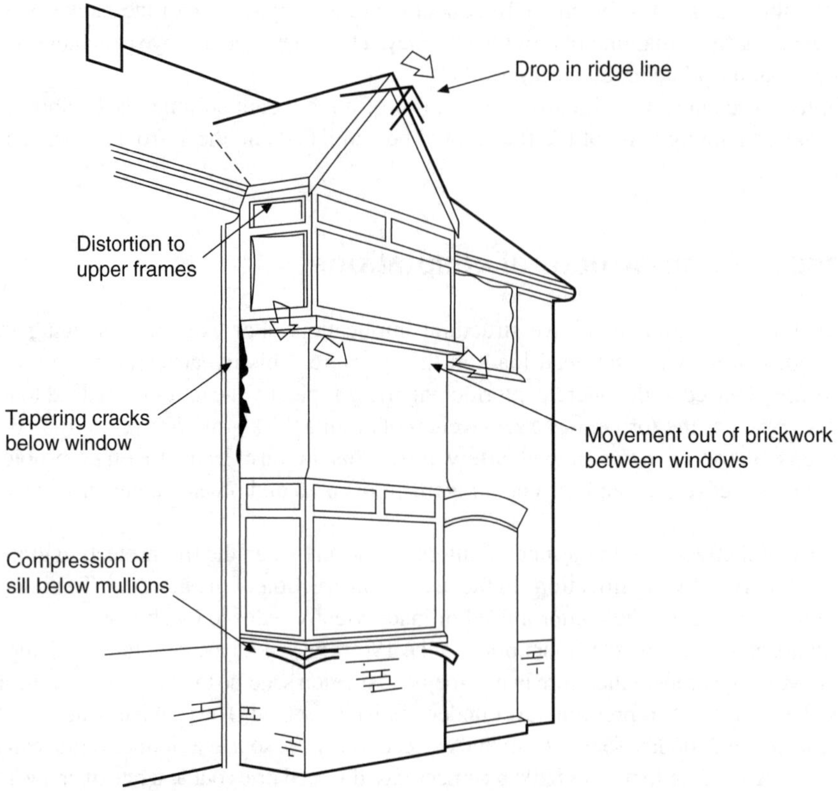

Figure 7.1 Points to watch out for when replacing bay windows (Nicholson, 1994)

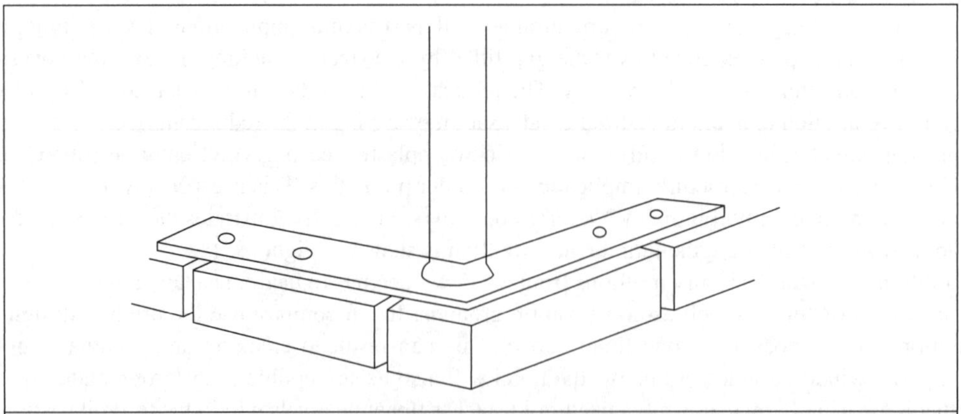

Notes

1. 65 × 8 mm galvanized mild steel plates bolted to wallhead to provide lateral restraint as well as spread load.
2. 50 mm minimum diameter galvanized steel pole (one at each corner of bay), which must go through the window sill and head to provide firm support for the structure. This can be hidden within the mullion plastic cover strips.

Figure 7.2 Design for welded post and spreader plate (Nicholson, 1994)

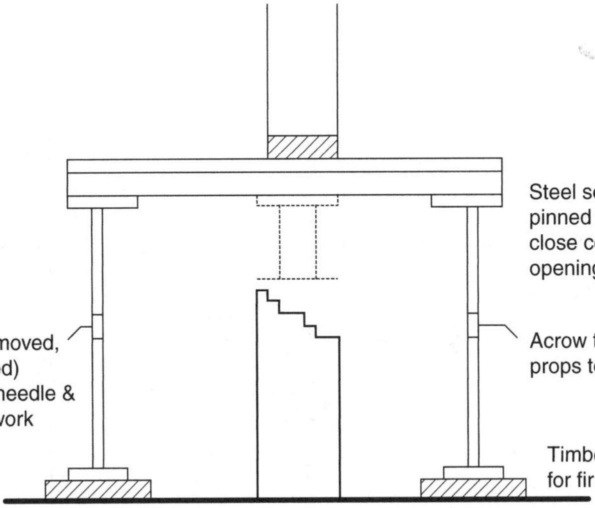

Figure 7.3 Typical dead shoring systems

Shoring

Shoring as well as permanent support is often required when undertaking structural alterations. Like adaptation work generally, however, no two shoring operations are the same. Each one must be considered on its own merits in terms of duration of use, loading conditions, risk of structural failure, and the need for protection of people and the surrounding construction.

Before any structural adaptation works begin, therefore, it is important to establish the need for and extent of any temporary as well as permanent support. Removing or altering load-bearing elements such as floors, roofs and walls is very likely to necessitate some form of shoring. As shown in the section on façade retention and façade replacement in Chapter 3 major structural alterations involving partial or full demolition of a building require temporary support. For example, flying shores are required for mid- or end-terraced properties that are being dismantled to prevent the adjoining buildings from becoming unstable.

Building Research Establishment (BRE) Good Building Guides (GBG) 10, 15 and 20 should be followed when forming new openings in existing walls and floors. BS 5975: 1991, the code of practice for falsework, should be followed. Highfield (1987; 2000) covers the traditional types of shoring as well as more complex support systems used in façade retention work.

These are summarized as follows:

- Dead shores: Comprising two vertical props and a horizontal member called a needle to support masonry and floors above when forming openings in existing walls (see Figure 7.3).
- Raking shores: Comprising inclined to buttress laterally unstable walls, such as the exposed gable of a terraced block (see Figure 7.4).
- Flying shores: As shown in Figure 7.5 comprising an arrangement of horizontal framework to support the exposed walls on either side of a mid-terraced property that has been demolished (see also Chapter 11).
- Cantilever shoring: Comprising an exo-skeleton or internal prop system (see next bullet point).
- Exo-skeleton systems: Consisting of a framework of steel box/I-sections for temporary support in façade retention schemes (as described in Chapter 3).
- Internal prop system: Comprising a box steel framework and/or series of box girder flying shores and towers as an alternative to the exo-skeletal system when undertaking façade retention work (again, see Chapter 3).

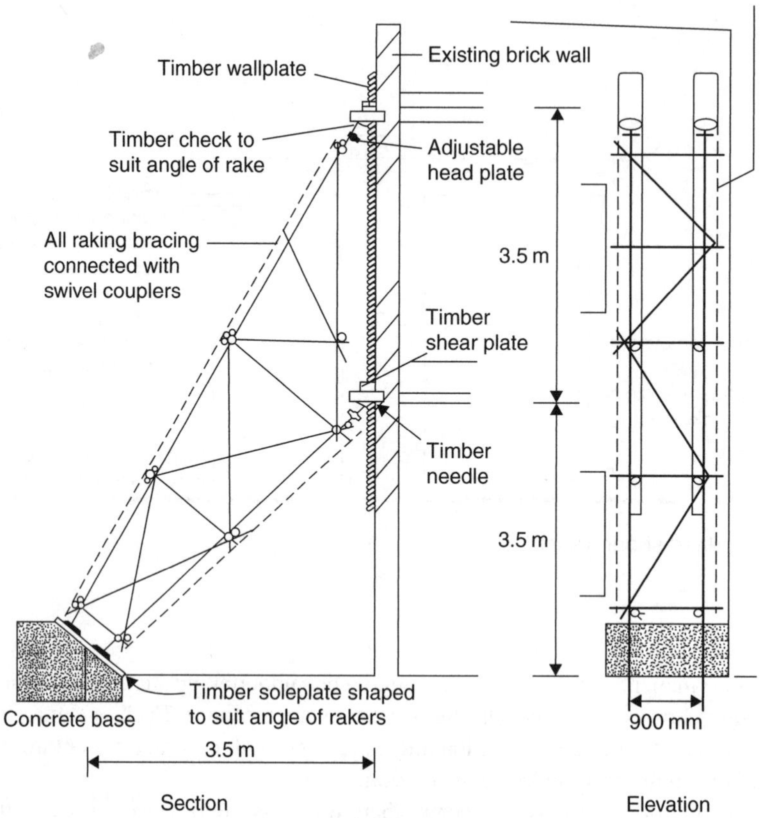

Figure 7.4 Typical example of a raking shore using scaffolding (Murray, 2002)

- Strutting: Comprising timber struts wedged between timber bearers in window and door openings to stiffen the surrounding masonry before façade retention work or prior to forming or enlarging an opening (see Figure 7.5).
- Internal propping: Comprising adjustable steel props (e.g. 'Acrows') to support a floor where the defective joist ends are being replaced (see Figure 7.6); they may also comprise props to support floors when adjacent walls are being removed or new openings are being formed.

Structural and fabric repairs

Generally

Remedial works to address a variety of structural and fabric problems often form part of major adaptations, such as large conversion schemes, extensions and modernization programmes. The advantage in combining improvements and repairs is that duplication and disruption of temporary installations and operations such as scaffolding and opening up can be minimized. These and other economies of scale can save on overall adaptation costs as well as help optimize maintenance and repair costs.

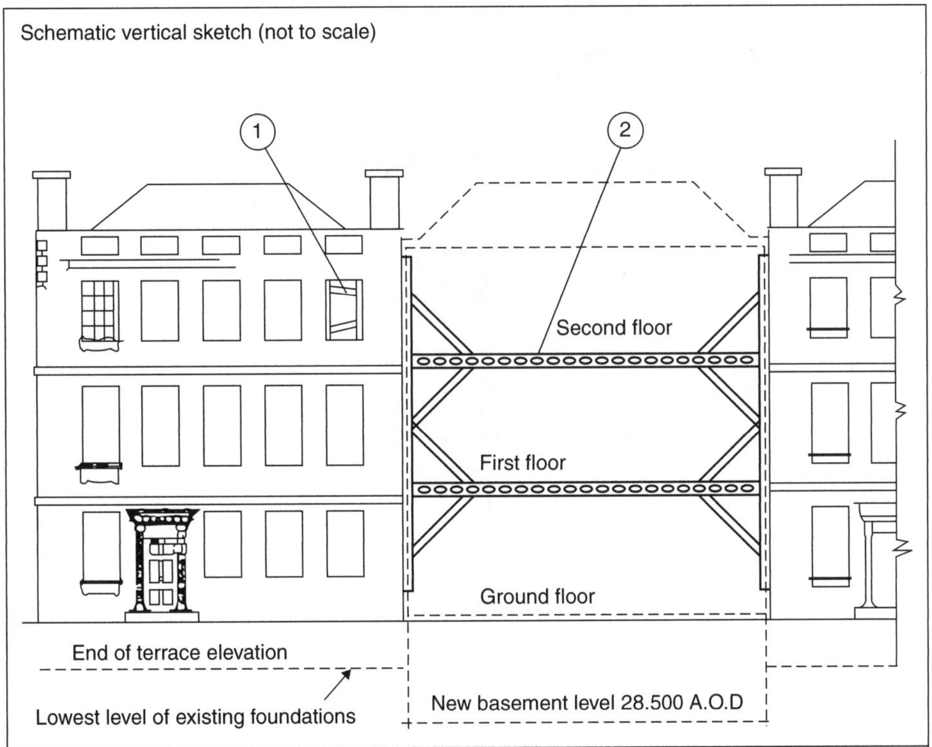

Notes
1. 100 × 500 mm timber strutting to openings around where temporary support works are being undertaken.
2. Castellated steel sections comprising main elements of flying shore system on 200 × 75 mm vertical timber walings (spacing of which is dependent on loading conditions). Alternatively, a tubular scaffolding system or heavy structural timber framework could be used as flying shores, however, these usually take up too much space so that they interfere with the new construction work.

Figure 7.5 Typical flying shore system

The following is a list of some of the main structural and fabric repairs that often form part of medium or large adaptation schemes. Such works may precede the main adaptation scheme but more usually form part of it to achieve economies of scale and minimize disruption to the occupancy of the building.

Timber decay

Preamble

Fungal attack and woodworm are the two main causes of timber decay in buildings. Table 7.1 compares these major forms of deterioration of timber.

Fungal attack

Dwellings affected by condensation and penetrating dampness are highly susceptible to timber decay. Fungal attack requires timber to have a moisture content exceeding 20 per cent. This is equivalent to

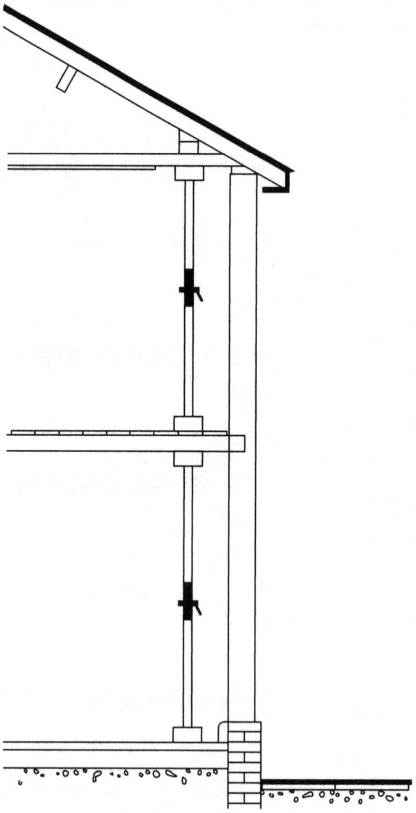

Figure 7.6 Internal propping (BRE GRG 15, 1992)

Table 7.1 Differences between fungal attack and woodworm (Coggins, 1980)

Decay caused by fungi	*Decay caused by woodworm*
No well-defined tunnels within wood structure.	Well-defined tunnels – often constant shape in transverse section.
No comminuted wood particles present, that is, no bore dust or frass.	Tunnel partly or entirely filled with frass – comminuted wood particles.
Change of colour of wood – usually darker in the case of Brown rots.	Colour of wood unchanged.
Loss of strength throughout whole mass of wood, which becomes friable.	Loss of strength of wood affected by tunnels; affected wood slightly friable if attack severe.
No circular or oval flight holes on external surfaces.	Circular or oval flight holes on external surface on completion of one life cycle.
Considerable loss in weight of whole wood mass.	Considerable loss in weight of wood tunnelled.
Transverse and longitudinal fissures present – often breaking into cubes.	No transverse and longitudinal fissures. No breaking into cubes.

85 per cent equilibrium relative humidity (Oxley and Gobert, 1994). The most serious form is *Serpula lacrymans*, the true dry rot fungus, because of its adverse effects on timber and ability to spread beyond the source of the outbreak. It is a Brown rot that thrives in the presence of warm (i.e. >22°C), damp (i.e. >85%RH), unventilated conditions (Ridout, 1999).

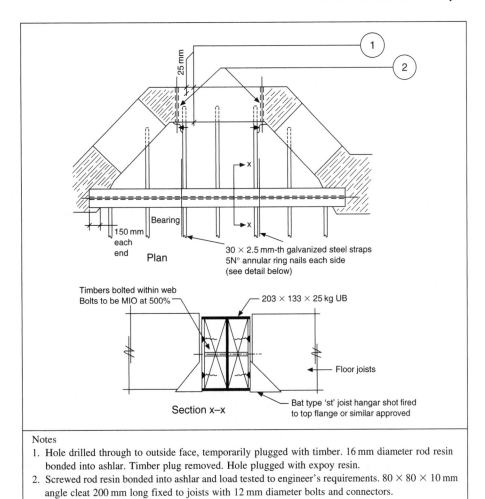

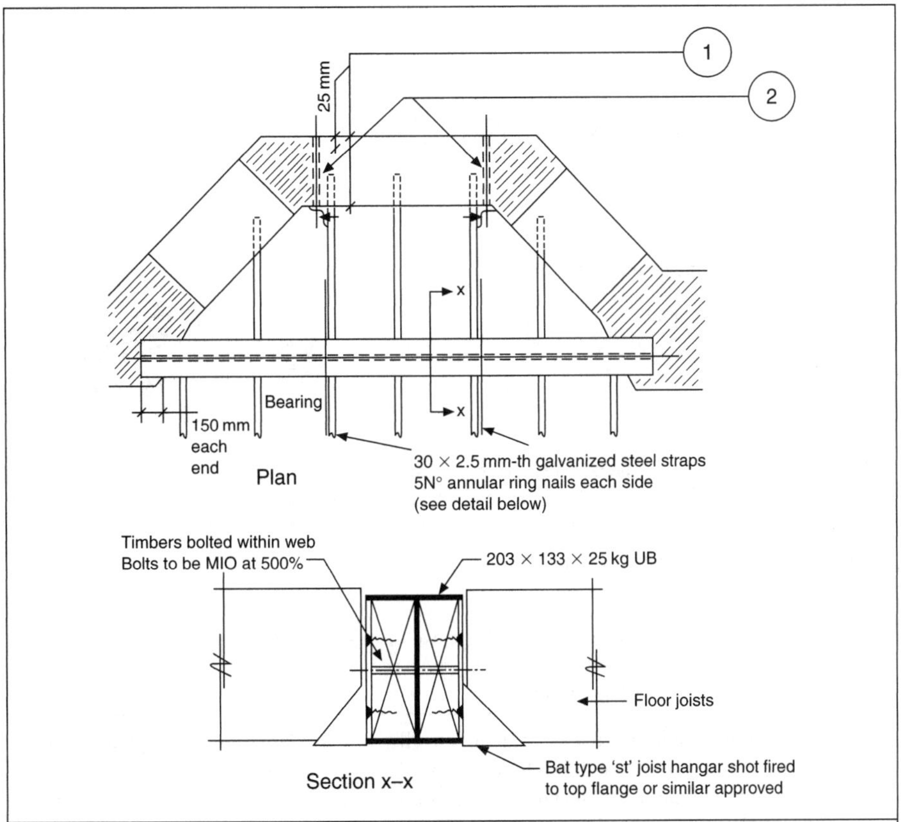

Notes

1. Hole drilled through to outside face, temporarily plugged with timber. 16 mm diameter rod resin bonded into ashlar. Timber plug removed. Hole plugged with expoy resin.
2. Screwed rod resin bonded into ashlar and load tested to engineer's requirements. 80 × 80 × 10 mm angle cleat 200 mm long fixed to joists with 12 mm diameter bolts and connectors.

Figure 7.7 Steel bressummer beam replacement

The timbers most vulnerable to dampness and fungal decay in buildings are:

- floorboards, joists and wall plates in ground floors, because of inadequate subfloor ventilation, penetrating dampness or lack of dpc;
- upper floor structures including timber staircases next to external walls or services;
- ends of built-in joists and bressummers in solid masonry (see Figure 7.7);
- 'safe' lintels above windows and bonding timbers in solid walling (see Figure 7.8);
- rafter feet, wallplates and sarking at eaves (especially if hidden by 'beamfilling', the crude masonry that closes off the eaves in the roof space);
- fascia boards, window sills, bottom rails and jambs of window sashes, and other exterior joinery.

Wet rots, such as *Coniofora puteana* (the cellar fungus), are still troublesome but do not spread beyond the source of infection. External joinery, roof timbers and woodwork in basement areas are most at risk of this form of fungal attack (Table 7.2).

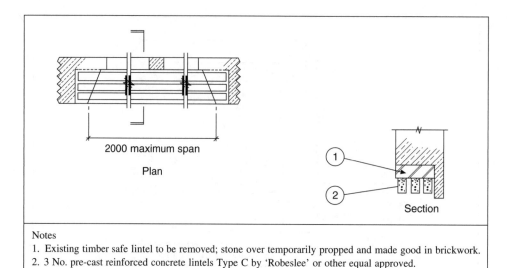

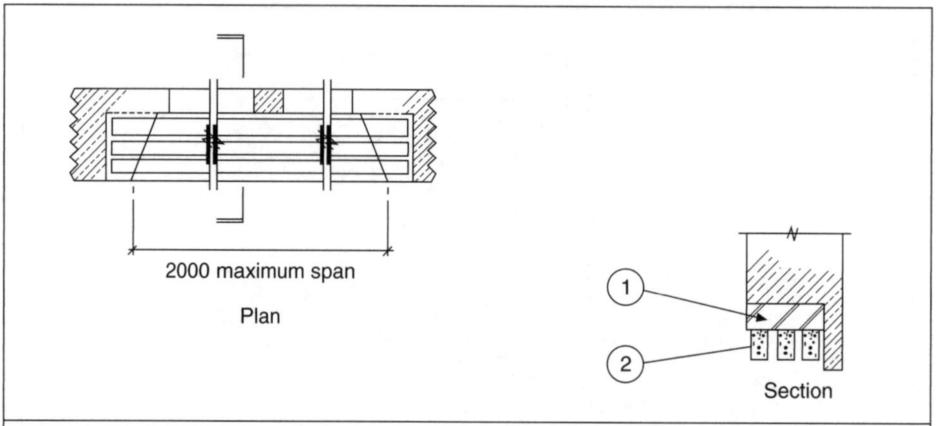

Notes
1. Existing timber safe lintel to be removed; stone over temporarily propped and made good in brickwork.
2. 3 No. pre-cast reinforced concrete lintels Type C by 'Robeslee' or other equal approved.

Figure 7.8 Replacing defective timber 'safe' lintel

Table 7.2 Comparison of characteristics for dry rot with those of wet rot (RICS)

Characteristics	*Dry rot (Serpula lacrymans)*	*Wet rot (Coniophora puteana)*
Mycelium	Damp conditions: masses of tears on silky white surface, with bright lemon patches. Drier conditions: thin skin of silver grey in colour, with deep lilac tinges.	High humidity: yellow to brownish in colour.
Decaying wood	Deep cuboidal cracking associated with differential drying shrinkage. Reduction in weight. Dull brown in colour. Resinous smell gone.	Cuboidal cracking on smaller scale. Thin skin of sound wood. Weight loss. Localized infection.
Strands (rhizomorphs)	3 mm in diameter. Brittle when dry. Off-white/dark grey in colour.	Thinner than dry rot. Flexible when dry. Creamy white in colour.
Sporophores (fruiting bodies)	Tough, fleshy pancake or bracket-shaped, varying from a few centimetres to a metre across. Ridged centre: yellow-ochre when young, darkening to rusty red when mature. Lilac/white edged. Distinct mushroom smell.	Not very common in buildings. Musty smell, rather than mushroom smell associated with an active growth of dry rot.

The main strategy for eradicating fungal attack is outlined in Table 7.3.

Insect attack

Woodworm and termites also pose a great threat to timbers in service. This type of infestation can seriously undermine their structural integrity. It is often found in damp timbers affected by fungal attack, especially in older buildings that usually contain unprotected sapwood.

As a general rule if there are more than 20 boreholes in any 100 mm section of timber, the infestation is considered severe (BRE Digest 307, 1986). Less than that, the timber may be treatable provided that it is not showing any other signs of deterioration or distortion.

Table 7.3 Measures for controlling fungal attack

Types of rot	*Control measures*	
	Primary	*Secondary*
Dry rot	Locate and eliminate source of moisture (e.g. leaking pipe, overflowing gutter, penetrating dampness?). Promote rapid drying out – use dehumidifiers (with the windows and doors closed) or hot air heaters (with the windows and doors partially open).	Remove decayed wood and other material (e.g. plaster) affected by fungus 300 mm past last sign of infection (1 m might be required if it is not dormant or is considered serious). Clean and affected masonry surfaces and sterilize them with a biocide. Irrigation of the walls with a fungicide is usually neither necessary nor effective. Moreover, it results in the build-up of potentially noxious chemicals into the fabric of the building. Ensure adequate dpcs and joist gloves are provided to built-in and other vulnerable timbers. Provide dry rot monitoring sensors (see below).
Wet rot	Locate and eliminate source of moisture. Identify the primary moisture source (e.g. rainwater penetration). Ascertain the nature and extent of decay. Determine if timbers behind or concealed by rotten wood are similarly affected.	If localized, cut out decayed section of timber back to sound substrate; apply wood hardener to substrate; insert copper nails to provide key for filler; apply epoxy resin acrylic filler to cut out section; repaint using a suitable protection system. If extensive, renew complete section (e.g. window sill or whole component).

Nowadays, spray-applied chemical treatments consisting of more environmentally friendly substances such as Boron diffusion fluid or Sovereign Chemicals' 'Flurox' should be used to preserve existing structural timbers. Injection of insecticide using a syringe or pressure system may be required if the infestation is more than slight. However, their use should be limited to the zone of infection. The use of traditional insecticides with containing lindane, DTT or other hazardous chemicals must be avoided.

Alternative approaches to timber preservation and monitoring

The 'environmental approach' to timber treatment is favoured in conservation work because it is less invasive and disruptive to the structure and fabric of historic buildings. It involves the two primary control measures listed in Table 7.3. In addition, this approach minimizes if not obviates the use of toxic chemicals. Rather, it relies on resolving the root cause of the moisture problem and ensuring the interior of the building is aerated, dried and kept free from dampness.

Monitoring moisture levels and timber decay is now a common strategy for important buildings. Installing moisture sensors in vulnerable locations (as highlighted in the above list) linked to a data logger is one way of achieving this. Hutton and Rostron's *H+R Curator*™ system (Hutton et al., 1993) is an appropriate technique for this kind of monitoring.

A more recent and specific monitoring approach is the use of dry rot sensors. The 'FUGENEX' Dry Rot Sensors (see www.fugenex.co.uk) are an innovative product that can save surveyors' time and money. The sensors are easy to install and use as part of the refurbishment or renovation of an old building containing vulnerable but valuable woodwork. They provide a new and effective way to conduct dry rot investigations but without the invasive disruption that often accompanies such work.

One of the key problems with dry rot is that the conditions for growth and development of the fungus are usually present in locations within the building that are not immediately accessible. Hitherto, the detection of rot was difficult and assessment of the extent to which it had developed would involve disruptive investigation.

Now these sensors can tell surveyors:

- when conditions for dry rot are present;
- when dry rot itself is present;
- if the dry rot is active;
- how far the dry rot has spread;
- if the dry rot is dead;
- if the treatment for dry rot has been effective;
- … all *without* the usual disruption to the floors, ceilings and walls.

The traditional (or 'slash and burn') approach is often more drastic. It usually consists of extensive stripping-out of all infected timbers and sterilization of surrounding construction. Many damp-proofing companies insist on irrigating the walls with a fungicide in the vicinity of an outbreak of dry rot as a condition for giving a guarantee for the work. Even then, however, irrigation should only be permitted in exceptional circumstances because this cannot ensure that all of the residual fungal spores are killed. In addition, it imposes extra costs to the remedial work and introduces toxic chemicals as well as more liquids into the building (Ridout, 1999).

Repairing delaminated claddings

In the second half of the 20th century small tile or mosaic panels were often used as a cladding to the façade of multi-storey buildings such as offices, educational buildings and blocks of flats. These cladding materials were often vulnerable to detachment because of failure in the bonding layer. Delamination of mosaic finish and spalling of renders is most acute at expansion joints. Moisture penetration at these locations exacerbates the problem. Loose mosaic or tile sections can sometimes fall off resulting in a major safety hazard.

Minor sections of these mosaic or tile façades can be repaired using an appropriate vacuum/low pressure injection resin system (see below in notes on repairs to cracks in concrete). This would avoid the problem of further delamination if a low-viscous resin grout under low pressure were used. If the grout is too viscous it can penetrate too deeply causing the mosaic to lift. The grout should have the consistency of mastic for maximum effectiveness in such cases.

If the delamination is widespread, however, a decision to overclad or replace rather than repair the mosaic or tile façade may be taken. This would be based on the grounds of enhancing both the aesthetic and safety characteristics of the façade. Chandler (1989) shows how this should be done.

Wall-tie repairs and renewal

Traditional masonry

Wall-tie failure is primarily caused by premature corrosion of the galvanized mild steel ties (see Figure 7.9). The BRE (IP 12/90) reported that this problem could eventually affect virtually all of the cavity-walled buildings constructed before 1981. After that time wall ties were designed to have much higher levels of resistance to corrosion. This means that some 10 million dwellings are potentially at risk of this problem (Good Repair Guide (GRG) 4, 1997).

Schematic cross-section (not to scale)

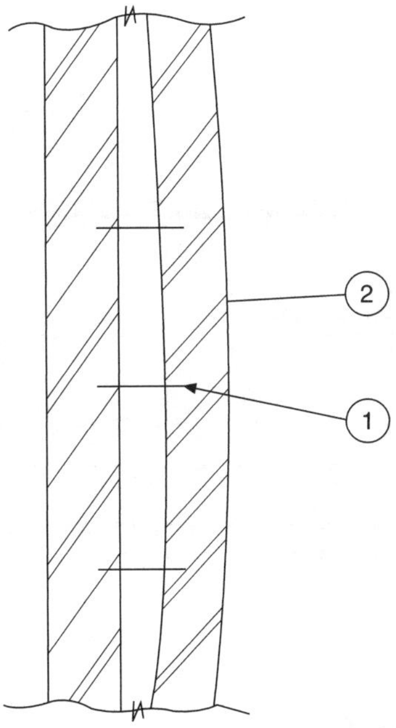

Notes
1. Corroded end of wall tie expands and causes cracking in mortar bed. Defective galvanized wall ties should be removed. Wire wall ties may remain in position after being bent downwards.
2. Outer leaf will tend to bulge outward or show distinct horizontal cracks every four courses.

Figure 7.9 Wall-tie failure

Accordingly, some housing modernization schemes may have to include cavity wall-tie replacement as part of the programme (see Figure 7.10). Two main methods of replacement can be used:

(i) stainless steel expanding bolts inserted into the both skins of brickwork from the outside; or
(ii) plastic or stainless steel reinforcing rods inserted into both skins but grouted with a thixotropic resin.

However, it is recommended by the BRE (GRG 3, 1996) that whatever method is used with the twisted type wall tie, the metal that is embedded in the outer leaf should be removed. If not, the original wall ties will continue to corrode and expand. As this problem is not so acute with butterfly wall ties, they can usually be left in place.

There are many proprietary retrofit wall-tie installations available these days – 'Hilti' and 'Red Head' are two typical examples. BRE GRG 4 and Digests 329 and 401 provide sound guidance on the methods of dealing with wall-tie failure.

In external conditions of low exposure an alternative to retrofit wall ties is to inject the cavity with a heavy-duty polyurethane foam. As well as enhancing the thermal efficiency of the cavity wall, this method ties both skins of walling. 'Tyfoam', manufactured by ICI, is one such system, which again specialist approved contractors must install.

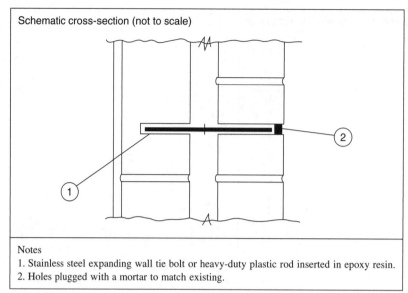

Schematic cross-section (not to scale)

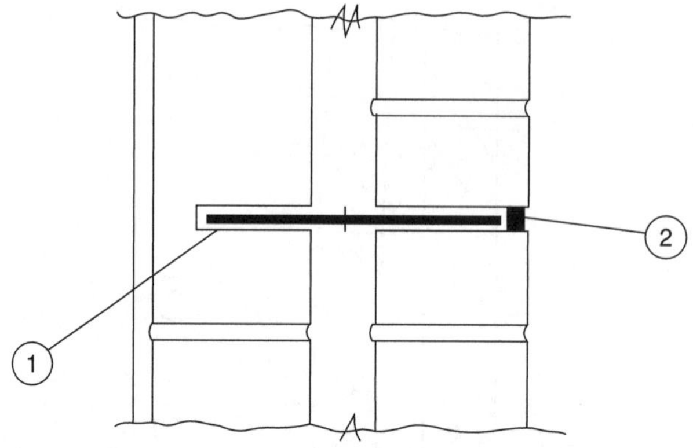

Notes
1. Stainless steel expanding wall tie bolt or heavy-duty plastic rod inserted in epoxy resin.
2. Holes plugged with a mortar to match existing.

Figure 7.10 Wall-tie replacement

Timber-framed construction

It would be easy to assume that timber frame construction is not affected by problems such as wall-tie failure. Shipway (1987) reported a number of structural problems found during a survey of some 266 low-rise timber frame houses in an estate in the north of Scotland. According to that investigation practically all the houses had areas of missing wall ties and that many of the existing ties were not adequately fixed or of recommended form. No movement joints had been provided in terraces up to 46 m long.

The remedial works to those dwellings involved inserting some 30 000 new wall ties and the cutting of new movement joints vertically at maximum 12 m centres. The retrofit wall ties for timber frame construction have a screwed end for insertion into the vertical studs. All of this was done without disturbance to the occupiers of the dwellings. Care should be taken to ensure that all the fixings are properly located in the centre of each stud.

Steel-framed construction

A popular choice of material in many non-traditional housing constructions, such as B.I.S.F. and Blackburn, was mild steel (Harrison et al., 2004). Bi-metallic corrosion as well as normal rusting is a common problem in the cladding fixings in these dwellings.

Repairing and reinforcing cracks in masonry

Every building has cracks. The stresses and dimensional changes caused by temperature, moisture and structural movements in a wall are sometimes distributed evenly throughout its thickness but these usually do not exceed 5 mm. Lack of suitable movement joints in brickwork, though, can lead to cracks exceeding the 5 mm 'safe' limit.

The type of repair to cracked masonry will depend on the extent of damage and the element affected. The main repair methods that can be used are as follows:

- Crack stitching (see next section).
- Repointing non-structural cracks with a weak mortar: with lime or a low cement content.

Sketches (not to scale)

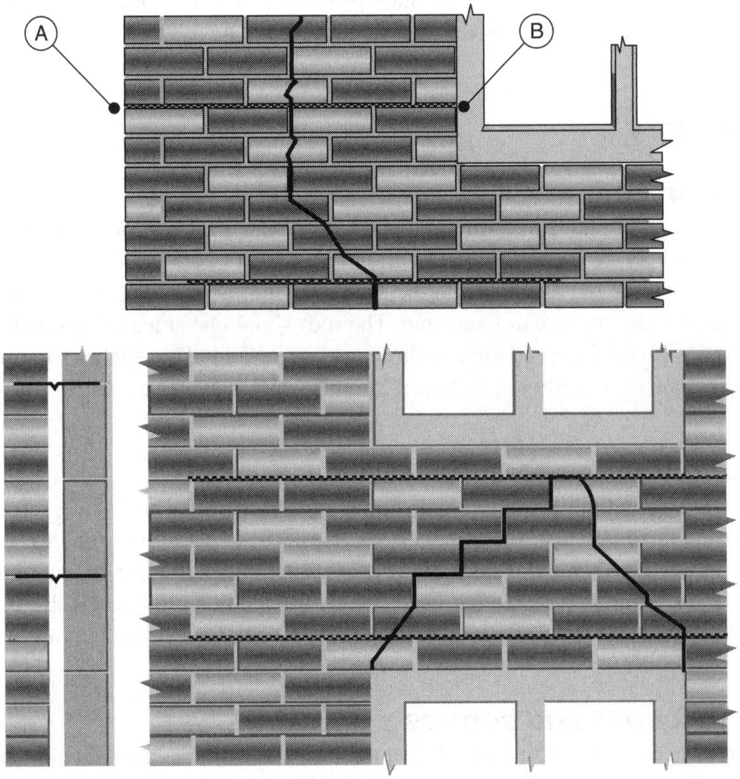

Notes

1. Repairing cracks near corners and openings using 'HeliBar': where cracks are less than 500 mm from an external corner (A) or an opening (B) at least 100 mm should be bent round the corner and bonded into the return wall or bent and fixed into the reveal, avoiding any dpc membrane.
2. Points to note:
   - Where two or more cracks are close together these may be stitched using one continuous length of HeliBar which must be long enough to extend 500 mm beyond the outer cracks. For example, if there are three cracks, each 250 mm apart, then the overall length of HeliBar required would be 1.5 m. The horizontal slot, normally the mortar bed, can be cut using a twin diamond blade chaser with vacuum attachment, an angle grinder or a mortar chisel.
   - All mortar must be removed, together with any loose debris, to ensure a sound bond.
   - Prepare the slot with HeliBond Primer or thoroughly wet the slot with clean water.
   - As standard, slots should be 10 mm wide to accommodate the 6 mm diameter HeliBar. For thin joints use 4.5 mm HeliBar.
   - HeliBond grout (non-shrink, pumpable, thixotropic, cement-based grout) is the recommended bonding agent. PolyPlus resin is used for smaller jobs and where loads are to be rapidly applied.

Figure 7.11 Stitching of defective brickwork (www.helifix.co.uk)

- Replacing cracked section/s of masonry, with compatible brickwork/stonework in a suitable mortar and with appropriate movement joints and restraint ties.
- Crack bonding: injecting structural cracks with an epoxy resin mortar. It may be possible to use vacuum pressure similar for concrete to achieve a better degree of fill.

Injected epoxy resin can also be used to reinforce masonry walls and columns. 'PC® Structo-Inject 1380' supplied and installed by 'TRADECC' is one such product. It's a solventless epoxy injection resin and was used to increase the load-bearing capacity of the large stone pillars in Antrwerp Cathedral in 2003.

Stitching masonry

Cracked masonry

Brick arches that have failed can be repaired using stainless steel reinforcing rods set in an epoxy resin (Robson, 1999). Brickwork or stonework with certain cracks up to about 5 mm wide in can be repaired using helical stainless steel rods embedded into the masonry joints across the cracks. 'Helifix' is one of several companies that provides this repair technique. The rods should be at least 500 mm long on each side a crack and are inserted into the prepared joint as shown in Figure 7.11. The stitches should not be any closer than four brickwork courses or further apart than 12 courses to achieve an effective 'brick beam'.

Stitching and grouting cores

Many old 'solid' walls were built using two outer skins of stonework or brickwork separated by a core of loose fill material and linked by occasional bonding stones. These walls are prone to bulging, because of a lack of bonding between the two outer skins of masonry. Stitching a loose core using stainless steel dowels embedded in a resinous grout in the affected walls can overcome this problem (Robson, 1999). It should, however, be done in a discrete manner so that the stitching holes are not conspicuous.

Stitching 'T' junctions and corners

Stitching may also be necessary to knit walls at 'T' junctions and corners. Anchor bolts can be used to stitch fragile masonry structures at 'T' junctions. Expanding stainless steel bolts 9 mm in diameter and 500 mm or 750 mm long (depending on the circumstances) can be inserted in such a junction (e.g. 'Clanstitch') at approximately 450 mm centres horizontally or downwards at a 45° angle. Alternatively resin injected stainless steel 'Helifix' anchors can be installed in a similar manner.

The corners of load-bearing masonry walls can also suffer cracking and distortion as a result of structural movement. Stitching can be used for both internal and external reinforcement of fractured masonry as indicated below:

- External corner stitching: Galvanized mild steel or stainless steel straps (about 75 × 5 mm) can be wrapped around the external corners of a building containing vertical cracking. Figure 7.12 shows how this can be done to an external corner of a building that is affected by structural movement.

  A modern way of stitching internal and external corners of brickwork is by using the 'HeliBar' method described in Figure 7.11. Its a more discrete method as it involves inserting these helical stainless steel rods, which are bent at the corner in an L-shape, into the bed joints of masonry. Only good quality stainless steel (e.g. Grade 304–316 austenitic stainless steel) should be specified for this type of repair.
- Internal corner stitching: Traditionally, expanded metal lath strips about 700 mm wide were used to stitch corner junctions vertically. The 350 mm wide strips of expanded metal lath are used over cracks in other parts of the walls (Robson, 1999). The expanded metal lath is fixed directly to the brick or concrete substrate, which is then plastered over. In more severe cases heavier gauge stitching consisting of stainless steel straps approximately 75 × 5 mm and about 600 mm long and bent to form an L-shape can be bolted to the bare masonry. These are normally fixed at 600–1000 mm centres using two 16 mm diameter bolts through each leg. The fixings are then plastered over.

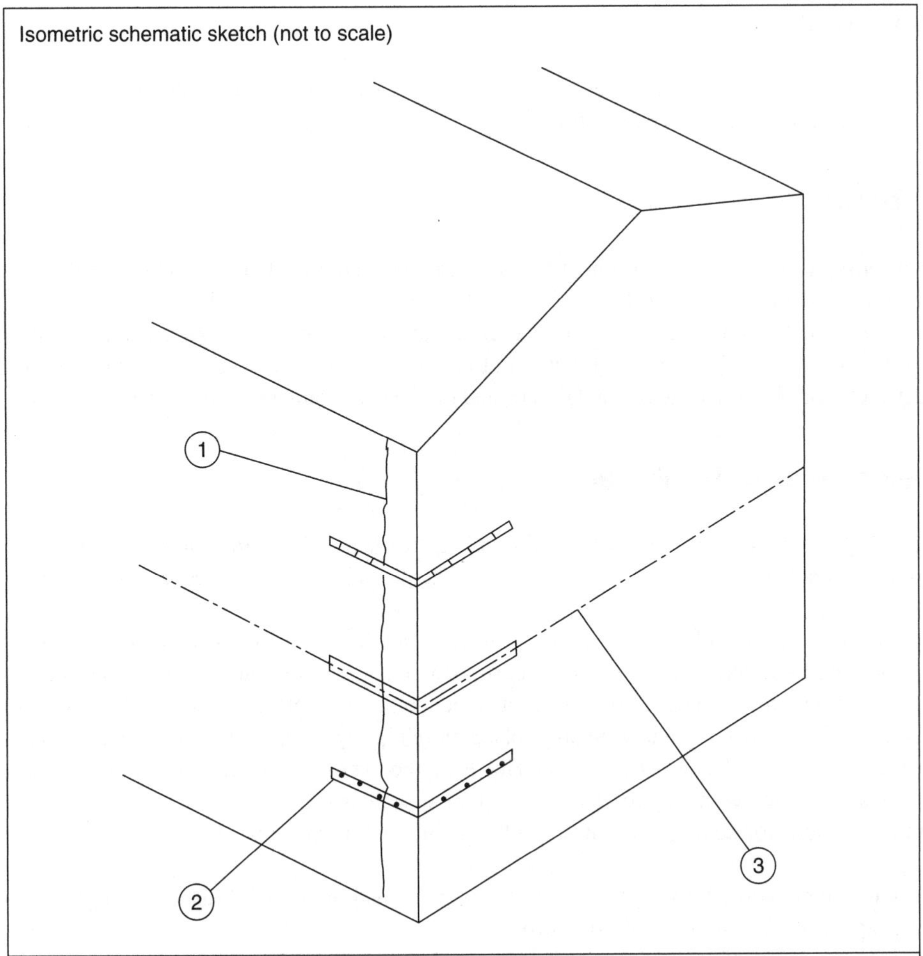

Notes

1. Vertical crack requiring attention. Structural cracks of this kind can be injected with an epoxy resinous grout.
2. 65 × 5 mm thick stainless steel or galvanized mild steel 'L' strap fixed to masonry using 105 mm long expanding bolts at 200 mm centres. Each leg of the strap should be at least 1.2 m long and at about 750 mm centres vertically or to other lengths and spacings as determined by a structural engineer.
3. First-floor level.

Figure 7.12 Stitching of an external corner

Girdle straps

Detached, semi-detached and short terraced bungalows in mining subsidence areas may have girdle straps installed around the entire building at eaves level. Two-storey dwellings may have a strap at first-floor level as well as at the eaves. Only buildings having a square or rectangular plan shape can be restrained using this technique. It is not suited for L-shaped, T-shaped and other irregular plan buildings because of the difficulty of achieving restraint at internal corners.

Restraint ties

Inserting these ties may be required in gables or other main walls that are affected by excessive bowing or leaning because of inadequate lateral restraint. Figure 7.13 illustrates how this can be achieved.

Roof spread

Defects in ceiling ties or lack of restraint at the wallhead can cause roof spread. This manifests as leaning and horizontal cracking of walls at the eaves as well as sagging of the ridge-line of the roof.

Inserting restraint ties similar to the process described in the next previous section can arrest this problem. Alternatively, Robson (1999) describes a method of arresting severe roof spread by using a combination of plywood gussets and plates fixed to ties and rafters, and mesh reinforcement fixed to the inside face of the wall.

Corrosion of structural steel

Moisture and aggressive compounds such as salts and acidic chemicals are the main enemies of steelwork in buildings. Rust and degradation can affect other metal elements in buildings – especially fixings, claddings and pipework.

Ferrous metals such as steel corrode when exposed to water and oxygen. Oxidation is the name for ordinary corrosion (chemical corrosion, electrolytic corrosion and stress corrosion are the other main types). In chemistry oxidation is the term used to describe any reaction in which electrons are lost from a substance and so become ions. The process of oxidation takes place more rapidly the higher the temperature. Thus as temperature increases so does the rate of oxidation. However, the rusting of iron and steel is not a case of simple oxidation. It is associated with the presence of moisture as well as air.

The main methods for dealing with these problems in structural steelwork are:

- Preparation: removal of all loose rust and scale by grit blasting or wish brush grinding;
- Priming: application of a rust-inhibiting primer;
- Protection: see anti-corrosion measures below.

Cathodic protection of steelwork

Steel-framed construction covered with stone or brick cladding has been a popular method for office and public buildings in Britain since about the late 19th century. In those early years little if any consideration was given to protecting the steelwork against corrosion. The columns and beams were often erected untreated. Even steel members that were encased in concrete or brickwork for fire protection often lacked any initial protection.

Treating corroded steelwork of a framed building is not straightforward. Many members such as columns and beams are hidden behind cladding or fire protection. This makes it difficult to obtain access to the sections affected by corrosion.

Cathodic protection was originally used to treat ferrous metals on boats. It used the sacrificial anode method. However, this version of cathodic protection is not suitable for steel-framed buildings (Historic Scotland, 2000).

Impressed Current Cathodic Protection (ICCP), on the other hand, can be used for halting but *not* reversing the corrosion of steelwork. According to Historic Scotland (2000), it is 'system in which an electric current is applied from an external source to oppose the natural flow of electric current that gives rise electrochemical corrosion'.

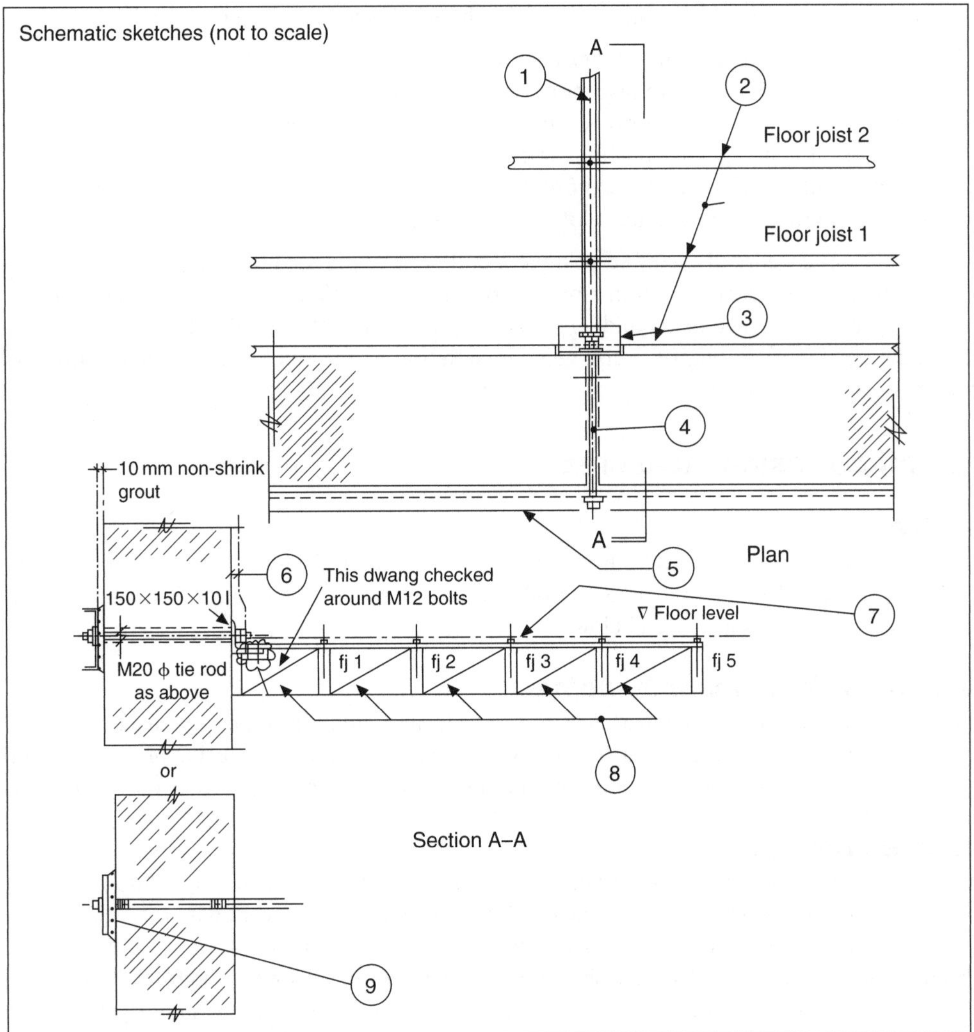

Notes

1. 60 × 5 mm thick mild steel flat tie connected to notched floor joists using 10 mm diameter × 75 mm long square-headed coach screws in 12 mm diameter hole in flat plate. 8 mm diameter pre-drilled holes in joists. Notch in existing floor joists to be 15 mm deep.
2. 60 × 5 mm flat mild steel plate connected over 5 No. floor joists as shown numbered.
3. 150 × 150 × 10 mm angle cleat 200 mm long.
4. 20 mm diameter 'Sherardized' screwed rod wrapped in 'Denso' tape and c/w mild steel washer, hex nut and lock nut assembly to each end.
5. 152 × 89 mm × 23.84 kg/m rolled steel channel (galvanized) gaps filled with non-shrinkable grout.
6. Skirting board and plasterwork removed to provide positive fixing against wall.
7. 60 × 5 mm flat mild steel plate connected over 5 No. floor joists as detailed above.
8. 1 No. line of full depth noggings skew nailed to existing floor joists.
9. External plate option – 300 mm diameter × 12 mm thick galvanized mild steel plate bedded on non-shrinkable grout – would be less ugly than the channel section described in (5).

Figure 7.13 Inserting steel restraint ties to gable

There are, therefore, a number of drawbacks with the ICCP and other similar techniques:

- initial costs are usually high, with savings achieved only in the medium to long term;
- voids around the steelwork lead to discontinuity in the current;
- if earthing is incorrectly installed or other metal components are nearby, stray current accelerated corrosion could occur, making the problem worse;
- some invasive work is still needed to expose the steelwork at connection points to the ICCP system;
- the system has a limited life expectancy of between 20 and 50 years.

Only suitably experienced and qualified designers and contractors should undertake ICCP design and installation work. The guidance given in Annex B of Historic Scotland's work on corrosion in masonry clad early 20th century steel-framed buildings (Historic Scotland, 2000) should be followed when dealing with this technique. See also English Heritage's research transactions on metals for more information on ICCP (English Heritage, 1998).

Other anti-corrosion measures

Preamble

Moisture and aggressive compounds such as salts and acidic chemicals are the main enemies of steelwork in buildings. Rust and degradation can affect other metal elements in buildings. The main methods for dealing with these problems are given below.

Separation of incompatible metals

Avoid dissimilar metals such as copper and mild steel from coming into direct contact with each other because in the presence of some moisture this can trigger electrolytic action causing bi-metallic corrosion. If contact is unavoidable, a neoprene gasket or other medium should be used to separate different metals.

Protective coatings

These are required to protect metals, especially those exposed to aggressive conditions such as high humidity levels or salt-laden water (British Steel, 1996). Priming the steelwork with rust inhibiting primer, such as epoxy aluminium, and coating it with a two/three coat protection system is the first requirement. Bituminous coatings can be used for suitably prepared exposed metalwork in high humidity conditions. Alternatively, a Micaceous Iron Oxide or 'Anode' paint could be used as the finishing system. Other protective paint systems for steelwork are cementitious coatings, and organic coatings such as an epoxy topcoat. In any event, preferably only paints with low volatile, organic compound (VOC) emissions should be employed to comply with the requirements of the Environmental Protection Act 1990.

Deterioration of reinforced concrete

Degradation mechanisms

Concrete as well as steel frames need protection against moisture and aggressive agents such as acidic gases like carbon dioxide and sulphur dioxide. The reinforcement in porous or other forms of poor quality concrete is highly prone to corrosion. Even good quality reinforced concrete, which was once thought to be a relatively maintenance free material, can deteriorate prematurely because of one or a combination of the following mechanisms:

- corrosion of steel reinforcement, especially with inadequate cover;
- carbonation – resulting in a loss of alkalinity in the concrete, which makes it more acidic and porous;
- chemical reactions that degrade the concrete such as alkali aggregate reaction (AAR), alkali silica reaction (ASR), sulphate attack, high alumina cement or calcium chloride reaction;

- impact and abrasion;
- biofouling and bacterial attack (which can release organic acids);
- frost damage;
- salt recrystallization (exfoliation);
- fire damage (see below).

The extent of carbonation, for example, can be easily determined by spraying the surface of the exposed concrete with Phenolphthalein solution. If the sprayed area of concrete retains its grey colour, carbonation of the concrete has occurred. If it turns a purple-red colour it is not affected by excessive carbonation.

Repairing degraded concrete

Re-alkalization may be required as a preventative measure. It is an electrochemical technique developed for the prevention of reinforcement in carbonated concrete (Historic Scotland, 2000). Another similar technique is electrochemical chloride extraction, which is a fast and simple method for removing salts from concrete structures exposed to de-icing salts or marine conditions (CRA, 1999). This can be combined with fibre-reinforced polymer (FRP) wrapping to strengthen in-situ concrete columns (see strengthening measures below).

Proprietary epoxy resin repair techniques are very effective for making good concrete elements damaged by spalling and carbonation (see Figure 7.14). The basic procedure for undertaking this work is as follows:

- Preparation:
  - Marking up: ascertain and mark up extent of damage or deterioration. A yellow wax crayon is useful for this exercise.
  - Cutting out: carefully remove loose, spalled and other defective concrete back to sound substrate.
  - Cleaning out: ensure all surfaces and exposed reinforcement are clean and free from dust and loose scale.
- Treatment:
  - Protection of the steel reinforcement: apply coat of rust inhibiting primer to exposed sections of reinforcement (and, where necessary, install additional bars if some steel is badly corroded).
  - Apply coat of epoxy bonding agent to exposed surfaces of concrete.

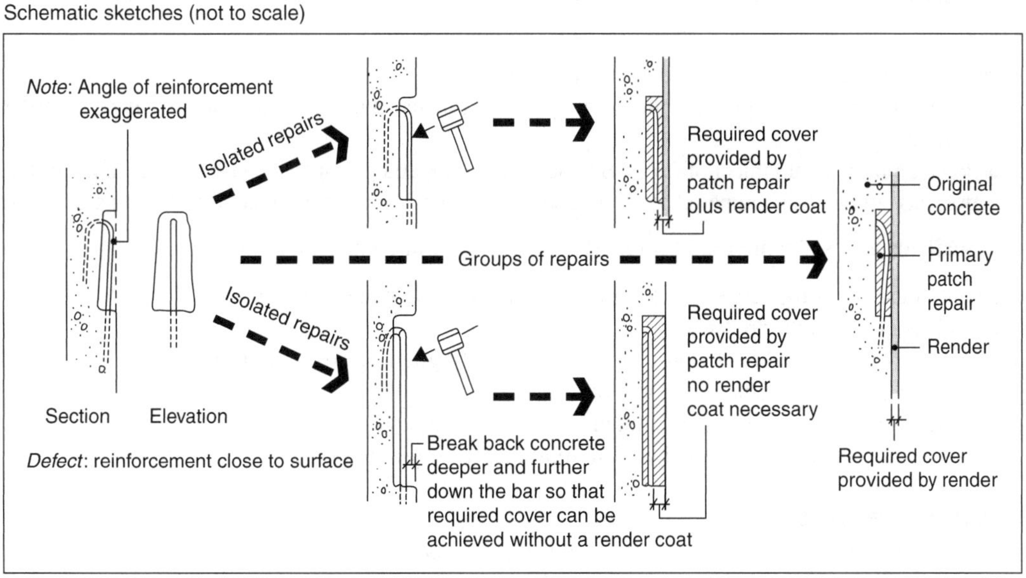

Figure 7.14 Typical repairs to reinforced concrete (Source: *Architects Journal*, 22 January 1986)

- Fill/mortar:
  - For localized areas of deterioration, apply epoxy resin filler to reinstate removed sections of concrete.
  - In more severe cases spray-applied concrete (e.g. 'Gunite') is required (see next section).

Repairs to fire damaged concrete

Fire in a building can cause extensive damage to its concrete elements. Building surveyors and architects with experience in dealing with damaged concrete can conduct the initial appraisal. In more serious cases, however, a structural engineer would be required to assess the degree of damage and determine which elements are repairable.

Concrete damaged by fire changes colour (see Table 7.4, in report by Concrete Society 1990). For example, at 300°C it turns a pinkish colour. Other symptoms such as spalling, deflection and exposed reinforcement may manifest themselves in the concrete.

Provided the element is structurally sound, fire damaged concrete can be repaired with 'Gunite'. This is a cement/sand mix spayed under pressure. Sand is used or an aggregate with a maximum size of 10 mm. A minimum thickness of 10 mm can be applied to the affected area. Twenty-five millimetre is the maximum thickness of one coat. Before this work is carried out, however, all pink coloured concrete and friable material is removed. The surface of the concrete and any exposed reinforcement is then grit blasted.

'Gunite' may be used for large areas – especially soffits of slabs or tunnel linings. A steel fabric reinforcement mesh should be used with thicknesses of about 40 mm or more. See BS8110: Part1: 1985 for details of repairs to concrete using 'Gunite'.

Repairs to cracked concrete

These are required to prevent ingress of aggressive agents such as moisture and carbon dioxide as well as to maintain continuity of the structure. The repair method selected primarily depends on the crack dimensions, particularly width. Other factors such as structural integrity, presence of water and reduction of chlorides and other deposits from cracks can also influence the selection of repair method.

The method of crack repair for the various widths of cracks in concrete slabs, beams and columns is as follows:

- Cracks 5–15 mm: use injection under gravity feed.
- Cracks 2–5 mm: use low or high pressure injection assisted with vacuum injection if appropriate to optimize level of fill.
- Cracks $<$ 2 mm: vacuum injection (e.g. 'Balvac' technique) and pressure injection; vacuum injection can achieve a 93 per cent fill compared to 75 per cent achieved using pressure injection according to 'Balvac'.

The main materials that are used for crack injection work in concrete and masonry are:

- Cementitious: good for larger cracks (i.e. $>$2 mm), but their bonding properties are generally lower than polyesters and other resins.
- Epoxy resins: traditional injection resin with generally good viscosity, but high level of mix control required when mixing the hardener and binder.
- Polyester resins: low viscosity, but suitable for vacuum injection and versatile under most conditions.
- Acrylic resins: such as Methylmethacolate mortars are similar to polyesters – low viscosity.

External coatings

Exposed surfaces or in-situ or pre-cast reinforced concrete members may require upgrading of their finish for aesthetic and protection reasons. This can be achieved either using an overcladding system (see Chapter 9) or painted finish to cover the exposed surfaces.

A two-coat paint system can be used to protect exposed concrete surfaces once they have been repaired. Either cementitious or plastic coatings can be used. The coating, however, should be vapour permeable to allow any moisture within the concrete to escape. For example, a 'Decadex' plastic paint system can provide good protection to as well as a degree of breathability for external surfaces of concrete. A range of colours is available with this material. A grey finish may be specified to retain the original appearance of the concrete.

Repairs to reinforced autoclaved aerated concrete

Reinforced autoclaved aerated concrete (RAAC) was a popular choice as structural material in educational, commercial and industrial buildings between 1950 and 1980. It was primarily used for pre-cast wall panels and flat roof planks in factory and warehouse blocks. 'Siporex', for example, was a common proprietary brand of RAAC wall panel.

However, the term autoclaved aerated concrete (AAC) is a slight misnomer in that it is not a true form of concrete. AAC is not concrete in its constituent materials or in its physical properties (Noy and Douglas, 2005).

AAC was also used for blocks in blockwork walling as well as for pre-cast wall and roof panels in low-rise residential properties. It is made under high-pressure steam-curing conditions by introducing bubbles of gas into a cement or lime mix. The finished product is a uniform cellular material that could be classed as a 'foamed mortar' – although it is sometimes described erroneously as 'foamed concrete' (Noy and Douglas, 2005). In one sense it is analogous to a 'no-coarses' concrete, in contrast to 'no-fines' concrete. As a result RAAC is relatively lightweight and has good thermal insulating properties.

RAAC, though, like ordinary Portland cement (OPC) is susceptible to water-induced degradation. Interstitial condensation and rainwater penetration are its main moisture-related deterioration mechanisms (Noy and Douglas, 2005). These can lead to corrosion of the reinforcement. Along with creep this can cause elements such as roof planks to sag by over 50 mm – depending on the span.

Structurally AAC blocks and planks are susceptible to following main problems:

- Cavity walls containing AAC blocks may have insufficient flexural strength to transfer wind loads or be poor at resisting impact loads, all of which would be exacerbated by poor condition of the masonry or the lack of ties between the leaves or the inadequacy of restraint fixings.
- As their modulus of elasticity is low AAC planks are not as strong as reinforced concrete slabs and are therefore more prone to sagging. When they are used as the structural deck on flat roofs this results in ponding.
- 'Siporex' flat roof planks may have a lower factor of safety against uplift than required by the current British Standard owing to inadequate holding down straps.
- There is a risk of shear failure occurring at the bearing of roof planks on the wallhead.

The extent of distortion in RAAC panels found during the initial survey of a building will determine the required response. Generally, the following actions would apply in the circumstances outlined:

- Deflections causing significant ponding, replace the roof.
- Deflections greater than 1 in 100, replace the roof.
- Deflections greater than 1 in 150, monitor annually.
- Deflections greater than 1 in 200, monitor every 5 years.

Traditionally, the repair method would involve replacing the defective deck. This is of course an expensive, time-consuming and disruptive option.

However, 'Metsec Building Products', have devised a suitable repair method that obviates the need for re-roofing these decks. It involves installing under the soffit of the RAAC planks a lightweight arrangement of steel castellated and lattice beams. The castellated beams are about 175 mm deep (depending of course on the span) and positioned at 2.4 m centres. A sub-grid of lattice beams 100 mm deep is positioned at

800 mm centres between the castellated beams. Special tubing is installed between the 50 mm nominal gap between the top of the beams and the soffit of the planks. It is then inflated to jack up the defected planks using a patented process called 'precise air lifting'. Non-shrink grout is then forced into the gap between the top of the beams and soffit to keep the planks in position once they have been lifted.

Strengthening existing buildings

Preamble

The structural elements of some older buildings may not adequately resist the load imposed by modern use. Some floors in a refurbished building, for example, may require upgrading to take 3–7 kN/m^2 loading from an original design of 2.4–2.9 kN/m^2.

Strengthening of buildings and other structures, therefore, may be required for a variety of reasons:

- To compensate for damage of the structure, such as after a fire/explosion or as a result of impact.
- To overcome deficiencies in the original design.
- To increase the load-bearing capacity by a changed use of the construction.
- To compensate for changes in the structure, such as drilling through the floor plate to take new services.
- To remediate errors made during construction (e.g. insufficient number of reinforcing bars in concrete members).

Installing an additional storey is sometimes another consideration for the building being adapted. In such a case some form of strengthening work to the structure may be required. The foundations, for example, may need upgrading to cope with the additional dead and imposed loads. In these circumstances underpinning is usually necessary (see separate section on underpinning later in this chapter).

Strengthening of a building structure may also be required to prevent overstressing, distortion or disproportionate collapse. Overstressing can occur because very heavy fittings such as safes and large bookcases have been installed in the building. It can result in cracking or distortion of structural elements such as beams, slabs and columns.

Dealing with strengthening measures to buildings is complex work. In most cases it is likely to necessitate the expertise of a structural engineer to design the required measures. However, it is important for other construction professionals such as architects, building surveyors and construction managers to understand when strengthening may be required and what measures are available. The main methods of strengthening specific load-bearing elements such as floors, walls and roofs are described briefly below.

Disproportionate collapse

Disproportionate collapse is said to occur if an accident or misuse removes one supporting member causes damage beyond the locus of the failure. The structure, in other words, must not be damaged to an extent disproportionate to the cause of the damage. It is sometimes also referred to 'accidental damage' or 'progressive collapse'.

Part A3 of the Building Regulations 2000 (as amended) for England and Wales is relevant only to:

(a) building having five or more storeys (each basement level being counted as one storey);
(b) a public building, the structure of which incorporates a clear span exceeding 9 m between supports.

Accordingly, A3 will apply if another level is added to a four-storey building or a five-storey (or more) building is to be used in a way that is not exempt from the requirements of the Regulations. In the context of adaptation, however, this regulation only likely to be used in existing buildings where very extensive structural alterations or additions to the number of storeys are being considered (Haverstock, 1998).

Typical strengthening measures

Load-bearing masonry buildings

Older buildings of solid load-bearing masonry construction built before the early 20th century sometimes require strengthening. For example, Georgian and Victorian tenements are often subject to substantial

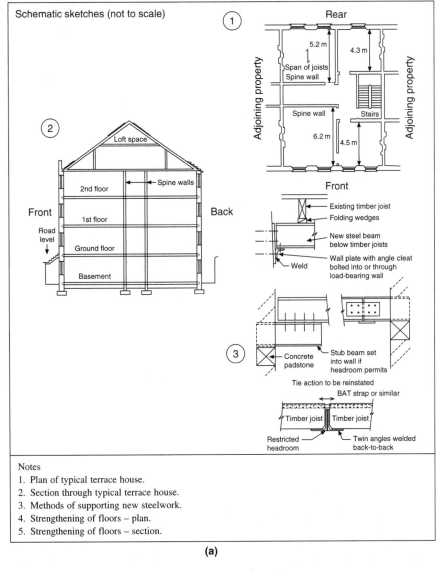

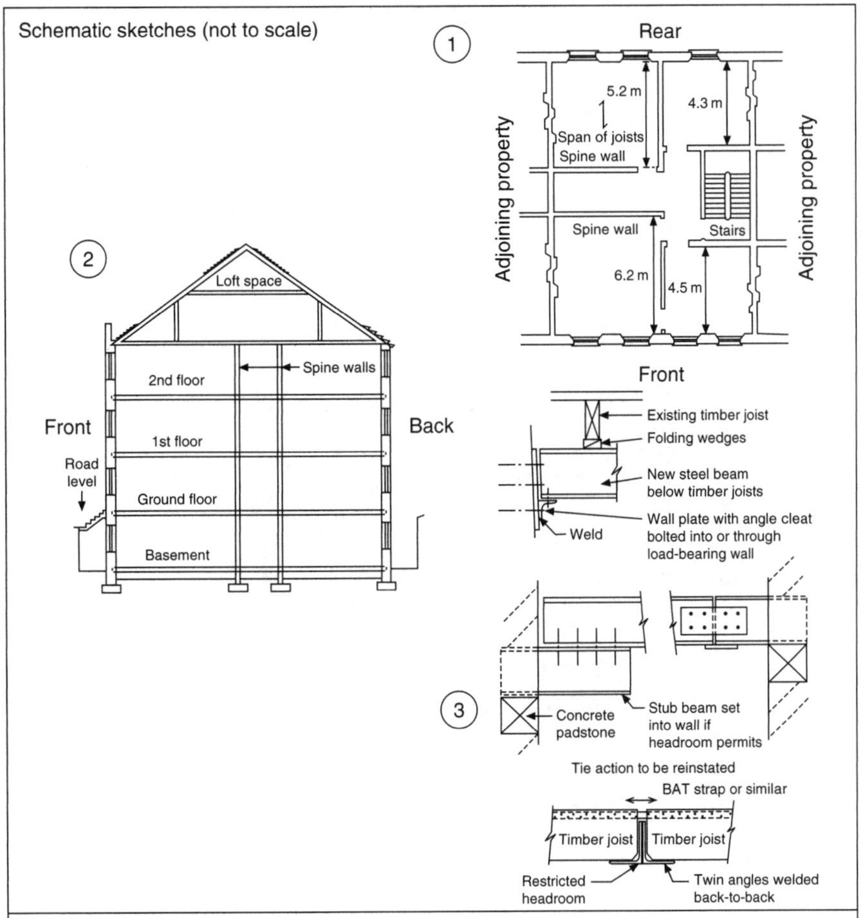

Notes

1. Plan of typical terrace house.
2. Section through typical terrace house.
3. Methods of supporting new steelwork.
4. Strengthening of floors – plan.
5. Strengthening of floors – section.

(a)

Figure 7.15 (a, b) Typical strengthening measures to a Georgian tenement

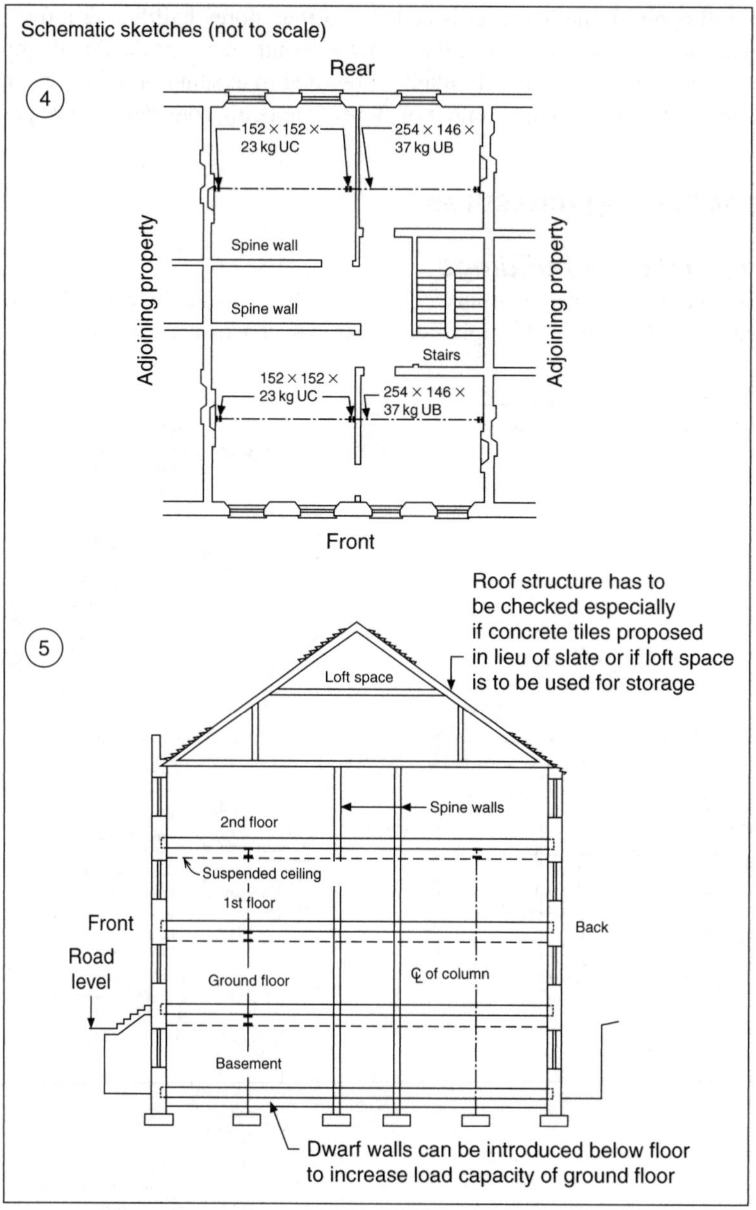

(b)

Figure 7.15 (a, b) (Continued)

adaptation. Such work can entail either a change of use or modifications to the layout, both of which may require an increase in the imposed loading that the existing structure was not designed for.

In many cases it may not be feasible to reinforce the existing elements of a building. They may be so defective, for example, that they cannot be repaired or strengthened without additional elements. Typical strengthening measures associated with the superstructure of load-bearing masonry buildings are shown in Figure 7.15.

Table 7.4 Main methods of strengthening framed buildings

Element	*Method*	
	Traditional	*Modern*
Floor slab	Steel plates/straps (c. 5 mm thick) glued and bolted to either top surface or soffit of concrete at critical locations. Inserting steel support beams under the slab at midpoints. Layer of reinforced concrete (c. 75 mm) over existing slab – although the effects on loadings as well as floor levels and staircases will need to be taken into account.	Preformed carbon fibre (1) composite (ribbed) plates/strips (c. 1 mm thick; 100 mm minimum width) or Fibre-Reinforced Polyester (FRP) sheet fixed to the concrete soffit using a polyamine-based thoxotropic adhesive (such as 'Sikadur 30') (2).
Beam/Joist	*Steel*: encased in reinforced concrete or new steel bottom flange plate welded onto existing or new steel channel sections bolted or welded to bottom flange. *Concrete*: steel channel sections bolted or glued to beam with epoxy resin.	*Steel*: carbon fibre strips (c. 1 mm thick) or FRP wrapped around exposed surfaces of beam using thixotropic epoxy adhesive for easy of application. *Concrete*: ditto or extending depth of beam using composite ply/balsawood box covered with carbon fibre wrapping. *Timber*: carbon fibre strips glued to the underside and sides of members.
Column	*Steel*: encased in reinforced concrete or new side-web plates or flange plates welded onto existing. *Concrete*: steel angle plates tied around column with steel straps.	*Steel or concrete*: carbon fibre strips (c.150 mm wide and 1 mm thick) wrapped around each column using thixotropic adhesive (3). *Concrete*: carbon fibre or FRP wrapping – also enhances the impact resistance of columns. This technique is usually used to improve the capacity of bridge supports to resist impact loads. Aramid (Kelvar) fibre wrapping can be used around columns and other structural members requiring extra impact protection particularly at corners.

Notes
1. Multidirectional carbon fibre laminates must be used.
2. Carbon fibre strip must be anchored using stainless steel bolts at the shear zone of the beam to prevent problems such as 'concrete rip-off', a form of shear failure.
3. Whilst carbon fibre is non-flammable, the adhesives used to attach it may not have good fire resistance. Their performance tends to reduce when the temperature exceeds 200°C. Where fire resistance is a requirement therefore the structural element strengthened using carbon fibre should be coated with a suitable intumescent coating, such as 'Ameron Steelguard FM 580', which is a thin film site-applied water borne intumescent coating that's environmentally friendly with low VOC, low odour and chlorine free.

Framed buildings

Older framed buildings, too, may require strengthening because of inherent defects or to take additional loads. As they were built before the present era of limit-state design their structural capacity may have been underestimated (see Bussell, 1997).

The traditional and modern methods of strengthening steel and concrete-framed buildings are summarized in Table 7.4. Sketches of some of these are shown in Figures 7.16 and 7.17.

Traditional strengthening methods are problematic because of the difficulty of installation. The floor surfaces or soffits would need to be exposed and flat enough to allow adequate coverage of the plates. In addition, the steel plates increase the dead load of the building and require corrosion protection. However,

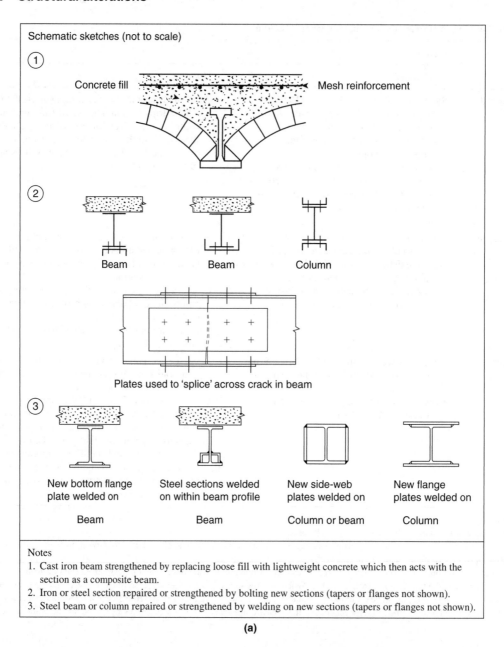

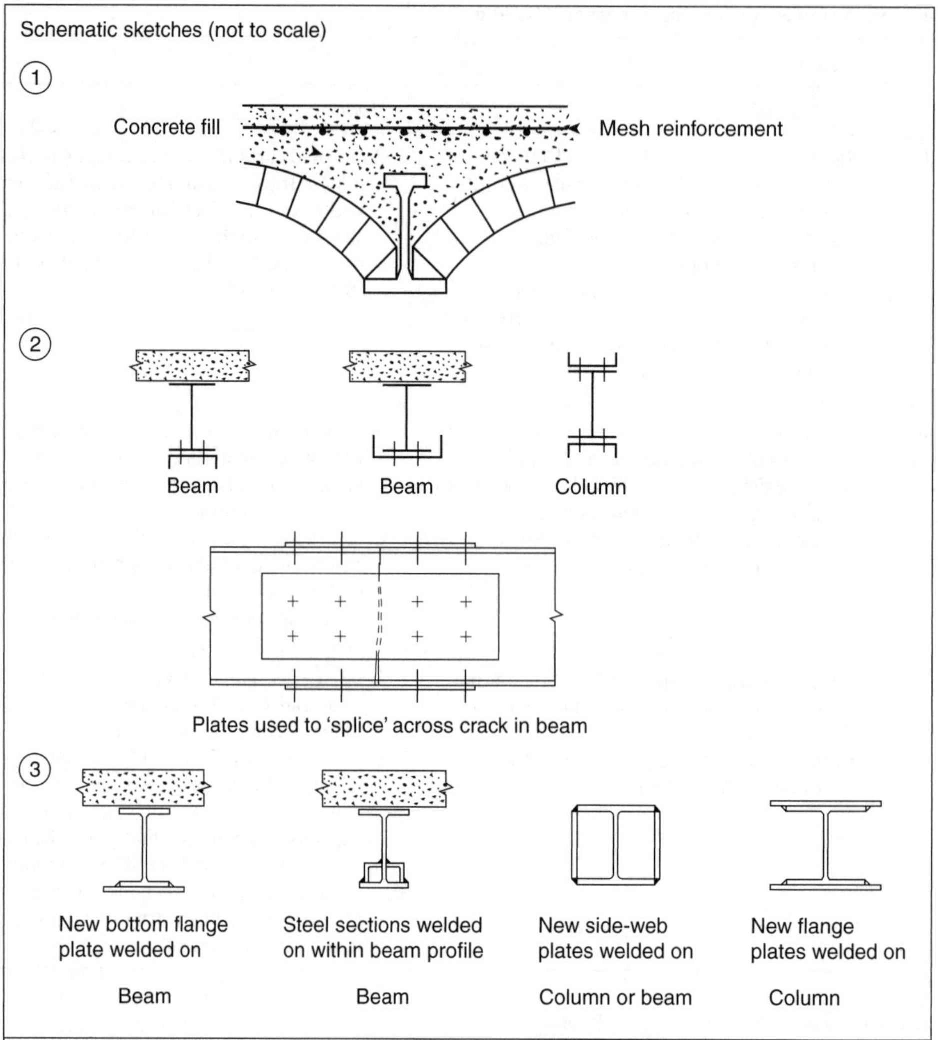

Notes

1. Cast iron beam strengthened by replacing loose fill with lightweight concrete which then acts with the section as a composite beam.
2. Iron or steel section repaired or strengthened by bolting new sections (tapers or flanges not shown).
3. Steel beam or column repaired or strengthened by welding on new sections (tapers or flanges not shown).

(a)

Figure 7.16 (a, b) Some typical strengthening measures to iron and steel frames (Bussell, 1997)

firms such as 'Balvac Whitley Moran Co Ltd', a subsidiary of Balfour Beatty, can undertake this specialist work.

The carbon fibre composite method, on the other hand, provides a light, very strong, and durable repair. It is highly flexible to apply and its strength-to-weight ratio is much better than steel. The 'Carbodur' system is one proprietary form of carbon fibre plate bonding (Barton, 1997).

TREDACC have devised an effective carbon fibre system to strengthen structural concrete elements. It uses cross ply carbon fibre wrapping or strips bonded to the concrete substrate using epoxy resin. When

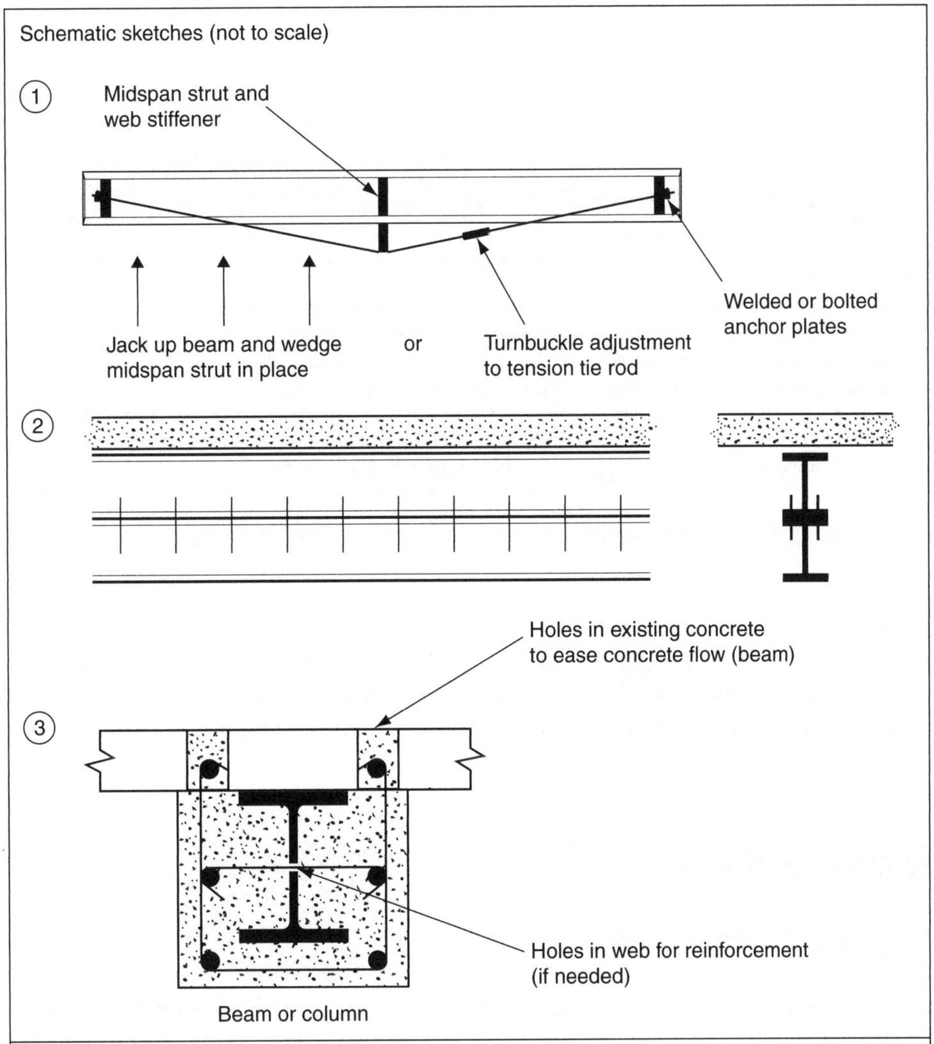

Notes
1. Beam strengthened by trussing.
2. Beam strengthened by addition of new member below existing beam.
3. Beam or column strengthening by concrete encasement.

(b)

Figure 7.16 (a, b) (Continued)

used on concrete or timber beams the carbon fibre strips have to be bolted near the ends of the members to prevent failure called 'forced stress'.

Large panel system buildings

In the early 1980s a spate of premature cladding failures occurred in some multi-storey flats of large panel system (LPS) construction. As a result local authorities and development corporations throughout the country embarked on an extensive repair programme to all of their LPS housing blocks.

Notes

1. Carbon fibre or FRP laminate fixed with an epoxy resin adhesive to bottom and sides of cleaned concrete beam. A balsa wood spacer can be placed to the underside of the beam and covered with carbon fibre laminate to increase the beam's lever arm and thus increase its strength and stiffness significantly.
2. Stainless steel bolts 25 mm diameter to anchor the carbon of slab.
3. Carbon fibre strips (100–150 mm wide) can be laid on top (or soffit) of slab to repair and strengthen defective concrete. A criss-cross grid arrangement can be used to achieve maximum strengthening. The treated surface can then be coated with a suitable epoxy resin floor coating.

Figure 7.17 Some typical strengthening measures to concrete beam and floor

Essentially, the work involved inserting 12 mm diameter expanding stainless steel bolts at a 45° angle upwards near the corner of each panel. The holes were then filled with a stiff cement grout.

Prior to any refurbishment work to LPS buildings such as overcladding a check would need to be made to on the adequacy of these fixings. This could be done by undertaking a sonar or impulse radar analysis of panels on a selective basis for each block.

Alterations to floors

Notching and holes in joists

In Britain, the ground and upper floors in residential buildings are predominantly of suspended timber construction. Timber joists under tongued and grooved boarding provide the most common and versatile form of suspended floor. This form of floor construction is especially appropriate on sloping sites, as it avoids the need for large volumes of cut-and-fill associated with ground-bearing floors.

Timber floor structures can be relatively easily repaired, strengthened and modified. Upgrading the fire resistance and sound insulation of these floors is also reasonably straightforward. This is dealt with in the next chapter.

Over their service life, however, suspended timber floors are often subjected to illicit or ill-considered alterations. Notching of timber joists is a typical example of one of the most common of such practices. In the writer's experience, some electricians and central heating installers are notoriously cavalier in their approach to inserting cables or pipes under the decking of timber floors. The easiest way to do this is to lift the floorboards and cut notches into the top of the joists to take the pipes or cables. Ideally, it is structurally more efficient to drill holes in the neutral axis of each joist to feed piping or cabling through. In practice, though, this is easier said than done. It is particularly difficult trying to feed rigid piping through a continuous series of small holes in tightly spaced joists.

The basic Building Regulation requirements for inserting notches and holes in joists are shown in Figure 7.18. The depth of the notches and their position within the joists should be within these strict parameters to avoid the problems indicated above.

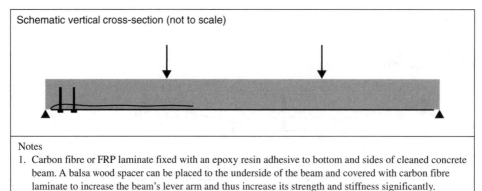

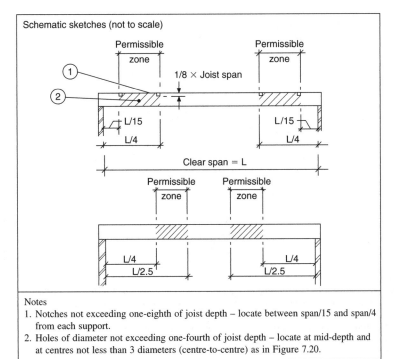

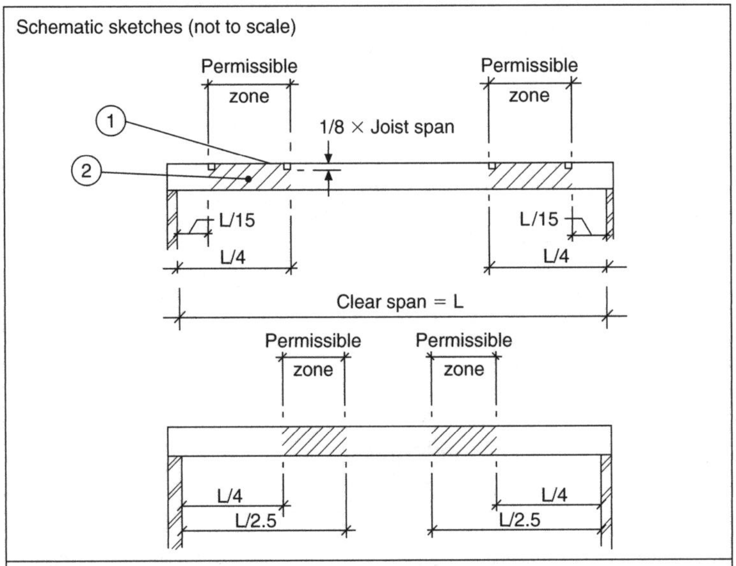

Notes
1. Notches not exceeding one-eighth of joist depth – locate between span/15 and span/4 from each support.
2. Holes of diameter not exceeding one-fourth of joist depth – locate at mid-depth and at centres not less than 3 diameters (centre-to-centre) as in Figure 7.20.

Figure 7.18 Notching requirements for joists

Existing floors that already contain excessive notching can be made good in one of two ways:

(i) If the deflection is present but not excessive (i.e. is less than 0.003 of the span), fix a mild steel cleat plate on either side of the joist under the notched section (Figure 7.19).
(ii) If the deflection exceeds the 0.003 × span limit, strengthen the entire floor structure (see below).

Even apparently innocuous changes such as removing minor load-bearing or ostensibly non-load-bearing walls under a floor can have adverse structural effects. It was not uncommon for some Victorian tenements in Scotland, for example, to contain a 'box bed recess' in nearly every flat. This was a small rectangular alcove in the kitchen area, which was just big enough to accommodate a bed for one of the servants. However, during the 1950s through to the 1970s the removal of the partition forming part of the enclosure to the box bed recesses to increase the size of rooms was common. In this type of property, the floors joists would normally run from front to back to provide lateral restraint to the main walls. Taking away the box bed recess partition effectively increased the span of nearly half the joists in the room by at least 1 m. This, coupled with excessive notching of the joists, can cause these suspended timber floors to sag as much as 50 mm or more.

Figure 7.20 shows the layout of a room in a Victorian tenement containing a box bed recess. The sketch shows the extent of sagging encountered in the floor of the kitchen/dining room.

Strengthening existing floors

Existing floors in a building being modernized or converted may require strengthening. This could also be due to a number of factors:

- excessive deflection caused by over-notching of joists;
- timber decay caused by woodworm or fungal attack;

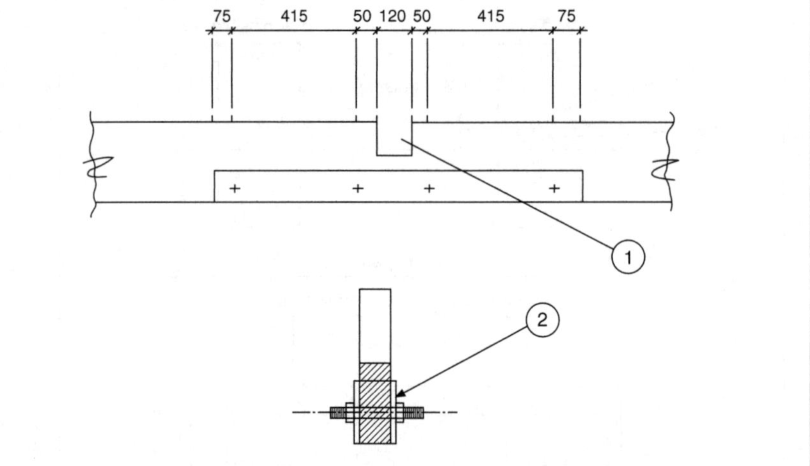

Notes
1. Where existing joists are notched to a greater extent than shown in Figure 7.19, this detail may be applied for notches up to 120 × 120 mm subject to a structural engineer's approval of location.
2. 2 No. 100 × 12 mm thick mild steel flat plates with 12 mm diameter bolts in 14 mm diameter holes + 2 No. 76 mm diameter toothed plate connectors.

Figure 7.19 Repairing an excessively notched joist

- the installation of a partition running parallel to the joists (see Figure 7.21);
- under-designed load-bearing members;
- overloading.

Overloading can result in a phenomenon called 'creep'. The degree of sag in a floor may not be such that its structural integrity is threatened. However, deflection in a floor can adversely affect its functionality or serviceability.

Thus, even if a deflected floor is structurally sound, it may not be comfortable to use. For example, users may feel that the floor uncomfortable to walk over, furniture placed on the floor will lean or appear lopsided, and the ceiling finish in the room below is likely to suffer extensive tensile cracking along its middle where deflection is greatest.

Changing the use of the building from, say, residential to commercial will in all likelihood incur higher imposed floor loadings from 1.5 kN/m^2 to 3 kN/m^2 (per BS 6399). Suspended ground and upper floors in dwellings were usually not designed to accommodate this increased loading. Unless some strengthening work is done to such a floor, it will sag over time. This affects not only the stability of the floor but also its serviceability.

The decision to strengthen or replace a deflected floor structure will be influenced by its state of repair and suitability for the upgraded use. If the deterioration of the floor structure is extensive, complete replacement may be the only feasible solution.

On the other hand, if a suspended timber floor structure is generally sound, it may only require strengthening to take increased loadings. This is usually done in one of the following ways:

(i) Inserting a steel or composite beam under the floor structure.
(ii) Inserting steel angle sections within the timber floor structure.
(iii) Inserting additional joists between the existing joists.

Schematic plan (not to scale)

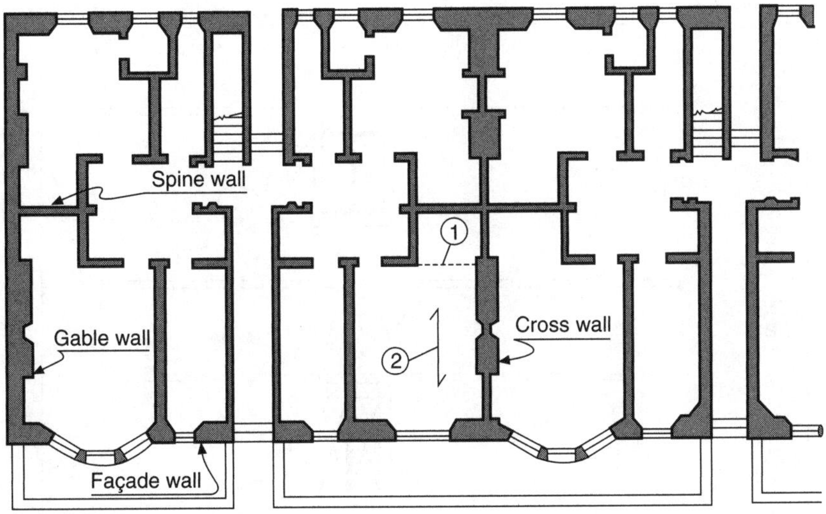

Notes
1. Partial partition wall with wide opening providing some support to the floor structure above.
2. Removal of partition increases the span of the joists, which are usually placed from front to back to give lateral restraint to main walls.

Figure 7.20 Removal of a box bed recess partition in a Victorian tenement

Vertical cross-section (not to scale)

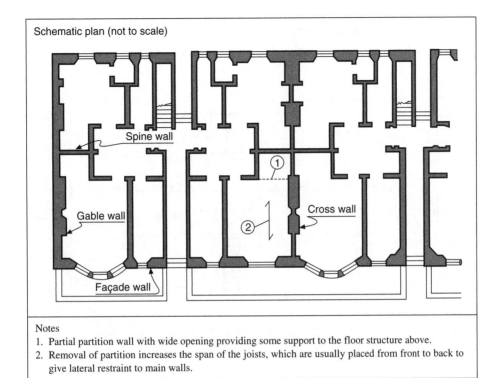

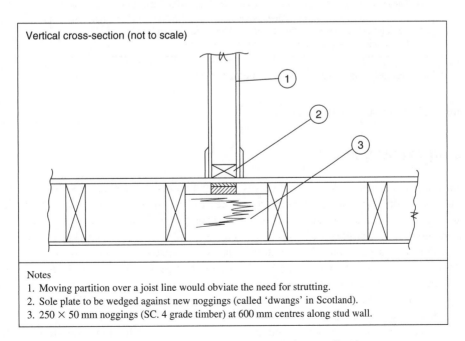

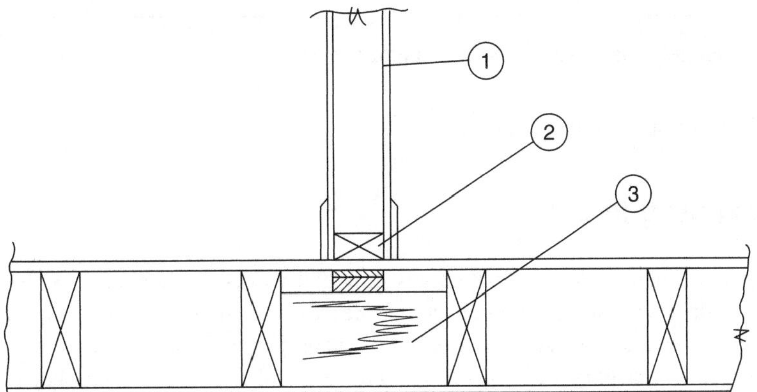

Notes
1. Moving partition over a joist line would obviate the need for strutting.
2. Sole plate to be wedged against new noggings (called 'dwangs' in Scotland).
3. 250 × 50 mm noggings (SC. 4 grade timber) at 600 mm centres along stud wall.

Figure 7.21 Detail where stud wall unsupported after removal of rotten flooring

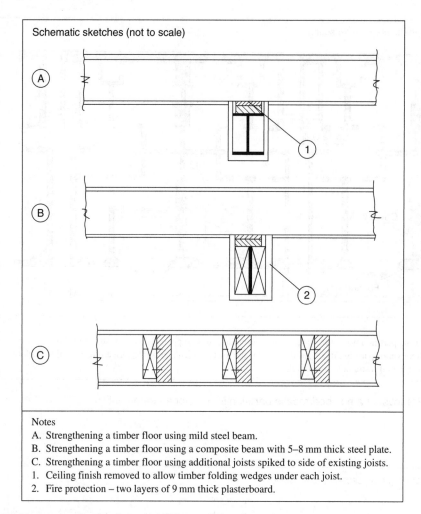

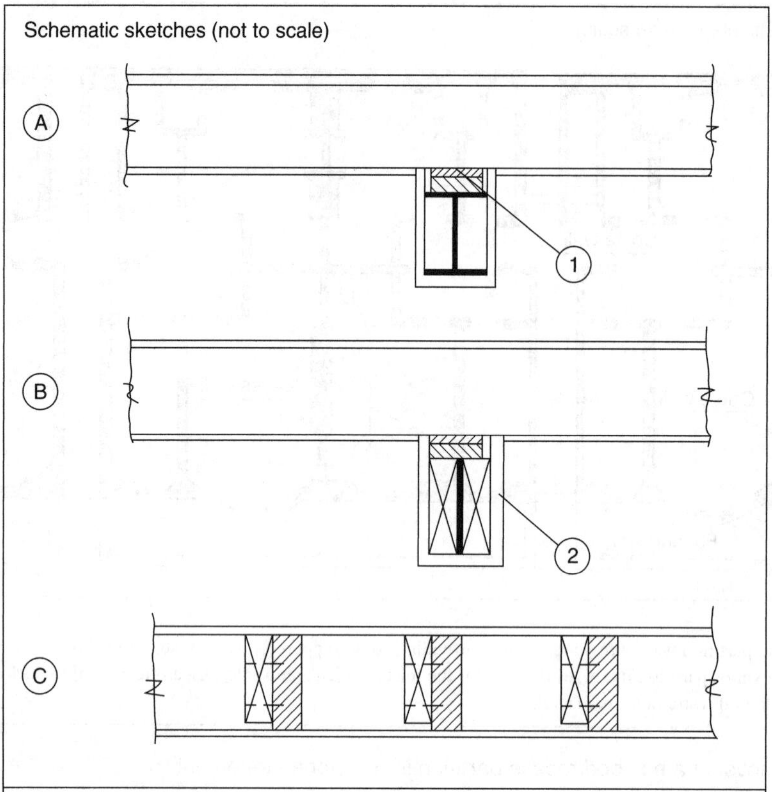

Notes

A. Strengthening a timber floor using mild steel beam.
B. Strengthening a timber floor using a composite beam with 5–8 mm thick steel plate.
C. Strengthening a timber floor using additional joists spiked to side of existing joists.
1. Ceiling finish removed to allow timber folding wedges under each joist.
2. Fire protection – two layers of 9 mm thick plasterboard.

Figure 7.22 Various methods of strengthening a timber floor

See Figure 7.22 for examples of these methods.

Repairing existing floors

Part of an adaptation scheme sometimes involves remedial works to elements affected by defects both natural and man-made. For example, as indicated earlier, the built-in ends of joists in older buildings are highly vulnerable to fungal attack. This is more likely in solid external walls that are regularly exposed to wind driven rain. Porous masonry will absorb the rainwater and any hygroscopic material such as timber in contact with it will become wet. This wetting increases the moisture content of the material well above the 20 per cent wood moisture equivalent (WME) safe limit. If persistent, such an increase in the WME inevitably leads to deterioration of the wood.

The type of timber decay is dependent on the conditions present. For example, the degree of saturation, the extent of background ventilation, and ambient temperature conditions will all be important influences. Built-in timbers are more likely to be affected by wet rot because of the high moisture levels encountered in solid walls that are exposed to the elements.

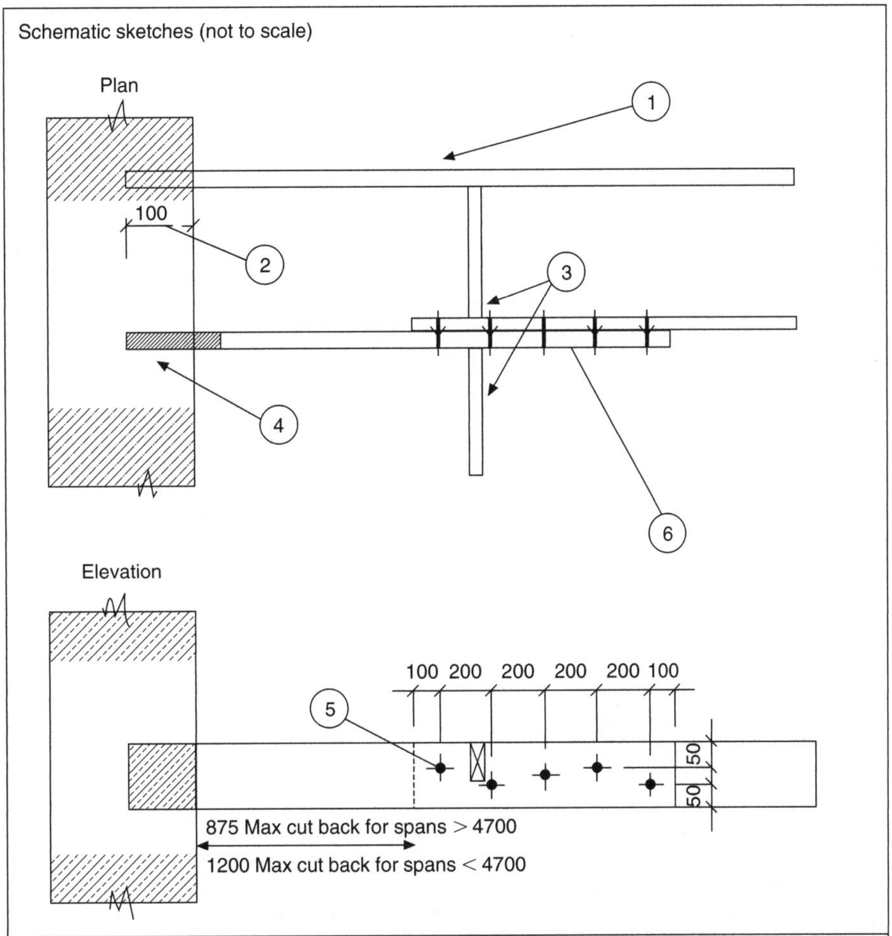

Notes

1. Floor joist (pockets around built-in ends of joists, pugging, floorboards and ceiling finish omitted for clarity). Temporary support required before rotted joist ends are cut back.
2. Minimum end bearing 100 mm. Bearings to be packed tight with slate after removal of any old wallplates.
3. 47 × 145 mm noggings/strutting.
4. New joist ends to be fitted with plastic shoe or dipped in approved bituminous preservative. Old joist pockets and gaps around remaining ones should be filled in with a weak mortar or expanding foam filler.
5. 5 No.12 mm diameter black bolts, grade 4.6. c/w flat washers, hex nuts and 51 mm diameter double-sided tooth plate connectors.
6. Splices are to be staggered where four or more are adjacent. No more than three splices to be carried out simultaneously. See Robson (1999) for alternative methods of replacing defective joist ends.

Figure 7.23 Typical joist-end splice for rotten ends of built-in timbers

As a rule of thumb, over two-thirds of the joist ends in properties built prior to the 20th century may be affected. Robson (1999) illustrates the requirements for repairing timber joist ends affected by fungal attack. The determining factors in most cases, however, are as follows:

- The exposure incidence of the wall in question.
- The quality of timber and the protection it is given.
- The extent of maintenance the property has received.

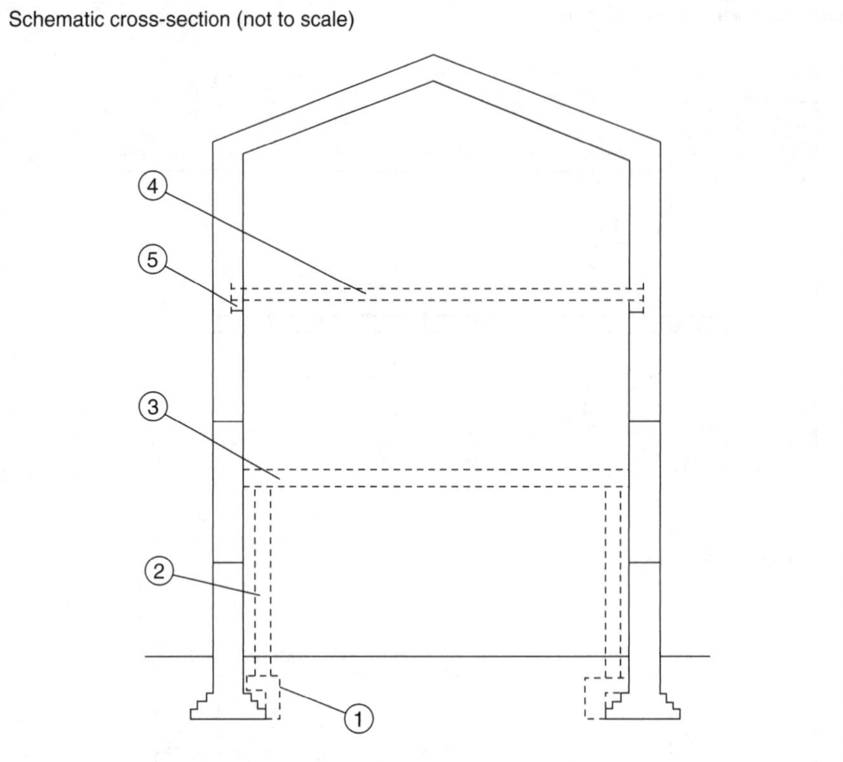

Notes
1. New reinforced concrete cantilevered/stepped footing for steel stanchion or masonry pier.
2. New steel stanchion or masonry pier supporting steel cross beam – size and spacings to be determined by a structural engineer.
3. New first floor structure – timber/steel beams supporting concrete planks or timber joist. Depth influenced by the span. See Figure 4.13 for detail of floor/window junction.
4. Timber/steel beams supporting concrete planks or timber joists would be an alternative method of supporting new first or second floor structure – provided the masonry is structurally stable to take the imposed loads.
5. New reinforced in situ concrete spreader beam in existing wall to support alternative first/second floor structure.

Figure 7.24 Inserting new floors in a building with a high floor to ceiling space

Georgian and Victorian tenements are especially prone to dry rot or wet rot in their suspended timber floors. This is because the floor joists are usually built-in to the masonry, without any protection. Ideally, built-in timbers such as bressummer beams and floor joists should be encased at each end in a 'shoe' made from water-resisting material such as plastic (see Figure 7.23).

Rafter feet are also susceptible to timber decay, particularly if the roof is poorly maintained or the eaves of the roof internally are closed off by 'beamfilling'. It is a feature that is found in the roof space of many Victorian villas.

Defective joist ends or rafter feet are usually repaired using spliced ends. These can be made from new, treated timber cleated and bolted to the existing sound sections (see Figure 7.24). Alternatively, resin splice ends fixed to the existing with dowels, or spliced ends in structural steel bolted to the existing, can be used (Highfield, 1987).

Replacing existing floors

Replacing suspended timber ground/upper floors affected by fungal attack or old, defective concrete/stone slab floor sometimes forms part of an adaptation scheme. In such instances the choice is to reinstate a replacement timber or pre-cast concrete suspended floor or if at ground level install a new solid ground-bearing concrete floor. The selection will be based on:

- The topography of the site. Suspended ground floors are more suitable on sloping sites, as they reduce the need for excessive cut and fill under the floor structure.
- Use of the room. Solid concrete floors are better at resisting larger imposed loads, provided that they are adequately reinforced on well-consolidated hardcore base.
- The type of floor structure elsewhere at ground/upper level. Feasibility of ventilating the subfloor void effectively if a suspended floor is selected; or maintaining subfloor ventilation for adjoining rooms if new ground-supported floor is being proposed.
- The relative cost of each option.
- The type of temporary support system required.

Inserting additional floors

In many older redundant buildings the floor to ceiling height is often well over 6 m. This feature is often encountered in single-storey properties, such as churches. A high floor to ceiling is a standard feature of ecclesiastical buildings, because of the need to have large spaces for giving a sense of grandeur and openness. However, in a redundant church building this high room size is a waste of space for most alternative uses. A room with a high ceiling is difficult and expensive to heat and decorate, and awkward to clean.

As indicated in Chapter 4 the potential to insert extra floors in such a building as part of, say, a conversion scheme is quite good. Ideally the new floor to soffit height should be at least 3 m. This is required to accommodate a floor to ceiling height in each level of about 2.3 m – leaving around 0.7 m for the new floor structure and ceiling void. There are, though, a number of issues that affect the choice of floor design:

- the span/s involved;
- the means of support for the new floor/s;
- the extent of coverage of the new floor/s;
- the anticipated floor loads (any crowd loading conditions?);
- the degree of separation involved, if multiple use is envisaged;
- the last point will therefore determine the fire and acoustic performance of the new floor/s;
- junction of floor with existing fenestration.

The proposed span will obviously influence the construction form of the new floor (see Figure 7.24). Timber joist floor structures are usually neither economical nor suitable for spans around 4.2 m or more. In such cases it may be more suitable to split the span in half by inserting a steel 'I' beam across the shortest width of the room. The new timber joists can then be installed across the longer span now reduced by half by the steel beam.

The form or method of support for the new floor structure depends on the suitability and condition of the wall it is abutting. Inserting pockets to accommodate joist or beam-ends in thick walls is sometimes acceptable. However, inserting the beam into the two pockets can be difficult in confined spaces. In such cases, it may be necessary to splice the beam in half to allow for easy installation. This is easily achieved using a steel beam, which can come in two sections if necessary to facilitate ease of installation. The flanges at each of the abutment side are then bolted together.

Schematic vertical cross-section (not to scale)

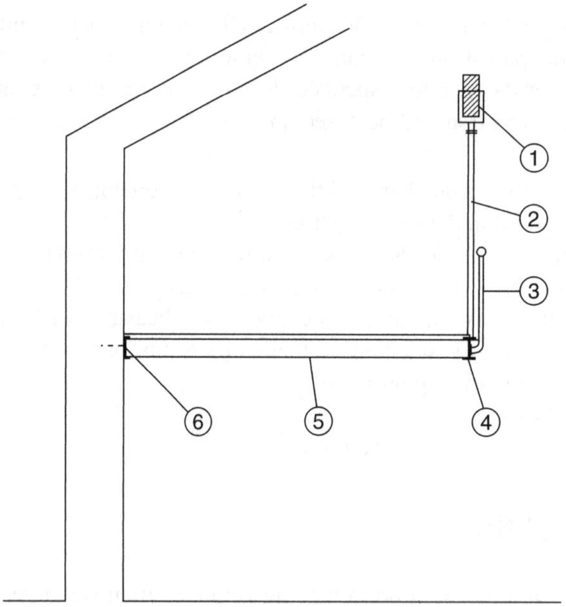

Notes
1. Mild steel channel bracket 6 mm thick, 100 mm wide and 150 mm long fixed to existing truss or roof beam using 10 mm diameter coach screws.
2. 50 mm diameter mild steel hanger pole welded to channel bracket at top and welded to mezzanine level 'I' beam. Spacing of hangers should be not more than about 2 m, but this will depend on the span and loadings.
3. Mild steel balusters welded to plate, which is bolted to web of mezzanine level 'I' beam.
4. Steel 'I' beam – depth determined by span and load.
5. Timber mezzanine floor structure resting on flange of steel sections. 20 mm thick tongued and grooved decking on 50 mm wide joists at 450 mm centres – depth as per values in Table A1 of the Building Regulations 2000 (as amended).
6. Steel channel section bolted to main wall using 150 mm long × 20 mm diameter expanding bolts at maximum 900 mm centres. Packing behind the channel web may be required if the wall surface is irregular.

Figure 7.25 Method of supporting a mezzanine floor using hangers

Installing joist hangers in existing walling may prove adequate for new floors with short spans and where the brickwork can easily accommodate them. However, this method of support is not suitable for large spans and where the brick or stone courses are so uneven for allowing the installation of hangers in a straight line. The alternative is to bolt either a 100 × 50 mm treated timber batten or 100 × 50 mm mild steel angle plate to the wall.

A novel way of supporting a mezzanine or additional floor level in a converted building is to hang the floor from the roof or floor structure above. Figure 7.25 illustrates how this can be done to achieve an attractive, lightweight solution.

In large volume buildings with high floor to ceiling heights such as Victorian churches inserting several new floor levels may be required to maximize the use of the interior space. This is probably best achieved by installing a steel framework within the building (see section on this later in this chapter).

Removing an existing floor

Taking out an existing floor in a building is a much rarer operation than inserting a new one. In the case of the former, this may be because the client wishes to increase the floor to ceiling height or otherwise enlarge a room's capacity in some part of the building. This may be particularly advantageous with a basement flat in a Georgian or Victorian townhouse to create a large, open-plan room facing the garden. Removing the floor structure of the ground floor above would allow the fenestration to discharge natural light to the enlarged basement room without the need to form a light well.

This form of floor alteration, however, has structural and non-structural implications. It may, for example, affect the foundations because of the changes in load paths induced by the removal of the floor. Structurally, floors provide lateral restraint to walls. If the former are taken away, this could result in instability in the latter, causing them to bow or crack. The extent of floor area being removed and the arrangement of the surrounding floor and wall construction will determine the need if any for installing retrofit steel tie rods to provide lateral restraint. On the other hand, retaining part of the floor to form a mezzanine floor may avoid the need for such ties.

The non-structural implications of removing a floor are:

- Accommodation: Reduces the available habitable floor space.
- Approvals: May not be obtainable if the property is listed or in a conservation area.
- Doorways at original upper level: Block off or provide landing and staircase. (Access enlarged basement room may be available via adjacent room.)
- Downtakings: Removal of skirtings, floorboards, ceiling finish, including cornice; joists, electrical cables or pipes in floor.
- Maintenance: Access for cleaning and decoration more awkward.
- Making good: Filling in joist pockets; patching plaster at old skirting level; plasterwork around filled-in doorway/s.
- Services: Modify heating, lighting, power and plumbing to suit enlarged room.

Alterations to walls

Generally

Adaptation schemes often involve removing or adjusting the size of structural and non-load-bearing walls (see Figure 7.26). Inserting openings in such walls are also a common feature of many conversions and extensions. Such work, however, is highly disruptive because of the noise and dust disturbance that it inevitably entails. In addition, there is usually some collateral damage to the masonry immediately surrounding the opening. Making good the reveals and surrounding wall finishes of the newly formed openings must also be taken into account.

Forming openings in existing walls

The location of any new opening in an existing building should be chosen with care. If this is not done the result could overstress the surrounding masonry. It may result in some distortion or excessive cracking of brickwork or stonework walls. This is especially the case near the junction with another wall.

As a general rule therefore the formation of a new opening in an external wall should leave at least 600 mm of walling between any reveal and an adjacent opening or wall to which it is perpendicular. Strengthening of the existing masonry may be required if the distance is any less than this (see below).

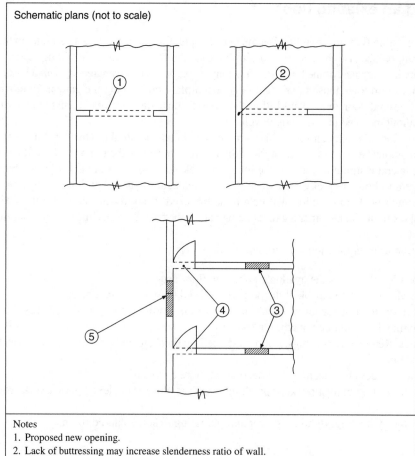

Schematic plans (not to scale)

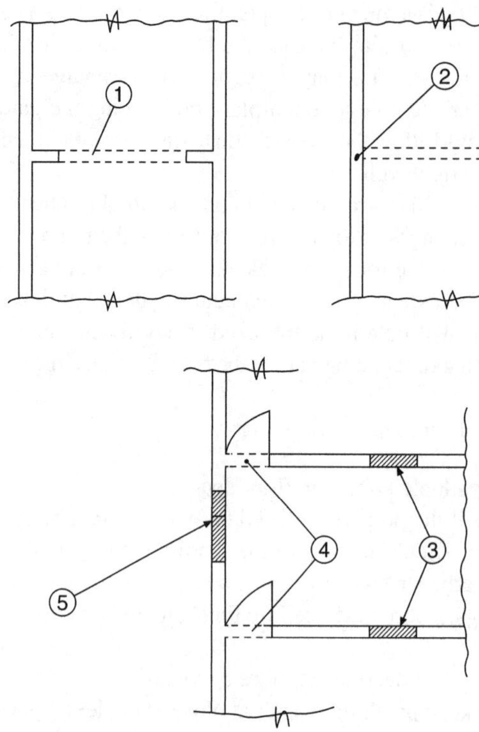

Notes
1. Proposed new opening.
2. Lack of buttressing may increase slenderness ratio of wall.
3. Existing doorways blocked up.
4. Formation of new doorways can have a detrimental effect on the stability of the surrounding wall construction.
5. Enlargement of existing or formation of new openings in external walls must not over-stress adjacent masonry.

Figure 7.26 Problems of forming new openings in existing walls

Once a suitable position has been determined, a number of factors need to be taken into account when forming a door or window opening in walls:

- protection or temporary removal of finishings that are being retained within the room;
- temporary support before and during the cutting of the opening;
- nature and extent of loading to be taken by the new lintel;
- insertion of permanent support;
- making good damage to surrounding masonry and adjacent floor;
- redecoration of affected wall and ceiling surfaces;
- adjusting any mismatch in floor levels either side of wall.

It is critical to bear in mind when forming an opening in an existing wall to determine the loading implications. Inevitably the load paths in such a wall will be changed. This can over-stress parts of the wall that are taking additional loading.

Forming the new opening has more than its fair share of difficulties. It is usually noisy, very messy and potentially hazardous. It is for these reasons that sealing off of adjacent rooms and adequate temporary support are required. The illustrated guidance given by the BRE in its excellent series of GBG 1, 10, 15 and 20 and GRG 25 should be followed when undertaking this work.

Removing existing walls

Taking down partitions or load-bearing walls is often done in conversion schemes to increase the size of an internal space or modify its layout. This will have a number of implications:

- Temporary support during the demolition work.
- Redirect dead and imposed loads from above, which may require strengthening of the affected elements.
- Making good of floor, wall and ceiling finishes.
- Disconnection and removal of any electric cables, power points or other services lines in the wall being taken down.
- Modifying existing lighting and power supplies to cater for the enlarged area.

In any event, the BRE's reminder, which can apply to any structural alterations, is well worth following:

> If you are considering the removal of a load-bearing wall, you must consult a chartered structural engineer, or similarly qualified person.

Simple worked example

Preamble

Assume that a semi-detached factory unit is being converted into single-occupier use. This will involve creating a new opening in the separating wall, which is 225 mm thick, to link the previously separate units.

Select a suitable pre-cast concrete beam from the manufacturer's tables (see Appendix C) to span the 2 m wide opening 2 m high in the party wall.

Procedure for determining a suitable pre-cast concrete beam

- Assess the load above the proposed opening. GBG 10 provides the best advice for assessing loads above ordinary openings in external and internal walls. It should not be used where the loads are suspected to be unusual or high.
- In summary, GBG 10 involves the following:
  - Establish the dead load above the span of the opening. This involves multiplying the volume of wall concerned by the density of common brickwork (say, 2300 kg/m^3 – taken from BS 648: Schedule of weights of building materials).
  - Ascertain any live loads above the span of the opening (i.e. determine if there are any floor or roof loads being taken by the party wall). If there are, these are assumed to act over the portion of wall spanned by the new opening (see BS 6399).
  - In this case, it is assumed that there are no floor or roof loads imposed on the party wall.
  - The minimum dead load is the weight of brickwork within the 45° load triangle. For a 2 m wide opening the working span would be about 2.75 m (i.e. 2 m + 0.375 m + 0.375 m). The load triangle would therefore be 1.38 m high × 2.75 m wide and have an area of about 1.9 m^2. This means that the weight of the brickwork within the load triangle is 1.9 × 0.225 × 2300 which is 983 kg = 9.6 kN (i.e. 983 kg × 9.81).

- Once the total load has been calculated (i.e. dead load + live load), reference should be made to an appropriate pre-cast concrete beam manufacturer's table of beam sizes. For example, Richard Lees Ltd provide such a table, as shown in Appendix C, which can be used to establish a suitable size of pre-cast concrete beam for varying spans and loading conditions. The nearest available section above the safe load of 9.6 kN is 225 mm wide by 150 mm deep, which has a maximum safe uniformly distributed load (in addition to self-weight) of 15.33 kN (for a 2.1 m span).

Procedure for determining a suitable steel 'I' joist or 'UB'

This procedure involves working from first principles of structural design. It necessitates obtaining a section that has a permissible deflection and load greater than the allowable deflection and load (multiplied by a suitable safety factor – say, 2):

- Determine the loadings above the new opening as before. It is assumed that the bearing at each end of the beam would be at least 150 mm and formed with in-situ or pre-cast concrete pad stones.
- Calculate an approximate size of steel beam using the rule of thumb (for a heavily loaded beam): depth = span/18. For a 2 m wide opening this requires a steel beam at least 111 mm deep. This figure should be rounded up to the nearest stock section depth (e.g. 127 mm). A 225 mm wide by 127 mm deep steel beam to support an opening in a 225 mm thick wall, though, is not a standard size. Two 127 × 76 × 13 kg steel 'UB' sections, encased in a two layers of 9.5 mm thick plasterboard to provide 1-h fire protection, would probably suffice for this opening.
- A more detailed structural calculation from first principles, of course, would be required to confirm the suitability of this section in terms of deflection and stress – the two key beam design factors. The maximum permissible deflection for a steel beam is 0.003 of the span – in this case 6 mm. The maximum permissible stress for ordinary grade mild steel is 150 N/mm^2.

Closing openings in existing walls

External walls

In major adaptation schemes, such as the conversion of a redundant whiskey bond to high-quality flats, many of the existing openings in the external walls may be superfluous for the new use. As a result they would normally be blocked up flush with the existing façade. In some cases, recessing the new masonry slightly may be favoured as a conservation design feature to avoid masking the original opening.

Where the finish is fair-faced brickwork the infill walling should match the surrounding masonry in terms of colour, bond and mortar strength. Normally, there is no need to tooth-in the infill brickwork to the existing. But jointing the new brickwork to the existing should suffice.

The infill brickwork should be of cavity or solid construction to match the surrounding masonry. However, if the opening is larger than, say, 1 m^2, the new brickwork should be tied-in to the reveals using stainless steel proprietary angle ties bolted to the existing brickwork. Structural engineering advice would be required in such circumstances.

Where the surrounding masonry has a rendered finish, the infill walling material can comprise blockwork, which is then rendered to match existing. It would be prudent to fix a strip of expanded metal lath over the joints between the old and new wall finishes to minimize cracking in the rendering at these positions.

Internal walls

These can usually be built-up using blockwork and plastered over to match the surrounding wall finish. Again, it may be prudent to fix a strip of expanded metal lath over the joints between the old and new wall finishes to minimize cracking in the plastering at these positions.

Strengthening existing walls

Flank walls and gables are often susceptible to failures such as bowing and cracking. Wall-tie corrosion and inadequate or defective lateral restraint can cause these problems. The removal of an old lean-to extension, which provided some form of lateral restraint, can also trigger these failures.

Existing walls can be strengthened in one of the following two main ways:

(i) Insert new steel tie rods, with protected steel pattress/cross plate or angle section (see Figure 7.13). This is also required of course to provide lateral restraint to a gable.
(ii) Increase the effective thickness of the wall by tying-in a new brick skin or buttressing it by installing brick/concrete piers at between 1.2 and 2m centres, which are anchored to the existing masonry using galvanized mild steel angle ties at least 5 mm thick and 60 mm wide – length, fixing and spacing to be determined by a structural engineer.

Alterations to roofs

Roof configuration

The type of roof structure will influence the extent of any proposed alteration of the loft area. Modern trussed rafter roofs do not lend themselves to internal adaptation. Anchoring and restraint in such roof structures is critical.

Changing the configuration of the structural members to accommodate additional useable space in the roof loft has to be dealt with carefully. If not handled properly failure in the roof structure could be the result.

Increasing the thermal efficiency of a roof may also require alterations to the roof structure. It depends, though, on what method is being used. For example, if the rafters are not deep enough to accommodate insulation between them, they may require deepening. Nailing a thick enough batten on the underside of each rafter could achieve this.

Strengthening ties

In older properties, the ceiling ties were usually not meant to take floor loadings. They were designed primarily to give lateral restraint to the roof structure to prevent roof spread, and provide a fixing medium for the ceiling finish. In any loft extension in such circumstances, these members have to be either replaced or strengthened using deeper ties bolted to one side of each tie.

Details of the various methods of strengthening the ceiling joists in a roof space to accommodate a loft conversion are shown in Figure 7.22.

Framework for accommodating over-roofing

In Chapter 6, the various options for converting flat roofs to sloping roofs were outlined. The structural condition and adequacy of the existing flat roof deck, however, will determine the form of the roofing framework. For example, different framing arrangements, usually in either metal or timber, are required for various types of flat roof decks as previously shown in Figure 6.16.

Closing or modifying existing openings

Over-roofing may occur on flat or pitched roofs with skylights. Light-wells, as indicated in Chapter 6, from the new roof to the existing skylight opening (with the original skylight removed) would maintain some of the previous levels of daylighting.

Closing off redundant light-wells in an existing roof or under a new roof may be required for security purposes. The visible parts of such work (i.e. the ceiling) will require careful detailing to hide joint. A suspended ceiling under the blocked of skylight would be a more effective way of masking it.

Inserting new openings

Inserting openings in a roof structure for a new dormer or rooflight can adversely affect the roof's structural integrity. The trimming rafters on either side of the opening would become overstressed if they do not receive any strengthening. Deflection of the rafters is the likely result in such instances. As well as distorting the profile of the roof slope this in turn can adversely affect the weather resistance of the roof coverings.

Alterations to frames

Changes in layout

Large factory or warehouse buildings are more often that not of some form of framed construction. Structural steel is the most popular material for the frame of low-rise structures. Pre-cast concrete frames because of their robustness and durability are common for portal frames in agricultural buildings. High-rise office buildings may, on the other hand, consist of cast in-situ reinforced concrete skeletal frame construction.

Because of changes in work practices there may be a need to increase the free floor space in these structures. Depending on the structural system used (whether of skeletal or portal frame) the combination of columns at 5–12 m centres, supporting steel trusses or lattice beams, is the usual configuration. This may, for example, necessitate the removal of one or more columns to increase the uninterrupted floor space.

Figure 7.27 illustrates the work involved in removing a column in a framed building to increase the free working space within a factory area. Some strengthening work to the remaining elements affected by this work may be required.

External additions

Typical additions to a framed building may consist of:

- a steel glazed or unglazed canopy for a new entrance foyer;
- mock or useable balconies at window openings;
- a stainless steel or aluminium open mesh walkway at inside floor level of each storey to facilitate easy cleaning and maintenance to the fabric, and provide a degree of solar shading.

These extra features can be easily added to the load-bearing members of a concrete- or steel-framed building. The cladding at the connection points would have to be temporarily removed to expose the existing framework for bolted or welded connections (see Figure 7.28).

Schematic cross-section (not to scale)

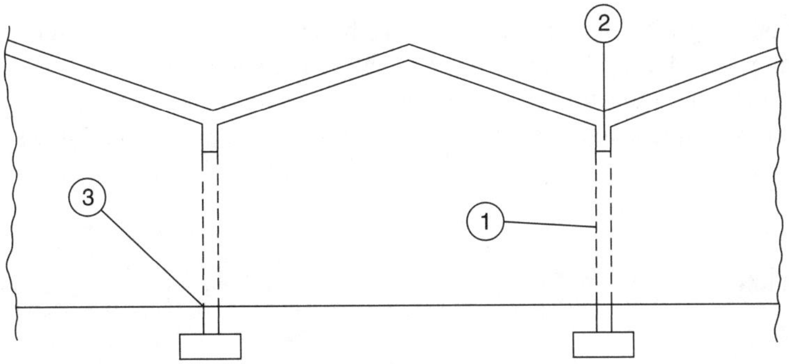

Notes
1. Columns to be removed – temporary support may be required.
2. New permanent support beam to span between retained columns perpendicular to widened span. These columns may need strengthening.
3. Floor slab made good.

Figure 7.27 Removing columns in a portal frame building to increase its free floor space

Schematic cross-section (not to scale)

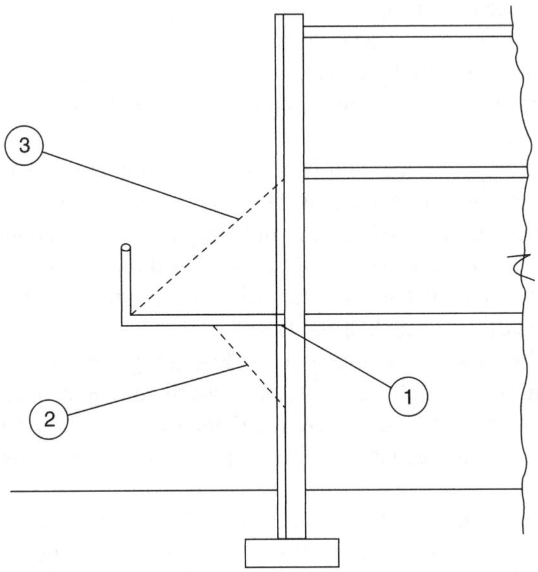

Notes
1. Cladding will need to be temporarily removed to allow cantilever brackets to be fixed to framework.
2. Bracket strut fixed to frame using welded plate and high-strength friction bolts.
3. High tensile 50 mm diameter steel tie as a possible restraint fixing for canopy – but not of course for balcony.

Figure 7.28 Adding a balcony or canopy to the outside of a framed building

Internal additions

The conversion of building with high internal spaces such as a large redundant church building is not feasible without creating more habitable areas. This is achieved either by inserting a skeletal steel/concrete framework or building cross walls topped by pre-cast concrete slabs at every storey.

For a new internal structure comprising two or more storeys in an existing building it is more likely that a steel frame would be used.

Underpinning

Rationale

A major repair of a building's substructure is termed underpinning. It is a term to describe where the existing foundation of a building is both excavated and supported, so that it transfers the loads to a lower bearing stratum. A fundamental principle of this process is that support to the construction above is temporarily removed in short lengths and new support installed. To ensure stability of the structure support should not be removed from more than about one-fifth of the building or wall being underpinned at any one time (Smith, 1994).

There are five main reasons why underpinning may be required to an existing building:

(i) to arrest excessive subsidence or progressive settlement;
(ii) to increase a building's load-bearing capacity;
(iii) to provide new or deeper basement accommodation;
(iv) to replace deteriorated or potentially defective substructure work;
(v) to make buildings with shallow foundations more 'future-proof' against ground erosion and flooding.

Any of these reasons could require underpinning to a building undergoing adaptation. This is especially the case with older buildings of load-bearing masonry construction on soils with low or poor bearing capacity. Such buildings may also require underpinning as part of a conversion to another use, because of the poor condition of their foundations. For example, farm buildings were often built with shallow foundations (i.e. less than 600 mm deep). The local authority may require underpinning of the foundations as part of proposed the change of use to such properties.

Partial underpinning is likely to be required to a building being adapted that requires repairs to the substructure of some but not all of its main walls. In such a case there is an increased risk of differential settlement in adjoining structures, and renewed localized settlement owing to the extra load imposed on the soil.

In rarer cases the whole building may require underpinning. This is usually restricted to low-rise detached dwellings.

In any event, underpinning can add substantially to the adaptation costs. As a rule of thumb, depending on the type and depth of underpinning used, at current prices it can cost between £600 and £1000 per linear metre to underpin a low-rise property (Smith, 1994).

Types of underpinning

Traditional

The main traditional forms of underpinning are brick with concrete footings, and mass concrete. They usually have a maximum depth of about 3 m and when complete are generally *continuous* over the length

Table 7.5 Main types of modern underpinning methods

Method	*Characteristics*
Grouting	This involves injecting a viscous cement or resin grout into the ground to 'stiffen' and/or expand the soil under the wall affected by subsidence. It is more suited for use in non-cohesive soils such as gravely and sandy soils.
Needle beams ladder frame	Usually used in conjunction with piles under the main ring beams. The needles link the ring beams to form a ladder frame under the building. This forms a continuous underpinning system.
Pier and beam	Suitable for shallow underpinning work (i.e. 1.5–3.5 m). The piers are installed at about 2–3 m centres – depending on the loadings and depth of beam. See type 4 underpinning in Figure 7.29.
Concrete piles	Pre-cast reinforced concrete jacks, installed in sections under the affected foundation. Jack-down piles can be used if the scarcement of the existing footing is wide enough to take the pile cap head.
	Bored in-situ piles (can be cantilevered if piling directly next to existing building may disturb the structure) with reinforced concrete needle beam.
	'Palo radice' underpinning uses 'a series of small-diameter piles, rotary drilled through the existing masonry and taken to an adequate depth in the ground below' (Lizzi, 1993). These raked mini-piles are installed in pairs at an angle through drilled holes in the existing substructure, typically at 1–1.5 m centres.
Metal piles	75–100 mm diameter cast aluminium helical piles drilled into the scarcements of the foundation in 1 m long sections to the required depth, such as 'Brutt Helical Pile'.
'Pynford' system	Pre-cast concrete stools inserted into the substructure above the foundation. This can be used in conjunction with in-situ reinforced concrete beams to link the stools within the body of the wall.
Raft floor	A reinforced concrete raft slab with downstand edge beam can be formed within part of the depth of the defective wall to provide a floor and some degree of underpinning.
Rehydration	This is a modern method of using water pumped into the soil so that it expands back to its near original level. Suitable for some low-rise buildings on clayey soils. The controllability of this technique, however, is uncertain.
'Safehoop' system	This involves installing a prestressed reinforced concrete girdle around the affected building. Usually suitable for relatively small- or low-rise dwellings of square or rectangular plan shape.

of wall being underpinned. The concrete is poured in staggered bays about 1.2 m wide and is taken up to within about 50 mm of the underside of the existing foundation – the gap being filled latter with a lean mix grout in a process called 'pinning-up'. This form of underpinning is, however, very expensive. It is also time-consuming and disruptive to install.

Modern

Table 7.5 lists the main forms of modern underpinning, some of which involve specialist proprietary techniques. Four common versions of modern underpinning techniques are shown in Figure 7.29.

Modern piles are either *continuous* or *discontinuous*. Continuous underpinning involves supporting the affected walling throughout its entire length. With discontinuous underpinning, on the other hand, a space is left between the new substructure supports.

The pier and beam system is continuous under the length of the underpinned wall. A more recent method using a continuous ladder frame of concrete needle beams may be required in very old buildings with thick walls. In contrast, the pile and beam method is discontinuous in that it does not cover the complete length of the foundation being underpinned.

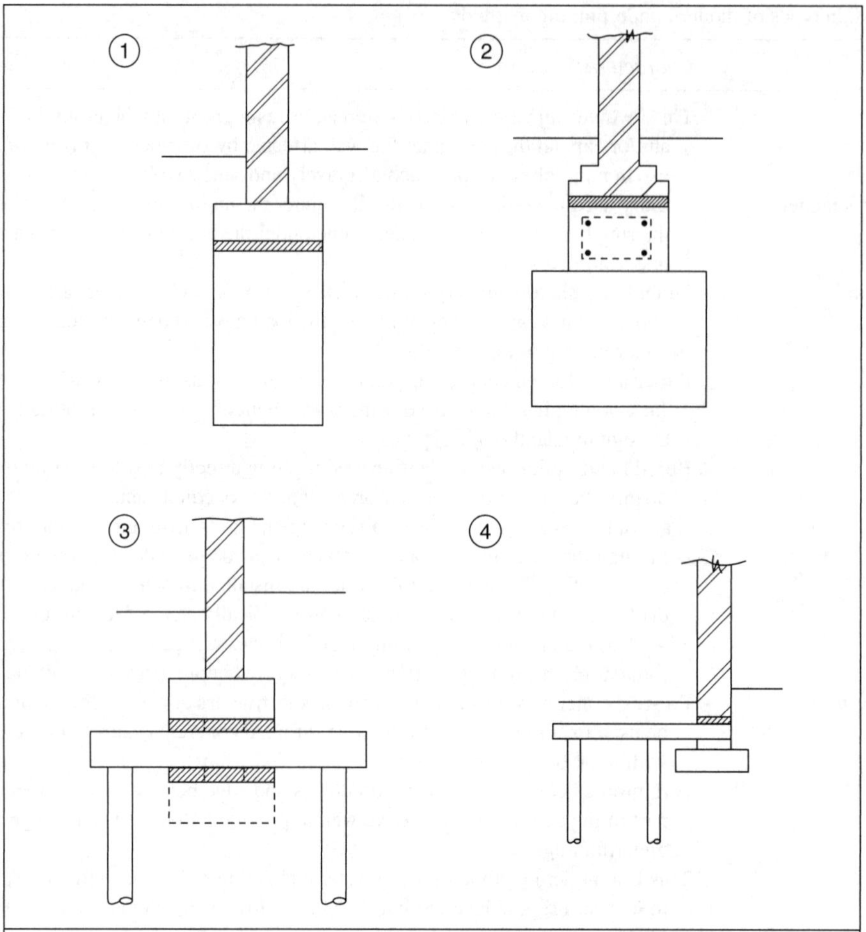

Notes

1. Continuous mass concrete legs cast in bays about 1.2 m wide under existing footing in 'hit and miss' sequence.
2. Shallow reinforced concrete beam and mass concrete pier underpinning.
3. Deep piles and needle beam underpinning. The needles may be installed either directly under the existing footing or through existing wall. They may also be cantilevered if there are problems of access or disturbance on either side of the wall.
4. Cantilevered to avoid internal work.

Figure 7.29 Four common types of underpinning

Helical metal piles can provide a quick and effective treatment for minor subsidence. The piles are fixed to the existing structure using a variety of methods and configurations. The capping generally involves using 8 mm stainless steel rods (such as 'Brutt Bar') grouted in to provide a shear and tensile connection to the existing structure. The 'Brutt Bar' cage is then usually connected back to the HeliPile by encasing it all in concrete.

Soil stabilization involving injection materials is becoming increasingly popular as an alternative to piling and traditional underpinning. For example, 'Uretek' is a proprietary soil stabilization technique using a deep injection non-compressible structural resin that provides up to 500 per cent increase in ground-bearing capacity. Multiple injections to 3 m in depth can be made in the identified weak soil layer beneath strip and rafted foundations and flooring. The resin expands about 30 times its liquid volume, exerting in excess of 50 tonnes/sq. m expansive pressure on surrounding soils and compresses any weak soil layer around it to prevent further subsidence. It expels ground water with lightweight material reducing weight on lower strata.

According to the manufacturer the URETEK system can be applied in either radial or linear form to create columns with similar characteristics to piled foundations. It's verifiable by penetrometer testing and other techniques. At face value it offers a non-disruptive solution – house foundations and floors can be treated within 1–2 days by the installer.

The lack of full control during installation and uncertain or limited durability of soil stabilization techniques are their main drawbacks. Injecting a stabilization substance into a soil can be difficult to control in terms of depth and extent of penetration. As yet, the long-term durability of soil stabilization materials remains unproven. There are unlikely to have the same service life as traditional masonry or modern concrete underpinning systems. The duration of any guarantee relating to a soil stabilization technique therefore is unlikely to exceed 30 years. Moreover, it's unlikely to be effective in cohesive soils.

Installation

Whatever underpinning system is used it is best that it only be undertaken by installer who is a member of the Association of Specialist Underpinning Contractors (ASUC). A 12-year Defects Insurance Guarantee should be obtained for such work.

Partial demolition

Rationale

Some adaptation schemes involve selective demolition of parts of a building. This consists of taking down specific elements to facilitate the adaptation work. Old or dilapidated lean-to extensions, as well as attached garages and water tank enclosures are typical examples of constructions that may require removal. This may be because of redundancy of the affected part of the building or to make way for another extension or additional access.

Problems

As with forming new openings in existing walls, care has to be taken to minimize collateral damage associated with partial demolition. This may consist of minor downtakings of ceilings or floors to more major work such as the demolition of load-bearing elements. Again, it is vital to consider the structural implications of removing parts of a building. Lean-to extensions and other additions as well as internal spine walls may be providing some lateral restraint to the wall they are abutting. Removal of such an extension can cause bowing or other similar failure in the exposed wall. The insertion of lateral restraint ties or the construction of masonry piers may be required to avoid these problems. A structural engineer should design these support mechanisms.

Any demolition work is messy and dangerous. Disposal of the waste generated in this process is also costly and difficult. The waste may have to be disposed of to special dumps and undertaken by a licensed contractor. Partial demolition is no different; in some instances it could be more troublesome than complete demolition of a building. Dust and noise control measures may need to be considered – such as the use of temporary screens, evacuation of the building and out-of-hours working.

Decapitating a building

As discussed in Chapter 6 in a few rare cases, it may be prudent to consider removing the top storey or more. This may be required as part of a major rehabilitation scheme to residential buildings. In some blocks the top storey is in a very poor state of repair and/or superfluous for the intended re-use. It may be worthwhile to consider removing this storey in such cases rather than either rehabilitating or demolishing the whole block (see Figure 6.11).

Granted, decapitating a building is not a common form of adaptation. It is only done where the result enhances the building and is beneficial to the intended occupiers and owners concerned. This option, though, does offer a novel way of radically altering a building without the need to remove it in its entirety.

It is normal practice to over-roof the decapitated building. This often results in a radical improvement in the building's appearance. It can transform the property beyond all recognition.

Naturally the remaining storey/s should be structurally sound for such work to be feasible. Moreover, because of the disruptive and dangerous nature of this work, decanting the occupants is mandatory – especially for those occupying the floor immediately below the storey to be removed.

Vertical reductions at the base of a building are virtually unknown. There is no reasonable scenario that would necessitate the removal of, say, a cellar or basement in an existing building. Even if such a part of the building were unusable, it would be easier to block it off than remove it.

Lateral reductions

In Chapter 5, we dealt with how buildings could be extended laterally. In rarer cases the size or capacity of buildings may require reducing. This could be to remove an unsafe or unused part of a building – such as an old lean-to or redundant wing of the building.

Demolition procedures

Preliminaries

Demolition, even when partial, is a dangerous and complex activity because of the higher level of uncertainty involved in this process. It requires a sound knowledge of various construction techniques, each of which has their own demolition method. It also requires an understanding of the legal and procedural issues involved. The Construction (Design and Management) Regulations 1994, for example, now require that the demolition of a building should be given consideration at its design stage. This should then be fed into the maintenance manual of the property (see Chapter 10).

Building professionals sometimes have to deal with and manage buildings that have to be demolished in part or in full. In such cases, those concerned need to have knowledge of demolition procedures and controls, as well as an understanding of basic techniques involved in such work. It must, for example, be carried out in accordance with BS 6187 and Health & Safety Executive Guidance Notes GS29/1, 3 and 4.

The criteria for any successful demolition project are summarized in Table 7.6.

Demolition plans

Plans required for full or partial demolition should contain the following:

- A block plan (to a scale of not less than 1:500) showing proposed demolished sections in red.
- The boundaries with land in different occupation.
- Particulars, appropriate to show that the operations involved will be conducted in accordance with the building operations regulations, code of practice and CDM regulations.

Table 7.6 Demolition criteria

Criteria	*Examples*
Safety	Avoid sudden or other unexpected total collapses of elements – even under controlled conditions. Demolition may need to be preceded by decontamination works (e.g. removal of asbestos, insect or vermin infestation, pigeon guano). Keep flying debris and dust to a minimum – use protective screens if the risk of this is high.
Thorough investigation	The building or part of the building being demolished should be thoroughly inspected beforehand to ascertain potential problem areas such as weak sections of the structure, hazardous materials, salvageable materials, etc.
Good neighbour policy	Adjacent occupiers and owners should be notified at all stages of the demolition process. A Schedule of Condition of remaining properties should be undertaken and agreed by the parties beforehand. The timing of the demolition work will be significant here.
Avoid damage and operational problems	Provide adequate protection to remaining parts of the building and adjoining properties. This could involve the erection of temporary plastic screens or sheeting during hazardous demolition procedures such as blasting.
Proper disposal	Waste disposed in proper bags or containers to an appropriate dump. In particular, hazardous waste requires special considerations (e.g. COSHH Regulations, and disposal to a licensed site).
Maximize recycling	Materials directly reusable (usually small %, require little reprocessing other than cleaning) if in good condition: windows, roof slates/tiles, doors, blinds, fireplaces, stone, sanitary ware, timber, carpets. Materials indirectly reusable (usually large %, require significant reprocessing): glass, ferrous/non-ferrous metals and crushed concrete/brick/stone for hardcore.

- Method statement to be provided. This summarizes the nature and sequence of the proposed demolition operations.
- Period of demolition to be indicated.

Initial procedures

Any partial demolition of a building will normally be undertaken in the same way as a full demolition work. The main difference of course is that the preparations are likely to involve the following:

- Obtain the necessary statutory consents, licenses, etc.: Permission to erect scaffolding once the demolition warrant has been issued must be obtained from the local authority.
- Disconnect or cap off redundant services: These should be made safe, whether dead or live.
- Protection of remaining elements: This could involve erecting screens over windows or whole façades nearby.
- Safety/security check: To ensure no unauthorized person is in the building during and immediately after the demolition work.
- Soft stripping: Non-load-bearing parts of the section of building being removed should be tackled first, such as partitions, windows and doors. Therefore, in any decapitation exercise the roof coverings should be tripped first.
- Pre-weakening: Cutting holes or chases in walls or other elements prior to using a pusher arm or other method to topple walls, columns and slabs.

Table 7.7 Demolition methods and their use in adaptation work

Category	*Method*	*Uses in adaptation work*
Piecemeal	'By hand' using a sledgehammer	This is the least damaging and disruptive option. It is the one most likely to be used in adaptation work.
	Pusher arm	Uses 'back-acter' or 'front shovel' machine. Suitable for low-rise redundant extensions.
	Mechanical demolition	Crane with ball and chain – limited use even in decapitation projects because of the risk of collateral damage.
	Hydraulic splitter	Used to pre-weaken or remove masonry or concrete elements.
	Thermal lance	May be used to help remove difficult concrete elements.
Deliberate collapse	Controlled using non-explosives	Pre-weakening walls and burning timber wedges and other temporary support.
	Controlled using explosives	Very limited use in adaptation work because of the danger and risk of collateral damage.

Basic techniques

The general sequence of demolition even in adaptation schemes is in reverse order of construction. In other words, demolition from the top down is the usual procedure.

The main techniques are listed in Table 7.7.

Procurement requirements

A contractor that is a member of the National Federation of Demolition Contractors (NFDC) should undertake awkward or complex demolition projects. Ideally, the NFDC standard form of contract should also be used for this type of work.

Problems with demolishing buildings

Partial demolition can pose a variety of problems in an adaptation scheme. It can be just as hazardous as full demolition:

- Collateral damage: Remaining parts of the building (wallheads, walls, floors, openings) can easily be damaged by careless demolition work.
- Estimating costs: This can be difficult and depends on the extent of deleterious or salvageable materials. The latter could increase costs substantially, whilst the latter may be sold off to help reduce costs.
- Salvaging materials: Limited scope for reuse? The extent of potentially valuable materials such as lead and copper should be ascertained and recorded.
- Deleterious materials: Identify, handle, treat or remove carefully (i.e. bagged and labelled and dispose to an approved dump) by a licensed contractor.
- Existing services: Blank off and seal?
- Disturbance/nuisance: Timing of demolition work? Additional or special protection measures need?
- Pests/wildlife: Bats or other protected species affected by the demolition?
- Archaeological finds: Any significant discoveries or previous uses identified? Notify appropriate authorities (e.g. English Heritage).
- Post-demolition survey: Check on the success of the operation and identify extent of areas of collateral damage.

Statutory approvals for structural alterations

Basic requirements

As previously indicated where adaptation works involve any structural alterations, the design of new or modified load-bearing elements will have to be approved by the local authority. This is normally done through the building warrant process. Approval can be achieved in one of two ways.

For minor work to buildings not exceeding three storeys in height

Structural calculations A chartered structural engineer or other qualified construction professional (e.g. building surveyor) with the appropriate skill and experience can prepare these for submission along with the warrant application. Calculations to determine sizes of members for lintels or new strengthening beams and other simple alterations can be based on the rule of thumb, which shows that the depth of beam can be determined as follows:

- Lightly loaded steel 'I' beam: 1/20 of span (100 mm wide flange).
- Heavily loaded steel 'I' beam: 1/18 of span (150 mm wide flange).
- Timber joist 50 mm wide: 1/24 of span + 50 mm (50 mm wide joists at 450 mm centres).

Several typical worked examples of the format and content of structural calculations are presented in Gauld (1996). It shows the procedure for determining the size and type of support members as a result of alteration work. For small-scale or straightforward jobs, such as a new beam under an existing floor to increase its load-bearing capacity or halt excessive deflection, this is often an adequate and convenient procedure.

Authoritative tables Reference in the adaptation drawings to authoritative or published tables of approved joist sizes (such as Table A.1 of the Building Regulations 2000 (as amended); or the Table in the Small Buildings Guide relating to the Scottish building regulations) may suffice where a new floor or flat roof structure is required.

This approach can be used when a concrete lintel over a new opening in an existing building is required. The appropriate size of a pre-cast (pc) concrete beam as the lintel in such cases can be determined reasonably easily. A table supplied by a pc reinforced concrete units manufacturer such as Richard Lees Ltd gives a variety of beam sizes for a range of spans and loading conditions (see Appendix D).

For buildings over three storeys in height and complex structural alterations in low-rise properties

A design certificate A chartered structural/civil engineer confirming the suitability of the proposed design must sign this certificate. Such an approach must be used because of the safety and professional liability implications involved in major structural alterations.

However, it cannot be overemphasized that in any cases of uncertainty or potential danger general building consultants such as architects and surveyors should not undertake complex structural calculations. The golden rule in such circumstances, therefore, is *if in doubt, consult an appropriate structural specialist.* Knowing the limits of one's own competency is a skill in itself.

In any event, the following procedure should be adopted when undertaking simple structural calculations for building warrant approval to install an additional structural element such as a new support beam:

- Use a *standard schedule format* for calculations (as shown in Gauld, 1996).
- Show existing and highlight as proposed *conditions* on sketch drawings.
- Determine the *loadings* (dead loads and live loads) and *load paths.*

- Determine the maximum *bending moment* of the member required.
- Determine permissible and actual *deflections and stresses or loads.*
- If necessary determine permissible and actual *loads and bearing and shear stresses.*
- If necessary determine suitability for lateral restraint.
- Select and check trial section of beam.

Essentially the fundamental objective of structural calculations is to ensure that the permissible stresses and loads exceed the actual stresses and loads by the appropriate factor of safety. This must of course be greater than 1 to prevent collapse or catastrophic failure of the structure.

The use of loads rather than stresses when dealing with factors of safety is to be preferred. This is because the latter only presupposes a reasonable margin against failure based on assumed stress values.

Thus the (load) factor of safety can be determined as follows:

$$\text{Factor of safety} = \frac{\text{Ultimate load (load to cause failure)}}{\text{Working load (actual load on structure)}}$$

The size of factor required depends on the predictability of both the loading conditions and properties of materials used in the structure. It can, however, range from about 1.5 for steel structures to as high as 5.5 for masonry (Seward, 2002). The quality and performance of the former material is more consistent. Brickwork and stonework, on the other hand, are less predictable in these terms. In addition, the tensile strength of masonry is variable, which makes it difficult to acquire a precise indication of its strength.

More complex structural designs are concerned with the probability of a structure's failure. In such cases, the services of a structural specialist should be commissioned.

Lintels for new openings

Preamble

Even small-scale adaptation schemes such as making two separate semi-detached units into one will involve forming a new opening in an existing wall. A new support beam or lintel will be required to support the masonry above the opening.

Worked example 1

Assume that a semi-detached factory unit of solid loadbearing masonry construction is being converted into single occupier use. This will involve creating a new opening in the party wall, which is 225 mm thick, to link the previously separate units.

Select a suitable pre-cast concrete beam from the manufacturer's tables (see Appendix B) to span the 2 m wide opening 2 m high in the party wall.

As an alternative option, design a suitable set of two steel beams to support the same span using the tables in Appendix C.

Procedure for determining a suitable pre-cast concrete beam

- Assess the load above the proposed opening. GBG 10 provides the best advice for assessing loads above ordinary openings in external and internal walls. It should *not* be used where the loads are suspected to be unusual or high.

- In summary GBG 10 involves the following:
  - Establish the dead load above the span of the opening. This involves multiplying the volume of wall concerned by the density of common brickwork (say 2300 kg/m^2 – taken from BS 648: Schedule of weights of building materials).
  - Ascertain any live loads (as per BS 6399) above the span of the opening (i.e. determine if there are any floor or roof loads being taken by the party wall). If there are, assume that these are acting over the portion of wall spanned by the proposed opening.
  - In this case it is assumed that there are no floor or roof loads imposed on the party wall.
  - The minimum dead load is the weight of masonry within the 45° load triangle. For a 2 m wide opening the working span would be about 2.75 (i.e. 2 m + 0.373 + 0.375). The load triangle would therefore be 1.38 m high × 2.75 m wide and have an area of about 1.9 m^2. This means that the weight of the brickwork within the load triangle is 1.9 × 0.225 × 2300, which is 983 kg = 9.6 kN (i.e. 983 × 9.81).
- Once the total load has been calculated (i.e. dead load + live load), reference should be made to an appropriate pre-cast concrete beam manufacturer's table of beam sizes. For example, Richard Less Ltd provide such a table which can be used to establish a suitable size of pre-cast concrete beam for varying short spans and ordinary loading conditions. The nearest available section above the safe load of 9.6 kN is 225 mm wide by 150 mm deep, which has a maximum safe uniformly distributed load (in addition to its own weight) of 15.33 kN (for a 2.1 m span).

Procedure for determining a suitable 'I' joist (RSJ) or 'UB'

- Determine the loadings above the proposed opening as indicated before. It is assumed that the bearing at each end of the beam is 150 mm and formed with in-situ or pre-cast concrete pad stones.
- Calculate an approximate size of steel beam using the rule of thumb (assuming it's a heavily loaded beam): depth = span/18. For a 2 m wide opening this requires a steel UB at least 111 mm deep. This figure should be rounded up to the nearest stock section depth (e.g. 127 mm). A 225 mm wide by 127 mm deep steel beam, though, is not a standard size. Two 127 × 76 × 13 kg steel UBs encased in two layers of 9.5 mm thick plasterboard (e.g. 'Gyproc Fireline') to provide 1-h fire protection would probably suffice for this opening.
- A more detailed structural calculation from first principles, however, would be required to confirm the suitability of this section in terms of deflection, stress and bearing, the three key beam design factors. The maximum permissible deflection for a steel beam is 1/360th of the span, in this case 6 mm. The maximum permissible stress for ordinary structural steel is 150 N/mm^2.

Worked example 2

Appendix H shows the schedule of calculations for a new lintel over a proposed enlargement of an existing opening.

New support beam for existing floor

Preamble

In older properties it is not unusual to encounter excessive sagging in the upper floors. Alternatively, with a proposed change of use from residential to commercial it may be necessary to strengthen the existing floor structure.

Worked example

The timber floor in the kitchen/dining room of a first floor flat in a Victorian tenement contains a 50 mm deflection. A new support beam under this floor is required to arrest this sag.

Determine the size of a suitable steel/concrete/flitch beam to span between bearings X and Y as indicated in the schematic layout plan in Appendix E.

Summary

Structural alterations form part of many adaptation schemes, from the smallest to the largest. They must be designed and constructed with care if collateral damage or structural failure is to be avoided.

Many buildings that are adapted need to have their structural capacity improved or altered in some way. The knock-on effects of these must be taken into account to prevent any danger to the building and its occupants as well as the public. Indeed, safety and stability are the two key criteria for any structural alterations.

Protection of structural elements often forms part of an adaptation scheme. The long-term durability of the structure may necessitate anti-corrosion measures to the steelwork or anti-fungal treatment to the load-bearing timbers. This work can also achieve an improvement in the appearance of these exposed elements if they are covered with an appropriate cladding.

Construction professionals such as building surveyors and architects should always consult an appropriate structural specialist before embarking on any work involving alterations to load-bearing elements.

8

Principles of refurbishment

Overview

This chapter outlines the options and processes of modernizing residential and commercial buildings and enhancing their overall performance. It focuses predominantly on the non-structural improvements that can be made to housing and non-residential buildings. The principal aim of such work is to bring properties up to modern standard in terms of thermal efficiency, weather-tightness, durability and appearance.

Background

The need for refurbishment

The principles relating to the refurbishment of buildings generally is the principal focus of this chapter. For example, some of the issues covered here, such as condensation control and indoor air quality, apply to non-residential as well as residential refurbishment work. Chapter 9 deals with more specific aspects of refurbishment that apply to most types of buildings.

As seen in Chapter 1, all buildings eventually are affected to a greater or lesser degree by some form of obsolescence or inefficiency. Deficiencies in the fabric and services occur because of their inability to satisfy current requirements and handle technological change. Sooner or later they fail to meet some if not all of the user needs or statutory requirements. This occurs for three main reasons regardless of whether the building is fully or partially occupied or wholly vacant. Firstly, construction standards and requirements are continually improving because of Government policy to enhance energy efficiency and building performance. As the demands and expectations of property users tend to increase over time, this is also having a major impact on the building regulations. Secondly, wear and tear as well as exposure to the elements results in ongoing deterioration or other adverse change in the building structure and fabric. Thirdly, advances in technology and rising demands by consumers and workers for better and more comfortable internal environments have prompted the need for ongoing modernization of buildings. This is especially the case with commercial and industrial properties. Many of these were built to an inferior standard up until the 1980s when tighter building regulations were introduced and user expectations increased.

'Better' buildings are more cost-effective to run, more suited to the needs of the user, and contribute to a more sustainable built environment. 'More comfortable' buildings have good indoor air quality, which means that the interior climate contains clean, fresh air with low levels of pollutants and other hazardous substances.

As indicated in Chapter 2 the vast majority of the total building stock in the UK (i.e. >95%) was built between the 19th and 20th centuries. Few if any of these buildings are able to meet 21st century needs and services without some physical improvement. The primary way of addressing this problem, other than redevelopment, is refurbishment.

Failure to match current building requirements, however, is not a problem that affects just old buildings. Even relatively new buildings can quickly fall below standard or fail to meet the increasingly exacting needs of their occupiers. The introduction of new or improved regulations can easily cause a recently constructed building to fail to comply with the regulations completely. It has to be emphasized, of course, that the building regulations are not normally applied retrospectively to all buildings, unless there are exceptional circumstances (such as danger to the public). They are only enforceable on existing buildings that are subjected to adaptation work, such as alterations, extensions and conversions. Even modifications to the layout of a kitchen, for example, may require a building control approval. When in doubt, of course, the local building control office should be consulted (see Chapter 11).

There are, though, some exceptions to the requirement for obtaining statutory approvals. The following types of modernization work, for example, do not normally require the permission of building control:

- window replacement (see next chapter);
- painting or repainting the façade of a building;
- rendering or re-rendering the façade of a building;
- recovering a flat or pitched roof (though the opportunity to improve its thermal efficiency should be taken as well);
- small porches not exceeding 8 m^2;
- the provision of internally applied insulation to or within a wall, ceiling roof or floor.

There are, however, some important provisos with all of these and other property improvements. Firstly, the building should be neither listed nor in a conservation area (see Chapter 11). Secondly, the work should not involve any structural alterations. Thirdly, the materials and workmanship must comply with the appropriate British Standards specifications and codes of practice. In addition, an often-overlooked consideration is that only a contractor who is a member of the appropriate trade organization should undertake the work. This is advisable but not mandatory. For example, in the case of window replacement, the installer should be a member of the Glass and Glazing Federation. This should give the client a degree of protection in the event of the contractor going bankrupt or where there is a dispute over the quality of the installation.

Health benefits of refurbishment

The impact of poor housing on health is significant according to many researchers (see e.g. Burridge and Ormandy, 1995). Preventing dampness and hypothermia is another reason over above applying energy saving measures why modernization of housing should include thermal efficiency. The Office of National Statistics (Anon, 2000b), for example, reported that premature deaths in the UK from cold-related illnesses such as respiratory and cardiovascular diseases have exceeded 50 000 per annum. Cold-related illnesses in other words account for around 10 per cent of all deaths in the UK, which has average annual mortality figure of about 580 000 according to the www.statistics.gov.uk. The elderly living alone and householders on low-income are especially susceptible to this modern-day scandal, which of course ought not to happen on this scale in any part of the world.

Health scientists have found that sudden drops in temperature can trigger the two main forms of cardiovascular problem: strokes and heart attacks. As temperatures fall this increases the body's adrenaline, which causes the blood to thicken and blood pressure to rise. The additional strain on the circulatory system resulting from hypothermia exacerbates any pre-existing ailment. Hypothermia occurs when the body's core temperature is at or below 35°C (Potter and Perry, 1999). The elderly, children, people with a heart condition, those who have taken excessive levels of drugs or alcohol and the homeless are at a higher risk of being affected by this condition.

Cold dwellings also increase occupiers' vulnerability to illnesses such as chronic bronchitis and emphysema. This is because low temperatures can easily lead to infection and other inflammatory processes in the pulmonary system.

Dampness in buildings is another major health risk (Committee on Damp Indoor Spaces and Health, 2004). Respiratory ailments such as colds and flu and rhinitis are more likely in damp homes. In particular, dwellings with these conditions can aggravate the following: asthma, angina, arthritis, rheumatic pain, blood flow problems, chest infections, coughs and mental health problems. Mould spore and bacterial growth, encouraged by dampness, can produce an allergic response on the skin or act as an irritant in the respiratory tract (Singh, 1993). Moulds such as *Penicillium chrysogenum* and *Aspergillus fumigatus* are toxic, whilst *Alternaria alternaria* and *Cladosporium cladosporioides* are moulds that are typically allergenic (Bech-Andersen, 2001).

Adverse health effects of dampness are also possible in non-residential buildings. For example, a workplace subjected to water-damage may result in a dramatic increase in the rate of asthma and other breathing problems in employees, and could be a substantial source of sick days, new research suggests.

In a study of workers at one leaky, mould-contaminated office building, scientists from the US government's workplace safety research body NIOSH found that the rate of adult-onset asthma among employees was more than three times that for the general population (Thompson, 2005).

Two-thirds of these cases were diagnosed after the employees had started working in the building. The researchers estimate that up to 12 per cent of employees sick days in a year could be attributed to the health effects of the building.

Recent research has also found that those living in damp housing have a higher risk of insomnia (Janson et al., 2005). This supports the author's contention that living or working in damp or otherwise inadequate building conditions can have adverse mental as well as physical effects (see psycho-social considerations below).

It is therefore clear that bad housing as well as poor quality office or hospital accommodation can exacerbate if not trigger ill health. Naturally, factors such as the level of housekeeping, hygiene standards, the intensity and type of use, and the quality of facilities are also influential in preventing building-related ailments, especially in children and the elderly.

Properties housing these vulnerable occupiers would nevertheless benefit from energy efficiency improvements and other modernization measures to their homes. Another important advantage of such work is that it can help improve the internal environment. Buildings that have been modernized are usually less prone to dampness problems as well as being warmer. This has obvious social and health benefits by reducing the ailments associated with building use. In addition, for health service related facilities it would help cut mortality rates and hospital admissions.

Higher temperatures associated with global warming of course may help to reduce the number of cold-related deaths annually in the UK – by about 20 000 (www.dh.gov.uk). However, warmer weather will bring its own adverse health effects, such as hyperthermia, for example, which is said to occur when the body's temperature is at or above 40.5°C (Potter and Perry, 1999). According to the Department of Health (2002) heat-related deaths are expected to rise by 2000 a year from the current rate of about 800. This will increase the need for buildings to have energy efficient cooling systems (see Chapter 10).

Granted, hot weather ailments such as skin cancer and heat stroke are associated with tropical outdoor environments. Harmful bacteria and other micro-organisms, however, can also flourish in warmer indoor climates in temperate regions, increasing the risk of food poisoning. Better ventilation and more stringent hygiene control therefore is required to combat these problems. Retrofitting passive stacks or other passive ventilation measures as part of an adaptation scheme, for example, would help reduce the level of airborne pathogens indoors.

Office buildings, too, may require modernization to overcome sick-building syndrome (SBS) and building-related illness (BRI). It includes measures dealing with indoor air quality, which is dealt with later in this Chapter.

Other impulses for refurbishment

Preamble

As with any adaptation or new-build scheme the needs of the client are paramount. Other considerations, though, may impact on the modernization of a building. In general, the impulses that drive this form of adaptation are summarized as below.

Environmental

Improving the quality of the urban landscape and obtaining a more sustainable environment is a goal of central and local governments. Tackling urban blight is not only possible through slum clearance schemes and major new developments. Major rehabilitation programmes to housing estates and refurbishment schemes to industrial and commercial premises also have a role to play in this area.

Modernization not only may form part of an urban renewal strategy, which can enhance the appearance and character of the building stock. It can also provide an opportunity to upgrade the overall performance of both the residential and office property. This form of adaptation is the most effective strategic way of maximizing the service life of these buildings.

It was noted earlier that sustainable construction is now an important part of the political and environmental agendas. Modernizing existing buildings can go some way to achieve a more sustainable environment. This means increasing energy efficiency and reducing wastage of non-renewable fuels and materials (see Chapter 10 especially).

Psycho-social

Modernization of buildings enhances not only the physical well being of their occupants. It may also have a positive effect on their mental health and overall performance. Thus not only does a 'healthy' commercial/industrial building maximize productivity, it can also reduce the incidence of absenteeism amongst the workforce.

A refurbished residential property, too, can provide a more comfortable and comforting environment for its occupants. It can help to reduce stress levels as well as minimize the incidence of BRI's mentioned above.

Achieving comfortable and efficient places to live and work, therefore, is a goal in any conscientious housing or commercial development. It helps to build social stability as well as economic prosperity. Creating a more attractive and pleasant habitat through such measures can have a positive knock-on effect on the health and morale of residents or staff.

There has been a great deal of research over the past 30 years on the impact of housing conditions on community attitudes and well-being (Coleman, 1989; Hillier and Hanson, 1989). Coleman (1989) propounded the view that anti-social behaviour of residents is affected by their living conditions. Although she has her critics (e.g. Hillier and Hanson, 1989), Coleman at least attempted to show the relationship between poor housing conditions and the behaviour of residents.

It is conceded, however, that merely improving the physical environment for the occupants in a previously run-down housing estate will not in itself eliminate these problems. Adequate welfare facilities to foster a good community spirit are also important. A subtle mix of these and other social factors such as accessible welfare provision and sufficient job opportunities for the local inhabitants is needed. These initiatives as well as crime prevention measures will play a vital role in preventing any refurbished housing area from degenerating back to its previous blighted state.

It is also generally acknowledged that the internal environment has an impact on human behaviour and performance in many other ways (see IP 13/03 Parts 1–3). The following is a summary of the areas where modernization can help to improve the quality of the indoor environment for its occupants:

- Residential: To enhance general comfort living conditions and reduce anxiety levels for occupants, especially for vulnerable people such as children, the poor and the elderly.
- Education: To enhance the learning environment for pupils/students and staff (see also section on school refurbishment later in this chapter).
- Healthcare: To maximize comfort conditions and recovery rates for patients and improve working conditions for staff.
- Hotel: To maximize comfort conditions for guests and provide an acceptable working environment for staff.
- Industrial: To enhance productivity by creating a pleasant and comfortable workplace for operatives and managers.
- Offices: To enhance productivity by creating a pleasant and comfortable workplace for staff and visitors.
- Retailing: To make the shopping environment more conducive to business for buyers and staff.
- Assembly: To attract visitors and users to the premises and encourage them to return.

Legal

Achieving a higher level of compliance with the building regulations and other statutory provisions such as fire regulations, energy efficiency, disabled access and soundproofing is obviously beneficial to all concerned (see Chapters 10 and 11 especially). Meeting these requirements makes buildings safer, more comfortable and efficient as well as user-friendly. The Home Energy Conservation Act 1995, for example, local authorities are now duty bound to assess the energy efficiency of their stock of dwellings. This clearly has implications for upgrading the thermal performance of housing generally.

Moreover, the Local Government Act 2000 allows local authorities to promote and participate in any scheme that would improve local economic and social well-being. The aim is to enable authorities to start 'to think globally and act locally'. Modernization of housing involving partnership schemes with the private sector is one way of doing this.

Most importantly, prevention of BRI and SBS will minimize the risk of a negligence claim or even criminal proceedings against the building's owners or maintenance personnel. A sobering example of this is the case of architect/maintenance manager Gillian Beckingham. At Preston Crown Court in April 2005 she was found guilty of charges laid against her under Section 7 of the Health and Safety at Work Act 1974, following an outbreak of Legionnaires' Disease at the Forum 28 Centre in Barrow-in-Furness in August 2002. Ms Gillingham also faced manslaughter charges because seven elderly people died as a result of the outbreak. Clearly this case has implications for everyone who manages health and safety at work, whether for maintenance or refurbishment. The legal fees as well as the emotional costs involved in such cases are often enormous.

Technical

Modernization offers an ideal chance for improving the overall appearance, durability and thermal efficiency of a building or group of buildings. It also provides a good opportunity to undertake major structural repairs to the stock. Overcladding or recladding schemes, for example, can enhance the overall performance of buildings (see below as well as Chapter 9).

Some older commercial and industrial premises may require the renewal or upgrading of their lightning protection. Given the predicted rise of fluctuations in weather patterns owing to global warming the risk of lightning strikes on high-rise buildings is likely to increase. Thus, allowance for additional or improved lightning protection may have to form part of the refurbishment of a property if it is of multi-storey construction.

Economic

Enhancing the value and extending the service life of the building stock and reducing its energy consumption can yield considerable financial benefits. A refurbished building, for example, will be more attractive to potential tenants or buyers. This helps to maintain if not increase its rental value or capital value.

Market influences

Many factors can trigger the modernization of commercial and industrial buildings. Internal organizational pressures such as changes in the size and composition of the workforce can drive the need to modify the layout and facilities within business properties. In addition, employees may demand or need better working and support facilities within their offices or work areas.

Companies are becoming increasingly aware as to the extent to which the quality of the workplace can have on the performance of their employees. It is of course not surprising to find that a comfortable, attractive and spacious workplace results in a happier, more content, less stressed, and thus more productive workforce.

External pressures such as the need to enhance the company's corporate image may prompt the modernization of its headquarters. The refurbishment of adjoining properties may also prompt a company to instigate a similar treatment to its building.

An office or factory building that is below current building regulation standard is not only likely to be unsuitable for the modern needs of its occupants in terms of amenities and comfort. It is also probably very inefficient to heat and illuminate. The running costs of such a building will be excessive. All of these features could reduce the building's marketability, whether for rent or sale. Commercial occupiers may take a holistic view of any available property by looking at operational expenses as well as rental levels and capital values.

A common characteristic of most industrial and agricultural buildings constructed during the last century is that their overall quality was given too low a priority. Capital expenditure for industrial buildings especially was minimal resulting in low construction standards. This was exacerbated by the degree of neglect or minimal maintenance suffered by these properties.

Refurbishment of non-residential buildings can also take place alongside major extension schemes. The opportunity to upgrade the main building whilst it is being enlarged can result in economies of scale.

The rapid rate of change affecting commercial buildings, of course, is most apparent in retailing. Shopping patterns and needs, like business attitudes generally, have become more customer-focused. Shop units and supermarkets nowadays have to adapt regularly to cope with these demand-side changes. This in turn has had an impact on the supply side. Many large supermarket outlets are open 24 h a day. In addition, it is common these days for many retail chains to refurbish their outlets about once every 5 years. This is not only to replace worn out or tarnished components and finishes but also to improve the layout or radically alter the image of the shop.

The short-life approach to retail construction is most noticeable in the fast-food outlets. These are often detached, single-storey structures that have a service life of about 20 years. The adaptation response in such facilities is usually restricted or directed to renewing the internal finishings such as flooring, wall coverings and fitments such as counters and furniture.

Refurbishment strategy

Preamble

In undertaking a major programme to improve, refurbish and upgrade buildings it is important to adopt an appropriate strategy. This will help ensure that the correct measures are implemented at the appropriate time to avoid duplication of effort or un-necessary works. For example, it is critical that any modernization measures are co-ordinated with the ongoing planned maintenance work to the building/s in question. This can help avoid wasteful exercises such as repainting old timber windows shortly before they are scheduled for replacement.

It's vital in any refurbishment strategy to set priorities. The needs of the user as well the owner of the building/s being considered for refurbishment should be addressed. Residential landlords, for example, should consult their tenants to avoid conflicts and achieve maximum user satisfaction with housing rehabilitation proposals. This could be done in a variety of ways:

- ward consultation meetings on capital programmes – the programmes from which major expenditure schemes like rehabilitation are funded (whereas maintenance is funded from the revenue budget),
- liaison with tenants groups,
- through feedback from housing surveys,
- other ways of determining tenants views.

According to the EST (2005) setting priorities is usually facilitated by:

- Targeting properties in greatest need – such as those in worst condition or those with the highest heat loss, inadequate heating or expensive heating systems such as space or ceiling heating.
- Initiating an energy efficiency programme to insulate and draught-proof all of the stock up to a set standard over, say, a 5–10 year period (such as the Decent Home Standard).
- Reviewing refurbishment practice, as well as repair and maintenance programmes, in the light of the energy efficiency targets, and upgrading the specification if necessary.
- Reviewing the window replacement programme.
- Reviewing any heating improvement schemes with a view to replacing them with heating/insulation/ventilation packages.
- Installing cheaper heating systems to help offset the cost of insulation, allowing more properties to be upgraded within a set budget.

Understanding typical life cycles of major renewals in or refurbishment of buildings forms an important part of any modernization strategy. This is required to achieve more effective replacement cycles. Table 8.1 shows typical life cycles for the elements of various building types.

The following, then, is a summary of the five key factors that must be considered when proposing to modernize buildings:

- *Context*
  Number and size of buildings – one or a whole stock? Distribution of the stock?
  Age and construction of the buildings?
  Condition and overall performance of the buildings?
  Prioritization of works required?
  Residential, Non-residential or Mixed?
  Finance required and allocated. Grants available?
  Statutory approvals required?

Table 8.1 Life cycles for major renewals or refurbishment of commercial buildings (adapted from Lawson, 1995)

Renewal/replacement			*Period (years)*
Generally			
	Roofs:	Flat-roof coverings	20–30[a]
		Pitched roof coverings	30–60[b]
		Rainwater goods	30–50[c]
	Walls:	Decoration (interior)	5–7[d]
		Re-rendering (exterior)	30–50[d]
	Windows and Doors		30–50[d]
	Major engineering plant		10–15[e]
	Electronic, communication and computer equipment		5–8[f]
Offices			
	Public rooms: renewal of carpets, fittings and furniture		5–8[h]
	Staff room/s:	Décor, furnishing fabrics	2–4[i]
		Carpets, electrical fittings	5–8
		Furniture	7–10[j]
		Bathroom fittings	10–15[j]
	Capitalized leased equipment		5–8[f]
Hotels, Restaurants and Bars			
	Public rooms: renewal of carpets, fittings and furniture		5–8[g]
	Guestrooms:	Décor, furnishing fabrics	2–4[i]
		Carpets, electrical fittings	5–8
		Furniture	7–10[i]
		Bathroom fittings	10–15[i]
	Capitalized leased equipment		5–8[f]
	Food service, kitchen and laundry equipment		7–10[j]
	Hotel buildings		20–25[k]

Notes
(a) For high performance polymeric systems.
(b) For slate and tiled roofs.
(c) 30 for uPVC and 50 for cast iron.
(d) To ft in with other refurbishment work.
(e) Major overhauls and replacement. Servicing and maintenance carried out regularly.
(f) Affected by obsolescence, introduction of new systems.
(g) Major refurbishments planned to fit in with the concept of life-cycle of the restaurant, bar, etc.
(h) Two-year decoration usually combined with fabric and carpet cleaning.
(i) May require renovation instead of replacement.
(j) Ten years for heavy-duty equipment.
(k) For amortization of loans. Building renovation likely.

- *Objectives*
  Extend the building/s service life: 30 years minimum.
  Energy efficiency and other measures to achieve better sustainability.
  Enhance appearance.
  Improve marketability.
  Functional efficiency improved – better layouts and enhanced accessibility.
  Achieve maximum code compliance.
  Improve comfort and health conditions for living/working.
  Contribute to urban regeneration of the area.

- *Key measures*
  Roofs: re-cover; overclad; replace; over-roof?
  Walls: re-point; paint/re-paint; overclad; re-clad?
  Windows and doors: overhaul/replace?
  External areas: new canopies; play areas and parking facilities?
  Floors: improve performance in terms of acoustic, fire and structural characteristics; and, improve appearance and durability of floorings?
  Dampness eradication: e.g., positive input ventilation (PIV) (see later in this Chapter)?
  Services: new heating systems; improved controls; rewiring electrics to current standards?
  Internal layouts: enlarge kitchens and bathrooms?
  Access: disability improvements?
  Common areas: redecorate and provide new floor coverings?
- *Plan*
  Investigation: desktop and field survey work to check accuracy of drawings with as-built conditions.
  Condition survey to determine the extent of necessary/required works.
  Establish priorities.
  Budgeting: calculate the extent and source/s of finance.
  Determine the timing and extent of works agreed works.
  Use of prefabrication possible/maximized?
  Procurement system: JCT 98 or Small Works Contract?
  Phasing of work: by element or by block; timing (one block at a time?).
  Decanting of occupants: partial or total?
  Programme: 5/7/10 year rolling programme?
  Consultation with occupiers: tenants/staff, on extent and timing of works.
- *Execution*
  Agree priorities with stakeholders (i.e. occupants/tenants, consultants, contractors, etc.).
  Phasing of operations: block by block or operation by operation (e.g. window replacement); if any breaking through work involved, kept till end if possible.
  Scaffolding requirements: moveable or fixed.
  Safety controls: CDM requirements.
  Security considerations: on-site security (i.e. night watchman); security patrols.
  Supervision of works: clerk of works, full-time on-site.
- *Aftercare*
  Defects liability period.
  Monitoring effectiveness of improvements.
  Revise existing programme of planned preventative maintenance.
  Undertake quinquennial review of maintenance and refurbishment requirements.

British housing stock

Context

The total British housing stock consists of about 23 million dwellings (DETR, 1998; Social Survey Division (SSD, 1998); Scottish Homes, 1997; Welsh Assembly, 1999). As indicated in Chapter 2, nearly two-thirds of the stock was built before 1965. This large proportion of the stock will require modernization as well as remedial works to bring them up to current standards owing to its increased ageing/dilapidation and obsolescence.

Table 8.2 Key deficiencies in the British housing stock (adapted from DETR, 1998a; Scottish Homes, 1997; Welsh Assembly, 1999 and Revell and Leather, 2000)

Dwelling details	*'000 dwellings England*	*Percentage*	*'000 dwellings Scotland*	*Percentage*	*'000 dwellings Wales*	*Percentage*
Lacking 1 or more basic amenities.	207	1	5	0.2	2.5	0.2
Unfit dwellings.*	1260	6.7	21	1	98	8.5
Vacant dwellings.	798	3.9	109	5	62	5
Disrepair costs.	Over 6 m dwellings required £1000 worth of urgent repairs (c. one-third of the stock).		Over 630 000 dwellings needed repairs costing over £3000 (30% of the stock).		287 500 dwellings had repair costs over £1000.	

*Dwellings classified as not fit for human habitation (or below the equivalent 'tolerable standard' in Scotland).

The case for refurbishment of the British housing stock is compelling. In 1997, the UK Government inherited a £19 billion backlog of renovation and improvement work to local authority (LA) housing, and 2.1 million homes owned by councils and registered social landlords (RSLs) that were below the decent homes standard (ODPM, 2005). Around one-third of homes in the private sector are failing to meet basic needs, especially as regards heating. Table 8.2 summarizes the main deficiencies in UK housing.

Clearly it would be neither financially feasible nor operationally practicable to spend billions of pounds in one go in what would be a massive housing refurbishment and maintenance programme. As with the multi-billion pound investment required for the railways, a long-term programme therefore would be needed to implement the comprehensive repair and modernization of the British housing stock. Optimally, a phased approach involving the expenditure of between £2 and £5 billion per year over a decade would be one realistic means of attaining this goal. Even then there is no guarantee that the political will necessary to adopt such a strategy with a radical expenditure programme, unless of course the private sector could be encourage to participate more in PFI/PPP schemes.

Basic amenities and fitness

As defined in the various Housing Acts, all dwellings are meant to have five basic amenities:

1. a kitchen sink;
2. a fixed bath or shower in a bathroom;
3. a wash hand basin;
4. hot and cold water to each of these;
5. an indoor WC.

As indicated in the previous English House Condition Survey (Research, Analysis and Evaluation Division, 2003) considerable progress has been made over the last 25 years in providing these five basic amenities. The same is also true in Scotland and Wales (Revell and Leather, 1999).

Fitness standard

The Local Government and Housing Act 1989 prescribes the Housing Fitness Standard in England and Wales. Houses not complying with this are classed as 'unfit for human habitation'. In Scotland the term

Table 8.3 The three phases of the non-traditional building boom in the UK (adapted from Chandler, 1991)

Phase	*Period*	*Extent*
1	1919–1944: The inter-war non-traditional stock.	Approximately 52 000 non-traditional dwellings were built during this period. Although they were built using many traditional features and internal floor plans, the external walls were replaced by a combination of concrete blocks or in-situ and pre-cast concrete frames and cladding components.
2	1945–1955: The Post-War Low-Rise Stock.	During this period 305 256 non-traditional dwellings were completed in England and Wales, and 100 648 in Scotland. These comprises a mixture of steel framed houses (such as the BISF house), pre-cast and in-situ concrete (such as no-fines).
3	1955 to c. 1980: The Second Post-War Building Boom.	Approximately 31 000 dwellings were built during this period.

Total stock of non-traditional housing in the UK: approximately 388 000 dwellings (about 2% of the total stock).

'Tolerable Standard' as defined in the Housing (Scotland) Act 1987 is used to describe a similar set of requirements. Houses not complying with this are classed as 'below tolerable standard'.

From the findings shown in Table 8.3, the degree of unfitness appears to be a greater problem in England and Wales than in Scotland. A dwelling is said to meet the requirements of the standard if:

- it is free from disrepair;
- it is structurally stable;
- it is free from dampness prejudicial to the health of the occupants (if any);
- it has adequate provision for lighting, heating and ventilation;
- it has adequate piped supply of wholesome water available within the house;
- it has an effective system for the draining of foul, waste and surface water;
- it has a suitably located WC for the exclusive use of the occupants;
- it has for the exclusive use of the occupants (if any) a suitably located bath or shower and wash-hand basin, each of which is provided with a satisfactory supply of hot and cold water;
- there are satisfactory facilities in the dwelling home for the preparation and cooking of food, including a sink with a satisfactory supply of hot and cold water.

The current fitness standard for England and Wales was introduced through the Local Government and Housing Act 1989 which inserted a new Section 604 in the Housing Act 1985. According to the DETR (1998) 'a dwelling is unfit if, in the opinion of the authority, it fails to meet one of the requirements set out in paragraphs (a) to (i) of Section 604 (1) and, by reason of that failure, is not reasonably suitable for occupation. The requirements constitute the minimum deemed necessary for a dwelling house (including a house in multiple occupation) to be fit for human habitation'.

Tolerable standard

According to the Scottish Executive (2003) 'The Tolerable Standard (which is equivalent to the Fitness Standard in England) as amended by the Housing (Scotland) Act 2001 was introduced in the 1969 Housing (Scotland) Act following recommendations made in the 1967 Cullingworth Report. Other than the incorporation of the "basic/standard amenities" (listed above) by the Housing (Scotland) Act 2001, it has remained largely unchanged.'

The Scottish Executive (2003) 'emphasizes that the standard is not intended to be a measure of acceptable housing conditions. It is distinct from the Building Regulations for example, which provide minimum standards for new construction and reflect modern expectations of the facilities and amenities to be provided in modern homes. The standard sets the base line below which houses should not be allowed to continue in occupation.'

A house meets the tolerable standard for the purposes of this Act according to the Scottish Executive (2003) if it:

- is structurally stable;
- is substantially free from rising or penetrating damp;
- has satisfactory provision for natural and artificial lighting, for ventilation and for heating;
- has an adequate piped supply of wholesome water available within the house;
- has a sink provided with a satisfactory supply of both hot and cold water within the house;
- has a water closet available for the exclusive use of the occupants of the house and suitably located within the house;
- has a fixed bath or shower and a wash-hand basin, each provided with a satisfactory supply of both hot and cold water and suitably located within the house;
- has an effective system for the drainage and disposal of foul and surface water;
- has satisfactory facilities for the cooking of food within the house;
- has satisfactory access to all external doors and outbuildings;
- any reference to a house not meeting the tolerable standard or being brought up to the tolerable standard shall be construed accordingly.

Decent homes and quality housing initiatives

It is the UK Government's aim that 'By 2010, bring all social housing (in England) into decent condition with most of the improvement taking place in deprived areas, and increase the proportion of private housing in decent condition occupied by vulnerable groups.'

A home is classified as decent (ODPM, 2005) if it:

- meets the current statutory minimum standard;
- is in reasonable repair;
- has reasonably modern facilities and services;
- provides a reasonable degree of thermal comfort.

Similar schemes are being proposed for Wales, Northern Ireland and Scotland. For example, the Scottish Executive (2003) set out proposals for a national standard based on a minimum set of quality measures for all houses in the social rented sector in Scotland. In February 2004 the Minister for Communities launched The Scottish Housing Quality Standard (here referred to as 'the Standard'). The announcement set out a range of measures that LA and RSL stock had to reach by March 2015 and required all social landlords to draw up Standard Delivery Plans (SDPs) to show how they were going to reach that target. This is similar to the Decent Home Strategy for social housing in England.

In Scotland the Government's intention has been 'to define a standard that is relevant to the 21st century and is consistent with views on what constitutes acceptable, good quality housing. It differs from the statutory Tolerable Standard (which is considered a very basic standard of acceptability) and the Building Standards as they apply to new housing' (Scottish Executive, 2003).

As initially proposed by the Scottish Executive (2003) the Standard is based on a number of broad quality criteria. To meet the Standard the house must be:

- compliant with the tolerable standard (as described above);
- free from serious disrepair (such as dilapidation or structural instability);

- energy efficient (i.e. has a National Home Energy Rating [NHER] of at least 5);
- provided with modern facilities and services (e.g. indoor WC, disabled access, etc.);
- healthy, safe and secure (e.g. contains no faulty electrical or gas installations, and is free from mould).

The new Housing legislation proposed by the Government will require sellers of dwellings to supply a standard set of information referred to as a 'Home Information Pack' (HIP) (Noy and Douglas, 2005). The HIP is deemed necessary for three main reasons (ODPM, 2003c):

1. To improve the home purchase process.
2. To enable compliance with EU Directive 2002/91/EC, Article six of which requires that Member States ensure an energy performance certificate is made available to the prospective buyer/tenant when buildings are constructed, sold or let.
3. To improve household energy efficiency.

Non-traditional housing

Origins

The 20th century saw a variety of major developments in the provision of housing in Britain. The impact of the two World Wars on the demand for and supply of housing was considerable. The existing stock of traditional dwellings in those days was unable to meet the housing needs of the country. Moreover, the rate of replacement of the stock by traditional means was slow because of the industry's lack of capacity. This was primarily caused by shortages both of skilled labour and traditional materials (especially timber) resulting from the two World Wars of the 20th century.

As a result, the government embarked on a programme of non-traditional housing. (See Glossary for a definition of this and other related terms such as 'traditional', 'industrialized' and 'system' forms of building.) This new type of residential construction was aimed at replacing the war damaged and slum quality stock quicker and more cheaply. It was also designed to help tackle the problem of overcrowding caused by a shortage of suitable housing.

Non-traditional construction was developed in three phases during the 20th century. These are summarized in Table 8.3.

Characteristics of non-traditional housing

Non-traditional construction for residential buildings has the following main characteristics:

- Modern techniques that rely heavily on the latest technology by maximizing the use of prefabrication and mechanization in the construction process. Ostensibly, these can produce better quality products under factory control conditions.
- It uses innovative design and materials: composites such as insulation-backed plasterboard, and lightweight metals such as aluminium for the framework of 'Prefab' dwellings.
- Framed construction obviates the need for primary load-bearing masonry walls. Relatively thin (i.e. 50 mm) precast concrete panels were often used for the external walls of proprietary single and two-storey dwellings such as Dorran and Tarran (Scottish Executive Building Directorate (SEBD), 2001). These would be finished internally with plasterboard on timber battens. The original window frames were either casement timber or mild steel.
- Unframed methods such as no-fines concrete (see below) or pre-cast concrete blocks are reasonably fast to construct and reasonably thermally efficient.

- Much of its construction process requires only semi-skilled labour on and off site.
- It offers faster speed of construction than conventional methods because of pre-assembly and standardization. Pitched roofs, for example, usually consisted of light steel or timber trusses covered with felt, battens and slates or corrugated steel sheeting. The typical construction of flat roofs was pre-cast concrete planks covered with a three-ply felt system.
- It maximizes the advantages of using modern technology – quicker production and consistency in product quality.
- Unusual features such as Queen closures at corner junctions.

The guides by the NCHA (1988) and SEBD (2001) describe and illustrate a wide range of typical examples of non-traditional housing. For detailed guidance on the identification of and defects in system-built dwellings and other forms of non-traditional construction, however, the reader should consult the extensive CD-ROM package of reports produced by the Building Research Establishment (BRE, 2002).

No-fines concrete housing

Preamble

This form of construction is a non-proprietary material that is used as the solid wall construction of dwellings (BRE leaflets BR 156 and 160, 1989). As its name indicates no sand or other 'fine' aggregate are used in the mix. It uses only cement and course aggregate (such as clinker) mixed with potable water. This mix produces a cellular reinforced concrete with relatively large voids uniformly spread throughout its mass (SEBD, 2001).

No-fines concrete was employed for housing construction in the late 1940s as a response to the shortage of bricks and bricklayers in the aftermath of World War II. Weir and Wimpey were two of the main British contractors who developed no-fines construction for dwellings, and implemented particularly in Scotland during the inter war period (SEBD, 2001). By the 1960s it was also being exploited as an infill-walling component to the in-situ reinforced concrete frame of some multi-storey blocks of flats. Since then it has been mainly used as the principal cast in-situ walling element in low-rise non-traditional housing built by organizations such as the Scottish Special Housing Association (now called 'Communities Scotland').

The solid walls of no-fines construction are generally between 200 and 225 mm thick (BRE leaflet BR 160, 1989). They are rendered externally with a two-coat render and usually finished internally with plasterboard fixed to timber battens attached to the no-fines walling using cut-nails. Up until the late 1960s this form of walling was considered to be relatively efficient in thermal terms. With increasing energy efficiency demands following the Oil Crisis of the early 1970s, however, many of these original no-fines dwellings are now considered thermally inefficient (see typical problems listed below). Some form of rendered 'raincoat' overcladding system to the no-fines blocks would be needed to rectify this deficiency (see Chapter 9).

Problems

Noy and Douglas (2005) highlighted the main problems that are often encountered in refurbishing no-fines concrete dwellings (see also Reeves and Martin, 1989; Williams and Ward, 1991). These can be summarized as follows:

- lack of energy efficiency,
- condensation and mould,
- rainwater penetration,

- rotting windows and external doors,
- deterioration of external render,
- drilling and fixing problems,
- plan modification difficulties,
- bulges in brick cladding,
- flat roofs and other inappropriate forms.

Improvements to non-traditional housing

Despite its attractions non-traditional construction has its drawbacks. These have been well documented elsewhere (Chandler, 1991). Suffice it to say, even today many dwellings of non-traditional construction that have not been demolished now require substantial upgrading.

The main modernization measures undertaken on non-traditional housing are summarized as follows:

- Renewing or over-roofing existing flat roofs (see Chapter 9).
- Overcladding of existing walls and pitched roofs (see below and Chapter 9).
- Fire-stopping of external wall cavities (at every storey and 10 m horizontally) and where services penetrate through floors, etc.
- Repairs to concrete elements affected by cracking, spalling, and corroded reinforcement – mainly caused by carbonation and lack of cover (see Chapter 7).
- Upgrading the fire resistance and acoustic performance of the upper floors.
- Double glazing and high quality timber or uPVC units for windows and external doors.
- Replacement of warm air and other inefficient domestic heating systems with low-energy heating systems (see Chapter 9).
- Installing solar energy and other energy efficiency measures in refurbishment (see Chapter 11).

Indoor air quality

Preamble

It is generally accepted that people in the developed world spend the vast majority of their time indoors – between 80 and 90 per cent. Moreover, ironically, the indoor environment is now in some cases considered to be more polluted than outdoors (Maroni et al., 1995). This means that in the West especially building users may be being exposed to unhealthy environmental conditions because of potentially dangerous levels of airborne contaminants.

There are a variety of factors that have made buildings nowadays more susceptible to indoor air pollution. The increasing use of double-glazing, draught proofing and insulation coupled with the reduction of open-hearth fireplaces in houses has created a more sealed indoor environment. All of this has resulted in a drastic reduction in the natural breathability of our homes and offices. As a result air trapped inside a building easily becomes stale and humid, which brings with it a whole host of problems. For example, if left unchecked contaminants such as dust, pollen, gases, smoke, fumes, smells, moisture, mould, toxins, airborne grease can all increase to harmful levels. It is also now well established that the house dust mite, which thrives in such conditions, is a major cause of asthma, eczema and allergies, especially in children (Singh, 1993).

Moreover, the incidence of asthma among teenagers in the UK (particularly Scotland) is the worst in the world, according to a report by the Global Initiative for Asthma in February 2004. Although no one as yet has been unable to pinpoint the cause of this problem, experts have identified a number of possible different reasons for asthma gripping western countries. For example, central heating, wall-to-wall carpeting as well as genetic predisposition and lifestyle are linked to the disease. As indicated earlier, however, asthma can be exacerbated in a water-damaged building.

Adaptation schemes, therefore, should give due consideration to indoor air quality. The primary goal is to ensure that the comfort conditions in the internal environment in the adapted building are kept at an acceptable level if not improved. If there has been a problem with the air quality in the building under consideration, this must be taken into account in any adaptation proposal. The principles relating to indoor air quality outlined in this Chapter apply equally to commercial and residential buildings.

There are many sources of indoor pollution: people, building materials, furnishings and coverings (e.g. especially thick piled carpets, which also act as niches for airborne contaminants), cleaning materials, food preparation, computer printer or photocopier use, hobbies and DIY work, and even ventilation systems themselves (Oseland and Raw, 1993). To minimize the adverse effects of these in any adaptation scheme attention should be given to the required:

- type and intensity of building use;
- ventilation rates (in l/s);
- air change rates (in air change rate per hour (ach));
- humidity range (in relative humidity (RH) percentages);
- temperature range (in degrees Celsius);
- air quality sensors (for measuring CO_2 in the indoor environment to highlight the need for increased ventilation when concentrations exceed a certain level).

The recommended rates and ranges for these comfort conditions are outline below.

Comfort factors

Generally

Determining suitable air quality levels is very complex because of the number of variables involved. In other words, there are several factors that affect indoor air quality. Table 8.4 summarizes the main influences in this area.

Although there are no universally agreed safety limits, the guidelines established by the World Health Organization (WHO) offer a good starting point. Table 8.5 lists some typical air quality guidelines.

Moisture

One of the most common influences in this context, however, is moisture. This is an important factor in adaptation work as well as in new-build schemes. It also acts as a catalyst for all sorts of problems such as mould growth, fungal and insect attack and water damage to the building's structure and fabric. These are dealt with in more detail below.

Temperature

The ideal temperature range in most buildings is 18°C to 22°C. Low temperatures in winter periods (i.e. $<$16°C) can lead to cold-related illnesses (i.e. hypothermia) and condensation. High temperatures increase the risk of bacterial and fungal infections as well as heat stroke within a building (i.e. hyperthermia). Moreover, buildings with warm ($>$21°C), dry ($<$30%) indoor environments can give rise to discomfort problems for their occupants. In such conditions the risk of electrostatic shocks and respiratory tract irritations is higher.

Table 8.4 Primary factors affecting indoor air quality

Factor	*Examples*	*Problems*
Moisture	Dampness from above ground sources – leaking gutters, downpipes and water supply, rain and snow. Dampness from below ground sources – ground water and leaking drains and supply pipes.	Soiling and softening of materials. Efflorescence. Corrosion of metals. Encourages the development of biocontaminants – with their subsequent adverse building and health effects. Reduces efficiency of thermal insulation.
	High internal humidity levels (i.e. >70%).	Biocontaminants or Microorganisms.
Biocontaminants or microorganisms	Fungi (moulds, dry rot and wet rots) Woodworm.	Bio-decay of constructional, finishing and furnishing materials. Fungi and moulds emit MVOCs. Organic toxic dust syndrome in buildings containing extensive fungal contamination.
	Bugs (House dust mites, cockroaches, etc.) Moulds and Pollen (microspores) Bacteria. Viruses.	Allergic reactions causing health problems such asthma and rhinitis. Increased risk of health problems usually of a respiratory nature, particularly in children and the elderly. Anthrax poisoning resulting from a terrorist attack.
Gases	Sulphur dioxide (SO_2). Carbon monoxide/dioxide. Nitrogen dioxide. Methane. Radon (releases a form of ionising radiation – see below).	Gaseous pollutant emissions from combustible appliances or from the subsoil can cause discomfort or even fatality if in sufficient concentrations. SO_2 can cause deterioration of parchment and other fragile materials. Excessive levels of gases such as carbon dioxide can adversely affect human performance. Unpleasant odours from noxious or unhygienic conditions or from people. Carcinogenic properties.
Volatile organic compounds (VOCs)	Various aromatic hydrocarbons in paints and related products. Alcohols, ethers and esters in consumer and commercial cleaning products. Ketones and various hydrocarbons in pesticides. Formaldehyde in cavity insulation.	Building materials consumer products, furnishings, pesticides and fuels release VOCs into indoor air. These substances can exhibit toxic, irritant and unpleasant odourant properties, and may be carcinogenic.
Radiation	Electromagnetic (non-ionizing) radiation (EMR). Extremely low-frequency EMR – from power lines. High-frequency EMR – from radar, microwaves and other electrical appliances.	Tissue and cellular damage if in sufficient concentrations. Possible cancer risks – but these have not been proven conclusively.

(*Contd*)

Table 8.4 (*Contd*)

Factor	*Examples*	*Problems*
Tobacco smoke	Cigarette smoking. Cigar smoking. Tobacco burning products (e.g. pipes).	It is a source of ammonia and many other toxine Nicotine staining on wall and ceiling surfaces. Lung cancer – even from passive smoking – and heart disease.
Other physical pollutants	Asbestos and man made mineral fibres (MMMF). Particulate matter (e.g. non-organic dust, silica dust, soot from combustible appliances and radiators). Vegetable/Organic dusts (e.g. grain, flour, wood dusts). Danders from domestic animals and parasites (e.g. dust mites).	Adverse heath effects – lung disease, possibly carcinogenic. Staining or soiling of wall and ceiling surfaces (e.g. pattern staining). Dust inhalation leading to respiratory problems. Lung disease. Allergens causing allergic reactions such as asthma, rhinitis, etc.

Table 8.5 Typical methods of measuring air-quality limits

Contaminant	*Measurement*	*Examples of reference levels (1)*
Fungi/Moulds	CFU/m³ air (2)	<200 CFU/m³ air is considered low for homes.
Bacteria	CFU/m³ air (2)	<500 CFU/m³ air is considered low for homes.
Radon	Bq/m³	200 Bq/m³ in Britain.
Gases	mg/m³ (3)	Acceptable long-term exposure range (ALTER) is ≤63 000 mg/m³ (≤3500 ppm) (4) for carbon dioxide. ≤63 000 mg/m³ (>3500 ppm) for carbon monoxide.
	μg/m³ (5)	<30 mg/m³ for carbon monoxide over 15 min average time. 400 μg/m³ for nitrogen dioxide over 1 h average time. Acceptable short-term exposure range (ASTER) for carbon dioxide is 1000 ppm.
Chemicals	mg/m³	0.3 mg/m³ for total VOCs having specific limits (i.e. excluding formaldehyde).
	μg/m³	60 μg/m³ for formaldehyde (0.05 ppm).
	ng/m³ (6)	10–20 ng/m³ for cadmium over an average of 1 year in urban areas.
Fibres	F/m³ (7)	No definite limit, but 40–200 F/m³ of air is a typical range for MMMF
	F/m³ (8)	0.01 F/ml for asbestos.
Dust	μg/m³	Acceptable long-term exposure range (ALTER) is ≤40 μg/m³ for fine particulate matter. Acceptable short-term exposure range (ASTER) is 100 μg/m³ – 1 h concentration.
	μm	Dust particles <2.5 μm in diameter.

Notes
1. Based primarily on WHO figures, which are subject to continual improvement and modification.
2. Colony forming units per cubic metre.
3. mg/m³: milligrams per cubic metre.
4. ppm: parts per million.
5. μg/m³: micrograms per cubic metre.
6. ng/m³: nanograms per cubic metre.
7. F/m³: Fibres per cubic metre.
8. F/ml: Fibres per millilitre.

Mould

There have been increasing health concerns in both the UK and the USA relating to mould (Committee on Damp Indoor Spaces & Health (2004). It occurs mostly in parts of a building where the moisture content is above 70 per cent RH, such as bathrooms. According to Reyers (2002) the health problems associated with moulds are:

- Immunological reactions: Exacerbate asthma, bronchitis, hay fever, etc.
- Toxic effects: See problems associated with Sick Building Syndrome (SBS) below.
- Infectious diseases: Rarer, but ailments such as *Aspergillosis* resulting from high concentrations of mould in buildings affected by bird droppings might occur in some cases.

Odour

Pungent or unpleasant smells in a building can be a warning that the air is unhealthy or that there is a problem with the building's indoor environment or drainage. Foul odours are often indicative of a build up of stale or bad air, often resulting from inadequate or inefficient ventilation. High carbon dioxide concentrations in the air (e.g. $>0.25\%$) are linked to bad odour conditions (Wanner, 1981). Strong smells in buildings can emanate from people, equipment, vegetation, materials, working processes such as painting and gluing, cooking and food preparation. However, tobacco smoke and body odour are the main sources of everyday foul smells in buildings, especially in commercial properties. Now that smoking is banned in many public and office buildings, of course, tobacco smoke should no longer pose a general problem in this regard.

Defects such as dry rot, dampness and faulty drains or defective sanitation can also of course trigger unpleasant odours in a building. Dry rot, for example, gives rise to a distinct mushroomy smell (Ridout, 1999). Mildew caused by condensation can also emit pungent mould spores. These micro-organisms emit microbial volatile organic compounds (MVOCs), which in turn can trigger allergic reactions and other respiratory problems in vulnerable people.

A broken or leaking foul drainpipe or manhole will inevitably lead to a distinct smell of sewage. Faulty or blocked interceptor traps and WC S-bends can also cause foul odours indoors. The resultant malodour is repulsive as well as hazardous to health. In extreme cases defective foul drainage can lead to cholera, which is a well-known health hazard associated with inadequate sanitation.

Therefore, the importance of odour as a factor affecting comfort should also not be overlooked. This has led to the use of two relatively new units of air quality: the 'olf' and the 'decipol'. Both of these rely on trained human beings as the detector to assess air pollution. One decipol is the pollution caused by one standard person (one olf) ventilated by 10 l/s of unpolluted air (a decipol is one-tenth of a pol).

The olf and decipol help achieve a useful but crude measure of air quality by assessing building users perception of indoor pollution. It is therefore highly dependent on the precision of the judgements made by individual subjects, and the number of subjects used (Oseland and Raw, 1993).

ASHRAE (1989) has recommended a decipol level of 1.4 ($=20\%$ dissatisfied) for some commercial premises. This can then be used to determine ventilation rates. For example, 1.4 decipol roughly equates to 7 l/s for the required outdoor airflow. However, this research in still in its infancy. According to Maroni et al. (1995) 'too little is known at present about the intensity of sources indoors, measured in olf, for the olf theory, as a whole, to be applied'.

Adaptation work can involve the installation of self-levelling screeds containing ammonia, or floor coverings using solvent based adhesives. This can easily lead to high concentrations of these substances in the indoor air.

Ammonia, for example, is a colourless gas with a strong characteristic smell. It is corrosive and an irritant, and can also be classed as a Volatile organic compounds (VOC). If the use of such a product in an adaptation scheme were unavoidable, the building would need to be thoroughly ventilated during and immediately after the work to avoid any obnoxious or toxic off-gassing in the affected and adjoining rooms.

'Off-gassing' is the steady release of fumes into the indoor environment. It can come from VOCs as well as from electrical equipment, furnishings and fittings.

Noxious odourless gases

Not all harmful gases in buildings emit noticeable odours. Carbon monoxide (see later under Gas Installations) and carbon dioxide (CO_2) are two odourless, tasteless and invisible gases found in every building. The former is emitted from faulty gas-fired appliances such as heaters and boilers. All humans of course emit CO_2 as a natural by-product of exhalation.

Traditionally the primary way in which air quality has been measured is in terms of CO_2 level. Haddlesey (2002) reported that research has shown that people's mental performance begins declining at levels over 1000 ppm. Inadequate ventilation is the primary cause of this.

Unhygienic atmospheric conditions and malodours are more prevalent when CO_2 levels are high. Concentration levels and illness rates in other words are directly related to indoor air quality.

Ventilation

Background ventilation must be provided even where draught-proofing measures have been implemented. It will help combat the build-up of noxious odours and gases as well as combat mould growth and help remove allergens and other dust-borne organic particles.

Trickle vents in windows can help achieve the 1 ach target in most rooms. Bathrooms and toilets should have a ventilation rate of around 3 ach. Light-switch operated mechanical extractor fans operating at a minimum capacity of 15 l/s for bathrooms/toilets and 60 l/s for kitchens (30 l/s if there is a cooker hood) are required.

Indoor air quality guidelines

At a basic level indoor air quality is usually measured in terms of concentrations of contaminants per unit volume of air or per unit mass. Table 8.5 summarizes some of the main measures that are used with an indication of recommended exposure levels. However, it must be emphasized that these guideline values are selective and only indicative. They can vary from building to building as well from country to country.

The guidelines highlighted in Table 8.5 have some use in adaptation work. They can provide a benchmark for converted or refurbished buildings where air quality is of vital importance – such as offices, hospitals, nursing homes and other public premises.

Moisture control

General measures

Moisture control is critical for many aspects of good indoor air quality (see ASTM, 1994). It can be maintained if not improved in an adapted property by adopting an appropriate control strategy. The strategy to be employed depends upon the macroclimate in which the building is located. Buildings in cold climates require different moisture control measures to those in warm climates (see below).

ISIAQ (1996) have, however, provided the following useful guidance on methods of moisture control that can apply to adapted buildings generally:

- RH of the indoor air should be maintained at a suitable level for the comfort of the occupants. Between 40 and 55 per cent RH is considered to be the optimum range (Maroni et al., 1995). If too low ($<30\%$) this will increase the incidence of skin irritation, dry eyes and throats as well as encourage electrostatic shocks (especially if carpeting or furnishings containing synthetic fabrics have been used in the refurbished building). If too high (>65%) this will encourage corrosion in metals, mould growth on wall and ceiling surfaces, and house dust mite infestation. Timbers exposed to RH levels over 85 per cent are at risk of deterioration by fungal attack.

- Elimination of excess moisture (and nutrients such as dirt, dust and other soiling materials) by adequate ventilation.
- Elimination of sites of water accumulation in or on the adapted building respectively by sufficient drain sumps (i.e. to prevent ponding or pools of water on floors in shower rooms and tank rooms) or adequate outlets and overflows in gutters (i.e. to prevent flooding by rainwater ingress).
- Ensure that wet or moist materials used in the adaptation scheme such as concrete screeds, plasters, mortars and saturated wood are allowed to dry thoroughly. This reduces the risk of construction water acting as a moisture source.
- Where envelope surfaces are below ground:
  - grading ground levels away from envelope surfaces;
  - installing free draining material next to the envelope surface and linking this material to a drainage system through a filter medium that prevents fines building up in the drains;
  - covering the free-draining material next to the envelope with a clay cap or other surface layer resistant to water flow, so that surface water does not drain into the material;
  - coating the envelope with a damp-proofing layer that provides an effective capillary break between it and the adjacent material.
- Installing continuous impermeable ground cover in sub-floor crawlspaces to act as an air barrier and vapour diffusion layer.
- Having a clear plan of action in the maintenance regime of the adapted building for dealing with water spills, leaks and other such contingencies. This requires the facilitation of a preventative maintenance system to identify potential sites of water accumulation accessible for inspection and service.
- Avoid using moisture-sensitive materials in locations where there are regularly high humid conditions (e.g. shower rooms). Rather, use moisture-resisting materials in such environments.

Cold climate moisture control measures

Regions such as the UK and the upper part of North America are typical cold climate areas. Buildings in these parts of the world need to have what is known as 'warming climates' (Ltsiburek and Carmody, 1993).

The following actions can lessen the likelihood of moisture/microbial problems in cold climates (ISIAQ, 1996):

- Reduce moisture input to the atmosphere by activities such as clothes washing and avoidance of drying clothes indoors, and change away from the use of portable kerosene and low-pressure gas heaters.
- Prevent entry of groundwater by providing damp-proofing as indicated above.
- Prevent wind-drive snow and rain from entering the building envelope.
- Provide envelope cavity drainage paths to the outside for condensate and leakage.
- Prevent internal surfaces of external walls from becoming too cold. This can be achieved by providing better insulation and better thermal resistance in the envelope.
- Reduce the entry of room moisture into the envelope. This may be accomplished by placing vapour diffusion retarders (vapour control layers) correctly, that is, towards the warm (humid) interior of the envelope assembly and not towards the cold exterior.
- Ensure heating system is efficient and controllable.

The ISIAQ (1996) have also indicted that in existing properties a change of mode of operation in the building is often the only practicable way of overcoming the problem of mould growth in the envelope.

The following additional strategies, along with reduction of humidity by source control, may be required in adapted buildings:

- ventilating the envelope cavity with air of relatively low humidity (i.e. $<60\%$);
- depressurizing the envelope cavity (e.g. by incorporating weepholes at 900 mm centres in the perpends of external skin of brickwork);
- isolating mould-susceptible cavity materials from any potential contact with cavity moisture;
- pressurizing the building (e.g. by using PIV);
- using dehumidifiers but not humidifiers;
- ventilating the building with air of low humidity (i.e. 40–50%).

Warm climate moisture control measures

The subtropics and tropics such as Singapore and Australia and in southern parts of the USA can be classed as warm humid climates. Thus buildings in these regions require 'cooling climates'.

The following actions can reduce the likelihood of moisture/microbial problems in buildings climates (ISIAQ, 1996):

- Avoid overcooling of internal spaces, especially the internal surfaces of external walls.
- Avoid overcooling which occurs when cold, air-conditioned air is directed against localized room surfaces.
- Dehumidify outdoor air prior to its introduction into the building. Air-conditioning systems should be of correct size for adequate dehumidification capacity. Prevent indoor RH from consistently exceeding 65 per cent.
- Maintain a net positive pressure by supplying outdoor air at a rate that exceeds the rate at which air is mechanically exhausted. Make the building envelope as tight as possible so that it is easy to maintain a positive pressurization. In this regard, an air barrier system should be installed typically toward the exterior of the envelope.
- Install a vapour diffusion retarder typically toward the exterior of the envelope. Avoid the use of low permeance materials on internal surfaces of the envelope. If the building envelope gets wet, it should be able to dry out toward the interior.
- Prevent moisture associated with rain from entering the building envelope. Providing robust water-shedding/waterproofing coverings along with adequate rainwater disposal goods minimizes this risk.
- Provide adequate insulation on air supply ducts and chilled water piping to prevent condensation.
- Avoid using wet or moist construction or finishing materials. If any materials are wet or moist, dry them adequately so that moisture is not trapped in wall assemblies.

Unhealthy buildings

Health effects

A key sustainability goal of any construction project is good indoor air quality (Andersson and Setterwall, 1996). Any adaptation scheme therefore should overcome or prevent adverse physiological and psychological reactions in the users of the building in question. If the existing property has had a bad reputation in this regard, this should alert the adaptation team to ensuring that the proposed scheme neither repeats nor intensifies the problem. Even if the building never had any such troubles, a poorly designed and executed adaptation scheme could inadvertently trigger health problems in its occupants.

'SBS' and 'BRI', are the primary examples of how a property can have an adverse impact on the health of its occupants. These problems have tended to manifest themselves predominantly in air-conditioned buildings built or adapted after the 1960s. However, potentially they can occur in any new or recently adapted building if certain circumstances are present – such as inferior comfort conditions, contaminated water supply pipes, poor air quality, inadequate environmental controls, inappropriate materials used in the construction, and poor siting of noisy or disruptive building activities and equipment (e.g. plant room) next to habitable areas.

Definition of SBS

SBS is a term used to describe adverse medical symptoms with an unclear aetiology (i.e. non-specific cause/s), but with a possible relationship to the indoor environment (Maroni et al., 1995). Not everyone that works in a building with an alleged SBS problem, however, is necessarily affected. It is very much related to the type of work activities undertaken in the property. If for example, the workers are highly stressed because of the demands of their job, they are more likely to suffer the consequences of SBS. Indeed, it is important to recognize that SBS is essentially a 'people problem' (Rostron, 1996).

SBS, therefore, being a contributory factor is likely to exacerbate poor health conditions in the building's workforce. It presents a variety of non-specific symptoms such as eye, skin, nose and throat irritations, headache and fatigue. Indeed, ongoing research into SBS has not conclusively identified any direct cause and effect relationship. It is caused by a subtle combination and interaction between several variables: the quality of the indoor environment, the extent of services within the building, the layout and shape of the workspace, and the type of work being undertaken.

Definition of BRI

BRI on the other hand, was a term originally referring to illness brought on by exposure to the building air, where symptoms of a diagnosable illness are identified and can be directly attributed to environmental agents in the air (Maroni et al., 1995). The aetiologies (i.e. causes) of these diseases, in other words, are usually known or identifiable.

However, BRI can also be caused by water-borne contaminants found in HVAC (heating, ventilating and air-conditioning) and domestic water systems. Harmful bacteria such as Pseudomonas (see below) can result from biofilm growths in pipework. In adhering to the inside of pipes or vessels biofilms are also instrumental in promoting corroding metal pipework.

The following is a summary of some examples of the health problems associated with BRI:

- nauseousness or irritation associated with the application of finishings such as oil-based paints or adhesives containing VOCs (e.g. aromatic hydrocarbons);
- humidifier fever or Legionnaires Disease from defective or poorly maintained HVAC plant and cooling towers;
- asthma triggered by allergen sources such as house dust mites and pets – especially in damp residential properties containing wide use of fabrics such as heavy curtains and thick pile carpeting;
- other allergic reactions such as rhinitis (i.e. inflammation of the nasal mucous membrane) as a result of bio-aerosols emitted from contaminated air-conditioning plant;
- Pseudomonas bacteria can increase the risk of eye, respiratory and skin infections in vulnerable people (e.g. the sick, elderly, children) if their immune system is compromised.

Other control measures

General indoor air control measures

Naturally moisture control is not the only measure that needs to be taken to achieve a satisfactory indoor air quality in an adapted building. The other factors that are relevant in this regard are:

- The ideal ambient temperature range within a building is between 16°C and 19°C. However, if the work undertaken inside the adapted building is primarily of a sedentary nature, the required temperature may have to be increased to 21°C, especially for older people.

- The ach should generally be between 0.5 and 3 ach, depending on the use of the facility.
- Install hard-wired detectors to monitor the presence of carbon monoxide and other harmful gases, particularly near combustible appliances.
- Ban smoking in public areas within the building and severely restrict it elsewhere around the property.
- Naturally ventilate the adapted building before re-occupancy to reduce VOC emissions from new furnishings, paints, floor coverings, sealants and adhesives. The use of water-based as opposed to spirit-based paints and adhesives should be encouraged to minimize any noxious odours from these products. Manufacturers such as F. Ball & Co, for example, are now promoting ultra low-emission polymer and resin adhesives; Dulux Trade supply 'Aquatech Gloss', which is a water-based paint that avoids the use of white spirit and still provides the performance of a traditional solvent-based gloss.
- Minimize if not eliminate the use of VOC-containing cleaning materials for maintenance operations.
- Use non-odorous and non-irritating materials in any adaptation work.
- Consider the need to install ionizers with adequate dust filters in areas where electrical appliances are used extensively. This will replenish the negative ions in the air depleted by electrical fields from those appliances, and help to reduce occupants' lethargy.
- Install interior foliage plants to serve as bio-indicators and for absorbing carbon dioxide (i.e. act as carbon sinks) and VOCs.
- Air-conditioning ducts and humidifiers should be serviced regularly to ensure that they do not contain any significant fungi levels.
- Air-conditioning plant should have high quality filters (i.e., that can retain pollen as well as dust and other contaminants out of the incoming air) and ultraviolet microbe-killing lights. Electrostatic precipitators can be used for air dust and fibre control.
- Only 'HEPA' (high efficiency particulate arrestor) filters should be used in air-conditioning and vacuum cleaning where control of bacteria, mould and other airborne contaminants is required. A HEPA filter is a filter that removes 99.97 per cent of particulates 0.3 μm or larger in size (Thompson, 2005). A conventional vacuum cleaner removes particulates down to about 35 μm or larger in size. Particles below 10 μm are invisible to the human eye while the most common airborne particle size is 2.4 μm. In comparison other particulates have the following sizes:
  - Asbestos (fibres): 3–20 μm
  - Bacteria: 0.3–50 μm
  - Dust mite faeces: 0–24 μm
  - Human hair: 60–80 μm
  - Mould: 4+ μm
  - Pollen: 10–40 μm.

Control of SBS and BRI in adaptation schemes

SBS is more difficult to eliminate than BRI because the former is influenced by psychosomatic factors such as stress levels in the workforce and the type of work that they undertake. However, the following control measures in conjunction with the precautions indicated above can be taken to reduce the possibility of both SBS and BRI occurring in adapted buildings:

- For mechanically ventilated buildings a minimum fresh airflow of 8 l/s per person is recommended by CIBSE. Where heavy smoking is permitted, an airflow rate of 32 l/s per person is suggested. In many buildings nowadays, of course, a strict no-smoking policy is enforced. This not only lowers the level of noxious fumes within a building but also helps to reduce its fire risk.
- Air velocities should be in the region of 0.10–0.15 m/s, rising to 0.25 m/s, in the summer. If too low, this can create stuffiness in the building. If too high, this can cause a wind-chill effect as if there were draughts in the building.

- Intakes for air handling plant should be located in such a way that they do not draw air contaminated by the likes of traffic pollution or cooking smells.
- Alterations in the layout of a building can affect the efficiency of both naturally and mechanically ventilated buildings. Normally aspirated buildings should be designed so that natural air movement is not adversely affected.
- Lighting should, wherever possible, be designed to give individuals control, with an emphasis placed on maximizing natural daylight. Diffusers to reduce glare should be used on all light fittings.
- Air-conditioning should also be designed to give individuals a degree of control at their desks and other work locations.
- Forms of intrusive noise that can be avoided by careful design or re-routing services include the sound of air passing through diffusers or ductwork, water in pipes, and machinery such as lift motors or air-conditioning. Anti-vibration mounts and noise-dampers should be fitted to ducts, fans and other equipment. Ideally, the sound level from air-conditioning and other noise-generating parts of the environmental services should not exceed about 40 dBA.
- Office equipment, such as photocopiers and printers, should ideally be located in a closed room with a separate air extraction system.
- Furnishings such as desks and cabinets should have a non-glare finish.
- Blinds should be fitted where sunlight might increase the amount of glare entering the workspace. These also give the occupants a degree of control over their working environment. Alternatively, the building could be re-clad with electrochromic glazing or the glass coated with solar reflective film (see below).
- Minimize the use of heavy fabrics such as curtains and thick piled carpets to reduce niches for allergens, dust mites and other potentially harmful micro-organisms. The use of laminated high-density fibreboard flooring for rooms and corridors can help reduce dust niches in dwellings as well as offices.
- Ensuring that adequate moisture control measures to air-conditioning plant and cooling towers are undertaken can prevent Legionnaires' Disease. Ponding and the possibility of stagnant water sources should be monitored and eliminated when discovered.
- The minimum recommended hot water storage temperature is 55°C to avoid the risk of Legionnaires' Disease. The minimum temperature at draw point should be 45°C (Palmer and Rawlings, 2002).
- Cleaning and maintenance are two of the most important factors influencing SBS. Therefore, the maintenance regime must ensure that floor surfaces are cleaned daily, and filters are replaced regularly. This work may include cleaning or sterilizing air-conditioning ducts and electrostatic precipitators. Chlorination can be used to sterilize water supply pipelines and water reservoirs in HVAC plant and cooling towers – to reduce the risk of Legionnaires Disease and other bacteriological infections.

Condensation and other dampness control

Symptoms and problems of condensation

Condensation has been a problem in British housing for many years. The various house condition surveys (Research, Analysis and Evaluation Division, 2003, for England; House Condition Surveys Team, 2004, for Scotland) indicate that around 1 in 5 dwellings in this country is affected by this problem. Basically, it occurs when warm moist air hits a cold surface. Affected buildings have an internal environment with an equilibrium relative humidity (ERH) that is persistently or regularly above 70 per cent (Garratt and Nowak, 1991).

The primary symptom of condensation is, of course, mould growth on walls and ceilings, and sometimes on floors (on or under the floorcoverings). Not only is this unsightly, it is also detrimental to the health of the occupants and the surfaces affected (Communities Scotland, 2004a). It also encourages the development of harmful microorganisms. Moulds emit fungal spores into the atmosphere, which along with bacteria contain mycotoxins and reduce the quality of the indoor air (Committee on Damp Indoor Spaces & Health, 2004). People who inhale these pathogens in sufficient quantities for prolonged periods are more susceptible to respiratory problems (Singh, 1993; Committee on the Assessment of Asthma and Indoor Air (2000)).

Sweating on the inside surface of single glazing or other cold, smooth surfaces is another manifestation of condensation. This causes pools of water to form on the sills or other ledges. It is also unsightly and in the case of glazing can lead to deterioration of wood and metal window frames.

Condensation is the main source of the moisture needed for the growth of fungi on the internal surfaces of residential building (Maroni et al., 1995). For example, it can cause a build up of multiple specks of organic growths such as *Aspergillus, Penicillium* and *Cladosporium*, three of the most common genera of moulds found in UK dwellings (Oreszczyn and Pretlove, 1999). In the USA *Stachybotrys Chartarum* is one of the main fungi causing health problems associated with toxic mould such as cold-like symptoms, rashes, and aggravation of asthma (Reyers, 2002).

Mould growth occurs on walls and ceilings when surface temperature is at or below dewpoint and ERH levels are near 80 per cent or more. These organic growths can also appear on the frames and sills adjacent to the glass of single glazed windows.

Aspergillus versicolor, for example, is a common species of mould in British buildings. It may, of course, be considered academic from a building adaptation perspective whether the mould is from one genera or species or another as the action taken to abate the problem is usually the same. However, as indicated above some moulds have allergenic effects whilst others are toxic. From a medical viewpoint, therefore, it may be important to identify the type of fungus in a building as this will influence the clinical treatment required for the affected individuals.

Poor or inadequate heating is the main contributory factor of condensation in dwellings (Garratt and Nowak, 1991). Lack of ventilation, inadequate fabric insulation and excessive moisture generation exacerbate this problem. These factors help to explain why dampness is often related to poverty and substandard housing. Overcoming dampness therefore is a major goal in housing modernization schemes.

Traditional condensation remedies

Preamble

Condensation remedies often form an integral part of a housing or commercial modernization scheme. Essentially these can be classified into two groups: short-term and long-term remedies.

Short-term remedies

These organic staining problems can be alleviated by sterilization and dehumidification. Applying a proprietary fungicidal wash (such as Dulux's 'Fungiwash') with a stiff brush to clean off the mould growths is normally needed. However, one should specify an approved biocide rather than diluted bleach to remove mould from walls and other surfaces. According to one expert (Thompson, 2005) bleach should not be used to tackle mould for the following reasons:

- It's too diluted, and therefore too weak to permanently get rid of mould. (It merely changes the mould's colour and does not kill it completely.)
- Bleach loses its cleaning strength over time. In fact, it has a 50 per cent loss in cleaning power in just the first 90 days inside a never-opened container. Chlorine constantly escapes through the plastic walls of its containers.
- Bleach won't penetrate porous materials such as drywall and wood. This means that it can't get to the roots of mould.

A more effective treatment for dealing with mould infected surfaces requires the following measures:

- Identify and control/eliminate the source of moisture. This will require a thorough investigation followed by an adequate therapy (Burkinshaw and Parrett, 2003).
- If the mould growth is extensive use a 'fogger' or other steam cleaner with fungicidal fluid to treat areas affected by mould (Thompson, 2005). A fogger machine is slightly larger than an electric kettle. It has barrel-shaped nozzle about 75 mm in diameter from which the steam-laden fungicide is emitted. 'Shockwave' is a typical concentrated, EPA-registered, quaternary ammonium chloride cleaner that disinfects hard, nonporous surfaces, and also sanitizes porous and semi-porous surfaces subject to microbial contamination. The fungicide should be sprayed on mouldy surfaces and mouldy building materials.
- Apply two coats of an appropriate anti-fungus paint to the affected wall and ceiling surfaces. 'AfterShock' is one such treatment designed to provide a durable, 100 per cent acrylic white coating that can be applied to common building materials including wood, OSB, drywall, concrete, and metal surfaces. It's ready-to-use, and requires no dilution. Five litres sprayed wet will cover about 20 m^2 of treatment area. According to the manufacturer:

> 'AfterShock' is an EPA-registered antimicrobial coating designed to kill residual mould and mildew remaining after pre-cleaning contaminated surfaces. 'AfterShock' also inhibits the future growth and spread of mould and mildew on the cured film surface in residential and institutional buildings. This extremely durable, easy to apply, 100% acrylic sealant offers the ultimate in durability in combination with excellent fungicidal characteristics to address mould growth from water damage. 'AfterShock' is recommended for use on interior wall surfaces such as plaster, wallboard, drywall, concrete, masonry block, wood, primed metal and galvanized metal. 'AfterShock' is also recommended for use on interior wood framing, primed metal, concrete, and wallboard inside the wall cavity. Do not use for HVAC system applications.

Along with sterilization of surfaces some form of dehumidification in the affected areas is required especially if the mould condensation is severe and using a dehumidifier to help dry out the wet indoor atmosphere. The aim is to get the ERH to between 50 and 60 per cent.

Modern dehumidifiers are reasonably quiet and economical to run. These units, however, take up space, consume energy and, unless plumbed in, require regular emptying of the collected water. They should only be seen as a means of alleviating rather than curing condensation problems.

Therefore, if the underlying causes of condensation are not addressed, short-term measures are likely to prove either ineffective or very costly to sustain. Usually, the symptoms mentioned above will return within a few weeks, depending on the extent of dehumidifier use, the prevailing ambient conditions and the time of year.

Long-term remedies

The conventional long-term strategy for overcoming condensation in housing and offices adopts a variety of measures. Each of the following used on its own is unlikely to cure a condensation problem:

- Heating: Upgrade the heating system (e.g. replacing small panel electric radiators with suitable size radiators placed in appropriate positions, such as below windows, and fed by small-bore hot water pipes) to ensure that the appropriate temperature levels are maintained (i.e. 20–23°C).
- Insulation: Improve the thermal insulation of the fabric (e.g. by installing loft insulation 200 mm thick with appropriate background ventilation, and/or cavity-fill insulation to the walls). This is required to 'warm up' the fabric so that its surface temperature is well above dewpoint.
- Ventilation: Increase the ventilation in key rooms such as bathrooms (see Table 3.6). An ach of 0.5–1.5 is the ideal range for dwellings (Garratt and Nowak, 1991). However, a higher ach range will be required for some non-residential premises where smoking is permitted. Over-ventilation, of course, is not energy efficient and may make matters worse by bringing in cold wet air in winter months.

- Education: Educate the occupants: – not to close off vents (because of the mistaken belief that by doing so heat losses are greatly reduced);
  - not to minimize the level or duration of heating/ventilation;
  - not to take off kitchen doors (for convenience, but will allow warm moist air to travel to colder parts of the dwelling);
  - not use inappropriate heating appliances (e.g. paraffin or portable gas heaters – which emit about 1 l of water for every litre of fuel consumed).

It is important to recognize that each of the above measures on its own is unlikely to cure a condensation problem. Condensation requires a subtle combination of improved heating, ventilation and insulation to ensure its control or eradication in a building. Even then, the co-operation of occupiers is required to ensure that they do not undertake any of the measures described in the last points in the above list.

Moreover, in some cases these conventional remedies will not solve endemic condensation in certain forms of housing. Medium and high-rise flats in deck access blocks or system-built construction, especially, are very prone to this problem. Inappropriate or sub-standard construction coupled with poor levels of heating and ventilation are the main contributory factors condensation in these properties. That is why innovative techniques such as PIV offer a solution to the condensation problem in dwellings.

PIV

This form of ventilation (sometimes called 'Positive Pressure Ventilation') is one of the most innovative method for dealing with condensation in dwellings. It has become so successful in many housing schemes that it is likely to be the main way of tackling this form of dampness for the foreseeable future.

The hardware used in the system is fairly straightforward. It consists of a high efficiency centrifugal fan and heat exchanger with twin filters, which is linked to a ceiling diffuser via a large concertina duct. The unit and diffuser are located in the loft space directly above the top landing. See Figure 8.1 for a schematic cross section illustrating how the system operates.

PIV works by a simple principle. Cold, relatively drier external air is drawn through the loft by a single fan unit in each dwelling. The heat that exists in all lofts then tempers this incoming air. Such heat

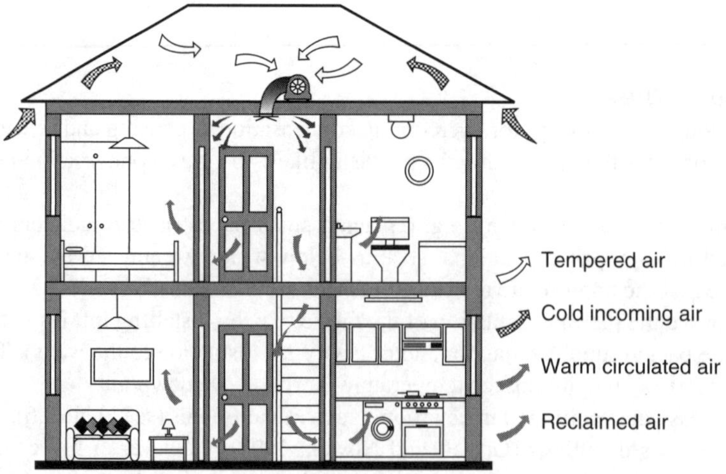

Figure 8.1 The use of input ventilation in a two-storey dwelling (Home ventilation NuAire, 2005)

comes from the solar gain combined with heat that is conducted through the ceiling in the dwelling. As the air passes through the diffuser it re-circulates the lost heat that gathers at ceiling level. It is effectively a three-way heat recovery system. One PIV system, 'NuAire Drimaster 5 + 5', has a British Board of Agrement (BBA) Certificate, No. 00/3727.

There are two versions of this system. The main one is designed for use in single to three-storey dwellings with a loft. The second is designed for use in flats and other dwellings without a loft (e.g. 'NuAire Flatmaster'). It uses a box ventilation unit with heat exchange capacity in the middle of a 75–100 mm plastic duct, which is connected to a fresh air inlet grille in the external wall. In all other respects it operates on the same principle as the home with loft version. The former costs about 20 per cent more than the latter because of the ducting required and the need to core the external wall for the intake (see indicative prices below).

PIV has the following advantages:

- Widespread successful use reported in problem housing estates in areas such as Glasgow, Fife, West Lothian, Inverness and Edinburgh. It is generally acknowledged that condensation is more of a problem in Scotland because of its colder wet climate compared with than other parts of Britain (Scottish Homes, 1997).
- Old contaminated vapour laden air in the home is continuously diluted and displaced through the leakage points (gaps around doors and windows, trickle vents, etc.) that are found in all dwellings. This helps to improve the air quality in dwellings.
- By replacing moist air in the home with fresh, filtered, tempered air it creates an environment in which condensation dampness is eliminated.
- High level of tenant/occupier satisfaction with the results of the system.
- Low capital cost – between about £500 and £600 (at prices in the year 2005) to supply and fit in each dwelling.
- Low running cost (2p/day, again at prices in the year 2001).
- Fully automatic heat recovery.
- Very low maintenance requirement (filters need changed only every 5 years).
- Complies with F1 and K1 respectively of the English and Scottish building regulations.
- Potential savings in building repairs, maintenance and decoration associated with condensation problems.
- Can help reduce radon levels within a building (see below).

PIV is good at tackling the symptoms of condensation, such as mould growth, sweating on surfaces, etc. In some instances it may need to be supplemented with additional or improved trickle vents and fabric insulation to ensure its complete success.

There are, though, three potential drawbacks with PIV. First of all, it may encourage freezing of unlagged or poorly insulated pipework in the loft space. Secondly, there is a risk that it can encourage condensation in well-sealed buildings. This is particularly the case in dwellings that have received over-efficient draught-proofing measures. Thirdly, the occupier may tamper with the settings on the system on the mistaken assumption that the unit is consuming too much energy.

Nevertheless, if the necessary precautions are taken (e.g. lagging pipework and ensuring new or existing windows have adequate trickle vents) these risks should be comparatively low. Installing tamperproof units with a plug-in electronic monitor chip to record run time of system can be used to combat illicit switching off or unauthorized re-setting by tenants. These monitoring devices can be checked remotely, without the need to obtain access to the dwelling to inspect the unit. This can be a very effective way of proving the abuse or misuse of the system in the event of tenants seeking a compensation claim against their landlord under the Housing Acts.

Penetrating and rising dampness

Preamble

Damp proof courses (dpc) in dwellings did not become mandatory until after the introduction of the Public Health Act 1875. Still, the proportion of extant dwellings in the UK built before that date is small – about 2 per cent of the total stock (SLASH, 1985), and is decreasing as more new stock is built.

It is not always essential, however, to install a retrofit dpc in an old building. Provided the five measures listed below are carried out, a building without a dpc is unlikely to have a severe rising dampness problem. It can, for instance, take 9 months to a year or more for damp walls to dry out – even after retrofitting a dpc.

Empirical and remedial work undertaken over many years by Massari and Massari (1993), though, has shown that precautions against direct rising dampness still may require a capillary barrier in extreme cases. The insertion of a retrofit dpc is still the only way to guarantee freedom from rising dampness where the substructure or soil is persistently wet and the masonry has a sufficiently high suction capacity. A high water table or a cohesive soil in such circumstances can act as a wick for ground moisture.

In many properties built before the emergence of national building regulations in the 1960s, lack of a suitable damp proof membrane (dpm) under the ground-floor slab is not uncommon. The installation of a dpm should form part of the renewal or improvement of the ground-floor slab in older properties undergoing adaptation. However, as Massari and Massari (1993) have pointed out, such a remedy runs the risk of diverting moisture below the slab to the main walls. This should not be a problem if such walls are protected against excessive ground moisture by carrying out the precautions listed below.

In any event, not all walling materials need a dpc. Cob and other earth walls, for example, require a higher level of moisture than conventional masonry. Therefore, a dpc inserted in an earth building could result in its walls becoming too dry and friable, leading to premature deterioration.

Forms of dampness

Apart from condensation penetrating rather than true rising dampness is the other main source of moisture problems in British houses. Research by Howell (1995) found that true or direct rising dampness is relatively rare in this country. Burkinshaw and Parrett (2002) have shown that sources of ground moisture other than true rising dampness are more prevalent.

Indirect rising dampness is more common than its direct form because the majority of buildings contain a dpc. The former is primarily caused by bridging moisture rather than by the substructure sucking up water from the soil. The source of moisture in such a case is usually either excessive surface water or leaking underground services waterlogging the ground. This gives rise to lateral moisture penetration, which can bridge the dpc or saturate the substructure masonry. The moisture can then wick up a wall by capillary action giving the impression that true or direct rising dampness is occurring.

Distinguishing between the various forms of dampness is not easy. Building surveyors and other investigators need to take care to avoid mistaking the actual source of the moisture. Table 8.6 gives guidance on how to tell the difference between the main forms of dampness.

Basic remedial measures

It should be remembered that whatever solution is adopted, whether it be the installation of a retrofit dpc system or basic measures described below, a period of at least 9 months is required to allow the affected part of the building to dry out. This will inevitably delay any decoration work to the affected walls. Occupiers must therefore be warned of this to avoid any disappointment or dispute over the time taken to solve dampness problem.

Table 8.6 How to distinguish between the various forms of dampness (adapted from Oxley and Gobert, 1997)

Evidence	*Source of dampness*			
	Air moisture condensation	*Rising/bridging dampness*	*Penetrating dampness*	*Other (e.g. Leaking drain or supply pipe)*
Volume of moisture	Low	Low	Sometimes high	Medium/High
Frequency of occurrence	Persistent during colder periods	Persistent	Periodical – during and just after heavy rainfall	Periodical
Moisture readings* at margin	Gradual change from wet to dry	Sharp change from wet to dry	Usually sharp change from wet to dry	Usually sharp change from wet to dry
Moisture readings* in skirtings, etc.	Low readings	High readings	High readings	Very high readings
Moisture readings within wall	High at surface, low at depth – unless interstitial	High all through	Generally high all through	Generally very high all through
Mould growth	Highly likely – spread evenly	Rare/Highly unlikely	Sometimes – but patchy	Sometimes – but patchy
Wall temperature	Near or below dew point	Usually above dew point	Usually above dew point	Usually above dew point
Contaminant salts	Absent	Nitrates present	Absent	Chlorides usually present

*Wood moisture equivalent (WME) readings taken using a 'Protimeter' or other proprietary electrical conductance moisture-reading instrument. Care should be taken when using this instrument as foil-backed plasterboard, salts and carbonaceous materials can give deceptively high WME readings.

The following measures are usually sufficient to reduce if not eliminate indirect rising dampness:

- Lower the outside ground level to below either the dpc or inside floor level if the building contains no dpc.
- Remove any plinths or rendering at ground level that may be bridging the dpc.
- Install a field drain (between 0.5 and 1 m away) in a suitable granular backfill along the affected outside wall, and drain it to a soakaway if necessary.
- Replace defective internal plaster with an appropriate salt retardant plaster up to a height of at least 0.3 m above the dampness tide mark or 1 m from ground-floor level, whichever is greater.
- Check the integrity of any drains or water supply pipes in the vicinity of the affected wall. A leaking service pipe can easily saturate both the surrounding soil and adjacent masonry. This can also lead to subsidence in severe cases (IStructE, 2000).

In cases of occasional seeping or persistent water penetration in cellars or basements, one of the various drained or un-drained tanking systems outlined in Chapter 6 may be required.

General housing improvements

The specific improvements relating to residential buildings are described in the following sections. The more general upgrading works that can apply these as well as to non-residential properties are dealt with after the section on commercial and industrial modernization.

Internal arrangements

Kitchens and bathrooms

The layout of many dwellings built before the 1940s are inappropriate for modern use. Kitchen and bathroom sizes were often too small and not properly sited in relation to each other. It is now not allowable to have a toilet directly off a kitchen, for example. Any such alteration to the internal layout to these areas may also entail some asbestos removal or encapsulation (see Chapter 2).

Kitchen layouts may also require remodelling, particularly those in tenements. The common features of these are their internal location (i.e. windowless), galley shape (long, narrow profile), insufficient preparation surfaces, and lack of cupboard space. In older, un-modernized properties, they may even lack adequate background ventilation – particularly if the open-hearth fireplace has been blocked up, thus the improvements that could be included in a housing refurbishment programme.

The following is a list of points to consider when refurbishing domestic kitchens:

- New mechanical ventilation, ducted to outside, with a large pad on-switch, and axial fan giving 60 l/s performance for kitchens, 15 l/s for toilets, with a 10 min time-delay cut-off.
- Filtered hood over cooker, which should be located as near to the door or ceiling vent grille as possible.
- Tube or lamp-size fluorescent light fittings should be installed.
- New kitchen fitments, laminated worktop with stainless steel recessed sink. New water-supply branches with easy-fit connections for washing machines.
- Thin plate laminated hardwood tongued and grooved floorboards for kitchens but not bathrooms. If the existing floor surface is irregular or uneven, a 2–3 mm thick layer of polystyrene insulation can be laid under the boards. The floorboards should be fitted with a 5 mm perimeter gap, which is covered by a moulded bead, to allow for expansion.
- A heat-detector wired to the mains should be installed on the ceiling, away from the cooker or other fume-generating appliances to minimize false alarms.

New internal surface finishes and doors

In some modernization schemes, the opportunity to provide new internal surfaces such as worktops and tile surrounds to the bath and wash-hand basin may be taken. At the same time, warped and damaged panelled doors could be replaced. Overhead closers should be fitted on kitchen doors to help reduce the risk of condensation in dwellings.

Lead pipework and lead-lined tanks

In tenements and other residential buildings built before 1900 lead was usually used for water supply pipework. The water supply, particularly if it is 'soft' (i.e. relatively acidic), in such properties can cause 'plumbo solvency' in lead pipes. This erodes the internal surface of the pipe allowing the released lead molecules to contaminate the supply.

The WHO have known for a number of years that a high lead content in drinking water increases the risk to health particularly for children and pregnant women because of its potential neurotoxic effect. It can also result in a low IQ as well learning and behavioural problems in children who consume such water.

Any housing improvement or large-scale modernization scheme must therefore make allowance for the replacement of all lead pipework including the replacement of any lead tanks. As shown in Chapter 2 a lead replacement grant may be available from the LA in certain individual cases. In any event, polybutylene (such as Hep_2O) or copper pipes and high-density polyethylene tanks should be specified for such work. All pipework exposed to low temperature conditions such as in roof spaces should, of course, be protected with pre-formed polyethylene lagging.

The replacement of lead-lined (and galvanized mild steel) water tanks with polypropylene tanks should form part of any major housing refurbishment scheme. All water tanks should be suitably lagged on all sides except the base to ensure that the water is kept well above freezing by the warm air rising from below.

Heating systems

Central heating did not become common in housing until the early 1960s (Harrison and Trotman, 2000). Prior to that period, an open-hearth coal fire in the living room and master bedroom was the standard type of heating in dwellings.

Warm-air heating was used as the central heating system in high-rise flats and in some low-rise detached dwellings in the 1960s and 1970s. It consisted of an electric/gas/oil-fired boiler that generated and forced warm air through a network of galvanized mild steel ducts. The ducts terminated at a grille located either in the floor or wall of each room.

Gas-fired warm-air heating can be especially dangerous owing to the risk of carbon monoxide poisoning (see next section) or other hazardous leakages. These contaminants can be discharged from a faulty or poorly serviced boiler. Broken duct connections caused by corrosion can lead to air leaks from the system directly into the rooms being heated. Dirty ducts and grilles can also adversely affect the indoor air quality.

As a result of these deficiencies, it is now normal practice in housing modernization schemes to replace forced warm-air heating systems with hot-water-fed radiator systems or electric night-storage heaters. The choice of heating system will be dictated by the installation cost, running costs, space requirements and availability of fuel.

In any event, replacing a warm-air heating system entails some disruptive work. This usually involves removing the redundant ducts and blanking off the voids they created. Some collateral damage to wall surfaces is inevitable, which prompts the need for complete redecoration of the affected rooms.

Nowadays, however, forced warm-air heating is more suitable for heating in buildings with large volumes. It is still used in the refurbishment of many non-residential buildings such as factories, warehouses, supermarkets, schools and offices.

Hot water

In many existing dwellings there is always a risk of scalding by hot water from taps. Water hotter than 45°C can scald sensitive skin. This is especially a problem in children and the elderly because their skin is thinner than the rest of the population.

Ideally, hot water from taps should be more than 40°C in temperature. To ensure that this limit is not exceeded it is now a requirement for all new and adapted dwellings for thermostatic mixing valves to be fitted to hot taps serving sinks, wash hand basins, baths and showers.

Gas installations

The Gas Safety (Installation & Use) Regulations 1998, which came into force on 31 October 1998, have imposed strict requirements for installing and using gas appliances and fittings. Safety is paramount to avoid the possibility of a gas explosion and reduce the incidence of gas-related carbon monoxide poisoning.

Carbon monoxide is a potentially lethal gas. Initially, for occupants it can trigger aggression, headaches, dizziness, nausea and vomiting, but may lead to a fatality if the exposure is prolonged. The main sources of this form of gas poisoning in buildings are blocked chimney flues, smoke from fires and emissions from defective gas or paraffin heaters.

The gas regulations will be especially relevant to adaptation work in a wide range of domestic and commercial premises. The main alterations to gas installations involving refurbishment work are:

- the refurbishment of existing masonry chimneys and flues;
- making or remaking an open fireplace and decorative fireplace surround;
- refitting or replacing a combustion appliance (e.g. boiler, built-in cooker, fire).

Generally, the gas regulations require that only a fitter registered with 'Council of Registered Gas Installers' (CORGI) must undertake gas installation work. This includes alterations to and replacements of existing gas appliances and fittings as well as undertaking new installations.

The following indicates the requirements in conjunction with any gas installations during adaptation work to housing:

- Provision of carbon monoxide detectors wired to the mains. These should be located on a suitable surface (e.g. ceiling) in the vicinity of gas-fired appliances.
- Adequate background ventilation to windows (i.e. 8000 mm^2).
- Combustible appliances must be fitted with a safety cut-out device on pilot lights.

Chimneys and flues

The masonry of many chimneys in older properties can be susceptible to sulphate attack. This causes the stacks to bend in the direction of the prevailing wind. In severe cases, where the stability of the stacks is in doubt, these should be carefully taken down and rebuilt. The installation of insulated metal flue liners within the new chimneys usually forms part of any improvements to the heating system. New flashings and cowls to the chimney are also required.

Combustible appliances such as gas fires must have adequate natural ventilation to comply with Part J of the Building Regulations for England and Wales and Sections 3 and 4 in Scotland. The diameter of the flue and other requirements depend on (1) the kW output of the appliance and (2) whether the appliance is room-sealed (i.e. requiring a balanced flue) or open-flued.

Ideally, balanced flues should be used for any new boiler installation. A standard horizontal balanced flue exiting straight out from the boiler should be used. However, in situations where the boiler cannot be positioned against an outside wall, an alternative method of proving a flue has to be adopted. For example, balanced flues involving: vertical flueing to ridge tile terminal; vertical flueing to roof slope; extended horizontal flueing (in duct); and extended mini terminal flue system (in duct). For gas fires, an insulated flue could be fed through the subfloor void of a suspended ground floor and exiting to the outside via an outlet grille.

Electrical rewiring

The minimum service life of an electrical installation in residential properties is about 25 years (NBA, 1985). Thus, rewiring and upgrading of the electrical system is likely to form part of a comprehensive modernization programme in older dwellings.

Nowadays, circuit contact breakers are used in the consumer unit instead of cartridge fuses. This enables the system to be reactivated easily in the event of a minor circuit failure.

Ideally, rewiring should be undertaken without the need to install surface conduits. The existing conduits or new conduits chased into the walling should be used wherever possible. Cabling using PVC-free materials is another desirable requirement.

Since 1 January 2005 tighter controls have been imposed by government legislation on the standard of electrical installation work in residential properties across England and Wales (Noy and Douglas, 2005). These controls are analogous to those required for gas installations by the CORGI.

The controls covering the standard of electrical installation work dwellings in are incorporated in Part P of the Building Regulations 2000. The Building Standards (Scotland) Regulations 2004, which have been in force since 1 May 2005, are likely to follow this development (in Section 4: Safety). The primary effect is to make electrical safety in housing a legal requirement (Noy and Douglas, 2005).

Approved Document P (Electrical Safety) applies to all fixed installations after the distributor's meter in buildings or part of buildings comprising dwellings and related property. It is therefore a legal requirement for anyone undertaking electrical work in residential buildings to comply with Part P of the Regulations by adhering to the fundamental principles for achieving safety delineated in BS 7671: 2001. Unlike this British Standard, however, Part P does not cover the inspection and testing of existing electrical installations (NICEIC, 2004).

Sanitation and drainage

The gutters, downpipes and brackets of the rainwater disposal system in many dwellings are often defective or inadequate. Lack of maintenance or vandalism can result in rainwater-penetration problems. These rainwater goods can be replaced with plastic or, preferably, matching cast iron components if the property is not listed or in a conservation area.

Replacement of defective or inefficient sanitary appliances may be worthwhile considering. This may be particularly beneficial where water conservation is an achievable sustainability requirement (see Chapter 10).

Accessibility provisions

To facilitate access to dwellings by disabled users, a number of provisions should be made in any residential adaptation scheme. Chapter 9 discusses the main design features that should be borne in mind to comply with these requirements. Other modifications, however, such as the position of electrical sockets and switches, and the height and width of the worktop should be considered if wheelchair users are likely in the rehabilitated dwellings.

Security installations, entry phones and call systems

These systems have become very common in multiple-occupancy dwellings such as blocks of flats and tenements. Such facilities, however, have to be very robust to resist the abuse and heavy wear and tear that they often are subject to by occupants and visitors.

Radon

Preamble

Radon is a natural product of decaying uranium. It is a radioactive noble gas that occurs naturally in the ground. The main sources of this gas are igneous granite, sedimentary sandstones and most other types of rock.

The existence of radon has been known since about the middle of the 20th century. The main parts of the UK where the amount of radon coming from the ground may present a risk are Cornwall, Devon, Northamptonshire, Derbyshire, Somerset, Manchester, Sheffield, Birmingham and the Highlands of Scotland.

Regular or ongoing exposure to radon can potentially have serious health implications. The ionizing radiation emitted by the products of the decaying uranium can give a high dose to lung tissue. The resulting tissue damage can cause lung cancer. Houses absorb this radiation because the pressure within them is less than that outside owing to wind and temperature effects. This means that radon is drawn into the dwelling from its surroundings, including through the soil beneath the property.

Sources of radon

Radon in a building may therefore emanate from three main sources:

1. the materials used in the building fabric (which account for about 25% of the radon);
2. the air that ventilates the building (about 25%);
3. the ground beneath the building (about 50%).

Radon is measured in becquerels per cubic metre (Bq/m^3). The recommended level for radon exposure is now 200 Bq/m^3. This was reduced recently from the previous 400 Bq/m^3 level. The design level for new dwellings has been set at 100 Bq/m^3. This is also applicable to houses that have been converted or otherwise substantially adapted.

If the property being adapted is in one of the risk locations referred to above, a number of actions may have to be taken. The first step is to instigate a survey of the property to detect the extent of any radon present. The National Radiological Protection Board (NRPB) can carry this out free of charge.

In the event of radon being detected, and if it is at or above the recommended limit, it is usually a combined operation by the building and electrical contractors to reduce the harmful effects to a 'safe' level. The way in which this can be done is described below.

Tracking down precisely where the radon is entering a building is more difficult. However, guidance on this and other aspects of radon treatment can be obtained from the local Environmental Health Department.

Remedial measures

Measures to minimize exposure to radon in an adapted building may include:

- the installation of a PIV system;
- sealing suspended floors and solid floors with a geotextile membrane containing a radon-resisting layer;
- sealing around all ducts and openings with a suitable mastic;
- providing more efficient or artificial ventilation in risk areas – particularly to the subfloor voids;
- avoiding the use of materials containing radium (see below);
- the construction of a depressurization sump.

The whole business could be costly, and it pays to remember that masonry materials are extracted from the earth. If these originated from an area of high activity, radon will be released by the structure itself.

Some plasterboard used as wall or ceiling linings may be manufactured using phosphogypsum. This form of gypsum contains a higher level of radium than natural gypsum, and it is the radioactive decay of this radium that forms gaseous radon and its by-products. It is this gaseous radon that is the main radiological risk to health. It is generally thought, however, that the radiation dose from phosphogypsum plaster is small (Watts and Partners, 2005).

Non-residential refurbishment

Rationale and strategy

The primary objective of any refurbishment programme is to improve a building's overall performance and enhance productivity (see Figures 8.2 and 8.3). Office buildings are no exception in this regard. Property investors know that the security of their investment is highly dependent upon the quality of the building they own. Giving the building a facelift is a distinctive way of enhancing its appearance and durability.

Traditionally, most office buildings were refurbished, if at all, only once every 15–20 years on average. The recession in the 1990s, however, led to a collapse of the commercial property market. This resulted in a glut of vacant office space in many cities.

Neither mothballing nor redevelopment is always the most suitable option for commercial premises. The former option is a potential waste of a rent-earning asset, even in an unrefurbished state. Similarly, redevelopment may not be viable because the market levels of rents could be insufficient to justify this option. The owner, for example, may hold the property on a ground lease with an uncooperative freeholder. The latter either may not permit redevelopment or will only do so upon very onerous conditions. In such circumstances, it becomes more profitable to refurbish the building, rather than opt for redevelopment.

Figure 8.2 Recladding of an office block – before and after (www.cape.co.uk/)

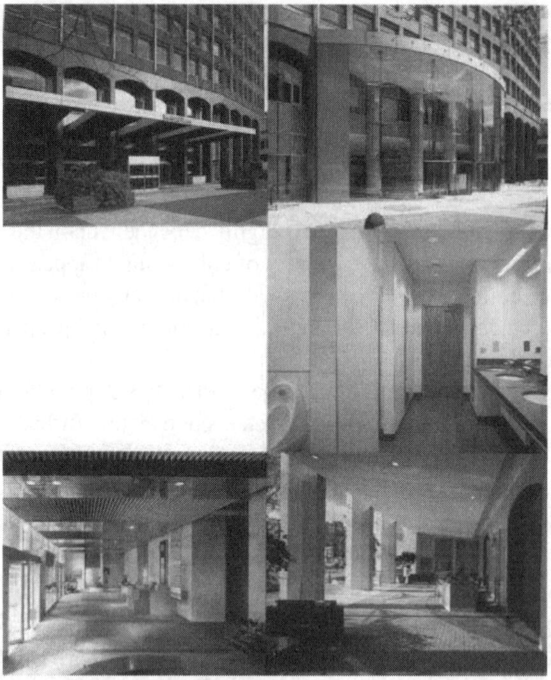

Figure 8.3 Some photographs of typical office refurbishment works (Source: 'Building Economist' supplement to *Building* magazine, 1994)

An adaptive reuse may not be possible owing to planning restrictions or other constraints. The alternative to conversion is to refurbish the building in its current use. The aim of this latter option is twofold:

1. To attract new tenants to improved premises previously unoccupied during times of recession.
2. To encourage existing tenants to stay by offering improved existing space rather than moving out as new office premises becomes available.

Property professionals such as chartered surveyors dealing with commercial premises know that occupiers want good-quality, flexible space that can cope with developments in information technology and is economic to maintain. The trend is towards more diversity in the workspace so that the needs of peripatetic workers as well as their sedentary colleagues are being met. Refurbishing redundant or inefficient offices to meet such requirements is the only way to prevent these premises from dereliction and eventual redevelopment.

Office refurbishment may be seen, however, as a compromise solution in some circumstances. Giving the building a makeover will by itself not guarantee its full occupancy. Indeed, even the best adaptation scheme will not transform a tight-fit building to suit the current and future needs of some commercial users. Big, modern businesses require large buildings with extensive, column-free 'floorplates' and responsive air-conditioning and power systems. Many old buildings cannot provide such facilities because of their restricted internal arrangement.

Any strategy for refurbishing an office building should consider:

- What are the most desirable features of the building to retain or enhance?
- What are the potentials for improvement? A cellular layout changed to a mixed layout consisting of open plan workplaces and some cellular offices provides the optimum combination.
- What repairs to the structure and fabric are required? Any strengthening work needed?

- What improvements are required to the external environment? Is a new entrance foyer desirable or permissible?
- Are the services adequate and efficient to cope with modern requirements?
- Are new or improved facilities required internally (e.g. shower rooms, etc. for staff)?

Cleaning and repairing building exteriors

Problems

Since the early 1970s, it has become fashionable in some areas to carry out vigorous cleaning programmes to many stone-faced buildings for aesthetic or conservation reasons. Such treatments, however, have not always been successful. In some cases, stone-cleaning schemes have made matters worse, either by bleaching the façade or by accelerating the rate of deterioration of the masonry.

The façade of a brick or stone-clad building in an urban area is likely to be soiled by both non-biological and biological contaminants. Dirt, graffiti, grime and soot are typical examples of the former. Algae, mould and lichens are common types of the latter.

Methods of cleaning masonry

There has been a great deal of debate and some controversy since the early 1990s about the need for, and benefits of, cleaning masonry. Some previous stone-cleaning exercises using chemicals or abrasive techniques have caused more problems than they have resolved.

Atmospheric pollution and lack of proper maintenance can often spoil the façade of masonry buildings. In particular, such properties in urban areas are even more susceptible to various aggressive agents such as dirt and grime, acidic gases and emissions from car fumes and other sources. The overall effect of these is to darken or soil the appearance of the stonework or brickwork.

To maintain a good appearance, minimize the likelihood of decay and to prevent defects becoming concealed by soot deposits, masonry that is exposed externally may have to be cleaned regularly. In addition, defective joints should be raked out and repointed as necessary with a compatible mortar (e.g. a lime-based mortar for stonework). However, the guiding principle of 'repairing like with like' should always be adopted wherever possible.

The frequency of cleaning required depends upon the degree of atmospheric pollution, the exposure and the type of masonry. Thus, limestone which is freely washed by rain, may be 'self-cleansing'; in protected positions, limestones and sandstones may require cleaning at 5–10 year intervals and brickwork at 10–20 year intervals, while in polluted atmospheres, polished marble may have to be wiped clean every month.

The appropriate method of cleaning masonry will depend on the factors listed in Table 8.7.

A fine spray of clean water can usually soften deposits on limestones, and sometimes, marbles can be cleaned in this way, but rarely sandstones or granites. *Caustic soda and soda ash are very damaging and must never be used on any stone or brick.*

Brickwork can be cleaned by water spray or by chemical cleaning using a dilute hydrofluoric acid solution similar to that used for stone-cleaning (see below). The use of steel brushes or abrasive wheels can easily damage the surface of brickwork.

Methods of cleaning the exterior of buildings can be divided into the following two main categories: traditional and modern.

Traditional methods

Traditional methods fall into two main groups: physical and chemical. Table 8.8 illustrates the differences between these two methods.

Table 8.7 Factors influencing the cleaning of masonry façades

Factor	*Example*	*Implications*
Type and condition of substrate.	Brick, sandstone, marble, render.	Sound? Friable? Patchy? Hard or soft?
Extent of detailing and moulding.	Cornices and string courses. Ornamentation. Plaques. Statues.	Loss of detail likely?
Degree and type of soiling.	Extensive? Patchy? Biological: algae, lichen, moss. Non-biological: soot, dirt, graffiti, efflorescence.	Disfigures appearance? Leaching of organic acids? Accelerated deterioration by slat crystallization?
Cost, time and disruption.	Equipment required. Specialist input needed? Extent and type of waste?	Expensive to hire? Few contractors available? Mess and disposal of waste?
LA's policy.	Some councils will not approve certain methods (such as chemical cleaning).	May delay or postpone renovation work or affect funding.
Environmental considerations.	Caustic or other hazardous chemicals used in the process?	Pollution of surfaces or drainage system.

Table 8.8 Traditional masonry cleaning techniques

Method	*Examples*	*Risks*
Physical	Dry techniques such as grit or sand blasting; abrasive wheel; wire brush.	Causes erosion of detailing. Cannot eliminate ingrained soiling. High degree of dust generation requiring extensive dust control.
	Wet techniques such as steam cleaning, high or low pressure water washing.	Very messy, and may cause water damage to other parts of the building.
Chemical	Acid or alkali wash (e.g. 5% hydrofluoric acid solution).	Can result in or encourage salt deposits and staining as well as erosion of the face of the masonry.
	Application of alkali-based paste poultice.	Should be restricted to heavily soiled areas to minimize costs and toxic side effects.

Modern methods

In more recent years, a variety of new techniques have been developed to clean the exterior of buildings. Their main advantage is that they minimize the level of abrasion and chemical damage to stonework.

- Dry ice (cryogenesis or 'drice' method): This method operates by firing grains of dry ice at the contaminated stone surface. It is a cleaning system developed in the USA and works by blasting the face of masonry with a jet of dry ice, which evaporates, leaving the dirt as waste.
- Laser: As with the 'drice' method, there is no need to dispose any gritty deposits or dirty water when using laser cleaning. The process may be time-consuming and expensive, however.

- Polystyrene pellets: Another method involves firing polystyrene pellets under pressure onto the face of the contaminated stonework. Like the previous two techniques, this is less damaging than conventional chemical or physical methods.

Graffiti removal

Most of the foregoing methods are not suitable for removing graffiti. Traditionally graffiti was removed either using high-pressure water or by application of a paint remover to eradicate the unwanted scrawl. Both these techniques have their drawbacks. They are messy and may not remove the graffiti completely. Smudging or ghost marks of the graffiti may still be conspicuous. Furthermore, the use of chemicals is not environmentally friendly.

In more extreme case where the graffiti is difficult if not impossible to remove, repainting is often seen as the best solution. However, it is neither appropriate nor easy to paint over graffiti on awkward surfaces such as in monuments.

More recently, a Mr Nigel Farrow has developed a more effective method of removing graffiti without damaging stone using a revolutionary sandblasting machine. He originally invented the machine as a means of cleaning barnacles off yachts. It gets rids of graffiti by a grit-blasting process using a blend of sand, hot water and air at low pressure. The machine removes graffiti quickly, cleanly and is environmentally friendly.

As a precautionary measure surfaces vulnerable to unwanted scrawls can be coated with a clear anti-graffiti coating. This is based on the water-repelling coatings described in the next section.

Waterproofing masonry

The use of clear silicone or alkylalkoxysilane-based water-repellent coatings to enhance the moisture resistance of brick and stone wall surfaces has gained some popularity in recent years. However, there are two major drawbacks with these coatings. First, their durability is limited. As a result, they normally have to be recoated every 5 years. Secondly, they could trap harmful salts within the body of the masonry, thus accelerating any erosion of the brick or stone.

Electrochromic glazing

Nowadays, photochromic or thermochromic capabilities can be incorporated within glass for windows. Glass with these qualities offers a form of passive solar control, but may not be very responsive to rapid changes in lighting or temperature conditions outside.

Electrochromic glass, however, offers a more active form of solar control. It can be used to reclad the façade or reglaze the windows of prestigious office blocks. This is an innovative, high-tech form of glazing that prevents glare and heat gain while letting sunlight flood into a building. It is a solution that obviates the need for using blinds, which are difficult and expensive to install and maintain (see Figure 8.4).

This form of glazing works by transmitting a low electric voltage across a panel consisting of a series of panes containing very thin coatings. When powered up, the glass changes colour from clear to blue, within about 12 min. According to Pearson (1999):

> The electrochromic film is sandwiched between two 4 mm layers of Pilkington's K-glass, a low emissivity glass designed to reduce heat loss through windows. One sheet has a tungsten oxide coating applied to it, the other a coating of vanadium-based oxide. Holding the sheets together is a polymer that is conductive to ions. An electric voltage applied across the film causes ions to be transferred out of the vanadium layer, through the polymer, to settle on the tungsten oxide film, which causes the oxide to turn blue.

One important drawback with electrochromic glass is that it creates more heat losses compared with normal glass. This in turn increases a building's heating load by about 6 per cent (Pearson, 1999). However, the energy efficiency potential of electrochromic glazing is still good because cooling a building is more

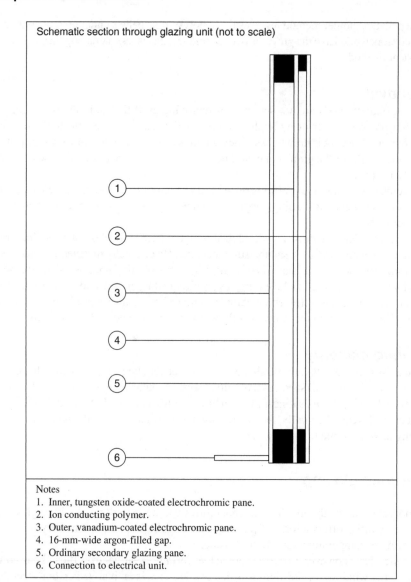

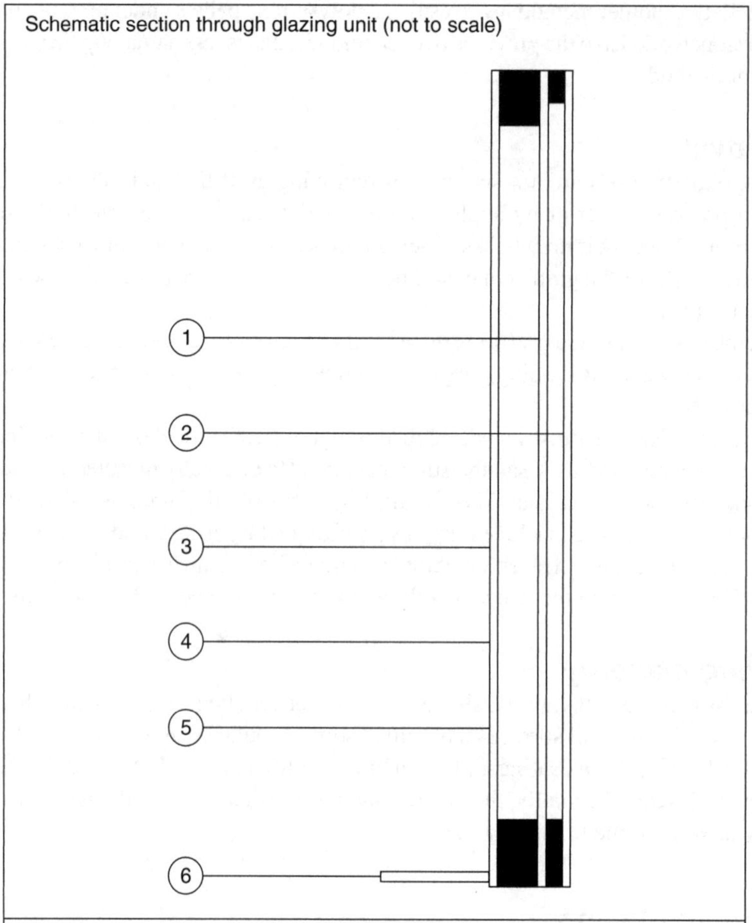

Notes
1. Inner, tungsten oxide-coated electrochromic pane.
2. Ion conducting polymer.
3. Outer, vanadium-coated electrochromic pane.
4. 16-mm-wide argon-filled gap.
5. Ordinary secondary glazing pane.
6. Connection to electrical unit.

Figure 8.4 Electrochromic glazing (Pearson, 1999)

expensive than heating it; any temperature gain can be used in winter to reduce the heating load; passive cooling can compensate for this in summer (see Chapter 9).

Window films

A cheaper option than electrochromic glazing is the use of window films on existing glass. The latter has other advantages over the former:

- it reduces solar heat gain and glare into the building;
- it can be applied to the interior or exterior of existing glazing;

Table 8.9 Strategic options for improving space (Stansall, 1997)

Building variable	*What to do*	*Effect on space*	*£Risk/Return*
Shell	Adapt	Improve efficiency net-to-gross, usable-to-net.	High cost High leverage*
Services	Refurbish	Ease changes in pattern of use.	High cost Variable leverage
Fitting-out	Refit	Improve flexibility in use and in replanning.	Variable cost High leverage
Furniture	Replace	Ease staff moves manage space footprint.	High cost Variable leverage
Interior Layout	Replan	Optimize footprint density and circulation pattern.	Low cost High leverage
Space-use	Reprogramme	Improve space use, ease new working patterns.	Low cost High leverage

*Strategic advantage in the use of space.

- it can improve the fabric's safety and security by holding glass together in the even of breakage;
- it can be designed to give one-way or two-way viewing for privacy;
- it can be used to upgrade glass to current safety standards (BS 6206, Reg. 14, 1981);
- its surface can come with scratch resistant qualities;
- some window films are guaranteed up to 10 years.

Nevertheless, the use of window films will tend to maximize the running costs associated with interior lighting. There is also a risk that they can be easily damaged and so may require more maintenance than ordinary glass over their service life.

Modifying the layout

After salaries, office space represents the highest operational cost for commercial organizations (McGregor and Then, 1999). Maximizing the space of an office building through refurbishment and interior replanning, therefore, could enable a company to rationalize its property requirements. This could result in considerable rental savings if it can accommodate its entire staff in one building rather than two or more properties.

The churn rate is also a factor influencing the regularity and nature of office refurbishment. A high churn rate may necessitate regular remodelling of the office layout – with all the cost implications and disturbance involved. The best strategic improvement in space use can be determined by considering the combination of the variables listed in Table 8.9.

A space planning exercise of an existing office block may reveal that its efficiency ratio (net-to-gross ratio or NGR) is poor. The NGR is a useful measure of a building internal space efficiency. The closer the NGR is to 1.00, the better or more efficient is the building's internal space. It can be calculated as follows:

$$\text{NGR} = \frac{\text{Net Lettable or Usable Area (NLA)}}{\text{Total (gross) Floor Area (TFA)}}$$

The following is a list of the expected NGR targets:

- Office buildings: 0.75–0.80
- Flats: 0.67–0.80
- Hotels: 0.62–0.70

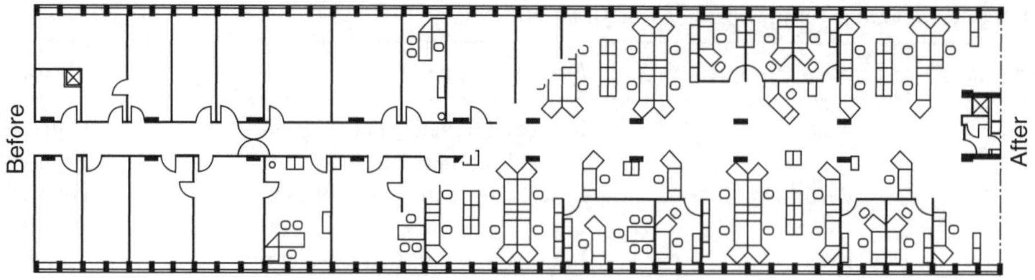

Figure 8.5 Typical reconfiguration of a 1950s office layout (Stansall, 1997)

- Schools: 0.55–0.70
- Hospitals: 0.55–0.67.

The proportion of circulation space to total floor areas is another useful measure of office space efficiency. Generally, circulation space should not be more than about 25 per cent of the gross internal floor area. A functional benchmarking exercise can help determine this, as indicated in Chapter 4. This uses benchmarks such as functions/facilities, space standards, floor areas, plan and service volumes, staff/occupiers numbers, equipment, and materials and output.

The office plan shown in Figure 8.5 contrasts the before-and-after layouts of a building constructed in the late 1950s.

Typical internal refurbishment

Common areas

Common parts of an office block such as entrance halls, stairs, as well as statutory fire escape requirements and landings lend themselves easily to refurbishment. These areas are the most conspicuous and most heavily used parts of commercial properties. Thus, their refurbishment rate may be higher than for other parts of the building.

Internal refurbishment of a commercial building can entail more than redecoration of the wall finishes. It can include stripping out existing finishes and installation of new wall panelling and doors. The doors would need to be attractive but robust and have adequate fire resistance.

The floor finishes in office corridors can be replaced with hardwearing sheet vinyl with welded joints. Alternatively, existing marble or terrazzo finishes can be cleaned by soft wet wiping to bring them back to their near original condition.

Main reception areas

Companies often require the reception area of their building to be of a high quality in terms of finish and colour. This is required to promote the corporate image as well as present an inviting and attractive focal point for newcomers as well as everyday users of the building. An open, spacious and well-lit reception area is also seen as appealing to visitors.

A refurbishment of the main reception area to achieve these objectives will normally involve stripping out existing finished and services. It will also include the installation of new suspended ceilings, wall finishes with wood panelling or high-quality wallpaper, as well as new floor coverings, and a resilient dirt-trapping matt at the front entrance.

Office floor (excluding tenants' fit-out)

Renewing the floor and ceiling finishes of office floor space can give it a rejuvenated look. The flooring can comprise close-pile anti-static carpeting in common office areas and cushioned sheet linoleum or vinyl for the circulation spaces. The installation of suspended ceilings with low-glare integrated light fittings can transform a previously tired looking space.

Conference rooms

These facilities are often a focal point for important meetings or presentations. They therefore usually demand a superior level of finishings and acoustic performance. This may require the stripping out of existing partitioning, ceilings and floor screed and finish. These can be replaced with new soundproof demountable partitions, raised access floor (if floor-to-ceiling heights permit), insulated perforated suspended ceiling with low-glare halogen lighting. The furnishings should comprise high-quality conference seating and facilities for making slide or video presentations.

Toilets

The lavatories in many older office buildings are often drab and unattractive. They are therefore usually ripe for refurbishment (see Figure 8.3). Giving office toilets a makeover can considerably enhance the ambience of otherwise unappealing facilities. New fittings such as wash-hand basins mounted flush on bright resilient worktops, as well as white WCs within light-coloured laminated partitioning, can rejuvenate common toilet areas. New ceramic or vinyl tile flooring, melamine panelled wall linings and suspended ceilings finishes along with recessed or indirect lighting can also help in this regard.

Kitchens and rest rooms

These areas can be transformed by stripping out existing cupboards and worktops and replacing them with bright, attractive, hardwearing finishings. The sink and tile surrounds should match the worktop surfaces. New, more attractive furniture can help transform facilities such as rest rooms.

Leisure facilities

Additional or improved facilities such as lounges, fitness rooms for office staff can help reduce their stress levels. Installing shower facilities for cyclists and lunchtime joggers in an office building can also help in this regard.

Refurbishing services

Heating

The opportunity to upgrade the heating systems within a building may form part of a major refurbishment scheme. Under-floor heating could be used where there is a need to maximize space and avoid exposed heating panels (as described in Chapter 5). This can be installed within a gypsum screed, but may not be easy with existing floor slabs and levels as it raises the floor level by at least 50 mm.

Lighting

The lighting system can comprise halogen lamps for high-quality illumination. Long-lasting sodium bulbs can be used in light fittings in fire escape stairs and less well-used areas of the building as part of an overall energy efficiency strategy. The various measures to improve the energy efficiency of lighting are discussed in more detail in Chapter 9.

Air-conditioning

Some office refurbishment involves installing improved mechanical and electrical services such as more efficient boilers, and new wiring and controls. They may also include the installation of a new variable refrigerant volume (VRV) comfort cooling air-conditioning system, provided there is sufficient space in the ceiling void. Some ceiling tiles are perforated, but others are of the dummy type to achieve better thermal diffusion within the office space.

Lifts

Major commercial modernization schemes often entail refurbishing the lift installations. This would involve overhauling the motors, guide-rails, cables and controls. It would also include refurbishing the lift car's internal surfaces as well as providing new doors and replacement of landing entrance surrounds. Providing controls that can be easily seen and used by disabled people is another improvement that could be incorporated in such work.

Commercial fire precautions

A new fire precaution system comprising both active and passive measures (see Figure 4.17) is also likely in an office refurbishment scheme. The requirements of the Loss Prevention Council (2000) should be followed in the case of commercial refurbishment projects.

Space for services

Horizontal space for service ducts and cables may be located within either a suspended ceiling void or raised floor space. Ideally, a minimum of 300 mm of deep space is required between the structural soffit and the suspended ceiling to accommodate the ductwork and associated equipment such as fans, heating coils, filters, etc.

The minimum depth of void for a raised platform floor is 100 mm. If this form of floor is used, consideration must be given as to how to accommodate changes in levels. Ramps may be required to link the raised floor areas with landings. Door openings will need to be raised. Window sills may also require raising to ensure that the floor-to-sill height does not fall below 1.05 m for windows at the first-floor level and above. If this were not feasible, a safety rail at the required height across the window should suffice.

Drainage

Below-ground drainage systems in buildings rarely need improving unless the pipework is so old as to require renewal. Above-ground drainage, however, may benefit from an upgrade, especially in terms of the surface water provision. For example, the outlets in gutters may require renewal or an increase in number because of deficiencies in the original design. Moreover, sufficient overflows should be installed in gutters serving all adapted buildings with large roof areas to avoid the problem of flooding caused by choked or defective outlets. Discharging the overflows at conspicuous locations, such as near to a front doorway and onto a gully, which will drain the water away safely, can act as an early warning indicator and can help to avoid the extensive water damage that is often caused by flooding.

Retail refurbishment

Preamble

Shops and shopping centres are among the most prominent features of the built environment in towns and cities. They have a major impact on the urban fabric because of their prevalence and conspicuous nature.

For commercial reasons, shop owners need to present inviting and well-presented premises. Competition from rival outlets as well as the need to keep up with current market trends means that retail premises are refurbished at a much higher rate than other properties.

Commercial considerations, therefore, are the main influences on the refurbishment of retail premises. The ongoing need to attract and retain customers is a critical requirement. However, the influence of the various regulations controlling retail premises cannot be overlooked (e.g. fire, management and other health and safety regulations; see Chapter 10).

Retail units can be categorized into two primary groups:

1. Traditional: Main-street premises at ground level (some in multi-storey premises).
2. Modern: Shopping-centre complexes (in the heart of town or at the periphery of an urban area).

Main-street shops

The typical range of refurbishment work for these retail premises is as follows:

- Renewal of shop front with new glazing and metal shuttering, entrance doors and shop sign (possibly illuminated). This should include the installation of barrier matting at the entrance threshold.
- Creating a clear and inviting entrance – with adequate width, illumination and signage.
- Mechanically operated revolving or sliding doors are popular for the larger retail outlets – with ordinary side-hung doors in case of an emergency.
- Replacement of floor coverings with laminated timber sheeting on foam underlay.
- Redecoration of walls with bright, attractive wallpaper or melamine finished interior wall boarding.
- New or replacement suspended ceiling with recessed light fittings.
- Modifying the layout to suit fire regulations and customer use.
- Improving the support facilities at the rear – such as upgraded toilets, kitchen and accessibility.

Shopping centres

Shopping centres are normally within an existing inner urban setting or as a stand-alone complex on the edge of the town. In any case, they will contain a variety of different shop types as well as a mall. The principal owner of a shopping centre would normally commission its maintenance and adaptation.

In the refurbishment of a shopping centre, however, a number of issues would require consideration. The extent of any upgrade, to the external fabric and internal finishings would need clarification. Attention could be given at the design stage of the adaptation scheme to reconfiguring the internal layout. This could include changing the retail mix, increasing or decreasing lettable space, creating additional units, extending or breaking up of existing units, or introducing totally new elements into the scheme. Each option should be examined for its cost and management implications and matched against financial criteria. Typical financial criteria could include:

- rate of return,
- expected rental levels,
- payback period,
- whole life cycle costs.

An analysis of the adaptation potential of the shopping centre would focus on external and internal improvements and fire safety.

The position of the existing centre has an impact on the surrounding environment. It can, for example, act as a magnet for adjoining property uses. Thus, the following external improvements can enhance its appeal and overall efficiency:

- Fabric: An upgrade of the fabric could include either overcladding or recladding (depending on the substrate) to make the external envelope more distinctive and attractive. It may also include recovering the roof/s or over-roofing scheme or installing new skylights on recovered flat-roof areas.
- Entrance and approach: Canopy at front and side entrances; ramps to facilitate disabled access; revolving doors.
- Delivery, servicing and disposal: New rear or side access doors and landings to cope with type and intensity of delivery loads; bays for refuse containers, their disposal and collection.

The following internal improvements should be considered:

- Improving mall layout and finishes by including ponds and water fountains as well as statues or other large ornaments and artwork.
- Installing large mechanically operated revolving or sliding entrance doors.
- Improving the mall environment by installing new heating and air-conditioning in the ceiling space.
- Providing enhanced amenities for shoppers – such as new seating, phone/internet booths; better colour scheme and lighting.

Retail fire safety

Like most buildings, it is unlikely that shopping centres fully comply with current regulations. Fire safety is one of the most important requirements in enclosed shopping centres. In particular, means of escape and smoke control requirements have become stricter, owing to developments in UK legislation influenced by directives from the EU.

The key technical considerations for fire safety in shopping centres are adequate means of escape, proper smoke evacuation, and sprinklers:

- Means of escape: The requirements for the design of fire exits in enclosed shopping complexes have been changed by the BS 5588: Part 10: 1991: code of practice dealing with fire precautions in the design and construction of buildings. The maximum floor space factor for shop units is 2 m^2/person. Using the simple formula shown below, this enables the number of occupants of a room or storey to be calculated. This in turn helps to determine the capacities of escape routes within a storey and of an exit leading therefrom. Thus, if the total area of a shop unit is, say, 420 m, the minimum width of escape route would be 900 mm (from Table 3 of BS 5588: Part 10: 1991).

$$\text{Number of occupants per storey} = \frac{\text{Area of room or storey } (\text{m}^2)}{\text{Floor space per person } (\text{m}^2)}$$

- Smoke evacuation: Since smoke is the main killer in fires, it is imperative that an effective exhaust system be provided to remove excess smoke without fanning the fire itself Thus, smoke vents in the roof of a shopping mall or atrium can alleviate this problem (Morgan, 1994). A BRE report by Morgan and Gardner (1991) highlights the main requirements in this regard.
- Sprinklers: These form a vital part of the smoke ventilation design in shops by preventing a fire spreading beyond the design fire size. According to Morgan and Gardner (1991) 'sprinklers should be installed in malls if they contain sufficient combustibles to support a fire larger than the design fire size of 5 MW, 12 m perimeter, during their operational lifetime'.

Hotel refurbishment

Hotel refurbishment and replacement programmes are dependent primarily on market conditions. The competition for hotel accommodation in urban areas is fierce. The best way to keep existing customers and attract new ones is to ensure that the facilities on offer are up to modern standard.

Hotels are subject to a great deal of wear and tear. This is because they are occupied 7 days a week all year round. Moreover, According Lawson (1995), the frequency and timing of such work will depend on:

- changes in market conditions,
- phasing of work,
- hotel performance,

- legal and insurance standards,
- difficulties in obtaining alternative facilities.

Table 8.10 summarizes the key factors to take into account in hotel refurbishment projects.

Industrial refurbishment

Preamble

The refurbishment of industrial buildings is probably less common than for other use categories. This is ironic in some respects because of the general low standard of construction of factory and warehouses compared to other categories of buildings.

Because of their low thermal mass and poor levels of insulation, industrial buildings are vulnerable to rapid swings in temperature and humidity. Any re-roofing therapy to these buildings, therefore, requires a high-quality roof covering system (e.g. EPDM, Ethylene propylene dien monomer, single ply membrane) to resist high thermal stresses induced by a well-insulated fabric.

The refurbishment of industrial premises is normally undertaken to improve operational performance and/or energy efficiency. The latter is dealt with in more detail in Chapter 10.

Improving operational performance

- Daylighting: New or improved domed or pyramid skylights or sunpipes can be installed to enhance natural lighting within the flat-roofed areas of the building. Stainless steel grilles below each skylight should be fitted to ensure adequate security. Rooflights or sunpipes can offer a similar benefit in the pitched-roof sections.
- Increasing capacity: Modifying the structure to cater for taller vehicles and working processes may be required to make the building more suitable for a wide range of distribution companies. Increasing the structure's load-bearing capacity may also help to enhance an industrial building's marketability. This may necessitate installing new machine bases with anti-vibration mountings where an additional plant is required.
- Access: Improving site access and loading bay facilities – this could include the installation of 150 mm diameter bollards in the hardstanding and steel and plates on landing edges and door reveals to protect vulnerable parts of the building.
- Fire protection: Installing sprinklers in offices as well as all storage and work process areas, to improve the premises fire safety. It may also include upgrading the fire warning system.

Public refurbishment

The extent of uses under the generic title 'public buildings' is wide. It includes hospitals, schools, swimming pools, council offices, as well as assembly buildings and other properties used by the general public. The refurbishment of each of these building categories could merit a chapter or even a book by itself.

The aim here, therefore, is to highlight some general refurbishment principles for public buildings by briefly describing the specific requirements of two main types: schools and swimming baths. The specific modernization of other categories of public buildings such as hospitals and assembly buildings is not addressed in this book. New build rather than adaptation schemes are more common for hospitals. This is because as these facilities are used 24 h a day, decanting the occupants to temporary or new premises is the only way to do refurbishment work to each building in the complex.

Table 8.10 Major hotel refurbishment (Lawson, 1995)

Area	*Construction changes*	*Engineering systems*	*Furnishings, etc.*
Rooms	Replanning divisions. Extensions. Conversions – suites. Bath/shower rooms. Fitting new windows, doors, locks, safes.	AC/vent/heating. Lighting, power. Telephone, video. Plumbing. Room management systems.[1]	Bathroom fittings. Furniture, mirrors. Carpets, furnishings. Lamps, equipment. Decoration, signage. Terminal equipment.
Circulation	Replanning of corridors, stairs, lobbies, elevators, service rooms, chutes, protection of evacuation routes.	Elevator (guest service goods), AC, lighting, power, vacuum systems, emergency telephones Fire safety,[2] sprinkler and security systems.[3]	Linings, decoration, carpeting, light fittings, signage. Fire resistant enclosures.
Lobby and front office	Changes in layout, Redesign of lobby, toilets, rental spaces, front desk, safe deposits, offices and business centre.	As above. House telephones. Telephone exchange. Public address and hotel management systems.[4]	As above. Lounge furniture, furnishings, piano bar, uniforms, graphics.
Public rooms	Changes in food concept, replanning interior design, bars, counters, kitchen planning.	AC, lighting, power. Energy management. Background music. Restaurant/bar. Management systems.[5] Fire safety.[2]	Furniture, carpets, furnishings, lamps, decoration features. Food service equipment, tableware, menus, graphics uniforms.
Meeting rooms	Division of areas. Movable partitions. Foyers, cloakrooms. Direct access. Service circulation storage.	As above. Computerized control. AVA projection systems, interpretation systems.	Furniture – meetings, banquets, seminars. Stage PA and AVA equipment. Food/bar service.
Recreation areas	Extension, conversion or renovation to form swimming pool, surge pools, sauna, changing rooms, gymnasium, bar.	AC/vent/heating, lighting, power, PA, TV monitoring, energy management, safety systems.	Changing room furniture, fitness equipment, signage. Lighting TV cameras.
Back-of-house	Replanning of loading docks, kitchens, laundries, stores. Staff facilities. Plant rooms.	AC/vent/heating, lighting, power. Refrigeration. Energy management.[6] Security, safety and maintenance systems.[7]	Wall and floor linings, ceilings. Laundry, and kitchen equipment. Stores, carts. Waste compactors. Employee toilets, showers and lockers. Dining room furniture.

Notes

Computerized systems and associated work:

1. *Room management*: Room status, energy adjustment, security monitoring, telephone metering, guest services, facsimile, in-room accounting in-room movies interactive information.
2. *Fire safety*: Detection, sprinkler systems, portable extinguishers, smoke door closure, alarm and indicator panels, emergency generator, pressurization of stairs, smoke exhaust systems.
3. *Security*: Electronic card access, internal fasteners, opening sensors for emergency doors; television monitoring of entrances, goods yards, lobbies, alarms in cashiers in and safe deposit areas, room areas.
4. *Hotel management*: Reservation, guest history room allocation, accounting, invoicing employee records.
5. *Restaurant-bar management*: Computerized ordering, recording, point of sale accounting, stock take and reordering menu and recipe costing.
6. *Energy management*: Monitoring of internal and external conditions, optimization of energy flows, zone controls, peak load shedding, energy cut-out for unoccupied rooms, improved insulation, energy reclaim systems from equipment.
7. *Maintenance*: Work scheduling, costing, programming inventory listing, stock taking and re-ordering.

Nevertheless, the five main requirements for the modernization of public buildings, most of which also apply to other building types, are summarized below:

1. Energy efficiency (see Chapter 10).
2. Enhanced comfort conditions in terms of acoustics, security and air quality, by installing better finishes and more responsive services (see below under Refurbishing schools).
3. Improved fire safety (see Chapter 4).
4. Increased durability (wall and/or roof upgrades – see Chapter 9; tighter security – see below under Refurbishing schools).
5. Accessibility (see Chapter 9).

Refurbishing schools

Context

Britain has around 25 000 primary and secondary schools (National Audit Office (NAO), 1987). As far back as 1987, the Department for Education estimated that £2 billion needed spent to bring school premises up to standard, including nearly £1 billion on structural repairs and maintenance (CSM, 1991). Allowing for inflation and further deterioration of the stock, the total figure is probably well over £5 billion at 2001 prices.

The extent of work required depends on the form of construction as well as the over condition of the stock. The types of school buildings are discussed in the next section.

Types of school buildings

Although there are a large variety of school buildings, they can be classified into two main groups: heavyweight and lightweight. Their characteristics are compared in Table 8.11.

Fluctuating school rolls and developments in teaching practices have had a major impact on the use of educational buildings. This has meant that some of these buildings have become either redundant or, overused. Those that are still in use, therefore, may require attention because of inadequacies in their capacity, layout and their poor overall physical performance.

The chronic under-investment in the maintenance and repair of school buildings is another factor influencing the need for refurbishment of these buildings (NAO, 1991). As indicated in Chapter 2, however, some local authorities are addressing this problem by embarking on multi-million-pound refurbishment and extension programmes to their school buildings using the Private Finance Initiative.

Table 8.11 Categories of school buildings (based on DES, 1991)

Category	*Era*	*Characteristics*
Heavyweight	Victorian/Edwardian schools, Schools of the inter-war period and 1970s.	Solid load-bearing masonry construction. Slated duo-pitched roofs. Large fenestration High floor-to-ceiling heights Slow thermal response – because of their dense, thick fabric.
Lightweight	System-built schools of the 1950s to 1970s (and beyond). Modern cladded, framed buildings.	Prefabricated pre-cast concrete panels over a steel frame using proprietary systems such as CLASP (see below). Traditionally finished with flat roofs covered with standard three-ply feltsystem. Fast thermal response – because of their thin fabric.

CLASP school buildings

CLASP is an acronym for the Consortium of Local Authorities Special Programme, which is an unincorporated association of public sector authorities. It developed a building system that was originally designed in the early 1950s for educational use. The early success of the CLASP system, however, meant that it was soon developed for a wide variety of other building types including hospitals, police stations, airports, laboratories and offices.

Because of its articulated foundation design, a CLASP system was often used for schools and other similar buildings in areas with mining subsidence problems. However, older versions (e.g. Mark 1 and Mark 2), especially, are likely to require extensive energy efficiency upgrades as indicated in Figure 10.7.

Educational benefits of refurbishment

Fairs (2002) reported that poor school accommodation has been found in an OfSTED report to have an adverse affect on the delivery of the curriculum. Refurbishment can help tackle this problem in the following key areas according to Fairs (ibid):

- Teacher motivation: Poor building conditions lead to higher teacher absenteeism, higher staff turnover, lower job satisfaction and reduced effort. Capital investment was seen by headteachers as one of the most powerful influences on teacher motivation, mainly through the boost to morale which teachers get from working in a professional physical environment.
- Pupil motivation: A depressed physical environment is believed by pupils (and parents) to reflect society's lack of priority for their education and is therefore detrimental to morale and effort. Headteachers viewed capital investment as having a strong positive impact on pupil behaviour, motivation and truancy. They reported that pupils care about the facilities they work in, and this affects their willingness to learn. Inner city heads in particular, observed a close correlation between the quality of the facilities and a sense that education is important, and that pupils are being valued by the system.
- Academic attainment: Secondary schools that spend relatively large amounts on capital improved their 'A' level performance by 9 per cent compared to those in which there was no capital spending. Primary schools in which there was significant capital spending improved their Key Stage 1 performance by 5 per cent compared to those with no capital spending.

SENDA requirements

The Special Needs and Disability Act (SENDA) 2001 came into force in September 2002. It addresses the rights for disabled students not to be discriminated against in education. Certain physical adjustments inn educational buildings are needed to comply with SENDA, which is now Section 4 of the DDA 1995). It requires institutions to provide auxiliary aids and services where these would help to prevent substantial disadvantage to disabled students, such as installing induction loops into classrooms.

Acoustic improvements in classrooms

There is a growing recognition of the importance of noise control in classrooms. Excessive noise can interfere with the learning and teaching environment. The modern classroom layout mirrors today's offices in that an open-plan layout is common. However, problems such as visual distractions and noise in multiple classroom zones can disturb activities in adjoining areas. The installation of the following acoustic and visual controls can help to overcome these problems:

- Acoustic quality suspended ceiling tiles (see Figure 8.6).
- 1.2 m high moveable screens used as acoustic breaks as well as dividers in semi-open plan classrooms.
- Double-glazing to main windows and, if present, rooflights.

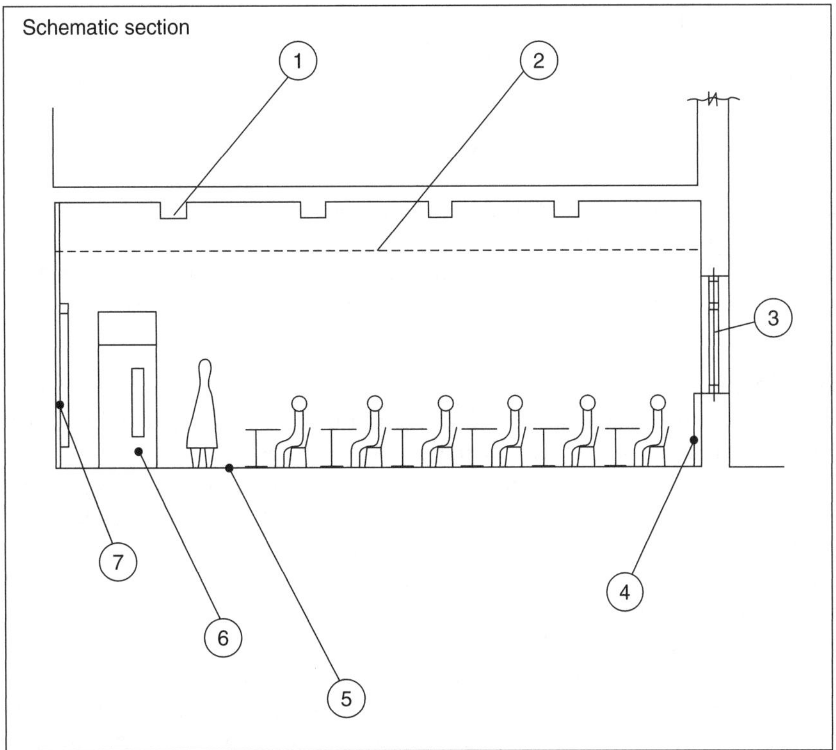

Notes
1. Existing uneven ceiling surfaces can deflect sound, causing acoustic 'dead zones'.
2. New suspended ceiling can reflect sound more efficiently to avoid occurrence of 'dead zones'.
3. New double glazing for acoustic as well as thermal efficiency. Low-emissivity glass used.
4. New radiators for more efficient heating.
5. Cushioned sheet vinyl or short-piled carpeting for acoustic and anti-static qualities.
6. Self-closing entrance door with lower viewing panel glazed Georgian wired glass.
7. Acoustic panels on walls.

Figure 8.6 Refurbishment works to a typical school classroom

- Resilient easy clean acoustic panels on walls.
- Hardwearing close-pile anti-static carpeting or welded sheet vinyl in classrooms to reduce impact noise and electrostatic shocks.
- Hardwearing epoxy resin or seamless vinyl floor coverings in corridors and stairs.

Improvement to school toilets

One of the main indirect indicators of a school's overall performance is the condition of its toilets. In many primary and secondary schools, the toilets for pupils are highly prone to misuse and abuse. Vandalism, graffiti and untidiness are common problems in these facilities especially in secondary schools. A school that is blighted by unruly behaviour amongst its pupils is more susceptible to these problems.

Unless carefully considered beforehand, any improvements in toilets on schools may be a waste of time and money. The following measures, therefore, should be considered for refurbishing school toilets:

- Stainless steel taps that operated by push-button controls to conserve water and avoid flood damage from running tap water should be installed.

- WC cisterns and other parts of the plumbing system should be located, if possible, behind wall panels to minimize the chances of damage.
- Sanitary ware such as basins, WCs and urinals should be of stainless steel or other robust material to minimize the chances of breakage. Trough urinals at floor level are preferable to knee height urinals as the latter can be used as a platform by pupils to stand on. Low-water-use fittings should be used.
- Decorations should be bright and resilient, with easy-clean surfaces, to give an attractive and comforting ambience to the toilets. The use of laminate panels for walls and non-asbestos mineral fibre tiles for the ceilings can offer attractive, hardwearing surfaces.
- Light-coloured ceramic or other hard tile finishing, rather than vinyl, should be used for the floorings.
- Anti-check strips fitted along hinge jamb of entrance and cubicle doorframes.

Fire protection

It is estimated that about one school a week in Britain is affected by a serious outbreak of fire. The most common cause of this problem in educational and similar buildings is arson.

Adequate security measures as described below along with detection sensors in all rooms and vulnerable areas (e.g. toilets, stores, corridors, etc.) will help to reduce the risk of fire occurring within the school and college buildings. Suitable management as well as physical measures, though, is also required to deal with a fire outbreak. For example, a fire-safety plan combined with explicit fire-drill procedures should be organized by the school's administration.

Emergency lighting is critical to highlight all escape routes and exit doorways. These can be aided by the use of fluorescent strips taped onto the floor coverings of key routes.

By the early 1970s open plan layouts became very popular in many primary school buildings. Unfortunately this obviated the need for fire barriers in pitched roof spaces. Potentially therefore in those buildings without such a barrier there is a risk that a fire started in one room could spread to other parts of the building via the roof space. As with over-roofing schemes, consideration should be given therefore to the installation of fire barriers, using reinforced mineral fibre blankets, every 8 m along the roof void (see Chapter 9).

Any prudent refurbishment scheme to a school therefore should include the provision of a suitable sprinkler system in accordance with BS 5306: Part 2 (1990). Retrofitting a sprinkler system in a building that has previously not had one is not always straightforward. The main pipework can be concealed within a suspended ceiling. Anti-vandal sensor heads should be used throughout the school to minimize the chances of accidental or deliberate operation of the sprinklers.

The fire-safety principles outlined in Chapter 4 can also apply to schools.

Security

The Dublane and Columbine atrocities in the late 1990s highlighted the need both in the UK and USA for improved security in education buildings (see NSSC publication). Therefore, in any proposed school-refurbishment programme, the following measures should be considered to improve security:

- Establish one main entrance point to the school, and maintain the remaining entrances as fire exits. A main door security system with receptionist should be provided. The entry doors should be made one-way (outwards, for fire escape purposes), so that once the school day has commenced, access can only be obtained via the reception.
- Closed-circuit television (CCTV) surveillance cameras installed to provide views of the main grounds and entrance point. Internal CCTV monitoring and recording equipment should be considered for main corridors and assembly points.

- Installation of good-quality security locks on all external doors and on key internal doors. Entrance to the reception area and into the school proper, for example, should be restricted to doors with coded button operated locks on the doorframe.
- Installation of window straps or catches to limit the opening of windows to 100 mm. This is especially important for obvious safety reasons at first-floor level and above. They should also be installed on windows at ground level for security as well as safety reasons. For example, ground-floor windows that open outward over a footpath are potentially hazardous for passers-by.
- Anti-climb paint on downpipes up to first-floor level and guard rails on parapets to prevent access to roofs.
- Installation of security lighting around the school buildings.
- The installation of a new or improved security fence at least 2 m high should be considered along vulnerable boundary perimeters.
- Consider installing security locks with wick release latches on the inside to all doors. Fire doors can be opened using quick-release pads on the inside.

General internal improvements

The main parts of the interior of school buildings that may require refurbishing are:

- Layout: Remodel the layout of rooms, if necessary, to enhance the spatial efficiency of the premises.
- Finishings: New suspended ceilings; redecoration of walls using brighter, more resilient finishes; panelling walls in vulnerable locations with impact-resistant boards such as fibre-reinforced polymer; new floor finishes – sheet vinyl in corridors and stairs, and close-pile carpeting in staff lounge.
- Services: Replace deficient heating installation (see Chapter 9) and upgrade lighting system, especially in classrooms, as indicated earlier.

Water consumption can be reduced by over 50 per cent by installing sanitary fittings with low-water-use features and recycling 'grey' water (see Chapter 9). For example, urinals account for a large proportion of water consumption in educational and other public buildings. Fitting urinals with infrared sensors or regulating valves can limit flushing to only after they have been used rather than every 20 min automatically. Vacuum WCs can also be installed to reduce water usage.

General external improvements

Re-roofing, overcladding and window/door replacements form the main constituents of common refurbishment works to the external fabric of school buildings. These works are described in more detail below and in the next chapter.

Refurbishing public baths

Many towns and cities throughout Britain contain public swimming baths built during the second half of the 19th century. These Victorian buildings were built to last many years. However, in recent times, many have had to undergo major refurbishment programmes to bring them up to modern standards.

Normal wear-and-tear, high-humidity conditions, and poor levels of maintenance have taken their toll on these buildings. Some have been left to become derelict. The more prestigious buildings have been refurbished.

Victorian baths were usually of a standard form of load-bearing masonry construction (e.g. thick solid sandstone walls topped with a slated pitched roof). The roof construction usually consisted of a heavy timber hammer beam trusses or mild steel trussed rafters. The energy-efficiency measures for school buildings having a high thermal mass described earlier, therefore, can apply to these properties.

In summary, the main precautions or measures that should be borne in mind when refurbishing public baths are:

- The use of moisture-sensitive materials should be avoided in the refurbishment of wet areas and other facilities that have conditions of regularly high humidity. For example, fibre-based suspended ceiling tiles should not be used in the main swimming pool or in adjacent areas such as the shower rooms, toilets and changing rooms. However, slatted timber can be used as a ceiling and wall finish in swimming pool and changing room areas. This can help to reduce noise levels and gives a warm quality to these spaces. The main precaution in using timber finishes such as slatting and tongued-and-grooved 'vee' boarding is that it should be finished in a fire-protective lacquer. Ideally, only hardwood or high-quality softwood should be used. The slats should be fixed on top of a black 'breather-paper' (i.e. moisture barrier) background.
- As an alternative to slatted timber, melamine-faced moisture-resisting tiles could be used for the suspended ceilings in 'wet areas'.
- Consideration should be given to applying internal insulation to walls, floors and ceilings in shower areas to avoid cold bridging (see Chapter 9). Special care, of course, has to be taken when installing insulation on floors because of the risk of leakage. A drained system below the insulation could be installed to discharge any leaked water to a suitable drainage point in the floor.
- Ceramic tiles with joints pointed with high-performance polysulphide mastic should be used for floor and wall finishes. Smooth finishes should be used for walls; dimpled finishes for floor surfaces to give them anti-slip properties in 'wet areas'.
- Curved internal and external corners (at least 30 mm in radius) should be used in floor and wall finishes to avoid traps for water and dirt.
- Any void areas such as ceiling spaces and services shafts should have adequate background ventilation.
- Stainless steel fittings and pipework should be used in all 'wet areas'. Showerheads should be self-cleansing or easy cleanable to avoid build-up of bacteria and limescale.
- Shower supply piping should be fitted with flush automatic on/off controls to conserve water and avoid accidental or deliberate damage. Anti-scald valves should be fitted to showers. Low-water-use urinals should be used.
- Shower traps should be made accessible for easy cleaning/sterilization.
- Improve the method of treating water (e.g. move from chlorine to ozone).

Summary

Refurbishment is an opportunity to enhance and modernize not only a building's appearance but also its overall technical performance. Given that nearly two-thirds of the existing housing stock in Britain was built prior to 1965, the need for ongoing improvement and upgrading of dwellings will continue for the foreseeable future. This will be driven by environmental influences as well as by the increasing demands for better-quality housing.

It is unlikely that new build will be able to meet the demand for additional dwellings completely. Refurbishment of poor-quality housing as well as the conversion of redundant buildings will be required to supplement the total supply of dwellings.

The increasing importance of sustainable construction will also drive the modernization of commercial and industrial buildings. Energy consumption of factories and offices needs to be reduced if companies are to remain competitive as well as cost conscious. This also applies to assembly buildings such as hospitals and schools (see Chapter 10).

The use of ESTs or RETs can help achieve a more sustainable building. Solar refurbishment can potentially reduce the annual energy consumption in a building by between 10 and 75 per cent, depending on the extent of the measures undertaken.

Some older commercial and industrial premises may require the renewal or upgrading of their lightning protection. Given the increasing fluctuations in weather patterns owing to global warming the risk of lightning strikes on high-rise buildings is rely to increase. Thus, allowance for additional or improved lightning protection may have to form part of the refurbishment of a property if it is of multi-storey construction.

Heavily used facilities in commercial premises such as restaurants, kitchens and toilets are often refurbished on a regular basis. They may have their internal layouts, fittings and fixtures and finishings renewed as frequently as every 5–10 years. This of course primarily depends on the degree of use and misuse as well as the standard of facility to meet either statutory requirements or the demands of the owners and occupiers.

Greater competition between businesses and higher expectations of employees will prompt a similar response for the modernization of commercial and industrial buildings. These influences, along with the need to combat fabric deterioration, are likely to increase refurbishment generally.

The next chapter considers more specific options for refurbishing these and residential properties.

9

Further aspects of refurbishment

Overview

This chapter outlines the refurbishment options and processes of upgrading specific aspects of all types of buildings to enhance their overall performance. Like the previous chapter, it focuses predominantly on the non-structural improvements that can be made to properties. Again, the principal aim of such work is to bring them up to modern standard in terms of appearance, durability and function. Although the focus here is on non-residential buildings, some of the issues such as indoor air quality can also apply to housing.

External improvements

Preamble

Various improvements to the external fabric normally form part of any major modernization scheme to many housing estates. The use of bright colours and intricate patterns on flank walls and gables, for example, can enhance the appearance of multi-storey blocks of flats. However, the main reason for doing this is to enhance the buildings' thermal performance. See section below on upgrading walls using external cladding systems.

Commercial properties, too, can benefit from an external upgrade. This may take the form of a recladding or overcladding and over-roofing scheme. These and other means of adapting the external fabric of a building as part of a modernization scheme are discussed below.

Recladding

Façade replacement or recladding is a relatively rare practice, especially in thick-walled masonry buildings. It can occur, however, if the external walling either is badly damaged as a result of, say, a bomb/gas blast or has prematurely deteriorated because of deficiencies in the original façade materials. This is more likely in lightly clad buildings (e.g. with curtain walling). It may also take place if the owner wants to replace, rather than overclad, the existing envelope finish with something more attractive and durable.

Recladding, therefore, may be the best option because the owner of an office or factory block wants the building to have a new look and a higher overall performance rating. The existing façade, though, may not lend itself to an overcladding system. If it is already of curtain walling, overcladding is unlikely to be feasible because of the difficulties of attaching the new cladding to the existing glazed substrate. The best solution in such circumstances would be to remove the curtain walling in its entirety and replace it with a new double-glazed cladding system comprising materials described in Table 9.1.

Many older industrial premises are often finished externally with corrugated asbestos cement sheeting. These buildings may require recladding to improve their appearance. This may also be done on health and safety grounds because of the presence of extensive asbestos or other deleterious materials within the cladding. Even more modern profiled metal sheeting may require replacement because of deficiencies in the external coating or fixings. Cladding damaged by inadvertent impact or vandalism may also require replacement with a more robust product (e.g. titanium – see below).

Therefore, recladding can provide a good opportunity to transform a building's appearance as well as improve its resistance to damage. A form of recladding using electrochromic glass, described below, can enhance both the energy efficiency and the internal comfort conditions of a building.

In any event, recladding is more likely to involve a 'jointed' system. 'Seamless' cladding versions such as insulated render would only be suitable if it were attached to a firm, continuous substrate such as brick/block infill panels. The relatively heavyweight materials such as stone or concrete panels may be

Table 9.1 Range of jointed lightweight multi-component cladding materials

Component	*Key performance criteria*	*Material*
Glazing	Anti-glare. Solar/heat reflective. Colour. Light/heat responsive. Impact/shock resistance. Durability.	Toughened glass to BS 6206. Low-emissivity 'k' glass. Electrochromic glass. Photochromic glass. Thermochromic glass Multi-layered: double or triple glazed, with argon filled cavities. Prismatic glass.
Panels(1)	Durability. Strength. Solar reflectivity. Finish(2). Thermal insulation. Weather-tight joints(3).	Glass-reinforced plastic (GRP). Glass-reinforced cement (GRC). Colour-coated aluminium lightweight profiled sheeting or sandwich panels, with integral urethane insulation. Polycarbonate translucent cladding. Stainless steel. Fibre-reinforced polymer. Zinc (but requires a minimum 12–15 mm air space behind it for ventilation to minimize the risk of corrosion). Structural glass. Titanium (see below).
Framing	Durability. Strength. Rigidity. Compatibility. Fire stopping.	Colour-coated aluminium. Hardwood. Stainless steel. Titanium. Mineral fibre battens every storey and at every 8 m horizontally.

Notes

1. All the materials used as panels require insulation backing, either installed separately or as an integral part of the cladding itself.
2. The designer's choice of colour, texture and style of panel will dictate this.
3. Open joints can be made weather-tight by using a pressure equalized rainscreen system. Closed joints would need to be sealed using a high-performance mastic or compatible cover strip.

used in recladding schemes to older buildings where a traditional finish is required. Generally, however, lightweight-cladding systems would be used to most other non-residential buildings.

Nowadays, the roof and walls of many industrial premises are clad with factory-engineered insulated panel-cladding systems (FEIPCS). These usually comprise aluminium-covered sandwich panels with an insulated core of fire-rated rigid urethane insulation. Each FEIPCS panel is about 65 mm thick, depending on the U-value required and the spacings between the fixing positions, and may be connected directly to the structural frame. The insulating sandwich panels are tongued and grooved, and have neoprene seals, to provide an easy, air- and water-tight fit. This system could be used as a recladding option for an industrial building undergoing refurbishment. One major requirement would be the use of fire-rated rigid urethane or other insulation with low-combustible properties in the sandwich panel.

'Kingspan', for example, is one of several manufacturers of factory-engineered fabricated cladding systems. They can produce lightweight sandwich panels in a wide range of colours and profiles. The panels contain a polyurethane insulation core.

When recladding is being carried out it will be necessary to erect screens between floors at the façade. These are required to provide shelter against wind and rain during the recladding work. The 100 × 50 mm softwood framing covered with translucent heavy-gauge polythene reinforced with polyester mesh can offer the necessary protection from the elements whilst still allowing some daylight into the building. Such screening should be easily demountable as well as robust.

Overcladding

Lightweight materials

In contrast to recladding, overcladding does not replace the existing external walling of a building. Rather, it retains the existing substrate but covers it with a new, usually lightweight, skin. Overcladding can comprise 'jointed' or 'seamless' versions. It can often accompany an over-roofing scheme (Lawson et al., 1998).

A wide selection of materials is available for the jointed, multi-component, lightweight versions of overcladding. These built-up cladding systems comprise three main components as listed in Table 9.1. In particular, fibre-reinforced polymer and titanium (see below) are two relatively new materials used for cladding. The former is a composite lightweight material that can be used as blast-resistant cladding (CIRIA, 1998).

Titanium

There is every chance that titanium will become one of the most popular innovative construction materials for the new millennium. Titanium was discovered in the late 1700s, but it was not until the last half of the 20th century that it began to be used primarily for a variety of military and industrial applications (Hochman, 1998). Nowadays, it is being used for aeroplane skins, golf-club heads, skis, spectacle frames, hip and knee replacement joints, and jewellery.

Titanium's main architectural application, however, is a cladding material. For this, it is supplied in panels up to about 1 m in width. The sheets can range in thickness from 0.016 mm gauge (for tent-like coverings) to 0.4 mm for roof and wall cladding (Santini, 1999).

Titanium has a number of advantages over other alloys. Its weight is 60 per cent that of steel, half that of copper and 1.7 times that of aluminium. Titanium's low coefficient of thermal expansion is half that of stainless steel and copper is strong, light and has excellent corrosion resistance (Hochman, 1998).

The cost of titanium is one of its main drawbacks. It is currently about 5 times more expensive than stainless steel. This is because of the extensive work required to process the raw material. However, the cost of titanium may reduce over time as its production efficiency increases.

From a sustainability perspective, the expensive and non-renewable characteristic of titanium's manufacture is obviously another shortcoming. Its use as an overcladding/recladding material in adaptation projects, therefore, is likely to be reserved for high-quality, top-class refurbishment schemes to prestigious assembly or commercial buildings.

Brickwork

Detached and semi-detached bungalows with external walls of solid 225 mm thick brickwork were a common form of construction for dwellings built before the 1940s, especially in rural and suburban areas. During the 1950s and 1960s dwellings of precast concrete-framed non-traditional construction, too, were occasionally used in rural and suburban areas as a cheap and quick form of housing for agricultural workers (Chandler, 1991).

In terms of weather-tightness and thermal insulation, however, these traditional and non-traditional constructions are inefficient. As well as having a relatively high U-value (i.e. $>1\,W/m^2K$) the fabric is usually not very well sealed, which renders them cold, draughty dwellings.

One method of upgrading the walls of these dwellings is to use an insulated brick-slip overcladding system (see Figure 9.1 below). This is most suited to one-brick thick, solid wall dwellings.

An alternative option is the construction of a full outer skin of facing brickwork or rendered blockwork complete with 50 mm wide cavity from 35 mm thick polyurethane insulation batts around the exterior of the building. This method is more suitable for thin-walled dwellings such as 'Dorran' and other forms of non-traditional construction (see Figure 9.1). It would also offer a more durable and robust solution than insulated brick slips or rendering on the external fabric of these buildings.

In any event, of course, the eaves need to be wide enough to accommodate any new wall cladding. In cases where there is only a few centimetres of overhang at the eaves, a special metal flashing trim can be fixed to the wallhead to allow the installation of an insulated render or brick-slip overclad. A 'clipped eaves' would obviate the construction of a full half-brick thick leaf unless the roof edge was extended at the wallhead. Alternatively, a suitable metal flashing trim was installed at the eaves to accommodate the new brick skin but the aesthetic impact of this might be deemed negative. To minimize disruption and cost this work could be done at the same time as the roof coverings are being replaced and thermal efficiency of the roof structure is being upgraded (see section later in this chapter on upgrading pitched roofs).

Furthermore, the width of the existing strip foundation would require widening by about 200 mm to support any new outer leaf of brickwork. The extended scarcement should be taken to at least the same depth and thickness as the existing footing and butt jointed against it. The local authority may insist on the insertion of dowels to link the new to the old foundations. These could consist of steel bars 16 mm in diameter at least 300 mm long inserted in the middle depth of the foundation (i.e. 150 mm into both the existing and new footings) at about 600 mm centres around the entire building (see Figure 9.1). As indicated in Figure 5.21, the end of each dowel bar within the extended part of the scarcement should wrapped in debonding tape to accommodate shrinkage in the new concrete and allow for some differential movement.

Installing a concrete kerb over the scarcement is an alternative to extending its width by forming a simple butt joint. In such a case there should be no need to insert dowel bars to link with kerb to the scarcement. However it would be prudent to install a 5 mm diameter steel mesh strip neat the top of the kerb to prevent it from shear failure.

It is possible that the formation level of the existing footing is too shallow (i.e. <600 mm). In such a case the new foundation should be taken down to at least 600 mm to avoid problems associated with frost heave or erosion. This may require some minor underpinning under the existing footing using mass concrete if the depth of the new foundation is any greater than about 100 mm below the original formation level.

Double-skin façade

This is a relatively modern form of overladding, which involves installing a curtain wall-like skin about at least 600 mm away from the structural face of a building (see Figure 9.2). Ideally the gap should be about

1 m to facilitate good convection airflow. The space can also be used as an access void with walkways for undertaking maintenance work to the fabric.

The main benefits of double-skin façades are the following:

- Lowers the energy demand for the building.
- Reduces the building's heating, ventilation and lighting costs.
- Lower CO_2 emissions from the building.
- Utilizes passive stack ventilation principles.
- Improves the overall appearance of the building.

Vertical cross-section (not to scale)

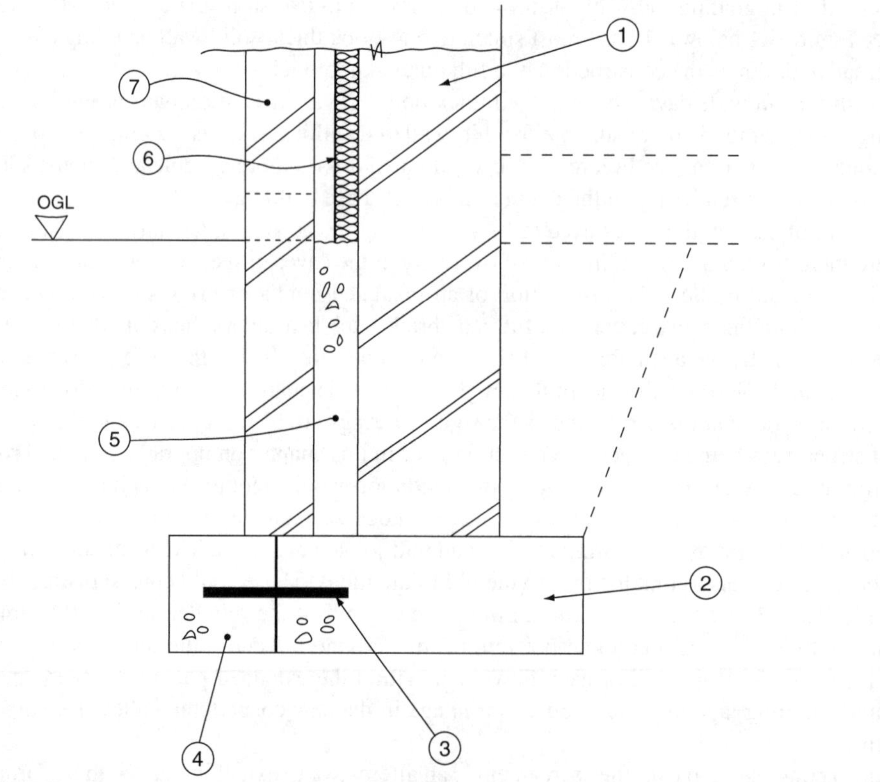

Notes

1. Existing solid one-brick thick wall or walling of reinforced concrete non-traditional construction.
2. Existing concrete strip foundation.
3. 16 mm diameter steel dowel bars 300 mm long at 600 mm centres (or as per engineer's instructions).
4. In-situ concrete nib providing 150 mm wide scarcement.
5. Lean mix 1:10 concrete infill in cavity up to ground level. Weep holes formed in brick perpends at 900 mm centres in course at ground level.
6. 50–75 mm thick polyurethane insulation batts with vapour control layer on warm side of construction.
7. New outer leaf of facing brickwork fixed to existing solid walling using stainless steel L-shaped wall ties, and forming 25–50 mm cavity.

Figure 9.1 Overcladding dwelling of solid wall or non-traditional construction using brick-skin, cavity and insulation batts

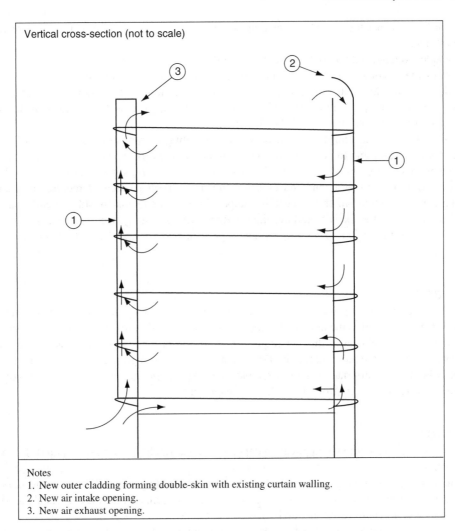

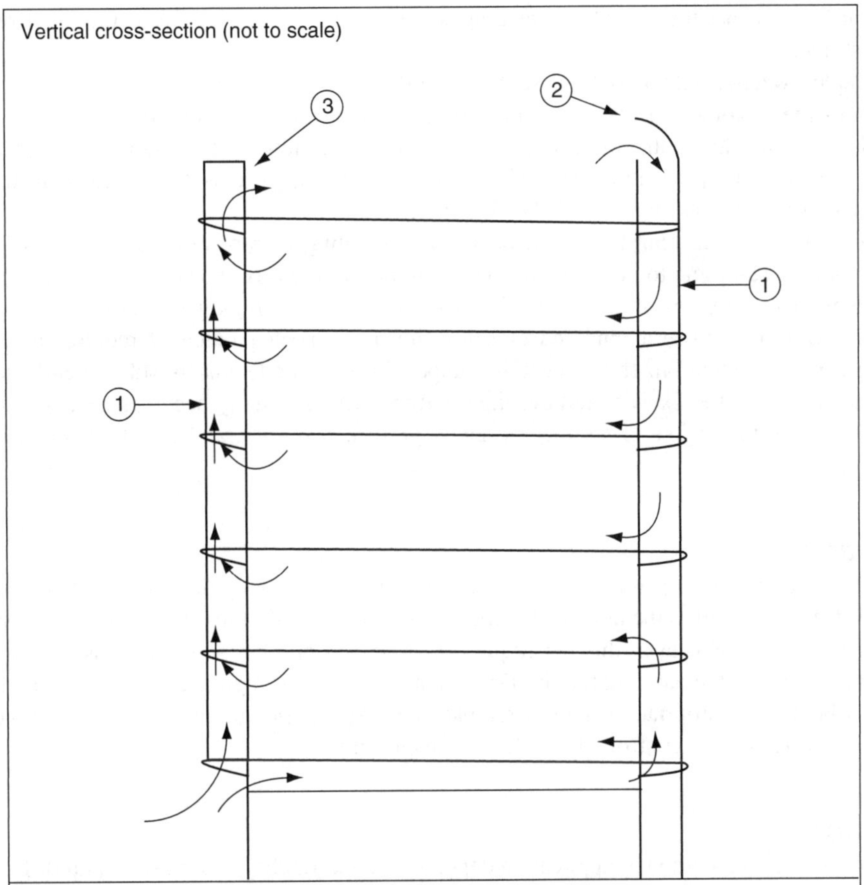

Notes
1. New outer cladding forming double-skin with existing curtain walling.
2. New air intake opening.
3. New air exhaust opening.

Figure 9.2 Double-skin façade

Double-skin façade cladding is of course more common on and suited to new buildings. However, in some cases it can be installed on the façade of an existing building. The main requirements to facilitate this are the following:

- External space (particularly at ground level) to accommodate extended width of building at set elevations.
- Existing substrate capable of receiving the required framework to support both the new curtain walling and walkways within the cavity thus formed.
- Planning restraints do not prevent this overcladding option.

Repainting façades

Preamble

Painting or repainting the exterior of a building is a relatively cheap and easy way of enhancing if not rejuvenating its appearance. Such a treatment can transform a previously dull or drab façade into a bright, attractive finish. It also has the benefit of providing some breathability and increased weather resistance to the coated surfaces.

For example, water-repellent coatings such as TEGOSIVIN® HL, which is a colourless ethoxy functional solvent-free oligomeric siloxane, can improve the appearance and durability of external walling. They are designed for the treatment of natural stone, bricks and low-alkaline substrata, such as artificial limestone. However, the precautions for using waterproof coatings to walls described in the previous chapter still apply in this case (see also BS 6477, 1992).

There are, of course, a number of drawbacks with painting or repainting a building's façade. The cost of scaffolding a building to facilitate this work can run into tens of thousands of pounds. Scaffolding itself brings with it problems of compromising security and encouraging vandalism during the period of the work. Moreover, once a rendered or brick finish has been painted, it requires a greater level of ongoing care and attention. If moisture is trapped in a substrate that is subsequently painted any new coating applied to it is likely to fail eventually. Repainting it usually becomes necessary every 5–15 years depending on the degree of soiling, extent of graffiti, exposure, and quality of paint system and workmanship.

Preparation

The extent of preparation for painting or repainting a building externally depends on the type and condition of the substrate as well as the degree of soiling. High-pressure washing of rough surfaces such as rendering may be needed. However, this technique brings with it problems. It is a messy process and will introduce more water to the building fabric. 'Dry' techniques such as high-pressure pneumatic hose or grit blasting may be more appropriate in some circumstances. Again, though, dust generation and disposal of waste, respectively, must be controlled and dealt with carefully.

Application

The form of coating and method of application depend on the durability and finish required. Table 9.2 lists the main options.

The colour of the new coating should be selected to suit the building context. For example, dark colours should be avoided wherever possible. Not only do dark coatings increase the fabric's heat absorption, they are also result more likely to result in a drab, depressing finish.

One company has launched a state-of-the-art silicate paint, ideally suited to the decoration and protection of buildings of historical and architectural importance. It's called 'Glixtone Silicate Paint'. The product is guaranteed for 15 years before first maintenance is required and has been tested in accordance with BS 476 Parts 6 and 7, achieving the optimum class '0' rating.

Window and door replacement

Precautions

It is important to ensure that existing openings are not enlarged or otherwise disturbed to accommodate new windows and doors. For example, some installers may be tempted to widen existing openings slightly to accommodate standard size frames. This can compromise the bearing of the lintels. Obtaining the precise frame size is the best solution in the circumstances. If for some reason this is not possible, it is better to use a slightly narrower/shallower frame; the side/top gaps can then be filled with a coated metal or polymer-cladded dead panel.

Other precautions include maintaining the existing vertical and horizontal damp proof courses and possible strengthening of the mullions at bay windows. This latter point was discussed in Chapter 7.

Table 9.2 Options for painting or repainting various exterior surfaces

Substrate	*Coating*	*Precautions*
Brick	Cementitious paint system (e.g. 'Sandtex'). Masonry paint (e.g. 'Powercote'). Polymeric paint system (e.g. 'Decadex'). Resinous paint system (e.g. 'Protex'). Silicone or siloxane clear coating (e.g. 'TEGOSIVIN® HL').	Surfaces must be clean and free from contamination and loose material. Defective joints must be repointed with a mortar using an appropriate mix. Delaminated or friable brick faces should be cut out and patched using a cementitious or resinious mortar repair paste. Should not be used on substrates contaminated with salts.
Concrete	Any of the coatings used for brick.	Ensure any concrete defects are repaired as described in Chapter 7.
	Bituminous paint. Multi-purpose epoxy coating (e.g. Amercoat 235).	Useful as a water-resisting coating covered by a durable cladding.
Metal	Micaceous oxide or zinc-rich paint systems. Aliphatic polyurethane topcoat (e.g. Amercoat 450E). Acrylic epoxy gloss coating (e.g. Amercoat 229).	Any loose scale and rust must be removed and bare surfaces treated with a rust-inhibiting primer before application of undercoat/s.
	Polysiloxane coatings (e.g. Ameron International's PSX® 700 siloxane epoxy – two-coat system).	These are based on pure inorganic siloxane binder are generally recognized as the newest generic class of high-performance protective coating.
Rendering	Any of the coatings used for brick.	Wet-dash (harling) substrates best suited to being painted – but even then it must be clean and free of contaminants such as algae, dirt, grime, etc. Dry-dash pebble render less suitable for repainting. A thicker paint coating should be used in such circumstances.
Stone	Any of the coatings used for brick.	Should only be used on clean, stable backgrounds. Salt crystallization if present can be trapped by the coating, causing damage (i.e. through crypto-efflorescence) to the substrate of the stone.
Timber	Microporous acrylic stain or varnish.	Ensure any treatment for fungal attack, mould staining or other wood damage is carried out before application.
Plastic	Any of the epoxy/polymeric coatings used for metal.	An etching primer should be used to achieve good paint adhesion.

Table 9.3 New thermal requirements for windows to satisfy the building regulations

U-value (W/m^2K)	*Material*	
	Timber or uPVC frame	*Metal frame with 8 mm thermal break*
1.8	High-performance, soft-coated triple glazing.	High-performance, soft-coated triple glazing.
2.0	High-performance, soft-coated double glazing.	Triple glazing.
2.2	Soft-coated double glazing.	High-performance, soft-coated double glazing.
2.5	Hard-coated double glazing.	Soft-coated double glazing, or argon gas-filled glazing.

The type of glass required should also be chosen with care to avoid using expensive, inappropriate as well as ineffective glazing for any window and door replacement programme. Table 9.3 lists the main types of glass available for window as well as cladding systems.

Any window and doorset replacement work should be undertaken in accordance with the appropriate codes of practice (British Plastics Federation, 1996; BS 8213–4, 1990).

Double and triple glazing

Replacing defective metal or timber windows and doors is a common item in non-residential as well as housing modernization programmes (see BS 8213, 1990). When this is being considered, the opportunity to replace these components with double glazed high-quality timber or uPVC units is normally taken. Double glazing helps to reduce heat loss from a building's fabric. Ordinarily its U-value is about 3.0 W/m^2K compared to 4.7 W/m^2K for single glazing.

The main requirements for such work, however, are the following:

- Minimum 4 mm thick panes of glass. Reductions in solar gain can be achieved, though, by using 10 mm thick 'anti-sun' outer pane and 6 mm thick Pilkington 'K' glass inner pane.
- 12 mm minimum cavity between panes of glass (20 mm is preferable for optimum thermal performance). The cavity can be filled with argon gas for better thermal insulation.
- Consider using sustainable products such as high-quality softwood/hardwood or pultruded glass fibre as an alternative to uPVC.
- Use a design of window appropriate for both the building and user (see below).
- Avoid installing full dead-light windows on all upper-floor rooms – thus eliminating the possibility of rapid ventilation and preventing a means of escape in case of fire.
- Adequate trickle ventilation (see Table 3.8).
- Windows on first floor and above must be cleanable from the inside (see Figure 9.3).

The U-values for windows have been reduced from the current 2.2 W/m^2K to 1.8 W/m^2K. The ways in which windows will need to be glazed to meet these changes in the building regulations are outlined in Table 9.3.

Ironmongery

The ironmongery of replacement windows and doors needs to be selected carefully to reflect the types of occupants. The elderly, because of their infirmity, require ironmongery that is light and easy to use. Window ironmongery at first-floor level and above should be designed with childproof latches to prevent toddlers opening the sash beyond 100 mm.

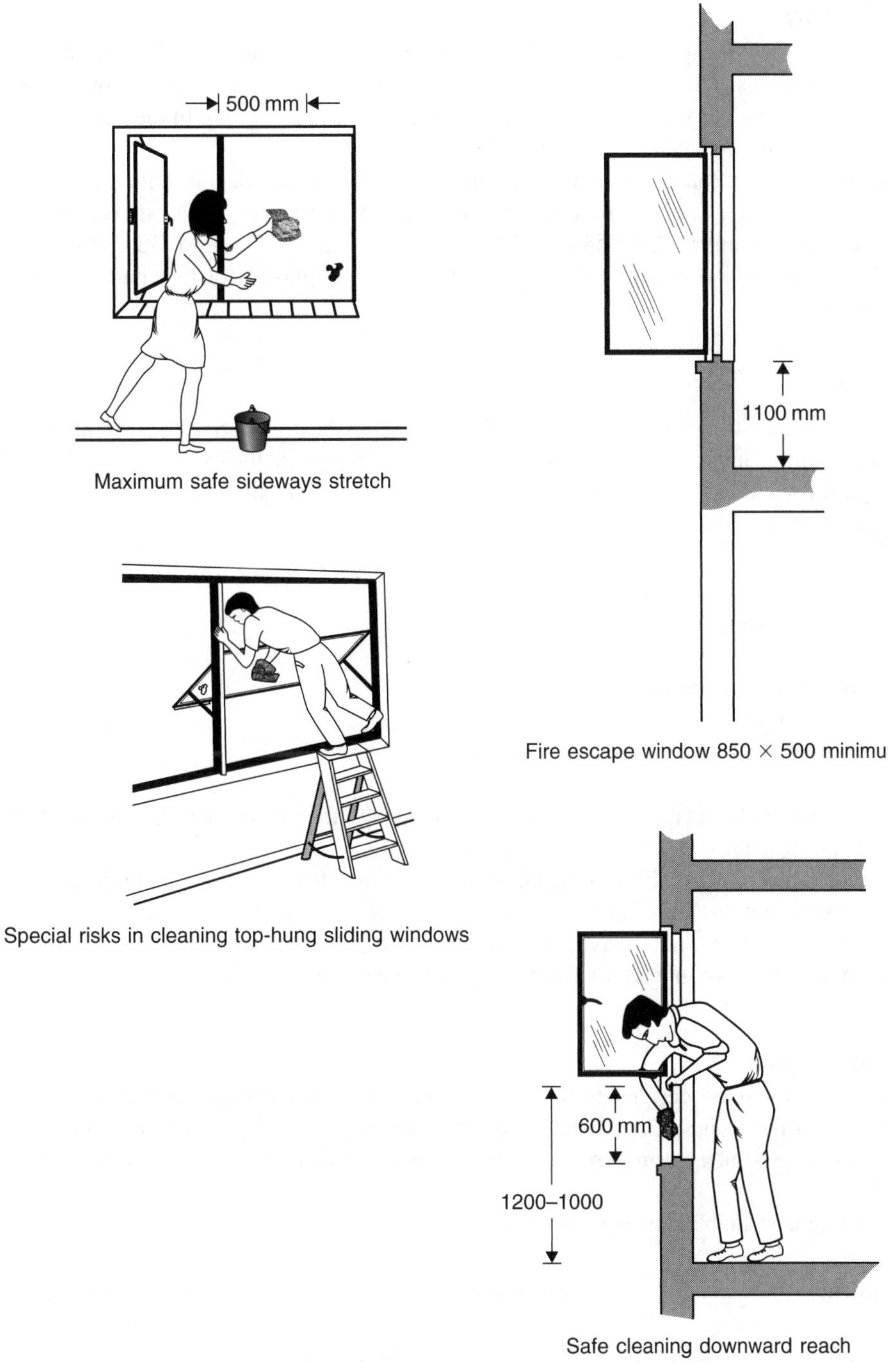

Figure 9.3 Design requirements of new windows for cleanability (BS 8213, 1990)

Safety glazing

Since January 1996 the Workplace (Health, Safety and Welfare) Regulations 1992 require safety-glazing measures in all buildings. This means that this type of glazing is required in areas between the floor and 1500 mm above in doors and side panels that are within 300 mm of the doors. In the case of partitions and windows, it is necessary between the floor and 800 mm above.

Safety glazing to BS 6202, 1981 helps to minimize the risk of accidental human impact against the glass. In adaptation work to facilities requiring greater security or resistance against explosions then either security/anti-bandit glazing to BS 5544, 1978 or bullet-resistant glazing to BS 5051, 1994 may be required. The 12 mm thick toughened glass can be used in fully glazed corridor partitions as part of an office refurbishment.

Materials

The fenestration of industrial and commercial buildings may be the only part of the external fabric of these properties that requires upgrading. Nowadays it is common to replace the existing defective softwood or steel single-glazed windows with one of the following materials containing double glazing with 20 mm minimum gap between panes:

- Colour-coated, aluminium-framed units.
- Stainless steel framed units.
- uPVC.
- Stained hardwood or good-quality softwood.
- Pultruded glass fibre.
- Composite system (e.g. timber and uPVC/metal).

Sustainability factors will dictate the choice as much as whole life cost influences. The finish, repairability and robustness of the material will also affect the choice.

The sills, whether of timber, uPVC or metal, should have an adequate bevel and projection to throw off any rainwater. Drips and throatings in the sills are also essential in this regard.

Only high-quality polysulphide mastic should be used to point around the windows frames. Any gaps or voids behind the frames should be filled with polyurethane expanding foam.

Window design

There are a variety of window designs that can be used. The selection will depend on the nature of the building and the needs of the occupier. Listed buildings, for example, would require a window design in keeping with the original style – such as small paned timber sashes for a Georgian property, whether residential or commercial.

The main designs available nowadays are the following:

- Pivot (vertical or horizontal – may lack weather-tightness if seals are not properly installed or of poor quality).
- Vertical sliding sash and case (for Georgian and Victorian buildings).
- Side hung casement (for more modern dwellings).
- Tilt and turn (not suitable for elderly users because of the complexity and awkwardness of operating this type of window).

Regardless of the style selected, the window design must take into account for safety and cleanability requirements. These are illustrated in Figure 9.3. It should also consider the controllability of the opening of windows. This can give occupants a degree of control over their environment, which may help in reducing the incidence of sick building syndrome (see below).

The ease of operation of windows and doors is another consideration, particularly where the occupants are likely to be infirm or elderly. Such users may not have the strength or agility to open, close or lock window sashes and doors properly. Heavy doors with awkward handles and stiff locks can also prove difficult to operate for older people.

The width of any dead light below opening frames should not be more than about 1 m. This is to avoid sagging of the transom rail, which will tend to become overstressed if the span is excessive. On openings wider than 1 m, a mullion taken down to the main sill should be installed to separate the two sashes.

Access for maintenance

Preamble

Many multi-storey buildings have little or inadequate access provision for maintenance. This can make repair and other maintenance work to the building both awkward and hazardous as well as expensive. The opportunity to rectify this deficiency, therefore, can be taken during a major refurbishment scheme.

Temporary access

For low-rise buildings ladders and hoists provide the main forms of temporary access. These are of course limited in height and may not always provide total access to the exterior of a building. In such circumstances and for buildings taller than two storeys more substantial access is required. Mechanical scissor hoists, hydraulic jib lifts with platforms or 'cherry-picker' hoists may be appropriate up to three storeys. 'Simon Snorkel' hydraulic jibs can reach to a height of 30 m if necessary, but these are expensive to hire and require adequate vehicular access. Improved access may need to be provided in any adaptation scheme.

Hiring portable access equipment can of course be expensive and disruptive. Where regular use of hoists, etc. would be inevitable, permanent access is likely to prove more cost-effective in the long term.

Permanent access

Permanent access can be formed in a variety of ways, not all of which can be easily installed as part of an adaptation scheme. The main examples of this form of access are summarized as follows:

- Face-mounted cradle system.
- Soffit fixed supports.
- Cantilevered track system.
- Cantilevered trolley system.
- Fixed ledges.
- Moveable ladders.

Figure 9.4 shows typical access provisions that can be made following the refurbishment of a multi-storey commercial building with either a flat roof or pitched roof. The opportunity to install a new cradle system for the maintenance of the main elevations will save on scaffolding and temporary hoist costs. However, the type of roof structure as well as the aesthetic implications of the access provisions on the building will determine the appropriateness of installing such equipment. Moreover the equipment itself will require maintenance, which will add to the building's aftercare costs. A cost–benefit analysis of the main options would help to determine whether or not providing permanent access would be cost-effective.

Schematic vertical cross-section (not to scale)

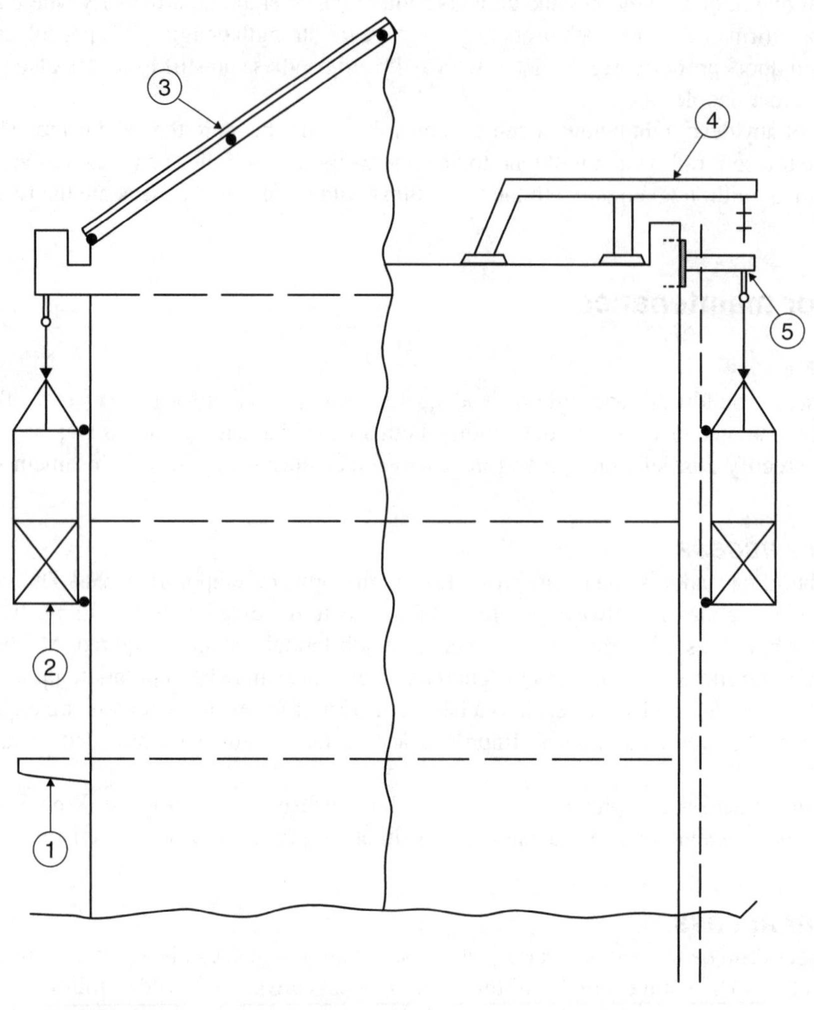

Notes

1. Galvanized/Stainless steel (G/SS) bracket supports under 600 mm wide open grid G/SS ledge at every storey.
2. Cradle system hung from 90 mm diameter G/SS tubular track support and runway rail fixed to G/SS square plates bolted to soffit. Stabilizers and ball wheels provided on cradle to ease its travel over face of cladding.
3. Aluminium sliding access ladder on rails over glazing or other vulnerable sloping roof surface.
4. Cantilevered jib and monorail track cradle system comprising G/SS I-sections bolted to concrete upstand bases anchored to structural deck of flat roof. Alternatively, a cantilever trolley system could be installed on a twin-track arrangement fixed to flat roof deck in a similar manner as above. The equipment should be housed in a garage on the flat roof where possible for protection and to make it more discrete.
5. Face-mounted cradle supports comprising rectangular hollow section welded to G/SS plate securely bolted to stable substrate. Weathering detail required where jib section passes through new overcladding.

Figure 9.4 Typical access requirements for maintenance

Environmental improvements to residential buildings

The overall appearance of a housing estate can be dramatically improved by incorporating some basic environmental works at the same time as the modernization programme is being undertaken. These could include the following:

- Providing better and more convenient recreational facilities – such as play areas and enclosed drying greens near the dwellings. The surfacing of play areas should be finished with soft rubberized coverings to minimize accidents from children falling.
- Providing safer and more convenient road layouts and car parking facilities – narrowing roads to create parking spaces and reduce car speeds, and traffic calming measures such as road bumps and chicanes.
- Installing more secure and discrete storage areas for 'wheely' bins or multi-compartment refuse containers for separating waste into distinct sections – for glass, metal, plastic, paper and organic materials.
- Installing bicycle parking bays next to dwellings to encourage more use of this 'green' mode of transport.
- Providing better access to and around the buildings for disabled and elderly people (see notes below).

Environmental improvements to non-residential buildings

Entrance foyer

The construction of a new canopy over the main entrance can transform an otherwise plain front elevation. It can give the building a distinctive new look (see Figure 8.3). This may include the installation of revolving doors as well as the provision of a very durable aluminium entrance matting system. Respectively these latter features will prevent draughts and dirt accumulation at the entrance.

As in housing, improved access for disabled people is another important requirement for commercial properties. This is especially the case with buildings that are used by the public. Buildings whose entrances are at ground level pose little trouble in providing access for disabled users. The same cannot be said about buildings that have steps leading up to the main entrance. A side ramp with handrail can be installed to overcome this problem (see Figure 9.16).

Car parking

Car parking facilities may also require upgrading along with the building. At least 1 bay in 10 should be reserved for disabled users. It could entail, for example, the use of electronically controlled traffic barriers to prevent unauthorized access to the car park. One form of these barriers can be inserted into the access road at the car park entrance. When operated a steel wedge-shaped plate spanning the road entrance to the car park is raised sufficiently to prevent any car from entering. The plate is lowered when deactivated to become flush with the road surface to facilitate car entry or exit. Alternatively, retractable stainless steel bollards can be used for protected parking and forecourt security.

Surrounding areas

The opportunity to upgrade the surrounding areas of a commercial building may be taken as part of its overall refurbishment. This could include works such as follows:

- Adjustable entrance bollards or thresholds: fully retractable or hinged stainless steel or other durable posts/plates to provide access control to the building, for parking and forecourt protection.
- Anti-bandit/crash barriers: galvanized or stainless steel bollards and rails at bumper height.

- Hardstanding: new surfacing to driveway and surrounds of premises using brick paviors or tarmacadam. Road markings should be done using chlorinated rubber paint for durability.
- Planting: new vegetation to enhance the soft landscaping around the premises.
- Bicycle canopies: covered stations with wheel rests to allow cyclists to lockup their bikes in a sheltered area.

Canopies

Ready-to-tile prefabricated roof sections for bay windows or storm porches at the front entrance to common stair entrances and individual dwellings are sometimes included in housing modernization schemes. This can help to improve the appearance of the building and reduce heat loss from the front entrance.

Canopies can also be added to commercial buildings as a way of highlighting the entrance and making the building more striking. They are usually much bigger than their residential counterparts. Such canopies are more likely to be either mono-pitched or barrel vault in overall profile, and require more substantial support – see Chapter 7.

There are several styles of small canopies but the main designs for dwellings are:

1. mono-pitched,
2. duo-pitched with gable,
3. duo-pitched with hipped end,
4. barrel vault.

Unlike the abutments of a proper 'extension' canopies do not normally require the installation of a cavity tray. The flashing, however, should be 150 mm deep, with the top edge chased into the main wall.

Polyurethane is a popular modern material for these canopies. They are prefabricated and come in a variety of colours and designs.

Upgrading floors

Requirements

Modernizing a building or changing its use can mean new requirements for its floor structures. The methods for improving the structural capacity of a floor have already been addressed. In this section the option for upgrading other aspects of the floor's performance is considered. The three other main performance requirements for floors are moisture resistance, fire resistance and sound insulation. These are especially important when a building is being converted to another use or where a multiple use scheme is being proposed (see BRE Digests 208, 1985 and 334, 1988a, b).

Hygrothermal performance of floors

Preamble

Ground floors are more problematic than upper floors as far as dampness is concerned. As discussed in the previous chapter, the main problem with upper floors in this regard is fungal attack resulting from unprotected built-in ends of joists. Floor timbers generally, though, are also vulnerable to woodworm.

Both suspended and solid ground floors are susceptible to moisture and heat loss. In any modernization programme, it would be prudent to consider the need for some upgrading of these ground floors to overcome these problems (see Figure 9.6). However, given the degree of disruption involved, the work described below would require the affected rooms to be vacated.

Suspended ground floors

Aeration of subfloor voids is essential to minimize the risk of dry rot in this part of a building. If necessary, therefore, the subfloor ventilation must be increased to 1500 mm^2/m run of external wall to comply with the building regulations. The opportunity should be taken to eliminate debris in the subfloor void and remove any blockages in dwarf walls or fresh air inlets (FAIs).

Installing ground cover on the solum will reduce moisture levels in the subfloor space. This is because moisture can evaporate from the bare earth, especially in damp soils. A 1200-gauge polythene sheet (e.g. 'Visqueen') under a 75 mm thick slab of oversite concrete would suffice. However, installing this effectively may be difficult owing to access problems, particularly if the crawlspace is shallow.

Ideally, the opportunity should be taken to install insulation between the joists. The 150 mm thick mineral fibre insulation laid on meshing for support between the joists could be used to achieve this. Uplifting of the floorboards to facilitate installation of the insulation would of course be required. A heat reflecting membrane (HRM) could be incorporated within the construction to increase its thermal efficiency (see Chapter 10).

Solid ground floors

It is sometimes necessary to replace an existing defective ground floor structure of either solid concrete or suspended construction. This is usually done with a new 150 mm thick concrete slab on a damp proof membrane (dpm) on a properly compacted inert base. In such cases a 'surface dpm' may be needed on the top of the slab to facilitate early laying of the floor covering (Pye and Harrison, 2003).

A surface dpm usually consists of an epoxy resin liquid applied in two coats each about 0.15 mm thick. 'TREMCO Treadfast ES3000' is a typical example of a proprietary surface dpm. It can be applied to screeds or slabs that have a moisture content of up to 90 per cent equilibrium relative humidity (ERH) – which is 15 per cent more than the limit set by BS 8204.

The opportunity to improve a solid concrete ground floor's thermal performance should be taken during any replacement work. This can be achieved by installing a layer of resilient insulation such as polyurethane or 'Neopor' about 75 mm thick under the new slab to achieve a U-value of 0.25 W/m^2K.

Fire resistance of floors

Separating floors need to have a minimum level of fire resistance to comply with the building regulations. Ordinary upper floors may become separating floors in the conversion of a building from single- to multiple-occupancy. This usually requires upgrading of their fire resistance to at least 30 minutes (see BRE Digest 208, 1985).

According to Haverstock (1998), the points that need to be covered when assessing fire resistance in existing floors are the following:

- Adhesion of plaster. Any extensive bossing or spalling of plaster?
- Plaster thickness. Too thick plaster on laths drops off, too thin breaks up; 16 mm is the best compromise but do not expect much from it (probe thickness and investigate bulges).
- Fixing and spacing of laths (especially if the property is pre-1900).
- Thickness, type and condition of floor boarding.
- Quality and condition of floor finish (a layer of hardboard may be required if it is uneven).
- Joists size, spacing and condition, especially if severely notched for services.

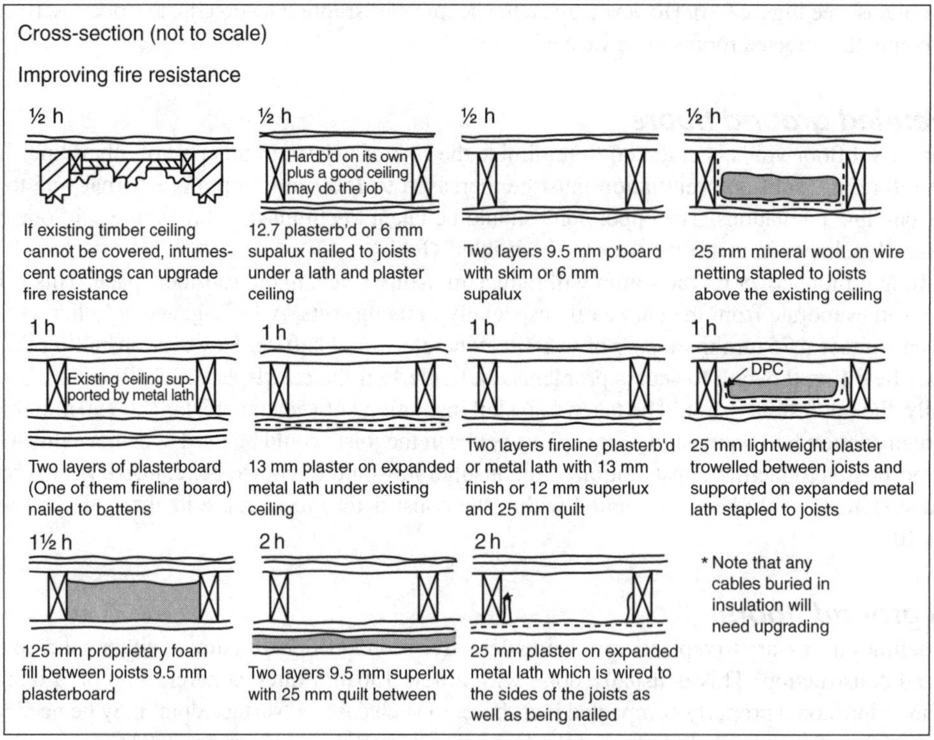

Figure 9.5 Options for upgrading the fire resistance of suspended floors (Haverstock, 1998)

- Unusual constructions (such as pugging, cast iron trimmers) present?
- Value or quality of existing ceiling and floor (Can they be covered over?).
- Gaps and passages for smoke – which is what usually kills people in a fire and therefore must be double-checked.
- Fire-stopping services through walls and floors, and other hazards such as gas pipes in voids.

The various methods of upgrading the fire resistance of floors are illustrated in Figure 9.5.

Sound insulation of floors

Effects of noise

Simply put, noise is unwanted sound. Sound transmission in buildings is airborne and/or structure borne. It has five main adverse effects on users of buildings:

1. Annoyance. Load music from hi-fis and televisions is particularly disturbing and stressful to adjacent occupiers if exposure is prolonged.
2. Hearing damage (operatives using loud tools or occupiers playing load music are more vulnerable to this problem).
3. Sleep disturbance.
4. Interference with communication or to hear other, more important sounds.
5. Lack of speech privacy.

Measures

It is important therefore in any adaptation, brief to address the client's expectations and statutory requirements on sound insulation (BRE report BR 238, 1993). In particular, adequate sound insulation will be critical in multiple-occupancy residential properties or in buildings where privacy from excessive noise is essential. This is especially important for party floors and walls (BRE Digest 334, 1988).

Meeting the statutory requirements may not be easy to achieve without some major alterations to the floor structure. It could also involve changes to the ceiling and floor finishes.

Some of the points listed in the previous section may need to be considered when assessing the sound resistance of existing floors. In addition, the quality and extent of fixings of the boards would influence the type and extent of sound insulation required in a floor (see BRE Digest 293, 1985). Figure 9.6 shows the traditional measures for improving the sound insulation of suspended floors.

To a certain extent there has to be a trade-off in adaptation schemes between achieving adequate levels of both sound and thermal insulation. As indicated in Chapter 3, measures to improve airborne sound insulation do not provide adequate resistance to impact sound transmission. Adding mass to the structure mitigates airborne sound transmission. Isolating or decoupling the mass from the structure, on the other hand, controls impact sound transmission.

There are currently a variety of proprietary sound insulation systems for improving the acoustic performance of floors (and walls). Apart from the infill techniques referred to above, these sound insulation enhancement systems comprise one of the following two methods, depending on the floor structure involved:

- Timber decking and joists: Plating the floor with a layer (or layers) of decoupling fibre/foam, overlaid with variable thicknesses and densities of sound barrier, in turn overlaid with tongued and grooved decking of various thicknesses and weights to suit specific conditions. This inevitably raises the floor level, which will necessitate refitting doors as well as adjusting skirtings and treads of stairs affected by such work. There is a range of standard systems to provide the combination of performance qualities required, depending on the existing floor and ceiling constructions (see Figure 9.7).
- In situ or precast concrete: An acoustic flooring system consisting of either timber battens on rubber pad cradles (which are manufactured from recycled products) or 10 mm thick neoprene strips under timber battens overlaid with chipboard or plywood in a range of standard floor heights and finishes can be used. Mineral fibre-acoustic insulation is placed between the battens. Patented versions such as the 'InstaCoustic' cradle and batten system solves the problems of sound transmission through concrete floors, whilst providing room for services. British Gypsum, for example, have developed a 'Gyproc SI Floor System' comprising 21 mm thick softwood tongued and grooved floor boarding with 'Gyproc Plank on SIF' channels; 100 mm thick 'Isowool 1000' glass mineral fibre between the joists and 'Gyproc Resilient Bars' at 450 mm centres.

In some older properties such as tenements the presence of timber decay within separating floors often results in patch repairs or wholesale replacement. If these works are not done properly this can lead to a marked reduction in the sound insulation performance of the floor (CIRIA, 1986). The methods for improving sound insulation in such cases are outlined in BRE IP 6/88.

Anhydrite screeds

Part of the refurbishment of a building may involve refurbishing the concrete ground or upper floors. Installing an anhydrite screed over an existing sound concrete substrate is one way of doing this. Moreover, anhydrite screeds work well with underfloor heating.

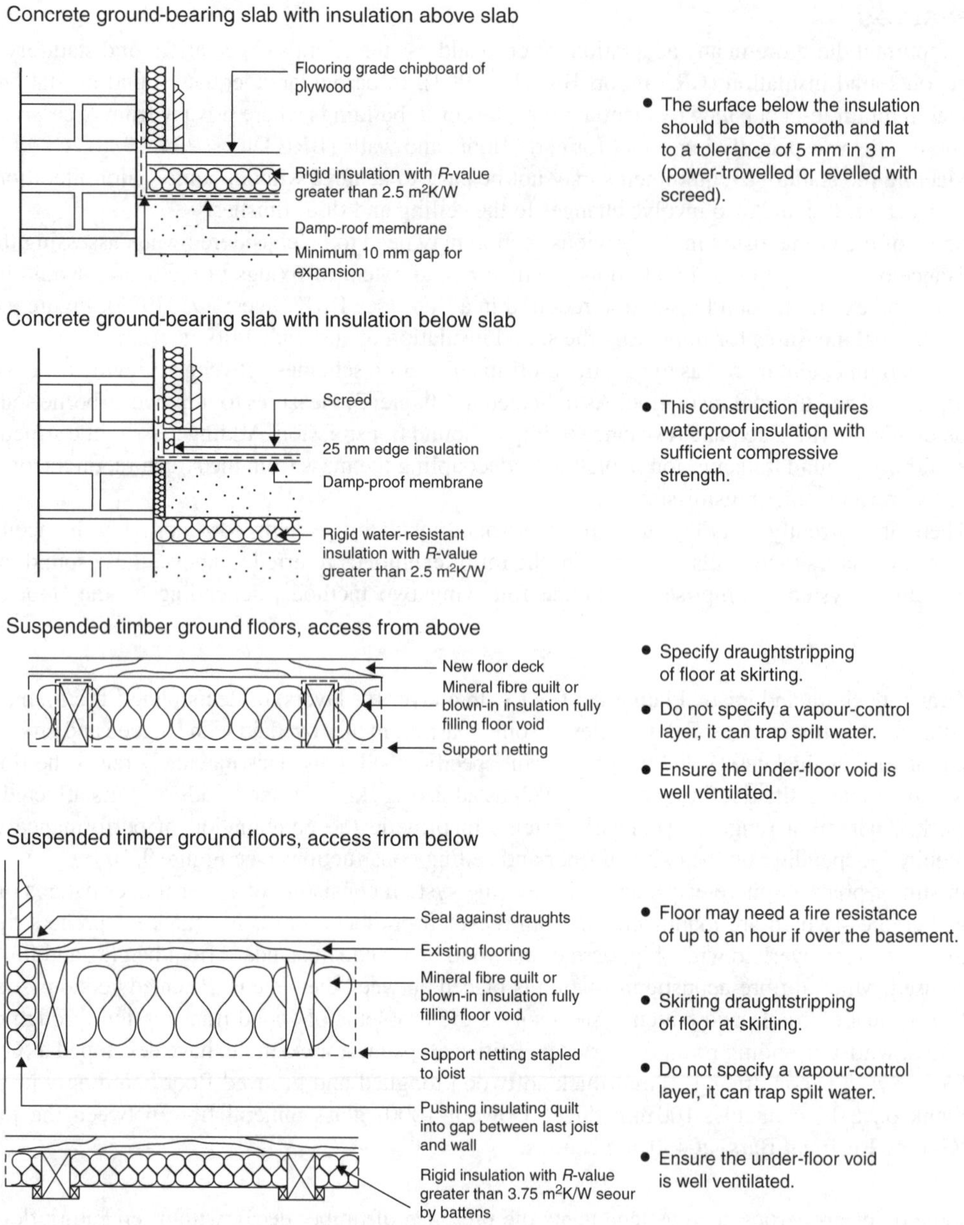

Figure 9.6 Options for upgrading the thermal insulation of ground and upper floors (GPG 155, 2001)

Anhydrite screeds, such as Isocrete's 'Gyvlon', comprise essentially calcium sulphate mixed with a synthetic binder. They are pump-applied and laid to an average thickness of about 40 mm. The other components of the screed system are the following:

- Polythene sheeting (1000 gauge) between the substrate and the anhydrite screed.
- An epoxy primer consists of a two-part epoxy mixture applied in two coats. The first coat is diluted with 1:1 water to facilitate penetration into the anhydrite screed. The second coat is applied neat.

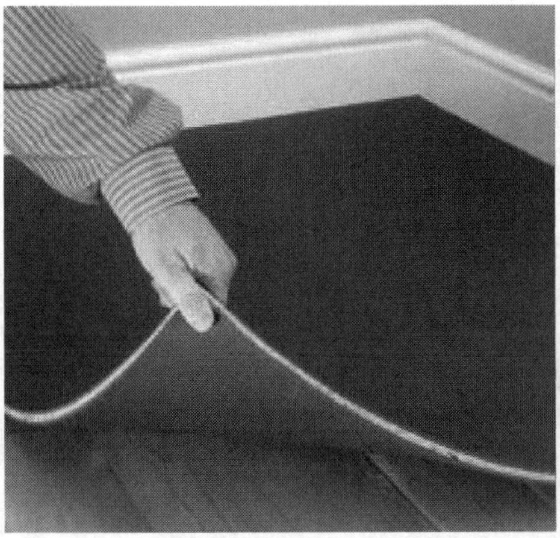

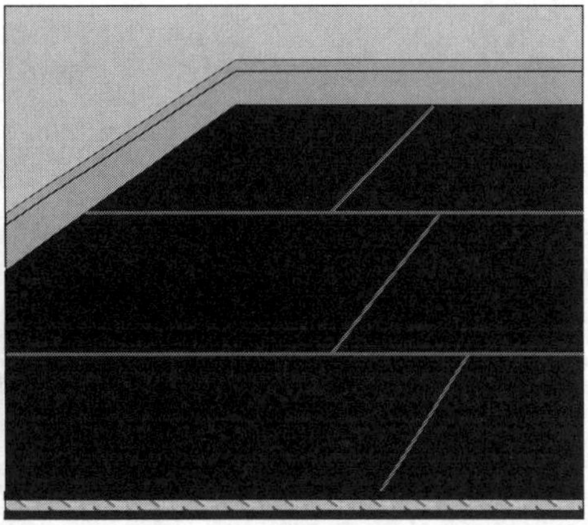

	Airborne improvement (dB)	Impact improvement (dB)
Acoustilay 3	0	30
Acoustilay 8	3	30
Acoustilay 15	5	30
Acoustilay 30	8	20

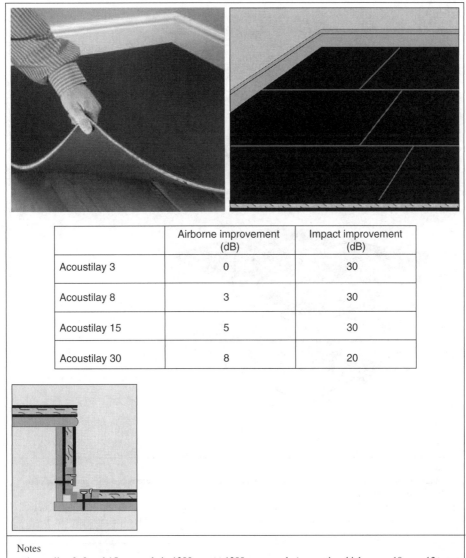

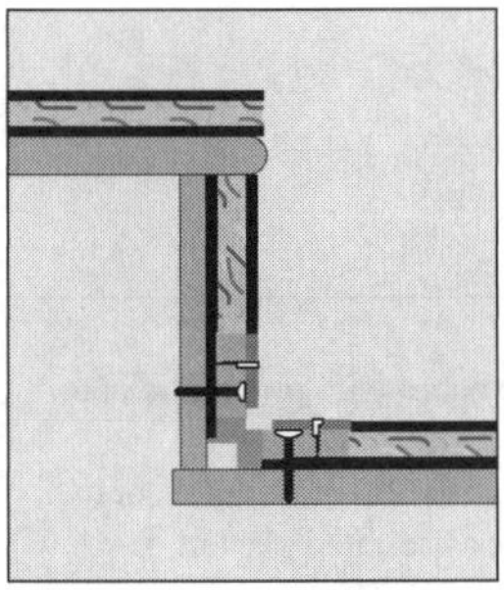

Notes
- Acoustilay 3, 8 and 15 are made in 1200 mm × 1200 mm panels (respective thicknesses 10 mm, 12 mm and 15 mm) whereas Acoustilay 30 is only made in a 1200 mm × 600 mm panel because of its weight (thickness 22 mm).
- Data on 'acoustilay' performance.
- 'Acoustilay' installed on stairs with perimeter strip.

Figure 9.7a Proprietary measures for upgrading the sound insulation of suspended timber floors and stairs (Courtesy of Sound Reduction Systems Ltd – www.soundreduction.co.uk/)

- Self-levelling screed (such as Laybond's 'Screedmaster 1') on top of the primed anhydrite screed. The former comprises a blend of fine aggregate, cement and organic binder, which is designed to produce a self-levelling screed when combined with water on a ratio of 1:5 (i.e. 1 of water to 5 of Screedmaster 1). It can be applied by trowel or by mortar-screw pump to a thickness range between about 3 mm and 100 mm.

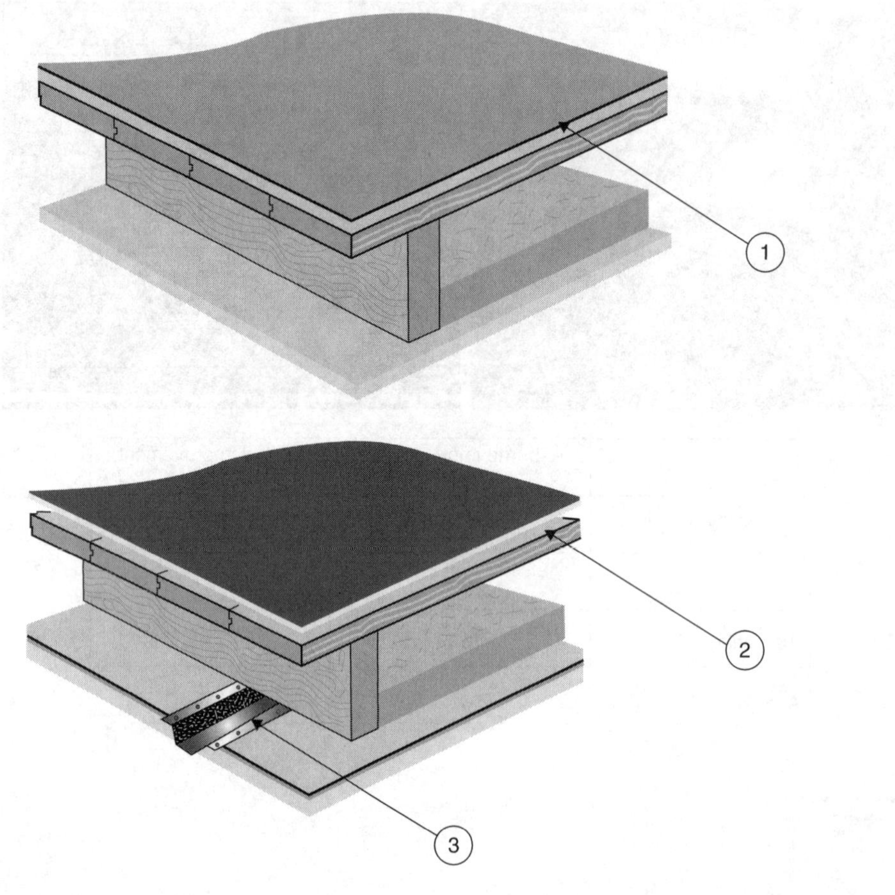

Notes
1. Plating of floor with 17 mm thick 'Maxiboard' cement-gypsum composite board with polymeric core.
2. 'Acoustilay 8' underlay 12 mm thick on floor.
3. Plating ceiling with Maxi 60 system comprising galvanized 'top-hat' section at 300 mm centres onto which 'Maxiboard' 17 mm thick is fixed and then plated with 12.5 mm thick fire-rated plasterboard to give 1h fire protection.

Figure 9.7b Proprietary measures for upgrading the sound insulation of timber suspended floors and stairs (Courtesy of Sound Reduction Systems Ltd – www.soundreduction.co.uk/)

However, there is a word of warning about the installation of these anhydrite screeds. After installation they must be protected against excessive abrasion and impact damage. The screed must be sanded down to remove surface laitance otherwise there is a high risk of debonding between the anhydrite and the self-levelling screed.

Dry 'screeds' or dry floor plating

Installing special plasterboard or composite gypsum fibreboard sheets instead of a screed is a relatively new way of improving an existing floor. These comprise insulation/foam-backed sheets laid 'loose' onto a prepared substrate. The base of course should have no irregularities and be free of dampness to avoid damaging or

Vertical cross-section (not to scale)

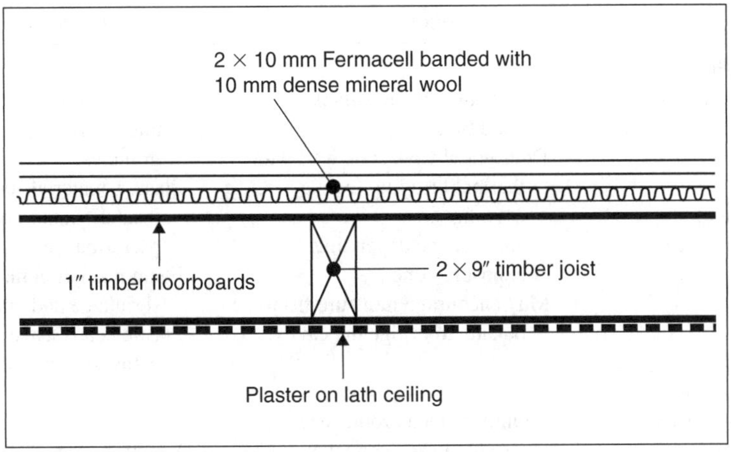

Notes
1. Resilient foam can be used instead of mineral wool to enhance the floor's acoustic performance.
2. Stair treads and risers should be treated in the same way.
3. If laminated timber or thin floor coverings such as lino are to be laid onto the fibreboard it must be plated with 5 mm thick plywood to minimize ridging.

Figure 9.8 Proprietary measures for upgrading the overall performance of floors and stairs (www.fermacell.co.uk/)

degrading the sheets. The joints of the sheets are then taped and scrimmed in the same way as wall plasterboard (Figure 9.8).

As with plating existing floors to increase their sound insulation, installing plasterboard sheets will have knock-on effects on the floor level. Doors and skirtings will need lifting to accommodate the raised floor surface. Special care will be needed to cope with any stairs and landings affected by this change in floor level.

Upgrading walls

Thermal insulation

In many modernization schemes, upgrading the thermal performance of walls often forms one of the main objectives. This is required not only to improve the building's energy efficiency. It is also done to arrest deterioration in the fabric as well as to enhance its appearance and provide weather protection.

It is estimated that there is around 7 million dwellings in the UK with solid walls about 225 mm thick (EST, 2001a). Up until the early 1930s, most British dwellings were built using this form of construction. Cavity walls had only begun to be used to any great extent a few years earlier. The thermal performance of both of these early forms of walling is poor because of their high conductivity value (see Table 9.4).

Improving the thermal efficiency of walls is obviously one of the main methods of increasing a building's energy efficiency. The other is draught-proofing doors and windows. The aim is to curb heat losses and energy consumption by lowering the U-value of the fabric. This can be done in one of three ways:

1. Internally applied insulation.
2. Cavity-fill insulation.
3. Externally applied insulation (see external cladding systems).

Table 9.4 Advantages and disadvantages of wall insulation methods (BRE GBG 5, 1990)

Advantages	*Disadvantages*	*Points to watch*
Cavity wall insulation		
Save up to 80 per cent of heat loss. Cost advantage over internal and external options. Not disruptive. Reduced condensation. Use existing wall thickness.	Cavity must be in suitable condition. Does not always cope well with thermal bridging (e.g. at openings). Limited scope depending on exposure zone. May encourage moisture from the outside to bridge the cavity.	Existing walls must be inspected. Choice of fibre, foam, beads or granules. Plastics materials provide better insulation, but not necessarily better performance. Foam insulation may suffer from shrinkage and emit gases. Plastic beads leave interconnected air paths so must include an adhesive.
Internal wall insulation		
Good for solid walls where external systems are not possible (e.g. listed buildings). Good where cavity cannot be filled. Good for adding insulation locally (e.g. to one room). Reduces thermal mass and improves responsiveness.	Slightly reduced room sizes. Does not always cope well with thermal bridging. Occupants must vacate rooms being treated during work. Strict on-site quality control essential. Skirtings, radiators and electric points need refixing. Five-year payback period.	Face of wall should be well sealed. Vapour barrier essential. Seal edges and service penetration points. Seal joints. Turn linings into reveals and soffits of openings.
External wall insulation		
Upgrades overall performance of wall. Overcomes thermal bridging. Moderate internal temperature swings. Little disturbance to occupants. Internal services not affected.	New weatherproof finish required. Wall thickness increased. Cost (most expensive option). Detailing around openings is critical. Rainwater goods need refixing.	Insulation must fit snugly. Finishes must be detailed. Cladding must include ventilated cavity. Render should be light coloured to prevent overheating in summer.

As these methods are considered to involve 'a material alteration' to an existing building they would fall under the definition of building work. Thus such improvements would be subject to the requirements of the building regulations – especially Part B (Fire Safety) and Part L (Conservation of Fuel and Power) of the Building Regulations 2000 (as amended).

A summary of the benefits and drawbacks of these three insulation methods is presented in Table 9.4. The decision on which system to use will depend on the age, construction and type of building as well as the cost involved.

Internal insulation

This method is most suited to upgrading the thermal performance of the external fabric of historic buildings (see Figure 9.9). Such properties are normally of solid masonry construction and because of listing restrictions have façades that cannot be replaced or covered over. Figure 9.10 shows the main precautions that would to be undertaken when installing this method of wall insulation.

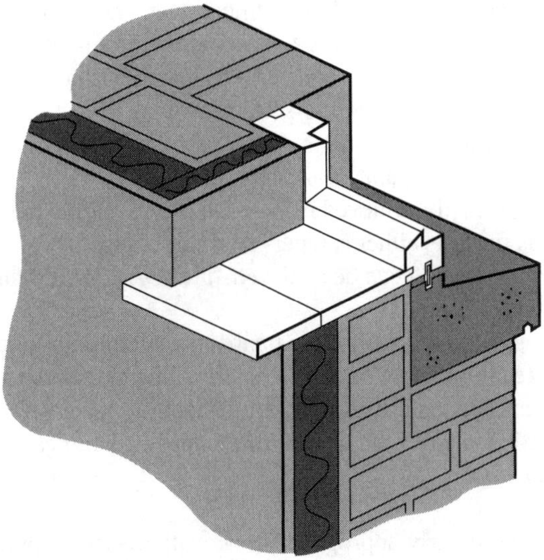

Figure 9.9 Internally applied insulation to walls (BRE GBG 5, 1990)

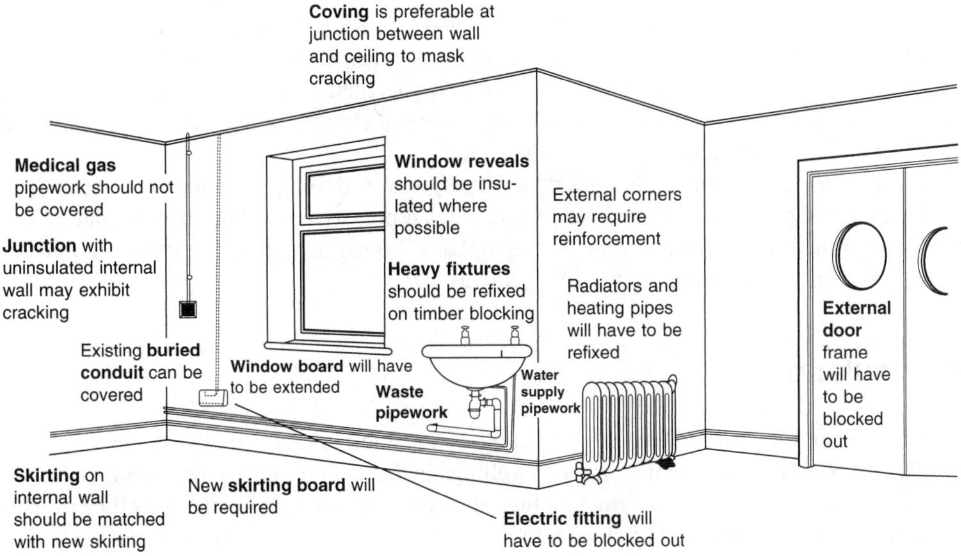

Figure 9.10 Precautions when installing internal insulation (NBA, 1980)

Ideally, a dry-lining technique should be used with any internal insulation method. This obviates the need for any wet-trade work such as plastering the walls. Composite insulation-backed plasterboard such as Gyproc 'Fireline' sheets fixed to the substrate using plaster dabs is one example of this method.

A more robust method of plating walls (as well as floors) is to use gypsum fibreboards. 'Fermacell' high-performance composite dry-lining board containing cellulose fibres is one typical example. It has the following advantages:

- Environmentally friendly (80% gypsum, 20% recycled paper and water).
- Excellent screw holding capabilities eliminates the need for pattressing or noggins to take brackets, etc. Direct fixings into the board can take 30 kg loads (50 kg if wallplug fixings are used).

- Very good fire resistance (F60 from single 10 mm thick layer, and Class 0 certification).
- Inherent moisture resistance makes it suitable for humid environments (up to 80% ERH) and can be installed before envelope is complete. This is because the material is not adversely affected by temporary wetting (unlike plasterboard, which is very moisture sensitive).
- Good sound insulating qualities.
- Ready to accept decorations such as paint, wallpaper, tiles.
- Fine surface treatment can be easily achieved using a Dolomite marble dust and latex paste mix, which eliminates wet trades. This finish is either scraper- or spray-applied.
- Rigid structure when used in framing can be achieved. TRADA certified raking board capabilities – up to 10 m high partitions on standard stud.
- Simple butt jointing system – glued, square edge boards produce continuous membrane. All four edges of boards can come with feather-edging to eliminate crowning at awkward joints.
- Good impact resistance, and damage tends to be localized. Gypsum fibreboard can withstand high point loads and heavy traffic loads. Its inherent stiffness means that it can support partition walls without the need for special support.

These composite boards are particularly appropriate in acoustic/thermal upgrades to the walls in historic buildings. Fixing the insulation on battens directly to the internal face of the wall is avoided to minimize damage to the fabric and the work is reversible. Boards can come in convenient 1.2 × 1.2 m sizes for easy one-man handling in refurbishment projects where access is limited.

According to the Energy Research Group et al. (1999), internal insulation should generally only be used if the façade cannot be changed, occupancy is intermittent, or, in the case of multiple-occupancy buildings, where not all the owners want to upgrade. In addition, this method precludes the use of the building's thermal mass as a heat store. As a result, it encourages thermal stresses in the outer skin, which increases the risk of interstitial condensation within this element.

The Germans, however, have made considerable progress in developing interior-lining materials for solid wall houses, such as 'Neopor', a silver-grey expanded polystyrene. These new insulation materials 'are 50 per cent more efficient than traditional "styrophone" panels as well as being thinner, thus reducing the amount of lost space', according to the EST (2001a).

Cavity insulation

Cavity walls were introduced in Britain in the early 1920s and became virtually mandatory after 1947 (Cairns, 1993). Nowadays, the vast majority of dwellings in this country have cavity walls (DETR, 1998). In the early years of this form of construction little thought was given to the thermal efficiency of cavity walls. Both skins usually comprised brickwork that was poor in terms of thermal resistance.

Cavity fill therefore can be an effective way of enhancing the thermal efficiency of old cavity walls. Figure 9.11 illustrates this technique. It costs, without a grant, about £380 to install cavity wall insulation in a typical three-bedroom semi-detached dwelling (EST, 2001a).

Urea formaldehyde (UF) foam was once popular as the injected cavity-fill material. However, as this is classed as a volatile organic compound (VOC), other cavity-fill materials have overtaken UF foam. In addition, UF foam in BS 5617 is only approved for buildings up to three storeys and of moderate exposure. Moreover, like other cavity insulation materials, it is not appropriate for use in timber-framed construction.

Heavy-duty polyurethane foam, for example, is one alternative. As indicated in Chapter 7 it can also act as a tie-foam to combat wall-tie sickness if the high-density version is used (Cairns, 1993). In such cases it adheres strongly to the inner surfaces of both skins in the cavity as a substitute for corroding wall ties. Again, though, this method would not be suitable on walls regularly exposed to wind-driven rain because of the risk of moisture bridging the cavity.

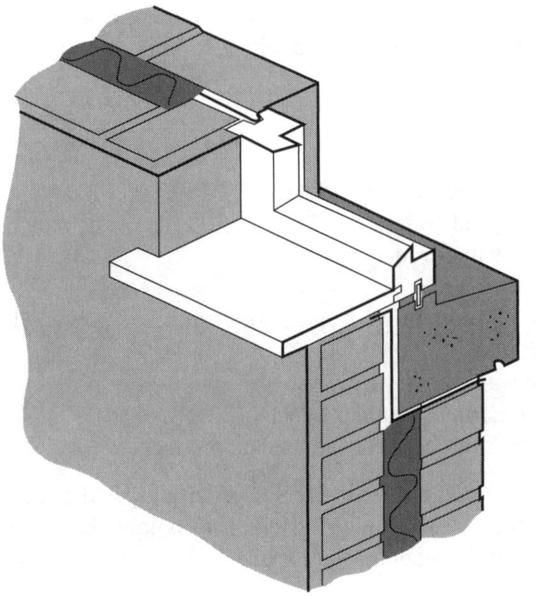

Figure 9.11 Cavity wall insulation (BRE GBG 5, 1990)

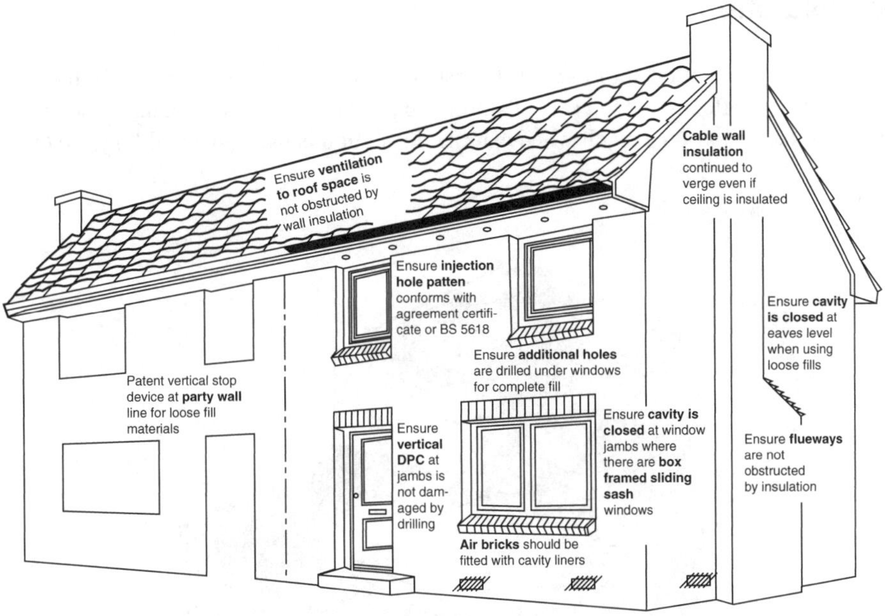

Figure 9.12 Precautions when installing cavity-fill insulation (NBA, 1980)

Plastic pellets or mineral wool beads are the most effective materials for injected cavity fill nowadays. The main drawback with these materials is that they tend to cost more (Cairns, 1993). However, as polyurethane and phenolic foams contain chlorofluorocarbons (CFCs), these should be avoided if possible (Johnson, 1993).

Injected cavity fill is normally installed by drilling the external skins at approximately 1.5 m centres in a five-of-clubs pattern (see Figure 9.12). The centre hole is filled first. The other holes are loosely plugged with small pieces of wood which act as tell-tales when the foam reaches the four corners of the pattern (Cairns, 1993). The holes should then be plugged with a 1:4 mix mortar.

The main advantages of using blown mineral fibre cavity insulation are that it:

- is non-combustible when tested to BS 476 Part 4 1970;
- allows the walls to breathe;
- will remain effective for the life of the building;
- will not transmit the passage of water;
- will not sustain mould growth or rodents;
- has no added chemicals or water;
- has no adverse effect upon wood, brick, wall ties or unsheathed polyvinyl chloride (PVC) cables;
- will neither settle nor shrink;
- will not escape from the wall;
- is manufactured from the Earth's natural products;
- is ozone friendly – no CFCs or chemical vapours are produced.

Cairns (1993) gives a detailed comparison of foam, pellet and bead materials used in injected cavity-fill systems. As with any specialist adaptation system or technique, however, only an approved/qualified contractor should install cavity insulation. The installer, therefore, should be a member of the National Cavity Insulation Association.

External insulation

Thermally this is the most efficient of the retrofit insulation methods (see Figure 9.13). It is known in the trade as the 'Tea Cosy' effect – the tea is kept warm and a decorative cover enhances the appearance and safety features of the teapot. As a result when external insulation is used on a building, the temperature of

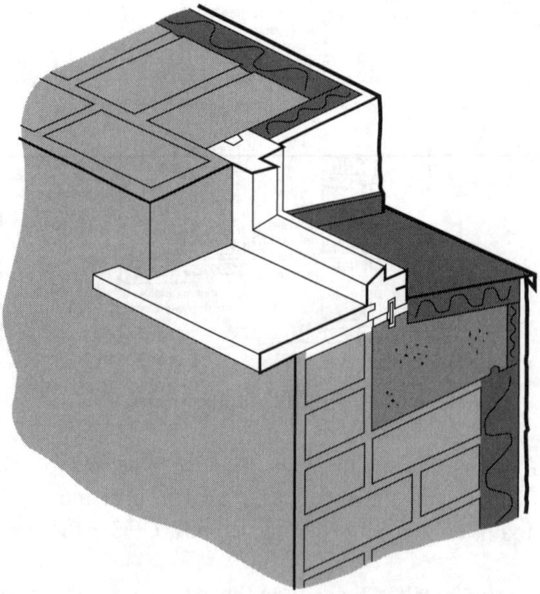

Figure 9.13 Externally applied insulation to walls (BRE GBG 5, 1990)

the main walls remains reasonably constant. Insulating existing buildings in this manner allows the installation of the system without decanting or disturbing the occupants. It also achieves a renewal of the existing external brick or rendered finish.

A major drawback of externally applied insulation to blocks of multiple occupancy is where only partial installation has been carried out. As well as being thermally inefficient, this gives the building a patched or incomplete look. The same effect can occur even in re-rendering schemes where only some of the dwellings in a block have been upgraded.

The following points, based on a technical article (Anon, 2000a), should be borne in mind when specifying an external insulated wall system:

- Check system is applied by an approved and trained contractor (i.e. who is a member of the External Wall Insulation Association). In addition, the insulation specified should have a current third-party certificate.
- Store insulation in a dry location, protected from direct sunlight, in its manufacturer's wrapping. Keep insulation materials flat, away from solvents and possible sources of fire.
- Insulation should be fixed with an approved mechanical fixing through the mesh into the substrate. At corners and around openings additional fixings should be used. A staggered pattern should be used for fixing the insulation panels.
- To minimize cracking, reinforce the render with a mesh suitable for and in accordance with the insulation system specified.
- A suitable aluminium/plastic/stainless steel stop bead, in accordance with the system specified, should be fixed to the structure above the damp-proof course. The damp-proof course must not be bridged by this work.
- Insulation should be cut cleanly using the correct tool and straight edge. Saw cut ends will be too ragged for a proper butt joint or edge finish.
- Provide adequate movement joints in accordance with the insulation system specified. These would normally be between 10 and 15 mm in width and filled with a polysulphide or other high-performance mastic. The horizontal joints are usually positioned at every level or alternate floor; the vertical joints are spaced about every 8 m.
- Where an impermeable cladding is used (e.g. rainscreen), ensure that the cavity is ventilated and drained. Install cavity barriers at all compartment floors and walls (see Figure 9.16).

Rendering and re-rendering

In many cases rendering or re-rendering external walling will form part of an external insulation system in a modernization scheme. It may also occur on its own where the existing external brick, stone or rendered finish has degraded and requires protection. 'Grinning through' occurs where old rendering is so eroded that the outline of the brickwork substrate can be easily seen.

Rendering external walls finished fair faced or with facing bricks may form part of a modernization scheme. In some cases, however, such walls may be re-rendered in isolation.

Adequate preparation of the substrate is essential to avoid failure of the rendering. The type and condition of the substrate will dictate the type and quality of preparation needed. Dense, low-suction-quality backgrounds, such as glazed brick, smooth concrete, require treatment to provide an adequate key. The main methods of providing a key for the rendering are:

(a) Raking out joints of brickwork to a depth of about 10 mm.
(b) Scarify the surface of the background (if dense/nonporous) using a toothed pneumatic chisel.
(c) Apply a coat of styrene butadiene rubber (SBR) bonding agent (3 mm thick) using a hair roller.
(d) Fix plastic or metal lath sheeting to the background using stainless steel fixings.

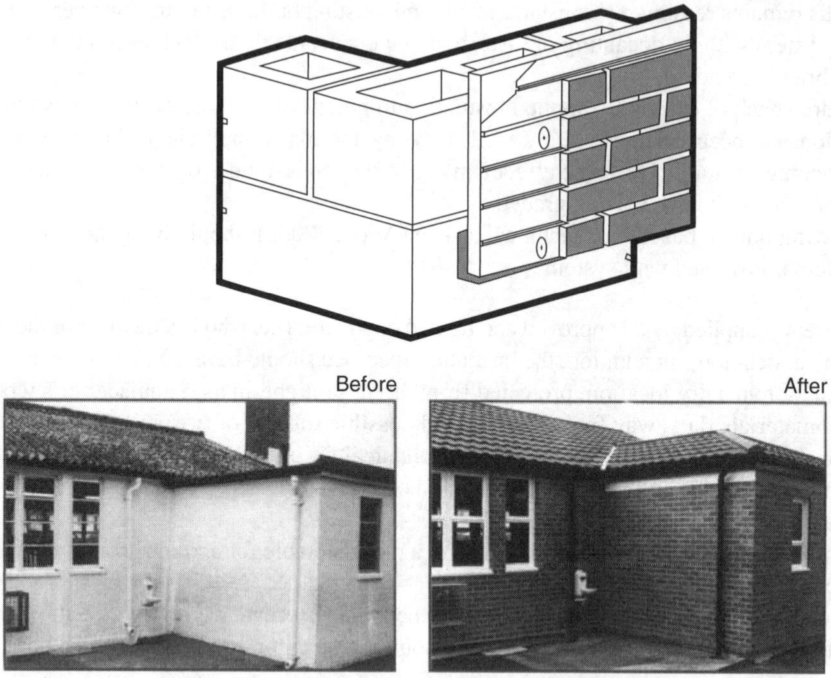

Figure 9.14 Typical insulated brick-slip overcladding system (www.eurobrick.co.uk/)

Methods (a) and (b) are messy and time-consuming. They can also cause irreversible damage to the substrate. For these reasons should only be used on small areas (i.e. say, $<10\,m^2$) and in exceptional circumstances.

Method (c) is probably the most common way of maximizing the bond between the render and the background. Polyvinyl acetate (PVA), however, should not be used for external works as it is water-soluble.

Method (d) is ideal where a degree of differential movement in the background is anticipated. Expanded metal lath strips should be used in any event across joints between different materials. For example, this should be done over the joints between a concrete lintel and clay brickwork to minimize cracking due to differential movement.

A fundamental principle of good rendering practice is that each subsequent coat should be thinner and weaker than the preceding coat/s. The total thickness of the rendering system should not exceed 20 mm. The following rendering specification should be used:

- First coat: 12 mm thick, using a 1:1:5 mix.
- Second coat: 8 mm thick, using a 1:1:6 mix.

Façade design

Preamble

The choice of overcladding or other improvements to walling will have aesthetic as well as constructional implications. Should the finish be plain or patterned, soft-tone or striking, for example? Judicious choice of pattern, if any, and colour or mixture of colours will impact on the success of the scheme. In some housing refurbishment schemes to multi-storey blocks of flats, the tenants/owners have had a say in determining the design of the cladding.

Table 9.5 Common façade design options

Style	*Characteristics*
Vertically accentuated	Emphasises mullions and other vertical aspects of the façade.
Horizontally accentuated	Emphasises exposed floor slab or balconies at each level or window sills.
Grid pattern	Accentuates the square or rectangular grid formed by the overcladding panels.
Random patterned	Panels of different colours or finish used to give distinctive or ornate appearance to the façade.
Mural	Large coloured painting reflecting an event, scene, object of local importance or portraying significant people. This style of façade design is mostly used if at all on conspicuous gables (as some might see it as a form of licensed graffiti).
Plain	Finish to match the existing or in a lighter colour for solar energy purposes.

Styles

A variety of styles can be used for different emphasis to give a distinctive appearance to the external envelope. The main designs are summarized in Table 9.5.

Acoustic performance

One-brick or even half-brick thick brickwork may require upgrading as regards their acoustic performance. This is especially the case if they become party walls as a result of a conversion scheme, to comply with statutory requirements. Party walls need to have good resistance to airborne sound transmission, which requires the element to have a relatively high mass. BRE Digest 293 (1985) outlines the main ways of improving the sound insulation of separating walls.

Cladding one side of such a wall with 'Gyproc gyplyner' wall-lining system with 25 mm thick 'Isowool 1200' glass mineral wool in the gap between the studs can achieve an airborne R_w rating of 60 dB. These techniques inevitably involve increasing the thickness of the wall being upgraded.

External cladding systems

Preamble

A fatal fire in a block of flats in Irvine, Scotland in June 1999 prompted a government review of external cladding systems. A fire in an apartment on the fifth floor of a 14-storey block of flats spread across the surface of the cladding to the upper floors. An elderly occupant of one of the flats above died from the effects of the fire.

As a result of this incident, the Select Committee on Environment, Transport and Regional Affairs carried out a parliamentary enquiry into external cladding systems. The Committee's report (1999) found that external cladding systems fall under three broad headings:

1. External wall insulation (brick-slip/render) systems (sometimes known as 'raincoat' systems).
2. 'Rainscreen' (sheet boarded) systems.
3. Preformed 'infill' systems.

The first two can be further classed as 'overcladding' systems. The third does not encapsulate the whole façade but only certain areas such as those incorporating fenestration.

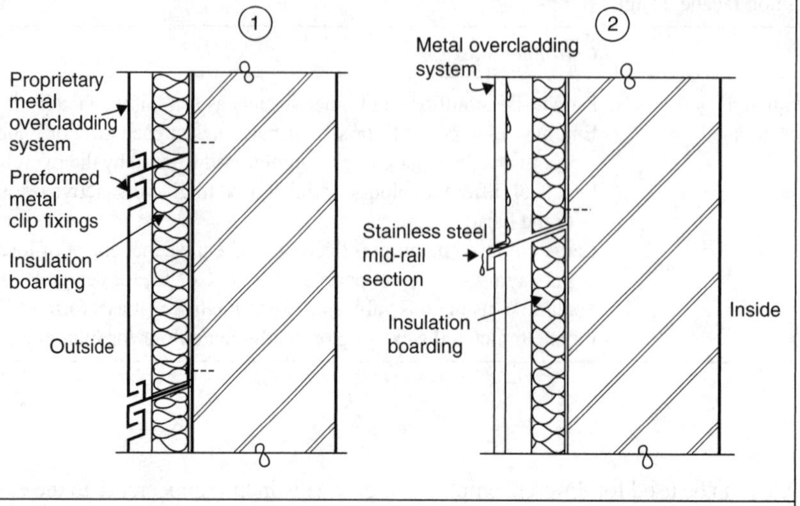

Notes
1. Pressure-equalized rainscreen cladding.
2. Drained and back-ventilated rainscreen cladding.

Figure 9.15 Types of rainscreen cladding (Oliver, 1997)

Preformed infill claddings

A preformed 'infill' system was the type of cladding involved in the Irvine tower block fire. This system does not cover the whole of a building's façade. It is usually fitted to small areas of a building to give a decorative effect. Unlike overcladding, preformed infill systems incorporate a window in each panel. It was the fire spread between the glazed and unglazed part of the unit that was the cause of the problem in the Irvine case. As a result in future it is likely that more stringent pre-installation tests will be required on the proposed cladding system to ensure that it does not encourage spread of flames across its surface.

There is, however, not much scope for applying this form of cladding to the refurbishment of buildings. It could be used as part of the partial recladding of high-rise blocks of flats constructed using the plate and column method. In this form of construction the floor slab edge beams are exposed.

Raincoat cladding

This is basically a term used to describe any insulated render system (see Figure 9.15). It can be classed as 'jointless' – even though construction joints are normally inserted at every storey. According to one source (Pearson, 2000) this is the most common type and accounts for more than half of cladding in the UK. Most systems use either a polyester or galvanized steel mesh or lath as the main reinforcement on the insulation to provide a key for the render.

A variation on the insulated render system is the insulated brick panel system. Instead of a rendered finish the insulation is covered with brick slips to give a traditional appearance to external walls. The brick slips are about 15 mm thick and are made from genuine kiln fired clay bricks. These are then attached to the insulated panel substrate using an epoxy resin adhesive. Figure 9.14 shows a typical proprietary example of this system. This method of overcladding is especially appropriate in older masonry buildings where a brick finish is desirable. According to various manufacturers it has an expected lifespan of at least 30 years.

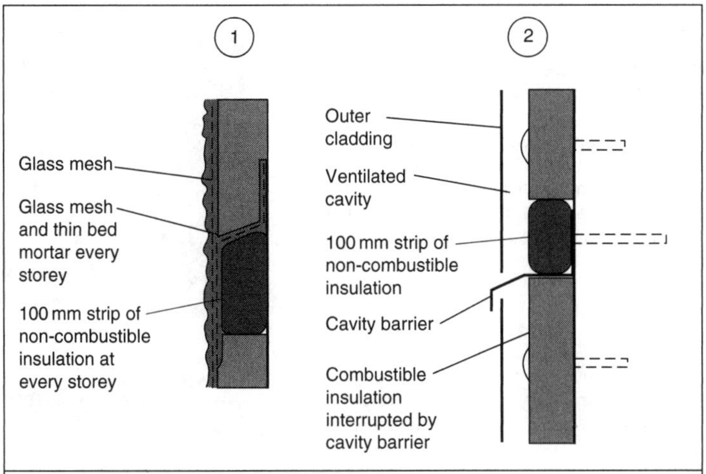

Notes
1. Render system incorporating cavity barriers.
2. Rainscreen cladding incorporating a non-combustible cavity barrier.

Figure 9.16 Cavity barriers

Rainscreen cladding

This type of cladding is the second most common in the UK, and is used in both residential and commercial refurbishment projects. Unlike the raincoat type, it consists of a jointed panel system. There are two versions of this form of cladding, both of which use a structural frame that supports an external panel and insulation. One is called the 'drained and back-vented system', which is the traditional version. The other is the 'pressure equalized' system (see Figure 9.15).

Fire stopping

Installing cavity barriers is the best way to prevent fire spread within an external cladding system. Fire-rated rigid urethane insulation batts can serve this purpose without creating a cold bridge. Figure 9.16 illustrates how this can be achieved.

Upgrading roofs

Generally

As shown in Chapter 6 over-roofing is becoming a common way of replacing existing flat roofs. In this section we shall look at ways of upgrading flat and pitched roofs that do not involve changing their basic profile. Such upgrading works entail one or a combination of the following three measures:

1. Replacing the existing roof coverings with similar or new material to enhance its appearance and weather resistance.

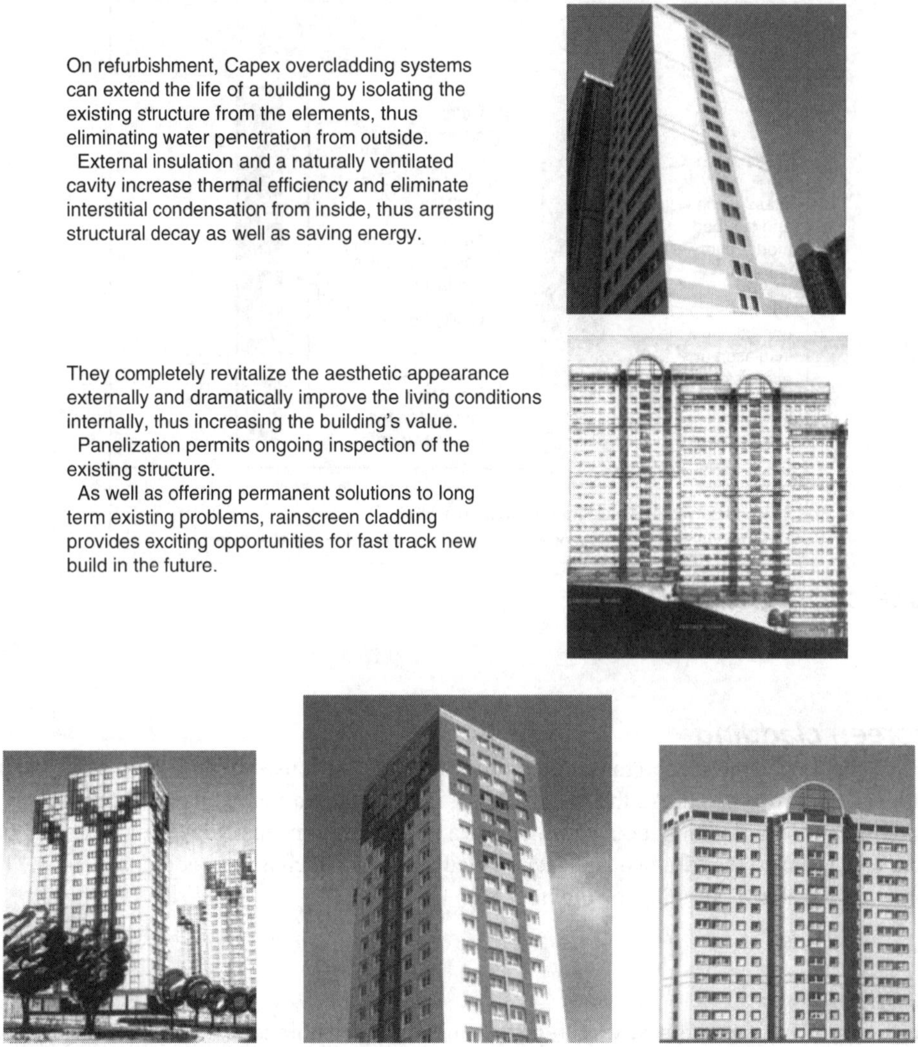
On refurbishment, Capex overcladding systems can extend the life of a building by isolating the existing structure from the elements, thus eliminating water penetration from outside.
External insulation and a naturally ventilated cavity increase thermal efficiency and eliminate interstitial condensation from inside, thus arresting structural decay as well as saving energy.
They completely revitalize the aesthetic appearance externally and dramatically improve the living conditions internally, thus increasing the building's value.
Panelization permits ongoing inspection of the existing structure.
As well as offering permanent solutions to long term existing problems, rainscreen cladding provides exciting opportunities for fast track new build in the future.

Figure 9.17 Photographs of typical overcladding systems to blocks of multi-storey flats (www.cape.co.uk/)

2. Increasing the roof's thermal insulation to upgrade its energy efficiency.
3. Improving the background ventilation to the roof void to obviate condensation problems.

The first can occur on its own without any need for improving the roof's thermal performance. This may be because the roof coverings have deteriorated prematurely through environmental degradation or vandalism. The opportunity can also be taken to provide roofing felt where none previously existed or where the existing felt has failed.

However, it is less likely that a roof would have its thermal efficiency and ventilation improved without changing or enhancing its coverings. Renewal of other related parts of the roof such as ridge and verge cappings, rooflight glazing, frames and flashings should also be considered.

Flat roofs

Roof profile

In many modernization schemes, upgrading the performance of flat roofs is an important consideration. This is required to improve the roof's thermal performance as well as its rainwater disposal efficiency.

Flat roofs prior to the early 1970s were normally laid to a fall of 1:80 as recommended in CP 144. In many instances this degree of slope was adequate to ensure disposal of rainwater. However, if there are any slight deflections or depressions in the roof slab or deck, ponding can easily occur despite the 1:80 fall. Ponding is a common problem with many flat roofs because standing water acts as a reservoir for leaks and encourages the formation of ice or organic growths such as moss on the roof surface.

Nowadays a minimum fall of 1:40 is required for flat roofs to avoid ponding and inefficient rainwater discharge. This can be achieved in a variety of ways as shown in Table 9.6.

Membranes

Basically flat roof membranes can be classified into two broad groups: seamless and jointed. Each of these are discussed as follows:

- Seamless membranes: These are either spray-applied or poured on. They especially suited to complex roof shapes or small, awkward areas such as balconies. The main materials available nowadays are:
  - mastic asphalt: traditional two-coat application on fleecing layer;
  - liquid-applied coatings: glass-reinforced polyester comprising a roller/spray-applied epoxy resin onto glass-fibre matting base or spray-applied polymeric coating on a polyester substrate.
- Jointed membranes: These are either felted or sheet metal. The former can come in single-, two- or three-ply systems, depending on the material selected. The more resilient the membrane the fewer the layers needed. Sheet metal coverings on the other hand are always single layer.

Rainwater disposal

The opportunity to improve the roof's rainwater disposal should also be taken at the same time the coverings and insulation are being replaced. This can consist of enlarging the size or increasing the number of outlets serving the roof. A 1 m × 1 m sump should be formed at each outlet to minimize the build-up of silt and debris at this position.

Conversion to pitched roof

As indicated in Chapter 6 existing problematic flat roofs provide the ideal opportunity for conversion to pitched solutions. This is usually achieved using lightweight galvanized steel frames where there is no intention of providing habitable space within the roof void.

Roof insulation

The new roof covering system should be of either the 'warm deck' or 'inverted' type. A warm deck system has the insulation above the deck. A vapour must be located on the warm side of construction.

In contrast the insulation in an inverted or upside down roof is above the membrane. The membrane also acts as a vapour check. However, an inverted warm deck flat roof must have a layer of ballast (rounded pebbles) about 150 mm thick or 50 mm thick pre-cast concrete slabs (laid on corner pads to spread the weight) on the insulation to shield it against damage from impact and the elements. This system has the added advantage of protecting the membrane from damage as well.

The BRE, following the ban implemented in Scotland for many years, now advise against the use of 'cold deck' flat roofs generally across the whole of the UK. This type of roof has the insulation below the deck.

Table 9.6 Factors influencing the method of achieving the falls on a flat roof

Deck	*Method*	*Precautions*
Timber or profiled metal	For large spans use tapered polystyrene or cork insulation board laid to the fall of 1:40. (Do not use a screed or other 'wet' solution on this type of deck.) Alternatively, joists or truss beams could be installed to achieve the required slope. A suspended ceiling would hide the sloping soffit. For short spans (e.g. garage roof or small lateral extension) firring pieces on joists.	Remove any existing 'cold deck' roof insulation. Make good all defective roof timbers or ceiling linings. Spray-apply boron diffusion preservative to all structural and vulnerable timbers. Use compatible materials for vapour control layer, insulation and membrane. Parapets and related flashings are likely to need raising.
Concrete	Lightweight concrete or 'Perlite'/Bitumen-bound screed to required fall of 1:40 might be considered. However, given that this is a 'wet' solution it is best to avoid a method in adaptation work that introduces more construction moisture into the fabric. Tapered insulation board laid to the required fall. This is the more modern (i.e. dry) method of achieving a fall on a flat roof. Precast planks laid on opposite wallheads at different heights to achieve the required fall. Again, if necessary a suspended ceiling could be installed to hide the sloping soffit.	Screeds were traditionally used under a mastic asphalt roof covering. However, 'perma-vents' may have to be installed at regular intervals on the deck to provide background ventilation to the screed to remove any build-up of construction moisture or water vapours from below. Replace any saturated or defective areas of existing insulation or decking. Ensure moisture-resisting integrity of seals around rainwater outlets, particularly in the centre of the roof. Replace any defective rainwater pipework through slab. The existing substrate must be checked to ensure it is flat and level prior to installing tapered insulation board. Timber firring pieces should not be used to provide falls on a solid substrate such as concrete as they create void areas between them that would require venting. Again, the height of parapets and related flashings are likely to need raising.

If used in a timber flat roof structure, the insulation is placed between the joists. A minimum 50 mm deep continuous vent gap between the top of the insulation and the underside of the deck must be maintained. This is to prevent interstitial condensation occurring above the insulation.

When re-covering a flat roof the opportunity to upgrade its thermal efficiency should be taken. The aim is to achieve a U-value of 0.25 W/m^2K or lower. Whatever system is used, however, it is critical that the vapour control layer is placed on the warm side of construction to avoid any interstitial condensation problems in the roof system.

Table 9.7 Options for upgrading flat roof coverings

Element	*Examples*	*Comments*
Solar protection	Two coats of bituminized aluminium paint.	Needs re-coating every 3/5 years.
	White mineral chippings.	Although it gives good resistance against fire spread, this finish encourages moss growth and makes it difficult to locate leaks.
	Paving slabs	Concrete paving slabs offer a durable resilient finish but this increases the dead load of the roof.
	Ballast	Rounded washed river pebbles should be used as ballast.
	Aluminium/copper foil finished felt.	Should be properly stored to prevent premature damage to the finished foil.
	Self-finished membrane (e.g. sheet metal, PVC 'felt').	
Membrane	Single-ply PVC felt. Single-ply EPDM membrane.	White, grey or green finish and can attain a 25–30-year life expectancy.
	Single-ply polyolefin membrane.	Like EPDM, polyolefin is a more environmentally friendly option than PVC.
	Two-ply high-performance elastomeric/polymeric felt.	Green mineral or metallic finish to provide solar protection.
	Two-coat mastic asphalt.	Solar reflective coating (bitumized aluminium paint) required.
	Liquid-applied coatings such as epoxy resin/polyurethane/acrylic/ modified bituminous coatings and other jointless roofing materials.	Reinforcing polymeric fibre underlay and solar protection is normally required for asphalt finish. Self-finish.
	Welded or welted sheet metal – stainless steel; lead; zinc; copper.	
Insulation	Polyurethane board. Polystyrene slab. Cellular glass board. Mineral fibre slab. Cork board. Perlite/vermiculite screed laid to falls.	The type of insulation used will be chosen on the basis of costs, availability, its tapering ability, U-value achieved and compatibility with both the substrate and roof covering. See Table 10.10.

EPDM: ethylene propylene dien monomer.

When adding insulation on top of an existing deck the height of the perimeter flashings and parapets will probably need increasing. This also applies to skirtings of skylights and other projections through the roof. Each vent pipe, for example, will require a new collar at its junction with the roof finish.

The main choices available for upgrading flat roofs are summarized in Table 9.7.

Access

Access is often inadequate to flat roofs for maintenance and inspection purposes in many flat roofs. Installing in the roof a stainless steel or aluminium access hatch can improve this with counter-balance for ease of use.

Pitched roofs

Recovering using traditional materials

Replacing the coverings of a roof may be required for two main reasons: to enhance the appearance of the roof and replace defective or worn coverings. Ideally, stainless steel or copper nails should be used to fix the slates or tiles.

The pitched-roof coverings of much of the mid-20th century stock of local authority housing may have by now reached the end of their useful life expectancy. The same problems are common in mid-terraced offices built in the Georgian and Victorian eras. Delamination and nail-sickness are two common deficiencies with slate and tile roof coverings. Storm damage, vandalism and neglect can exacerbate these problems.

In such cases it is necessary to replace the roof coverings. On the other hand, the opportunity may be taken to recover the roof with a new material rather than on a like-for-like basis. Any selection should, however, take cognisance of:

- the loading implications on the existing roof structure;
- the service life and maintenance characteristics of the replacement roof covering material;
- the aesthetic impact of the new roof finish will have on the building and surrounding properties;
- the need to improve the roof's thermal insulation (see section below);
- local planning authority guidelines on changes in the external appearance of buildings in their area;
- the roof coverings' whole life costs.

Recovering using modern materials

The health and safety risks associated with cutting, grinding, and drilling concrete and clay tiles (see Chapter 11) could prompt the specification of alternative coverings for the re-roofing part of an adaptation project. For example, instead of traditional slates or concrete tiles lightweight roof tiles made from granular/mineral-coated mild steel could be considered.

Typical proprietary pitched roof coverings of the metallic kind are 'DECRA' tiles or 'Tufftiles' (BBA Certificate No. 89/2267). The former comprise mild steel sheeting protected with an aluminium-alloy zinc coating, acrylic priming system and finished with stone granules with an acrylic overglaze or lightweight epoxy-coated metal.

'Tufftiles' are made from 'extra strong galvanized steel profile, covered with an acrylic basecoat and ultraviolet (UV) stabilized topcoat plus the hardwearing granular surface', all of which 'helps to reduce maintenance costs and keep up appearances'. Each 'tile' can be made to have a cover length of 1.6 m, and has a depth of 371 mm. This means that each square metre contains only 1.72 tiles and 12 fixings. Installation time can thus be dramatically reduced.

Metallic lightweight roofing tiles have the following advantages:

- Weigh 8 kg/m^2, which is up to 85 per cent less than concrete tiles, thus reducing the dead load on the building.
- Provide a strong, robust, vandal- and wind-resistant roof finish.
- Have a minimum 30 years guarantee.
- Quicker installation than conventional tiles or slates.
- Can provide a tough, impact resistant finish on vertical surfaces.
- Dry-fixing system obviates the need for mortar to fix perimeter cappings.
- Lower initial cost than clay tiles.
- Minimal maintenance.

Coating existing roof coverings

'Turnerization' is a trade name for a process involving the application of a black or rust red-coloured bituminous coating to a slated or tiled roof (Hollis, 2005). The rationale behind this brush- or spray-applied repair method is that it is intended to provide a watertight seal to the roof. It is sometimes considered as a means of refurbishing old or 'nailsick' roof coverings.

Applying a bituminous coating to a roof, however, is a risky venture. It effectively traps any moisture in the roof space, which could lead to condensation problems if there is inadequate background ventilation in the loft area. Black- and other dark-coloured coatings increase the thermal absorptivity of the roof in very hot summer conditions. This could cause an excessive temperature build-up in the roof space as well as cause the coating itself to melt. The coating has a lifespan of between 5 and 10 years, which necessitates its re-application on a regular basis. Moreover, 'Turnerization' obviates the possibility of recycling the slates at a later date because of the difficulty in removing the bituminous coating.

A similar technique to 'Turnerization' uses a 'Gortex'-like aqueous-based acrylic coating instead of bitumen. The advantage with this spray-applied coating is that it can offer a degree of breathability to the roof. It also inhibits the growth of algae and lichen on the roof coverings. Like the traditional bituminous coating, this technique can come in a variety of colours, such as rust red, orange and light green.

The various options for upgrading a roof's thermal performance are as follows.

Conventional thermal upgrades for pitched roofs

Normally additional or new insulation in unoccupied roof spaces is easily installed above the ceiling. Unfortunately, placing the insulation between the ceiling ties creates a 'cold roof' space (see Figure 9.17). This requires two major precautions when using traditional construction practices: background ventilation at eaves and ridge to avoid condensation problems in the roof space as indicated above, and lagging of all exposed pipes and tanks (see below).

The recommended minimum thickness of insulation is now 200 mm, using mineral fibre or blown cellulose fibre. Mineral fibre, for example, can come in a polythene wrapped coating for ease of installation and to minimize skin irritation when during handling. It should be laid in two 100 mm thick layers; the first laid between the joists; the second layer across the joists and butted up to the gable and separating walls. Care must be taken, however, to ensure that the eaves ventilation is not compromised, and that electrical cabling is placed above the insulation to reduce the risk of the wires overheating.

A U-value of around 0.2 W/m^2K can be obtained using this method (BRECSU GPG 12, 1992). The use of 200 mm thick loft insulation with a British Agreement Board (BBA) Certificate gives a payback period of about 2–3 years according to GPG 12 (1992).

Condensation problems in the roof space can manifest themselves in a number of ways. This can cause efflorescence on the roof timbers and softening of the fibreboard or plasterboard sarking, as well as trigger fungal attack and corrosion.

Pipes and tanks that are not lagged in an insulated loft have a high risk of freezing in cold weather. The water damage to the rooms below following a burst pipe is often extensive and extremely disruptive and damaging. This can be avoided by encasing all exposed pipes with preformed lengths of polyethylene foam lagging.

The sides and top of the water tanks should also be insulated. A 75 mm thick board of sealed mineral fibre or polyurethane insulation is usually sufficient for this purpose. The space under the tank should be kept free of insulation to allow heat rising from the room below to keep the water from freezing.

According to Appendix D and clause 9.4.7.1 of BS 5250, 1989, for roofs with pitches of greater than 15°, a 10 mm continuous vent gap at the eaves should be provided (25 mm if the pitch is below 15°). This is also the requirement stated both in Section 7.2-D11 of the National House Building Council (NHBC) Standards and Part G4.1/G4.2 of the Building Standards (Scotland) Regulations 1990. 'Redland' or 'Marley' eaves vent slates/tiles can be used to achieve this. An alternative to these is the 'Klober' *Inline Plain Tile Ventilator*, a proprietary uPVC tile that provides 9000 mm^2 of free airflow.

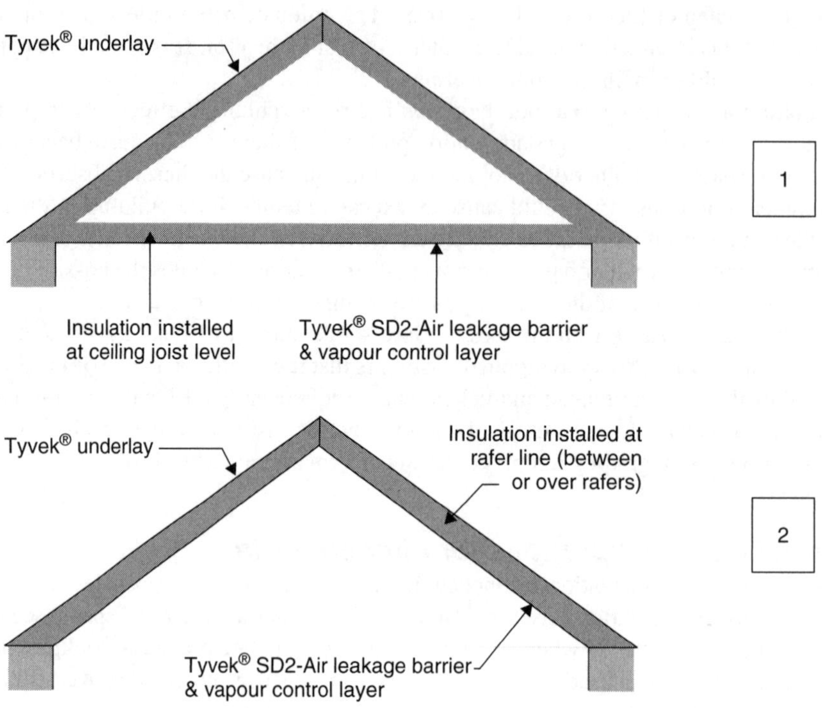

Notes
1. 'Cold roof' (unventilated because of the 'breathable' underlay).
2. Warm roof.

Figure 9.18 Pitched roof applications for breathable underlays (Dupont™ Tyvek®, 2004)

Ridge ventilation can be provided via 'Ubbink' or other proprietary ridge vents, which give an open area equivalent to continuous opening of at least 5 mm over the length of ridge. Clause 9.4.7.1 of BS 5250 states that this is particularly recommended at pitches of 35° and over.

There is, however, no guarantee that installing retrofit eaves or ridge ventilation will prevent condensation problems in every type of pitched roof. On the contrary, introducing cross ventilation in a sealed loft may increase the risk of dampness in 'cold' pitched roof-spaces. As external air often has a higher humidity than exists in the interior, this can ironically make matters worse (see cold 'breathing' roof in Figure 9.18).

Modern thermal upgrades for pitched roofs

In Chapter 6 installing insulation between the rafters was described as a means of improving the thermal efficiency of a loft extension by forming a 'warm pitched roof'. Instead of using a ventilated air space between the rafters, however, a more modern approach to achieve a 'warm roof' is to use insulation that is integral with the roof structure itself. This can be done in one of two ways:

1. A deck comprising a rebated polyurethane insulation board (approximately 85–100 mm thick) fixed over and between the rafters. This technique obviates the need for a ventilated air space within the roof

structure. It would of course involve uplifting and refixing or replacing the existing coverings including providing new roofing felt, battens and counterbattens.

2. A sprayed polyurethane foam applied to the underside of the exposed slates/tiles, battens and rafters. 'Renotherm' (BBA Certificate 93/2939) is one proprietary version of this technique. It can help to prevent slippage of tiles as well as reduce ingress of airborne noise, dust and water. The material's thermal quality helps to reduce the property's heating costs and CO_2 emissions. This method does not require any scaffolding for the installation, which involves little if any disruption, mess or inconvenience. However, as with 'Turnerizing' a roof externally, this internal method of coating the roof coverings runs the risk of trapping any residual moisture with the roof structure. It should therefore only be used for refurbishing existing uninsulated roofs that are in very good condition (i.e. structurally sound and weather-tight).

Investigations into modern roof constructions by the Building Research Establishment (BRE) have indicated that conventional ventilation is not the only solution to condensation in loft spaces. Warm breathing roof construction (alternatively called 'sealed pitched roof construction') offers a dry, draught-free loft area, and gives better energy efficiency. This type of roof design requires minimal background ventilation.

An example of this modern pitched roof design, using minimal ventilation (i.e. 1 mm wide continuous vent gap at the eaves and ridge) can be used in re-roofing works but usually only if it comes with a BBA Certificate. It relies on an integrated air barrier and vapour control layer to achieve a 'sealed roof', which is breathable. For example, vapour permeable high-tech membranes made from high-density polyethylene (HDPE) such as 'Tyvek® SD2' (BBA Nos. 99/3635 and 94/3054) can be installed under the insulation of both 'cold' and 'warm' pitched roofs. The entire roof structure adheres to the principle of 'vapour-open and airtight' when used with a standard 'Tyvek'® breather membrane on top of the rafters and insulation. This allows moisture in the roof void to escape easily to the outside, but keeps draughts out of the loft.

According to its manufacturer Tyvek® is 'made from very fine, HDPE fibres. Tyvek® brand protective material offers all the best characteristics of paper, film and fabric in one material. This unique balance of properties, not found in any other material, makes Tyvek® lightweight yet strong; vapour-permeable, yet water-, chemical-, puncture-, tear- and abrasion-resistant. Tyvek® is also low-linting, smooth and opaque'.

In some non-residential buildings with a pitched roof having a flat top (i.e. a half mansard) the original metal (e.g. lead or zinc) coverings can be replaced with a polymeric single-ply membrane. A ribbed profile can be used on the sloped faces of the roof covered with this newer material to give a sheet metal effect – in green or grey to mimic copper or lead, respectively.

The opportunity to replace 'wet fix' ridge, verge and hip cappings with 'dry-fix' components should be taken when re-covering a pitched roof. These can comprise cloak verges with secret fixing method. The traditional method of fixing such cappings on a bed of mortar is no longer suitable because of its poor resistance to wind damage. In such cases special fixing clips have to be used. Moreover, mortar is prone to cracking and spalling over time. This obviously reduces the weather-tightness of the verge or ridge.

Lead roofs

Underside corrosion

The external surface of lead is exposed to CO_2 and pollutants both from air and rainwater. These react the lead to form a protective grey patina, which gives the material its traditional durability (English Heritage, 1998).

Deterioration, though, can quickly occur to the underside of the lead. Upgrading works to the roof can easily trigger this. For example, it is not uncommon these days to renew the lead flat/pitched roof of older buildings as part of their general refurbishment. The opportunity to improve the roof's thermal insulation is usually taken in such circumstances. However, this can drastically increase the risk of underside corrosion to the lead. It is a problem usually caused by condensed moisture, which being relatively acidic can accelerate deterioration of this otherwise reasonably durable material. Organic acids from timbers such as oak near or in contact with the leadwork can exacerbate the corrosion.

Precautions

The joint English Heritage/Lead Development Association advice note for specifiers should be followed when undertaking major refurbishment work to a lead roof. In summary, the general guidance below is recommended (English Heritage, 1998):

- A ventilated warm roof design should be used. Adequately sized air spaces, ventilation inlets and outlets should be provided. This involves forming a 50 mm continuous vent gap above the warm deck insulation but below the lead sheeting and its deck. The 50 mm air space should provide full ventilation from eaves to ridge with no dead spots or obstructions.
- The existing roof structure must be dry before and during installation of lead coverings. For example, the timbers and other wooded parts of the roof system should have a moisture content below 18 per cent when the new covering and insulation is being installed.
- The underside of the lead should be coated with a protective passive film such as a slurry of chalk powder dispersed in distilled water.
- A sealed air and vapour barrier should be installed below the insulation.
- The substrate should be of low chemical reactivity. Any new timbers or decking material should be free from acidic preservatives and have a pH value above 5.5. Installing an inert fleecing layer or underlay between the existing timbers and the new covering can prevent these other aggressive chemicals attacking the lead.
- Avoid using substrates comprising manufactured wood-based boards such as plywood, blockboard, chipboard, hardboard and orientated strand board, because of their chemical and moisture reactivity.
- The lead should ideally be installed in warm dry weather – say, May to July.
- Consider installing sensors (e.g. *H+R Curator*™ system) to monitor moisture and temperature conditions within vulnerable sections of the roof – (e.g. under outlets, gutters, etc.).

Modular construction

Preamble

Prefabrication methods and other aspects of system building such as mechanization of the production process have been around since the early 1950s. As with the problems associated with industrialized construction in the 1960s and 1970s, however, this method of building fell out of favour in Britain shortly after that period.

Up until the late 1970s most prefabricated buildings were used as site huts and for temporary accommodation purposes. Nowadays, though, these modular units can be tailored made for the building in question. They are especially advantageous for toilet and bathroom facilities in fast track schemes. Because they are made under factory conditions, the quality of prefabricated building units is much higher than can be achieved on site.

The main types of modular units that can be used in adaptation schemes are:

- roof pods;
- room pods;

Communal space added to the roofs in Rodowa, Copenhagen

Figure 9.19 Modular roofs unit being lifted into place in a Copenhagen refurbishment project (Lawson, 2001b)

- vestibule/porch pods;
- modular cassettes.

Roof pods

Pre-assembled roof units are being used in the refurbishment of period and modern buildings in several European cities (Lawson, 2001b). In this type of modernization prefabricated penthouse pods, complete with bathrooms and kitchens, are installed onto the top of the building. The existing roof is first removed and the exposed wallhead prepared to take the new element. A prefabricated roof pod complete with internal fittings and finishings as well as external coverings replaces the old roof (see photograph in Figure 9.19).

There are a number of advantages of installing penthouses this way:

- Although the preparation time is longer than for traditional loft extensions, overall the installation work is much faster.
- Disruption to existing residents is minimized – because of the speed of installation.
- Scaffolding is kept to a minimum – which reduces costs, lessens security problems and avoids disfiguring the façade with scaffolding.
- The units and their finishings, being manufactured under factory conditions, are of a high quality.

Room pods

In some cases modular units can prove useful even in the refurbishment of an existing property. Figure 9.20 shows how pods could be used in the renovation of a Victorian tenement building. The internal configuration

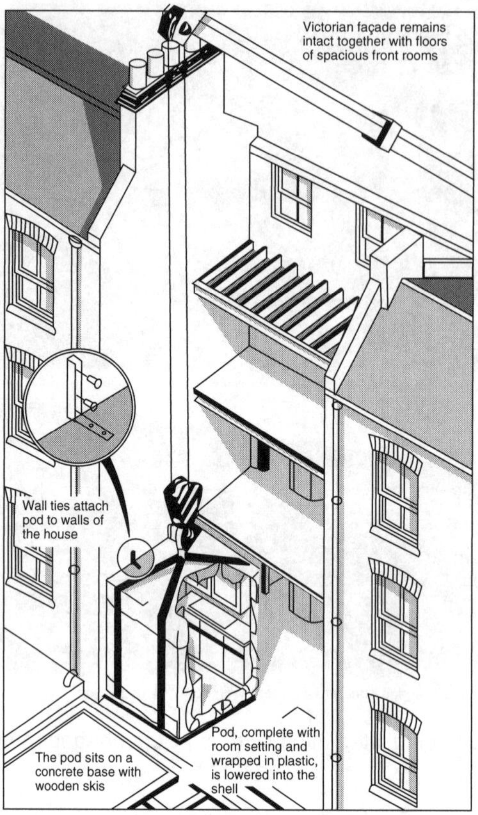
Victorian façade remains intact together with floors of spacious front rooms
Wall ties attach pod to walls of the house
The pod sits on a concrete base with wooden skis
Pod, complete with room setting and wrapped in plastic, is lowered into the shell

Figure 9.20 Use of room pods in housing modernization scheme (Source: *The Sunday Times*, 12 March 2000)

of the building must be able to accommodate these units without too much alteration. Some strengthening work to the existing structure may also be required (see Chapter 7).

The main difficulties associated with installing room or roof pods are the following:

- Large low-loader trucks are required to deliver the units.
- Large-span jib cranes are then needed to lift the modules onto position.
- The road at the front of the premises would have to be closed off temporarily during the installation process.
- Keeping the building wind and watertight during and after the removal of the existing façade or roof cannot be guaranteed.
- The areas of the existing building have to be prepared to receive the units. For example, this entails installing steel and concrete beams to fix the roof units onto the wallhead.

Vestibule pods

Prefabricated cubicles slightly larger than telephone boxes can be installed at the front entrance to ground-floor flats and other dwellings for enhanced security and comfort. These pods come complete with tiled duo-pitched gable roof, external light, security door and threshold. The opportunity to install a retrofit ramp to this new doorway should be taken to facilitate easy access.

These vestibule pods can be made from a polymeric cladding with a core based on recycled glass and refractory waste material. They are attached to the face of the building using bolted stainless steel angle brackets. The abutment at the sides is then pointed with a bed of polysulphide mastic, whereas the junction between the duo-pitched roof and the existing wall is sealed with a lead flashing step dressed into the masonry.

Modular cassettes

These may form part of a room pod or can be fitted individually into an existing space. They are especially suitable for toilets and bathrooms to facilitate the installation of pre-arranged units such as wash-hand basin complete with taps and fittings. These readily installed services are then connected to the building's main services.

Accessibility improvements

The Disability Discrimination Act 1995

Nowadays, as a matter of course, all new public buildings are required to be accessible to people with disabilities. Unfortunately, this is not yet the case with many existing properties. In Chapter 8 importance of disabled access was highlighted. This chapter assesses the provisions of the Disability Discrimination Act 1995 (DDA), as they are more likely to impinge on non-residential buildings.

The aim of the 1995 Act is to discourage those providing goods, facilities and services from discriminating against disabled people. This means that service providers have to ensure that their buildings are accessible to as many people as possible. The ultimate goal is to produce a barrier-free environment to all buildings used by the public.

This Act for the first time requires progressive improvement in the way that owners and property managers of thousands of public buildings and private offices made provisions for people who have a disability or special need to gain access to their premises. The DDA, which was introduced in December 1996, has three stages of implementation. The deadline for Phase one was December 1996; the degree of compliance required those concerned to recognize the problem. Phase two's deadline was October 1999, by which time those responsible had to start making adjustments (e.g. formulate a strategy to comply fully with the requirements). Following the October 2004 deadline for Phase three there is now a need to comply.

The above timetable places the onus directly on owners and managers of public buildings and private offices to make access provisions for the disabled. This has presented a major problem for consultants and statutory officers alike. Problems with the cost commitment, the logistics of planning, access audits and refurbishing many buildings are also significant.

The typical properties that will need to comply with the DDA are offices, restaurants, factories, public buildings, shops, hotels, hospitals, assembly buildings such as theatres and cinemas, and educational establishments.

The consequences of failure to comply are potentially serious. They could result in the owner being taken to an employment tribunal. It may result in a civil court proceedings or disciplinary action.

The provisions of the DDA are enforced by the Disability Rights Commission, which started work in April 2000. Its main areas of influence will be to:

- carry out investigations;
- assist disabled people to take cases to court;
- arbitrate in disputes.

Previously the main standard that building designers used for disabled access was Part M (or in the old, now deleted, Part T of the Scottish Building Regulations). The two main problems with this are:

1. The Regulations are concerned primarily with the building. The 1995 Act, on the other hand, is concerned with the needs of people.
2. The Regulations do not cater for all disabled people. In contrast, the 1995 Act contains a broader definition, which includes wheelchair users as well as deaf and visually impaired people.

It is important to note therefore that mere compliance with the Building Regulations does not mean that the provisions of the 1995 Act have been satisfied, and vice versa. However, adhering to the guidelines indicated in Figure 9.21 would go some way to satisfy these requirements.

Modifications

There are therefore a variety of modifications that may have to be made to an existing building for 2004 for compliance with the 1995 Act. These improvements (see Table 9.8) will inevitably have significant cost implications for service providers.

In properties of historic or architectural importance the implementation of some of the required modifications can prove problematic. Nevertheless, the installation of front entrance ramps behind balustrades can offer a discrete and effective way of overcoming access problems (Bone, 1996; Foster, 1997).

Level entry to buildings is an important requirement of accessibility. However, the majority of existing dwellings in particular have a 200–250 mm high step, which acts as a shallow flood guard. This could only be overcome by installing a ramp with a slope of not more than 1 in 12, which is deemed 'level'.

Nevertheless, 'level' entry requires the threshold to be not more than 15 mm high and have sufficient precautions against water ingress. This can be done by using an aluminium or stainless steel threshold plate with a drainage grating provided on the landing, and a weather bar on the bottom of the door, with heavy-duty neoprene weather strips underneath the bar.

Other basic modifications that can be made to most buildings to enhance their accessibility are summarized as follows:

- Automatic door operators installed (with battery back-up to ensure operation in the event of a power failure), activated via infra-red sensors or push-pad panels. These devices are fitted in a similar way to, and are about the same size as, ordinary overhead door closers.
- Installation or upgrade of induction loop system to assist those with hearing difficulties.
- Larger and more illuminated signage fitted around and within the building.
- More handrails and larger door handles fitted within the main access routes in the building.
- Installation of retrofit lifts or stair-lift (see next section).

Retrofit lifts

Preamble

In the modernization of some premises such as nursing or residential homes as well as offices and some dwellings generally vertical access via only a staircase is likely to be unsuitable for many users. Compliance with the DDA may make it worthwhile therefore to consider installing a retrofit lift as part of the adaptation work.

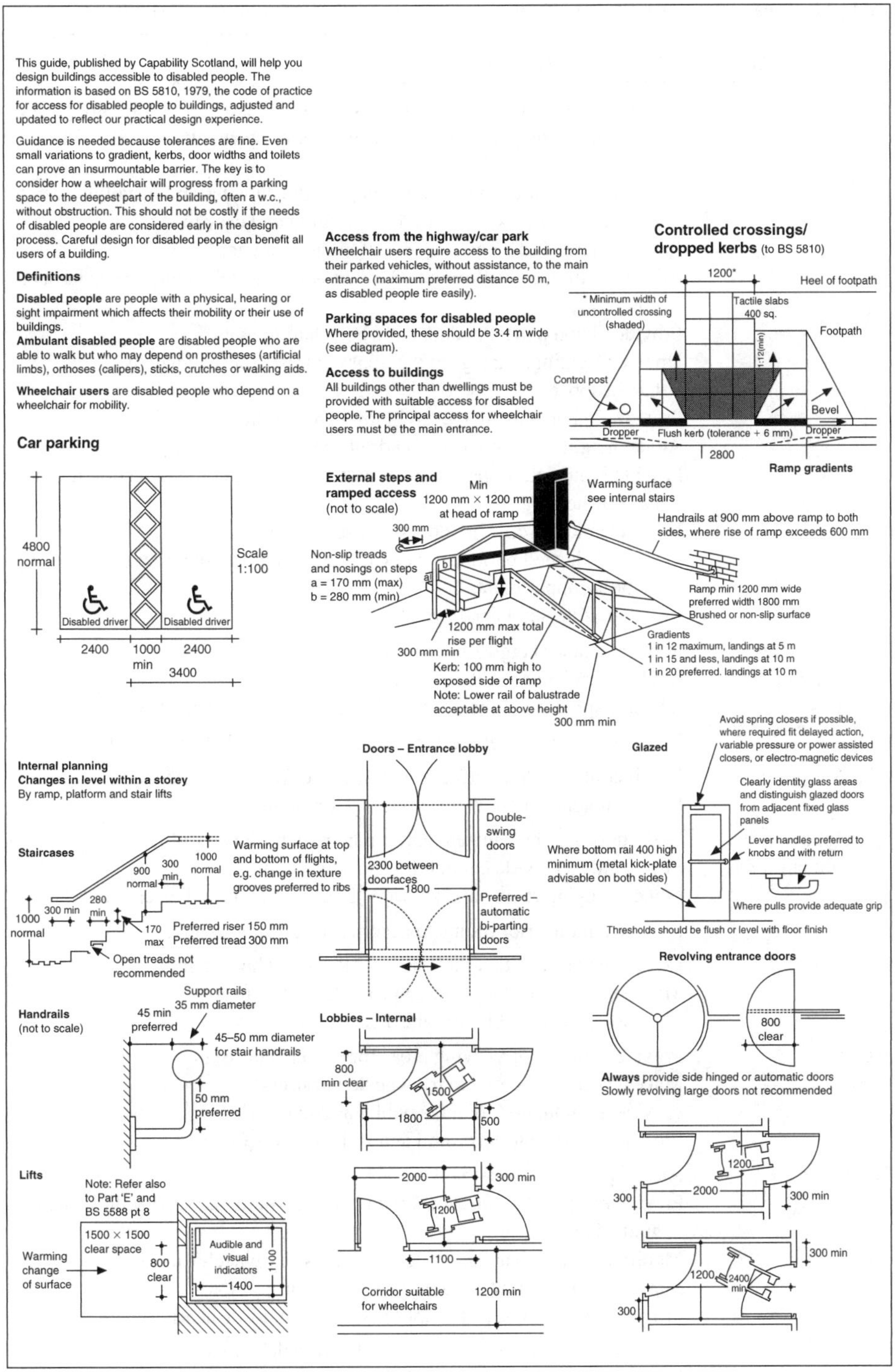

Figure 9.21 Disabled access guidelines (Capability Scotland, 1993)

Table 9.8 Types of accessibility modifications required after 2004 (based on CEM, 2000)

Type of building	*Accessibility improvement*
All	Widen doorways (900 mm), remodel entrances and access routes (e.g. to avoid steps) by providing level access or ramp at entrance (as discussed above). Alter toilet facilities, including enlarging the cubicles and installing easy to use taps. Cubicle doors should be folding and easy to use. Provide tactile indicators for visually impaired people. Adjust position of wall-mounted features such as telephones, fire alarm call points, etc. Provide vision panels in doors at wheelchair height. Upgrade lighting, signage and colour contrasts for the visually handicapped. Provide lifts or stair-lifts, even in low-rise buildings. Adjust height and colours of handrails and other guarding. Provide induction loops and other aids to improve communication for the hard-of-hearing. Fit mechanically opening doors on fire escapes. Provide dedicated parking spaces and dropping off points for people with mobility impairments. Install audible and visual fire alarms.
Retail buildings and shopping centres	Reduce counter heights for wheelchair users. Modify checkouts to allow wheelchair access. Lower displays to allow goods to be reached from wheelchairs.
Banks	Make counters and self-service machines accessible from wheelchairs.
Offices	Install controllable lighting, easily reached from a wheelchair. Reduce heights of counters and reception points.
Sports and leisure buildings	Make door widths wide enough for sports wheelchairs. Provide lockers wide enough for artificial limbs. Make changing rooms and showers large enough for wheelchairs.
Hotels	Provide a number of rooms to accommodate wheelchairs.
Museums	Lower displays, so that they are visible from wheelchairs. Make study spaces large enough for wheelchair users. Make toilets accessible for wheelchair users.
Stadia	Provide a choice of viewing areas for wheelchair users. Make turnstiles accessible for wheelchair users. Provide an induction loop for public address. Provide access to toilets, food and souvenir concessions.
Theatres	Allow space for wheelchairs throughout. Provide extra wide seats, seats with extra legroom and seats with space for a guide dog. Modify access to bars and toilets for access by wheelchair users. Make box offices low enough for wheelchair users. Provide sound loops, or headphones for the hard-of-hearing.
Restaurants	Make self-service counters suitable for wheelchair users. Provide moveable tables and chairs for wheelchair users. Make toilets and terraces accessible.

Schematic cross-section (not to scale)

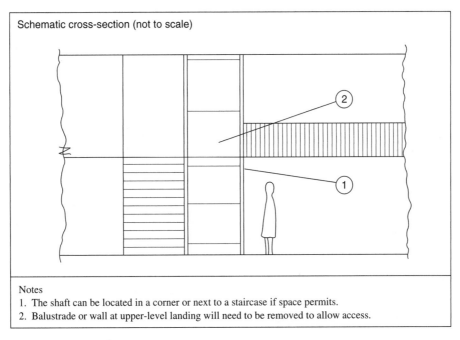

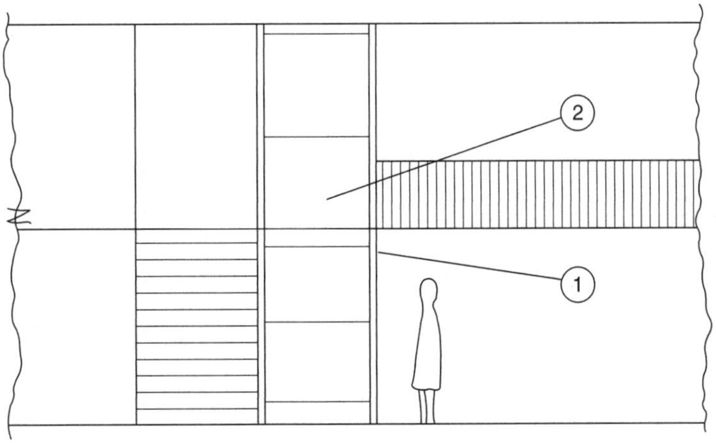

Notes
1. The shaft can be located in a corner or next to a staircase if space permits.
2. Balustrade or wall at upper-level landing will need to be removed to allow access.

Figure 9.22 Retrofit porch-lift

The main techniques that can be used to provide alternative vertical access in such circumstances are the following:

- Stairway lift: seat on glide rail fixed to stair wall. Suitable for two-storey domestic applications.
- Stair carrier lift: platform on rails fixed to stair and landing walls. More suitable for commercial applications not exceeding two storeys.
- Platform/porch lift: see below.
- Through-floor lift: see below.

Platform lift

In some adaptation schemes a modular lift servicing up to three levels could be installed. This option is more suited for nursing/residential homes and commercial buildings where vertical access is a problem because of the intensive use of the premises by wheelchair or other disabled users. It usually comprises a free-standing, self-contained lift unit complete with integral shaft (shown schematically in Figure 9.22). The unit has about a 1.6×1.4 m 'footprint', and is made from steel or aluminium with toughened glass panels.

There are a number of advantages in using a platform or porch lift:

- relatively easy to install – no lift core required;
- no pit at the base of the lift core is needed;
- can be sited next to staircase or in corner position;
- no dedicated motor room is required (the lift mechanism is built into the shaft).

Single-phase electric systems are available with quiet hydraulic or ball screw drive from companies such as 'Access Options', Edwin Road, Manchester; 'Axess 4 All', Venture House, 7 Leicester Road, Loughborough; and 'Movement Management', Abbey Lane, Leicester. The call/send controls are integrated into the gate and doorframes.

Schematic cross-section (not to scale)

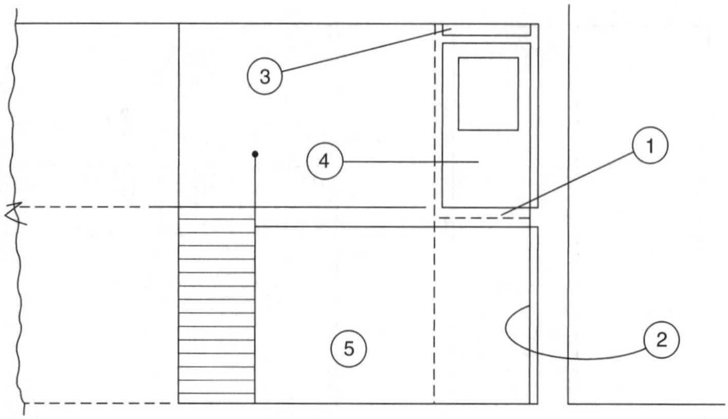

Notes

1. Existing floor structure trimmed to form aperture to accommodate lift car.
2. Guide for lift car.
3. The motor and control circuitry box is located at top of the guide rails, near the ceiling on the first floor.
4. Lift car.
5. A floor-to-ceiling safety pole at both levels may be required to minimize the chances of someone or something interfering with the car as it travels up and down the rails.

Figure 9.23 Retrofit through-floor lift

Through-floor lift

This lift is for transporting one standing passenger or wheelchair user from the ground level to the first floor (see Figure 9.23). It usually features a total enclosure box or capsule, unlike other personal vertical lifts, which are basically platforms in a shaft. The user is completely protected during travel without feeling 'closed in' by its steel car and 'plexiglass' or other safety glass front and sides. The car rides in two steel guide rails, which are mounted directly to the wall. 'Access Options' also supply and install this type of lift, which is more suited for domestic and nursing/residential home use.

Another, more basic, version of the through-floor lift is the manually operated platform for servicing a maximum of two levels (e.g. ground floor to first floor). This is suitable for one standing passenger who operates it by turning a lever wheel, which in turn drives a chain to lift the platform. Typically such a system costs around £5000 + value-added tax (VAT) at 2001 prices to supply and install. Given the need to undertake alterations to the layout and structure to install such a lift, this work would require building control approval.

Summary

Modernization of non-residential buildings is prompted primarily by economic and legal considerations. For landlords it can be seen as a beneficial marketing measure to attract new users as well as retain existing tenants. For occupiers it can result in a more comfortable and energy efficient building. The main energy efficiency measures for most building categories are, however, addressed in Chapter 10.

The extent of the refurbishment work undertaken is a reflection of the owners' future intentions for the building. The long-term aim may be to sell the property at a suitable date. A refurbished building will be a more attractive option for potential purchasers.

External refurbishment often involves a major change to the finish of the fabric. This not only enhances its appearance but also extends the building's service life.

Internal refurbishment is primarily aimed at achieving a better use of the space whilst at the same time maintain if not enhance comfort conditions. However, improvements to comply with the provisions of the DDA are driving the need to ensure that all buildings used by the public are accessible to as many people as possible.

10

Sustainable adaptation

Overview

Sustainability is the key theme of this chapter. In particular it focuses on measures for improving energy efficiency as well as reducing pollution and waste in the refurbishment of existing buildings. Measures to both residential and non-residential building types are dealt with in this chapter.

Background

Climate change

The hypothesis of adverse change in the global climate is now generally accepted (UNEP, 2001). It is mainly the precise causes, scale and significance of this phenomenon that are in some dispute. The UK government in late 2000, however, unveiled a clear strategy aimed at reducing the country's contribution to climate change by cutting emissions of carbon dioxide by 20 per cent below 1990 levels by 2010. Carbon dioxide is emitted during the combustion of fossil fuels. It accounts for over 80 per cent of all greenhouse gas emissions and is generally thought to be the main human cause of global warming these days (Graves and Phillipson, 2000).

One of the consequences of global warming over the last several decades is that tropical sea temperatures have been increasing slightly – by around 1 or 2°C. This has triggered the spate of intense hurricanes emanating from the Gulf of Mexico that affected several southern states of the USA in recent years. For example, the Category 5 hurricanes (i.e. wind strength 155 mph+ on the Saffir–Simpson scale), Katrina and Rita caused substantial flooding and storm damage to large areas around America's Gulf Coast during September 2005. These extreme weather conditions can be attributed to natural climate cycles as well as being exacerbated by human industrial activities.

Inevitably, therefore, climate change has had and is continuing to have an impact on buildings (Graves and Phillipson, 2000). According to various press reports the severe gales that battered Britain in 1987 resulted in a total repair bill of around £1.4 b. Two major storms in 1990 caused an estimated £3b worth of damage. The initial estimated total cost of rectifying the severe flooding and storm damage in late 2000 was around £1b.

It is estimated that the average insurance claim for a wind-damaged roof of a residential property in the UK is between £1000 and £2000 [Building Research Establishment (Graves and Phillipson, 2000)]. In some cases where the internal damage has been caused and the whole roof requires replacing, the cost could be as high as £20 000 for each dwelling.

The British government-sponsored Climate Impacts Programme (CIP) has identified a number of implications of climate change for the UK by the second half of the 21st century. The following changes will necessitate building adaptation responses:

- An increase in the average temperature of London by over 2°C. However, paradoxically, overall temperatures in the northern parts of the UK may actually decrease owing to the cooling of the Gulf Stream by the melting of the polar ice caps, triggered by global warming.
- A rise in sea levels causing an increased risk of flooding, particularly in low-lying areas of the country such as East Anglia.
- More fluctuations in weather conditions resulting in more storms and wind damage. This means that for Britain winters will get wetter, summers will get drier – respectively leading to floods and droughts occurring within short periods of each other.
- Not only will weather fluctuations generate more flooding in winter but will also probably result in subsidence and heave problems. As summer months hand are more likely to become drier and warmer, subsidence is likely to be a problem, particularly in areas with shrinkable clay soils such as the south east of England. Warmer summers will increase the demand for energy consuming facilities such as air conditioning for cooling.
- Rising ground temperatures can result in more active soil contamination and allow the ingress of gases such as radon into buildings more easily.

Government policy changes

One of the main policies emerging from the CIP that has a bearing on building owners/users has been to update the relevant parts of the Building Regulations. Since 1 April 2002 major changes to Part F (equivalent to Section 3 in Scotland), Part J (Section 4), Part L (Section 6) and Part M (Section 4) of the Building Standards have come into force. They are more complex than earlier versions and for the first time elements apply to changes to existing buildings.

These amendments to the Regulations are a response to the government's aim to increase energy efficiency and reduce carbon dioxide emissions. The new requirements are more complex than earlier versions and for the first time elements apply to changes to existing buildings.

In mid-September 2005, the Office of the Deputy Prime Minister (ODPM) and Department for the Environment, Farming and Rural Affairs (DEFRA) announced new measures to make buildings more energy efficient to save one million tonnes of carbon per year by 2010. This is equivalent to emissions from more than one million semi-detached homes.

According to the Workplace Law Network e-bulletin No. 260:

> The changes to Parts F and L (ventilation and fuel conservation) of the Building Regulations two years ahead of schedule from April 2006 and the implementation of the Energy Performance of Buildings Directive, will make a major contribution to the UK's commitment to combat climate change. Buildings will need to be better insulated and make use of more efficient heating systems.
>
> The revised Part L will also make air pressure leakage testing of buildings mandatory, improving compliance with the regulations by showing where there is unacceptable leakage, which can reduce the energy efficiency of buildings.
>
> The ODPM believes these measures, alongside changes to condensing boilers, will deliver increased energy standards for new buildings, including around 27 per cent in non-dwellings, 22 per cent in houses and 18 per cent in flats. On average the increase in dwellings will be 20 per cent, which reflects the growing proportion of flats being built with more people now living alone. The new measures taken together with changes to strengthen Building Regulations in 2002 will improve standards by 40 per cent, cutting fuel bills by up to 40 per cent for new homes built from 2006.

> Part L of the Building Regulations sets out standards for building work in order to conserve fuel and power and minimise heat loss, raising energy efficiency standards through the use of more energy efficient materials and methods. The measures are performance-based which allows housebuilders' flexibility about how the new standards are met.
>
> To ensure a high level of compliance and understanding of the new regulations ODPM is introducing nationally recognised qualifications for surveyors and will be promoting the development of self-certification schemes for Part L schemes to improve regulation. ODPM has already put in place a training and information programme.
>
> To maximise the impact of Building Regulations on climate change these measures are being brought forward by two years from 2008 to April 2006. This includes a deferral of three months from the implementation date set out in the Energy White Paper, to give the building industry sufficient time to prepare.
>
> In addition, from April 2006 all new residential development receiving government funding will need to meet a new national Code for Sustainable Buildings. The Code will go beyond Building Regulations covering not just fuel and power but also the efficient use of water, ensuring much higher sustainability standards.

The Climate Change Levy and Enhanced Capital Allowance Scheme are two other climate change policies but these relate to non-domestic buildings. These measures are aimed at encouraging or discouraging companies where necessary to adopt more sustainable processes.

In summary the measures in climate change policy deal with:

- air tightness standards through robust construction and pressure testing new buildings;
- the carbon performance rating;
- efficient air conditioning and mechanical ventilation systems and appropriate controls;
- maximum carbon intensities for heating systems;
- special requirements for replacement of boilers (with the more efficient condensing boilers) in both residential and non-residential buildings;
- increased thermal insulation standards – involving significantly improved *U* values;
- resistance to solar overheating;
- performance levels of lighting fittings;
- commissioning, operation and maintenance of building engineering services;
- waste water treatment systems.

Significance of sustainability

Sustainability is defined as a set of processes aimed at delivering efficient built assets in the long-term (Department of the Environment, Transport and the Regions (DETR), 1998). It is about taking a strategic view of enhancing the impact of human development on the environment. The fundamental aim is to satisfy the requirements of people today without undermining the ability of our descendants to meet their own needs (Brundtland Commission, 1987). In other words it is about social and economic progress and technological advance, which recognizes the needs of everybody, as well as the environment in which they live and work. Or to put it another way:

> Sustainable practices are an investment in the future. Through conservation, improved maintainability, recycling, reduction, reuse and other actions and innovations, we can meet today's needs without compromising the ability of future generations to meet their own. (US Air Forces Facilities Guide 2005, cited in http://renovation.pentagon.mil/)

Two aspects of sustainability are relevant to the built environment: sustainable development and sustainable construction. The former has wider applications than the latter aspect of sustainability. Sustainable development deals with urban and regional issues and local issues such as development densities and public transport as well as the relationship between land-use and transport. It includes social planning issues such as creating workplaces and housing near each other to reduce waste and minimize transportation problems. It also means design for adaptation by making new buildings loose-fit or more flexible to allow for easier reuse, which relates it to sustainable construction.

Sustainable construction, in other words, whether in relation to new or existing buildings, deals with a variety of proactive processes. If a building can continue to function effectively for an indefinite period, it is considered sustainable. For example, sustainability in this context is primarily concerned with matters such as minimizing construction waste and pollution, saving energy, increasing the use of recycled and locally produced materials and relying less on toxic chemicals. It is also about using whole-life cycle costing in the design of new build and adaptation schemes to help determine economic maintenance cost levels.

The primary goals of sustainable construction, therefore:

- cause minimum damage to natural and social environments through effective protection and prudent use of natural resources;
- reducing both embodied energy (the energy required to make or adapt a building) and operational energy (the energy required to use a new or adapted building);
- minimize the use of scarce resources such as fossil fuels and non-renewable materials, and maximize the use of renewable energy sources;
- enhance the quality of life through the maintenance of high and stable levels of economic growth and employment;
- make outputs (adapted as well as newly constructed buildings) acceptable to future generations (DETR, 1998).

Sustainability has become increasing important in adaptation as well as in new build work. The oil crisis in the early 1970s triggered a growing awareness of environmental issues. Since then environmentalists have been continually warning of the adverse effects of pollution, ozone depletion, the greenhouse effect and global warming. This in turn prompted a number of responses at both national and international levels.

The United Nations framework convention on climate change held in Kyoto, Japan, in December 1997, addressed the issue of sustainability on an international basis. This was followed in 2002 by the UN convention on sustainable development in Johannesburg, South Africa. Both these conventions have committed member states to set targets in reducing pollution and carbon dioxide emissions, as well encouraging them to adopt a host of other initiatives, such as enhancing sanitation and using environmentally sound technologies (ESTs).

According to Strong (1999) the main ESTs in use today are:

- Biomass: using the energy stored in plants and organic matter, such as woodchip-powered heat and power plant.
- Geothermal: tapping thermal energy in the soil to warm buildings using heat pumps.
- Hydropower: using waves or large mass of stored water to generate electricity.
- Solar thermal: capturing sunlight onto a receiver dish which converts the light into heat.
- Photovoltaics (PV): converting light energy to electricity using photosensitive plates.
- Wind: using wind power to generate electricity, pump water and grind grain and wind-driven ventilation systems.

Of these ESTs solar thermal and PV have the greatest potential for use in adaptation schemes. The main features and drawbacks of these technologies for use in refurbishment work were briefly discussed in Chapter 8.

An important concept in relation to environmental issues affecting adaptation is 'permaculture'. This is a term that is becoming internationally recognized as a key environmental criterion. Its leading tenets are ethically principled, sustainable, sound and creative design for the care of the earth and its people (Mollison and Slay, 1991).

'Green' issues such as ecology and 'permaculture' therefore have become fashionable nowadays. Sustainability is an important policy response to these issues and has become a significant factor in construction as demonstrated by a government paper on the subject (DETR, 1998). This has led to the development of ESTs, sometimes called 'renewable energy technologies' (RETs). These are designed to generate electricity for passive heating of buildings and heating of domestic and commercial process water supplies.

It is no wonder then that the environmental impact of the construction industry is gaining increasing attention in government and other official circles. The construction and use of buildings are among the most significant contributors of greenhouse gas emissions. It is estimated in UK about half of all carbon dioxide emitted is directly related to our use of buildings (Graves and Phillipson, 2000). This has contributed to global warming as well as to acid rain. The extensive use of timber and wood-based products in buildings has been one of the major causes of depletion of the tropical rainforest. The extraction of aggregates and other minerals for construction has resulted in super-quarries, which can cause aesthetic as well as environmental blight to an area.

All of this results in the following consumption and emission factors:

- Initial embodied energy (measured in $GJ/m^2/year$ – over, say, 60 years). This is the amount of energy generated in the construction of the building.
- Embodied carbon dioxide is the amount of CO_2 emitted during production of unit quantity of a material.
- Energy consumed in use (this is the energy consumed for heating, cooling, etc., as represented by the fuel bills for the building).
- Embodied energy of buildings and their maintenance and adaptation (over each building's predicted life).
- Transport energy (the total daily commuter travel and business travel associated with the building).
- Consequent gaseous pollutant emissions (e.g. carbon dioxide and other acid gases from chimneys, and volatile organic compounds (VOCs)).

Sustainability, then, is about achieving a greener or more environmentally friendly approach to adaptation and maintenance as well as to new build work. The aftercare and reuse of old buildings is one of the key goals of sustainable construction. It obviates if not minimizes the need for wholesale demolition, which is a dangerous, polluting and waste-producing and energy-consuming process.

One way to achieve optimal sustainability in construction is by using as many ESTs or RETs as possible in any adaptation scheme. The aim is to maximize the efficient use of an existing facility and minimizes any negative impact on the environment. Another goal is to make new and adapted buildings 'greener' by using local, renewable materials (Anink et al., 1996) and considering maintenance and whole-life cycle costs at the design stage. Assessing the adapted building using the BREDEM 98 rating method, for example, could assist this.

It is important to appreciate, however, that there is no end-point in this process – sustainability means continuous improvement. Not surprisingly, then, building adaptation is considered one of the most effective strategies for sustainability. Compared with new build it involves lower costs in relation to materials, transport, energy and pollution. It also minimizes on the new services and infrastructure that are often needed for a greenfield site (Energy Research Group [ERG], 1999).

Importance of sustainable adaptation

Adapting a property as opposed to constructing a new building not only helps to reduce energy consumption, pollution and waste. As pointed out by Edwards (1998) 'recycling buildings and giving them new uses is as important as recycling bottles,' mainly because 'the UK construction industry generates about 70 m tonnes of waste a year, which is nearly a quarter of all waste'.

Adaptation also saves valuable resources as well as reduces the volume of material being sent to landfill. That is why environmentally it is more sensible to refurbish than to demolish and rebuild. In this context, then, sustainability consists of some if not all of the following:

- Re-using old buildings that might otherwise have remained empty or underused (i.e. by undertaking an adaptive reuse – in particular see Chapter 4).
- Improving the physical performance of existing buildings (i.e. through modernization – see Chapters 8 and 9). This includes improving environmental performance – thermal insulation to reduce energy demand.
- Retaining heritage features of buildings being maintained and adapted for conservation purposes (see Chapter 3).
- Environmental and ecological issues included in the adaptation brief and maintenance policy to minimize any adverse effects of such work.
- Maximizing access to the building for all users – disabled people as well as bicycle/pedestrian access to and around the building.
- Designing extensions and conversions for easier reuse/recycling/demolition.
- Retaining trees and other ecologically important vegetation on the site to preserve if not enhance bio-diversity.
- Using indigenous materials wherever possible to support local suppliers and avoid generating excessive transportation usage.
- Achieving community links by providing multi/flexible-use facilities in adapted public buildings.
- Maximizing use of second hand/recycled/salvaged products (e.g. reusing old slates, old wooden floors and flagstones and using waste wood to make chipboard).
- Maximizing the use incinerator ash, blast-furnace slag or other by-products of industrial processes. These can be used for example in the concrete mix of ground floor slabs to replace existing defective suspended timber floors.
- Maximizing the use of environmentally sound/green materials (e.g. renewable products such as non-imported softwood timber).
- Minimizing or controlling the use of environmentally unfriendly materials such as ammonia, urea formaldehyde and polyvinyl chloride (PVC) (see below).
- Avoiding the use of substances containing toxic chemicals such as 'Lindane', pentachlorophenols (PCPs) and polychlorinated biphenyls (PCBs), the latter in timber preservation, electrical equipment, flame-retardants, paints and plastics.
- Minimizing the use of non-renewable energy such as fossil fuels, and scarce materials in construction such as marble.
- Reducing CO_2 and other toxic emissions from buildings to a minimum (e.g. to help combat the greenhouse effect) by making building more energy efficient. This means using insulation materials, such as mineral fibre, that do not contain chlorofluorocarbons (CFCs) or hydro-chlorfluorocarbons (HCFCs), the main ozone-depleting chemicals. The use of Halon, a CFC-containing fire-fighting gas in fire extinguishers, should also be avoided.
- Using natural lighting as much as possible to minimize energy-consuming lighting.
- Using low energy lighting with appropriate controls to reduce energy costs.
- Achieving a highly energy efficient fabric – super-insulated and with excellent air-tightness, but adequately ventilated to prevent excessive moisture build-up within the building.
- Reducing water use in buildings, which accounts for half of all such consumption in Britain. Installing low water-use sanitary ware in the adapted building can assist this. For example, twin-flush water closets (WCs) can use 2 or 6 l instead of water compared to 9 l with conventional single flush units.

- Reducing the amount of toxic and noxious chemicals used in adaptation and maintenance works. This means moving away from spirit-based materials to aqueous-based ones, such as paints, adhesives and chemical water-repellents, with low/no VOCs.
- Incorporating more responsive and energy efficient services – with sensors and automatic controls to regulate the use of energy consuming services to where and when they are needed.
- Making buildings more futureproof.
- Eliminating fuel poverty and tackling hard to treat homes (see EST, EC 21, 2003).

Futureproofing buildings

Generally

Adapting buildings isn't just about making them suitable for today's uses and conditions. It's also concerned with making them suitable for occupancy requirements and weather conditions in later years. Achieving a loose fit in a building will facilitate its adaptive reuse if necessary in future.

Protecting buildings against adverse environmental influences that may occur in future as a result of global warming, for example, is an increasingly important sustainability objective. Depending on their location, existing as well as new buildings will need to be made more robust to resist more intense levels of rainfall, extremes of cold and hot weather, flooding and stronger winds triggered by climate change. The following measures can help in this regard:

- Buildings can be made more weathertight and better ventilated by replacing wooden windows with double-glazed unplasticized polyvinyl chloride (uPVC) frames containing trickle vents. However, this has to be balanced against the environmental problems posed by uPVC as discussed in Chapter 1 and below.
- Strengthening fixings on roofs of buildings by using dry-fix verge and ridge cappings. Using lightweight metal roofing tiles that are mechanically fixed will also improve the wind resistance of the roof finish (see Chapter 9).
- Improve the design of natural ventilation and passive cooling in buildings to reduce the need for air conditioning or other forms of active cooling in hot summer months. Installing retrofit windows with hopper lights, and passive stacks ducted to the roof to facilitate rapid natural ventilation in hot weather periods, will help in this regard.
- Render or overclad buildings to make them more resistant to rain and wind penetration.
- Underpin existing buildings that have shallow foundations or are otherwise vulnerable in flood/subsidence-prone areas. More underpinning measures are required to combat subsidence problems associated with warmer, drier summers, particularly in areas with shrinkable clay soils. However, as indicated in Chapter 5 this option should be considered with care as installing partial underpinning can create 'hard spots', which might in turn lead to differential settlement.
- More condensation control measures (such as positive input ventilation) required in housing because of increasingly mild but damp winters in certain regions (see Chapter 8).
- Install flood control measures in and around buildings in flood risk areas.

Dry-proof flood control measures

These involve measures to improve flood resistance of external walls and floors to prevent water ingress to the building. They comprise the following:

- Flood barriers for doorways and other openings at or near ground level.
- Covers for airbricks and other vents at low level.
- Valves in drainage lines to prevent backflow.

Wet-proof flood control measures

These involve means of improving the resistance of internal walls, floors and fittings to improve the ability of materials to withstand the effects of internal flooding. They comprise the following two main measures:

- Raise electrical sockets above predicted flood line.
- Use water-resistant materials for wall and floor finishes in basement and ground floor rooms.

Requirements for sustainable adaptation

Profitability

For any business, profitability is an important criterion of sustainability in relation to new build. It is a valid objective for construction generally but is much harder to attain in adaptation work, particularly in the context of building conservation. An adapted building can rarely achieve the same rate of return as a newly constructed facility. This is because, inevitably, the former cannot match the latter in terms of overall performance.

Flexibility

As with new build, an adapted building ought to be seen as an asset for life. It should be so designed that it will be cherished to increase the chance of re-use in future. Flexibility needs to be incorporated in the adaptation design but, again, this is more difficult to achieve than in new build. Adapted buildings should be designed so that they are cheap to run, by paying particular attention to water, heat and power systems. They should also be designed and constructed with health and safety in mind. Adopting a coherent safety policy and undertaking risk assessments where appropriate both before and after occupancy can achieve this (see Chapter 11).

Energy efficiency actions

In an adapted building this can be best achieved by reducing energy consumption and minimizing heat losses. Lighting, for example, accounts for the majority of energy consumption in commercial buildings (see *THERMIE Maxibrochures*, 1992). Adaptation schemes should therefore attempt to maximize natural daylighting (e.g. by installing light-wells or sun-pipes) if possible and provide energy efficient lighting where necessary.

Global warming is likely to raise the demand for active cooling systems in buildings. Air conditioning in a building increases its energy consumption. In many cases it is more expensive to cool a building than it is to heat it. More reliance therefore will need to be placed on passive cooling measures to combat this problem.

Draught proofing and insulation measures to the external fabric can help minimize heat losses. However, excessive draught proofing can increase the risk of condensation occurring by trapping vapour in roof spaces. These measures and precautions are dealt with in more detail below.

Environmental performance indicators

Setting sustainability goals is one thing; meeting them is another matter. The best way to measure as to whether or not sustainability targets have been met is to use established indicators. The Sustainability Working Group of M4i has identified six key indicators:

- Operational energy ($kgCO_2/m^2/year$).
- Embodied energy ($kgCO_2/m^2$).
- Biodiversity (the area of flora and fauna habitat preserved and or created as a percentage of total site area as well as taking into account the area of wildlife habitat existing prior to development).

- Transport (km per total on-site working hour, or transport movements per total development area).
- Waste (m^3/100 m^2 floor area).
- Water (m^3/person (equivalent)/year).

Eco-friendly materials

The environmental and health implication of using certain materials should be considered in any adaptation scheme. Thus, ideally, only benign materials should be specified. In addition, materials containing known allergens (e.g. formaldehyde) or other disease triggers should be avoided. Materials with hygroscopic properties should be specified that would aid moisture management and help maintain low relative humidity levels in which moulds and mites (two main asthma triggers) cannot thrive.

Therefore, only environmentally acceptable materials such as timber (treated with breathable rather than solvent-based coatings) and ceramics should be used in preference to substances such as PVC and solvent-based glues and paints.

Alternatives to PVC in adaptation schemes

Background

Plastic (or 'polymer' as it is more accurately called) was developed for use in the construction industry on a large scale after the mid-1940s (Baumann, 1966). It is therefore a relatively modern material and has been promoted because of its relatively low cost, rot-free, low maintenance, weather-resistance and wide colour range qualities. For these reasons the performance of uPVC, for example, is still considered better than timber, particularly in helping to make buildings more 'futureproof' as described earlier.

There are two main categories of polymers namely thermoplastics and thermosetting plastics and their uses are listed in Table 10.1. Despite the potential environmental problems with using some polymers like PVC, however, thermoplastics such as the 'Stevens EP' polyolefin single-ply waterproofing membrane offer a sustainable, safe and durable flat roof covering.

Thermoplastic polyolefins (TPOs) come in two main forms: Polypropylene-based (PP) and Polyethylene-based (PE). According to Stevens Europe (2001), PP- and PE-based materials have markedly different performance characteristics. PP-based TPOs have been used for over 15 years in the roofing industry in the USA. Unlike PE-based membranes they 'exhibit a wide melting range allowing for ease in welding, good elevated temperature performance particularly useful with dark colours and good dimensional stability over a broad temperature range'. In contrast, the long-term performance of PE-based membranes is not yet known (Stevens Europe, 2001).

The main benefits of TPOs over other polymeric membranes are:

- strong, reliable seams;
- fire resistance (using non-chlorine flame-retardants);
- wind resistance;
- impervious to a wide range of chemicals;
- weather resistance (UV stabilizers);
- energy efficient (using white and light-coloured roof finishes);
- environmentally friendly (lower percentages of halogen than PVC sheets, no chlorine or plasticizers).

PVC is one of the most commonly used plastics in buildings today. It also comes in uPVC/PVCu/PVC-U (unplasticized polyvinyl chloride) form, which is now used for windows, doors and cladding. uPVC is slightly more environmentally friendly than PVC. This is because the former does not contain the plasticizers that are normally added to the latter. Leaching of plasticizers is one of the main pollution effects of PVC. Moreover, uPVC tends to have better durability than ordinary PVC because it lacks these chemical additives.

Table 10.1 Common plastics and their uses in buildings

Types	*Examples*	*Typical uses*
Thermoplastic	PVC (polyvinyl chloride)	Drainage pipes, windows and doors, cladding, cable sheathing, etc.
	PVA (polyvinyl acetate)	Interior grade wood adhesive
	Polypropylene (PP) and CPVC (Chlorinated PVC)	Cold water supply pipes
	Polystyrene	Insulation beads/boards
	Polyethylene (PE)	Vapour control layers and damp-proof courses
	Polycarbonates	Transparent sheeting with heat and impact resistant qualities
	Acrylonitrile butadiene styrene (ABS)	Above ground drainage and cold water supply pipes
	Polyolefin	Reinforced waterproofing membrane 2 mm thick that is chloride- and plasticizer-free
	Ethylene Propylene Diene Monomer (EPDM)	Roofing membrane
Thermosetting	Polybutylene	Hot and cold water supply pipes
	Polyurethane	Insulation foam/board
	Phenolics	Insulation foam, laminate boards, cisterns
	Polyesters	Roofing felt
	Urea formaldehyde	Insulation foam
	Nylon	Carpeting
	Epoxy resins	Adhesives and filler
	Silicones	Mastics, water-proofing membrane/film
	Elastomers	Waterproof membranes
	Pultruded glass fibre	Doors and windows

Problems with PVC

Environmental pressure groups such as Greenpeace (1997) have, however, pointed out the serious risks associated with the manufacture of plastics such as PVC and its use in buildings. Several local councils throughout Europe are now restricting the use of this material in construction. Some countries such as Sweden are even committed to a complete phase-out of soft PVC and rigid PVC with harmful additives, which applies to refurbishment as well as new build projects. Indeed, the European Commission (EC) issued a Green Paper in July 2000 pointing out the adverse environmental effects of PVC.

In summary, then, the main problems with PVC are as follows:

- Because of its chemical make-up PVC has carcinogenic and mutagenic properties. Dioxin and Phthalates, for example, are two of its constituents, both of which have known hormone-disrupting properties. The environmental effects and impact on human health of PVC, therefore, cannot be ignored.
- The manufacture of PVC generates highly toxic, tarry wastes as well as pollutants such as vinyl chloride monomers, cadmium and mercury.
- Proper disposal of PVC products and wastes is difficult, even when used as landfill or underground dumping. For example, leaching of PVC chemicals such as phthalate additives or modifiers can contaminate groundwater.
- The fire safety performance of PVC is very poor because of its very low fire resistance. Moreover, in a fire it produces toxic smoke and gases such as hydrogen chloride, which releases corrosive hydrochloric acid on contact with moisture. Electrical fires are particularly dangerous in this regard.

Table 10.2 Examples of non-PVC building materials

Element or component	*PVC alternative*
Roofing membranes	EPDM. Glass reinforced polymers (GRPs) seamless membranes with epoxy resin. Polyolefin single-ply waterproofing membrane. Rubberized felt.
Geomembranes	Butyl rubber sheeting.
Vapour checks and HRMs	Aluminium foil.
Tensile structures	Silicone coated glass-fibre stressed fabric.
Windows	Aluminium and wood combination. Pultruded glass fibre. Timber – softwood or hardwood.
Doors	Hardwood, better still, pine pitch or other recyclable timber reclaimed from a demolished property. Pultruded glass fibre.
Cladding and profiles	GRP sheeting. Glass reinforced cement (GRC) sheeting. Timber, strand board.
Flooring	Linoleum sheeting or tiles, cork tiles or boards, laminated timber boarding.
Above ground drainage	Stainless steel, cast iron, copper, aluminium, HDPE (High density polyethylene).
Below ground drainage	Vitrified clay, HDPE.
Electrical cables and wiring	Zero halogen low smoke ('HLS) or low smoke and fume (LSF) cabling. Ethylene propylene rubber cables for low voltage use.
Hot and cold water pipes	Polybutylene.

Adapted from Greenpeace (1997).

The alternatives to PVC in adaptation and new build schemes are listed in Table 10.2. One of the most exciting developments in this regard is pultruded glass fibre. Using Fibre Reinforced Thermosetting (FRT) technology offers better durability and overall performance than PVC and other window materials. As a result, it's been branded the 'next generation' window profile.

There is of course a dilemma with the use of plastics in buildings. On the one hand, polymers can be considered as environmentally unfriendly because of the pollution implications associated with their production. On the other hand, however, their durability and thermal qualities make them a good material for windows or insulation. A balance, therefore, has to be struck in using plastics as opposed to natural or environmentally friendly materials for certain components in adaptation and new build schemes.

Energy efficiency

Rationale

According to the BRE and the Energy Saving Trust (EST) buildings in Britain consume up to 50 per cent of the country's energy. Twenty-eight per cent of the UK's carbon dioxide emissions come from domestic energy use. Ninety per cent of total energy consumption is building energy use – the remaining 10 per cent is related to energy in manufacture. This coupled with the estimated £10 billion of energy wasted in the UK every year makes energy efficiency a major sustainability criterion. The British Government responded in the mid-1990s by issuing several pieces of legislation to tackle this problem. For example, the Home Energy Conservation Act 1995 and the Energy Conservation Act 1996, deal specifically with this issue. These two Acts required all local authorities with housing responsibilities to prepare, publish and submit to the Secretary of State (for the then DETR) an energy conservation report identifying energy conservation measures for residential accommodation in their area.

The principles for both energy efficiency and water conservation are basically the same: reduce demand, recycle and exploit renewable supplies (Edwards, 1998). The ways in which some of these measures can be implemented in adaptation work are addressed in this chapter.

Government-backed guidance on energy efficiency in buildings is extensive and deals with refurbishment as well as new build schemes. *The Energy Efficiency Best Practice Programme* publications indicated in this book cover a wide range of building types, both domestic and non-residential. A whole series of case studies as well as good practice guides related to that Programme are available from the EST.

The EST was established after the 1992 Earth Summit in Rio de Janeiro. It's is a non-profit organisation funded by government and the private sector, to help reduce CO_2 emissions in the UK. Its website contains an extensive range of informative articles on energy efficiency matters (see Bibliography).

It is essential therefore that cost-effective energy efficiency is integrated with any refurbishment programme, particularly in housing, to maximize economies of scale as well as reduce fuel consumption and pollution. Housing, after all, accounts for one-quarter of all UK CO_2 emissions (Anon, 2002). At least half of energy savings must be achieved among priority groups in housing – such as those on low income or disability benefits or the elderly.

Heat losses

Heat loss through a building's fabric has a major impact on its energy efficiency. According to Cairns (1993), the approximate percentages of heat losses from an uninsulated dwelling are as follows (with the revised figures as a result of increasing the insulation levels shown in brackets):

- 25% escapes through the roof (8% after insulation);
- 35% escapes through the walls (10–12% after insulation);
- 15% escapes in air leakage through the doors, etc. (9% after insulation);
- 15% escapes through the ground floor (9% after insulation);
- 10% escapes through the windows (5% after insulation).

Accordingly, following refurbishment, all housing should be as energy efficient as cost-effectiveness allows (GPG 82, 1992), in order to:

- reduce tenants' fuels bills and provide affordable warmth (as well as combat condensation dampness);
- minimize management and maintenance costs;
- maximize rental value;
- reduce global and local pollution;
- conserve fossil fuel resources.

There are, of course, a variety of ways of attaining better energy efficiency in existing buildings. Solar energy schemes comprise one group and these are examined below. At a basic level, however, energy efficiency goals can be attained in a refurbishment scheme by:

- achieving good levels of thermal insulation in the external fabric;
- conserving water use in buildings;
- using energy efficient lighting;
- providing an efficient heating and hot water system;
- substituting controlled ventilation for existing (uncontrolled) draughts.

Benefits of energy efficiency

The benefits of greater energy efficiency therefore are not hard to identify. They can be summarized as follows:

- lowering fuel costs;
- reducing demands on non-renewable fuels;
- enhancing comfort conditions within buildings, and thereby improving health and productivity;
- curbing environmental pollution;
- delaying if not mitigating the greenhouse effect;
- creating jobs in energy conservation.

Barriers to energy efficiency

Despite the best efforts of government and others there are still a number of barriers that inhibit if not prevent the attainment of greater energy efficiency. Such barriers can be summarized as follows:

- Economic: The capital costs of schemes are often perceived as a disincentive. The payback period for some measures may exceed 10 years.
- Legal: The delays in implementing any measures may undermine their effectiveness. Statutory constraints such as obtaining approvals may inhibit the implementation of energy efficiency measures.
- Human: Excessively sceptical or hostile attitudes towards the greenhouse effect hypothesis, coupled with problems such as environmental pollution and the depletion of fossil fuels may hamper the promotion of energy efficiency. Ignorance of the benefits and costs of undertaking energy efficiency is another factor that can inhibit the implementation of these measures.
- Technical: The difficulty (in terms of access, compatibility or fixing) of installing energy efficiency measures to the fabric or services. Installations such as solar reflectors or PV panels may compromize the appearance of a building.

Guidance on energy efficient refurbishment

As shown in the Bibliography, BRECSU have published an extensive range of guides on energy efficient refurbishment for most building types. The following is a summary of some of the relevant guidance available:

- Factories, GPCS 192 (1994)
- Hospitals, GPG 206 (1997)

- Hotels, GPCS 205 (1995) and GPCS 244 (1995)
- Houses, GPG 155 (2001)
- Offices, GPG 35 (1993)
- Public houses, GPG 150 (1995)
- Shops, GPG 201 (1997)
- Schools, GPG 233 (1997).

Energy efficiency strategy

Under the Home Energy Conservation Act 1995 local authorities are required to prepare a strategy for improving the energy efficiency of their housing stock. This involves setting targets for a 10-year plan to improve the energy efficiency of residential accommodation by a set percentage (30.9%).

The following is a typical list of energy efficiency measures that local authorities may set to achieve within 10 years:

- all houses with hot waster tanks to have tank insulation;
- 85% of houses with lofts to have at least 200 mm of insulation;
- 85% of houses shall be draught stripped;
- 80% of houses with cavities shall have cavity wall insulation;
- 40% of old heating systems shall be renewed with non-condensing boilers;
- 40% of old heating systems shall be renewed with condensing boilers;
- 80% of houses with lofts to have at least 200 mm of insulation;
- two low-energy lights per household shall be installed in 80% of those without them;
- reflective foil shall have been fitted behind radiators in 50% of those homes with wet central heating systems.
- radiator shelves shall have been fitted above radiators in 50% of homes
- secondary glazing shall have been installed in 50% of properties with single glazing.

Table 10.3 Typical energy efficiency measures for domestic premises (Based on BRECSU guidelines – see various case studies)

Item	*Before refurbishment*	*After refurbishment*
Pitched roof	Uninsulated	200 mm mineral fibre insulation laid between joists, 60 mm across joists (background ventilation maintained in roof space or 'sealed roof' breather membrane used).
External walls	600 mm sandstone	75 mm semi-rigid mineral fibre insulation boards and 12.5 mm plasterboard fitted internally.
Close/Stair walls	Uninsulated	50 mm glass wool behind 12.5 mm wallboard.
Windows	Timber framed single glazing	Replacement timber frame windows with double glazing (20 mm air gap) and draught-stripping.
Ventilation	Uncontrolled	Draught-stripping windows, bathroom and kitchen extract fans (linked to light switch).
Heating	Coal and electric fires	Gas-fired central heating with combination boiler (balanced flue), programmer, room thermostat and thermally regulated valves.
Hot water	Electric immersion heaters	From combination boiler system.

Effects of energy efficiency measures

Typical energy efficiency measures to a Victorian tenement in Glasgow were described by BRECSU in its GPCS series. These are summarized in Table 10.3.

The results of these measures on the building in question are listed in Table 10.4. The successes in achieving energy efficiency and pollution reduction in this instance are obvious. The payback period is usually between 2 and 5 years, depending on the extent of improvements undertaken.

Energy audit

Preamble

Before fixing the brief for the adaptation work it may be worthwhile undertaking an energy audit of the building. This would be undertaken to identify building design dilemmas such as building usage; seasonal conditions; establish the building's thermal capacity. It would also help to identify energy appraisal yardsticks, such as fuel consumption in kWh/m^2 per annum (see RICS and EEO, 1992).

Table 10.6 list some typical performance yardsticks for various categories of buildings based on floor area. As an alternative the yardsticks can also be based on building volume (i.e. GJ/m^3). These energy benchmarks can be used to determine the extent of the required energy efficiency measures. They are normally found by calculating the Normalized Performance Indicator (NPI) using the following formula:

$$\text{NPI} = \frac{\text{Corrected Annual Energy Consumption}}{\text{Floor Area}}$$

Appendix G shows an example of how the NPI is calculated for a hotel.

Energy appraisal procedure

- Pre-inspection: consult previous reports or obtain information on the history of the building to determine any problems with heating system or energy system.
- Building fabric survey: undertake a condition assessment of the building; identify major defects that may impinge on its energy efficiency; pinpoint areas of heat loss or energy inefficiency.
- Energy usage: assess data on energy consumption per annum; compare with a benchmarked or similar building.
- Services: assess the efficiency and capacity of the heating, lighting and other major energy consumption installations.

Therapies

According to ERG (1999), these could involve the consideration of the following points:

- Increasing daylighting through additional or more efficient roof lighting (see Chapter 6).
- Reducing overheating through the use of passive cooling techniques (see Chapter 9).
- Reducing heating demand through installation of draught lobbies and by adding insulation to external walls and roof.
- Enhance envelope performance by installing better windows and doors (see Chapter 8).

Table 10.4 Appropriate energy efficiency measures

Improvement area	*Aim*	*Measures*	*Building categories**
Building fabric	Reduce air infiltration	Draughtproofing around windows and doors. Fit automatic door-closure devices. Install draught lobbies.	1–15
	Improve thermal insulation	Insulate roof voids. Insulate flat felt roofs during refurbishment. Cavity wall insulation. Add internal insulation panels. Reduce excessive glazing area/fit insulated infill panels. Double/triple glazing.	1–15
Boiler plant	Improve operating efficiency	Regular maintenance checks/maintenance contracts. Improve thermal insulation to boilers and pipework. Isolate inoperative boilers. Isolate space-heatingcircuits in summer months (combined heating and direct hot water plant). Replace old inefficient boilers with condensing type.	1–15
	Avoid overheating	Check thermostats are set correctly. Turn off radiators in unused rooms. Use frost protection thermostats outside occupation periods. Install zone controls for areas of extended use. Install weather-compensating controls. Install more accurate thermostats and thermostatic radiator valves (tamperproof).	1–15
Space heating systems	Reduce operating costs	Fit time controls to eliminate out-of-hours heating. Ensure frost protection operates during holiday periods. Install optimum start control to reduce preheating times. Consider BEMS.	2–15
	Avoid overheating	Same as for boiler plant.	2–15
Hot water service	Reduce heat losses	Insulate hot water storage tanks and pipework. Remove redundant pipework and avoid long pipe runs where water can stagnate. Install point of use water heaters in place of central plant.	1–14
	Reduce hot water demand	Check hot water thermostat settings are correct. Install spray taps.	1–15

(Contd)

Table 10.4 (*Contd*)

Improvement area	*Aim*	*Measures*	*Building categories**
Electrical services	Reduce costs	Check tariffs and maximum demand ratings are still appropriate.	1–15
Provision of own on-site electricity and heat	Reduce electricity and heating costs	Install CHP plant.	7, 11 and 15
	Reduce distribution losses	Install weather compensating controls.	2–15
		Improve thermal insulation of site distribution pipework, valves and flanges.	
		Isolate space heating circuits in summer months (combined heating and hot water service plant) or shut down main plant in summer and use local water heating systems.	
Lighting	Improve lighting efficiency	Regular maintenance and cleaning of lamps and luminaries.	1–15
		Install more efficient lighting source (e.g. fluorescent instead of tungsten lamps).	
		Decorate the building with lighter colours.	2–15
	Avoid unnecessary lighting	Switch off unnecessary lights.	1–15
		Install automatic lighting controls (time, daylight or occupant detection control) for public areas.	
		Install 'task' lighting where possible and reduce ambient artificial lighting.	
Mechanical ventilation	Optimize operation	Switch off kitchen and toilet extract fans out of hours.	2–15
	Recover reject heat	Install heat recovery devices in ventilation exhaust systems.	2–15
Mechanical extract ventilation	Reduce operating costs	Switch off kitchen extract fans when not required.	2–15
		Clean filters, grilles and fan blades regularly to prevent grease build up.	
		Ensure correct siting of extract fans.	
		Close external doors when operating fans.	
Air conditioning	Optimize operation	Reduce air volume handled where permissible.	2–15
		Set room sensor cooling temperature to 22°C or higher where possible.	
		Provide controls to prevent simultaneous use of heating and cooling circuits in air handling units.	
		Ensure system uses free cooling effect of outside air when possible.	
		Have plant regularly maintained.	

Dishwashing and hot water supply	Improve dishwasher utilization	Maximize dishwasher loads with correct stacking. Clean and maintain machines regularly. Consider using sanitized liquids and water softeners to reduce boost temperatures.	4 and 6
	Reduce hot water storage costs	Insulate hot water storage tanks and pipework. Install point of use water heater at appropriate places. Check hot water thermostats settings are correct and reduce where possible.	4 and 6
	Reduce hot water demand	Install spray taps for handwashing facilities. Ensure taps are switched off after use and leaks attended to.	4 and 6
Food storage	Reduce refrigeration costs	Locate refrigerators and freezers away from sources of heat. Minimize the frequency of opening appliances. Avoid putting hot food in refrigerators. Adopt a planned defrosting programme. Consider installing a heat recovery system for reject heat from refrigerator compressors.	6
Refrigeration (cold stores)	Maintain performance	Adopt planned maintenance procedures. Minimize defrost cycles. Check thermostat settings.	5, 11 and 15
	Reduce losses	Improve thermal insulation. Minimize the number and size of doors. Fit automatic closure devices to doors and minimize door opening times. Install heat recovery systems. Consider centralized refrigeration packs with heat recovery where practicable. Install cold air curtains on doors openings. Fit loading bays with dock seals and sensors.	5 and 11
Heating and ventilation (dining areas)	Reduce heat losses	Ensure kitchen extract ventilation system does not draw excessive outside air into restaurant. Consider heat recovery system for mechanical ventilation. Improve thermal insulation in walls and roofs. Consider double-glazing (reduces down draughts). Install draught lobbies at external doorways. Draught proof doors and windows. Fit automatic door closers.	6

(Contd)

Table 10.4 (*Contd*)

Improvement area	*Aim*	*Measures*	*Building categories**
	Avoid overheating	Check room thermostats are set correctly.	
		Install time controls for heating or intermittently occupied areas.	
		Optimize table arrangements to make use of occupant and lighting heat gains.	
Cellar cooling	Avoid excessive cooling	Check thermostats are set correctly.	5
	Reduce energy losses	Consider the installation of heat recovery and heat pump systems to utilize reject heating.	
Hydrotherapy and swimming pools	Reduce pool losses	Fit swimming pool cover.	4 and 7
		Check whether pool water and air temperatures are correct or could be reduced.	
	Reduce ventilation losses	Consider ventilation heat recovery systems.	4 and 7
Process energy	Recover reject heat	Install heat recovery and storage devices in industrial processes and compressor rooms, and use recovered heat for space or water heating.	11

*Building categories: 1. Dwellings, 2. Offices, 3. Shops, 4. Schools, 5. Entertainment, 6. Catering establishments, 7. Health care facilities, 8. Hotels, 9. Courts, depots and emergency services buildings, 10. Factories and warehouses, 11. Libraries, 12. Museums, 13. Art galleries, 14. Churches and 15. Public houses.
Adapted from various EEO guides.

Table 10.5 Energy effects before and after refurbishment (based on BRECSU guidelines)

Item	*Top floor 1-bed mid-flat*		*1st floor 2-bed gable flat*	
	Before	*After*	*Before*	*After*
SAP rating	27	79	59	94
Annual space and water heating costs (£)	338	101	279	114
CO_2 emissions (tonnes/year)	7.2	2.3	6.6	2.9

Table 10.6 Energy performance yardsticks for various buildings based on floor area

Type of building	*Energy efficiency rating (kWh/m² per year)*		
	Good	*Fair*	*Poor*
Cinemas	<650	650–780	>780
Dwellings	<350	350–580	>580
Nursery	<370	370–430	>430
Primary school, no indoor pool	<180	180–240	>240
Primary school with indoor pool	<230	230–310	>310
Secondary, no indoor pool	<190	190–240	>240
Secondary with indoor pool	<250	250–310	>310
Secondary with sports centre	<250	250–280	>280
Prisons	<550	550–690	>690
Police stations	<440	440–620	>620
Restaurants	<410	410–520	>520
Public houses	<340	340–470	>470
Small hotels and guesthouses	<240	240–330	>330
Medium size hotels	<310	310–420	>420
Large hotels	<290	290–420	>420
Library	<200	200–280	>280
Museum or art gallery	<310	310–420	>420
Church	<90	90–170	>170
Factories (with little or no process energy requirement)	<31	31–42	>42
Factories (with heat gains from plant)	<25	25–36	>36
Warehouses (heated)	<19	19–34	>34
Cold stores	<55	55–75	>75
Hangers	<15	15–55	>55
Hospitals (GJ/100 m³)	70	70–80	>80

Adapted from EEO (EST *Energy Efficiency in Buildings booklets*, 1993).

- Natural ventilation by adding opening sections to windows and rooflights.
- Controlling ventilation and casual infiltration.
- Performance of active systems through better controls (time clocks, thermostats, building energy management systems, BEMS) and more efficient fittings (lights, heat emitters).
- Improving indoor air quality by substituting natural for synthetic finishes (linoleum, water based paints, see Chapter 8).

A checklist for points to consider for energy efficient renovation is shown in Table 10.7. More specifically, the energy efficiency issues to consider in commercial refurbishment work are summarized in Table 10.8 (see also Tables 1.2 and 1.3).

Table 10.7 Checklist for energy efficient renovation

Element	*Objectives*	*Means*
Space heating and ventilation	Reducing demand	• Insulate fabric. • Reduce solar infiltration. • Utilize solar gain.
	Improving efficiency	• Consider efficiency of heating appliances – use condensing boilers. • Improve controls.
Space cooling and ventilation	Reducing overheating	• Reduce solar demand. • Improve efficiency of lighting and other heat producing equipment. • Use natural ventilation.
	Improving efficiency	• Ensure efficiency of pumps and fans. • Improve controls. • Specify efficient cooling plant (if cooling plant is unavoidable).
Lighting	Reducing demand	• Improve daylighting (use SunPipes). • Rationalize space usage.
	Improving efficiency	• Redesign artificial lighting layout. • Specify efficient lamps, luminaires and ballasts. • Improve controls.
Water	Reducing demand	• Use water treated for human consumption only where necessary. • Improve water storage and pipe layout. • Install water metres to reduce demand.
	Improving efficiency	• Specify water-conserving fittings – low flush WCs, etc.
Building management	Reducing demand	• Educate building users. • Ensure good 'housekeeping'.
	Improving efficiency	• Set targets and monitor performance. • Ensure effective maintenance and operation. • Consider a range of energy management systems.

Adapted from ERG (1999).

Thermal performance

Existing position

The *U*-values of the external enclosure of a building gives an indication of its thermal performance. These can be useful in helping to establish the extent of thermal upgrading required. Table 10.9 (see also Table 9.3) summarizes these.

Thermal efficiency of insulation materials

In upgrading the thermal performance of elements it is important to be aware of the relative properties of each of the various insulation materials in relation to their thickness. These are summarized in Table 10.10.

Table 10.8 Energy efficiency issues in commercial refurbishment

Issues to consider	*Examples*
Before fixing the brief for the work	• Identify energy efficiency targets. • Undertake an energy audit of the building (see guide by RICS and EEO, 1992).
Identify the building's potential for environmental improvement	• Increasing daylighting through roof/sky lighting and sun-pipes. • Reducing overheating through the use of external louvres or blinds or solar tinted glazing. • Reducing heating demand through installation of draught lobbies and by adding insulation to external walls and roof. • Enhance envelope performance by better quality windows and doors. • Natural ventilation by adding opening sections to windows and roof lights. • Controlling ventilation and casual infiltration. • Performance of active systems through better controls: time clocks, thermostats, BEMS, and more efficient fittings – lights, heat emitters. • Improve indoor air quality by substituting natural for synthetic finishes: linoleum, water based paints.
Consider the following when refurbishing	• Improve controls on active service systems. The following will often be cost-effective: ▪ solid-state programmable controllers for heating; ▪ automatic switching systems for lighting; ▪ individual thermostatic room and/or radiator control; ▪ weather compensating controls. • Improve air tightness in the external envelope. • Improve thermal insulation: not always easy, but where roof finishes are being replaced it may be possible at a modest extra cost to upgrade thermal insulation significantly. External wall insulation can enormously enhance thermal performance and increase thermal comfort. • If windows or external door sets are to be renewed, the best performing models available will generally be worth installing. • Secondary glazing can create small sunspaces, pre-heat ventilation air and reduce transmission of external noise. • The best available floor and wall finishes will increase service life out of proportion to cost. • Passive climate control devices, including draught lobbies at external entrances, external shading.

Adapted from ERG (1999).

The existing thermal performance of the fabric being upgraded will obviously have to be taken into account in determining the thickness of insulation required for the desired *U*-value.

However, with the ever-tighter controls on energy efficiency in the Building Regulations it is becoming increasingly difficult to achieve. For example, to attain a target *U*-value of 0.16 W/m^{2}K for a flat roof the mineral fibre insulation would have to be about 250 mm thick.

Fire performance will probably be another factor influencing the choice of insulation material. Mineral and glass fibre insulation boards have good fire resistance but their thickness tends to reduce in a fire. This can allow smoke to spread within a cladding system unless adequate fire stops are provided.

Table 10.9 Typical *U*-values of some main building elements (based on CIBSE Guide Part A, 1988)

Element	*Typical U-value (W/m^2K)*	*Refurbishment option*
Solid brickwork 225 mm thick.[1]	1.9	External insulation (see Chapter 9). Internal insulation (ditto).
Cavity wall of 105 mm brick, 50 mm air-gap and 105 mm brick.[1]	1.5	Ditto Cavity fill insulation.
Cavity wall of 105 mm brick and 125 mm block.[1]	0.6	Ditto Cavity fill insulation.
No-fines concrete 220 mm thick.[2]	0.6	External insulation (see Chapter 9). Internal insulation (ditto).
100 mm thick roof insulation.[3]	0.38	Increase to 200 mm (and ensure that pipes and tanks are adequately insulated and eaves ventilated).
150 mm thick roof insulation.[3]	0.25	Ditto
200 mm thick roof insulation.[3]	0.22	Ensure that pipes and tanks are adequately insulated and eaves have appropriate levels of ventilated.

Notes
1. Originally plastered internally, fair-faced externally.
2. Originally plasterboard on battens internally, rendered externally.
3. Roofs with horizontal insulation between/over joists.

Polystyrene, being a thermoplastic, tends to drip molten plastic when burned, as well as release toxic fumes. Fire-rated rigid urethane, on the other hand, because it is a thermosetting plastic, does not shrink but develops a hard charred finish when burned. As shown earlier, it is used as the insulation within some insulated sandwich cladding panels.

A highly sustainable insulation material is sheep's wool. It is a natural fibre from a fully renewable resource. The life cycle of this material, therefore, has an ideal energy balance, which makes it eminently suitable as an environmentally friendly product in adaptation schemes as well as in new build work. It can be used in a wide range of floor, roof and wall constructions.

Sheep's wool offers the following advantages:

- it keeps the fabric cool in summer and warm in winter;
- its naturally breathable;
- it offers good condensation control;
- its energy efficient (uses only 14% of the embodied energy that is used to manufacture glass fibre according to Second Nature UK Ltd, a supplier of sheep's wool insulation called Thermafleece™);
- its safe (as its non-toxic and non-irritating it can be handled without the need for gloves and protective clothing);
- its durable (life expectancy over 50 years);
- its 100% recyclable;
- its fire resistance is better than cellulose and cellular plastics (and it can be treated with a fire-proofing agent to improve its intrinsic fire resistance).

Its lack of inherent rigidity is one key drawback of using sheep's wool insulation. This would therefore limit its use as an insulation where stiffness is needed (e.g. under floor and roof decks).

Table 10.10 Some notional *U*-values for typical thicknesses of a range of insulation materials (partly based on thermal conductivity 'k' values in CIBSE, 1989)

Insulation material	*Thickness (mm)*	*U-value (W/m^{2}K)*
Mineral fibre slab and Expanded polystyrene slab (EPS)	50	0.70
	75	0.46
	100	0.35
	200	0.18
Cellular glass insulation (e.g. 'Foamglass')	50	0.60
	75	0.40
	100	0.30
	200	0.16
Polyurethane board	50	0.50
	75	0.33
	100	0.25
	200	0.18
Sheep's wool (e.g. 'Thermafleece™')	50	0.77
	75	0.52
	100	0.39
	200	0.19

As thermal efficiency requirements increase there is a greater need for better performance from insulation materials. This means that insulation materials should not need to be too thick to achieve the necessary thermal resistance. As indicated earlier 'Neopor' is one such new material.

Enhancing performance

One of the main ways in which the thermal performance of an existing building can be improved is to lower the *U*-value of its fabric. The main methods of doing this are shown later. At this juncture it is worth looking at the targets set by the government as part of its campaign to improve energy efficiency through the building regulations. Parts L and Section 6 of the English/Welsh and Scottish building regulations respectively are regularly being overhauled to provide tighter controls on energy efficiency and conservation.

In this regard, Table 10.11 summarizes the *U*-values for the main parts of the fabric of both domestic and non-residential buildings. Dwellings with low-efficiency heating systems would require slightly lower *U*-values to compensate for this deficiency.

Improvements to enhance the general as well as thermal performance of a dwelling are shown in the plans of a dwelling in Figure 10.1. Note the differences between thermally important items (e.g. wall insulation), thermally relevant items (e.g. full house heating) and general modernization measures (e.g. kitchen and bathroom enlargements).

Heat-reflecting membranes

The space programme has spawned many positive terrestrial applications. 'Teflon' the non-stick coating for frying pans and other cookeryware is one well-known example. Space scientists from National Aeronautics and Space Administration (NASA) also developed long-life heat-reflecting membranes (HRMs) to protect orbiting craft for solar radiation. Initially, though, these materials were very expensive.

Table 10.11 Existing and target fabric *U*-values

Exposed elements	*U values*	
	2005	*GIL 72 Advanced standard*
Pitched roof with insulation between rafters[1,2]	0.16	0.08
Pitched roof with insulation between joists	0.16	0.08
Flat roof[3]	0.16	0.08
Wall	0.25	0.15
Floor	0.22	0.10
Windows, doors and roof-windows (area weighted average), glazing in metal frames[2]	2.0	1.50
Windows, doors and roof-windows (area weighted average), glazing in wood or PVC frames[2]	1.8	1.50
Roof-lights[3,4]	1.8	Not stated
Vehicle access/similar large doors	Not stated	Not stated

Notes
1. Any part of a roof having a pitch of 70° or more can be considered as a wall.
2. Display windows, shop entrance doors and similar glazing are not required to meet the standards given in this table.
3. This standard only applies to the performance of the unit excluding any upstand. Reasonable provision would be to insulate any upstand or otherwise isolate it from the internal environment.
4. Roof windows may be considered as roof-lights.

However, mass production techniques are now available that can supply tough, cost effective HRMs for use within buildings. Figure 10.2 illustrates the use of an HRM in a loft conversion.

According to one supplier, Apollo Energy Research, its HRM system 'will help to keep a building warm, yet in summer prevents the sun from overheating its interior. In summer, infrared energy from the sun is easily absorbed by roof and wall structures. Through conduction and convection, excessive heat can penetrate the building's insulation layers, eventually radiating to the interior causing an unwelcome temperature rise. An HRM System blocks the infrared energy passing through structures so the insulation layers and the interior are kept cool.'

'In winter, the system has the reverse effect; it stops warmth escaping from ceiling, wall and floor structures. Thus sealing the building's envelope and deflecting infrared energy back into the interior do not waste fuel wasted in heating up the building's insulation layers.'

The primary function of an HRM is to block infrared energy radiating across air spaces. Infrared energy converts to conducted heat when it strikes a surface such as a wall or roof. The membrane works either by reflecting back most of the radiating infrared energy striking its bright non-tarnish surface (reflectance) or by not radiating heat (emittance).

To act as a radiant barrier the HRM is installed within a structure facing at least one air space, usually 19–25 mm. Heat conducting through the structure and crossing the air space as infrared energy is blocked by the membrane.

Energy labelling

Types of energy rating

Energy rating of existing buildings is a useful benchmark for determining their need for refurbishment. There are two main energy ratings in the UK: Standard Assessment Procedure (SAP) and National Home Energy Rating (NHER).

APOLLO
ENERGY RESEARCH

Application Drawing............C.A.001
Published.....................August 2003

Attic room – cold & temperate climates
Alternative detail

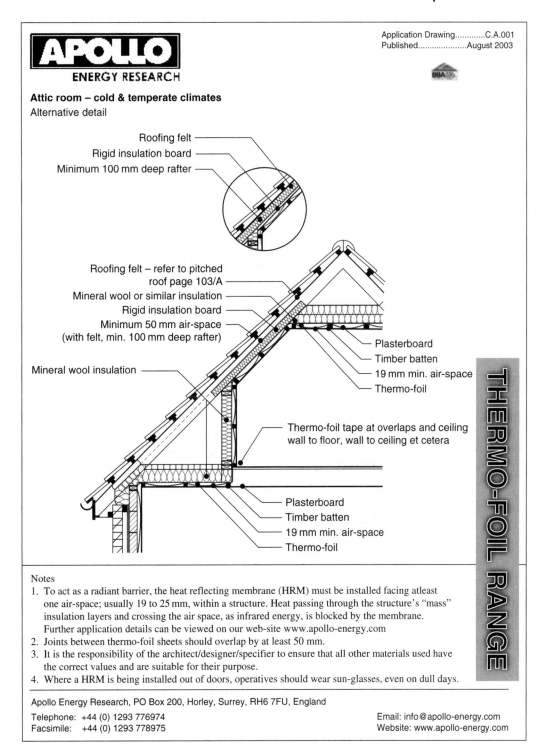

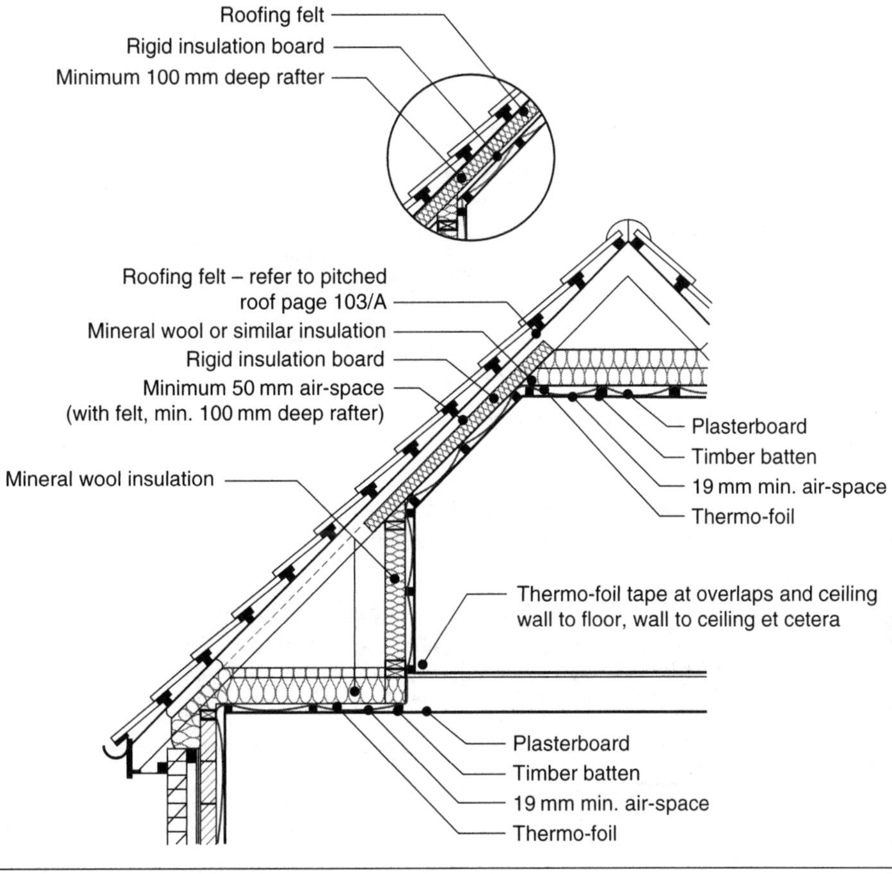

THERMO-FOIL RANGE

Notes
1. To act as a radiant barrier, the heat reflecting membrane (HRM) must be installed facing atleast one air-space; usually 19 to 25 mm, within a structure. Heat passing through the structure's "mass" insulation layers and crossing the air space, as infrared energy, is blocked by the membrane. Further application details can be viewed on our web-site www.apollo-energy.com
2. Joints between thermo-foil sheets should overlap by at least 50 mm.
3. It is the responsibility of the architect/designer/specifier to ensure that all other materials used have the correct values and are suitable for their purpose.
4. Where a HRM is being installed out of doors, operatives should wear sun-glasses, even on dull days.

Apollo Energy Research, PO Box 200, Horley, Surrey, RH6 7FU, England
Telephone: +44 (0) 1293 776974
Facsimile: +44 (0) 1293 778975
Email: info@apollo-energy.com
Website: www.apollo-energy.com

Figure 10.1 HRM used in an attic conversion (www.apollo-energy.com/)

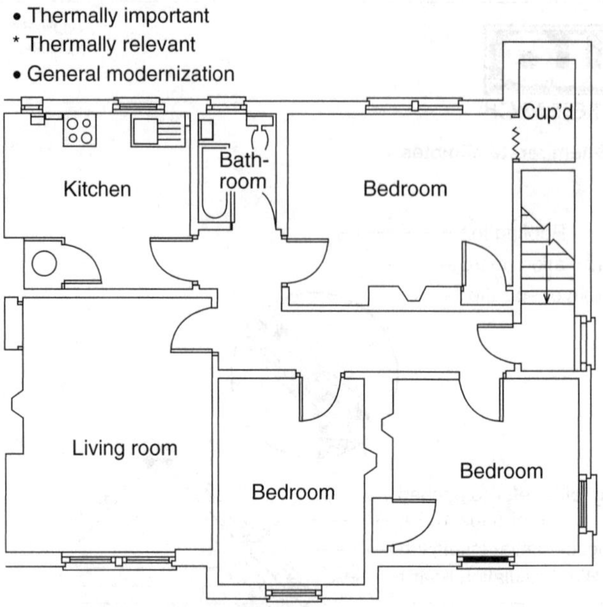

Ground-floor plan before upgrading

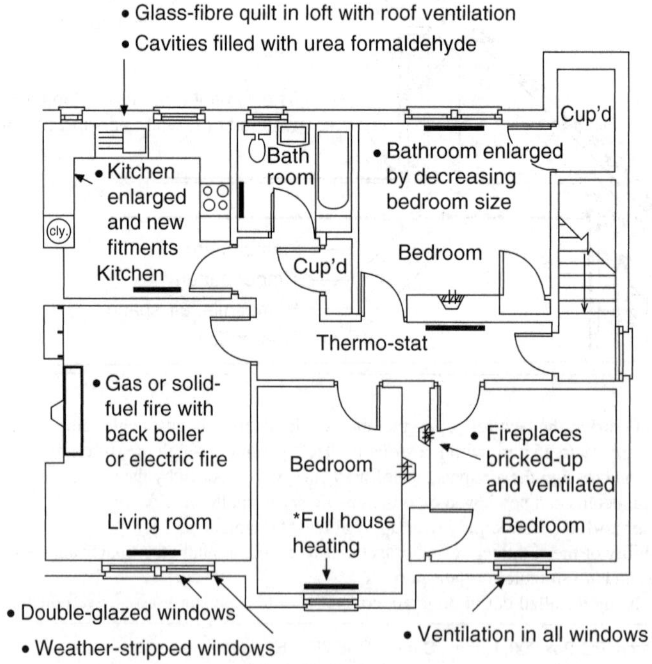

Ground-floor plan after upgrading

Figure 10.2 Ground floor plans of a dwelling before and after modernization (CRP, 1987)

Table 10.12 SAP rating scale (NHER, 2005)

Your new home has a SAP rating of at least: **89**

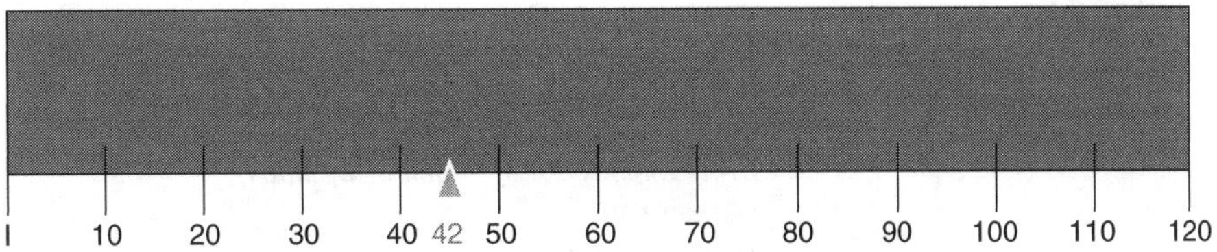

The Government standard for bench marking the energy efficiency of a home is called the Standard assessment procedure (SAP). The SAP gives you a score of between 1 and 120 (with 120 the best). The average SAP rating of homes in Britain is 42.

What is a SAP rating?

The UK Government's recommended system for the home energy rating is the SAP. According to NHER (2005),

> this energy cost rating is based on energy costs for space and water heating only. A SAP rating is required for all new build dwellings and those that are undergoing significant material alteration (such as the addition of an extension to the dwelling). Housing Associations and Councils which own stock are all required to submit average SAP figures for their regions so that Government can monitor the amount of energy used, and associated carbon emissions, from domestic dwellings in the UK.

The current version of SAP is 'SAP2005' (v9.8). It has a scale of 1–120. The higher the figure, the more energy efficient is the dwelling: 1 being very poor, 120 being excellent. A typical SAP for an average house in England is about 45. A 'SAP2005' rating on a house built to current Part L building regulations would be closer to 80–100 or more, SAP points.

The SAP energy rating method applies to all new dwellings formed by new construction, or by conversion through a material change of use. Buildings affected by these works are given a SAP rating. Its impact on major residential conversions of existing buildings therefore cannot be ignored.

A SAP rating of 60 or less indicates a strong need to upgrade the insulation standards of the building's fabric. Alternatively, consideration may be given to modifying the design details to improve the energy rating to a value of above 60. For example, this may be achieved by using a more efficient heating system boiler.

Initial calculations may show that a high-energy efficiency rating is achieved (i.e. a SAP rating of 100 or above). It may allow compliance with the fabric insulation requirements of the building regulations. This could be demonstrated using an appropriate domestic software Energy Rating programme (e.g. BREDEM, The BRE's domestic energy model, see BRECSU General Information Leaflet 31, or TICK Program, 42 Mitchell Street, Dalkeith EH22 1JQ).

The actual rating need not be declared to a building control authority until completion of the extension or building's conversion into residential use. However, if the energy rating method is used to show compliance with the building regulations, it may be prudent to submit the SAP calculations at the design stage.

The SAP rating method, though, is not foolproof. A sudden rise in fuel prices could potentially invalidate the rating. It has also been revised recently to provide for the optional calculation of CO_2 emissions, expressed in tonnes/year. According to the DETR's construction web site (2000):

> The proposed amendment introduces a new optional Carbon Index (CI) calculation, which could be used as a way of showing compliance with Part L. The CI is based on the CO_2 emission figure, but adjusted for floor area so that it is essentially independent of dwelling size for a given built form. It is expressed on a scale of 0.0 to 10.0, the higher the number the better the performance. The achievement of a specified level of the CI is proposed as one way of demonstrating compliance with the next revision of Building Regulations for the Conservation of Fuel and Power.

Table 10.13 Comparison of NHER and SAP ratings (NHER, 2005)

N/A	*Uses local weather conditions?*	*Accounts for the affects of cooking, and lights and appliances?*	*Accounts for the differences in room temperature between zones? (e.g. you may prefer that your living room is warmer than your bedroom)*	*Can calculate the costs and energy usage based on the actual number of occupants and their habits?*	*Quality assured rating?*
NHER	Yes	Yes	Yes	Yes	Yes, for NHER certificates (SAP too when issued on NHER certificate).
SAP	No	No	No	No	No – unless ISO 9000 and the SAP crown logo is displayed upon the certificate.

What is an NHER rating?

The NHER rating takes into account the local environment and the affect it has on the building's energy rating (NHER, 2005). The NHER calculates the costs of space and water heating as well as cooking, lights and appliances. This is why it is claimed that the NHER is more accurate than the SAP.

The NHER can only be calculated using computer software as supplied by National Energy Services (NES). It first emerged in 1983 at the Energy World Exhibition in Milton Keynes, and later the Government used a simplified version of this model that could be calculated by hand. This became known as SAP.

The NHER was conceived as a rating with as scale of 0.0–10.0. 0.0 is poor; 10.0 is excellent. An average dwelling in England would currently score between 4.5 and 5.5 (NHER, 2005). A building meeting current Part L1 Building Regulations would probably score higher, perhaps around 8.0 or more.

According to the NHER (2005):

> The two energy scales measure slightly different things: the SAP looks only at the fixed elements of the home and is the same wherever the property is located in the UK. All homes built to the same design should have exactly the same SAP. The NHER includes various location-specific elements (including whether the home is South facing or sheltered from wind by other buildings) and so reflects actual running costs. If two homes have the same floor area but different NHERs, then the home with the better (higher) NHER should cost less to run. All good Home Energy Labels should be prepared under a BS EN ISO 9002 quality system and authorised by the Department for the Environment, Farming & Rural Affairs (DEFRA).

General energy improvements

Preamble

As pointed out by BRECSU (GPG 155, 2001) the energy efficiency of a dwelling can be improved without waiting to undertake a full refurbishment package. Repair and improvement schemes provide many opportunities for energy measures. Indeed, the economies of scale involved normally mean that it is cheaper to combine energy efficiency measures with repair and improvement work. It is usually more expensive and disruptive to do these measures separately at a later date.

The main energy efficiency measures to buildings are summarized as follows:

- Wall insulation systems (see Chapter 9).
- Floor insulation methods (see Chapter 9).
- Roof insulation methods (see Chapter 9).
- Insulate hot water cylinder with at least 75 mm thick plastic-coated mineral fibre quilt cover.
- Insulate water tank (except at its base) and pipes – respectively with rigid 50 mm thick polyurethane board insulation and 20 mm thick PE pre-formed insulation.
- Provide insulated doors for external doorways.
- Draught stripping to external doors and windows.
- Extract ventilation to kitchens and internal toilets/bathrooms (with humidistat control and cut-off delay timer).
- Trickle ventilation to windows (8000 mm^2 for habitable rooms).
- Add a porch or vestibule at entrances to reduce heat losses at entrance.
- Use energy efficient lighting (see below).
- Install water conservation measures in the building (see below).
- Improve heating, lighting and ventilation controls to reduce use of these services when the space is unoccupied.
- Install 'suncatcher' tubes (with or without integral vent pipe) in the roofs to increase natural lighting (and ventilation) in rooms underneath (see below).
- Install roof vent ducts and cowls to provide passive stack ventilation (see below).

Figure 10.1 shows the schematic layouts of a dwelling before and after modernization comprising energy upgrading measures.

Not surprisingly, sustainable design measures are now an integral part of most if not all large public sector refurbishment and new build schemes. Sustainable design includes not only environmental considerations. It is also about how environment integrates with cost, schedule (time), operations, maintenance and worker/employee considerations.

The multi-million dollar Pentagon renovation project in Washington, DC mentioned in Chapter 1 includes the following design and construction practices:

- Sustainable site planning: Minimizing the impact of the construction project on the surrounding natural environment through limiting disturbance of soils, vegetation and wildlife throughout construction and operations and maintenance of the facility.
- Water efficiency.
- Energy performance and atmospheric disturbance: Maximizing use of renewable energy sources and minimizing fossil fuel use.
- Conservation of materials and resources: Using natural resources efficiently through maximizing reuse/recycling of material resources an accurately predicting quantities of raw materials needed – to include water, wood, etc.
- Indoor environment quality: Enhancing the indoor working environment for all who occupy the facility through promoting worker health and safety with design and construction innovations.
- Management of life cycle cost: To manage the project using long-term operations and maintenance considerations to increase cost effectiveness and maintain long-term value of the facility.

More specifically, the types of sustainability measures that form part of the above practices are varied. For example, according to the Pentagon Renovation Programme (2005), the various schemes in the Pentagon renovation project contain the following sustainable features:

- Wedge 1:
  - 90 per cent of all concrete and metal diverted from landfills.
  - High efficiency lights requiring less lumens.
  - Recycled gypsum wall board.
  - Highly sustainable recycled-content carpet.
  - Energy efficient fixtures.
  - Energy efficient heating and cooling systems.
  - High reflectivity of the Terrazzo floor surface and painted walls.
- Wedges 2–5
  - 90 per cent of all concrete and metal diverted from landfill.
  - Using high efficiency T-5 and T-8 lights.
  - Recycled gypsum wallboard.
  - Highly sustainable recycled content carpet.
  - Recycled content ceiling tile.
  - Universal space plan allows for easy space reconfiguration, reducing future construction waste.
- Remote delivery facility
  - Registered as Leadership in Energy and Environmental Design (LEED) pilot project.
  - Green vegetated roof.
  - Indigenous vegetation and water re-use.
  - Building control system for energy efficiency and indoor air quality.
  - Elimination of unnecessary materials.
- Metro entrance facility
  - At least 50 per cent of waste was diverted from landfill.
  - Electric vehicle outlets installed.
  - Vegetation covers half the open space restoring life back to the site.
  - High reflectance Energy Star roofing installed.
  - 20 per cent savings in energy consumption.
  - Over 50 per cent of the building materials were assembled within 500 miles.
  - Over 50 per cent of the materials contain recycled content.
  - 21 per cent of wood-based materials were FSC Certified.
  - Permanent CO_2 monitoring system installed.
- Pentagon athletic centre
  - Aggressive construction waste recycling programme.
  - Green vegetated roof.
  - Vegetation covers half the open space restoring life back to the site.
  - Rapidly renewable materials (bamboo and cork flooring).
  - Bathroom partitions made from 100 per cent post consumer recycled plastic.
  - 20 per cent of fly ash was added to concrete.
  - Composite wheat board for millwork panels.
  - Ozone system for the pool.
  - Energy reduced through heat exchanger technology, occupancy and light sensors.
  - Permanent CO_2 monitoring installed.
  - Low-emitting materials used.

Modern sun-catcher tubes

A contemporary and more efficient alternative to the traditional light-well is to use a proprietary circular sealed daylight duct. 'SunPipe', for example, is a silverized aluminium tube that comes in diameter ranges from 200–1000 mm. Larger diameter tubes are best used in commercial buildings. These can be used on flat roofs as well as pitched ones.

The super-reflective mirror-finished tube intensifies and directs daylight and sunlight down to the habitable room below where the light is evenly spread by a translucent polycarbonate ceiling diffuser. It produces the equivalent of up to 1200 watts of natural light even on cloudy days. A hemi-spherical cap of polycarbonate transparent glass complete with collar and flashing finishes the duct at the roof line.

A sun-catcher tube is more effective than a conventional light-well because it offers the following advantages:

- Highly sustainable because the unit consumes no energy.
- Can provide daylight to rooms without windows.
- Since the fitting provides a sealed column of air, solar gain and condensation associated with conventional rooflights are eliminated.
- Minimal heat loss in winter months because the sealed column of air acts as an insulator.
- Less disruptive to install in an existing building than a conventional skylight.
- It achieves about three times more natural light than a roof-light of the same size.
- Less space is taken up compared to a skylight giving the same level of natural light.
- Provides a neat, unobtrusive and relatively maintenance free finish.
- The dome is self-cleansing and any rainfall will usually wash off any surface dirt.
- On smaller diameter tubes no structural alterations to the roof structure are required.
- The adjustable bends in the tube allow daylight and sunlight to be taken down to virtually any part of the building.

Figure 10.3 shows how a modern SunPipe system is installed.

Modern roof vent stacks

Another passive energy efficiency measure is to install a roof vent stack, which acts as a 'windcatcher'. It is a form of passive stack ventilation that has a circular or square louvred cowl. Monodraught Ltd is one company that supply and fits this method of maximizing natural daylighting in a building.

This 'windcatcher' can be installed on its own or incorporated within a suncatcher tube. It is similar in style to the suncatcher pipe except that it does not require a reflective inner lining.

Figure 10.4 illustrates how a 'windcatcher' device can provide excellent background ventilation to a refurbished school building.

Energy efficient services

Rationale

Services account for most if not all energy consumption in a building. They also account for around 40–50 per cent of the capital cost of a new work and can form a substantial part of the cost of an adaptation scheme. Moreover, services can take up nearly 30 per cent of the space in a building. It is therefore crucial that attention is given to the energy efficiency of services within a building.

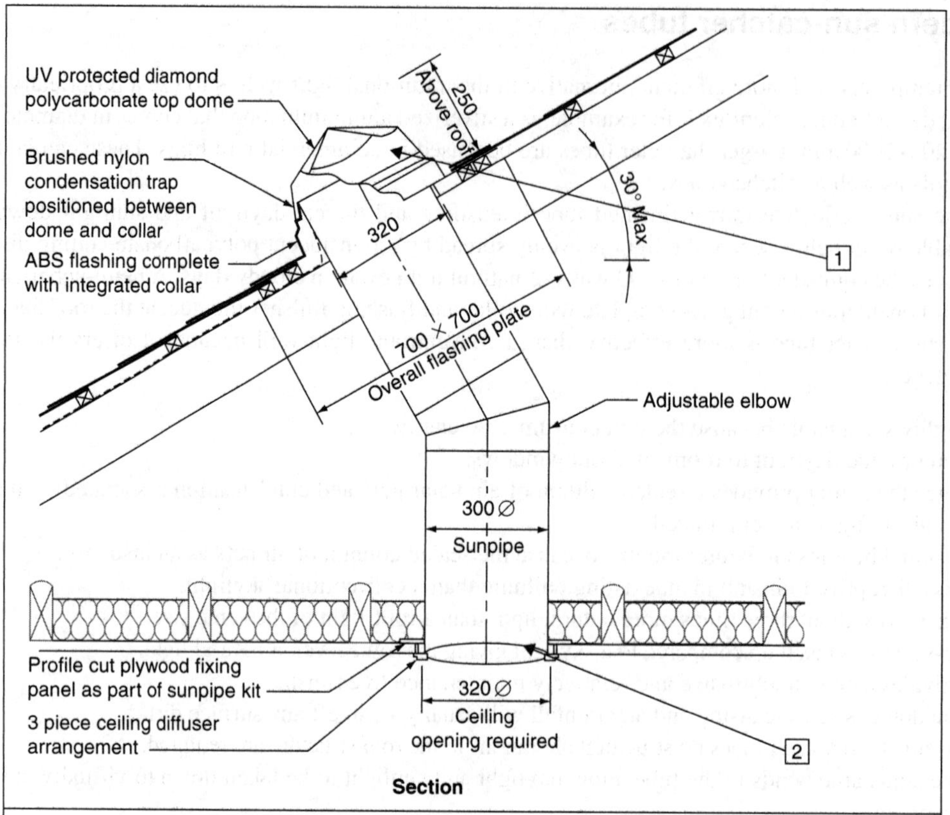

Notes

1. Polycarbonate top dome can come in a variety of styles (e.g. square, for conservation work; plain dome, for other schemes).
2. A ceramic based ceiling diffuser can be installed for use in conjunction with the complete SunPipe range, to provide up to 2 h fire resistance. Alternatively a fire-choke collar can be fitted above a plaster board ceiling and thereby leaving the 'SunPipe' ceiling diffuser unaffected. 'Sleev-it' fire choke collars are designed to prevent the spread of smoke and fire through a roof space by crushing the 'SunPipe', when the aluminium is subjected to fire giving up to 2 h protection. The 'Fire Choke' patented system contains the expanding intumescent material which crushes the aluminium SunPipe when the temperature in the room exceeds 150°C.

Figure 10.3 Example of the use of a 'SunPipe' tubed skylight system (Monodraught, 2005)

Space heating and hot water

Criteria

According to GPG 187 (1997), the criteria of an energy efficient heating system are that it should:

- be correctly sized to warm up the dwelling from cold within a reasonable time;
- use fuel as efficiently, and cheaply, as possible;
- be capable of control and thus provide heat only when and where needed;
- have controls that are easy to use and understand;
- be reliable and easy to maintain.

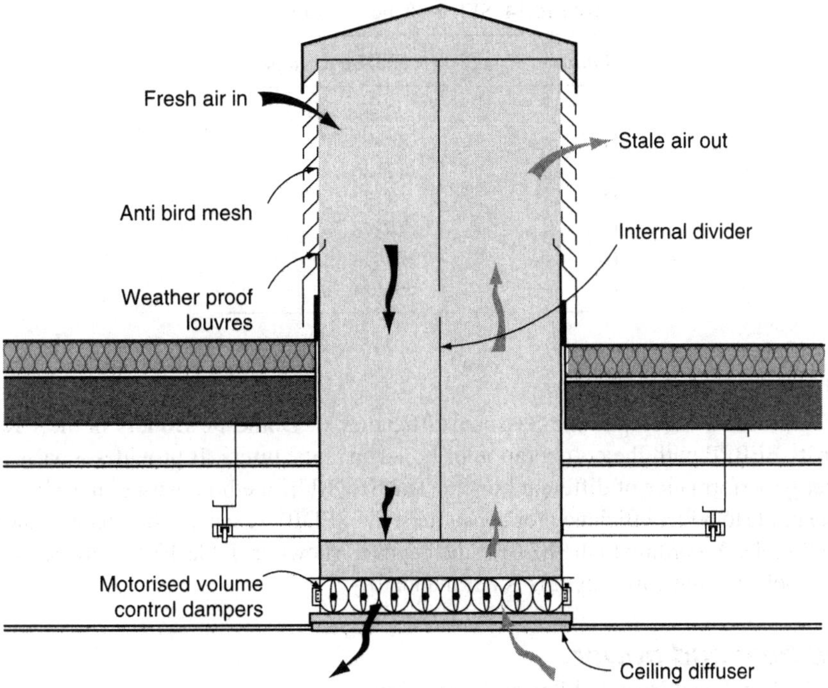

Figure 10.4 Example of the use of a 'Windcatcher' monovent (Monodraught, 2005)

Choice of fuel

Gas and off-peak electricity are the two major fuels for heating systems in dwellings. The latter is the more environmentally friendly because of its lower CO_2 emissions. For heating in two or more bedroom dwellings gas is more cost effective (Energy Consumption Guide (ECG) 6, 1992). Off-peak electricity costs, on the other hand, are comparable with those of gas for schemes with mainly one- or two-bedroom dwellings.

Boilers

Replacing old or inefficient central heating boilers is often necessary in residential and commercial refurbishment schemes. In such instances condensing boilers should be used because of their energy-saving potentials. In particular, condensing gas boilers operate at an average annual efficiency of 85 per cent, which is about 15 per cent more than the standard type (Harrison and Trotman, 2000).

The recommended basic package for boiler and radiators of a gas central heating system indicated in GPG 187 (1997) is summarized as follows:

- Fan-assisted boiler (condensing for large dwellings).
- High-recovery hot water cylinder with a minimum of 50 mm insulation (CFC- and HCFC-free).
- Insulation of primary pipework (i.e. from boiler to cylinder).
- Electronic programmer with separate space and water heating control must be readily accessible, easy to understand and operate.
- Room thermostat in the hallway, suitably positioned (e.g. not over a radiator).
- Both space heating and primary hot water circuits pumped.
- Unvented domestic hot water system (if mains pressure sufficient).
- Feed and expansion tank for space heating system.
- Thermostat on hot water cylinder.

Table 10.14 SEDBUK boiler ratings

Band	*SEDBUK range (%)*
A	90 and above
B	86–90
C	82–86
D	78–82
E	74–78
F	70–74
G	Below 70

Boiler efficiency database

The British Government developed the Seasonal Efficiency of Domestic Boilers in the UK (SEDBUK) project under its EEBPP with the co-operation of boiler manufacturers. It provides a basis for fair comparison of energy performance of different boilers. The SEDBUK method is used in SAP.

As a simple guide to boiler efficiency for consumers, the SEDBUK scheme has been created with bands on an 'A' to 'G' scale. According to the EEBPP, the bands as shown in Table 10.14 may be used on product literature and labels, though currently this is not mandatory.

Combined heat and power

According to the Combined Heat and Power Association:

> CHP plant is an installation where there is simultaneous generation of usable heat and power (usually electricity) in a single process. The term CHP is synonymous with 'cogeneration' and 'total energy', which are terms often used in the United States or other Member States of the European Community. The basic elements of a CHP plant comprise one or more prime movers usually driving electrical generators, where the heat generated in the process is utilised via suitable heat recovery equipment for a variety of purposes including: industrial processes, community heating and space heating.

However, combined heat and power (CHP) is only really suitable for high density urban housing (EST Guide CE 102, 2005). Its' not appropriate for low density urban housing, distributed urban housing or rural housing. The EST (EST Guide CE 102, 2005) considers that 'the most appropriate applications of CHP in existing housing are likely to be on single sites with large numbers of dwellings, such as estates that already have communal heating from central boiler houses. In such cases, when boilers are replaced by CHP plant, the existing heat distribution systems can often be re-used, with little or no modification. It is however necessary to modify on-site electricity distribution systems to use the locally generated electricity'.

Gas and off-peak electricity are the two most common fuels for heating systems in dwellings. A more sustainable CHP system, however, is one that is waste-fired or uses wood-fuel boilers. For individual dwellings or groups of dwellings clustered together, micro-CHP systems are now available. These are predominantly gas-powered and deliver between 1 kW and 3 kW of electric power (1–3 kWe), and between 4 kW and 8 kW of heat (4–8 kWth).

Lighting

Energy consumption

Lighting is a considerable consumer of energy in many buildings. It is also a major part of the cost of services. For example, according to THERMIE (1992) lighting accounts for the following:

- In offices about 50% of the electricity used is for lighting, and the lighting costs can exceed those for heating.

- In hospitals between 20% and 30% of the electricity consumption may be used for lighting.
- In factories, the proportion of energy used for lighting is typically around 15%.
- In schools, lighting accounts for 10–15% of the energy use.
- In dwellings, about 20% of the electricity used is for lighting (BRECSU, ECG 6, 1992).

Energy saving measures

Energy savings relating to lighting in residential and non-residential buildings can be improved by as much as 50 per cent, using a combination of the following measures:

- *Daylighting*: Making the most of natural light into a building will reduce the need for artificial lighting. Installing additional skylights or sun-pipes in the roof can do this (see below).
- *Lamps*: Replace ordinary tungsten light bulb lamps with compact fluorescent lamps; and replace 38 mm diameter fluorescent tubes with 26 mm diameter tubes. In addition, the use of 'low-loss' high frequency electronic ballasts can achieve savings of around 20 per cent in new installations. However, because of their bulk and heat build-up, they may be proved difficult to fit into existing luminaires unless these are replaced as part of the adaptation work.
- *Luminaires*: In a refurbishment scheme replacing existing light fittings using modern equipment can often result in substantial energy savings as well as improved visual conditions (THERMIE, 1992). Modern luminaires use reflector systems, which replace existing diffusers or prismatic panels.
- *Controls*: Controlling the use of lighting only when it is required using sensors and time-regulators. The use of localized switches can also help occupiers to have more control over their working environment. Automatic re-set switches wall/ceiling-mounted can be used as well as automatic central off-switching to control lighting use.
- *Direct lighting*: Lighting should be positioned where it is needed. This is termed task-directed lighting, and helps to minimize wasted light.

Water conservation

Rationale

Just over 50 per cent of all water used in the UK is attributed to buildings. Power generation accounts for around 36 per cent and industry and farming 13 per cent (Edwards, 1998). Given the uncertainties about the extent of global climate change, which is still likely to result in drier summers but wetter winters for Britain, and the increases in demand and charges for water, water conservation is becoming increasingly important.

Water conservation in refurbishment can be achieved, according to Edwards (1998), 'by specifying low water-using appliances, exploiting new technologies such as vacuum WCs and taking simple measurements to reduce waste (such as using self-closing taps)'. The benefits of waster conservation in buildings are summarized as follows:

- Lower water charges. Now that many residential as well as commercial and industrial users are charged directly for their water consumption, any reduction in this will have obvious financial benefits.
- Groundwater supplies conserved for future generations, which is clearly a major goal of sustainable development.
- Reduced stress on the current infrastructure.
- Reduced pressure to build new reservoirs, etc.
- Reduced use of hot water, so saving energy.
- Reduced energy use in waster and sewerage systems (Edwards, 1998).

Recycled water

Rainwater from roofs and hardstandings can be collected to underground tanks and recycled without much reprocessing. It's usually delivered on demand from such below ground containers by an in-tank submersible pump direct to toilets, washing machines and outside taps.

The term 'greywater' can be used to describe non-foul wastewater from sinks, showers and baths. It therefore requires a higher level of treatment than ordinary rainwater. According to Anglia Water on average 45 per cent of water supplied to domestic properties is discharged as 'greywater' ('freewaterUK').

Using greywater and rainwater for WCs, washing machines and outside taps helps to reduce the demand for potable water. They can potentially replace more than 50 per cent of mains water (Edwards, 1998).

In the short term, recycling greywater and rainwater may not be economically viable for all adaptation schemes. However, in the long term it can offer a cost-effective solution to tackling water demand, particularly in the refurbishment of larger buildings. For example, installing a large water tank in the basement of a block of 16 flats could be used to store, screen and redistribute the greywater back into the system.

Even recycling of rainwater, though, necessitates high levels of maintenance to the plant and pipework to avoid the adverse microbiological reactions that can pose health problems or damage drainage systems (Edwards, 1998). Thus proper filters and sterilization methods will require regular servicing and occasional replacement in any such installation.

Efficient sanitation

A conventional WC uses about 9 l of water with every flush. Reducing flush volume will therefore reduce water consumption. The British government's Environment Committee asserts that 'reducing WC flush volume is a priority… We understand this could save 10 per cent water usage… sufficient to mitigate any effects of climate change now predicted. It called upon the government through a Water Regulations Committee to adopt a 6 l flush standard for all new WCs and to encourage replacement. We cannot afford to wait 50 years for cisterns to wear out…' (Environment Committee, November, 1996).

The 1999 Water Supply (Water Fittings) Regulations introduced the following changes for plumbing systems:

- a maximum flush volume not to exceed 6 l;
- a secondary dual flush to remove fluid contamination not to exceed two-third of the full flush volume;
- the use of drop valves to replace the monopoly since 1935 of the siphon in the UK as a means of flush initiation;
- the introduction of an internal overflow to the bowl.

Solar energy in refurbishment

Advantages of solar energy

Refurbishment or other adaptation work can present an ideal opportunity to apply different solar energy options. It also presents challenges, however, because retrofitting such techniques in existing buildings can be difficult owing to the physical constraints and financial restraints involved.

'Solar energy is a clean and sustainable energy source and, therefore, has economic as well as environmental and social value' (ISE, 1997). It can be categorized into 'active' and 'passive' techniques. The former uses turbines, fans and pumps, some of which are often used in conjunction with passive systems. The latter uses the form and fabric of the building to admit, store and distribute primarily solar energy for heating and lighting (O'Sullivan, 1988). It has the following advantages:

- Energy is saved. For example, according to the ISE (1997) an annual energy demand before refurbishment of 200 kWh/m^2 to 100 kWh/m^2 after standard refurbishment was achieved; in one scheme a further reduction down to 50 kWh/m^2 was obtained after solar refurbishment.
- Thermal and visual comforts are improved.
- Maintenance problems are reduced.
- The architectural image is enhanced.
- The utility of the space is enhanced (better indoor comfort conditions are achieved).

The main passive solar techniques can be divided into two areas: solar thermal and PV (see below).

Disadvantages of solar energy

The main drawbacks with solar energy measures in refurbishment are related to cost and practicality. The high capital cost of many solar energy measures and their long payback period can prove a major inhibition for many property owners. The controllability of these measures, too, is not as efficient or responsive as non-solar energy systems. Some solar energy systems are so specialized that there may be a problem over their availability.

Solar thermal

The term solar thermal encompasses a variety of techniques that use the light and heat from the sun. These are summarized in Table 10.15.

PV

Basic principles

The use of solar energy using RET in buildings such as PV is gaining greater acceptance. PV technology uses silicone solar cell contact grids on a metal base plate to collect solar radiation (Sick and Erge, 1996). The collected energy from light is converted to electricity (see Figure 10.5) to satisfy some if not all of a dwelling's electrical needs.

PV can also be used, for example, to enhance the efficiency of the hot-water heating system in a dwelling. The effect of this is to raise the temperature from 6°C to >20°C over the winter half of the year. Applied solar power manufacturers such as 'Powertech' claim that this type of collector can increase hot water efficiency and flow by as much as 30 per cent and throughout the rest of the year temperatures from the collector can exceed 90°C.

According to the Energy Saving Trust (2005):

> In the UK, a 1 kWp system is expected to produce at least 750 electrical units – kilowatt hours – every year. The average household in the UK uses approximately 3,300 kWh every year. Therefore, a kWp system will produce nearly half of your yearly requirements and avoid around 650 kilograms of carbon dioxide emissions. With a life span of at least 25 years, a 2 kWp system will generate around 37,500 kWh and avoid approximately 16 tonnes of carbon dioxide emissions in its lifetime - enough to fill three hot air balloons…
>
> …The ideal roof pitch is approximately 35–40 degrees or less. PV can be successfully installed on a flat roof as there is much scope for ideal orientation. However, on a pitched roof in excess of 35–40 degrees the area exposed to light will be restricted and this will affect your system's performance.

Table 10.15 Typical passive solar thermal techniques and their use in adaptation

Technique	*Means*	*Potentials in adaptation*
Solar collectors	These can consist of building-integrated large reflector dishes on the roof or wall panels to capture solar radiation and convert it for heating and domestic hot water purposes. Thermosyphoning air panels form one type of solar collector.	Possible use in facade replacement schemes. They can be integrated in the roof or in a south-facing facade.
Glazed balconies	To act as a means of controlling heat gain/loss.	Limited to blocks of flats or offices with balconies.
Daylighting	Installing light-wells in pitched roofs via skylights to provide more natural light into the building (see Chapter 6).	Depends on extent of free roof surface and on whether there are any legal restrictions preventing this installation (e.g. listing).
'Trombe' walls	Named after a French inventor. Sometimes called 'solar walls'. Comprises double-glazed external wall panel in front of a heavy masonry/concrete wall, which has slots at top and bottom.	More appropriate in new build house- and school-building schemes.
Transparent insulation	Highly insulating glazings with light guiding properties.	As an insulation material for external walls. As a fill material for advanced glazing.

You will need at least 10 square metres of un-shaded, exposed roof area facing predominantly south, with a horizontal angle of up to 40 degrees. Chimneys, roof lights or nearby trees can all shade your system and need to be taken into consideration when deciding where to position your system.

Building-integrated PV

The integration of PV technology into the actual finished weathering skin of a structure is called BIPV for short (Building-Integrated PV). BIPV materials are manufactured with the dual purpose of producing electricity and serving as building materials. They can replace traditional building components, including curtain walls, skylights, atrium roofs, awnings, roof tiles, slates and shingles and windows (US Department of Energy, 2005).

BIPV panels are installed especially in south-facing elevations to maximize the exposure to sunlight). Such systems include:

- Roof panels in place of tiles/slates – crystalline modules integrated into roofing systems and as 'eyebrows' over windows.
- Glass-on-glass modules used in skylights and view walls.
- Amorphous silicon modules, both opaque and semi-transparent, used in curtain wall systems.

Traditionally, its high capital cost and relatively low service life coupled with relatively cheap mains-gas rendered this energy conservation measure unattractive for many users in Britain. PV panels will not become more popular in buildings generally until the technology has improved to reduce the installation and material costs.

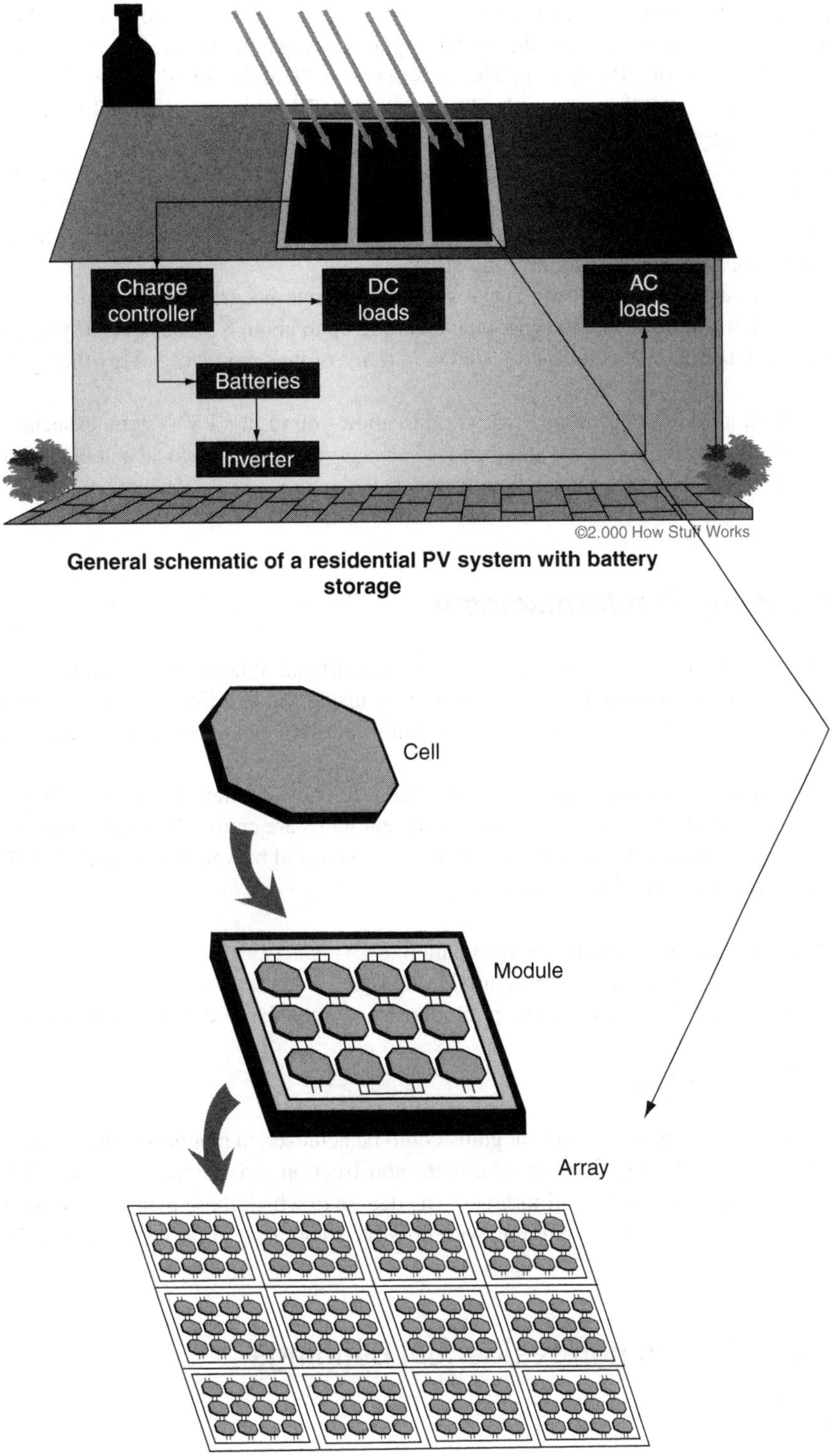

Figure 10.5 The photovoltaic effect (Sources: http://www.howstuffworks.com/solar-cell.htm and http://science.nasa.gov/headlines/y2002/solarcells.htm)

According to the US Department of Energy (2005), however, 'the technology holds the promise of extensive market penetration throughout the world, replacing conventional facade and roofing materials and "buying down" the cost of a PV system. The system cost is "bought down" simply by eliminating the need for ordinary shingles or other materials. The market for BIVP is growing at 30 per cent per year. Rising demand will increase production volumes and prices should drop'.

Retrofit BIVP or bolt-on measures can be done as part of an adaptation scheme. They can take advantage of the building's existing surface area, and help reduce the capital cost of the PV systems by the cost of slates/tiles or other building materials they replace (see Table 10.16). The biggest retrofit opportunities for this technology are with the existing housing stock.

PV is a highly sustainable technology. These systems generate no greenhouse gases, saving approximately 325 kg of carbon dioxide emissions per year – adding up to about 8 tonnes over a system's lifetime – for each kilowatt peak (kWp – PV cells are referred to in terms of the amount of energy they generate in full sun light).

Some local authorities require planning permission to allow you to fit a PV system, especially in conservation areas or on listed buildings. One should therefore always check with the local authority about planning issues before installing a PV system. Obtaining retrospective planning permission can be difficult and costly.

Passive cooling in refurbishment

Energy consumption relating to comfort conditions is not restricted to heating of buildings during winter. Even in temperate climates such as Britain some cooling inside the building will be required to combat overheating during summer. In some large office buildings cooling can account for a substantial proportion of their energy costs.

The aim therefore is to minimize heat gains in a building and modulate them when they occur. Heat gains originate either externally or internally. External heat gains are caused by solar radiation and ambient outside temperature conducted through the fabric and convected by ventilation and air infiltration.

The primary sources of internal heat gains are:

- metabolic heat (the heat produced by the occupants);
- artificial lighting – especially in non-residential buildings;
- appliances – such as cookers, televisions, photocopiers, computers – and other equipment in modern offices;
- cooking, bathing and washing.

The main ways in which protection from heat gains could be achieved in a building undergoing adaptation are summarized in Table 10.17. The intensity of use, the construction and orientation of the building and the extent of services within the property will influence the degree to which these measures could be applied.

A schematic sketch of a refurbished building with possible solar energy options is shown in Figure 10.6.

Energy efficiency in non-residential buildings

Preamble

Energy efficiency, like spatial performance, is an important design criterion for offices. The primary environmental effect of energy use is the emission of carbon dioxide, which is a major contributor to the

Table 10.16 Approximate costs for BIPV systems (ECOTEC, 1998)

System	*Installed Cost £/m² (2005 prices)*
PV curtain walling, glass/glass crystalline modules	780
PV curtain walling, glass/glass thin-film amorphous modules	280
Conventional wall systems (for comparison)	
Double glazing	350
Cavity wall (brick/block)	50–60
Stone cladding	300
Granite faced pre-cast concrete	640
Polished stone	850–1500
PV rainscreen cladding	600
Steel rainscreen overcladding (for comparison)	190
PV roofing tiles (housing estate)	500
Roofing tiles – clay or concrete (for comparison)	32
PV modules on a pitched roof (large office)	650
Aluminium pitched roof (for comparison)	44

greenhouse effect. If a building is made more energy efficient, therefore, by consuming less energy it will produce less pollution.

The following five key refurbishment options for achieving maximum energy efficiency in offices are suggested by BRECSU in GPG 35 (1993).

Office fabric and insulation

Criteria

The main requirements for office buildings are:

- Retain maximum opportunity for natural light and ventilation, even where the building will initially be air-conditioned. With deep plan, consider inserting courtyards and atria.
- Roof and wall insulation should be improved where possible. Opportunities are greatest in pitched roof spaces and when re-roofing and re-cladding.
- The thermal capacity of floor-slabs, internal partition walls and ceilings can help to stabilize internal temperatures, reducing the need for and capacity of any air conditioning. Try to avoid lightweight internal linings and suspended ceilings, particularly in buildings without air conditioning.
- Avoid unwanted air infiltration. Provide good door and window seals. Pay attention to construction joints as these often let in more air than the windows and doors.

Windows and natural ventilation

- Retain natural lighting where appropriate. To reduce solar gain, consider using shading and/or smaller windows. With tinted and reflective glass the lights will tend to be used more.
- New windows should normally be double-glazed, preferably with low emissivity glass. Window frames should be at least as well insulating as the glass.
- In naturally ventilated offices, windows should both seal tight and have means for controlled 'trickle' ventilation, to help avoid over-ventilation in cold weather and condensation, particularly while the building is drying out. Alternatively, background mechanical ventilation can be used (but there will be a penalty in terms of fan power).

Table 10.17 Typical passive cooling techniques and their use in adaptation

Technique	*Means*	*Potentials in adaptation*
Shading[1]	Using overhangs on pitched roof. Using retractable awnings. Using light-coloured internal blinds or shutters. However, internal blinds on their own will not necessarily reduce heat gain if used in conjunction with non-solar reflective glazing. Using automatic/adjustable timber or plastic louvres.	Part of an over-roofing scheme. Possible installation for retail outlet fronts. Gives occupants a degree of control. Could be fitted as part of an overcladding scheme.
Vegetation[1]	Planting trees or hedges to reduce solar glare.	Limited space available on urban sites. Suitable mainly for low-rise properties.
Insulation[1]	Thermal insulation in the building's envelope to reduce heat transmittance through its fabric.	Installing external insulation as part of an overcladding scheme.
Light colour finishes[1]	Using reflective or light-coloured wall and roof finishes.	Selecting light-coloured finishes for any redecoration work or overcladding scheme.
Efficient appliances[1]	Electrical appliances that give off low levels of waste heat.	Siting appliances in positions that encourage easy cooling to the outside.
Thermal storage[2]	Uses materials for the fabric with high thermal mass such as brick, stone and concrete.	Not very feasible in 'lightweight' buildings.
Natural ventilation[3]	Using wind induced ventilation via rapid vent openings or stack ventilation via roof vents or 'solar chimneys'.	Installing additional or larger diameter roof vents or stack vents.
Radiative cooling[3]	Using flat roof as an adjustable radiator. Moveable insulation panels cover plastic bags, filled with water, during the day and expose them to the sky dome during the night in summer.	Installing a 'Roof pond' or 'Skytherm system' on a large flat-roofed building. More appropriate for buildings in hot, dry, sub-tropical climates.
Evaporative cooling[3]	Using pools and fountains, combined with wind channelling by vegetation and natural ventilation.	These could be retrofitted in shopping centres and large open plan offices.
Earth cooling[3]	Using the earth or rock bed under the ground floor as a heat sink.	Replacing suspended timber ground floors with ground supported floors. Or increasing ventilation to subfloor void of suspended ground floors.

Notes
1. These techniques are used to minimize if not prevent heat gains.
2. This technique can be used for the modulation of heat gains.
3. These techniques are used for the rejection of heat from the interior of the building to heat sinks (by natural cooling). Hybrid cooling uses processes that are enhanced mechanically such as fans or pumps (i.e. active cooling).

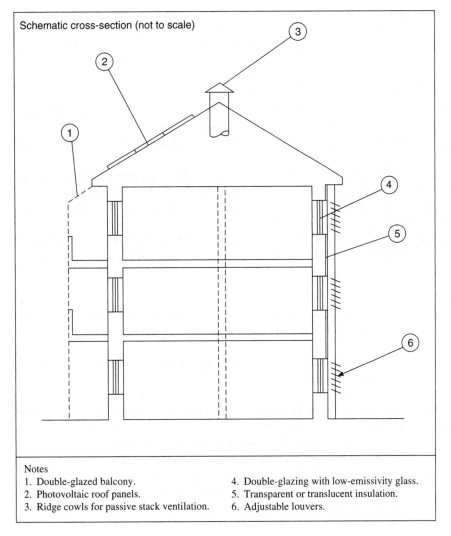

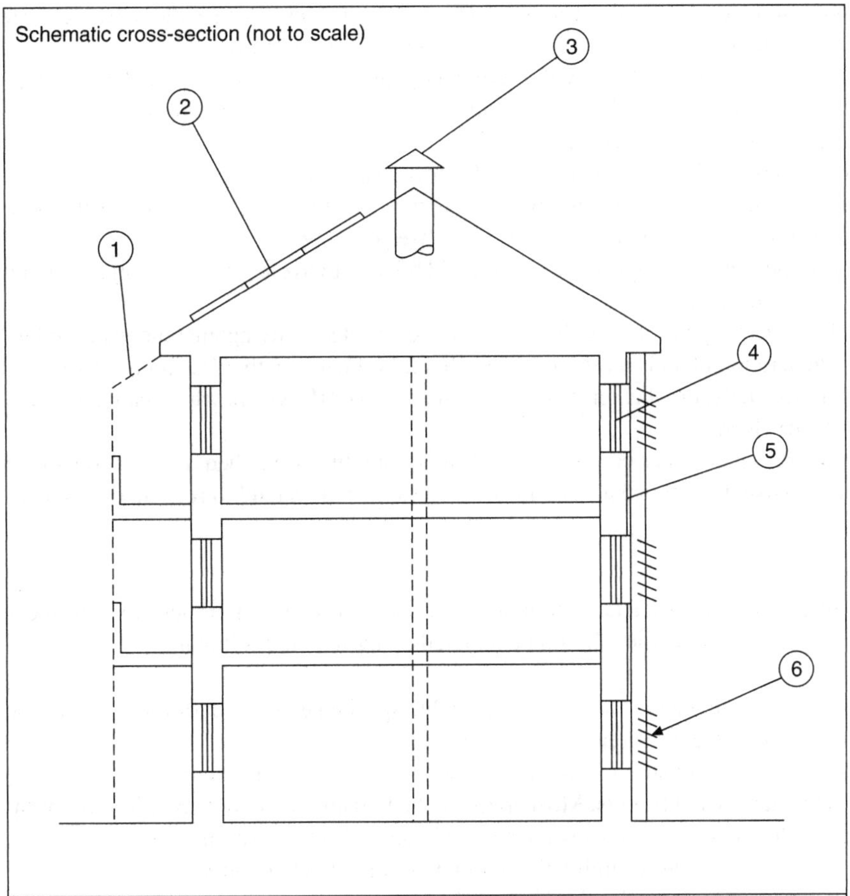

Notes
1. Double-glazed balcony.
2. Photovoltaic roof panels.
3. Ridge cowls for passive stack ventilation.
4. Double-glazing with low-emissivity glass.
5. Transparent or translucent insulation.
6. Adjustable louvers.

Figure 10.6 Some passive solar and related energy saving techniques in refurbishment

- Where natural ventilation is retained, review the design of the windows and the selection of window 'furniture' to give good control and performance.

Heating, ventilating and air-conditioning systems

Heating, ventilating and air-conditioning (HVAC) systems are major consumers of energy in buildings. Their use should be limited wherever possible:

- Choose efficient plant (e.g.: condensing boilers). See BRECSU's associated GPG listed in the Bibliography for further advice on this matter.
- Provide effective central, zone and room controls.
- Avoid the 'tails wags the dog effect' where large central systems run to meet small loads. Examples are central systems, which are used to provide hot water services in summer, local cooling overnight in

winter and ventilation air pre-heating on cool summer mornings. Provide separate systems to meet these small loads, such as independent water heaters for summer use.

- Avoid wasteful hot water systems with inefficient plant and labyrinthine circulation systems.
- Consider whether full air conditioning is always necessary. Natural ventilation may be a preferable option from a sustainability perspective.
- Consider 'free cooling' systems using outside air or evaporated cooling.
- With air conditioning annual fan energy consumption usually exceeds chiller consumption. Design for low fan power and hours of operation as well as 'free cooling'.
- Separate summertime and daytime loads from 24 h loads to avoid excessive fan and pump energy and low part-load efficiencies.
- Consider heat recovery but review the cost of any extra electricity against the value of the heat saved.
- If standby generation plant is required by the client, evaluate whether it can be run to help cut the peak demand for electricity, or whether it is worth installing a CHP system which can meet some of the base heat and electric loads.
- Use proprietary windcatcher cowls such as 'Monodraught' on pitched and flat roofs to minimize the need for expensive mechanical air conditioning systems which tend to recycle the same air.

Lighting

Lighting in offices accounts for a large proportion of their energy use. For these reasons the general guidance on lighting referred to earlier should be adopted along with the following:

- Choose appropriate standards but do not over-light. Special needs for additional lighting should be met locally and not for the entire area.
- Choose light coloured finishes where possible to improve inter-reflections.
- Select efficient lamps and fittings. Most areas can be lit using no more than 2.5 W/m^2 of installed lighting power (including control gear) per 100lux of standard service illuminance.
- Use electronic control gear for lights that will be on for extended hours.
- Where considering high intensity discharge lighting, remember the lamps take several minutes to warm up to re-start and so will tend to be left on unless special provision is made.
- Use tungsten or tungsten-halogen lighting very sparingly for essential access lighting only.
- Provide efficient lighting not only in the offices themselves but also in corridors, WCs, etc., where running hours may be longer.
- Consider automatic controls, particularly for lights in open area offices. Try adopt a policy manual ON/OFF but with automatic OFF providing a reminder/backup. Integrate automatic controls with daylight where possible. Provide local switches so that users can easily control their own lighting levels.
- Consider occupancy sensing controls in intermittently used areas where otherwise lights will tend to be left on.

Controls and monitoring

- Provide effective central, zone and room controls for health and safety, energy efficiency, prolonged plant life and responding to the needs of the occupants and management.
- Make sure that devices requiring regular re-setting and re-programming are readily accessible to the people responsible.
- Provide feedback devices, so that the status of the installation can be regularly monitored.
- Consider electronic BEMS, but as a management tool.
- Avoid excessive complication, beyond the normal capabilities of site staff and maintenance contractors.

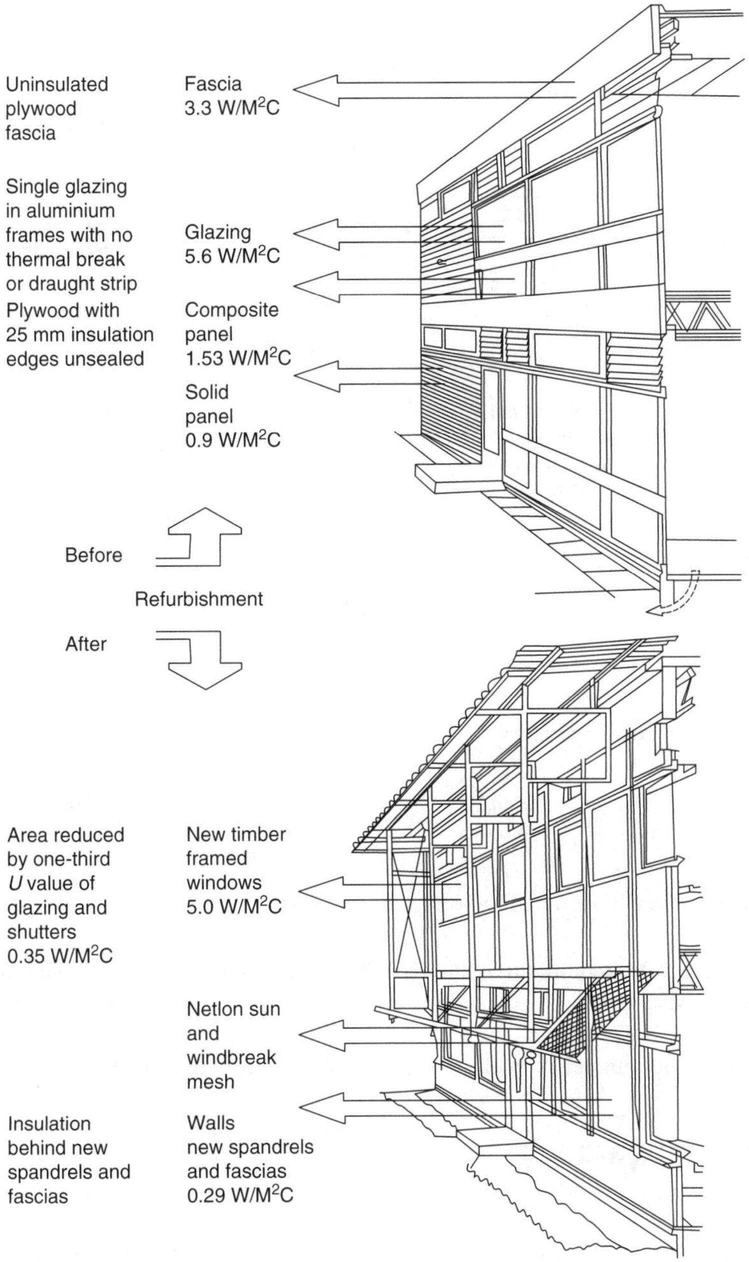

Figure 10.7 Façade of lightweight school before and after refurbishment

Schools

The measures required to improve the energy efficiency of the fabric of school buildings are illustrated in Figure 10.7–10.10. The other measures that can be taken are summarized as follows:

- Lighting: Replace tungsten filament lamps with fluorescent tubes and robust diffusers (to minimize breakage through vandalism); and provide effective luminaire control in classroom areas.

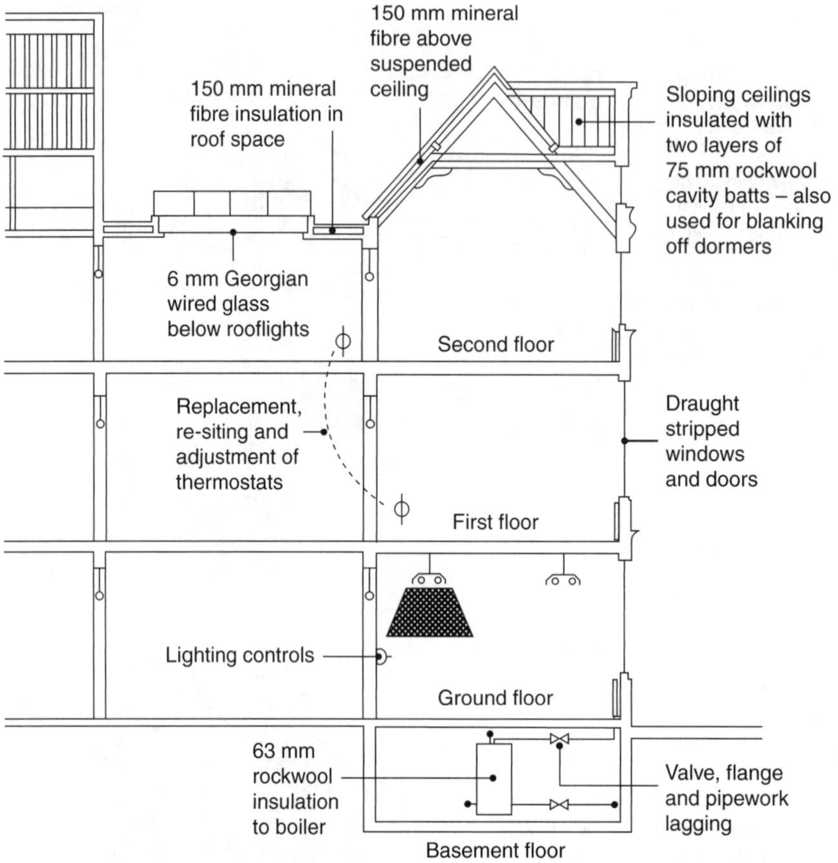

Figure 10.8 Typical section through heavyweight school building showing energy efficiency measures

- Boiler controls: These are required to reduce boiler cycling (on/off operation) to a minimum.
- Heating control: These should be situated in secure areas so that only authorized use is allowed (see BRECSU, GPG 233, 1997).
- Fit sun-pipe natural lighting tubes in classrooms below roofs to reduce the need for artificial lighting.

Industrial fabric upgrades

Preamble

Overcladding or recladding of the external walls and pitched roofs of an industrial building (as discussed in Chapter 9) can enhance its appearance as well as energy performance. The choice of overcladding or recladding will depend on the condition of the existing frame or substrate.

Insulated re-roofing of flat-roofed areas with high performance waterproof membranes to reduce heat loss and enhance the durability of this often extensive and vulnerable part of the building.

Strip curtains

In industrial buildings, strip curtains should be installed over openings between one area and another. These heavy-duty plastic strips reduce heat loss from, and ingress of cooler air to, the main factory areas.

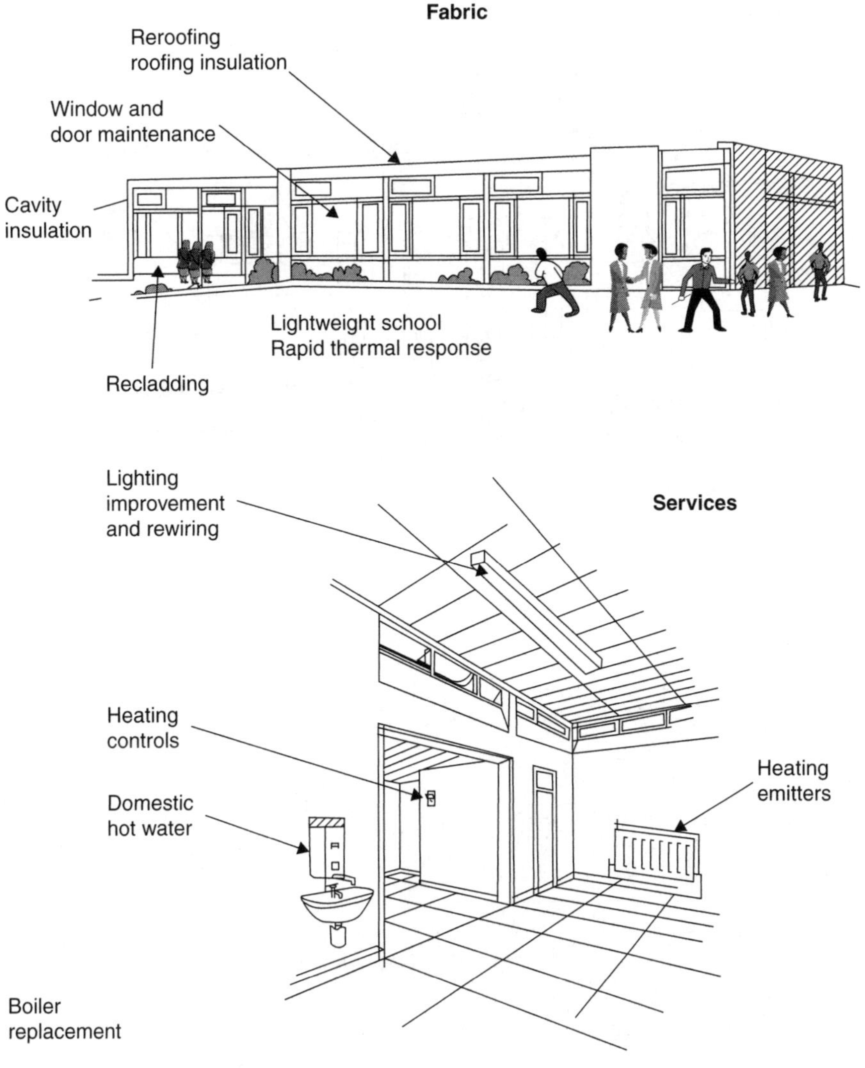

Figure 10.9 Refurbishment of fabric and services of lightweight school building

Heating controls

Tamperproof time/temperature controls for the space heating system should be sited in the factory Manager's office. This is to prevent unauthorized setting of the thermostats and time clocks.

Destratification fans

These fans should be used in large volume spaces such as warehouses, factories and hypermarkets. As warm air rises in these spaces a difference of over 5°C in the temperature between floor finish and head level can be created. Destratification fans (c. 1.2 m in diameter) should be installed in the activity/work-space to overcome this problem (see Figure 10.11). These are low-velocity sweep ('punkah') fans, which prevent a steep temperature gradient occurring in the building. They also reduce the heat loss through the roof by repelling the building's natural heat rise.

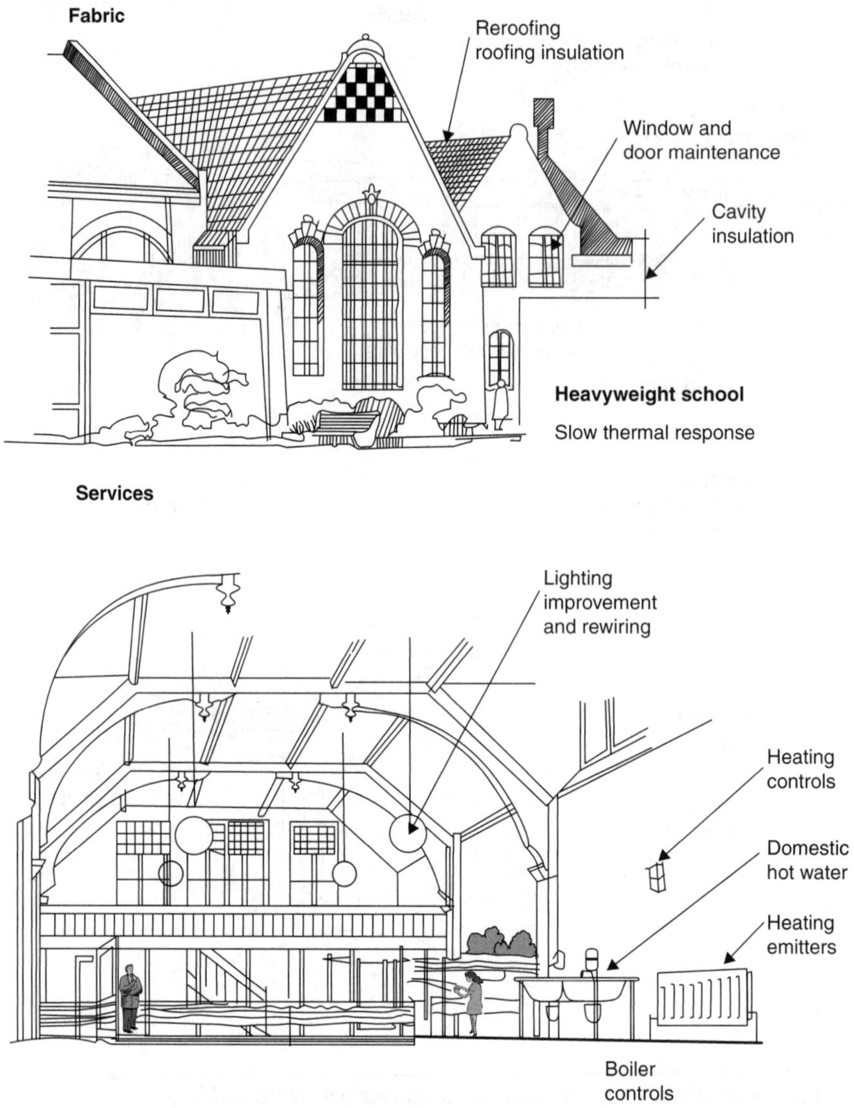

Figure 10.10 Refurbishment of fabric and services of heavyweight school building

Lighting

Low energy (sodium) light fittings should be fitted to the main factory space. The offices attached to the factory area should be lit using krypton-filled fluorescent light tubes. The other lighting efficiency measures referred to earlier should be adopted where possible.

Future energy efficiency requirements

The statutory requirements relating to energy efficiency are unlikely to stand still. Improvements will still need to be made to continue reducing non-renewable fuel consumption, waste and pollution. The extent of these improvements will be influenced by the rate of climate change and government responses to it.

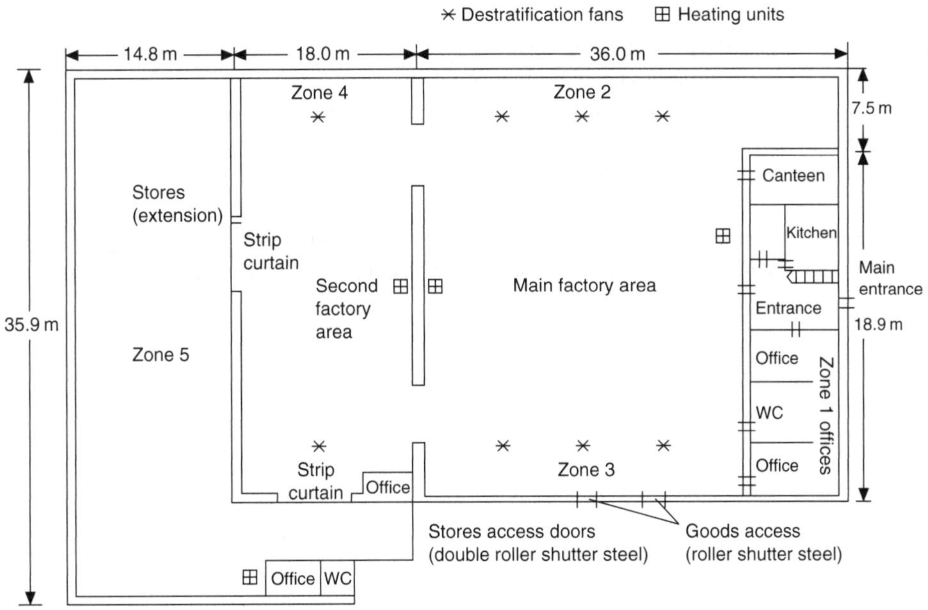

Figure 10.11 Layout of typical factory building with energy-efficiency measures

According to Gold and Martin (1999a,b), the likely aims of future legislation controlling energy efficiency and the environment are:

- use of recycled materials for construction;
- reduction in the amount of materials used (i.e. do not over-design);
- environmental life cycle assessments;
- reduction of energy used in construction (including, of course, adaptation);
- preference given in selection to contractors and suppliers who employ environmental best practice;
- reduction of energy used in operation and maintenance, by:
  - use of passive measures
  - consideration of shape and orientation of building
  - better controls to reduce un-necessary energy use.

Summary

The increasing importance of sustainable construction is likely to remain one of the key drivers of the modernization of commercial and industrial buildings. Energy consumption of factories and offices needs to be reduced if companies are to remain cost conscious as well as competitive. This also applies to assembly/institutional buildings such as hospitals and schools, for example (see BRECSU, GPGs 206 and 233, 1997).

The use of ESTs or RETs can also help achieve a more sustainable building. Solar refurbishment can potentially reduce annual energy consumption in a building by between 10 per cent and 75 per cent depending on the extent of the measures undertaken.

Some of these measures might include some if not all of the following initiatives:

- Structural and finishing materials selected for low embodied energy and minimal environmental impact.
- Displacement (passive stack) ventilation with 100% fresh air.

- Radiant heating and cooling.
- Extensive use of daylighting (using SunPipes, etc.).
- Air-conditioning system without cooling tower or CFCs.
- No CFCs and minimal HCFCs used in construction or in operation.
- Drought-resistant native landscape materials.
- Construction waste re-use and recycling.

Therefore the role of refurbishment in achieving the UK's sustainable construction targets should not be underestimated. As indicated in Chapter 1 it's a sector that accounts for nearly half of the building industry's output.

11

Implementation

Overview

This chapter looks at the requirements and procedures involved in implementing an adaptation scheme once a desired option has been determined. It outlines the principal administration procurement processes that are peculiar to such work. Although they contain many similarities with new-build projects, adaptation schemes require a unique range of technical and management skills.

Adaptation assessment procedure

Adaptation plan of work

Background

The plan of work for adaptation schemes can generally follow that for new-build projects. The version described below is based on the Royal Institute of British Architects (RIBA) Plan of Work. This procedure, of course, is primarily designed for new-build schemes. However, cognizance has to be taken of the subtle differences between adaptation and new build. There are likely to be greater levels of consultation between the client and contractor in the refurbishment of an occupied property. This is because since it already exists the client will have a greater understanding of the building's attributes.

Proper briefing with the client of an occupied building that is going to be adapted is essential if problems are to be minimized and raised expectations dashed. For example, a re-roofing exercise may form part of the major refurbishment of a commercial building. During the course of this work there is no guarantee that the roof can maintain its weather-tightness. This would mean that there is a risk that rooms on the top floor may be affected by rainwater ingress whilst this work is being undertaken. Temporary covers such as tarpaulins can help to reduce the incidence of rainwater penetration if they are correctly positioned and adequately anchored.

It cannot be emphasized enough, however, that each construction project whether new build or adaptation is unique. Every building has its individual characteristics and problems regardless of its similarity with other structures. Moreover, each scheme will not always have exactly the same group of participants, or have the same monetary requirements or the same property.

Establish brief (inception)

Determining the needs of the client is the most important initial task for the professional adviser. The consultant has to determine the extent of the client's requirements and establish the suitability and potential of the building for adaptation.

Investigations (feasibility)

The consultant will then need to assess the feasibility of the scheme by undertaking a cost analysis once a brief has been established. To do this the consultant conducts a survey of the building to form a preliminary opinion about its general condition and suitability for adaptation. The client's budget provides the benchmark against which the proposal and alternative options can be assessed (see Chapter 2).

Assessment of data and requirements (scheme design)

Briefing charts and bubble diagrams and other graphical devices can be used to plan out the proposed scheme. Outline sketch proposals are formulated at this stage and these are then tested to see if they meet the client's needs, budget and statutory requirements.

Formulation of proposals (detail design)

Once the general form and arrangement of the internal space and the extent of external changes have been established, firm working drawings and details can be produced. This will enable the consultants to submit the proposals to the appropriate local authority planning and building control sections for approval. Development of the specification and bills may also occur at this stage.

Implementation (solution)

Preparation of contract documents and selection of procurement route takes place at this stage. Following the granting of planning permission (PP) and a building warrant a single stage selective tendering process can be instigated. Only contractors who have a proven track record of undertaking adaptation work should be considered for such a contract.

Aftercare (occupancy)

Six months after completion of a major adaptation scheme, a building surveyor (BS) or architect could carry out a post-occupancy evaluation (POE) on the property. This will identify the successes and failures of the scheme and provide feedback on points for future projects. It will also give the maintenance manager information for formulating an aftercare strategy for the adapted building (see last three sections).

Data collection sources

Client

The first source of information on what is required for the adaptation is, of course, the client. The client may be the owner of the building in question but may not be intending to occupy it once the adaptation work has been completed.

Occupiers

If the client has no intention of occupying the property then feedback on the likely user requirements is necessary to ensure that the completed building fulfils the needs of the user. Feedback from people who use other buildings where similar activities are being carried out may also prove informative.

Where the user is unknown, as in the case of a speculative conversion scheme to a commercial building, flexibility is the key requirement. Open plan spaces to maximize this will be important.

The building

The property itself can yield a great deal of information on its suitability for adaptation. Data from the measured and condition surveys will prove useful. Material contained in archives will also yield helpful information.

Previous projects

The consultants involved in formulating the adaptation proposals for a particular building may consult recently completed projects undertaken by them or others previously. Although not every project is identical, there are likely to be similarities as regards statutory requirements and costs. Moreover, problems peculiar to certain types of adaptation that were previously undertaken by the team could be flagged up. There are always lessons to be learned from any project; adaptation work is no exception.

Consultant's previous experience or training

The consultant may have extensive knowledge of certain types of adaptation schemes, such as loft extensions. Such a person is likely to know most, if not all of the pitfalls associated this type of work. Alternatively other designers familiar with the kind of adaptation being proposed may provide useful feedback.

Local authorities

Building-control surveyors and planning officers may have 'local knowledge' of property problems. They may therefore, be in a good position to clarify statutory requirements or provide information on the history of the building being adapted.

Official publications

Statutory, governmental and other legal publications can be a vital source of information on a whole range of issues relating to construction and buildings. All government departments now have information sites on the Internet. The various sites contain a wealth of information on construction matters, from accessibility to sustainability. See the Bibliography for a list of some of the main British governmental sites.

Non-statutory sources of information on issues relating to adaptation and new build can also be helpful. Technical journals such as *Structural Survey* or *Building Appraisal*, magazines such as *Building* and the *Architects Journal*, books such as those listed in the Bibliography, as well as various digests, guides, information papers and reports produced by the Building Research Establishment (BRE), can all provide informative sources of guidance.

The number of web sites relating construction is extensive and growing rapidly. However, a short list of some informative web sites of non-statutory bodies in construction world-wide is also listed in the Bibliography.

Analysis of data and requirements

Generally

Once the main sources of information have been consulted, it is important for the building professional to analyse the data that has been collected. Priorities must be established.

Briefing chart

For small adaptation schemes a briefing chart could be devised to highlight all the key requirements of the client. Table 11.1 illustrates a simple briefing chart for a proposed kitchen extension.

Synthesis of data and requirements

Association charts

In large conversion or extension schemes involving several different activity spaces association charts can be used to determine an appropriate layout. The charts may be compiled in normal tabular form with all

Table 11.1 Simple briefing chart for kitchen extension

Requirements	*Details*
Arrangement	
Position	At rear of building. New back door needs to be formed.
Next to	Original kitchen, which is being enlarged.
Near to	Living room.
Floor	Suspended timber ground floor – to match existing floor construction.
Area	
Aspect	SW–SE.
Size	30 m^2 minimum.
Shape	Rectangular or 'L' shaped?
Features	
Light	500 lux. One 2 m long fluorescent tube.
Humidity	Between 40 and 60 per cent relative humidity (RH).
Temperature	Between 18 and 23°C.
Power	Three double-sockets positioned along worktop.
Equipment	Mechanical extract ventilation over cooker required to remove cooking fumes.
Work surfaces	Worktops (with cooker recess) finished with durable, light-coloured surface and rounded edges.
Floor surface	Dust free and hard – hardwood laminate sheeting or high-quality foam-backed vinyl sheeting.
Wall surface	White glazed tiles above worktop. Generally washable.
Ceiling surface	Plasterboard with painted 'Artex' finish. Or tongued and grooved 'vee' jointed boarding finished with a fire retardant lacquer.
General remarks	Lean-to or duo pitched roof preferred. Pitch will be determined by the fenestration of the rear elevation. Position of adjoining owners' boundaries needs clarifying.

the spaces written along both the top and the side. The resulting chart can be used to assist in the preparation of bubble diagrams relating to the proposed scheme.

The conversion of the redundant mill shown in Figure 3.11 could be planned using an association chart. Table 11.2 illustrates how this could be done.

Evaluation

This can consist of an inter-scheme evaluation and/or an intra-scheme evaluation. The former would involve a comparative assessment of several schemes using an adaptation analysis matrix described in Chapter 2. With an intra-scheme evaluation, critical or key factors in the selected solution could be compared against some desired standards. The assessment would consider performance indicators such as:

- Cost: The estimated cost of the work against an authoritative benchmark – such as the Building Cost Information Service (BCIS) prices.
- Value: The estimated value of the adapted building, and compared against current market values of similar properties.
- Thermal efficiency: The Standard Assessment Procedure (SAP) rating or *U*-values of the adapted building.
- Noise reduction: The degree of noise reduction achieved/required.
- Spatial efficiency: The net-to-gross ratio or proportion of office space to circulation space (see Chapter 9).
- Aesthetics: The overall quality of the design in relation to the surroundings.

Table 11.2 Association chart for proposed conversion of a mill into apartments

Floor Level	*Spaces*
Ground	Living Room
	Dining Room
	Kitchen
	Study
	Hall/Stair
First	Main Bathroom
	Double bedroom 1
	Double bedroom 2
	Single Bedroom

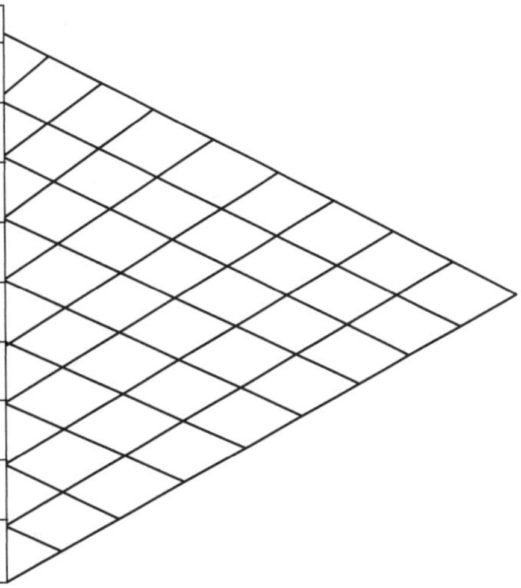

Legend
Functional relationship
- Very close connection
- Close connection
- Some connection
- No connection

Physical conditions
- Identical
- Compatible
- Different
- Opposed

Bubble diagram

A bubble diagram is a pictorial representation of the spatial arrangement of the building being designed or adapted. Each primary space is represented by a circle with the name of the space written in it. Lines with arrows are used to join the bubbles, showing the flow or circulation between the spaces. Lines of varying thickness are used to indicate amount or importance of the circulation required.

Figure 11.1 illustrates the application of a bubble diagram to the adaptation scheme referred to above.

Planning policy

Cognizance of national and local planning policy must be borne in mind when contemplating the adaptation of a building. The Office of the Deputy Prime Minister's web site contains downloadable copies of the main planning policy guidance notes and circulars.

As indicated in Chapter 3 PPG15 as amended by Circulars 14/97 and 01/01 must be heeded when planning to adapt a building of historic or architectural importance. These planning circulars and guidance notes articulate central government's policy for the identification and protection of historic buildings, conservation areas and other elements of the historic environment (PPG15/94).

Statutory consents

General requirements

It is essential for a variety of reasons to obtain the appropriate statutory approvals when proposing to adapt a building. Failure to do so will in all likelihood delay, or even cancel, any proposed sale of the property. Documents such as PP, building control approval, 'Letter of Comfort' (see below), or Lawful Development

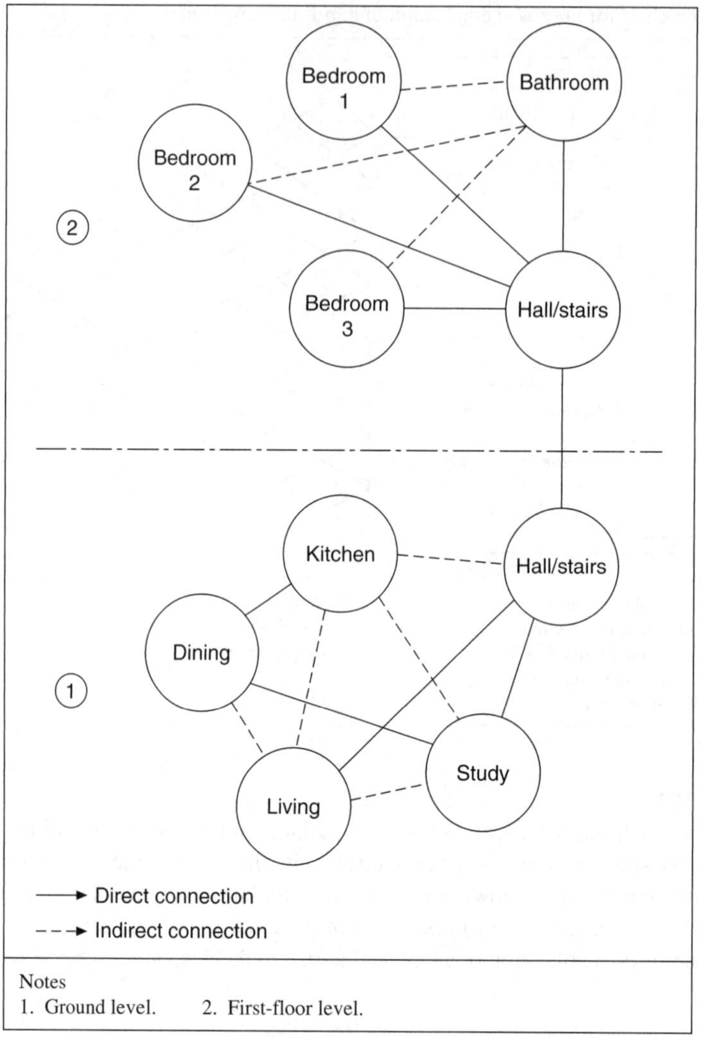

Figure 11.1 Bubble diagram for proposed conversion of a mill into apartments

Certificate, are important. They are needed to confirm any adaptation work by the lawyers handling any conveyance before the deal can be finalized.

External alterations and extensions can have a major impact on the appearance and massing of an existing building. This is particularly a problem in residential adaptation schemes. As the number of households altering their homes increase the cumulative effect of their alterations will change the character and appearance of the street. These may also include the erection of new boundary walls, fences and planting of trees or other vegetation.

Internal alterations can affect the layout and circulation within a building. They may also necessitate structural alterations or major changes to the original character or function of the building. There is also a need to control the removal as well as addition of space or elements within a building.

Before undertaking any adaptation work, therefore, it is important to establish the extent of any statutory approvals that may be required. The degree of adaptation and the category of building involved determine this. It will also have cost implications for the project and may delay the timing of the work.

However, as indicated earlier, the guiding rule regarding whether or not any statutory approval is required is:

> If in doubt, consult the appropriate authority (e.g. local planning office), and make sure any agreement or decision is confirmed in writing.

Planning permission

The purpose of planning law of course is to control the use and development of land. It is primarily concerned, therefore, with the proposed nature and density of any development – whether new build or work to existing property. As already shown with conversions and extensions a range of adaptation work may require conforming to these statutory requirements.

The various Town and Country Planning Acts determine the rights applying to the enlargement, improvement or other alteration to a property. Such rights can only be used once in relation to increasing the volume of a property. Previous extensions may therefore have used up these rights. This should be checked during the desktop investigation stage.

As a general rule, external alterations and extensions require PP. Some minor works (such as the erection of a single garage) to domestic properties, however, are classified as Permitted development (PD) as consolidated in Schedule 2 of the General Permitted Development Order (GPDO) 1995. The exemptions of these specific alterations change as the laws are amended. For example, certain rooflight and dormer installations (especially to rear elevations) now fall within the PD provisions.

If the character of the street is of high quality then there is a possibility that it is in a conservation area or the houses are listed as being of historic or architectural interest. The proposed adaptation work could involve replacing the original rainwater goods or repainting the front door or external rendering. In such cases an application to the planning authority will be required (see below).

The extent of PD for existing buildings is not clear-cut. However, as indicated by Haverstock (1998), PP is not required for adaptation works, such as a garage or house extension, in the following cases:

- For a terraced house, and for any dwelling in a conservation or special area, they do not increase the existing volume of the dwelling by more than 50 m^3 or 10 per cent, whichever is the greater; in any other case, whichever is the greater of 70 m^3 or 15 per cent (with a maximum of 115 m^3).
- The height of the property will not be increased. In the case of dormer windows or other developments affecting the roof, however, the projection of the proposed development from the roof should not be more than 100 mm.
- No alteration is proposed to the original roof – but a separate Class in the GPDO details conditions for some PD when:
  - work is kept below the highest part of the roof;
  - no part of the house would, as a result of the works, extend beyond any existing roof slope fronting any highway (i.e. no dormers);
  - it is less than 40 m^3 for a terraced house or 50 m^3 otherwise;
  - the cubic content of all the resulting work is within the figures given above;
  - above all, the shape of the roof will not be altered.
- No part of the garage or house will be in front of any wall of the original dwelling that fronts on to a road.
- The work is further than 20 m or the line of the original property from the curtilage highway/s.
- The work is within 2 m of the boundary curtilage of the dwelling and is less than 4 m high.
- The total area of ground covered by buildings within the curtilage (excluding the original house) is less than 50 per cent of the total area of the curtilage (excluding the ground area of the original house).
- The Change of Use is permitted within the Use Class Order (see Chapter 3).
- Any new walls or fences do not exceed 1.2 m or 2 m in height, respectively.
- Ordinary television aerials and BSkyB-type satellite dishes. Standard-size installations to listed buildings or larger aerial/satellite (e.g. communal/commercial) installations will require PP.

Building control

Compiling drawings to submit to building control for approval is not straightforward, even for simple adaptation schemes. A checklist of points that should be covered when undertaking this can help to maximize the chances of complying with the requirements of the building regulations. Such a checklist is shown in Appendix E.

As indicated earlier the building regulations are primarily concerned with the quality and safety of the proposed construction. Any adaptations (involving some non-structural as well as structural alterations) will normally require building control approval, unless the building is an exempted class (e.g. a temporary structure). The specific building-regulation requirements relating to conversions and extensions were addressed in earlier chapters. The property owners who either undertake such work without a building warrant or buy a property that had alterations done to it without the proper approvals can take one of two courses of actions to rectify this legal anomaly.

Firstly, some local authorities may allow an application to the local building control for a 'Letter of Comfort'. This follows what some local authorities call a 'Property Inspection' procedure by one of their building control surveyors if the alterations are minor and do not involve any major structural works. Such a letter from the local authority confirms that they have no objection to the alteration. For example, the removal of a non-load-bearing partition to enlarge, say, a lounge may be classed as a 'minor' alteration. If the work is more extensive, however, the next procedure must be followed.

Secondly, a retrospective application for a building warrant may be required, particularly if it involves structural or other substantial alterations. This would require the same set of drawings and other information had the application been made before the work was undertaken.

Failure to obtain the necessary statutory approvals at the very least can cause delays or problems in the conveyancing or re-mortgaging of a property. It may also expose the owner to being served a statutory notice requiring that the building be brought up to minimum required standard, especially in terms of safety. This must be started within a reasonable period (usually 28 days). Non-compliance could expose the owner to legal sanctions, such as a fine. Alternatively, in extreme cases, the council may be forced to initiate the work and charge the owner with all the administrative and construction costs involved. This is potentially a more expensive option because of the additional charges imposed.

Listed-building consent

Any property that is listed under the Planning (Listed Buildings and Conservation Areas) Act 1990 [or the Planning (Listed Buildings and Conservation Areas, Scotland) Act 1997] requires listed-building consent. This covers any proposed work that would affect the character of such a building internally as well as externally. If the proposed adaptation is restricted to the interior then only listed building consent may be needed. PP will also be required if external works such as an extension are proposed.

The application for listed building consent must include:

- a set of 'as existing' plans, sections and elevations of the building, to the appropriate scales;
- a set of 'as proposed' plans, sections and elevations coloured to highlight the location and extent of the new works;
- details of alterations and new works such as doors, windows, disabled ramps, new railings, new external or internal openings.

Applications for listed-building consent that are recommended for approval by the local planning authority are then referred to English Heritage via the Department of Culture Media and Sports (in England and Wales) or Historic Scotland (for Scotland). These conservation agencies have 28 days to ratify the approval, after which consent is issued.

Conservation-area consent

This is required for the demolition of any building or any part of a building in a conservation area. An adaptation scheme in a conservation area involving the demolition of, say, an old lean-to to make way for a new extension or more space around the main building would require conservation-area consent. In such a case the local planning authority may demand evidence to justify such demolition such as:

- a full building-survey report,
- details of demolition and adaptation costs compared with repair/restoration costs.

Party Wall Act 1996

The requirements of this Act cover only England and Wales. It updated the 1939 Party Wall Act, which only applied to London. Thus adaptation schemes in England and Wales may have to take cognizance of the 1996 Act, which applies where work is being proposed to a party wall or boundary wall, which can often occur with adaptation schemes in urban areas. For example, it applies where there is a need to excavate below the level of the adjoining building or structure that is within 6 m of the boundary (Billington et al., 2003).

Inner London

The Building Act 1984 made the Building Regulations apply to Inner London as well as the rest of England and Wales. Nevertheless, additional requirements in relation to building control still apply to Inner London through the London Building Acts 1939–1982. For example, the 1939 Act:

> requires special fire precautions in tall buildings (i.e. a building that has a storey at a greater height than 30 m, or 25 m if the area of the building exceeds 930 m^2, or is a large building, or warehouses over 7100 m^3). In addition, local authority consent is required if two buildings are united by forming an opening in a separating wall, or external wall if access is obtained between the buildings without passing into the outside (Billington et al., 2003).

Feu superior's consent

In Scotland anyone proposing to adapt his or her building must obtain the consent of the feu superior before work can proceed. In the context of housing this is likely to be the original developer or contractor involved in the construction of the dwelling. Whilst refusal for valid adaptation work is rare, there may be delays in obtaining consent if the payment of a nominal fee to the feu superior for giving consent is not forthcoming. The client's lawyer normally handles this matter.

This procedure, however, will cease when the Abolition of Feudal Tenure (Scotland) Act 2000 comes into force, probably within the next 3 years. In the meantime, householders and other property owners in Scotland will still require the feu superior's consent before embarking on any alteration, extension or conversion of their property.

Obtaining approval

The following stages are involved in obtaining the necessary building control approval:

- Application: Complete appropriate forms for PP and/or Building/Demolition/Change of Use. These can be obtained from the local authority's planning or technical services department, and can be completed by either the client or his/her agent or professional adviser.
- Fees: The level of fees is dependent on total cost of the proposed works, and is usually calculated on a percentage basis (per the Building [Prescribed Fees] Regulations 1994). The minimum fee is around £75.

Each local authority may work to its own fee schedule, but this is based on government guidelines. Fees are payable on submission of application.

- Copies: At least two if not three copies of the drawings must be submitted, one of which must be a 'permanent' copy such as linen or plastic coated paper. Either a design certificate signed by a qualified engineer or detailed calculations must be provided for works involving structural alterations (e.g. insertion of a lintel over a new opening in an existing wall, or additional support beam to strengthen an existing floor). (See notes on graphic presentation of proposals below.)
- Building notice: In England and Wales (but not in Scotland) the Building Notice procedure can be used. With this approach less information is presented than with depositing full plans. It enables the client to commence the work by giving notice before full approval of the proposals is obtained.
- Decision: One of three decisions is likely – Approved, Amendments required (most likely initial outcome) or Refused (uncommon, unless submission is very poor or proposed works are inappropriate). Local authorities aim to give a decision is given within 6 weeks of submission of a building warrant application. This timing also applies to decisions relating to planning applications. Therefore, if an adaptation scheme requires planning approval as well as a building warrant, a period of at least 3 months is needed before any work can commence.

Amendments if accepted by building control must be shown on the drawing. When finally approved, the works must be carried out within 3 years of date of issue of building control approval. If the local authority refuses to grant such approval, the applicant may appeal to the county/sheriff court within 21 days of the date of the decision.

Other legislation affecting adaptation work

Before undertaking any adaptation work it is important to establish the extent of any statutory approvals or requirements that may have to be obtained or satisfied once the building is in use. These can be summarized as follows:

- Building Act 1984 or the Building (Scotland) Act 2003.
- Civic Government (Scotland) Act 1982.
- Clean Air Acts 1956, 1968 and 1993.
- Disability Discrimination Act 1995.
- Environmental Protection Act 1990.
- Factories Act 1961.
- Fire Precautions Act 1971.
- Health and Safety at Work etc., Act 1974.
- Housing Act 1986 or the Housing (Scotland) Acts 1974 and 1984.
- Housing Grants, Construction and Regeneration Act 1996.
- Home Energy Conservation Act 1995 and the Energy Conservation Act 1996.
- Highways Act 1980 or the Roads (Scotland) Act 1984.
- Planning (Listed Buildings and Conservation Areas) Act 1990 or the Planning (Listed Buildings and Conservation Areas) (Scotland) Act 1997.
- Town and Country Planning Act 1997 or the Town and Country Planning (Scotland) Act 1999.
- Special Educational Needs and Disability Act 2000.
- Sewerage (Scotland) Act 1968.
- Water Industry Act 1991.
- Water Resources Act 1991.

Consultants involved

Very rarely will the appointment of one qualified person be sufficient to undertake all the professional work involved in adapting a building. The following is a list of some of the main players:

- Architect: To undertake design and supervision of work.
- Building surveyor: Analysis of defects and, for small–medium scale adaptation schemes; provides similar project services to the architect.
- Construction manager: To manage medium–large refurbishment projects.
- Party-wall surveyor: Some BS are also qualified to undertake the assessment under the Party Wall Act 1996.
- Quantity surveyor: Cost analysis and planning; possible project management of larger schemes.
- Structural engineer: To design any structural alterations to the building being adapted.
- Services engineer: To design more complex services installations to the building undergoing adaptation.

Professional fees

At one time professional bodies in the UK produced their own scale of fees for their services. However, this was deemed contrary to the principle of fair market competition by the British government in the 1980s. Professional bodies like the RICS were required by the Office of Fair Trading to discontinue their fee guidance in relation to professional services. As a result, for example, BS fee scales were abandoned in February 2000. The new edition of the documentation for appointing a Chartered Building Surveyor was published after this date.

The minimum fee for a small-scale adaptation project, such as a loft conversion, could be based on a fixed sum. A typical fee for such work would likely be about £1500 + value added tax (VAT) but exclusive of statutory charges such as building control and planning approval. For larger projects the fee level is usually based either on an agreed percentage of the total cost of the work or on an agreed hourly rate plus expenses. A range of 5–10 per cent is typical for most adaptation projects.

Spatial considerations

Sources of space data

The sources of space data can be divided into two main groups: statutory and non-statutory. For example, Part M (Facilities for Disabled People) of the 1991 English building regulations and Section 4 of the 2004 Scottish building standards provides some statutory sources of space data.

There is a large variety of non-statutory sources of spatial standards. For example, BS Codes of Practice (e.g. BS 5810: Code of practice for access for the disabled to buildings), The Parker Morris Standards 1961 (though these are seen as too generous nowadays), and the various planning and design data reference texts (e.g. Adler, 1999). Compliance with these non-statutory sources is not mandatory, but their inclusion may be desirable in any application for PP or building-control approval.

Space requirements

A guide to the space requirements for parts of buildings subjected to adaptation schemes is presented in Table 11.3. The figures indicated are only guides. Variations may be needed in these standards to suit specific situations and may have to be increased to meet certain statutory requirements.

Table 11.3 Summary of the main space standards (based on various sources)

Occupancy	*Suggested area (total or per person per m^2)*
Domestic kitchen-dining area	12 m^2 would be a useful minimum starting point.
Living room	18 m^2 would be a useful minimum starting point.
Double bedroom	14 m^2 would be a useful minimum starting point.
Single bedroom	10 m^2 would be a useful minimum starting point.
Bathroom	3.7 m^2 would be a useful minimum starting point.
Water closet (WC)	3 m^2 would be a useful minimum figure.
Circulation space	14.4 m^2 would be required for a single dwelling.
Storage general	12 m^2 would be a useful minimum starting point.
Workrooms in factories	Average 7 m^2. The nature and type of work being carried out will influence the minimum space required.
Work area in offices	Average 8 m^2. The Offices, Shops and Railway Premises Act 1963 requires a minimum volume of 9.3 m^3 per person, but no space over 4.27 m high is included in the volume.
Retail shops serving area	Average 10 m^2 per assistant.
Restaurant seating area	Average 1.25 m^2 per seat.
Assembly halls seating area	Average 0.75 m^2 per seat.

Graphical presentation of proposals

Format

Compiling drawings by hand is the traditional way of presenting architectural plans, elevations and sections. For simple adaptation schemes, A2-size paper is usually big enough. If the work is small-scale, such as forming an opening in a load-bearing wall, one A2 sheet will usually be sufficient to contain all of the following key information:

- Location plan, with property shown in red (Ordinance Survey scale).
- Project information panel.
- As existing plans and elevations.
- As proposed plans, elevations and sections.
- Notes and specification panel.

The pieces of key information that are usually presented on a drawing are shown in Figure 11.2.

All alterations should be highlighted using appropriate colours. Downtakings or formations of new openings, for example, must be shown in red. The conventional indication of materials using colours and shadings as prescribed in BS 1192 (Drawing Office Practice) should be followed.

Nowadays, many adaptation schemes are prepared using computer-aided drawing (CAD) packages. *Autocad LT* or its more advanced version *Autocad 2000* can be used to compile the necessary details. The main benefits of using CAD in this type of work are:

- Any technical drawing that can be produced by hand can be created in a CAD package.
- Drawing with the aid of CAD is much quicker than working by hand. A skilled operator can produce drawings as much as 10 times faster than the conventional method.
- There is less tedium when working with CAD. Features such as text, which can be very tedious when entered freehand, can be added to a drawing with minimum effort.
- Drawings can be inserted into other drawings without having to redraw them.

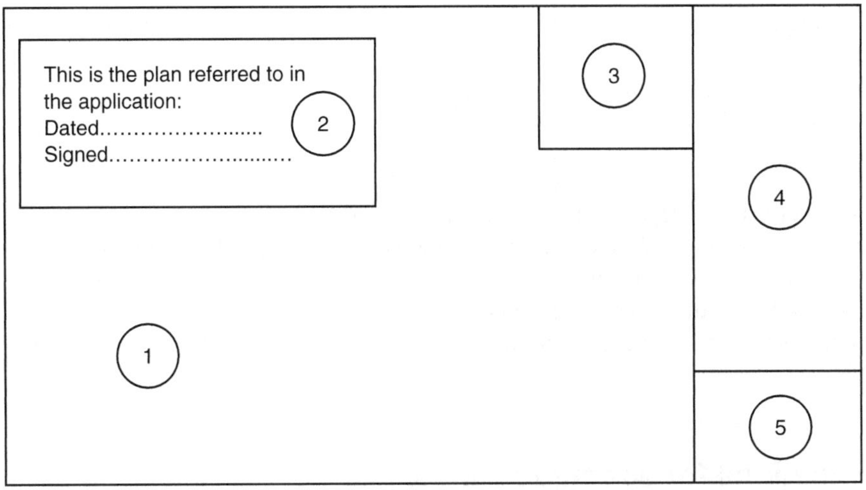

Notes
1. Main plans/elevations – preferably in two sheets:
   - Drawing 1: 'As Existing': block plan, main plans and elevations only.
   - Drawing 2: 'As Proposed': sections as well as plans and elevations. Small adaptation schemes (e.g. forming a new opening in a load-bearing wall), can show 'as existing' and 'as proposed' drawings on one sheet.
2. Docquet or label on each drawing and any copies confirming its application for building control approval. The actual size of the label should be not more than about 80 × 35 mm.
3. Location plan. The property in question should be highlighted in red for ease of identification.
4. Notes and specification (see example in Appendix D).
5. Project information panel: This should describe the nature of the adaptation work and its location. It should also incorporate the name and address of the client. The number of drawings involved must be indicated, as should when they were done and who compiled them, as well as an indication of any amendments.

Figure 11.2 Typical A2 size drawing layout

- Parts of drawings can be easily copied, moved, mirrored, arrayed (i.e. continuously repeated) without the need to redraw. In fact a basic rule when drawing with CAD is: *Never draw the same detail twice!*
- Adding dimensions to a drawing is very fast and when using associative dimensioning, this reduces the possibility of dimensioning error.
- Drawings created in CAD can be saved as files on disk, considerably reducing the amount of space required for storage of drawings.
- Drawings can be printed or plotted to any scale from the same drawing, reducing the need to make separate drawings for each scale.
- Use of layering can be helpful to highlight separate features such as walls, services, etc.
- Standard details and plans of common items such as furniture and sanitary ware are readily available. They can be either in one of the *AutoCAD* files or can be taken from CD ROM files of manufacturers' product databases. This saves a great deal of time in drawing standard items from scratch.

Requirements for plans

Typical local authority requirements for the particulars to be shown on the plans are as follows:

- Block plan shall be 1:500 where possible. It should show the size and relationship of the proposed adapted property to adjoining buildings and streets.

- Locality plan shall be 1:1250 where possible.
- The following minimum scales should be applied so far as is necessary to enable the local authority to determine e-compliance:
  - Plans of the affected floor/s } 1:100
  - Sections of the affected floor/s } 1:100
  - Elevations } 1:20 or 1:50
  - Details } 1:20 or 1:50
- Position, materials, line depth, inclination and dimensions.
- Alteration work should be clearly identified (coloured – e.g. red for downtakings).
- Where submitting an amendment to a building control application, the area affected by the change should be clearly identified (i.e. again coloured).
- The drawings should show the relationship and distance from any building to site boundaries and other properties.

Specifications for adaptation work

Types of specifications

Definition

Basically, a specification is a set of written instructions to the contractor. Its primary objective is to describe in a concise and accurate manner the nature of the materials, standard of workmanship and schedule of works required.

Even for small adaptation schemes clients should state explicitly their requirements. This is best achieved in an appropriate form of specification that gives a clear and itemized description of the work to be done. A firm price in writing, not an estimate, for the work should be requested, which should clearly indicate whether VAT is included. In addition the contractor involved should be a member of the appropriate trade association, have adequate liability insurance cover and give guarantees on the quality of their materials.

There are, however, two main types of specification used in construction: prescriptive and performance specifications.

Prescriptive specifications

These are the traditional, trade/works based documents, which are 'means' or input orientated. They place a heavy responsibility on the specifier. The contractor's responsibility, in contrast, is to meet the requirements of the specification. If he does so correctly but this results in a defect of the element or component concerned, the specifier (or client) may be held responsible.

Prescriptive specifications are still the most common type in use for adaptation projects. In small-scale work such as loft conversions or single-storey extensions, the specification can be included within the drawing under the Notes column (Panel B in Figure 11.4). A typical set of prescriptive specification clauses for inclusion in the Notes section of a drawing of a small (front) porch extension is presented in Appendix D.

In certain circumstances where drawings are not required, such as redecoration and remedial works, the relevant specification can stand-alone. Few adaptation schemes would entail this limited range of work.

For larger jobs, the specification is usually a separate document that supplements a set of drawings. Major extensions and conversion projects would require this format to amplify the information contained in the drawings. Moreover, a quantity surveyor may be required to prepare a bill of quantities, which incorporates the relevant information from the specification. As a general rule, in such a case the specification becomes excluded from the contract. However, the specification would be used on the adaptation works to indicate the location of any item of work that could be neither shown on the drawings nor indicated in the bills.

Performance specifications

These are the more modern type of specification and are output-orientated. In other words they are concerned with ends rather than means (JCT and BDP, 2001).

Performance specifications have some but limited uses in adaptation and maintenance works. This is especially true where matching like for like is a requirement of the contract.

These functional or output specifications rely on the expertise of the supplier. It means that the responsibility is shared between the specifier and contractor.

Compiling the specification

Wording

Once the client has decided to adapt his or her building, the consultant surveyor-architect is instructed to prepare a scheme, obtain both PP (unless the work is PD) and Building Consent, and invite tenders for the work.

The specification will form an integral part of the scheme design. Wherever possible it should be written in a style that is readily understood by the layperson as the contractor. In general terms, the following rules should be observed:

- The instructions must be in the imperative or commanding terms (e.g. provide and fix new skirting). To say 'A new skirting is to be fixed in …' is inappropriate and weak.
- The expression 'shall' ought to be used instead of 'will' (as the latter suggests a predictive intention). Thus 'The Contractor shall ensure that the premises are left clean and tidy on completion of the work …' is more appropriate than 'The Contractor will ensure that the premises are left clean and tidy on completion of the work …'.
- Ambiguity should be avoided when using certain terms. For example, 'Approved', if used should be followed by the words 'by the Consultant or Supervising Officer'.

Structure of a prescriptive specification

A prescriptive or traditional specification contains two main sections: 'Preliminaries' and 'The Works'. These main headings that follow are based on SMM7:

PRELIMINARIES

(a) Job Particulars:
- Project, parties and consultants.
- Description of the property.
- Drawings and other documents.

(b) Contract:
- Form, type and conditions of contract.
- Contractor's liability.
- Employer's (i.e. Client's) liability.
- Local authorities' fees and charges.
- Obligations and restrictions imposed by the employer.

(c) Works by nominated sub-contractors, goods and materials from nominated Suppliers and works by public bodies:
- Works by nominated sub-contractors.
- Goods and materials from nominated suppliers.
- Works by public bodies.
- Works by others directly engaged by the employer.

(d) General facilities and obligations:
- These include items such as staged inspections, construction, design and management (CDM) requirements.

(e) Contingencies:
- An allowance for contingent and unforeseen work is essential in any adaptation scheme. This may be as high as 10 per cent of the contract value and be expressed, for example, as follows: *The provisional sum of £5000 for contingent or unforeseen works to be expended or deducted in whole or in part as directed by the Consultant (or Supervising Officer).*

(f) Maintenance:
- Under this section the following clause could be inserted: *The Contractor shall maintain the new work executed in good order for at least 6 months after completion and shall, at his own expense, make good defects arising from faulty workmanship or defective materials as may appear during that time.*

(g) Materials and Workmanship:
- This section should contain typical clauses such as:
  The Contractor shall provide everything for the safe, proper and expeditious completion of the work including plant, ladders, materials, labour and superintendence.
  All materials and workmanship are to be the best of their respective kinds or as described herein, and shall be to the satisfaction in all respects of the client. The relevant British Standard specification for any material will be regarded as the minimum quality that will be acceptable.
  Workmanship shall comply with the relevant code/s of practice (BS numbers stated).

The works

This consists of a series of sections, preferably numbered using a decimal hierarchical notation format, covering the following:

(a) Preparation.
(b) Demolitions/Downtakings.
(c) Element by element on a chronological or sequential basis.

Proforma specifications

The National Building Specification (NBS) provides a set of model specification clauses in loose-leaf ring binder format that can be used on most construction projects. However, some of the clauses may have to be modified to suit adaptation work. The system is also available on CD-ROM format.

Procuring adaptation work

Characteristics of adaptation work

It should be apparent by now that adaptation work is not identical to that of new build. Granted, there are many similarities as regards the participants and operations involved in both these types of construction operations. Adaptations, however, usually involve fewer operatives on site at any one time and take much less time than new-build projects.

For the purposes of procurement adaptation work can be classified into two main groups: small/medium scale and medium/large scale. Although this is a crude dichotomy, most adaptation schemes fall into one of these two categories. The value, size, type and technical complexity of the adaptation scheme will determine its scale (Griffith, 1992). These and other criteria can be used to compare the general procurement

Table 11.4 Comparisons between procurement of various sizes of adaptation work

Criteria	*Medium/large*	*Small/medium*
Value	>£10 00 000	<£10 00 000
Extent	Whole building or several buildings affected.	Small-scale commercial extension. Large-scale domestic extension (i.e. over 30 m^2). Only part of one building affected (e.g. re-roofing work).
Characteristics	Complex or substantial works, involving nominated sub-contractors. Drawings, specification and bills of quantities used.	Minor works, not involving any nominated sub-contractors. Work described by drawings and/or specification or a schedule of work, but can be without bills of quantities.
Contract	Standard form usually required.	Informal agreement – but not advisable because of the difficulties in resolving disputes. Simple ad-hoc form can be used but a standard form (e.g. Minor Works Contract) is more preferable.
Examples	Major rehabilitation programme to a housing scheme (including window replacement). Multi-storey extension to an office building. Commercial/housing conversion such as a change of use of a large house or block of flats. Refurbishment scheme to a major office building (which may include overcladding/recladding and/or over-roofing).	Two-storey lateral extension to a dwelling (e.g. garage with room above). Domestic loft conversion. Window replacement to one house or several houses. Residential conversion of one dwelling into several flats. Modernization work, including wall and roof insulation, to one dwelling or a small number of dwellings.

requirements of adaptation projects. Thus, the differences between these two main groups are summarized in Table 11.4.

Unlike new build, adaptation work usually involves internal cleaning and clearing up. Debris can easily accumulate in void areas such as basements, cellars, previously unused or underused rooms and roof spaces. This may consist of builders' rubbish from past work or uncollected refuse left by previous or existing occupiers.

Debris within a building is undesirable for a number of reasons. First of all, it is a potential fire risk. Secondly, it may attract vermin by providing food or nesting material. Thirdly, rubbish may either hide a building defect, such as dampness, or act as a nutrient source for fungal attack. For example, paper and cardboard contain cellulose, which is the main constituent of timber attacked by *Serpula lacrymans*, the true dry rot fungus (Ridout, 1999).

Forms of contract

Basically, any contract is a legally binding agreement between two parties. It requires an offer and an acceptance. In the context of construction the two parties are the client (usually known as the employer in a contract) and the building contractor. But building contracts, because of their unique nature, are considered a special category of such legally binding agreements.

Building contracts for adaptation works can be classified into two groups: standard forms of contract and non-standard or ad-hoc contracts. The latter are sometimes used by large organizations such as development

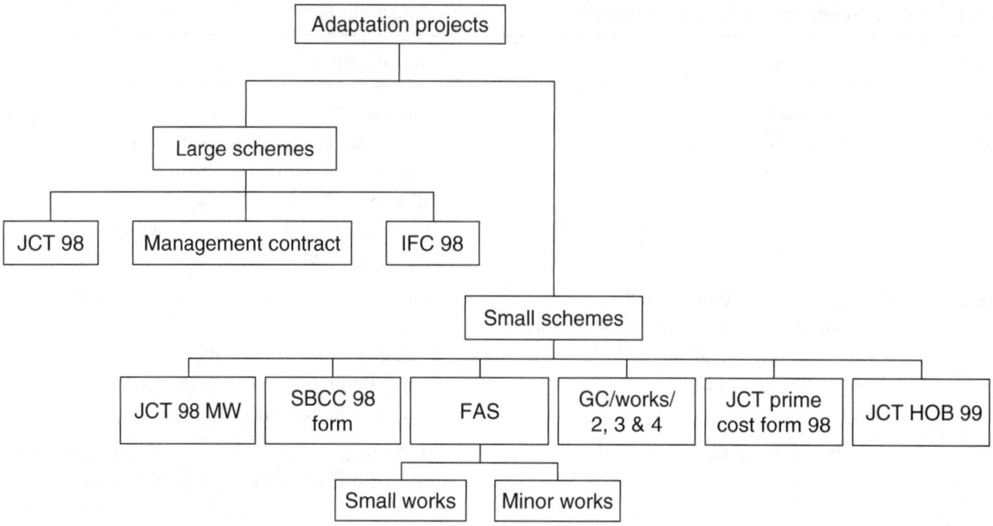

Figure 11.3 Typical forms of contract for the various sizes of adaptation projects

corporations or by individuals for undertaking specialized small-scale work of a unique nature. The clauses are normally tailored to reflect the client's particular building characteristics.

More usually, however, the parties to a building adaptation project will use a standard form of contract. There are numerous standard forms of building contract, each of which is suited to specific sizes and types of construction work (see Figure 11.3).

Standard forms of contract are produced on a pro-forma basis by a variety of organizations. The most commonly used standard forms of contract in Britain are the various types supplied by the Joint Contracts Tribunal Ltd (JCT). This is a limited company formed in the early 1900s. It is now sponsored by several of the main professional institutions and trade organizations in construction.

Some adaptation schemes such as large commercial extensions can be procured using a management contract because of the predominance of new-build work in the project. Many other schemes, however, particularly the smaller ones would be best suited to one of the various minor works contracts (Griffith, 1992). This of course would depend on the estimated total value of the contract and nature of the work involved (see Table 11.5).

The main types of standard forms of contract that can be used in adaptation work are:

- JCT Building Contract for Home Owner/Occupier 1999 HOB99 Without Consultant: According to the RICS this is 'the first of its kind, designed for use by homeowners and occupiers covering home improvement, extensions and general repairs. The contract is also designed for surveyors to use as a sales tool and as a user-friendly, plain-English contract to ensure confidence from their customers'.
- JCT 98/JCT 80: May be applicable in some instances for very large adaptation schemes such as large extensions or major refurbishment programmes to commercial properties.
- JCT Prime Cost Form 98: May be suitable for very specialized refurbishment schemes.
- JCT Management Contract 98: Suitable where the client wishes to take greater control and hence more of the risk by appointing a management contractor to provide the service of management for an agreed fee.
- Intermediate Form of Contract (IFC 98): Applicable for large or small adaptation schemes of simple content (e.g. major window replacement programme).
- JCT 98 Agreement Minor Works 98: Suitable for small–medium-size projects up to the value of £70 000, mainly in the private sector.
- Scottish Building Contract Committee (SBCC) Form for Minor Works: Suitable for small building projects in Scotland (because of the requirements of Scots Law).

Table 11.5 Typical risk bearers in adaptation projects

Risk bearer	*Examples of risks and hazards*
Client	Loss of use or interruption of use of building due to delays or overruns in the contract period. Increased repair costs or insurance premiums due to accidental or criminal damage. Lack of tenants or buyers for either the adapted building or units within the adapted building.
Consultants	Negligence suit for alleged breach of professional duty – e.g., failure to inspect, design, specify, supervise or communicate properly. Injured whilst on site as a result of exposure to some of the same hazards as operatives. Loss or devalue of reputation as a result of bad publicity following a botched refurbishment scheme. Underestimation of fees or time required. Failure to obtain the necessary statutory approvals.
Contractor	Loss of production due to delays or overruns in the contract period. Disruption of work due to accidents or other safety scares on site. Failure to hire appropriately experienced/skilled operatives. Failure to obtain materials of adequate quality or at the correct time. Under-pricing work. Underestimating the time taken to complete work.
Occupiers and the public	Falls from landings or other raised platforms with open or inadequately secured balustrades. Colliding with or bumping into unmarked and/or unprotected scaffolding. Collapse of crane or scaffold. Standing on or being ripped by protruding nails. Struck by dropped or falling objects from cranes, trucks or scaffolds. Tripping or slipping on irregular, untidy or inadequately covered surfaces. Exposures to pollutants such as hazardous chemicals, high levels of noise, smoke or dust. Disturbance of or damage to asbestos. Inhalation of dust, gases and other noxious substances.
Operatives	Falls from roofs, landings, balconies, scaffolding or other high point. Struck by dropped/falling objects. Collision with transport vehicles (e.g., forklift/dumper trucks). Tripping or slipping on irregular, untidy or inadequately cleared/covered surfaces. Collapse of structural elements or temporary supports (e.g. reckless formation of new opening in an existing wall or illicit removal of load-bearing wall). Electrocution from faulty equipment/wiring. Exposures to pollutants such as high levels of noise, smoke, and hazardous chemicals. Disturbance of or damage to asbestos. Improper handling of materials resulting in back injuries. Cuts, punctures and other injuries from hazardous working operations using saws, electrical drills and saws, generators and electrical other equipment, nail guns, etc. Burns from hot working processes (e.g. welding, flame cutting, laying asphalt, torching on bituminous felt, hot air paint-stripping gun, etc.). Standing on sharp objects (e.g. protruding nail, glass or other sharp implement). Dust inhalation from breaking, cutting, grinding or drilling cementitious and other particulate materials.
Building	Accidental damage to the structure/fabric during adaptation work (e.g. inadvertent or careless removal of a load-bearing wall causing structural movement and cracking). Impact damage from crane or materials being lifted during crane operation. Flooding due to burst pipes (particularly in winter), leaking services or defective roof coverings/drainage. Water ingress can cause extensive damage to valuable contents and could trigger fungal attack. Fire resulting from carelessly discarded cigarette, electrical fault or hot working process (e.g. improper or careless use of blow torches, which has caused major fire outbreaks in conservation works to historic buildings).

- Faculty of Architects and Surveyors (FAS) Small Works Contract: Applicable for use in both public and private sectors to works described by drawings, specification, but not using bills of quantities.
- FAS Minor Works Contract: Suitable for works carried out on a lump sum basis, for use in both public and private sectors.
- General Conditions of Government Contract for Building and Civil Engineering Works (GC/Works/2, 3 & 4): Suitable for small building and adaptation projects in the public sector.
- JCT Agreement for Renovation Grant Works: Suitable for use in renovation work to dwellings where a grant is receivable under the Housing Act 1974.
- JCT Form for Building Works of a Jobbing Character: Suitable for maintenance and repair works.

Developments in contract law

Contracts (Rights of Third Parties) Act 1999

The Act came into force on 11 November 1999. It applies to all contracts entered into from 11 May 2000.

This new Act reforms the rule of privity of contract, where a person can enforce a contract only if he is party to it. It may also apply to contracts entered into between 11 November 1999 and 1 May 2000 where the contract provides expressly for it to do so.

The Housing Grants, Construction and Regeneration Act 1996

The following main Parts of this Act impinge on adaptation work:

- Part 1: Covers grants for renewal of private sector housing, including repair schemes, group repair schemes, home repair assistance.
- Part II: Concerns construction contracts including the right to refer disputes to adjudication and payment conditions.
- Part III: Amends the law relating to architects' registration.
- Part IV: Concerns grants, etc. for regeneration.
- Part V: Is a miscellany of general provisions including some home energy efficiency schemes.

The Construction Act 1995

This Act came into force in May 1998. It introduced the concept of adjudication in construction contracts. One of its main features is that it provides a rapid dispute-resolution system that is aimed at making payments more equitable.

Tendering

Selection criteria

This part of the procurement process involves determining the number and identity of the contractors invited to price the proposed adaptation scheme. The selection criteria is likely to comprise some, if not all of the following:

- Reputation: Has the contractor a good track record in undertaking the kind of adaptation work involved?
- Reliability: Can the contractor deliver the project on time and within budget?
- Past experience: Does the contractor's past experience equip them for the proposed project?
- Capacity: Has the contractor the financial and operational capacity to undertake the project?
- Availability: Can the contractor commence the work when required by the client?
- Profitability or status: Is the company in a sound financial position?

- Value for money: Can the contractor deliver good value for money?
- Indemnity: Does the contractor have the adequate level of insurance?
- Guarantees: Is the quality of the contractor's materials/installations guaranteed?
- Trade membership: Is the contractor a member of an appropriate trade organization?

Tendering procedure

A single-stage selective tendering process would be appropriate for most types of adaptation project. Generally, the Code of Procedure for Single Stage Selective Tendering published by the National Joint Consultative Committee for Building should be adopted.

Assuming no tenders are subsequently qualified or withdrawn, the following procedure for an adaptation project could be adopted:

- Make preliminary enquiry – to determine the contractors' interest.
- Dispatch tender documents to the selected contractors.
- Select and adopt an appropriate standard form of contract.
- Allow at least four working weeks for the tender period.
- Open submitted tenders as soon as possible after the deadline and evaluate them on the basis of price and quality of submission.
- Make recommendation to decision-makers on which tender to select.
- Select successful tender on the basis of price and quality.
- Accept successful tenderer's offer in writing within a prescribed period.
- Once the successful party has been notified and confirms their intention to proceed, inform unsuccessful tenderers as soon as possible, preferably by through a debriefing meeting.

Debriefing

Unsuccessful tenderers for large-scale adaptation work may warrant having a debriefing meeting or being sent a brief explanatory letter. The project manager, who would explain to the unsuccessful contractors on an individual basis why they were not selected, would normally undertake such a task. This helps to ensure greater accountability, openness and transparency in the selection process, which are especially important in public sector projects.

Risk in adaptation work

Types of risks

In the context of building adaptation, risks are predominantly either technical or commercial in nature. These can each be further divided into five main contrasting groups, some of which overlap: Fundamental; Pure or Speculative; Particular; Downside or Upside; and Controllable or Uncontrollable.

Fundamental risks

These are risks such as war damage, nuclear pollution and sonic bangs (Latham, 1994). In relative terms their impact on adaptation projects is very rare.

Pure/speculative risks

Pure risk usually arises from the possibility of an accident or technical failure, such as damage caused by fire, flood or storm. It is thus one of the main types of risk relevant to an adaptation project. It should be

Table 11.6 Risk schedule estimating maximum risk allowance for a large public building refurbishment project

Refurbishment option: north/south phasing		*Average risk*		
Risk element (1)	*Types of risk (2)*	*Base value of risk element (3)*	*Probability factor (4)*	*Assessment (5)*
Rot repairs	V	25 000	0.95	23 750
Additional masonry repairs	V	30 000	0.50	15 000
Additional roof repairs	V	25 000	0.90	22 500
Asbestos abatement	V	50 000	0.50	25 000
Structural alterations	V	300 000	0.50	150 000
Fire protection	V	20 000	0.75	15 000
Fire officer requirements	V	100 000	0.40	40 000
Safety officer requirements	V	25 000	0.50	12 500
Funding body's requirements	V	10 000	0.50	5 000
Restrictive working methods	V	40 000	0.75	30 000
Penalty clause	F	10 000	0.25	2 500
Average risk allowance (6) £				341 250

Notes
1. Determined by the nature of the adaptation scheme and type of building.
2. F = Fixed (e.g. penalty clause), V = Variable (e.g. dry rot repairs).
3. Based on approximate cost of remedial works on similar size buildings.
4. A high probability is 95 per cent; this is represented as a probability factor of 0.95.
5. This is simply (3) × (4).
6. This figure can form the basis of the minimum contingency sum for the contract.

contrasted with 'speculative risk', which is concerned with the possibility of loss or gain. Activities such as gambling, oil exploration and building conversions for an unknown user are typical examples of speculative risk. However, other types of speculative risk such as inflation, and wage increases, changes in taxes, shortages of materials and skilled labour can impinge on adaptation work.

Particular risks

These are the risks specifically associated with the project. For example, subsidence, vibration, removal of support can easily occur in an adaptation scheme as well as in a new-build project. A fire outbreak is a common risk within the building during adaptation. The collapse of temporary supports such as dead shoring or scaffolding is another major risk area. These risks may be either fixed or variable (see Table 11.6).

Downside/upside risks

Risk is normally associated with negative connotations and thus most are regarded as 'downside risks'. These relate to where things may turn out worse than originally planned. They concern events that are connected to the concept of loss only. Risk management, for example, is concerned with reducing the possibility or controlling the effects of downside risks. The opposite, 'upside risks', where things may turn out better than planned are not always allowed for in adaptation work. They are more readily associated with activities such as gambling.

Controllable/uncontrollable risks

Controllable risks, as the term suggests, are those risks which can be managed and controlled. Moreover, their likelihood of occurrence in the form of damage or errors associated with adaptation work involving the contractors, the professional advisers and the client, can be estimated. In contrast, uncontrollable risks are outwith the control of these participants. Adverse weather conditions or occurrences such as flooding caused by excessive rainfall or burst water mains are examples of uncontrollable risks.

Controllable risks are unlikely to prove troublesome in conversion projects. Incidents such as flooding, on the other hand, can cause extensive damage during and after adaptation work.

Typical adaptation risks

General perils of adaptation

The project manager of any adaptation scheme should take cognizance of the common problems associated with this type of work to avoid financial as well operational disasters. As highlighted in a short article in the journal *Building* (November 2001), refurbishment (i.e. adaptation) projects are notoriously risky. It listed some of the main problems in this regard:

- The building can be different to the information on the original drawings.
- Access can be difficult, particularly for large pieces of plant and equipment.
- Non-standard sizes and imperial measurements mean special parts are needed.
- Even comparatively recent buildings are not suitable for modern requirements, with low floor-to-ceiling heights and lack of space for a raised floor.
- Poor quality construction and use materials below modern standards.
- Structural capacity is not suitable for modern equipment.
- Difficult to bring the building up to thermal and fire safety standards.
- Damp, insect and fungal attack.
- Listed buildings, where even the floor joists could be untouchable.

The other main problems are described in more detail as follows:

Leaks

Opening up the external fabric, particularly in the roof, to install new windows, chimneys or cladding, will inevitably make the building vulnerable to water penetration in periods of inclement weather. Precautions such as timing any opening-up work to coincide with periods of fair weather should be made. However, if this is impracticable, tarpaulins can be used to cover the exposed areas, but the risk of some rainwater penetration in such instances is still high.

Damage and mess

Debris and dirt are often by-products of adaptation work. They are awkward to clean up as well as pose health risks to workers. Adequate preparation and protection of adjacent areas by covering them with heavy-duty dustsheets may be needed to minimize such problems.

Temporary enclosure during breaking through work (see Chapter 5) can reduce the transmission of dust and dirt during such operations. It also provides a suitable safety zone provided it is well signposted.

Interruption or disruption of core business

By its very nature, adaptation work on an occupied or unoccupied building can easily interfere with the core business of the property. If the building has to be decanted during the bulk of the work, the potential loss of earnings, especially if there are delays, must be ascertained. Even when the business operations need to be continued within the building because of economic or other pressures, a degree of disruption for both these operations and adaptation work is inevitable. The aim therefore is to keep 'downtime' in the client's business activities in the building being adapted to a minimum.

Health and safety of occupiers and the public

Any building that is unsafe for those undertaking the adaptation work will be even more dangerous for its occupants and members of the public who may use or pass by it. In some respects, the required standards may be higher than for site operatives because of occupiers lack of building safety knowledge.

Scaffolding is frequently the most prominent feature or indicator of a building undergoing adaptation. It is important that uprights and projecting members at ground level are wrapped in pre-formed resilient fabric covered padding to protect the public and operatives from accidentally bumping into protrusions (see Figure 11.7). This is preferable to makeshift jute wrapping, which is often used on scaffolding in adaptation projects. Whatever type of wrapping is used, though, it should be made conspicuous by highlighting it with red and white striped tape.

Health and safety of construction operatives

Reducing the number of injuries and fatalities in the construction industry is vital if it is to achieve the decent and safe working conditions that the Egan Report (1998) demands. Regrettably, the building industry still has one of the highest records for accidents.

Working on fragile roof coverings, for example, is a high-risk activity for operatives and surveyors. Precautions such as harnesses, crawlboards or safety nets must be provided to prevent them falling through such vulnerable parts of an existing building.

Dust inhalation is one of the most serious health risks on any adaptation or new-build project. Appropriate mouth masks, safety goggles and gloves are required for operatives undertaking this work. The presence of cement in the air generates dust particles usually not much larger than 2 μm in size. Particles less than 10 μm in size can pass through the larynx and penetrate deep into the lung and trigger chronic respiratory ailments. Silica particles from stone cutting and grinding, etc. are especially hazardous to health because of their carcinogenic properties (Curwell and March, 1986).

Permit to work

This is an important requirement to limit work in confined or other dangerous areas. In some situations it is not permissible to work in certain locations or undertake potentially hazardous operations (e.g. using a blow torch or other hot working process). This is especially the case in adaptation works to historic buildings.

Fire

Some adaptation works involve hot working processes such as plumbing, laying asphalt and torching of roofing felt. Flames or sparks from these activities are especially susceptible to causing an outbreak of fire in buildings. Indeed the use of blowtorches in the renovation of old properties is now seriously curtailed because of the fire risk associated with this type of operation. Some of the risks associated with plumbing, though, can be minimized by off-site fabrication. A 'hot work' permit would normally be required in such circumstances.

It would be advisable for medium- to large-size adaptation projects (i.e. those costing over, say, £100 000) to have a fire safety plan. The Loss Prevention Council (2000) in its Fire Code recommends the formulation of such a plan. This requires that:

> All parties involved must work together to ensure that adequate detection and prevention measures are incorporated during the contract planning stage; and, the work on site is undertaken to the highest standard of fire safety thereby affording the maximum level of protection to the building and its occupants (i.e. employees, general public, contraction workforce, etc.).

The assessment of fire risks associated with buildings before or after adaptation is dealt with in Chapter 2.

Contingent or unforeseen works

Adaptation projects can generate considerably more headaches for consultants and contractors than new build. Hidden defects or unpredictable problems often occur during the contract – such as corrosion behind a stone façade or dry rot behind wall panelling. These unforeseen works can inflate the construction costs well above the original estimate.

In most major rehabilitation or renovation schemes a great deal of cutting out or opening up is usually required. This work can easily expose hidden defects such as dampness and fungal attack, its main adverse effect on timber. Underestimating the extent of these and other defects such as concrete deterioration can easily be made.

Remedial treatment of fungal attack, for example, is probably the most troublesome and expensive form of additional unforeseen work that can be encountered in an existing building of load-bearing masonry construction. In particular dry rot (*S. lacrymans*) is the most pernicious form of timber decay known to man (Ridout, 1999). It can spread to other parts of the building beyond the source of infection. The rate of spread in search of more timber and other cellulose-based materials to consume is dependent upon the prevailing environmental conditions. Generally, though, dry rot spreads at a rate of about 1 m a year – although 4 m a year is possible if the environmental conditions are right (Hollis, 2005).

A thorough pre-adaptation inspection of the building should ascertain the extent of any such defects. Unless an element of invasive survey work has been done on the building, however, there is no guarantee that all of the inherent or concealed defects have been identified. Built-in or hidden timbers such as 'safe lintels', joists ends and rafter feet are highly prone to such problems. The risk schedule in Table 11.6 shows how the potential cost of these unforeseen works can be assessed.

Problems with uncooperative occupiers

Not all occupiers welcome the refurbishment of their property. This is particularly the case in housing rehabilitation schemes. The journal *Building* (2002) reported that the refurbishment of the first of several block of flats in London was a disaster, partly because of the contractor's failure to deal with some awkward tenants who refused to co-operate with the refurbishment works. The second phase of the refurbishment programme in the same estate was a success because of the lessons learned from the first contract.

A proactive approach is needed to avoid or at least minimize problems with awkward tenants and consultation will go a long way to reassure occupiers of the property owners intentions.

Other potential risk factors

Obtaining grants

As indicated in Chapter 2 some grant aid may be secured for certain types of adaptation work. Obtaining partial funding from a grant awarding body is dependent on a number of factors. Most grant awarding bodies have limited funds. There may be problems in processing the application, which could delay or postpone the grant award. Moreover, such grants normally are only awarded on certain conditions, such as using only recognized repair methods, appointing only 'approved' or 'accredited' consultants, etc.

Programming

The timing of an adaptation project can be critical. This is particularly the case where the building is occupied during the adaptation work or can only be vacant for a limited period. Failure to meet any deadlines may result in more delays.

Contractual–financial

The form of contract chosen for an adaptation project will reflect the allocation of financial risk between the parties. The nature and scale of the work being undertaken will therefore have a bearing on this. In this

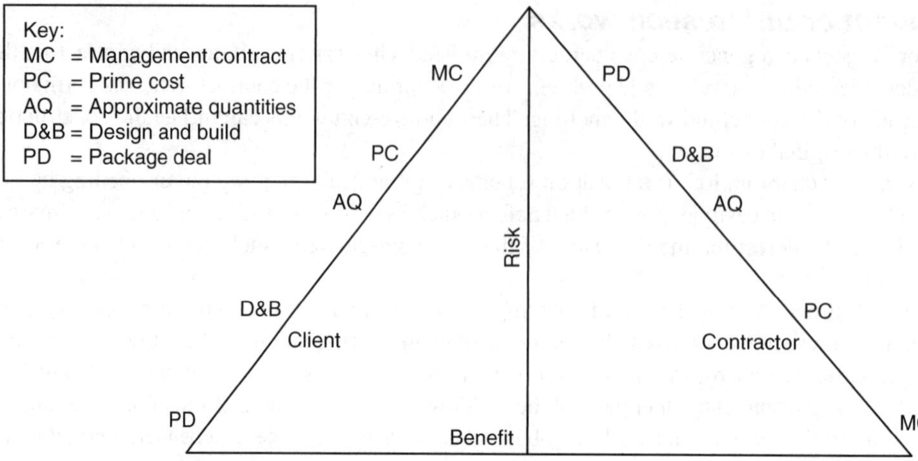

Figure 11.4 Risk–benefit triangle of building contracts

regard, the risk weighting between client and contractor is generally the same for adaptation work as it is with new build.

Selecting an inappropriate contract can result in excessive claims for additional payments. There is also the danger of the client or contractor going bankrupt. Moreover, an unscrupulous main contractor might attempt to undertake illicit subcontracting of the work.

Figure 11.4 illustrates the risk distribution of the main types of contract for adaptation work. It shows, for example, that a Management Contract (where a main contractor, using works sub-contractors, undertakes the work for an agreed lump sum) yields maximum benefit and minimal (financial) risk for the contractor and minimal benefit and maximum risk to the client.

Statutory requirements

The need to satisfy statutory requirements such as those imposed by the Disability Discrimination Act 1995 is a potentially extensive financial risk for many building owners. The costs of adapting a building to comply with this Act could run into tens of thousands of pounds for some service providers. A dispute over the interpretation of the regulations could delay the work.

Deleterious materials

As shown in the previous chapter the presence of hazardous materials in a building could increase health as well as financial risks for the occupier. Removing or treating such materials could be costly, disruptive and time-consuming.

Pest infestation

Any building that has lain empty for more than a few months is susceptible to infestation by vermin and other pests. The health and comfort of operatives undertaking the adaptation work may be adversely affected by such infestation. As indicated in Chapter 2 a specialist pest control contractor may be required to eradicate this problem.

Security

Undertaking work on an existing building often means that the occupiers have to work alongside the contractor's operatives. This can result in security problems for both parties. Occupiers' belongings or workers' tools can be stolen.

Risk management

Safety is a major concern for all those that are involved in construction. Accidents not only cause financial loss resulting from reduced productivity and damage to equipment. They also cause injury to human beings or loss of life.

Like any other construction project, adaptation work has more than its fair share of risks. That is why some form of risk management is needed to minimize or control potential problems arising from unforeseen works and occupation during the works, all which are common in building adaptation schemes.

Risk management is a generic term that covers a broad subject area. As with adaptation there is no single, universally accepted 'correct' definition of the term. However, according to one authoritative source (CUP, 1997), risk management is defined as 'the planned and systematic process of identification, assessment, monitoring and control of risk'.

Risk management is relatively new in construction, having emerged in the early 1990s (Flanagan and Norman, 1993; Raftery, 1994). The pressures on the building industry to improve its overall performance (e.g. as a result of the Latham Report 1994 and Egan Report 1998) have prompted more attention to risk management because of the cost implications of ignoring the financial and physical hazards in construction.

Moreover, competition from foreign contractors encouraged by the dismantling of international trade barriers, better transportation and more efficient communications have all contributed to the increased use of risk and value management in construction. Some have suggested that construction is an industry that is subject to more risk and uncertainty than any other (Flanagan and Norman, 1993). Naturally, industries such as mining and deep-sea oil explorations are just as onerous in this regard.

Risk analysis

Preamble

Risk analysis is the branch of risk management that concentrates on the measurement and assessment of hazards and is, of course, a vital component of any health and safety programme. It is a discipline that can be applied to any scheme, project or endeavour with potential problems, dilemmas, dangers and hazards. For example, in offshore engineering schemes, construction projects and manufacturing risk assessment is now standard practice, if not a mandatory requirement.

Before considering risk assessment in detail, as it relates to building adaptation, it would be pertinent to clarify the meaning of the terms and parameters used. According to a Royal Society study group report (1992), 'a general concept of risk is the chance, in quantitative terms, of a defined hazard occurring. … Risk is a combination of the probability, or frequency, of occurrence of a defined hazard and the magnitude of the consequences of occurrence'. Risk, therefore, is the likelihood or probability of some adverse event occurring during a specific time frame or results from a particular challenge (Royal Society, 1983).

Risk and uncertainty

Closely related to risk but not necessarily synonymous with it is uncertainty. Risk is defined in the Oxford English Dictionary as a verb as 'to expose to the chance of injury or loss'. It is something that is generally considered to be objective and thus quantifiable. Uncertainty, on the other hand, is more subjective but not quantifiable. It exists where there is no previous data or precedent or there is ignorance relating to the situation being assessed (Flanagan and Norman, 1993). Some writers have, though, asserted that in certain fields the distinction between risk and uncertainty serves little purpose (Raftery, 1994). This is probably because these two terms are essentially at different ends of the same spectrum (see Figure 11.5). However, the essential difference between risk and uncertainty is that there are consequences with the former but not with the latter.

Figure 11.5 The risk–uncertainty spectrum (Raftery, 1994)

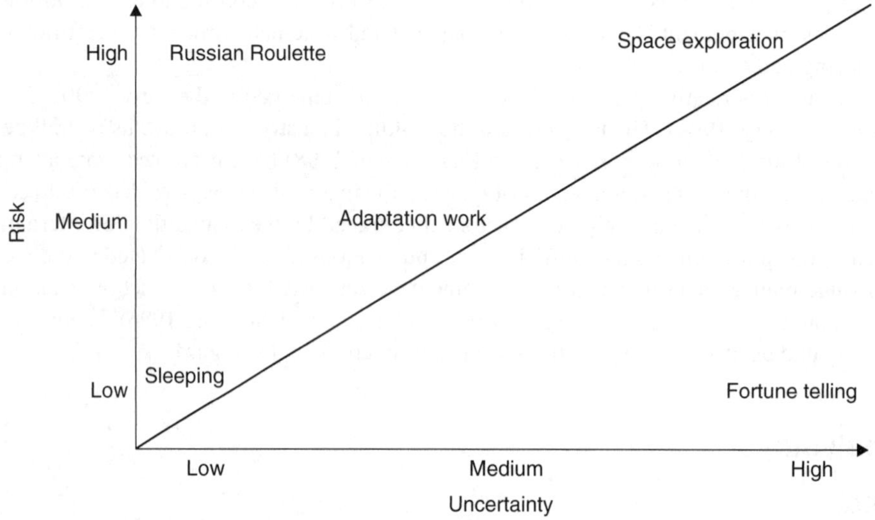

Figure 11.6 Risk and uncertainty relationship

The relationship between risk and uncertainty is not immediately apparent. Figure 11.6 attempts to illustrate this relationship more clearly. Not all situations that have a high degree of uncertainty will entail a high level of risk, and vice versa. For example, fortune telling has a high degree of uncertainty but involves little if any risk. In contrast, Russian Roulette, because of its limited possibilities (i.e. one gunshot out of a maximum of six may be fatal), is a highly dangerous and hence very risky activity. It is hypothesized here that adaptation work occupies the central location of the diagram because in relative terms it has a medium level of risk and uncertainty.

According to Smith (1994) risk is a term that actually covers two parameters: frequency analysis and consequence analysis.

Frequency analysis

This relates to the probability (or rate) of a particular event. In the case of an adaptation scheme, the probability of, say, discovering a major unforeseen defect might be one every year. Alternatively, the probability of abating a problem such as asbestos in the building being adapted may be as high as 0.95 (i.e. 95%).

Consequence analysis

The scale of consequence (perhaps expressed in degree of damage or fatalities) also has to be considered. For adaptation work it could range from a small patch repair of a defective element to partial or even complete renewal of the defective section of a building. Some cost figure may need to be applied to the risk (see below).

Policy

It is realistic to acknowledge that construction risks can never be entirely eliminated. All of the legislation relating to health and safety, therefore, has the primary objective of reducing and controlling if not avoiding risks to those at and around work. Companies will have different safety priorities depending on the nature and hazard level of their business. Those priorities are normally articulated in a set of safety objectives.

A contractor's site health and safety policy will usually state the safety objectives, such as:

- Prevent fatalities during the project.
- Reduce injuries and serious accidents to a minimum.
- Prevent damage to the structure and fabric of the building.
- Maintain good housekeeping on the site – keeping the building and its surroundings clean and tidy.
- Promote a health and safety culture among staff, operatives and any visitors to the site.
- All personnel must wear hard hats and any other personal protective equipment (PPE) where appropriate whilst on site.

Risk identification

Adaptation work involves business risks, which are primarily financial in nature. Minimizing underestimates of construction costs and reducing mistakes in tendering documents can reduce these risks.

The nature or type of adaptation project will determine the technical hazards and risks associated with the work activities. For example, an over-roofing project will involve operatives working from at least one storey in height. Materials and equipment will have to be delivered to this high location.

Falls and dropped objects from heights over 1 m are amongst the most common and predictable accidents that occur in building sites (see Table 11.5).

Risk evaluation

Either a qualitative or quantitative approach can be used to assess the significance of the risks identified. The risks could be divided into two main groups: institutional risks (organization) and operational risks (adaptation).

A risk profile could be compiled to rate the risks on a 'high, 'medium' or 'low' basis. For example, if the building has a high fire load (i.e. extensive amounts of material such as timber and flammable material to fuel a potential fire), the risk would be rated as high. Special precautions over and above standard safety measures would be required where such a risk has been identified. These may include the provision of additional fire detectors and extinguishers.

Table 11.6 shows how the anticipated risks in the refurbishment of a large public building can be quantified. The total figure is then used as the basis of the contingency sum for the contract.

Risk control

Insurance

Recent cases of accidental damage during residential adaptation work have highlighted the need for adequate levels of property insurance. The standard JCT contract requires the contractor to insure against collateral damage to adjacent properties. It does not, however, insure the existing dwelling itself from damage caused by the adaptation works. In such cases, therefore, the client should take out additional insurance cover via a special policy covering non-negligent damage for the period of the contract, which is known as Clause 21.2.1.

Site precautions

Once the likely hazards have been evaluated, the necessary measures to reduce if not eliminate the risks can be planned. Under the CDM Regulations a health and safety plan must be formulated at the tender stage and finalized at the construction phase. The planning supervisor must prepare this.

Project managers involved in large projects such as the Channel Tunnel found that a safe site is an efficient site. This also applies to adaptation as well new-build schemes.

In larger refurbishment and new-build schemes it is mandatory for all those on site to wear the following three pieces of personal protective gear:

1. Hard hat
2. High visibility vest
3. Safety boots/shoes.

Further typical safety precautions and facilities on a building adaptation project are:

- Adequate evacuation procedures and means of escape clearly identified. Dead ends when installing flammable materials, such as floor tiles, should be avoided.
- Dust control measures (see below).
- Strict controls and close supervision of hot working and other activities involving flammable processes. Appropriate portable fire extinguishers should be readily available nearby such work.
- Flammable materials and low-pressure gas cylinders should be kept in secure, adequately ventilated enclosures or stores.
- Timing hazardous operations (such as asbestos removal, hot or noisy and dusty work) outwith core business hours to minimize disruption and nuisance.
- Properly rated and earthed electrical equipment, with circuit breakers. A qualified electrician must check these at least once a year.
- Create separate access routes for occupiers and the public so that they do not intermingle with site operatives.
- Padding and red–white striped bands with conspicuous warning stickers should be fitted around scaffolding poles and at other dangerous or vulnerable installations.
- Other adequate personal protective equipment (PPE) over and above the three items referred to above must be worn by all operatives when undertaking hazardous operations such as cutting, drilling, scraping, working with toxic substances etc. – e.g., goggles or safety glasses/visors, face masks, safety gloves, overalls, etc.
- Adequate harnesses and crawlboards to protect operatives working on pitched roofs, particularly on buildings with fragile coverings.
- All operatives and staff on site are adequately briefed and regularly updated as to safety precautions and emergency procedures. To facilitate this large, durable safety notices should be placed at conspicuous points in and around the building.
- Waste chutes constructed from articulated plastic cone-pipes to transfer debris safely down scaffolding to a skip (see Figure 11.7).
- The building being adapted must be kept clean and tidy on a daily basis. Rubbish and construction waste should be removed and deposited in a suitable skip. Any rubbish containers such as skips need to be emptied regularly to an approved dump.

Risk monitoring

During the undertaking of the adaptation work, regular (daily or weekly) checks must be made by the Planning Supervisor, to ensure that the safety plan is being adhered to for compliance with the CDM Regulations. After the works are completed, a proactive (i.e. planned preventative) maintenance system, which would incorporate a health and safety file under these regulations, would be required.

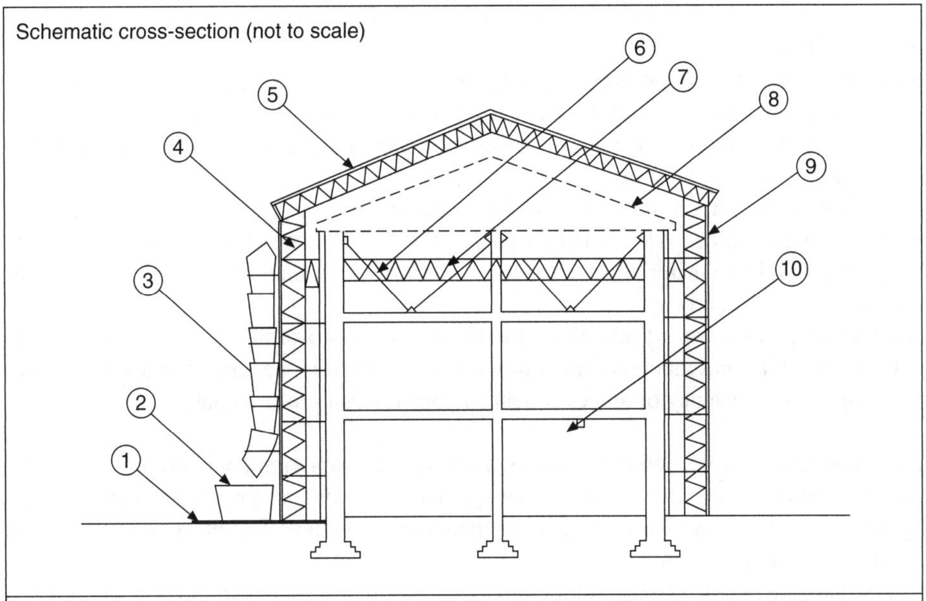

Notes
1. Main road at front of building.
2. Refuse skip.
3. Refuse chute formed using plastic pipes to discharge waste into skip.
4. Aluminium lattice framework to support temporary roof.
5. Aluminium lattice trusses covered with heavy-gauge polythene sheeting reinforced with polyester grid.
6. Adjustable raking props to provide temporary support to wallhead.
7. Temporary flying shore beam to provide lateral restrain to vertical lattice framework.
8. Proposed new roofline to match original.
9. Independent scaffolding.
10. Additional support beam to strengthen existing floor.

Figure 11.7 Temporary enclosure over a major restoration project

Audit, review and feedback

Having a safety plan is all very well, but if it is not checked, failings in the system may emerge or be repeated. The regular 'site' meetings would be the ideal forum for reporting on safety matters during the adaptation contract.

Managing adaptation projects

Preamble

Despite their differences adaptation work and new-build schemes share many similarities. These two forms of construction work are procured using building contractors employing tradesmen, as well as semi-skilled and un-skilled operatives. Both involve the expertise of consultant representatives such as surveyors, architects, engineers and other professionals. A project manager may be appointed for a large adaptation project such as a major refurbishment scheme to an office building involving millions of pounds of expenditure.

There are, however, a number of important differences between an adaptation scheme and a new-build project to properties of similar size. The former is normally of much shorter duration than the latter. It is on a smaller scale both in terms of cost and extent of contractor's input. Much more heavy plant and equipment will be used in a new-build scheme.

Accordingly, it is vital that those involved in the adaptation project have some experience in dealing with work to existing buildings. This is vital if problems such as inherent defects (such as dampness and fungal decay) and traditional materials (such as lime mortar) and problems in working in an existing building are to be dealt with properly.

Managing an adaptation project involves a variety of tasks to co-ordinate and control a wide range of activities. It consists of a combination of the following overlapping management functions, many of which should be within the competency of any experienced construction professional:

- Contract management: Selection and administration of the adaptation contract.
- Design management: This deals with the management of technical aspects of the project. It includes 'configuration management', which deals with changes to drawings. Technical audits of key details also forms part of 'design management'.
- Financial management: Identifying sources and types of finance. Also included under this heading are pricing, estimating, cost planning and cost control of adaptation work.
- Health and safety management: Formulation of safety policy and implementation of safety plan for compliance with CDM Regulations (see below).
- Operational management: This is concerned primarily with planning and control of operations on site, and uses specific techniques such as business continuity planning (BCP), just-in-time (JIT) and total quality management (TQM). It also includes 'interface management' (co-ordination of roles within an organization) as well as 'systems management' and 'programme management'.
- Project management: A general term that embraces many of the management functions listed.
- Resource management: This comprises 'personnel management', 'reliability management' and 'supply chain management'. It could also include 'conflict management'.
- Risk management: This involves the implementation and administration of a methodology on control and minimization of risk. It can include BCP, which may help in determining a suitable disaster recovery strategy to ensure minimal interruption to normal business operations in the event of a major accident (e.g. fire, flooding).
- Time management: This involves the application of specialist techniques such as JIT. 'Change management' can be included under this heading.
- Value management: Comprises value planning, value engineering and value analysis functions, to improve business efficiency. It is a structured method of eliminating waste from the brief and from the design before the contract is signed. It involves the functional analysis of proposed components and elements (Kelly and Male, 1993).

Security

The key issues of security and safety are, of course, paramount in any refurbishment scheme of a building whether occupied or empty. Security in adaptation schemes is important for a variety of reasons:

- to minimize theft of building materials, components, plant and equipment;
- to reduce breaches of security so that unauthorized personnel are prevented from entering the refurbished building; and

- to prevent non-essential personnel from entering the work-zone so that they do not endanger either themselves or others (and thus risk a negligence claim on the owners).

For larger schemes, this may require a permanent security guard on the premises during the adaptation scheme.

Benchmarking

Definition

Benchmarking is the process of comparing and measuring an organization (or an asset) with leaders in the field to gain information that will help it to take action to improve performance. It is a useful management tool for determining the overall level of performance of an asset against a known best standard. It can help construction firms to understand their performance measures up to their competitors' and force improvement up to the best in the class standard.

The benchmarking process for an adaptation scheme

- Identify and measure the gap. To what extent does the building being adapted compare against similar modern or recently adapted properties?
- Learn what makes the difference in performance (and instigate an action plan to close the gap). Compare the technical performance of the building with a similar property that is considered one of the best in its class.
- Measure success in closing the gap. Undertake a Post-adaptation Evaluation (PAE) of the adapted building (see below).
- Keep comparisons going for continuous improvement. Monitor performance regularly through a suitable aftercare system (see below).

Functional benchmarking

Generally the largest factor that forces up the cost of a commercial construction project, whether involving new build or adaptation, is the space in a building. Functional benchmarking is a technique that can help ensure that the correct area is provided for a specific function. If too little space is provided in an adaptation scheme, the property will be unable to accommodate the function for which it was designed. This could also lead to overcrowding. Too much space in the adapted building results in unnecessary capital expenditure and possibly underused areas.

Space allocation for a current adaptation project, therefore, can be compared with similar buildings or past projects. A functional benchmarking exercise may help a client decide that it needs, for example, less circulation space.

The four main phases in benchmarking

1. Planning:
   - Identify Benchmark Issues: Cost, quality, space, value and risk?
   - Identify Critical Success Factors: Which key performance indicator (KPI) is a priority?
   - Determine Data Collection Method: Condition or energy appraisal survey; evaluation of similar schemes using an adaptation analysis matrix as shown in Table 2.4; functional benchmarking exercise.
2. Data Collection: By a surveyor or other building professional, using questionnaire or other site survey technique.
3. Analysis: Assess the results to determine the extent of the differences, if any, between the building being adapted and the benchmark property.

4. Action: Measures to bring the building being adapted up to the standard of the 'best in the class' building. This could, for example, involve increasing the thermal performance of the property's roof and walls, or decreasing the circulation space to make room for more office accommodation.

CDM regulations

As seen in the section on risk assessment, construction is one of the most dangerous industries in the country. Adaptation work is no exception in this regard. In response to this the safety regulations for building work were tightened up through the introduction of the CDM Regulations 1994.

The CDM Regulations came into effect on 31 March 1995. Although they are primarily aimed at new-build projects, the regulations are also applicable in certain circumstances to adaptation work.

One of the most important implications of the CDM regulations is that both clients and 'designers' (which includes surveyors and project managers as well as architects) become liable for health and safety on site, along with the contractor. Thus, the burden of health and safety is not solely placed on the contractor, but everyone involved in the process.

The legislation is concerned not only with health and safety on site once work has begun. It also requires that the design of the project and the sequence in which it is carried out are such that health and safety risks are not compromised.

The CDM Regulations according to Regulation 3:

> do not cover house occupiers carrying out work on domestic premises which they use as a private dwelling, nor projects where a domestic client commissions an architect, BS or engineer and a builder to build, alter or extend a house. Commercial developers who sell houses before the project is complete will remain clients to whom these Regulations apply (Regulation 5). Where domestic premises are not used solely as a private dwelling, but are let out, opened to the public or used also for business purposes, then, depending on the nature of the use, those clients may be subject to the Regulations.

The directive is wide in its scope. It applies to all other construction work including excavations and construction operations, conversions and fitting out as well as, repairs, drainage, upkeep and maintenance. The CDM regulations even extend to works such as painting and cleaning, and demolition.

Under the CDM Regulations certain construction projects are 'notifiable'. These include projects that are of a non-domestic nature, of more than 30 days duration, or that involve more than four workers on site. In Britain the Health and Safety Executive (HSE) is the relevant authority that must be notified. The HSE has extensive powers and will readily use them if safety on site is being severely compromised. Failure to notify them or respond to their demands for changes in safety precautions may result in a conviction.

Adapting occupied buildings

Refurbishing an occupied property imposes a heavy burden on the contractor to undertake the work with the minimum nuisance. It can result in disturbance or danger to the building's occupants or those nearby as well as the public. The users of the building as well as the operatives may have to contend with working nearby one another. This can lead to disruption owing to noisy activities, access problems, etc.

Co-ordinating the adaptation work around the normal activities being undertaken within the building itself is not easy (McLennan et al., 1998). The building operations are often noisy and disruptive. Interference with the power supply to the occupied part of the building, for example, is a risk that must be taken into account.

Out-of-hours working may be a standard requirement or feature of the adaptation contract. This in itself also brings problems of security and degree of supervision to ensure operatives are in attendance at the correct times.

Adapting unoccupied buildings

Vacant buildings can also pose problems when subjected to adaptation. Such properties are more likely to be in a poor state of repair or even dilapidated. Ideally, before work commences all their windows and doors would normally be boarded up. If left open or unsecured vacant premises are more prone to damage by squatters and vandals.

Moreover, because they are not heated or cleaned, unoccupied buildings tend to deteriorate at a faster rate than occupied properties. Moisture and dust levels with these properties inevitably rise, bringing with them problems of dampness, dirt and grime. Infestation by pests such as pigeons and vermin is also likely, particularly if the building has been left unoccupied for any length of time. Debris is often left in such properties and this can act as a magnet for vermin. It may require removal by a specialist contractor prior to adaptation. This must be allowed for in the total adaptation cost.

If the degree of infestation is great, a major cleansing operation may be required in the building to disinfect it before work can start on its adaptation. Steam cleaning of areas contaminated with fleas and other such pests may be required. In more serious cases it may involve fumigating the building's interior with a smoke-bomb type of insecticide. These techniques can be hazardous and costly as well as time-consuming.

In large renovation projects involving renewal of the roof and other substantial works, it may be appropriate to erect an enclosure over the building. Figure 11.7 shows a cross section through such a building, with the use of a sectional refuse chute for disposing construction waste.

Dust control

Adaptation projects are more to generating excessive levels of dust than comparable new-building schemes. Dust is frequently generated in refurbishment schemes involving stone masonry work and façade renovations. Sandblast cleaning of buildings, demolition processes or dismantling elements and concrete scabbling, for example, can produce high levels of dust. Moreover, cutting, drilling and grinding cementitious and other particulate building materials are potentially very hazardous in this regard. Even sweeping up and other cleaning operations in workplaces carried out after refurbishment and maintenance can generate considerable dust levels.

Air extraction equipment may be required where large amounts of dust are generated. This is particularly important where natural ventilation is limited.

Wetting down the affected surfaces may help to minimize disturbance of dust. However, this can also be messy and may result in water damage to the building's fabric if carelessly applied. In these and other areas where hazardous work is being carried out a permit to work procedure is likely to be required.

Robust screens of polyester reinforced polythene sheeting with duck-taped seals over or around unused openings will reduce noise as well as dust transmission. For small areas these can comprise 50 × 50 mm softwood framing covered with heavy gauge plastic sheeting (see Figure 5.26). Reinforced plastic sheeting covering 100 × 50 mm section framing would be required for larger openings/enclosures.

Weather protection

Temporary screens comprising timber framework covered with heavy-duty translucent polythene or other reinforced polymer sheeting should be used to cover large openings. These screens should be translucent to allow some daylight into the building.

This screening provides wind and weather protection during operations such as façade replacement (i.e. recladding) work to an existing building. Raking struts installed behind the framework give rigidity to the screen (see Figure 11.8).

Schematic vertical cross section (not to scale)

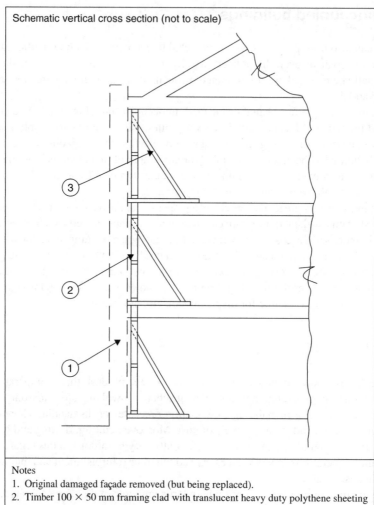

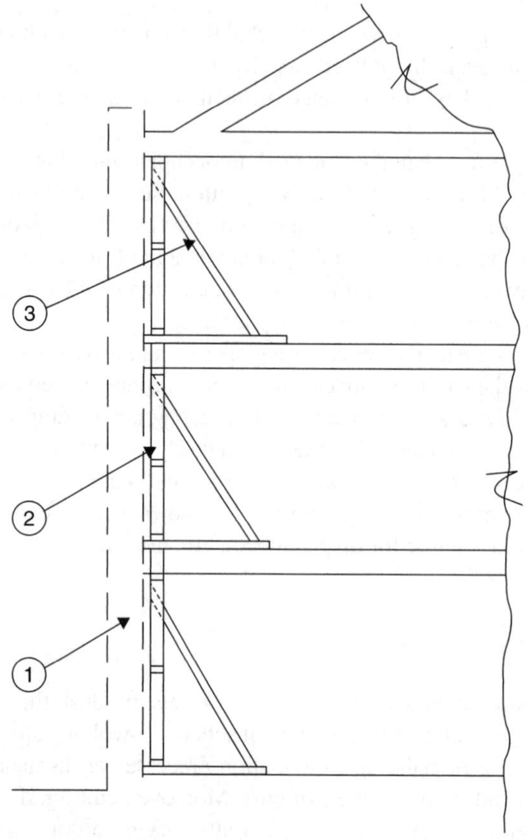

Notes
1. Original damaged façade removed (but being replaced).
2. Timber 100 × 50 mm framing clad with translucent heavy duty polythene sheeting installed prior to completion of new façade.
3. Timber struts at 1.2 m centres.

Figure 11.8 Temporary weather-protection screening

Operational problems

In particular, a number of problems of logistics can occur for contractors especially in the adaptation of properties in tight sites, particularly in inner urban areas. Some of the common problems in this regard are:

- Position, size and setting up of offices and other site accommodation may be difficult to achieve in one location.
- Access into property for building components and materials may be limited.
- Access to property for plant and equipment may be restricted as well as limited.
- Delivery restrictions may be imposed on the street or site itself.
- Egress from building for demolition waste may be awkward.
- Position and oversailing limitations of crane has to be taken into consideration as regards their impact on adjoining properties.

- Health and safety of the public, requiring extra protection measures, such as screens, fans on hoarding and scaffolding, padding around scaffolding poles, etc.
- Nuisance to public and adjacent property owners – disruption of access, dust and noise pollution, etc.

Using project management

Preamble

Any construction work, from the smallest adaptation job to the largest new-build scheme will require some form of project management if the critical success factors (time, cost, safety and quality) are to be achieved. This form of management has become the main methodology used to administer major projects from space programmes to new car designs as well as large building work. It can also be employed in medium–large scale adaptation schemes that require careful administration and control because of the importance or complexity of the work involved.

The three fundamental functions of project management and their application to adaptation work are:

1. project planning
2. project monitoring and control
3. project review.

Project planning

This stage involves several key objectives:

- Developing the client's brief by helping them to translate their requirements into a coherent and achievable form.
- Proactive measures – anticipating the problems and identifying the requirements (e.g. lack of access, disruptive work; tackled by scheduling some work outwith normal working hours).
- Calculating work and resource requirements, and defining levels of quality and standards.
- Communication – informing all participants and letting them contribute to the discussions.
- Preparing the overall programme of the proposed scheme.
- Target setting – agreeing and establishing yardsticks that can be measured. Table 11.7 illustrates the use of a simple bar chart for a medium-size extension project.

Project monitoring and control

This stage involves using a variety of measures to ensure that the project is on target through:

- Design and contract procurement – selection of design team members and type of contract appropriate for the proposed scheme.
- Managing and monitoring the design and construction.
- Establishing and controlling the budget. This involves cost monitoring – comparing the budget costs with the actual costs.
- Comparing planned to actual performance (see simple programme chart in Table 11.7).
- Cost control – adjusting payments in light of the monitoring results. Figure 11.9 shows a graph containing a planned cost envelope for the extension referred to above.
- Monitoring of operations – observing and recording progress of the work.
- Communicating with all members of the design team on behalf of the client.
- Analysing impact on project of changes resulting from modifications to client's brief, delays in materials delivery, unforeseen events such as a major accident on site.

Table 11.7 Example Gantt chart for the construction of a medium-size extension (Dark bars represent predicted time-scale; lighter equal actual time-scale)

Activities	*Weeks* 1	2	3	4	5	6	7	8	9	10	11	12	13	14	15	16
Preparation of site																
Substructure work																
Preparation of existing building																
Superstructure of extension																
Fitting out extension																
Tidying up and making good																
Progress meetings	↑			↑				↑				↑				↑
Milestones		1				2								3		4

Note
Up to twelve weeks prior to the commencement of the contract may be required to obtain the necessary statutory approvals.

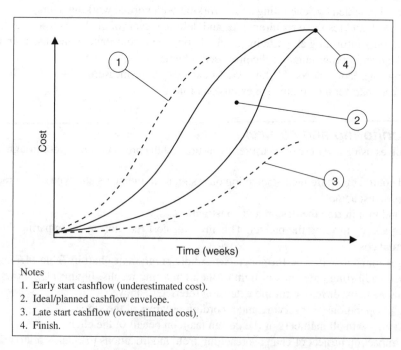

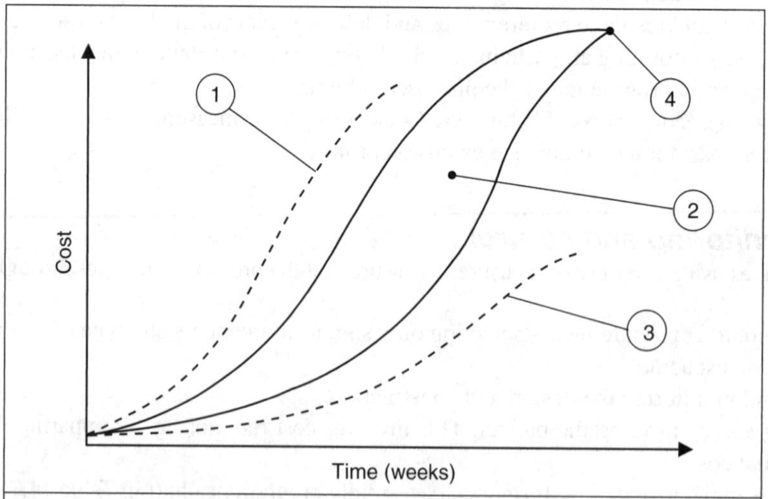

Notes
1. Early start cashflow (underestimated cost).
2. Ideal/planned cashflow envelope.
3. Late start cashflow (overestimated cost).
4. Finish.

Figure 11.9 Example showing the planned cumulation of costs for the above project

- Receiving feedback on cash flow and progress of work.
- Control of operations – adjusting the system (e.g. increasing or reducing the number of operatives on specific tasks).
- Problem-solving – acting as the client's fixer to resolve conflicts and difficulties associated with the proposed design or construction.

Project review

Regular briefings to assess progress as identified in the monitoring reports are important. Site meetings held once a week/month (depending on the size of the job and time-scale involved) would be the main forum to review progress of the project.

It would also be helpful for the project manager to review the successes and failings of the project once it is complete. Such feedback can then be used to help improve the knowledge base for future projects of this kind. Some form of post-occupancy appraisal exercise can also help in this regard (see below).

Project managers can also be involved in the final reconciliation of an adaptation scheme. They can advise on a range of diverse issues such as defects liability period, maintenance, landlord and tenant requirements, etc.

Dispute resolution

Not every construction project goes without a hitch. Dissatisfaction on the part of either the client or contractor can occur for a variety of reasons.

Adaptation schemes, no matter how small, can also entail a dispute between the parties. This can, for example, concern the quality of the work or occur as the result of problems relating to the timing and size of payments.

The following are the major causes of disputes in construction projects including adaptation schemes:

- Interpretation of contract clauses.
- Performance of product or quality of service poor.
- Failure to agree price for additional work.
- Delays or later deliveries.
- Changes in requirements.
- Disagreement over responsibility.
- Poor communication.

According to Watt (2004) the main avenues for dealing with construction disputes are:

- Arbitration
- Conciliation
- Expert determination
- Litigation
- Mediation.

Post-adaptation evaluation

Purpose

It is useful to evaluate the performance of the building once the adaptation work is completed and the occupier has settled in to using the premises. This exercise is likely to prove worthwhile for medium–large-scale adaptation schemes. It is unlikely to be cost-effective for minor adaptation works.

A PAE, therefore, is for all intents and purposes a form of POE. This is a diagnostic tool and evaluation system, which enables surveyors and other building professionals to identify and appraise critical aspects of building performance in a systematic manner (Preiser, 1989). It was developed in the USA in the late 1960s by architects interested in finding out how well their buildings were satisfying the needs of the user.

POE, a user-based appraisal technique that was one of the first systematic attempts by building professionals to achieve better buildings for their clients. In the words of Wolfgang Preiser, one of the originators of POE, it can be defined as follows:

> POE is an appraisal of the degree to which a design setting satisfies and supports explicit and implicit human needs and values of those for whom the building is designed.

There is no reason why this appraisal technique could not be applied to an adapted building. POE is a process of assessing buildings in a rigorous and systematic manner at least 6 months after occupancy. It involves assessing the overall functional performance of occupied buildings. POE does not therefore normally include a detailed technical appraisal of the building at the initial stages. However, the feedback obtained may point to deficiencies in the fabric that can affect performance.

POE is designed to assist in the quality control of both the building and its process. It can help complex or large organizations to improve and understand the significance of building performance. Thus the information obtained from a POE/PAE exercise would yield the following:

1. The overall quality of the building would be improved in terms of health/safety/security; functionality/efficiency; and social/psychological/cultural satisfaction.
2. Adds to the state-of-the-art knowledge, local experience and contextual factors.
3. Saves the cost of maintaining and operating facilities over life cycles.
4. Improves morale of the building users.
5. Helps generate better design guidelines.
6. Benchmarks successful concepts that can be used for future adaptation schemes of the same or similar type.
7. Helps establish standards and databases.

Useable POE/PAE tools

The tools formulated in POE (Preiser, 1989) can be applied, with modification where necessary, to adaptation schemes. Rating factors can be recorded in the format indicated in Table 11.8. The satisfaction of performance criteria is illustrated by the example shown in Table 11.9.

Success criteria for adaptation schemes

Generally

Traditionally the accomplishment of any adaptation or new-build scheme was judged on four key criteria: on time, within budget, as specified and done safely. If a project fulfilled these main requirements it was likely to be judged a success.

The recommendations of the Egan Report (1997), however, extended the list of success criteria and made them more explicit. Rethinking Construction was the focus of the Report for initiatives to improve relationships and processes in the UK construction industry as well as to enhance client value. The report highlighted the deficiencies in the construction industry compared to other sectors. It 'challenged the industry to measure its performance over a range of its activities and to meet a set of ambitious improvement targets'.

Table 11.8 Rating of performance factors

Performance met	*Yes*	*No*
Ambient temperature (18–20°C)	✓	
Air quality	✓	
Security (medium level required)	✓	
Access (for disabled users)		✓

Table 11.9 Measuring the satisfaction of performance criteria

Item number	*Requirement*	*Ratings*			
		Excellent	*Good*	*Fair*	*Poor*
1	Adequacy of space				
2	Lighting				
3	Acoustics				
4	Temperature				
5	Odour				
6	Aesthetic appeal				
7	Security				
8	Flexibility of use				
9	Accessibility for disabled – generally				
10	Other (specify)				

It is generally accepted that the construction industry's clients want their projects delivered:

- on time,
- on budget,
- free from defects,
- efficiently,
- right first time,
- safely,
- by profitable companies.

Regular clients expect continuous improvement from their construction team to achieve year-on-year progress through reductions in project costs as well as reductions in project times. Safety, too, has become an important goal in any construction project.

Time

Naturally, the work should be completed within the agreed or estimated time-frame. Over-runs caused by delays are very common in construction projects. Adaptation work can also take longer than expected because of unforeseen work, which can easily lead to disruptions and delays as well as overruns on time and costs.

Cost

The scheme should be within budget. Few clients are happy when their construction project exceeds the original estimate, unless provision for extras or contingencies were made in the original contact sum. In adaptation schemes, more than in new build ones, unforeseen work can often occur. That is why accurate and tightly controlled tendering as well as an adequate contingency allowance is a critical requirement in procuring adaptation work.

Quality

Quality in construction has two key attributes:

1. compliance with standards,
2. fitness for purpose.

The level of satisfaction, the client feels about the adaptation can determine the overall quality of a scheme. The standard of workmanship has a great bearing on the achievement of this goal. This is in turn influenced by the level and competency of the site supervision and control.

Zero defects

An increasingly important requirement of modern construction is that the finished product should have zero defects. This may seem a utopian ideal but it has brought to prominence the need to produce buildings that do not contain any major failures.

Safety

Any building project that incurs injuries or hazards for those involved has not been a complete success. Under the CDM Regulations 1994 safety is now a key factor in any building project employing more than five operatives.

Location

For commercial buildings location is a vital requirement. This will ultimately influence the level of demand for (and hence the value of) an adapted commercial property.

Key performance indicators

The purpose of the KPIs is to enable measurement of project and organizational performance throughout the construction industry. This information can then be used for benchmarking purposes, and will be a key component of any organization's move towards achieving best practice.

Clients, for instance, assess the suitability of potential suppliers for a project, by asking them to provide information about how they perform against a range of indicators. Some information will also be available through the industry's benchmarking initiatives, so clients can see how potential suppliers compare with the rest of the industry in a number of different areas.

The 12 KPIs prescribed (www.constructingexcellence.org.uk/) are:

- Client satisfaction – product: measured as a per cent scoring 8/10 or better; 72 per cent.
- Client satisfaction – service: per cent scoring 8/10 or better 63 per cent.
- Defects: per cent scoring 8/10 or better 53 per cent.
- Predictability cost – design: per cent on target or better 63 per cent.
- Predictability cost – construction: per cent on target or better 52 per cent.
- Predictability time – design: per cent on target or better 41 per cent.
- Predictability time – construction: per cent on target or better 60 per cent.
- Profitability: median profit before interest and tax 5.5 per cent.
- Productivity: mean turnover/employed £28 k/£36 k.
- Safety: mean accident incident rate 1088.
- Cost: change compared with 1 year ago +2 per cent.
- Time: change compared with 1 year ago +1 per cent.

Means of achieving success

These primarily involve fully comprehending the client's requirements and knowing the building to understand its construction, condition and structural arrangement. These will help in determining its suitability for adaptation. An adequate budget along with an appropriate procurement method is also necessary to ensure the success of any adaptation scheme.

Aftercare strategy

Facilities management

The facilities or maintenance management team for an adapted building should have been involved in the process from day one. Their input to the design and commissioning processes is essential to help ensure that the property and its systems are readily manageable.

A proactive maintenance regime is needed to smooth out maintenance costs in use over the building's service life. If a reactive (i.e. unplanned) approach to maintenance is adopted, the maintenance expenditure is much more likely to be erratic and unpredictable over the long term. These differences are shown in Figures 11.10 and 11.11.

The proactive approach means that maintenance budgeting is more realistic and uniform over the period. It makes it much easier to bid for maintenance funds because consistency of expenditure in the long term is more achievable with this method.

A key component of any building aftercare system is the maintenance of plant, equipment and other electrical/mechanical facilities. This can be delineated in a maintenance manual (or log book see below) for the building and should include, for example:

- servicing of fire safety equipment (including smoke ventilation systems) on a regular basis (say every 6 months) is critical;
- testing of alarms and sensors should be carried out on a more frequent basis (say every fortnight);

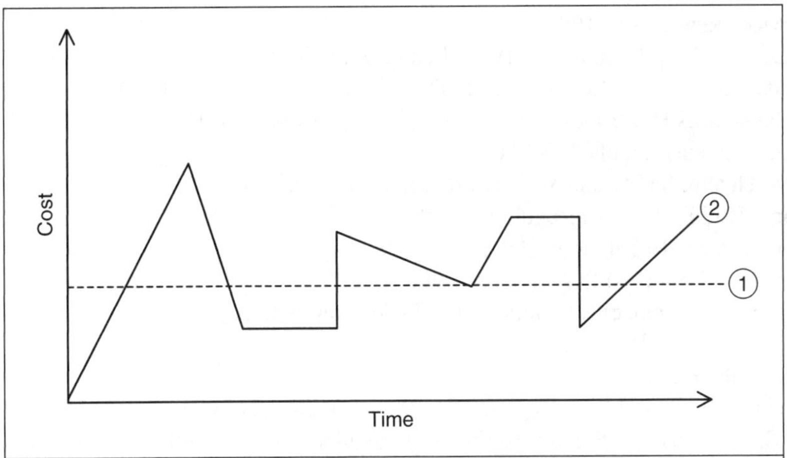

Notes
1. Original condition.
2. Erratic maintenance expenditure profile.

Figure 11.10 Notional reactive maintenance cost profile

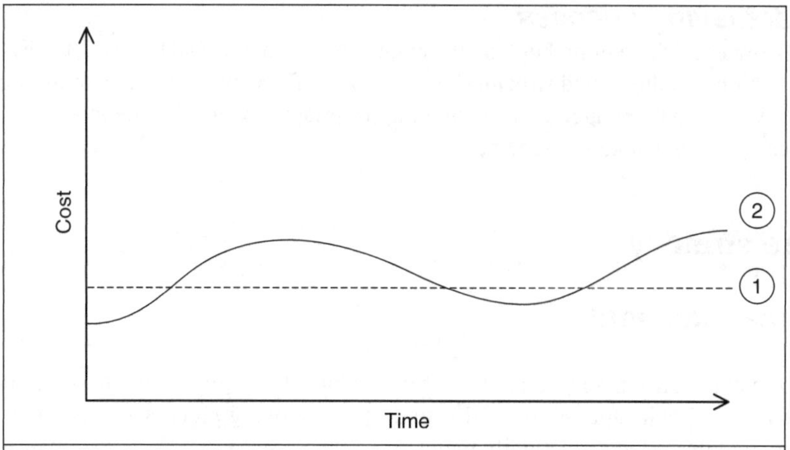

Notes
1. Average maintenance expenditure.
2. Smooth cyclical maintenance expenditure trend.

Figure 11.11 Notional proactive maintenance cost profile

- replacement of air conditioning filters at least twice a year;
- sterilizing humidifiers and cooling chambers annually if not more regularly by chlorinating and steam cleaning containers, pipes and other parts of the system that may become contaminated.

Health and safety regulations

The controls regulating health and safety requirements of property are extensive in scope. Any aftercare regime would have to take cognizance of some if not all of the following Regulations before, during and after any adaptation work:

- Confined Spaces Regulations 1997.
- Control of Lead at Work Regulations 1980 (Lead Regulations).
- Control of Asbestos at Work Regulations 2002 (Asbestos Regulations – CAWR).
- Control of Substances Hazardous to Health Regulations 1994 (COSHH).
- COSHH (Amendment) Regulations 2004.
- Construction (Health, Safety and Welfare) Regulations 1996.
- Construction (Head Protection) Regulations 1989.
- Construction Products Regulations 1991.
- Education (School Premises) Regulations 1999.
- Fire Precautions (Workplace) Regulations 1997 (Fire Regulations).
- Gas Appliances (Safety) Regulations 1995.
- Gas Safety Regulations 1991.
- Gas Safety (Installation and Use) Regulations 1998 (Gas Regulations).
- Health and Safety (Display Screen Equipment) Regulations 1992 (otherwise known as the Display Screen Regulations).
- Health and Safety (Safety Signs and Signals) Regulations 1996.
- Lifting Operations and Lifting Equipment Regulations 1998.
- Management of Health and Safety at Work Regulations 1992 (Management Regulations).
- Manual Handling Operations Regulations 1992 (Manual Handling Regulations).

- Noise at Work Regulations 1989 (Noise Regulations).
- Personal Protective Equipment at Work Regulations 1992 (PPE Regulations).
- Provision and Use of Work Equipment Regulations 1992.
- Reporting of Injuries, Diseases and Dangerous Occurrences Regulations 1985 (RIDDOR).
- Water Supply (Water Fittings) Regulations 1999.
- Workplace (Health, Safety and Welfare) Regulations 1992.
- Work at Height Regulations 2005.

Adaptation projects to offices should be carried out in accordance with developments in the various workplace and fire regulations enacted under the Health and Safety at Work Act 1974 and the Fire Precautions Act 1971, respectively. The Fire (Workplace) Regulations 1997, for example, came into operation on 1 December 1997. These regulations, which arise from two separate European Commission Directives, complement the existing legislation relating to fire, health and safety. In some respects they amend the current Management Regulations.

Highly specialized risk areas involving major hazards such as toxic waste processing, ionizing radiation, high-pressure vessels, genetic manipulation, etc. have their own regulations.

Building logbooks or maintenance manuals

Preamble

In order to ensure the ongoing success of any new or adapted building, a logbook or maintenance manual should be compiled. If none has previously existed for the building a large adaptation project should provide the impetus to correct this deficiency. The information compiled within the safety plan for such a scheme as required under the CDM Regulations could provide the basis for the manual. The operational and maintenance personnel responsible for the regular upkeep of the facility would then use it subsequently. Facilities managers especially have a key role to play in using the log book in passing on any strategic understanding of the building (CIBSE, 2003).

Purpose

The revisions to Part L of the Building Regulations 2001 require that a building logbook be provided to the owner/occupier of all new and refurbished non-residential properties. Such a manual is intended to provide a simple overview for operational and maintenance staff of how the building is meant to work (CIBSE, 2003). This is so that the building can run effectively (e.g. provides adequate comfort conditions) and efficiently (e.g. minimizes costs, pollution and energy wastage).

A building maintenance manual or logbook has similar functions to a car handbook. The building logbook, however, describes the nature and type of the original development or previous major adaptation work carried out. It is aimed at ensuring that everyone working on the building (premises manager, facilities manager, etc.) understands the basic design philosophy so that managing the premises is made as uncomplicated as possible. This is normally expressed in the logbook in several cross sectional schematic diagrams plus some concise explanatory text.

According to CIBSE (2003) a building logbook should provide:

- a summary of the building;
- a key reference point (where it exists at all – information is commonly found in a range of documentation);
- a source of information/training (giving all those working on the building a good understanding of the basic design philosophy;
- a day-to-day log (somewhere to log energy performance and changes to the building).

As suggested by CIBSE (2003) a typical contents list of a building logbook is as follows:

Summary: general information, historical background, etc. relating to the building.

1. Updates an annual reviews.
2. Purposes and responsibilities.
3. Links to other key documents (e.g. health and safety plan).
4. Main contracts – procurement options, types of contract – daywork/measured term; financial matters – budget; outsourcing.
5. Commissioning, handover and compliance.
6. Overall building design.
7. Summary of areas/occupancy.
8. Summary of main building services points.
9. Overview of controls and Building Management System.
10. Occupant information.
11. Metering, monitoring and targeting strategy.
12. Building energy performance records.
13. Maintenance information and review: maintenance of structure and finishes – requirements; maintenance of services – servicing and inspection requirements; cleaning – required frequency, methods and recommended products; and special requirements/features: fire extinguisher replacement; lift maintenance; solar energy requirements; security measures.
14. Major alterations and other adaptations.
15. Results of in-use investigations.

Appendices relevant certificates and schedules (see below).
Index (for easy reference).

The manual should contain in schedule format information essential for the effective long-term maintenance of the adapted building, such as:

1. Description of materials/products used.
2. Names and addresses of manufacturers/suppliers.
3. Location of materials/products in the building.
4. Comments on availability/installation.
5. Brief specification of system/installation.
6. Recommended maintenance/servicing/cleaning cycles.
7. Any other relevant textual information concerning the building (e.g. presence of deleterious materials).
8. Working details of as-built sections.
9. Safety requirements relating to materials, etc.

A proper maintenance manual for any new or adapted building can only best be prepared once the full extent of the project has been identified, agreed and completed. Where such a manual does not exist in the case of an existing property, it can be compiled using the same approach as for a new building. The main problem is that obtaining information on all aspects of the building such those indicated below may not always be possible.

A member of the design team should compile the manual. Architects, BS, maintenance managers and facilities managers are the most appropriate professionals to compile a building logbook. The consultants involved in the formulation of the adaptation brief and design can provide assistance on the preparation of a complete manual. The property manager or the person responsible for the day-to-day maintenance of the facility will then hold and use the final document.

Maintenance schedules

Schedules provide a useful way of presenting information in a succinct and readable form. They can be used to summarize building information (specification details) and Maintenance Requirements (i.e. inspection and cleaning cycles). Appendix F shows an example of each of these sections.

Maintenance audit

This form of audit is normally undertaken every 3–5 years. The three main functions of a maintenance audit are:

1. To report to the management on the existing situation as regards the efficiency and running of the maintenance regime.
2. To make recommendations which will help to remedy any existing anomalies with the aftercare of the property or properties.
3. To help in determining an appropriate policy as regards planning, procurement and operational strategy for the organization's property maintenance.

It involves checking the effectiveness and efficiency of the following key aspects of a maintenance regime:

- Organizational: An appraisal of the management structure should cover issues such as the composition and hierarchy of the maintenance team. It should also look at the decision-making process, and review the policy regarding the use and performance of Directly Employed Labour as opposed to outsourcing maintenance work.
- Operational: The day-to-day effectiveness of the maintenance work should be monitored regardless of the procurement system employed. This would include the quality of environmental conditions within the building. Client or user feedback is essential to determine this.
- Financial: This may involve an analysis of the pattern of maintenance expenditure over a 5- or 7-year period. Budget demands and requirements over that period may also be assessed. Sources of finance may have to be reviewed as well.
- Legal: Compliance with statutory requirements is essential for any organization. In addition, contractual matters associated with maintenance work need to be reviewed. This could, for example, take the form of an appraisal of the extent and effectiveness of various procurement methods – measured term, daywork term or lump sum contracts. It may highlight the need to change how the maintenance work is procured (see RICS, 1997).
- Technical: Condition and physical performance of the stock should be ascertained. Common or regular incidents of failures or breakdowns should be analysed. This could help in the formulation of a modernization strategy as part of the maintenance planning process.

Summary and conclusion

Adaptation work requires its participants, the consultants and contractors especially, to have a sound understanding of traditional as well as modern construction techniques. For example, they ought to have some knowledge of solid masonry walling, close couple pitched roof structures, lath and plaster and other forms of obsolete construction. In addition, they are also expected, when renovating or restoring historic buildings, to have sensitivity to conservation needs.

Some of the risks associated with conversions, extensions, refurbishment, restoration and modernization schemes are different from those in new build. Proximity to occupiers and the public using the premises being adapted can increase both safety and security risks. The requirements of the CDM Regulations place a heavy responsibility on those involved.

Building adaptation will play an increasingly important role in helping to obtain a more sustainable environment. It is not the only option however. Eventually every building will reach the end of its life expectancy after many adaptations. This should not be seen necessarily as a failure of the building. The time will come when every property has to make way for another structure. Adaptation along with maintenance can ensure that the efficiency, use and service life of a building is maximized.

It is hoped that this work will prove helpful for contractors as well as BS and other construction professionals with some guidance on the problems and methods of adapting buildings. Building adaptation is about responding to changes in the demand for and supply of property. It is a process that is likely to become more common as more attention is given to maximizing the service life of existing properties.

The implications of information technology in both the home and work environments are enormous. Intelligent properties which will involve responsive environmental controls, as well as interactive/internet-linked television and energy efficiency measures are likely to be amongst the most influential developments in building technology within the next 20 years. It is not only new buildings that will have these modern facilities. Existing properties, too, will also need to accommodate these advances to avoid obsolescence. Building adaptation is the process whereby this can be done.

So long as we have properties, we will need to adapt them, because of:

- Advances in technology, which potentially can make some buildings obsolete or prompt their upgrading or modification to make best use of such changes.
- Changes in market conditions can easily result in some commercial and industrial buildings becoming unattractive if not redundant.
- Changes in living habits and working environments requiring different building conditions and layouts.
- Demographic changes such as increasing proportion of elderly people, more single occupiers and single parent families resulting in a greater need for flats with fewer rooms.
- Conservation of the built environment necessitating the adaptive reuse or refurbishment of old buildings.
- Renovating dilapidated historic buildings to ensure their ongoing beneficial use.
- Restoring salvageable buildings damaged by floods, storms and impacts.
- Increasing wear and tear of the ageing building stock, which requires modernization.
- Climatic change requiring existing as well as new buildings to become more robust and energy efficient.
- The need to mitigate the increasing adverse effects of environmental pollution.

Building adaptation, therefore, can clearly make a substantial contribution to sustainable construction and minimizing the adverse effects of climate change on the built environment. This work has attempted to demonstrate that and gives an indication of how it could be achieved.

Appendix A

Checklist for assessment of habitability

Reproduced by permission of Building Research Establishment (BRE). BRE Good Building Guides are available from CRC, 151 Roseberry Avenue, London, EC1R 4GB. Tel.: 020 7505 6622.

Externally

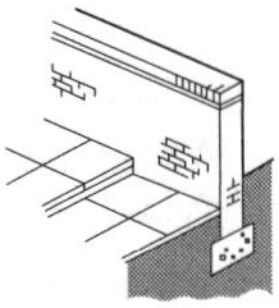

- Access, particularly for elderly or less mobile and for prams? (Give special consideration for basements.)
- Access to gardens, walkways?
- Condition of openings/paths, steps particularly for elderly, disabled, prams, bicycles?
- Safety of play areas and equipment?
- Refuse storage space and access for collection?
- Parking or turning space or access for emergency vehicles? (Particularly for elderly or less mobile.)

Roof spaces

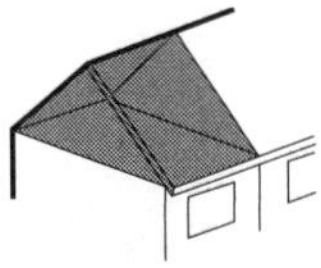

- Safe access for storage or maintenance?
- Birds, vermin, insects excluded? (*Note*: Bats already living in a roof space are protected and their access must not be restricted.)

Balconies

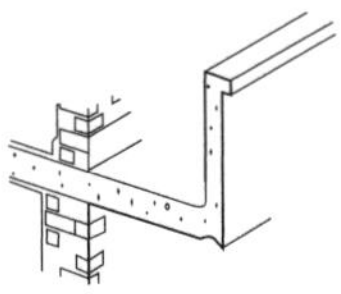

- Guardings of least 1100 mm high, unclimbable by small children and robust?
- Vertical gaps below guardings less than 100 mm?
- At least 2 m above the ground?
- Safe access, particularly for the elderly?
- Sufficient step/threshold to avoid room flooding?

Living, dining, bedrooms and halls

- Heating level comfortable and even?
- Ventilation adequate?
- Lighting adequate?
- Sound resistance adequate?
- Room size, layout and door position adequate for activities and circulation?

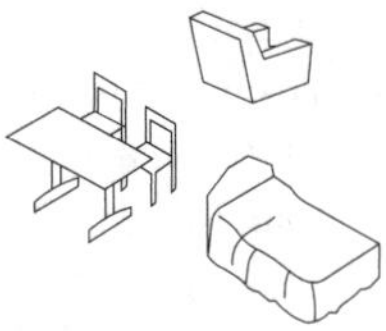

- Floors level and even enough for coverings? (solid floors not cracked or pitted, timber floors not cupped, springy or with gaps between boards)
- No awkward changes of floor level?
- Wall/ceilings in suitable condition for redecoration?

Kitchens

- Ventilation adequate?
- Lighting adequate?
- Services and fittings satisfactory and safe?
- Storage space adequate?
- Sufficient hygienic and easy to clean work surfaces?
- Floors hygienic, easy to clean and moisture resistant; level and even enough to receive new coverings?
- Layout suitable for sensible work pattern?

Bathrooms and WCs

- Ventilation adequate?
- Lighting adequate?

- Floors coverings hygienic, easy to clean and moisture resistant; level and even enough to receive new coverings?
- Services and fittings satisfactory?

Basements

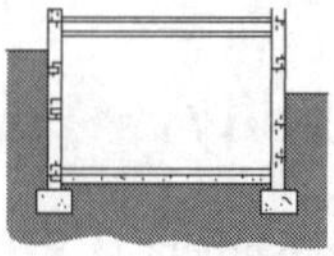

- Lighting adequate?
- Ventilation adequate?
- Damp proofing practical?
- Heating level comfortable, even?
- Headroom adequate; can it be improved by lowering floors?
- Does damp proofing system limit the type of wall fittings which can be used?

Stairs

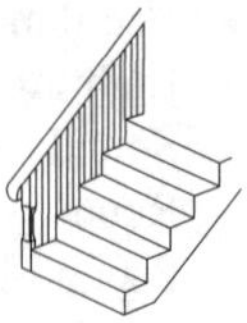

- Lighting adequate with no glare?
- Headroom adequate?
- Is width of stair adequate?
- Are steps uniform?
- Step rise and going within acceptable limits?
- Safe and secure balustrades/handrails along full length of stair light?

Doors

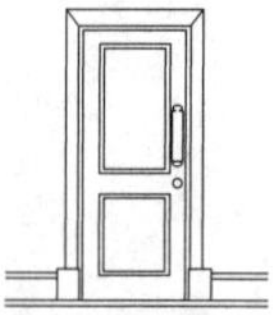

- Clear opening dimensions minimum 1900 mm × 700 mm? (wider for doors to gardens or yards or for less mobile)
- Safety glazing used?
- Thresholds shaped to avoid tripping?
- Handles at correct height and easy to operate? (spring return levers most suitable for elderly or less mobile)
- External doors to individual homes secure?
- Floor level changes, steps or stairs avoided near internal doors?
- Clashing door swings avoided?

Windows

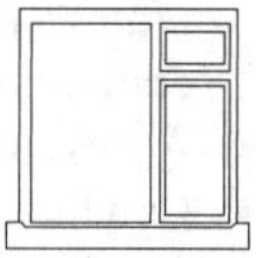

- Adequate size for daylight and ventilation?
- Trickle ventilation incorporated to ensure background ventilation when closed?
- Reasonably draught-free?
- Easy to open and clean, especially by the elderly or less mobile? (upstairs windows should be secure against opening by children)
- Not causing a hazard by opening onto paths?
- Safety glazing below 800 mm?

Appendix B

Key aspects of performance

Key aspects of performance

Assessment of the relevant performance considerations for each building element/service (all circles in the table below) should reveal most of the deficiencies which will require attention during rehabilitation. All of these considerations are important, but evidence from BRE surveys indicate that some (those marked with a star) are more frequently overlooked than others.

	Strength and stability	Dampness (rising)	Weather-tightness	Fire	Thermal	Condensation	Durability and maintenance	Sound	Habitability	Function	Health and safety
Basements	*	○		○	○	○	*		*		
Foundations	○						*				
Ground floors (solid or timber)	○	○				*	*		○		
Solid intermediate floors	*			○	*	*	*	○	○		
Timber intermediate floors	○			○	*		*	○	○		
Solid or cavity masonry external walls	○	○	○	○	○	○	○				
Doors and windows			○			○	○		○		○
Balconies and canopies	*		*				*				○
Bays and porches	*	○	○		*	*	*				
Pitched roofs	*		○	○	○	○	○		○		
Flat roofs	*		○	*	○	○	*				
Parapets	*		*				*				
Chimneys	*	*	*	○		*	*				
Separating and compartment walls	*	*		○				○			
Stairs and landings	*			*			○				○
Internal joinery							○			○	
Internal piped and wired services				*			*			○	
Above and below ground drainage							*			○	
External works	*						**		○		*

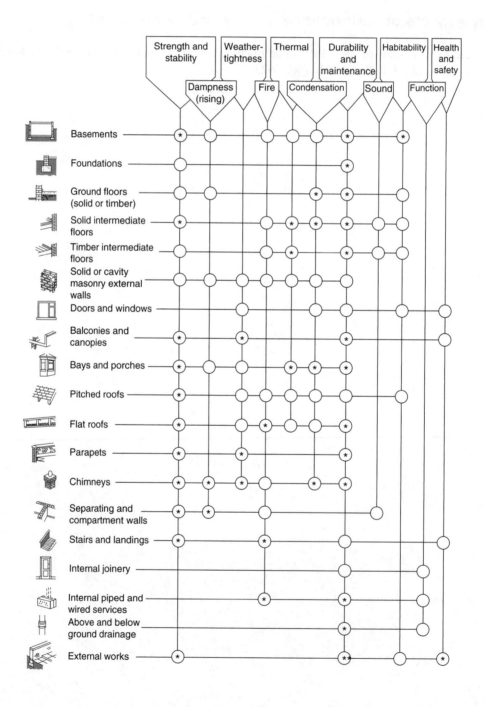

Appendix C

Table of suitable sizes of pre-cast (pc) concrete lintels for various spans and loadings (from Richard Lees Ltd catalogue, c. 1995)

Beam Lintels
Maximum safe uniformly distributed load (kN/m) (in addition to self weight)

Beam lintel	W × D (mm)	0.90	1.20	1.50	1.80	2.10	2.40	2.70	3.00	3.30	3.60	3.90	4.20	4.50
							Clear span between supports (m)							
☐	100 × 100	5.97 / **14.55**	4.54 / **8.51**	3.64 / **5.54**	3.86	2.82	2.13	1.65	1.30	1.04	0.84	0.69	0.56	0.46
☐	100 × 150	30.47 / **44.31**	17.88 / **26.08**	11.69 / **17.09**	10.23 / **12.02**	7.54 / **8.88**	5.76	4.52	3.62	2.95	2.44	2.03	1.71	1.45
☐	100 × 225	60.67 / **70.76**	37.25 / **51.07**	24.41 / **33.53**	23.42 / **23.63**	17.33 / **17.49**	13.30	10.50	8.47	6.95	5.79	4.87	4.15	3.56
☐	140 × 100	19.67 / **27.30**	11.51 / **16.02**	7.48 / **10.46**	6.54 / **7.32**	4.79 / **5.38**	3.64	2.83	2.25	1.82	1.48	1.22	1.01	0.84
☐	140 × 150	37.18 / **54.91**	21.80 / **32.29**	14.22 / **21.15**	12.12 / **14.85**	8.91 / **10.95**	6.79	5.31	4.24	3.44	2.83	2.35	1.97	1.66
☐	140 × 215	80.88 / **91.07**	52.19 / **69.89**	34.21 / **47.10**	30.35 / **33.20**	22.46 / **24.58**	17.23	13.59	10.96	8.99	7.48	6.30	5.36	4.59
☐	150 × 100	22.19 / **31.11**	12.99 / **18.27**	8.45 / **11.94**	8.36	6.15	4.68	3.66	2.92	2.37	1.95	1.61	1.35	1.13
☐	150 × 150	42.40 / **65.14**	24.87 / **38.36**	16.24 / **25.15**	15.50 / **17.68**	11.42 / **13.05**	8.73	6.85	5.49	4.47	3.70	3.09	2.60	2.20
☐	150 × 225	87.53 / **98.41**	59.12 / **75.52**	38.76 / **53.27**	34.15 / **37.54**	25.27 / **27.80**	19.39	15.29	12.33	10.12	8.42	7.09	6.03	5.17
☐	190 × 100	24.28 / **34.06**	14.19 / **19.97**	9.21 / **13.03**	9.11	6.68	5.07	3.95	3.14	2.54	2.07	1.71	1.42	1.18
▉	190 × 150	48.19 / **73.70**	28.24 / **43.34**	18.41 / **28.38**	17.22 / **19.93**	12.67 / **14.70**	9.66	7.57	6.05	4.92	4.05	3.37	2.82	2.38
▉	190 × 215	112.16 / **127.37**	86.05 / **97.75**	57.09 / **64.60**	44.15 / **45.53**	32.69 / **33.72**	25.10	19.81	15.99	13.13	10.93	9.22	7.85	6.74
▉	225 × 100	33.29 / **43.73**	19.48 / **25.66**	12.68 / **16.76**	11.41 / **11.73**	8.38 / **8.61**	6.37	4.97	3.96	3.20	2.62	2.16	1.80	1.51
▉	225 × 150	58.37 / **98.62**	34.22 / **58.26**	22.32 / **38.19**	20.82 / **26.85**	15.33 / **19.82**	11.69	9.16	7.33	5.96	4.91	4.08	3.43	2.89
⬓	*100/150 ×150	38.19 / **46.76**	24.07 / **34.37**	15.74 / **22.54**	14.99 / **15.86**	11.07 / **11.72**	8.47	6.66	5.35	4.38	3.63	3.04	2.57	2.19

Notes
Bold type indicates High Load Lintels – Please specify when ordering. All loads shown are unfactored.
*Safe loads also apply to steel fronted lintels.

Appendix D

Typical notes and specification column in a drawing for a small extension scheme

Notes and specification

General

1. These plans have been prepared for Building Control approval only and do not purport to be a full description of the work.
2. The local authority must be informed of any proposed deviation from these plans to ensure compliance with the appropriate legislation.
3. All work must comply with the relevant codes of practice, British Standards and Building Regulations.
4. All relevant statutory authorities and public utilities must be notified of commencement of the works where appropriate.
5. The drainage proposals shall be to the satisfaction of the local inspectorate and shall comply with the relevant regulations.
6. The electrical installation shall be carried out with polyvinyl chloride (PVC) sheathed and insulated cables with protective metal conduit where necessary and shall comply with BS 7671. The client shall determine the final location and full quantity of electrical points, although the minimum requirements shown on plan must be installed.
7. All dimensions and sizes shall be checked on site.
8. All divergences between drawn and actual site conditions shall be notified to the client.

Substructure

9. Block underbuilding built on mass concrete strip foundations. The strip foundations shall be at least 150 mm thick, and have 150 mm wide scarcements. The top of the foundations to be a minimum of 450 mm below ground level or at the same depth as existing – whichever is deeper.
10. 225 mm wide sold 'Turbo' substructure grade blockwork walling to be built below ground level.
11. New floor to be constructed of 20 mm 'Weyroc' T&G decking on high-density polyurethane board 35 mm thick on 1200 g 'Visqueen' DPM on 150 mm thick mass concrete slab on 150 mm thick layer of well consolidated Type 1 blinded hardcore. The concrete mix shall be 1:3:6.

Wall construction

12. 125 mm thick 'Turbo' blockwork inner leaf with 35 mm thick polyurethane insulation batts, 50 mm wide cavity and 102 mm facing brick outer leaf to match existing. Mortar shall be not less than 1:4 strength mix.
13. Outer skin tied through 50 mm wide clear cavity with stainless steel wall ties at 900 mm centres horizontally and 450 mm vertically (300 mm at reveals of openings).
14. New cavity wall tied at abutment using 'Furfix' or other equivalent stainless steel wall profile/starter plates bolted to existing wall.

Downtakings

15. Existing timber screen and lean-to roof to be removed.
16. Non-load bearing partition in the kitchen to be carefully removed and the ceiling made good.
17. New door opening between kitchen and dining room to be formed complete with 'Richard Lees Ltd' or other equivalent heavy-duty pc lintel inserted in existing wall.

Roof

18. Roof tiles to match existing laid to manufacturer's printed instructions on 20 × 40 mm treated w.w. battens on 14 × 35 mm counterbattens, on reinforced roofing felt, on 13 mm thick exterior grade sarking board on 200 × 50 mm lean-to rafters at 450 mm centres. 200 × 50 mm runner plate bolted to existing wall at maximum 900 mm centres.
19. Roof structure to be insulated with 65 mm 'Coolag' insulation board between the rafters. A 50 mm vent space shall be achieved above the insulation to the underside of the new sarking.
20. Eaves shall be ventilated using 'Redland Red-Vent' eaves vents giving an equivalent 20 mm continuous soffit vent gap.
21. New 'Velux Type GGL-4' rooflights to be formed in new lean-to roof on raised kerbing to achieve the necessary pitch.
22. New 18 mm thick exterior grade WBP plywood fascia to be installed with 100 mm diameter unplasticized PVC (uPVC) gutter connected to existing system.
23. Roof ceiling to consist of 9 mm thick taper-edged foil-backed plasterboard taped ready for decoration.

Windows

24. New casement windows in 'tanalised' softwood to match existing, with 8000 mm^2 trickle ventilation slots in each head rail.
25. New clay tile sub-sills to match existing, complete with all necessary damp proof courses.

Completion

26. The contractor shall leave the job clean and tidy on completion of the works.

Appendix E

Building control checklist

Overview

Surveyors and other building professionals can use a plan-checking technique to help ensure that details of their proposed adaptation work achieves compliance with the building regulations. No checklist, however, can cover all contingencies, as every building project is unique in its own way. Still, the following plan-checking schedule, adapted from the one presented in Building Engineer (May, 1997), covers the main issues that need to be addressed when drafting plans for building control approval.

Key Areas	*Applicable?*		*Comments*
	Yes	*No*	
1. Water/fire service comments/special premises (HSE involvement) • Fee/block plan-indicating boundaries. • Purpose group use/of rooms.			
2. Site problems • Flooding/land drainage/land slip/ filled ground/migrating gas. • Boundaries. • Trees/vegetation nearby. • Access. • Drains. • Utilities.			
3. Structure • Foundations – size/depth/existing basement/ adjacent drains/trees/expose existing foundations and lintels. • Cavity width/ties/continuous cavities/cavities closed/retaining walls, trays, land drains. • Wall connection required? • Strength of blocks/bricks (especially below ground level) – frost resistance? • Openings in walls/piers/buttress returns (550 mm)/parapets/glazed walling. • Expansion joints (c. 6 m centres in blockwork, 12 m in clay brickwork).			

(*Contd*)

Key Areas	*Applicable?*		*Comments*
	Yes	*No*	
• Lintel types/size/insulated/and if UDL. • Lateral restraint to floors, roofs, flat roofs, tie down straps (strap thickness, dwangs, etc.). • Trussed rafters; • longitudinal/diagonal/ceiling bracing (>8 m web) dwangs, etc.; • disproportionate collapse over compartment walls/support over wall offsets, etc. • Floor, ceiling and flat roof joists (sizes and centres) • Herringbone strutting to joists over 2.5 m. • Double up joists under baths. • Double up rafters on either side of opening. • Overloading existing floors, roofs or structure. • Rafter, purlin sizes, centres, adequate loading. • Height and ratio of chimneys. • Disproportionate collapse to floors > five storeys, and to portals (see BS). • Calculations sent to structures? • Roofs – suitability, pitch, firings, ventilation (> or <15°), straps, rating, slate lathe size, tiling specification for nailing and clipping.			
4. Fire • Ancillary groups (1/3 storage/shops; 1/5 other). • Fire resistance (steelwork/internal walls/ external walls/separating walls and floors/ protected structure/glazing/basements). • Compartment sizes/compartments walls and floors/fire stopping/fire resistance and non-combustibility. • Is any floor >4.5 m above GL or three storeys? • Cavity barriers to ceilings, roof voids, wall cavities, etc. • Protected shafts (only bathrooms, etc. allowed) fire doors/cavity barriers in floor void/roof void/glazed areas (1.2 m^2 maximum for ½ hour and 0.5 m^2 for 1 hour, non-combustible beading). • Unprotected areas, dormer cheeks, notional boundaries (one building to be residential or assembly). • Stairs (limited combustibility). • Protected stairs and corridors. • Fire doors (seals, intumescent strips, rebate sizes, letterboxes, fire resistant glazing, ironmongery, fanlights, etc.). • Surface spread of flame (SSF) walls internally and externally, ceilings, circulation spaces, voids.			

(*Contd*)

Key Areas	*Applicable?*		*Comments*
	Yes	*No*	

- Class I, Class O, TPA light diffuser panels.
- Roof finish, SSF rating in relation to boundaries, rooflights (thermoplastic, TP) and their designed layouts.
- Pipes, vents, flues fire stopped where passing through compartment walls/floors.
- Lift shaft provisions (1% ventilation area), fire resistance of lift doors/walls, etc.
- Smoke detection (17 m horizontal)
- Inner rooms (okay when kitchen, laundry, utility, bathroom, WC or shower – with mechanical ventilation), otherwise 850 × 500 mm wide and at 800–1000 mm high, or roof window minimum of 600 mm high or dormer 800 mm high, both 1.7 m from eaves. Vent pipes required for non-dwellings.
- Firefighting shafts (fire mains required where firefighting shafts provided).
- Fire vehicle turning facilities: to within 45 m and with 3.1 m between gates, 3.7 m between access kerbs, 16.8 m turning (between kerbs) and 19.2 m between walls for turning.
- Basement venting (except single family dwelling house, or where any floor less than 200 m^2 and not more than 3 m below ground level).
- Means of escape (MOE) existing being reduced or impaired especially during construction operations. MOE as proposed/emergency lighting/signs/exits/fire alarm applicability/alternative fire detection/travel distance/escape door furniture, etc.

5. Resistance to moisture
   - DPCs/DPMs/cavity trays/tanking/air grates/150 mm air space/maintain existing subfloor ventilation/existing ground levels/hard-core compaction, and blinding.
   - Damp penetration through walls/roof/floor (existing and proposed).
   - Weepholes every 900 mm at ground level and lintels.
6. Sound insulation
   - Stairs/separating walls/flanking/floors.
   - Refuse chutes/flues back to back.
7. Ventilation to all building types
   - Ventilation to existing and proposed rooms/toilet lobbies/height of ventilation 1.75 m.
   - Trickle ventilation – 8000 mm^2 to all habitable rooms, 4000 mm^2 elsewhere.

(Contd)

Key Areas	Applicable?		Comments
	Yes	No	
• Mechanical ventilation – kitchen 60 L/S or 30 L/S in hood (216 m^3/hour or 108 m^3/hour) or passive stack ventilation (PSV) – bathroom 15 L/S or PSV/utility 30 L/S – WC (only) 1/20 floor area or mechanical ventilation 6 L/S (15 min overrun) – and smoke rooms 16 L/sec/person.{AQ1}			
8. Hygiene			
• Provision of bathroom, and sufficient sanitary accommodation.			
• Provision of hot and cold water to wash-hand-basins (WHBs), etc. (dwellings).			
• Sufficient sanitary macerators.			
• Refuse collection and solid waste storage.			
9. Plumbing			
• Air admittance valve required for soil stack?			
• Sink waste pipe length – 2.7 m maximum for 30 mm pipe; 3 m maximum for 40 mm pipe; 4 m maximum for 50 mm pipe.			
• Bath waste pipe length – 3 m maximum for 40 mm pipe; 4 m maximum for 50 mm pipe.			
10. Drainage			
• Layout (existing and proposed), materials, sizes, jointing, access, falls, trapped, consent from adjoining owner, vented, depth of cover, protection, bedding, internally – screwed down and double sealed, inverts, position of sewer drains within buildings protected, undermining foundation, damage by trees.			
• Rainwater drainage – gutter/rwp sizes/falls/materials/outlets/weirs/soil pipe quality internally, expansion joints.			
• Soakaway trial hole inspection required.			
• Cesspools, septic tanks (see separate departmental guidance notes).			
11. Heating			
• Appliances, class/flue/hearth/combustion air/chimney height/debris catchment/distance to combustible material.			
• Hot water storage/heating system controls (exempt $<$15 L space-heating, or industrial processes).			

(*Contd*)

Key Areas	*Applicable?*		*Comments*
	Yes	*No*	

12. Stairways
    - Private stairs – rise 220 mm maximum/ going 225 mm minimum.
    - Access stairs – rise 170 mm maximum/ going 250 mm minimum.
    - Headroom 2000 mm minimum/unclimbable guarding/handrails/maximum 100 mm between balusters and treads, glazing including low level landings/ramps 1:12/floor levels/subdivision.
13. Thermal
    - SAP rating required?
    - Ideal *U*-values (W/m^2K) – 0.25 for walls, 0.16 for roofs, 0.22 for exposed floors, 1.80 for doors and windows.
    - Cold bridging to lintels, reveals, air leakage.
    - Heating controls.
    - Lagging to pipework and tanks.
    - Suitability of cavity wall insulation for the exposure.
    - Where building is $>100\,m^2$, then efficient lighting and controls required.
14. Access for disabled people, etc.
    - Dropped kerbs, tactile warning, 1:20 approach, ramps, stairs, landings (non-slip), handrail, access hazards, tread profile, door widths, visibility zones, lobbies, lifts, WC compartment, communication aids, access to all services, signage, MOE.
    - Level thresholds, weathering provisions.
    - Appropriate ironmongery.
15. Safety glazing
    - 1500 mm high to doors/side panels, 800 mm elsewhere, screen protection to glazing (maximum 75 mm) and restricted to 100 mm maximum window opening, manifestation.
    - Georgian wired glass required?
16. Generally
    - Notes/amendments on drawing(s)?
    - Location plan to scale of 1:1250 showing property in red?
    - Colours to show downtakings and additions.
    - Statement on drawing to building control signed.

Appendix F

Typical specimen maintenance schedules

1. Building information

Property: 2 Acme Road, Anytown				
Area: rear extension		*Section: roof finishes*		*Ref: (A)*
Item	*Material/product*	*Manufacturer/supplier*	*Location*	*Comments*
(A)a	The pc concrete paving slabs	Builders' Supplies Co, 12 Right Road, Busytown	Flat roof (A)	Trade quality only
(A)b	Ballust pebbles	Builders' Supplies Co, 12 Right Road, Busytown.	Flat roof (A)	Clean pebbles only to be used
(A)c	Single ply elastomeric membrane	'Bitumera' Ltd, 1 Any Street, Utopia	Flat roof (A)	This material is currently not marketed in the UK, but may be available by special order
(A)d	Cork insulation	'Corkyco' Ltd, 1 New Terrace, Busytown	Flat roof (A)	This material is tapered to give the required 1:40 falls.
(A)e	'Truvent' vapour control felt	'Bitumera' Ltd, 1 Any Street, Utopia	Flat roof (A)	
(A)f	'Metalco' vertical spigot gunmetal rainwater outlet with flat or dome grating to suit	'Metalco' Ltd, 3 Broad Avenue, Busytown	Flat roof (A)	

2. Maintenance details

Property: 2 Acme Road, Anytown			
Area: rear extension		*Section: roof finishes*	*Ref: (A)*
Item	*Description of system*	*Maintenance requirements*	*Servicing cycle*
	For the location of the roof see drawing RE/GEN/03 (A2 size)		
(A)1a	ROOF OVER MAIN EXTENSION 50 mm thick pc concrete paving slabs (with 20 mm nominal rounded washed pebble ballust to provide 200 mm margin at abutments and perimeter) on plastic base pads at each corner; single ply elastomeric membrane; cork insulation tapered to falls; reinforced vapour control layer on concrete deck.	Inspection required to: identify damage to felt upstands; and to ensure that flashings are secure and in good order;	Once per year minimum
		pebbles are still in place; concrete slabs and gutter grating are secure and undamaged; rainwater outlets are clear and in good order; wall copings are secure and in good order.	Four times a year

Appendix G

(Reproduced by permission of the Energy Efficiency Office)

Normalized performance index (NPI) calculation form

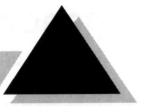

NPI Calculation Form for Hotels

1. Convert your energy use into kWh units

Add your quarterly or monthly use over one year for each fuel and enter below

Natural gas	Therms	× 29.31	=	kWh	
	Cubic feet	× 0.303	=	kWh	
Gas oil (35 sec)	Litres	× 10.6	=	kWh	
Light fuel oil (290 sec)	Litres	× 11.2	=	kWh	
Medium fuel oil (950 sec)	Litres	× 11.3	=	kWh	
Heavy fuel oil (3500 sec)	Litres	× 11.4	=	kWh	
Coal	Tonnes	× 7600	=	kWh	
Anthracite	Tonnes	× 9200	=	kWh	
Liquid petroleum gas (LPG)	Litres	× 7	=	kWh	
	Tonnes	× 13900	=	kWh	
Electricity	kWh	× 1	=	kWh	
Total energy use for the year			=	kWh	A

2. Find your space-heating energy use

If you can identify any of the fuels above used *only* for space heating, enter the total energy use in kWh

1.
2.
3.

Add these to give total kWh B

For fuels used for space heating *and* hot water, where these are not separately metered, take 60% energy used. This figure may also be used for all electrically heated buildings.

1.
2.

Total	× 0.60	=	kWh	C
Annual space heating energy	(B or C)	=	kWh	D
Annual non-space heating energy	(A-D)	=	kWh	E

3. Adjust the space-heating energy to account for weather

Find degree days for energy data year = F

The weather correction factor $= \frac{2462}{F} =$ G

Adjust the space-heating energy to standard conditions (D × G) = kWh H

4. Adjust the space-heating energy to account for exposure

Obtain the exposure factor from the booklet to suit the location of the building = J

Adjusted space-heating energy (H × J) = kWh K

5. Find normalised annual energy use

E + K = kWh L

6. Find floor area in square metres

Floor area = m^2 M

Or the number of bedrooms = N

7. Find the Normalised Performance Indicator (NPI)

Floor area $NPI = \frac{L}{M}$ kWh/m^2

Bedroom $NPI = \frac{L}{N}$ kWh/bedroom

Appendix H

(kind permission of The City of Edinburgh Council)

House extensions

City development planning

Objective

This guideline is supplementary to local plan policies on conservation and design, providing additional guidance on the erection of house extensions. The Council seeks to protect and enhance the character and amenity of dwellings and neighbourhoods.

Policy context

Local plan policies promote new development of the highest quality and seek to protect character and amenity. For example, Policy CD_{19} of the Central Edinburgh Local Plan states:

> 'The Council will permit alterations and extensions to buildings which in their design and form, choice of materials and positioning are compatible with the character of the original building, will not result in an unreasonable loss of privacy or natural light to neighbouring properties and are not detrimental to neighbourhood amenity and character. Particular attention will be paid to ensuring that such works to listed buildings and non-listed buildings in conservation areas do not damage their special architectural character.'

Scope of guidance

These supplementary guidelines apply to extensions to houses and flats on a city-wide basis.

Statutory requirements

Listed building consent will always be required for the erection of an extension on a listed building. In certain circumstances, an application for planning permission will also be required.

On unlisted buildings, planning permission may be required for the erection of an extension, depending on the location and type of building and the size and height of the extension. Planning permission will always be required for extensions to flatted property.

Proposals for house extensions will be considered against the relevant local plan policies and the design principles set out below, as well as the individual circumstances of the application. Compliance with these guidelines will not in itself guarantee that consent will be granted.

1. Quality

(a) It is not the Council's intention to prevent owners from adding extensions but rather to ensure the extensions are of a quality of construction and design appearance equivalent, or superior, to that in the existing development. The size and position of the extension should be such as to permit this to be achieved.

(b) High quality, modern, innovative design will be encouraged where it enhances the existing building and townscape.

2. Townscape

(a) Where an extension, of a similar height and/or massing as the original dwelling, could connect separate houses into a continuous terrace, planning permission will normally be refused. Where a separation distance of 5 m between the extension and the side boundary cannot be achieved, extensions of a

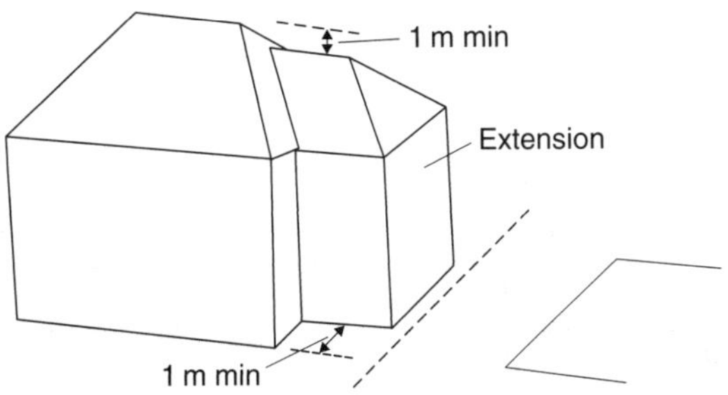

Figure 1.

similar height as the existing dwelling should be set back a minimum of 1 m from the existing front and rear building lines and be a minimum of 1 m lower than the existing at ridge (see Figure 1).

(b) In general, the position and design of an extension should allow for the possibility of neighbours adding similar or equivalent extensions. No extension should detract from the integrity of the original dwelling or particular character of the area.

(c) There is a general presumption against extensions which project beyond the front building line. It is frequently beneficial to set the extension behind the line of the existing dwelling to give clear definition between the new and existing design and materials. Single storey extensions, which project beyond the building line may be acceptable where a minimum of 9 m separation to the site boundary is achieved. Modest porches may also be acceptable where they do not detract from the design of the original dwelling or street scene.

(d) Where corner gardens contribute to the character of the area, their openness will be protected by resisting any significant intrusion into the corner ground.

3. *Materials*

Facing materials should normally match those of the existing building or alternatively make an attractive contrast with them. Where the existing dwelling is constructed from stone, it will be appropriate to use stone to form any additions. Where stone gables are built over by an addition, the new elevations should be formed in natural stone. Where modern materials are proposed, they should be of the highest quality and complement with the existing dwelling.

4. *Roof design*

(a) There is a general presumption against flat roofs on two-storey extensions. There is also a presumption against mansard roofs on all extensions unless complementary to the existing house. In general, the pitch and form of an extension roof should match that of the existing roof. There is a presumption against new eaves above the eaves of the existing dwellings.

(b) New gables, greater than single storey in height, will not normally be allowed to the rear of the original back wall.

(c) Development above the existing roof ridge will be resisted. A rear roof, terminating at a similar or equivalent height as the existing roof, should be no greater in length than 50% of the depth of the original dwelling (see Figure 2).

5. *Dormers*

(a) Modest individual dormers are generally more appropriate than large single box like dormers. Dormers should be of a size such that they do not dominate the form of the roof. It is not normally acceptable

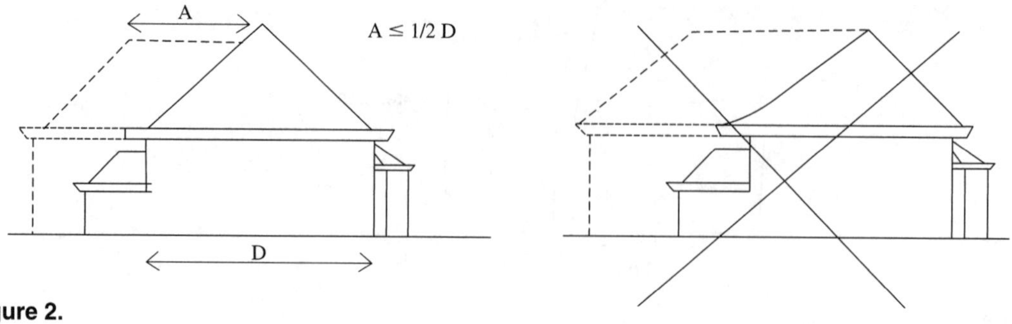

Figure 2.

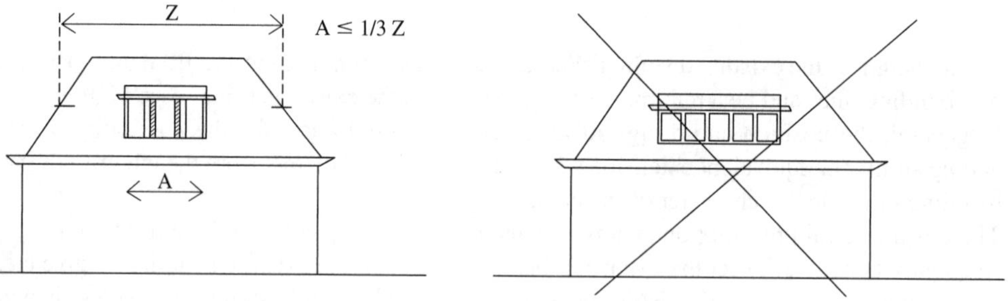

Figure 3.

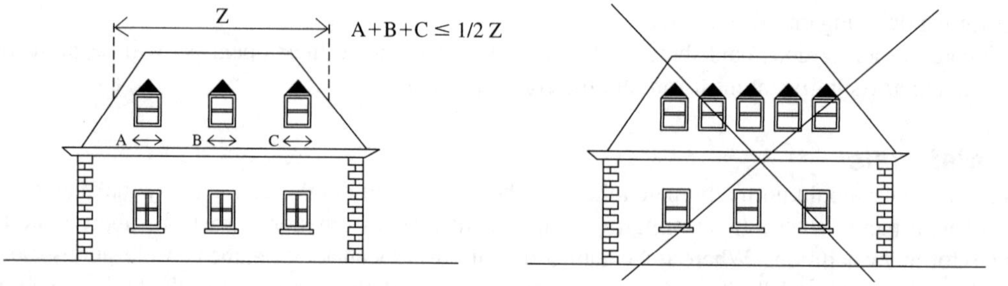

Figure 4.

for a single dormer to be greater than one third of the average roof length (see Figure 3). There is a presumption against two or more dormers which represent greater than 50% of the average length of a single roof plane (see Figure 4).

(b) Where a number of dormers are incorporated, on a single roof plane, the separation between the dormers should be carefully considered. It is normally more appropriate to follow the proportions and alignment of the window openings of the existing dwelling (see Figure 4). Where a dwelling has a number of individual dormers, infilling the area between these dormers will not normally be allowed.

(c) The relationship between the dormer and the eaves, ridge, gable and hip is particularly important. It is normally preferable to keep dormers a minimum of 500 mm clear of the roof ridge and hip, a minimum of 500 mm from the eaves and a minimum of 1 m from the gable (see Figure 5). If this cannot be achieved it is preferable to use velux roof-lights where acceptable. An exception would be where an appropriate traditional dormer is proposed on a traditional dwelling. There is a presumption against the extension of an existing ridge, to the front or side, to form a dormer (see Figure 6).

(d) Where dormers are proposed on an existing dwelling, there is a presumption against dormers on >50% of the original roof planes. It is preferable to locate the dormers on opposing roof pitches.

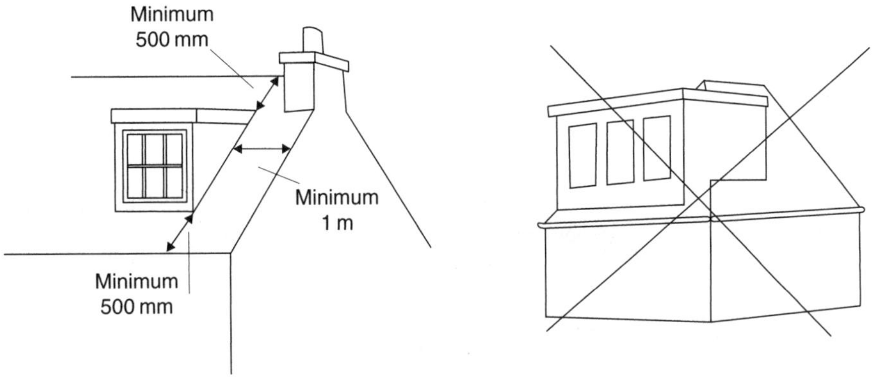

Figure 5.

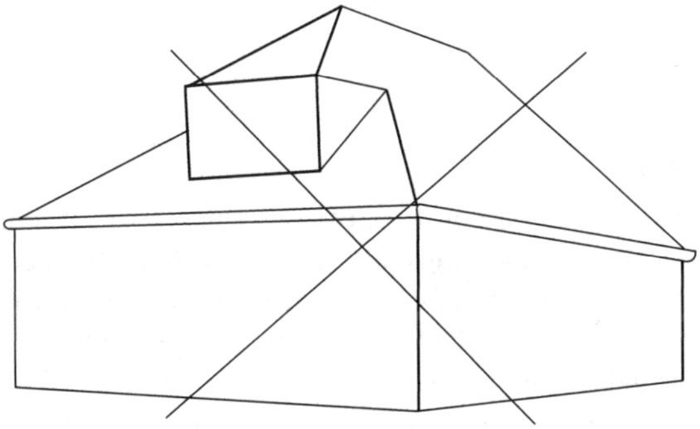

Figure 6.

6. Amenity

(a) Extensions to the rear should not extend beyond one third of the rear garden depth, or a distance equal to that of the main original dwelling depth, from the principal rear building line (see Figure 7). Where there is a traditional development pattern, such as villas with single storey outbuildings, this may determine the form and size of any addition.

(b) A minimum of a 9 m deep garden, within the ownership of the applicant, must be retained, immediately to the rear of any extension. In certain circumstances, such as four-in-a-block flatted accommodation, it may not be appropriate to carry out any form of extension due to the modest amount of garden ground.

(c) New development should be located hard on to a boundary or no less than 1 m from the boundary to prevent dead unusable space and allow access for maintenance.

7. Windows

Generally the scale and proportions of new openings should reflect those of the existing dwelling. All new windows should comply with the Council's guidelines on Window Alterations, where relevant.

8. Garages

The location of a proposed garage should be in harmony with the existing dwelling and street scene. There is a presumption against the construction of a garage beyond the front building line of the existing dwelling.

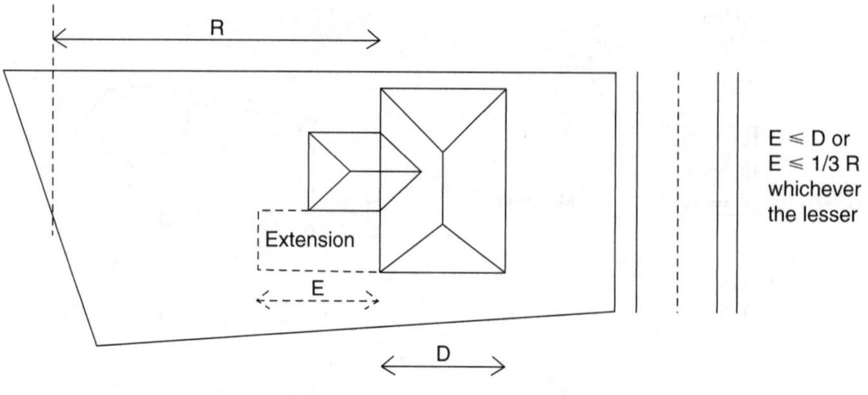

The rear garden dimension is the average dimension from the back wall of the main building mass to the rear boundary.

Figure 7.

Reasoned justification

The Council has operated a guideline on house extensions since 1976. It aims to reconcile the need to maintain the architectural character of dwellings and amenity with the frequent requirement of occupiers to maximize accommodation and to ensure a fair and consistent standard throughout the city. Other relevant guidelines cover the following topics: daylighting, privacy and sunlight; villa areas and the grounds of villas; window alterations; and conservatories.

The appointment of an architect is strongly encouraged in all cases. The Royal Incorporation of Architects in Scotland can provide a list.

A Building Warrant may be required for any addition, for which application should be made to City of Edinburgh Council, City Development, Property Management, 329 High Street, Edinburgh EH1 1PN.

Applicants are advised that there may be limitations on development in their title deeds: for example separate feu superiors' consent may be required and legal advice may be necessary. The Council is unable to provide legal advice on these matters.

Appendix I

(Reproduced by permission of the BRE)

Office scorer refurbishment/redevelopment evaluation form

Existing Building Details...

Please tell us about your existing building

Enter details in square meters or square feet?	○ Sq. Metres ○ Sq. Feet
Gross internal area of building ? (only include area of atria if classed as internal spaces.)	0 m^2/ft^2
Number of storeys including basement ?	0
Plan depth ?	▼
Plan shape of building ?	▼
Current occupant density ?	▼
Potential thermal mass of floors ?	◉ High ○ Low
What ceilings does your building have? ?	◉ Suspended ◉ Exposed
Ventilation / Cooling system ?	▼
Does the building overheat in the summer? ?	▼
Is the building heated with electric panels/storage heaters? ?	○ Yes ◉ No
Average glazing ratio ?	▼
Location ?	▼
Current rental value ?	£ 0 Per m^2/ft^2

Next

Refurbish or Redevelop Decision...

Please tell us about your existing building

Have you decided to?	Refurbish ▼
Main reason for decision if made	☐ Usability of space ⓘ ☐ To accommodate more staff ⓘ ☐ Location ⓘ ☐ Market forces in locality ⓘ ☐ Length of lease ⓘ ☐ Major works due ⓘ ☐ Risk of planning refusal ⓘ ☐ Length of construction phase ☐ Site ratio ☐ Provision of parking ⓘ ☐ Floor to ceiling height ⓘ ☐ No raised access floors ⓘ ☐ Floor loading ⓘ ☐ Building is too cold at times ⓘ ☐ Building overheats at times ⓘ ☐ Length of investment/payback period ⓘ
Expected Occupant Density	10–15 m^2 per person ▼

Next

Refurbishment Specification...

Please enter the following refurbishment details:

Is this a major or complete refurbishment? ?	☐ Refurbish Floors, ceilings, internal walls & services only ☒ Refurbish all services & fabric except floors & structure
Are you adding an internal atrium or additional storeys? ?	Neither ▼
Revised plan depth (if adding an atrium) ?	Less than 14 m ▼
Gross internal area of building. ? (only include area of atria if classed as internal spaces.)	0 m²/ft²
Total number of storeys (inc basement and any additional storeys) ?	0
Ceiling Type ?	☐ Suspended ☒ Exposed
Ventilation/Cooling strategy ?	Nat vent with natural night cooling ▼
Is the building heated with electric panels/storage heaters? ?	☐ Yes ☒ No
Have you considered renewable energy systems? ?	☐ Yes ☒ No
Average Glazing Ratio ?	Between 30–45% ▼
Are you refurbishing/providing new solar shading? ?	Blinds ▼
Have you considered efficient water systems? ?	☐ Yes ☒ No
Expected rental value ?	£ 0 Per m²/ft²

Next

Redevelopment Specification...

Please enter the following redevelopment details.

Gross internal area of building ? (only include area of atria if classed as internal spaces.)	0 m²/ft²
Number of new storeys (inc basement) ?	0
Plan shape of building ?	Square or Rectangular ▼
Plan depth ?	Less than 14 m ▼
Potential thermal mass of building ?	◉ High ☐ Low
External facade ?	☐ New ◉ Retain
Ceiling Type ?	☐ Suspended Ceiling ◉ Exposed floor structure
Ventilation/Cooling Strategy ?	Nat vent with natural night cooling ▼
Is the building heated with electric panels/storage heaters? ?	☐ Yes ◉ No
Have you considered Renewable Energy Systems? ?	☐ Yes ◉ No
Average glazing ratio ?	Between 30–45% ▼
Are you refurbishing/providing new solar shading? ?	Blinds ▼
Have you considered efficient water systems? ?	☐ Yes ◉ No
Are you relocating? ?	City centre ▼
Expected rental value ?	£ 0 Per m²/ft²

Results

Glossary of building adaptation-related terms

Adaptation

Any work to a building over and above maintenance to change its capacity, function or performance. In general terms adaptation means the process of adjustment and alteration of a structure or building and/or its environment to fit or suit new conditions (Chudley, 1983). However, more specifically it is also considered as work accommodating a change in the use or size or performance of a building, which may include alterations, extensions, improvements and other works modifying it in some way.

Adaptive reuse

Conversion of a facility or part of a facility to a use significantly different from that for which it was originally designed (Iselin and Lemer, 1993).

Alteration

Modifying the appearance, layout or structure of a building to meet new requirements (Watt, 1999). It often forms part of many adaptation schemes rather than being done on its own.

Beam-filling

A crude masonry infill about 600 mm high found in traditional pitched roofs of Georgian and Victorian buildings to provide a degree fire stopping and draughtproofing at the eaves.

Bressummer

A large timber beam found in older buildings. It spans a wide fireplace or a bay window. The built-in ends of this wooden load-bearing member are highly prone to fungal attack.

Carbonation

The transformation in concrete or cement mortar of the free alkali and alkali-earth hydroxides in the cement matrix into carbonates, due to attack by carbon dioxide in the atmosphere (CIB, 1993).

Conservation

Preserving a building purposefully by accommodating a degree of beneficial change. It includes any 'action to secure the survival or preservation of buildings, cultural artefacts, natural resources, energy or other thing of acknowledged value for the future' (BS 7913: 1998).

Consolidation

Basic adaptation and maintenance works to ensure a building's ongoing beneficial use.

Conversion

Making a building more suitable for a similar use or for another type of occupancy, either mixed or single use.

Cryptoclimate

The climate created by the building and influencing its fabric (Harper, 1978). Adapting an existing building can change its cryptoclimate.

Dampness

Condition of being slightly wet – usually not so wet that liquid water is evident – such as wetness caused by condensation on a porous substrate or water transmitted up a porous wall by capillary action (CIB, 1993). Dampness can be said to occur when an atmosphere or material is wetter than 85% RH (Oliver et al., 1997).

Defect

Non-conformity with standard or shortfall in performance (CIB, 1993). Also known as a fault, or deviation from the intended performance of a building or its parts (ISO 15686-1:2000).

Degradation

This usually refers to the deleterious effects of sunlight, in particular, on organic materials. Strictly speaking degradation is defined as the conversion of a complex molecule into simpler fragments.

Design life

The period of time over which a building or a building subsystem or component (e.g., roof, window, plumbing) is designed to provide at least an acceptable minimum level of performance (Iselin and Lemer, 1993). It is the period of use as intended by the designer – for example, as stated by designer to the client to support specification decisions (BS 7543).

Deterioration

A reduction in ability to perform up to the anticipated standard (CIB, 1993). It often results in decay beyond normal repair.

Durability

The ability of a building or its parts to perform its required functions over a period of time and under the influence of internal and external agencies or mechanisms of deterioration and decay (Watt, 1999). It is also a measure of a building's ability to resist deterioration.

Economic life

The period of time over which costs are incurred and benefits or dis-benefits are delivered to an owner; an assumed value sometimes established by tax regulations or other legal requirements or accounting standards not necessarily related to the likely service life of a facility or a subsystem (Iselin and Lemer, 1993).

Extension

Expanding the capacity or volume of a building, whether vertically by increasing the height/depth or laterally by expanding the plan area.

Failure

The termination of an item's ability of to perform a required function. It can also be classed as the consequence or effect of a defect.

Fault

An unexpected deviation from requirements, which would require considered action regarding the degree of acceptability. It is also considered as a departure from good practice.

Hot desking

Allocating desk space to staff in a large office on a daily or weekly basis when they need them.

Improvement

Bringing a building and/or its facilities up to an acceptable or higher standard as required by the building regulations or occupier. Beneficial improvement entails replacing something with a new item on a like-for-like basis. Substantive improvement, on the other hand, involves the replacement of an element or component with a new item having a higher performance rating.

Industrialized building

This is not necessarily synonymous with 'non-traditional' construction. Industrialized processes could be used with traditional building, although generally more extensively with more non-traditional forms. Industrialization is a wider term than non-traditional because it is primarily concerned with the rationalization of the building process itself. In building operations it implies the use of mechanical plant and the replacement of in situ work by prefabricated units (AMA, 1984).

Insured life

The period of time between the end of premature failure and average service life is considered the insured life of a component – that is, the point at which failure would cease to be regarded as premature (HAPM, 2000).

Kentledge

This a series of heavy blocks, usually reinforced concrete, to act as a stabilizing mass on scaffolding or as a means of testing the load-bearing capacity of structural elements such as floors and piles.

Life cycle

The sequence of events in planning, design, construction, use, and disposal (e.g., through sale, demolition or substantial renovation) during the service life of a facility; may include changes in use and reconstruction (Iselin and Lemer, 1993).

Life-cycle cost

The present value of all anticipated costs to be incurred during a facility's economic life; the sum total of direct, indirect, recurring, non-recurring and other related costs incurred or estimated to be incurred in the design, development, production, operation, maintenance, support, and final disposition of a major system over its anticipated life span (Iselin and Lemer, 1993).

Loss

The consequences of a defect or failure, expressed in terms of costs, injuries, loss of life, etc. (CIB, 1993).

Maintenance

A 'combination of all technical and administrative actions, including supervision actions, intended to retain an item in, or restore it to, a state in which it can perform a required function' (BS 3811: 1993). Maintenance involves routine work necessary to keep the fabric of a building, the moving parts of machinery, etc. in good order (BS 7913: 1992). In other words, it consists of regular ongoing work to ensure that the fabric and engineering services are retained to minimum standards (Ashworth, 1997). The word 'maintenance' comes from the French verb 'maintenir', which means to hold (Chudley, 1983).

Macroclimate

The typical climate of a region or country, based on meteorological records (Chandler, 1989). This should be taken into account when planning a major adaptation scheme such as refurbishment or large extension.

Mesoclimate

The climate of a district or part of a region, influenced by topography and exposure (Harper, 1978). Again, cognizance of the macroclimate should be taken at the adaptation design stage to reduce problems caused by topography or changes in exposure.

Microclimate

The site climate created by deviations from the macroclimate and microclimate resulting from local land usage (Harper, 1978). Adaptation such as lateral extensions can affect a property's microclimate by altering wind and other exposure patterns around the building.

Modernization

Bringing a building up to current standards as prescribed by occupiers, society and/or statutory requirements.

Non-traditional building

This form of construction may use unfamiliar products or the same basic traditional materials in new ways, employing new techniques in fixing and erection, which differ, for instance, from the traditional method of laying by hand one brick, concrete block, on top of another. It is a narrower concept than 'industrialized building' but also uses prefabricated elements (AMA, 1983).

Obsolescence

The condition of being antiquated, old fashioned or out of date, resulting when there is a usually rapid change in the requirements or expectations regarding the shelter, comfort, profitability or other dimension of performance that a building or building subsystem is expected to provide. Obsolescence may occur because of functional, economic, technical or social and cultural change (Iselin and Lemer, 1993).

Performance

The degree to which a building or other facility serves its users and fulfils the purpose for which it was built or acquired; the ability of a facility to provide the shelter and service for which it is intended (Iselin and Lemer, 1993). It is a quantitative expression of behaviour of an asset or product in use (BS 6019).

Physical service life

The time it takes for a building, subsystem or component to wear-out. It is the time period after which a facility can no longer perform its function because increasing physical deterioration has rendered it useless (Iselin and Lemer, 1993).

Preservation

Arresting or retarding the deterioration of a building or monument by using sensitive and sympathetic repair techniques. Preservation means 'the state of survival of a building or artefact, whether by historical accident or through a combination of protection and active conservation' (BS 7913: 1998). It also can be defined as 'the act or process of applying measures necessary to sustain the existing form, integrity and materials of an historic property (Weeks and Grimmer, 1995). Preservation focuses on the maintenance and repair of existing historic materials and retention of a property's form as it has evolved over time. It includes protection and stabilization measures.

Protection

The legal use of this term involves the provision of legal restraints or controls on the destruction or damaging of buildings, etc. with a view to ensuring their survival or preservation for the future. Physical protection may be either temporary (e.g., tarpaulins over an exposed roof surface undergoing refurbishment) or permanent (e.g., over-roofing scheme).

Rebuilding

Remaking, on the basis of a recorded or reconstructed design, a building or part of a building or artefact that has been irretrievably damaged or destroyed (BS 7913: 1999).

Reconstruction

The re-establishment of what occurred or what existed in the past, on the basis of documentary or physical evidence (BS 7913: 1999). Reconstruction, in other words, re-creates vanished or non-surviving portions of a property for interpretative purposes (Weeks and Grimmer, 1995).

Recycling

Transforming or re-utilizing a redundant or other underused/unused building or its materials for more modern purposes.

Redundancy

The result for a building when it becomes superfluous or excess to requirements. It is often triggered by obsolescence.

Refurbishment

Modernizing or overhauling a building and bringing it up to current acceptable functional conditions (Watt, 1999). It is usually restricted to major improvements primarily of a non-structural nature to commercial or public buildings. However, some refurbishment schemes may involve an extension.

Rehabilitation

Work beyond the scope of planned maintenance, to extend the life of a building, which is socially desirable and economically viable (Watt, 1999). It is a term that strictly speaking is normally confined to housing. Rehabilitation can also be defined as 'the act or process of making possible a compatible use for a property through repair, alteration and additions while preserving those portions or features which convey its historical, cultural or architectural values' (Weeks and Grimmer, 1995). It acknowledges the need to alter or add to a historical property to meet continuing or changing uses while retaining the property's historic character.

Reinstatement

Major repair and restoration works to put back a building to its condition prior to substantial damage such as fire, flood or earthquake.

Relocation

Dismantling and re-erecting a building at a different site. It can also mean moving a complete building to a different location nearby.

Remodelling

This is a North American term analogous to adaptation. It essentially means to make new or restore to former or other state or use.

Renewal

Substantial repairs and improvements in a facility or subsystem that returns its performance to levels approaching or exceeding those of a recently constructed facility.

Renovation

Upgrading and repairing an old building to an acceptable condition, which may include works of conversion.

Repair

This is the 'restoration of an item to an acceptable condition by the renewal, replacement or mending of worn, damaged or decayed parts' (BS 8210: 1993). It is associated with the rectification of building components that have failed or become damaged through use and misuse (Ashworth, 1997).

Restoration

To bring back an item to its original appearance or state (BS 3811). It is often undertaken to depict a property at a particular period of time in history, while removing evidence from other eras. This usually involves reinstating the physical and/or decorative condition an old building to that of a particular date or event. It includes any reinstatement works to a building of architectural or historic importance following a disaster such as extensive fire damage. Restoration may also be defined as 'the act or process of accurately depicting the form, features and character of a property as it appeared at a particular period in time by means of the removal of features from other periods in its history and reconstruction of missing features from the restoration period' (Weeks and Grimmer, 1995).

Retrofitting

The redesign and reconstruction of an existing facility or subsystem to incorporate new technology, to meet new requirements or to otherwise provide performance not foreseen in the original design (Iselin and Lemer, 1993). In other words, retrofitting is the replacement of building components with new components that were not available at the time of the original construction (Ashworth, 1997).

Revamp

An informal term used to describe overhauling a building to upgrade its appearance and facilities. It is sometimes used by laypeople as an alternative expression to refurbishment.

Revitalization

Extending the life of a building by providing new or improving existing facilities, which may include major remedial and upgrading works (Watt, 1999).

Service (or working) life

Actual period of time during which no excessive expenditure is required on operation, maintenance or repair of a component or construction – as recorded in use (BS 7543).

Stabilization

Substantial maintenance and adaptation works to ensure a building's long-term beneficial and safe use. It often includes major repairs and strengthening works such as stitching and underpinning.

Subsystem

Functional part of a system, and often used interchangeably with that term – for example, heating subsystem being part of HVAC system (Iselin and Lemer, 1993).

Sustainability

A set of processes aimed at delivering efficient built assets in the long term (DETR, 1998). 'Eco-renovation' occurs where sustainable issues are deliberately and explicitly incorporated into an adaptation scheme (Harland, 1998).

System

Collection of subsystems, components or elements that work together to provide some major aspect of shelter or service in a constructed facility (e.g., plumbing system, heating system, electrical system and roofing system). Also, a set of building components specifically designed to work together to facilitate construction, such as an integrated building system (Iselin and Lemer, 1993).

System building

This form of construction usually relies on both non-traditional and industrialized methods (AMA, 1985). It took these a stage further by producing a combination of materials and methods for a design and construction package. A system building provides both a design and technique (often with specially made plans and components), which is available from one design (and possible one construction) organization. System building therefore is synonymous with proprietary forms of construction, such as 'Airey', 'Bison', 'Orlet', 'Skarne', etc.

Technology

This can be defined simply as the systematic study and application of how artefacts are made and used. Building adaptation is one aspect of applied construction technology.

Teleworking

This is modern work practice in offices, which is intended to achieve more flexibility, using computer monitors and keyboards with phone links, such in satellite television call centres.

Terotechnology

This is a combination of management, financial, engineering, building and other practices applied to physical assets in pursuit of economic life-cycle costs (BS 3811: 1993). Terotechnology is aimed at achieving the best possible value for money for a user from the procurement and subsequent employment of a physical asset. It is concerned with total costs over a building's full life, and is derived from the Greek word 'tereo', *I care*).

Traditional construction

Traditional methods are basically on the principle of an on-site operation where all the materials traditionally required for the building are first gathered together, such as bricks, cement, sand, ballast, timber tiles and plaster. It predominantly involves the use of relatively small-scale units such as bricks and blocks, which are assembled or installed on site.

Turnerization

The term used to describe a proprietary system of applying a bituminous coating to slated or tiled roofs to enhance their weathertightness.

Upgrading

Enhancing the performance characteristics of a building's major elements, components and/or services.

Whole life-cycle cost

Generic term for the costs associated with owning and operating a facility from inception to demolition, including both initial capital costs and running costs (Watt, 1999).

Typical maintenance activities

Built-in maintenance
This is maintenance that is required in every building to meets the functional needs of the owner/occupier.

Built-on maintenance
This is a form maintenance performed on materials that have been included in the building, but are not necessary for its function. As an example, consider the need for carpeting on the office corridor floor of a warehouse area. A value-engineering analysis would probably eliminate the need to provide such a carpet.

Condition-based maintenance
This is a form a preventive maintenance. It is initiated as a result of knowledge of an item's condition from routine or continuous monitoring.

Contingency maintenance
This is similar to reactive maintenance, except that it contains an element of planning. It is sometimes referred to as 'casual maintenance'. Unplanned requisitioned or emergency maintenance can fall under this heading.

Corrective maintenance
Maintenance carried out to restore (including adjustment and repair) an item that has ceased to meet an acceptable standard.

Cyclical maintenance
This comprises those items of maintenance, which recur at regular intervals such as redecoration, changing filters or the resurfacing of roads or paths, etc. It is termed cyclical because the maintenance process will have to be repeated regularly during the life of the building.

Day-to-day maintenance
This deals with instances of essential repair, which cannot be left until the next routine maintenance cycle without serious consequences. Thus day-to-day maintenance is difficult to predict and organize.

Design-out maintenance

Other forms of maintenance may be inappropriate; therefore maintenance needs are designed out to achieve the required level of reliability.

Emergency maintenance

This is maintenance that is necessary to be attended to immediately. It aims to avoid serious consequences, usually in terms of safety and security. Unforeseen breakdown or damage may necessitate emergency maintenance.

Fixed time maintenance

Activities repeated at pre-determined intervals, within a planned maintenance system.

Just-in-time maintenance

This is maintenance derived from manufacturing and delivery industries in which the basic approach is to continually reduce (product) costs. This is achieved by stressing the elimination of waste, no rejects, no delays, no stockpiles, no queues, no idleness and no useless motion.

Maintenance-free

This is an ideal objective but is usually difficult to achieve owing to limited resources. Some materials and components are occasionally claimed to be 'maintenance-free' – such as stainless steel doors, etc. Maintenance-free products, however, can be damaged or disfigured, so may still require some upkeep.

Opportunity maintenance

Work done as and when possible within the limits of operational demand.

Periodic renewals

Regular changes of items, such as replacing carpets, painting or overhauling compressors (Iselin and Lemer, 1993).

Planned maintenance

Maintenance organized and carried out with forethought, control and the use of records to a pre-determined plan.

Planned preventive maintenance

Maintenance carried out at pre-determined intervals, or to other prescribed criteria, and intended to reduce the likelihood of an item not meeting an acceptable standard.

Preventive maintenance

This is work carried out in anticipation of failure.

Planned short service life

A decision that the service life of a facility should be shorter than might typically be expected; implies selection of components that have low first cost and low durability; similar to the term 'planned obsolescence' used in the automobile and consumer products industries (Iselin and Lemer, 1993).

Reactive or unplanned maintenance

Maintenance that is left until there is a major breakdown or a serious complaint from the user before action is taken.

Reliability centred maintenance

Reliability centred maintenance (RCM) is a method of maintenance designed to anticipate the mode and consequences of failure. It helps to select the appropriate tasks to prevent failure before it occurs based on risk and experience.

Running maintenance

Maintenance carried out while the item is in service.

Schedule maintenance

This is designed to cover the items that deteriorate at a more or less uniform rate and which have a high degree of urgency.

Shut-down maintenance

Maintenance that needs to be carried out on an item when it is taken out of use.

Statutory maintenance

Maintenance required by law to prevent serious injury or damage should failure occur. Examples of this type of maintenance are servicing of lifts and associated plant, electrical equipment, water treatment, steam boilers, pressure vessels and air conditioning.

Maintenance response phrases

Mean down time

Mean down time (MDT) applies to corrective maintenance. Down time is the total time from failure occurring till the failure has been rectified and inspected.

Mean maintenance time or mean time to repair

This applies to preventive maintenance. Mean maintenance time (MMT) or mean time to repair (MTTR) includes the repair phase and check out phase of MDT. It is often expressed in percentile terms such as the 95 percentile repair time shall be 1 hour.

Mean life

Mean life (ML) describes the average life of an item, taking into account wear-out – whereas mean time to fail (MTTF) is the average time between failures.

Mean time to fail and mean time between failure

This is defined as the ratio of cumulative time to the total number of failures for a stated period in the life of an item. It is applied to items that are not repaired, such as bearings and transistors. Mean time between failure (MTBF) is similar to mean time to fail (MTTF). The only difference between them is in their usage. MTBF is applied to items that are repaired.

Bibliography

This bibliography offers a comprehensive list of publications dealing with the adaptation of buildings and cognate constructional issues. It contains the sources referred to in this book as well as other references that the reader may wish to consult. Also included are many titles as well as guides and reports not directly mentioned in the text on various aspects of refurbishing and maintaining buildings.

Books and papers

Some of the books listed hereunder were originally published in the USA. Despite the environmental and constructional differences between Britain and America, though, the texts in both countries contain much that is relevant to all other developed countries.

Although several of the titles listed are out of print they should still be available for reference in any good construction library within academic establishments and professional institutions dealing with the built environment.

Addis, B. and Talbot, R. (2000) *Sustainable Construction Procurement, C571*. London: Construction Industry Research and Information Association.

Addis, W. and Schouten, J. (2004) *Principles of Design for Deconstruction to Facilitate Reuse and Recycling, C607*. London: Construction Industry Research and Information Association.

Addleson, L. (1992) *Building Failures: A Guide to Diagnosis, Remedy and Prevention* (3rd edition). London: Butterworth-Heinemann.

Addleson, L. and Rice, C. (1992) *Performance of Materials in Building*. Oxford: Butterworth-Heinemann.

Adler, D. (ed.) (1999) *Metric Handbook: Planning and Design Data* (2nd edition). London: Architectural Press.

Agocs, Z., Ziolko, J., Vican, J. and Brodniansky, J. (eds) (2004) *Assessment and Refurbishment of Steel Structures*. New York: Van Nostrand.

Alexander, D. E. (2002) *Principles of Emergency Planning and Management*. Herts: Terra Publishing.

ALGAO (1997) *Analysis and Recording for Conservation and Control of Works to Historic Buildings*. London: Association of Local Government Archaeological Officers.

Allen, W. (1997) *Envelope Design for Buildings*. London: Architectural Press.

Allinson, K. (1997) *Getting There by Design (An Architect's Guide to Design and Project Management)*. London: Architectural Press.

AMA (1983) *Defects in Housing Part 1: Non-Traditional Dwellings of the 1940s and 1950s*, July 1983. London: The Association of Metropolitan Authorities.

AMA (1984) *Defects in Housing Part 2: Industrialised and System Built Dwellings of the 1960s and 1970s*, March 1984. London: The Association of Metropolitan Authorities.

AMA (1985) *Defects in Housing Part 3: Repair and Modernisation of Traditional Built Dwellings*, March 1985. London: The Association of Metropolitan Authorities.

Ambrose, J. E. (1992) *Building Construction and Design*. New York: Van Nostrand Reinhold.

Andersson, H. E. B. and Setterwall, A. K. (eds) (1996) *The Energy Book: A Resume of Present Knowledge*. Stockholm, Sweden: The Swedish Council for Building Research.

Andrew, C. et al. (1995) *STONECLEANING: A Guide for Practitioners*. Edinburgh, Scotland: Historic Scotland and The Robert Gordon University.

Anink, D., Boonstra, C. and Mak, J. (1996) *Handbook of Sustainable Building: An Environmental Preference Method for Selection of Materials for Use in Construction and Refurbishment*. London: James & James (Science Publishers) Ltd.

Annesley, B., Horne, M. and Cottam, H. (2003) *Learning Buildings*. London: School Works Ltd.

Anon (1993) Housing Refurbishment Special Report, *Architects' Journal*, February 1993. London: Emap Business Publications.

Anon (1994) Building Renewal (Peninsula Barracks conversion), *Building* supplement, 9 December 1994. London: Tower Publications.

Anon (2000a) Insulating External Walls, *Building Engineer*, December 2000. Northampton.

Anon (2000b) Health through Warmth Pledge to People at Risk, *Green Government*, October 2000. Manchester: Partnership Media Group.

Anon (2002) Working Together to Get Results, *Green Government*, September 2002. Manchester: Partnership Media Group.

Anon and Building Research Establishment (1959 and 1962) *Principles of Modern Building*, Vols 1 and 2. London: HMSO.

Anstey, J. (2000) *A Practical Manual for Party Wall Surveyors*. London: RICS Books.

Aplin, G. (2002) *Heritage: Identification, Conservation and Management*. Oxford: Oxford University Press.

APM (2000) *Book of Knowledge*. London: Association of Project Management.

Appleton, N. and Leather, P. (1998) *Carrying Out Adaptations (A Good Practice Guide for Registered Social Landlords)*. London: The Housing Corporation.

Architects' Journal (2003) Housing Refurbishment, Special Report. London: Architects Journal.

Ashjack™ (2001) *Over-roofing, Company Brochure*. West Bromwich: Ash and Lacey.

ASHRAE (1989) *ASHRAE Standard 62-1989 Ventilation for Acceptable Indoor Air Quality*. Atlanta, Georgia: American Society of Heating, Refrigeration and Air-Conditioning Engineers Inc.

Ashurst, J. (1994) *Cleaning Historic Buildings, Vol. 1: Substrates, Soiling and Investigation*. Shaftsbury: Donhead Publishing.

Ashurst, J. and Ashurst, N. (1988) *Practical Building Conservation*, (Vols 1–5). *English Heritage Technical Handbook*. Aldershot: Gower Press.

Ashworth, A. (1997) *Obsolescence in Buildings: Data for Life Cycle Costing, Construction Paper 74*. Englemere: Chartered Institute of Building.

Ashworth, A. (1999) *Cost Studies of Building* (3rd edition). London: Longman.

Askham, P. and Blake, L. (1993) *The Best of Mainly for Students*, Vol. 1. London: Estates Gazette.

Askham, P. and Blake, L. (1999) *The Best of Mainly for Students*, Vol. 2. London: Estates Gazette.

ASTM (1994) *Manual of Moisture Control in Buildings, ASTM Manual 18*. California: American Society of Testing and Materials.

Atkinson, M. F. (2000) *Structural Defects Reference Manual for Low-Rise Buildings*. London: E & FN Spon.

Audit Commission (2001) *Improving School Buildings*. London: Department for Trade and Industry. Available online at http://www.dtigov./construction/kpi/

Auer, M., Fisher, C. and Grimmer, A. (eds) (1988) *Interiors Handbook for Historic Buildings*, Vol. I. Washington, DC: National Park Service/Historic Preservation Education Foundation.

Auer, M., Fisher, C., Jester, T. C. and Kaplan, M. (eds) (1993) *Interiors Handbook for Historic Buildings*, Vol. II. Washington, DC: National Park Service/Historic Preservation Education Foundation.

Austin, S. A., Robins, P. J. and Goodier, C. I. (2002) Construction and Repair with Wet-process Sprayed Concrete and Mortar, Technical Report No. 56. London: The Concrete Society.

Baird, G. and Gray, J. (1995) *Building Evaluation Techniques*. London: McGraw-Hill.

Ball, R. (1999) Developers, regeneration and sustainability issues in the reuse of vacant industrial buildings. *Building Research and Information,* **27**(3), 140–148. London: E & FN Spon.

Banks, O. F. and Tanqueray, R. (1999) *Lofts: Living in Space*. New York: Universe Publishers.

Banton, J. H. (1980) Management of Modernisation, *Site Management Information Service Paper No. 81*. Englemere: Institute of Building.

Barnett, N. P. (ed.) (2004) *Wessex Alterations and Refurbishment Price Book* (9th edition). London: Wessex.

Barton, R. (1997) Carbon fibre 'plate bonding'. *Structural Survey*, **15**(1). Bradford: MCB University Press.

Basement Development Group (1991) *Basements 2 – A Preliminary Assessment of the Design of Basement Walls, Options for Quality in Housing*. London: British Cement Association.

Basement Development Group (1993) *Basements 3 – Thermal Performance of Dwellings with Basements, Options for Quality in Housing*. London: British Cement Association.

Basement Development Group (1994) *Basements 4 – House with a Basement: Design Exercise, Options for Quality in Housing*. London: British Cement Association.

Basement Development Group (1999) *Basements 1 – Benefits, Viability and Cost, Options for Quality in Housing* (2nd edition). London: British Cement Association.

Baumann, J. A. (1966) In Skeist I. (ed.), *Construction Aids, Plastics in Building*. New York: Reinhold Publishing Corporation.

BBA, BRE et al. (1983) *Cavity Insulation of Masonry Walls – Dampness Risks and How to Minimise Them*. London: British Board of Agrément.

BCIS (2002) *Access Audit Price Guide*. London: Building Cost Information Service.

BCSC (1994) *Refurbishment of Shopping Centres*. London: British Council of Shopping Centres.

Beall, C. (1998) *Thermal and Moisture Protection Manual*. New York: McGraw-Hill.

Bech-Andersen, J. (2001) *The Dry Rot Fungus and Other Fungi in Houses* (5th edition). Vejdammen, Denmark: Hussvamp Laboratoriets Forlag.

Bech-Andersen, J. (2001) *Indoor Climate and Mould*. Vejdammen, Denmark: Hussvamp Laboratoriets Forlag.

Becker, F. (1990) *The Total Workplace (Facilities Management and the Elastic Organisation)*. New York: Van Nostrand Reinhold Inc.

Becker, F. and Steele, J. (1994) *Workplace by Design*. San Francisco: Jossey-Bass.

Beckmann, P. and Bowles, R. (eds) (2004) *Structural Aspects of Building Conservation* (2nd edition). Oxford: Butterworth-Heinemann.

Belle, J., Hoke, J. R. and Kliment, S. A. (eds) (1991) *Traditional Details for Building Restoration, Renovation and Rehabilitation (From the 1932–1951 Editions of 'Architectural Graphics Standards'*. New York: John Wiley & Sons.

Beneke, J. (2000) *Converting Garages, Attics and Basements*. New York: Sunset Books.

Benson, J. et al. (1980) *The Housing Rehabilitation Handbook*. London: Architectural Press.

Best, R. and de Valance, G. (1999) *Building in Value (Pre-design Issues in Construction)*. London: Arnold.

Bett, G., Hoehnke, F. and Robison, J. (2003) *The Scottish Building Regulations Explained and Illustrated* (3rd edition). Oxford: Blackwell Publishing.

Bickerdicke Allen and Partners (1996) *Design Principles of Fire Safety*. London: HMSO.

Billington, M. J., Simmons, M. W. and Waters, J. R. (2003) *The Building Regulations Explained and Illustrated* (12th edition). Oxford: Blackwell Publishing.

Binney, M. (1984) *Chatham Historic Dockyard: Alive or Mothball*. London: SAVE Britain's Heritage.

Binney, M. and Martin, K. (1982) *The Country House: To Be or Not to Be*. London: SAVE Britain's Heritage.

Binney, M., Machin, F. and Powell, K. (1990) *Bright Future: The Re-use of Industrial Buildings*. London: SAVE Britain's Heritage.

BMI (1997) *BMI Special Report on Rehabilitation* (Serial 287). London: Building Maintenance Information, RICS Business Services.

BMI (2001a) *The Economic Significance of Maintenance* (Serial 306, BMI Special Report), Building Maintenance Information. London: Building Cost Information Service.

BMI (2001b) *Life Expectancy of Building Components: Surveyors' Experiences of Buildings in Use – A Practical Guide* (Serial 307, BMI Special Report). London: Building Cost Information Service.

BMI (2004) *Building Maintenance Price Book 2004*. London: Building Cost Information Service.

BMI (2005) *Review of Rehabilitation Costs* (Serial 339, BMI Special Report). London: Building Cost Information Service.

Bone, S. (1996) *Buildings for All to Use*, (Special Publication No. 127). London: CIRIA.

Bonshor, R. B. and Bonshor, L. L. (1996) *Cracking in Buildings*. Watford: Construction Communications Ltd.

Boonstra, C. (1997) *Solar Energy in Building Renovation*. London: James & James.

Bradbury, D. (2002) *Loft Style: Styling Your City-Center Home*. London: HarperCollins.

Bragstad, J. and Fracchia, C. (1987) *Converted into Houses*. London: Thames and Hudson.

Brand, S. (1994) *How Buildings Learn*. New York: Viking.

Brantley, L. R. and Brantley, R. T. (1996) *Building Materials Technology (Structural Performance and Environmental Impact)*. New York: McGraw-Hill Inc.

Brereton, C. (1995) *The Repair of Historic Buildings: Advice on Principles and Methods*. London: English Heritage.

Bridger, A. and Bridger, C. (1998) *Altering Houses and Small-scale Residential Development*. London: Architectural Press.

Brierton, J. M. (1999) *Victorian: American Restoration Style*. Layton, Ohio: Gibbs Smith Publishers.

Bright, K. (2004) *Disability: Making Buildings Accessible – Special Report* (2nd edition). Cambridge: Workplacelaw Network.

British Plastics Federation (1996) *Code of Practice for the Installation of PVC-U Windows and Doorsets*. London: British Plastics Federation.

British Steel (1996) *The Prevention of Corrosion on Structural Steelwork*. Cleveland: British Steel Publications.

Brolin, B. C. (1980) *Architecture in Context: Fitting New Buildings with Old*. New York: Van Nostrand Reinhold.

Brooker, G. and Stone, S. (2005) *Re-Readings: Interior Architecture and the Design Principles of Remodelling Existing Buildings*. London: RIBA Enterprises Ltd.

Broomfield, J. P. (1994) *Assessing Corrosion Damage on Reinforced Concrete Structures Corrosion and Protection of Steel in Concrete* (ed. by Narayan Swamy). Sheffield: Sheffield Academic Press.

Broto, C. (2002) *Reborn Buildings*. USA: Gingko Press.

Brown, A. (1996) *Small Spaces: Stylish Ideas for Making More from Less in the Home*. Japan: Kodansha International.

Brown, S. A. (2000) *Communication in the Design Process*. London: Spon Press.

Brundtland Commission (1987) *Our Common Future: The Report of the World Commission on Environment and Development*. Oxford: Oxford University Press.

Bruntskill, R. W. (1994) *Timber Building in Britain* (2nd edition). London: Weidenfeld Nicholson.

Bruntskill, R. W. (1997) *Brick Building in Britain* (2nd edition). London: Weidenfeld Nicholson.

Bruntskill, R. W. (1999) *Traditional Farm Buildings of Britain*. London: Weidenfeld Nicholson.

Bruntskill, R. W. (2000a) *Illustrated Handbook of Vernacular Architecture* (2nd edition). London: Faber & Faber.

Bruntskill, R. W. (2000b) *Houses and Cottages of Britain: Origins and Development of Traditional Buildings*. London: Weidenfeld Nicholson.

Bruntskill, R. W. (2004) *Traditional Buildings of Britain: An Introduction to Vernacular Architecture*. London: Cassell.

BSCA (1984) *Historical Structural Steelwork Handbook*. London: British Constructional Steelwork Association Ltd.

BSI (2000a) *ISO 15686-1:2000 Buildings and Constructed Assets – Service Life Planning* (Part 1: General principles). London: British Standards Institution.

BSI (2000b) *ISO 15686-2:2000 Buildings and Constructed Assets – Service Life Planning* (Part 2: Service life prediction procedures). London: British Standards Institution.

Building Science Corporation (1999) *House and Barn Renovation*. Westford MA: Building Science Corporation.

Building Surveying Faculty (2005a) *Stock Condition Surveys* (2nd edition RICS Guidance Note). London: RICS Books.

Building Surveying Faculty (2005b) *Building Surveys and Inspections of Commercial and Industrial Property – RICS Guidance Note* (3rd edition). London: RICS Books.

Bullivant, R. A. and Bradbury, H. W. (1996) *Underpinning*. Oxford: Blackwell Science.

Bunn, R. and Roberts, D. (2000) *Screeds with Underfloor Heating: Guidance for a Defect Free Interface (Publication C150)*. London: The Concrete Society and BISRIA.

Burkinshaw, R. and Parrett, M. (2003) *Diagnosing Damp*. London: RICS Books.

Burman, P. and Stratton, M. (eds) (1997) *Conserving the Railway Heritage*. London: E & FN Spon.

Burns, J. A. (ed.) (1995) *Recording Historic Structures*. New York: John Wiley & Sons.

Burridge, R. and Ormandy, D. (eds) (1995) *Unhealthy Housing: Research, Remedies and Reform*. London: E & FN Spon.

Burton, S. (2001) *Energy Efficient Office Refurbishment*. London: James & James (Science Publishers) Ltd.

Bussell, M. (1997) *Appraisal of Iron and Steel Structures* (Publication SCI-P-138). London: Steel Construction Institute.

Bussell, M., Lazarus, D. and Ross, P. (2003a) *Retention of Masonry Facades – Best Practice Guide*, C579. London: Construction Industry Research and Information Association.

Bussell, M., Lazarus, D. and Ross, P. (2003b) *Retention of Masonry Facades – Best Practice Site Handbook*, C589. London: Construction Industry Research and Information Association.

Cadie, J. M. C., Stratford, T. J., Hollaway, L. C. and Duckett, W. H. (2004) *Strengthening Metallic Structures using Externally Bonded Fibre-reinforced Composites*, C595. London: CIRIA.

CAE (1998) *The Design of Residential Care and Nursing Homes for Older People* (Health Facilities Note 19). Leeds: NHS Estates.

CAE (1999) *Designing for Accessibility: An Essential Guide for Public Buildings*. London: Centre for Accessible Environments.

Cairns, A. H. (1992) *New Construction Materials in the Design, Maintenance and Refurbishment of Buildings* (Vol. 1). Edinburgh: Heriot-Watt University.

Cairns, A. H. (1993) *New Construction Materials in the Design, Maintenance and Refurbishment of Buildings* (Vol. 2). Edinburgh: Heriot-Watt University.

Calder, A. J. J. (1986) Repair of Cracked Reinforced Concrete: Assessment of Injection Methods, Research Report 81. Transport and Road Research Laboratory. London: Department of Transport.

Calder, A. J. J. and Thompson, D. M. (1988) Repair of Cracked Reinforced Concrete: Assessment of Corrosion Protection, Research Report 150. Transport and Road Research Laboratory. London: Department of Transport.

Cantacuzino, S. (1975) *New Uses for Old Buildings*. London: The Architectural Press.

Cantacuzino, S. (1989) *Re/Architecture: Old Buildings/New Uses*. London: Thames and Hudson.

Cantacuzino, S. (1990) *Old Buildings, New Uses*. London: Thames and Hudson.

Cantacuzino, S. (1994) *What Makes a Good Building?* London: Royal Fine Art Commission.

Cantacuzino, S. and Brandt, S. (1980) *Saving Old Buildings*. London: The Architectural Press.

Capability Scotland (1993) *Access Guide* (October 1993 edition). Edinburgh: Capability Scotland.

Cape External Products (1997) 'Overcladding', *Brochure*. Burnley: Cape External Products.

Carillion Services (2001a) *Defects in Buildings: Symptoms, Investigation, Diagnosis and Cure* (3rd edition). London: The Stationery Office.

Carillion Services (2001b) *A Guide to Measured Term Contracts: Value for Money in Property Maintenance and Improvements*. London: The Stationery Office.

Carter, L. and Skipper, S. (1992a) The Roof that Leaked…and leaked (Technical Case History). *The Building Surveyor*, **1**(5), February. London: RICS Journals Ltd.

Carter, L. and Skipper, S. (1992b) The Roof that Leaked...Part II (Technical Case History), *The Building Surveyor*, **1**(6), March. London: RICS Journals Ltd.

Cassell, J. and Parham, P. (2000a) *Repair and Renovate: Bathrooms*. New York: New Publishers.

Cassell, J. and Parham, P. (2000b) *Repair and Renovate: Kitchens*. New York: New Publishers.

Cassell, J. and Parham, P. (2000c) *Repair and Renovate: Doors and Windows*. New York: New Publishers.

Cassell, J. and Parham, P. (2000d) *Repair and Renovate: Walls and Ceilings*. New York: New Publishers.

Catt, R. and Catt, S. (1981) *Conversion, Improvement and Extension of Buildings*. London: Estates Gazette.

CEM (1993) Development (Refurbishment). *CEMicircular*, **1**(2), December. Reading: The College of Estate Management.

CEM (2000) Disability Discrimination Act 1995. *CEMicircular*, **7**(1), March. Reading: The College of Estate Management.

Central Unit on Procurement (1996) *No. 54 Value Management, C.U.P Guidance*. London: HM Treasury.

Central Unit on Procurement (1997) *No. 41 Managing Risk and Contingency for Construction Work C.U.P Guidance*. London: HM Treasury.

Cerver, F. A. (1999) *Lofts: Living and Working Spaces*. New York: Area.

Chandler, I. (1989) *Building Technology 2 – Performance*. London: Mitchell/CIOB.

Chandler, I. (1991) *Repair and Refurbishment of Modern Buildings*. London: B.T. Batsford Ltd.

Chanter, B. and Swallow, P. (1996) *Building Maintenance Management*. Oxford: Blackwell Publishing.

Chanter, F. W. B. (1985) *Conservation of Timber Buildings*. Shaftsbury: Donhead Publishing.

Chappell, D. (2004) *The JCT Minor Works Form of Contract* (3rd edition). Oxford: Blackwell Publishing.

Chappell, D. (2006) *The JCT Intermediate Building Contract* (3rd edition). Oxford: Blackwell Publishing.

Charles Roberston Partnership (1987) *Thermal Upgradings in House Modernisation*. Edinburgh: Scottish Development Department.

Chew, M. L. Y. (1999) *Construction Technology for Tall Buildings*. Singapore: Singapore University Press/World Scientific.

Ching, F. D. K. and Miller, D. E. (1983) *Home Renovation*. New York: Van Nostrand Reinhold.

Chitty, R. and Fraser-Mitchell, J. (2003) *Fire Safety Engineering: A Reference Guide*, BRE Report BR459. Garston: Building Research Establishment.

Chudley, R. (1983) *Maintenance and Adaptation of Buildings*. London: Longman.

Church Commissioners (1993) *Annual Report and Accounts*. London: Church of England.

CIB (1993a) Building Pathology: A State-of the-Art Report, CIB Report Publication 155, CIB Working Commission W86, June 1993. Holland: International Council for Building.

CIB (1993b) Some Examples of the Application of the Performance Concept in Building, CIB Report Publication 157, CIB Working Commission W060, June 1993. Holland: International Council for Building.

CIBSE (1988) CIBSE Guide, Volume A: Design Data. London: The Chartered Institution of Building Services Engineers.

CIBSE (2003) 'Building log books: A guide and templates for preparing building log books', *TM 31*. London: The Chartered Institution of Building Services Engineers.

CIC (1997) *Definitions of Inspections and Surveys of Buildings (An Explanatory Leaflet)*. London: Construction Industry Council.

CIOB (1987) Code of Estimating Practice. *Refurbishment and Modernisation* (Suppl 1). Englemere: Chartered Institute of Building.

CIP (1998) Refurbishment. *Construction Health and Safety Manual* (Section 27). London: Construction Industry Publications.

CIRIA (1986) *Structural Renovation of Traditional Buildings*. Report 111. London: Construction Industry Research and Information Association.

CIRIA (1997) *New Paint Systems for the Protection of Construction Steelwork*. Report 174. London: Construction Industry Research and Information Association.

CIRIA (1998) *Fibre-reinforced Polymer Composites for Blast-resistant Cladding, CON51*. London: Construction Industry Research and Information Association.

CIRIA (2000) *Sustainable Construction Indicators, C563*. London: Construction Industry Research and Information Association.

CIRIA (2002a) *A Simple Guide to Building, SP153*. London: Construction Industry Research and Information Association.

CIRIA (2002b) *A Simple Guide to Controlling Risk, SP154*. London: Construction Industry Research and Information Association.

Civic Trust (1971) *Financing the Preservation of Old Building*. London: HMSO.

Clark, K. (2002a) *Informed Conservation*. London: English Heritage.

Clark, K. (2002b) *Building Education – The Role of the Physical Environment in Enhancing Teaching and Research*. London: Institute of Education, University of London.

Clarke Associates (1989) *Loft Conversion: Inception to Completion*. Cambridge: Keith Clarke Associates.

Coates, R. (2001) Converting farm buildings to business use. *CEMicircular*, **8**(1), April. Reading: The College of Estate Management.

Coggins, C. R. (1980) *Timber Decay in Buildings: Dry Rot, Wet Rot and Other Fungi*. London: Rentokil.

Cohen, N. (2002) *Urban Planning, Conservation and Preservation*. New York: McGraw Hill.

Coleman, A. (1989) *Utopia on Trial: Vision and Reality in Planned Housing*. London: Macmillan.

Collins, J. (2002) *Old House Care and Repair*. Shaftsbury: Donhead Publishing.

Committee on the Assessment of Asthma and Indoor Air (2000) *Clearing the Air: Asthma and Indoor Air Exposures*. Division of Health Promotion and Disease Prevention, Institute of Medicine. Washington, DC: The National Academies on Health.

Committee on Damp Indoor Spaces and Health (2004) *Damp Indoor Spaces and Health*. Board on Health Promotion and Disease Prevention, Institute of Medicine of the National Academies. Washington, DC: The National Academies Press.

Communities Scotland (2004a) Housing and Health in Scotland, *Scottish House Condition Survey 2002*, SHCS Working Paper No. 4. Edinburgh: Scottish Executive.

Communities Scotland (2004b) 'Fuel Poverty in Scotland. *Scottish House Condition Survey 2002*. Edinburgh: Scottish Executive.

Communities Scotland (2004c) *Scottish Housing Quality Standard: Delivery Plan Guidance and Assessment Criteria*, July 2004. Edinburgh: Scottish Executive.

Concrete Society (1984) Repair of Concrete Damaged by Reinforcement Corrosion, Technical Report No. 26. London: The Concrete Society.

Concrete Society (1989) Cathodic Protection of Reinforced Concrete, Technical Report No. 36. October 1994. London: The Concrete Society.

Concrete Society (1990) Assessment and Repair of Fire Damaged Concrete, Technical Report No. 33. London: The Concrete Society.

Concrete Society (1991) Patch Repair of Reinforced Concrete – Subject to Reinforcement Corrosion (Model Specification and Method of Measurement), Technical Report No. 38. London: The Concrete Society.

Concrete Society (1995) The Relevance of Cracking in Concrete to Corrosion of Reinforcement, Technical Report No. 44. London: The Concrete Society.

Concrete Society (1999) Alkali-silica Reaction: Minimising the Risk of Damage to Concrete (Guidance Notes and Model Clauses for Specifications), Technical Report No. 30 (3rd edition). London: The Concrete Society.

Concrete Society (2000) Diagnosis of Deterioration in Concrete Structures: Identification of Defects, Evaluation and Development of Remedial Action, Technical Report No. 54. London: The Concrete Society.

Concrete Society (2003) Strengthening Concrete Structures using Fibre Composite Materials: Inspection, Acceptance and Monitoring, Technical Report No. 57. London: The Concrete Society.

Concrete Society (2004) Design Guidance for Strengthening Concrete using Fibre Composite Materials, Technical Report No. 55 (2nd edition). London: The Concrete Society.

Concrete Society and Corrosion Engineering Association (1989) Cathodic Protection of Reinforced Concrete, Technical Report No. 36. London: The Concrete Society.

Construction Audits Ltd and HAPM (1993) *Defects Avoidance Manual: New Build*. Garston: Building Research Establishment.

Cook, G. K. and Hinks, A. J. (1992) *Appraising Building Defects*. London: Longman.

Cooper, M. (1998) *Laser Cleaning in Conservation – An Introduction*. London: Butterworth-Heinemann.

Cooper, R. and Buckland, J. (1985) *A Practical Guide to Alterations and Improvements*. London: International Publishing Ltd.

Corbett-Winder, K. (1995) *The Barn Book*. London: Random Century.

Corke, M. (2000) *Repair and Renovate: Floors and Stairs (Repair and Renovation)*. New York: New Publishers.

Coupland, A. (ed.) (1997) *Reclaiming the City: Mixed Use Development*. London: E & FN Spon.

Coutts, J. (2006) *Principles of Loft Conversions*. Oxford: Blackwell Publishing.

CRA (1992) *The Application and Measurement of Protective Coatings for Concrete, Guidance Note*. Aldershot: The Concrete Repair Association.

CRA (1999a) *Electrochemical Repair using Re-alkalisation and Chloride Extraction Techniques, Guidance Note*. Aldershot: The Concrete Repair Association.

CRA (1999b) *The Route to Successful Concrete Repair, Guidance Note*. Aldershot: The Concrete Repair Association.

CSM (1991) *Schools are Deteriorating 'Faster than We Think'*, *The Building Surveyor*, **1**(2), October. London: RICS Publications.

Cuito, A. (2003a) *Lofts: Good Ideas (Good Ideas Series)*. USA: Harper Design International.

Cuito, A. (2003b) *Small Lofts*. USA: Harper Design International.

Cunningham, G. S. (2002) *The Restoration Economy: The Greatest New Growth Frontier.* San Fransisco: Berrett-Koehler Publisher Inc.

Cunnington, P. (1988) *Change of Use: The Conversion of Old Buildings*. Dorset: Alphabooks (A & C Black).

Cunnington, P. (1993) *How Old is that Church?* Dorset: Marsdon House Publishers.

Cunnington, P. (1999) *How Old is Your House?* (2nd edition). Dorset: Marsdon House Publishers.

Cunnington, P. (2002) *Caring for Old Houses* (2nd edition). Dorset: Marsdon House Publishers.

Currie, R. J. and Robery, P. C. (1994) *Repair and Maintenance of Reinforced Concrete*, BRE Report BR 254. Garston: Building Research Establishment.

Curtin, W. and Parkinson, E. (1989) Structural Appraisal and Restoration of Victorian Buildings, *Conservation and Engineering Structures*. London: Thomas Telford.

Curwell, S. R. and March, C. G. (eds) (1986) *Hazardous Building Materials: A Guide to Selection and Alternatives*. London: E & FN Spon.
Curwell, S. R., March, C. G. and Venables, R. (eds) (1990) *Buildings and Health; The Rosehaugh Guide to the Design, Construction, Use and Management of Buildings*. London: RIBA Publications.
Dale, D. (1992) *Renovate or Demolish?* London: National Housing and Town Planning Council.
Dallas, R. (ed.) (2003) *Measured Survey and Building Recording for Historic Buildings and Structures, Guide for Practitioners 4*. Edinburgh: Historic Scotland.
Darley, G. (1988) *A Future for Farm Buildings*. London: SAVE Britain's Heritage.
Davey, A. et al. (1995) *The Care and Conservation of Georgian Houses* (4th edition). London: Butterworth Architecture.
Davies, B. and Begg, N. (2004) *Converting Old Buildings into Homes*. London: The Crowood Press.
Dawson, S. (ed.) (1997a) *Architects' Working Details 3*. London: Emap Construct.
Dawson, S. (ed.) (1997b) *Architects' Working Details 4*. London: Emap Construct.
Dawson, S. (1998) *Architects' Working Details 5*. London: Emap Construct.
Dawson, S. (1999) *Architects' Working Details 6*. London: Emap Construct.
Dawson, S. (2000) *Architects' Working Details 7*. London: Emap Construct.
Dawson, S. (2002) *Architects' Working Details 8*. London: Emap Construct.
DCMS (1999) *The Disposal of Historic Buildings, Guidance Note for Government Departments and Non-departmental Public Bodies*. London: Department for Culture, Media and Sport.
Defra (RDS) (2004) *Bats, Buildings and Barn Owls: A Guide to Safeguarding Protected Species when Renovating Traditional Buildings (Guidance Leaflet)*. London: Department for Environment, Food and Rural Affairs.
DEFRA (2004) *Fuel Poverty in England: The Government's Plan of Action*. London: Department for Environment, Food and Rural Affairs.
Delafons, J. (1997) *Policies and Preservation (A Policy History of the Built Heritage 1888–1996)*. London: E & FN Spon.
Department of the Interior (2004) *The Preservation of Historic Architecture: The U.S. Government's Official Guidelines for Preserving Historic Homes*. Washington, DC: The Lyons Press.
Department for Culture, Media and Sport (1999) *The Disposal of Historic Buildings, Guidance Note For Government Departments and Non-departmental Public Bodies*. London: The Stationery Office.
DES (1991) *A Guide to Energy Efficient Refurbishment: Maintenance and Renewal in Educational Buildings*, Building Bulletin 73, Department of Science and Education. London: The Stationery Office Ltd.
DES (1996a) *Adaptable Facilities in Further Education for Business and Office Studies*, Building Bulletin 64, Department of Education and Science. London: HMSO.
DES (1996b) *Adaptable Facilities in Further Education for Accommodation Studies*, Building Bulletin 66, Department of Education and Science. London: HMSO.
Design Council (1979) *Designing Against Vandalism*. London: The Design Council.
DETR (1998) Sustainable Construction, Report. London: The Stationery Office Ltd.
DETR (2000) *Conversion and Redevelopment: Processes and Potential*. Department of the Environment, Transport and the Regions and the National, Assembly for Wales. London: The Stationery Office.
DETR and DTZ Pieda (2000) *Demolition and New Building on Local Authority Estates*, Department of the Environment, Transport and the Regions. London: The Stationery Office.
DfE (1993) *Security Lighting: Crime Prevention in Schools*, Building Bulletin 78. Department for Education (Architects and Building Branch). London: HMSO.
DfEE (1996) *Improving Security in Schools – Managing Schools Facilities Guides 4*, Building Bulletin 93. London: The Department for Education and Employment.
DfEE (1999) *Access for Disabled People to School Buildings – Management and Design Guide*, Building Bulletin 91. London: The Stationery Office.

DfEE (2000) *Asset Management Plans*. London: Department for Education and Employment.
DfES (2003) *Acoustic Design of Schools: A Design Guide*, Building Bulletin 93, Department for Education and Skills. London: The Stationery Office.
Dhir, R. K., Dyer, T. D. and Paine, K. A. (eds) (2000) *Sustainable Architecture: The Use of Incinerator Ash*. London: Thomas Telford.
Diamonstein, B. (1978) *Buildings Reborn: New Uses, Old Places*. New York: Harper and Rowe.
Di Leo, G. (1990) *Altering, Extending and Converting Houses: An Owner's Guide to Procedures*. London: International Thomson Business Publishing.
DOE (1985) *Guide to Energy Efficient Renovation of Housing*. London: HMSO.
DOE (1986) *Asbestos Materials in Buildings* (2nd edition). London: HMSO.
DOE (1992) *Houses into Flats: A Study of Private Sector Conversions in London* (Vol. 1. Report of Main Findings). London: HMSO.
DOE (1997) *Evaluation of Flats Over Shops*. London: The Stationery Office.
DOE, SDD and Welsh Office (1971) *New Uses for Old Buildings* (Aspects of Conservation 1). London: HMSO.
DOE, SDD and Welsh Office (1972) *New Life for Historic Areas* (Aspects of Conservation 2). London: HMSO.
DOE, SDD and Welsh Office (1977) *New Life for Old Churches* (Aspects of Conservation 3). London: HMSO.
Donaldson, R. J. and Donaldson, L. J. (1993) *Essential Public Health Medicine*. London: Kluwer Academic Publishers.
Douglas, J. (1997) The development of ground floor constructions – Part II. *Structural Survey*, **15**(4). Bradford: MCB University Press.
Driscoll, R. M. C. and Crilly, M. S. (2000) *Subsidence to Domestic Buildings*. Garston: Building Research Establishment.
Drysdale, D. (1994) Fire science. In Stollard, P. and Johnston, L. (eds), *Design Against Fire: An Introduction to Fire Safety Engineering in Design*. London: E & FN Spon.
Duell, J. and Lawson, F. (1983) *Damp Proof Course Detailing* (2nd edition). London: Architectural Press.
Duffy, F. (1993) 'Measuring building performance'. *Facilities*, **8**, 17–20. Bradford: MCB University Press.
Duncan, et al. (1998) *Tomorrow's World, Friends of the Earth*. London: Earthscan.
Dupont™ Tyvek® (2004) *21st Providing protection in Construction*. Clevedon: Dupont™ Tyvek®.
Dworin, L. (1996) *Renovating and Restyling Older Homes: The Professionals Guide to Maximum Value Remodeling*. Carlsbad, CA: Craftsman Book Company.
Earl, J. (2003) *Building Conservation Philosophy* (3rd edition). Shaftsbury: Donhead (in Association with the College of Estate Management).
ECC Building Control Services (1995) *Loft Conversions and the Building Regulations*. Exeter: Exeter City Council.
ECD Partnership et al. (1986) *A Designer's Manual for the Energy Efficient Refurbishment of Housing* (The Application manual to accompany BS 8211 Part 1). London: British Standards Institution.
ECOTEC, ECD and NPAC (1998) *The Value of Electricity Generated from Photovoltaic Power Systems in Buildings*, ETSU S/P"/00279/REP. London: ETSU.
Edmunds, I. R. (1991) *Alteration or Conversion of Houses*. London: Pearson Publishing.
Edwards, B. (1998) *Green Buildings Pay* (2nd edition). London: E & FN Spon.
Edwards, L. (1995) *Practical Risk Management in the Construction Industry*. London: Thomas Telford.
Edwards, B. and Hyett, P. (2001) *Rough Guide to Sustainability*. London: RIBA Publications.
Edwards, B. and Turrent, D. (eds) (2000) *Sustainable Housing: Principles and Practice*. London: E & FN Spon.
Edwards, R. (2002) *Handbook of Domestic Ventilation*. Oxford: Butterworth-Heinemann.

Egan, J. (1998) *Rethinking Construction*. London: Department of the Environment, Transport and the Regions.

Egbu, C. O. (1995) Perceived degree of difficulty of management tasks in construction refurbishment work. *Building Research and Information,* **23**(6), 340–344. London: E & FN Spon.

Egbu, C. O. (1996a) *Characteristics and Difficulties Associated with Refurbishment, Construction Paper 66*. Englemere: Chartered Institute of Building.

Egbu, C. O. (1996b) *Management Education and Training for Construction Refurbishment, Construction Paper 69*. Englemere: Chartered Institute of Building.

Egbu, C. O. (1997) Refurbishment management: challenges and opportunities. *Building Research and Information,* **25**(6), 338–347. London: E & FN Spon.

EKOS Ltd and Ryden Property Consultants (2001) *Obsolete Commercial and Industrial Buildings*. Edinburgh: Scottish Executive Central Research Unit.

Eley, P. and Worthington, J. (1984) *Industrial Rehabilitation: The Use of Redundant Buildings for Small Enterprises*. London: Architectural Press.

Emmitt, S. (2002) *Architectural Technology*. Oxford: Blackwell Science.

Emmons, P. H. (1994) *Concrete Repair and Maintenance Illustrated: Problem Analysis, Repair Strategy, Techniques*. Massachusetts: R.S. Means Co. Inc.

Endean, K. F. (1995) *Investigating Rainwater Penetration of Modern Buildings*. Aldershot: Gower.

Energy Research Group et al. (1999) *A Green Vitruvius – Principles and Practice of Sustainable Architectural Design*. London: James & James (Science Publishers) Ltd.

Energy Saving Trust (2005) *Solar PV: Your Guide to Creating Clean Electricity*. London: Energy Saving Trust.

Engler, N. (2000) *Renovating Barns, Sheds and Outbuildings*. New York: New Publishers.

English Heritage (1993) *The Conversion of Historic Farm Buildings, Statement*. London: English Heritage.

English Heritage (1995a) *Easy Access to Historic Properties*. London: English Heritage.

English Heritage (1996a) *London Terraced Houses 1660–1860, A Guide to Alterations and Extensions*. London: English Heritage.

English Heritage (1996b) *Sustainability and the Historic Environment, Report prepared by Land Use Consultants and CAG Consultants*. London: English Heritage.

English Heritage (1997) *Sustaining the Historic Environment: New Perspectives on the Future*. London: English Heritage.

English Heritage (1998) *Research and Case Studies in Architectural Conservation*, Vol. 1: Metals. *English Heritage Research Transactions*. London: James & James (Scientific Publishers) Ltd.

English Heritage (1999) Graffiti on Historic Buildings and Monuments: Methods of Removal and Prevention. Technical Note. London: English Heritage.

Enterprise Foundation (1991a) In Duncan, W. (manual ed.), *Substantial Rehabilitation and New Construction*. New York: Van Nostrand Reinhold.

Enterprise Foundation (1991b) In Ruckle, G. (manual ed.), *Multifamily Selective Rehabilitation*. New York: Van Nostrand Reinhold.

Enterprise Foundation (1991c) In Santucci, R. M. (manual ed.), *Single Family Selective Rehabilitation*. New York: Van Nostrand Reinhold.

Environment Committee (1996) *Water Conservation and Supply*. London: House of Commons.

EST (2001a) *Domestic Energy Fact File: England, Scotland, Wales and Northern Ireland*, November 2001. London: Energy Efficiency Trust.

EST (2001b) *Energy Efficiency Strategy*, November 2001. London: Energy Efficiency Trust.

Eternit Tac Ltd (1987) *Options for Urban Renewal* (Vol. 1: Large Panel System Buildings). Royston: Eternit Tac Ltd.

Euroroof Ltd (1985) *Re-roofing: A Guide to Flat Roof Maintenance and Refurbishment*. Northwich: Euroroof Ltd.

Evans, J., Hyndman, S., Stewart-Brown, S., Smith, D. and Petersen, S. (2000) An epidemiological study of the relative importance of damp housing in relation to adult health. *Epidemiol Community Health*, **54**, 677–686.

Fairs, M. (2002) 'Better by Design', Schools for the Future, PUBLIC SECTOR BUILDING Education Supplement, March 2002. Broomsgrove, England: Ascent Publishing Ltd.

Fawcett, W. and Palmer, J. P. (2004) *Good Practice Guidance for Refurbishing Occupied Buildings*, C621. London: Construction Industry Research Association.

Feilden, B. M. (2003) *The Conservation of Historic Buildings* (3rd edition). London: Architectural Press.

Feirer, M. (2000) *Attics: A Quick Guide (The Quick Guide Series)*. New York: New Publishers.

Fire Protection Association (1992) *Fire Protection in Old Buildings and Historic Town Centres*. London: The Fire Protection Association.

Fishlock, M. (1992) *The Great Fire at Hampton Court*. London: The Herbert Press.

Fitch, J. M. (2000) *Historic Preservation: Curatorial Management of the Built World*. Virginia: University Press of Virginia.

Fitchen, J. (1988) *Building Construction Before Mechanisation*. Massachusetts: MIT Press.

Flanagan, R. and Jewell, C. (2003) *The Risk of Mould Damage Over the Whole Life of a Building*. London: RICS Foundation and University of Reading.

Flanagan, R. and Norman, G. (1993) *Risk Management in Construction*. Oxford: Blackwell Science.

Foley, P. D. and Green, D. H. (1984) *A Practical Guide to the Conversion and Subdivision of Industrial Property*. Ilkley: D Howard Green Associates.

Foster, L. (1997) *Access to the Historic Environment – Meeting the Needs of Disabled People*. Shaftsbury: Donhead Publishing.

Foulks, W. G. (ed.) (1997) *Historic Building Façades: The Manual for Maintenance and Rehabilitation (Preservation Press Series)*. New York: John Wiley & Sons.

FRA (1999) *Maintenance and Refurbishment*, Information Sheet No. 11. London: Flat Roofing Alliance.

Friedman, D. (1995) *Historical Building Construction: Design, Materials, and Technology*. New York: W.W. Norton & Co.

Friedman, D. (2000) *The Investigation of Buildings: A Guide for Architects, Engineers, and Owners*. New York: W.W. Norton & Co.

Friedman, D. and Oppenheimer, N. (1997) *The Design of Renovations*. London: W.W. Norton & Co.

Garner, J. F. and Edmund, R. (1985) *Alteration or Conversion of Houses* (5th edition). London: Longman.

Garratt, J. and Nowak, F. (1991) *Tackling Condensation: A Guide to the Causes of, and Remedies for, Surface Condensation and Mould in Traditional Housing*, Report 174. Garston: Building Research Establishment.

Gauld, B. J. (1996) *Structures for Architects* (2nd edition). London: Longman.

Gause, J. A. et al. (1996) *New Uses for Obsolete Buildings*. Washington, DC: Urban Land Institute.

GB Geotechnics Ltd (2001) Non-Destructive Investigation of Standing Structures, *Technical Advice Note 23*. Edinburgh: Historic Scotland.

GGF (1999) *Conservatories – A Guidebook*. London: Glass and Glazing Federation.

Gibson, G. (1996) *Remodel! An Architect's Advice on Home Renovation*. New York: John Wiley & Sons Inc.

Gilbert, J. and Flint, A. (1992) *The Tenement Handbook: A Practical Guide to Living in a Tenement*, ASSIST Architects Ltd. Edinburgh: RIAS.

Gilbertson, K. and Richards, J. (2003) *Asbestos: Duty to Manage – Special Report*. Cambridge: Workplacelaw Network.

Glover, P. (2003) *Building Surveys* (5th edition). Oxford: Butterworth-Heinemann.

Gold, C. A. and Martin, A. J. (1999a) *Refurbishment of Concrete Buildings – Structural and Services Options*, Guidance Note GN 8/99. London: BSRIA.

Gold, C. A. and Martin, A. J. (1999b) *Refurbishment of Concrete Buildings – Designing Now for Future Re-use*, Guidance Note GN 9/99. London: BSRIA.

Gomez, L. (2001) *Lofts*. USA: Client Distribution Services.

Goodier, C. and Gibb, A. (2004) *The Value of the UK Market for Offsite*. Loughborough: Buildoffsite and Loughborough University.

Goss, R. (2001) *Roofing Ready Reckoner* (3rd edition). Oxford: Blackwell Science.

Gourley, C. (c.1910) *Elementary Building Construction and Drawing for Scottish Students*. Glasgow: Blackie & Son Ltd.

Gourley, C. (1922) *The Construction of a House* (2nd edition). London: B.T. Batsford Ltd.

Graining, J. (1999) *Compact Living*. San Francisco: Soma Books.

Grammenos, F. and Russell, P. (1997) *Building Adaptability: A View from the Future, 2nd International Conference on Buildings and the Environment*, June 1997. Paris.

Graves, H. M. and Phillipson, M. C. (2000) *Potential Implications of Climate Change in the Built Environment*. London: BRE and Construction Communications Ltd.

Gray, C., Hughes, W. and Bennett, J. (1994) *The Successful Management of Design – A Handbook of Building Design Management*. Berkshire: University of Reading.

Green, H. and Foley, P. (1986) *Redundant Space – A Productive Asset: Converting Property for Small Businesses*. New York: Harper and Row.

Greenpeace (1997) *Use of UPVC, Explanatory Brochure on the Dangers of and Alternatives to PVC*. London: Greenpeace.

Greer, N. R. (1998) *Architecture Transformed: New Life for Old Buildings*. Massachusetts: Rockport Publishers Inc.

Gregg, T. R. and Crosbie, J. (2001) Refurbishment of buildings for residential use. *Building Engineer*, **76**(4), April 2001. Northampton: Association of Building Engineers.

Griffin, C. (ed.) (1994) *No Losers – New Uses: New Homes from Empty Properties*. London: National Housing and Town Planning Council and Empty Homes Agency.

Griffith, A. (1992) *Small Building Works Management*. London: Macmillan.

Guy, C. (1994) The Retail Development Process: Location, Property and Planning. London: Routledge.

Haddlesey, P. (2002) *A Breath of Fresh Air, Schools for the Future, Public Sector Building, Education Supplement*, March 2002. Broomsgrove: Public Sector Building.

Hall, G. T. (1984) *Revision Notes on Building Maintenance and Adaptation*. London: Butterworths.

Ham, R. (1998) *Theatres: Planning Guidance for Design and Adaptation*. London: Architectural Press.

Handisyde, C. C. (1991) *Everyday Details*. London: Butterworth.

HAPM Ltd (1991) *Defects Avoidance Manual: New Build*. Garston: Building Research Establishment.

HAPM Ltd (2000) *Life Component Manual* (2nd edition). London: E & FN Spon.

Hargreaves, M. (1987) *Restoration and Refurbishment – Design Data, AJ Focus*, August 1987. London: Architects Journal.

Harland, E. (1998) *Eco-Renovation: The Ecological Home Improvement Guide* (2nd edition). Devon: Green Books.

Harper, D. R. (1978) *Building: The Process and the Product*. Lancaster: The Construction Press Ltd.

Harris, J. and Wiggington, M. (2000) *Intelligent Skin*. London: Architectural Press.

Harris, S. Y. (2001) *Building Pathology: Deterioration, Diagnostics, and Intervention*. New York: John Wiley & Sons Inc.

Harrison, H. W. (1996) *Roofs and Roofing: Performance, Diagnosis, Maintenance, Repair and the Avoidance of Defects*. London: BRE Building Elements series, Construction Research Communications Ltd.

Harrison, H. W. and de Vekey, R. C. (1998) *Walls, Windows and Doors: Performance, Diagnosis, Maintenance, Repair and the Avoidance of Defects*. London: BRE Building Elements series, Construction Research Communications Ltd.

Harrison, H. W. and Trotman, P. M. (2000) *Building Services: Performance, Diagnosis, Maintenance, Repair and the Avoidance of Defects*.London: BRE Building Elements series, Construction Research Communications Ltd.

Harrison, H. W. and Trotman, P. M. (2002) *Foundations, Basements and External Works: Performance, Diagnosis, Maintenance, Repair and the Avoidance of Defects.* London: BRE Building Elements series, Construction Research Communications Ltd.

Harrison, H. W., Hunt, J. H. and Thomson (1986) *Overcladding External Walls of Large Panel System Dwellings,* BRE Report BR 93. London: Building Research Establishment.

Harrison, H. W., Mullin, S., Reeves, B. and Stevens, A. (2004) *Non-Traditional Housing Types – An Aid to Identification.* Garston: IHS Rapidoc (BRE Bookshop).

Harvey, J. H. (1972) *Conservation of Building.* London: John Barker (Publishers) Ltd.

Hasluck, P. N. (2001) *House Decoration* (Facsimile of 1871 edition). London: Donhead Publishing.

Haverstock, H. (1998) *Building Design Easibrief* (3rd edition). Kent: Miller Freeman.

Heckroodt, R. O. (2002) *Guide to the Deterioration and Failure of Building Materials.* London: Thomas Telford Publishing.

Heldmann, C. (1987) *Manage Your Own Home Renovation* (rev. edition). Vermont: Garden Way Publishing.

Hendriks, C. F. (2001) *Durable and Sustainable Construction Materials.* Netherlands: Aeneas.

Hendriks, C. F. (2002) *Sustainable Construction.* Netherlands: Aeneas.

Henket, H. A. J. (1992) *Forecasting the Technical Behaviour of Building Components: A Model, Proceedings of CIB Symposium, Innovations in Management, Maintenance and Modernisation of Buildings*, October 1992. Rotterdam: CIB.

Henley, E. J. and Kumamatro, H. (1983) *Reliability Engineering and Risk Assessment.* NJ: Prentice-Hall.

Heritage Preservation Services (1995) *A Checklist for Rehabilitating Historic Buildings.* Washington, DC: National Park Service.

Herzog, T., Krippner, R. and Lang, W. (2002) *Façade Construction Manual.* Germany: Birkhauser Verlag AG.

Highfield, D. (1987) *Rehabilitation and Re-use of Old Buildings.* London: E & FN Spon.

Highfield, D. (1991) *The Construction of New Buildings Behind Historic Facades.* London: E & FN Spon.

Highfield, D. (2000) *Refurbishment and Upgrading of Buildings.* London: E & FN Spon.

Hillier, B. and Hanson, J. (1989) *The Social Logic of Space.* Cambridge: Cambridge University Press.

Hillier, M., Lawson R. M. and Gorgolewski, M. (1998) *Over-roofing of Existing Buildings using Light Steel*, SCI Publication P246. London: The Steel Construction Institute.

Hilton, D. O. and Oliver, M. A. (1985) *Timber Framed Housing: Repairs, Maintenance and Extensions.* London: E & FN Spon.

Hinks, A. J. and Cook, G. K. (1997) *The Technology of Building Defects.* London: E & FN Spon.

Historic Preservation Education Foundation (1995) *Preserving the Recent Past I.* Washington, DC: HPEF.

Historic Preservation Education Foundation (2000) *Preserving the Recent Past II.* Washington, DC: HPEF and National Park Service.

Historic Preservation Education Foundation/National Park Service (1997). *Window Rehabilitation Guide for Historic Buildings.* Washington, DC: HPEF.

Historic Scotland (1993) *Memorandum of Guidance on Listed Buildings and Conservation Areas.* Edinburgh: Historic Scotland.

Historic Scotland (1994) *Stonecleaning – A Guide for Practitioners.* Edinburgh: Historic Scotland.

Historic Scotland (1996) Biological Growths on Sandstone Buildings: Control and Treatment. *Technical Advice Note 10.* Edinburgh: Historic Scotland.

Historic Scotland (1997a) Access to the Built Heritage. *Technical Advice Note 7.* Edinburgh: Historic Scotland.

Historic Scotland (1997b) Fire Protection Measures in Scottish Historic Buildings. *Technical Advice Note 11.* Edinburgh: Historic Scotland.

Historic Scotland (1997c) Stonecleaning of Granite Buildings, *Technical Advice Note 9.* Edinburgh: Historic Scotland.

Historic Scotland (1998) The Installation of Sprinkler Systems in Historic Buildings. *Technical Advice Note 14*. Edinburgh: Historic Scotland.

Historic Scotland (1999a) *Fire Protection and the Built Heritage. Conference Proceedings*. Edinburgh: Historic Scotland.

Historic Scotland (1999b) *Rural Buildings in the Lothians – Conservation and Conversion*. Edinburgh: Historic Scotland.

Historic Scotland (2000) Corrosion in Masonry-Clad Early 20th Century Steel Framed Buildings. *Technical Advice Note 20*. Edinburgh: Historic Scotland.

Historic Scotland (2001) Fire Risk Management in Heritage Buildings. *Technical Advice Note 22*. Edinburgh: Historic Scotland.

Hochman, P. (1998) All Hail Titanium, Dicaprio Among Metals, *Fortune*, October 26 1998, USA.

Holdsworth, W. and Sealey, A. (1992) *Healthy Buildings* (A Design Primer for a Living Environment). Harlow, Essex: Longman.

Holland, R. et al. (eds) (1990) *Appraisal and Repair of Building Structures*. London: Thomas Telford Services Ltd.

Hollis, M. (2005) *Surveying Buildings* (5th edition). London: RICS Books.

Holmes, S. and Wingate, M. (1997) *Building with Lime: A Practical Introduction*. London: Intermediate Technology Publications.

Holt, A. (2001) *Principles of Construction Safety*. Oxford: Blackwell Science.

Homebuilding and Renovating Magazine (2005) *Book of Barn Conversions: 22 Inspirational Projects*. London: Homebuilding and Renovating Magazine.

Home Ventilation Ltd (2000) *Dri-Master 5 + 5 input ventilation system*. London: *NuAire* Home Ventilation Ltd.

House Condition Surveys Team (2004) Scottish House Condition Survey 2002 (Main Report). Edinburgh: Communities Scotland.

House Condition Surveys Team (2005) *Housing and Disrepair in Scotland: Analysis of the 2002 Scottish House Condition Survey*. Edinburgh: Communities Scotland.

Howell, J. (1995) Moisture Measurement in Masonry: Guidance for Surveyors, *COBRA 95, RICS Construction and Building Research Conference*, at Heriot-Watt University, 8–9 September 1995. London: The Royal Institution of Chartered Surveyors.

HSE (1984) *Health and Safety in Demolition Work Part 3: Techniques*, GS29/3. London: Health and Safety Executive Books.

HSE (1992) *Façade retention, Guidance Note* GS 51. London: Health and Safety Executive Books.

HSE (1993) *Asbestos in Buildings*, HSE186/2. London: Health and Safety Executive Books.

HSE (1997a) *Health and Safety in Construction*, HSG 150. London: Health and Safety Executive Books.

HSE (1997b) *Fire Safety in Construction Work*, HSG 168. London: Health and Safety Executive Books.

HSE (1998a) *Approved Code of Practice on the Gas Safety (Installation and Use) Regulations 1998*. London: Health and Safety Executive Books.

HSE (1998b) *Approved Code of Practice for the Control of Substances Hazardous to Health (COSHH) Regulations 1994*. London: Health and Safety Executive Books.

HSE (1999a) *Working with Asbestos Cement* (2nd edition), HSE186/2. London: Health and Safety Executive Books.

HSE (1999b) Collapse of a Three-Storey building. A Report on the Accident at Woodthorpe Road, Ashford, Middlesex that Occurred on 1 August 1995. London: Health and Safety Executive.

HSE (2004) *Investigating Accidents and Incidents*. London: Health and Safety Executive.

HUD (1999) *Innovative Rehabilitation Provisions: A Demonstration of the Nationally Applicable Recommended Rehabilitation Provisions*. Washington, DC: U.S. Department of Housing and Urban Development.

Hudson, R. (1984) *Corrosion of Structural Steelwork, London.* Swindon Laboratories: British Steel Corporation.

Hughes, G. (1985) *Barns of Rural England.* London: The Herbert Press Ltd.

Hughes, P. (1986) The Need for Old Buildings to Breathe, *SPAB Information Sheet 4.* London: Society for the Protection of Ancient Buildings.

Hurley, J. W., McGrath, C., Fletcher, S. M. and Bowles, H. M. (2001) *Deconstruction and Reuse of Construction Materials.* BRE Report BR 456. London: Construction Research Communications Ltd.

Hutchins, N. (2000) *Restoring Old Houses* (Revised and Updated). New York: New Publishers.

Hutchison, B. D., Barton, J. and Ellis, N. (1973) *Maintenance and Repair of Buildings and their Internal Environment.* London: Newnes-Butterworths.

Hutton, T. and Lloyd, H. (1993) Mothballing buildings: Proactive maintenance and conservation on a reduced budget. *Structural Survey,* **11**(4), 335–342. Bradford: MCB University Press.

Hutton, T., Lloyd, H. and Singh, J. (1993) The environmental control of timber decay, *Structural Survey,* **10**(1), pp. 5–20. London: Henry Stewart Publications.

Hymers, P. (1999) *Home Extensions: The Complete Handbook.* London: New Holland.

Hymers, P. (2003) *Home Conversions: The Complete Handbook.* London: New Holland.

Hymers, P. (2004a) *Home Renovations: The Complete Handbook.* London: New Holland.

Hymers, P. (2004b) *Home Alterations and Repairs: The Complete Handbook.* London: New Holland.

Hymers, P. (2005) *Planning and Building a Conservatory.* London: New Holland.

ICOMOS (1991) *Guide to Recording Historic Buildings.* London: Butterworth Architecture.

Iddon, J. and Carpenter, J. (2004) *Safe Access for Maintenance and Repair. C611.* London: CIRIA.

Illston, J. M. (ed.) (1993) *Construction Materials – Their Nature and Behaviour.* London: Chapman and Hall.

Imber, A. and Blake, L. (2004) *The Best of Mainly for Students* Vol. 3. London: Estates Gazette.

Inions, C. (2002) *One Space Living.* USA: Watson-Guptill Publications.

Innocent, C. F. (1916) *The Development of English Building Construction.* Cambridge: Cambridge University Press (1999 reprinted ed. by Donhead Publishing Ltd, Shaftsbury).

Insall, D. W. (1972) *The Care of Old Buildings Today – A Practical Guide.* London: The Architectural Press/SPAB.

IoH (2001) *Repair and Maintenance. Good Practice Briefing No. 22.* London: Institute of Housing.

IoH (2004) Turning Empty Properties into Homes. *Good Practice Briefing No. 28.* London: Institute of Housing.

Irwin, J. K. (2002) *Historic Preservation Handbook.* New York: McGraw Hill.

ISE (1997) *Solar Energy in Building Renovation, International Energy Agency Solar Heating and Cooling Programme, Task 20.* London: James & James (Science Publishers) Ltd.

Iselin, D. G. and Lemer, A. C. (eds) (1993) The Fourth Dimension in Building: Strategies for Minimising Obsolescence. *Committee on Facility Design to Minimize Premature Obsolescence*, Building Research Board. Washington, DC: National Academy Press.

ISIAQ (1996) Control of Moisture Problems Affecting Biological Indoor Air Quality. *ISIAQ-guideline TFI-1996.* Ottawa: International Society of Indoor Air Quality and Climate.

IStructE (1991) *Guide to Surveys and Inspections of Buildings and Similar Structures.* London: The Institution of Structural Engineers.

IStructE (1992) *Appraisal of Building Structures.* London: Institution of Structural Engineers.

IStructE (2000) *Subsidence of Low-Rise Buildings* (2nd edition). London: Institution of Structural Engineers.

James, N. (1992) *The Conversion of Agricultural Buildings: An Analysis of Variable Pressures.* Newcastle upon Tyne: Department of Agricultural Economics.

Janson, C. et al. (2005) Insomnia is more common among subjects living in damp buildings. *Journal of Occupational and Environmental Medicine*, **62**, 113–118.

JCT and BDP (2001) *JCT Guide to the Use of Performance Specifications.* London: RIBA Publications.

Jenrette, R. H. (2000) *Adventures with Old Houses*. Charleston: Wyrick & Company.

Jester, T. C. (ed.) (2000) *Twentieth-Century Building Materials: History and Conservation*. New York: National Park Service/McGraw-Hill Inc.

JM Consulting (2002) Study of Science Research Infrastructure, A Report for the Office of Science and Technology. London: The Stationery Office.

John, G. and Sheard, R. (2000) *Stadia: a Design and Development Guide* (3rd edition). London: Architectural Press.

John Pryke and Partners (1987) *Surveying Cracked Property: A Guide for Engineers and Surveyors* (2nd edition). Essex: John Pryke & Partners.

Johnson, A. (1988) *Converting Old Buildings*. London: David and Charles.

Johnson, S. (1993) *Greener Buildings: Environmental Impact of Property*. London: Macmillan.

Jokilehto, J. (2002) *History of Architectural Conservation*. Oxford: Butterworth-Heinemann.

Joyce, R. (2001) *The Construction (Design and Management) Regulations 1994 Explained* (2nd edition). London: Thomas Telford.

Kay, G. N. (1992) *Mechanical and Electrical Systems for Historic Buildings*. New York: McGraw-Hill Inc.

Kelly, J. R. and Male, S. P. (1993) *Value Management in Design and Construction: The Economic Management of Projects*. London: E & FN Spon.

Kendrick, C., Martin, A. J. and Booth (1998) Refurbishment of Air-conditioned Buildings for Natural Ventilation. *Technical Note TN 8/98*. London: BSRIA.

Kerzner, H. (1999) *Project Management*. New York: John Wiley & Sons Inc.

Kibert, C. J. (ed.) (1999) *Reshaping the Built Environment*. Washington, DC: Island Press.

Kidd, S. (ed.) (1995) *Heritage Under Fire: A Guide to the Protection of Historic Buildings* (2nd edition). London: The Fire Protection Association.

Kidney, W. (1976) *Working Places – The Adaptive Use of Older Buildings*. Pittsburg: Ober Park Associates.

Kincaid, D. (2003) *Adapting Buildings for Changing Uses: Guidelines for Change of Use Refurbishment*. London: Spon Press.

Kirk, S. J. and Spreckelmeyer, K. F. (1988) *Creative Design Decisions – A Systematic Approach to Problem Solving in Architecture*. New York: Van Nostrand Reinhold.

Kitchen, J. L. (1995) *Caring for Your Old House: A Guide for Owners and Residents (Respectful Rehabilitation Series)*, National Trust for Historic Preservation. New York: John Wiley & Sons Inc.

Knight, J. (1995) *The Repair of Historic Buildings in Scotland*. Edinburgh: HMSO for Historic Scotland.

Kubal, M. T. (1993) *Waterproofing the Building Envelope*. New York: McGraw-Hill.

Kwakye, A. A. (1994) Built Asset Management: Refurbishment and Optimum Land use. *Construction Paper No. 29*. Englemere: Chartered Institute of Building.

Lander, H. (1979) *House and Cottage Conversion*. Cornwall: Acanthus Books.

Lander, H. (1982) *A Guide to Do's and Don'ts of House and Cottage Interiors*. Cornwall: Acanthus Books.

Lander, H. (1992) *The House Restorer's Guide* (rev. edition). Devon: David and Charles.

Lander, H. (2002) *Do's and Don'ts of House and Cottage Restoration*. Cornwall: Acanthus Books.

Langston, C. (edition) (2001) *Sustainable Practices in the Built Environment*. Oxford: Butterworth-Heinemann.

Latham, D. (2000a) *Creative Re-Use of Buildings* (Vol. 1: Principles and Practice). Shaftsbury: Donhead.

Latham, D. (2000b) *Creative Re-Use of Buildings* (Vol. 2: Building Types – Selected Examples). Shaftsbury: Donhead.

Latham, M. (1994) Constructing the Team, Final Report. London: HMSO.

Lawson, B. (1994) *How Designers Think* (2nd edition). London: Butterworth Architecture.

Lawson, F. R. (1995) *Hotels and Resorts: Planning, Design and Refurbishment*. London: Architectural Press.

Lawson, R. M. (2001a) *Better Value in Steel: Light Steel Framing in Renovations*, SCI Publication P282. London: The Steel Construction Institute.

Lawson, R. M. (2001b) *Better Value in Steel: Modular Construction in Building Extensions*. SCI Publication P284. London: The Steel Construction Institute.

Lawson, R. M., Pedreschi, R., Popo-Ola, S. and Falkenfleth, I. (1998) *Over-cladding of existing buildings using light steel.* SCI Publication P247. London: The Steel Construction Institute.

Lazarus, D., Bussell, M. and Ross, P. (2003) *Retention of Masonry Facades – Best Practice Guide. C579.* London: Construction Industry Research and Information Association.

Leather, P. (2000) *Crumbling Castles? Helping Owners to Repair and Maintain Their Homes*. York: The Joseph Rowntree Foundation.

Lee, V. and Main, R. (2000) *Recycled Spaces: Converting Buildings into Homes*. San Francisco: Soma Books.

Leo (1991) *Altering, Extending and Converting Houses*. London: International Thomson Publishing Services.

Lim, W. B. P. (ed.) (1988) *Control of the External Environment of Buildings*. Singapore: Singapore University Press.

Lion, E. (1982) *Building Renovation and Recycling*. New York: John Wiley & Sons.

Liska, R. W. (1990) *Means Facilities Maintenance Standards*. Massachusetts: R.S. Means Co. Inc.

Litchfield, M. W. (2000) *Renovation: A Complete Guide* (2nd edition). New York: Prentice- Hall.

Lizzi, F. (1993) 'Palo radice' structures. In Thorburn, S. and Littlejohn, G. S. (eds), *Underpinning and Retention* (2nd edition). London: Blackie Academic and Professional.

Lloyd, D., Fawcett, J. and Freeman, J. (1979) *Save the City* (2nd edition). London: SPAB.

London District Surveyors Association (2003) *Extensions Toolkit*. London: London District Surveyors Association.

London Hazards Centre (1990) *Sick Building Syndrome (Causes, Effects and Control)*. London: London Hazards Centre Trust Ltd.

Long, H. (2001) *Victorian Houses and Their Details*. Oxford: Butterworth-Heinemann.

Loss Prevention Council (1993) *Asbestos Use in Buildings: The Hazards and Their Limitations*, Report SHE 9:1993. Borehamwood: Loss Prevention Council.

Loss Prevention Council (1995) *Joint Code for Protection from Fire of Construction Sites and Buildings Undergoing Renovation*. Borehamwood: Loss Prevention Council.

Loss Prevention Council (1996) *Code of Practice for the Protection of Unoccupied Buildings* (2nd impression). Borehamwood: Loss Prevention Council.

Loss Prevention Council (2000) *The Design Guide for the Fire Protection of Buildings 2000, A Code of Practice for the Protection of Business*. London: The Fire Protection Agency.

Loughborough University and Milan Polytechnic (2004) *Health and Safety in Refurbishment Involving Demolition and Structural Stability*. London: Health and Safety Executive.

Lowenthal, D. and Binney, M. (eds) (1981) *Our Past Before Us: Why Do We Save It?* London: Temple Smith.

Ltsiburek, J. and Carmody, J. (1993) *Moisture Control Handbook (Principles and Practices for Residential and Small Commercial Buildings)*. New York: Van Nostrand Reinhold.

Lunardi, I. (1999) Refurbishment – Façade Retention. *The Impact of Engineering on Buildings, Supporting the City Conference at the University of Strathclyde*. 16–17 September 1999, Glasgow, Scotland.

Luscombe-White, M. (2004) *Barns: Living in Converted and Reinvented Spaces*. London: Conran Octopus.

Macdonald, S. (ed.) (1996) *Modern Matters: Principles and Practice in Conserving Recent Architecture, Proceedings of the English Heritage Conference 1995*. Shaftsbury: Donhead.

Macdonald, S. (ed.) (2001) *Preserving Post-war Heritage: The Care and Conservation of Mid-twentieth-century Architecture, Proceedings of the English Heritage Conference 1998*. Shaftsbury: Donhead.

Mack, L. (2000) *The Art of Home Conversion: Transforming Uncommon Properties into Stylish Homes*. USA: Sterling Publishing Co. Inc.

Male, S., Gronquist, M., Kelly, J., Damodaran, L. and Olphert, W. (2004) *Supply Chain Management for Refurbishment*. London: Thomas Telford Ltd.

Mant, D. C. and Gray, J. A. M. (1986) Building Regulation and Health. BRE Report. Garston: Building Research Establishment.

Marks, S. (ed.) (1996) *Concerning Buildings*. London: Architectural Press.

Markus, T. (1993) Cold, condensation and housing poverty. In Burridge, R. and Ormandy, D. (eds), *Unhealthy Housing: Research Remedies and Reforms*. London: E & FN Spon.

Markus, T. A. (1988) *Old Buildings – New Opportunities: A Guide to the Possibilities of Converting* (2nd edition). Salisbury: COSIRIA.

Markus, T. A. (ed.) (1979) *Building Conversion and Rehabilitation*. London: Newnes – Butterworths.

Markus, T. A. et al. (1972) *Building Performance, Building Performance Research Unit*. London: Allied Publishers Ltd.

Maroni, M., Seifert, B. and Lindvall, T. (eds) (1995) *Indoor Air Quality (A Comprehensive Reference Book), Air Quality Monographs*, Vol. 3. The Netherlands: Elsevier Science BV.

Marsh, P. (1983) *Refurbishment of Commercial and Industrial Buildings*. London: Longman.

Marshall, D. and Worthing, D. (2000) *The Construction of House* (3rd edition). London: Estates Gazette.

Marshall, D., Worthing, D. and Heath, R. (1998) *Understanding Housing Defects*. London: Estates Gazette.

Martin, A. J. and Gold, C. A. (1999) *Refurbishment of Concrete Buildings – The Decision to Refurbish, Guidance Note GN 7/99*. London: Building Services Research and Information Association and British Cement Association.

Massari, G. and Massari, I. (1993) *Damp Buildings, Old and New* (Translated by C. Rockwell). Rome, Italy: ICCROM.

Matarasso, F. (1995) *Spirit of Place: Redundant Churches as Urban Resources*. Stroud: Comedia.

Matulionis, R. C. and Freitag, J. C. (eds) (1990) *Preventive Maintenance of Buildings*. New York: Van Nostrand Reinold.

Mayer, P. and Wornell, P. (eds) (1999) *HAPM Workmanship Checklists*, HAPM Publications Ltd. London: E & FN Spon.

Mazzolani, F. and Ivanyi, M. (2002) *Refurbishment of Buildings and Bridges* (CISM Centre for Mechanical Sciences Courses and Lectures). New York: Springer-Verlag Telos.

McDonald, S. (ed.) (2001) *Preserving Post-War Heritage (The Care and Conservation of Mid- Twentieth-Century Architecture)*. Dorset: Donhead.

McDonald, S. (ed.) (2002) *Building Pathology: Concrete*. Oxford: Blackwell Publishing.

McDonald, T. C. (1994) *Understanding Old Buildings (The Process of Architectural Investigation), Preservation Brief 35*, Heritage Preservation Services, National Park Service. Available online from: http://www2.cr.nps/tps/briefs/brief35.htm

McAfee, P. (2000) *Stone Buildings: Conservations, Restoration, History*. New York: New Publishers.

McGee, F. and Pawson, J. (1999) *Barns*. London: Booth-Clibborn Editions.

McGregor, W. and Then, D. S. S. (1999) *Facilities Management and the Business of Space*. London: Arnold.

McLennan, P. M., Nutt, B. and Walters, R. (eds) (1998) *Refurbishing Occupied Buildings (Management of Risk Under the CDM Regulations)*. London: Thomas Telford.

Melville, I. A. and Gordon, I. A. (1993) *Professional Practice for Building Works*. London: Estates Gazette.

Melville, I. A. and Gordon, I. A. (1998) *The Repair and Maintenance of Houses* (2nd edition). London: Estates Gazette.

Melville, I. A. and Gordon, I. A. (2004) *Inspections and Reports on Dwellings: Assessing Age*. London: Estates Gazette.

Melville, I. A. and Gordon, I. A. (2005a) *Inspections and Reports on Dwellings: Inspecting.* London: Estates Gazette.

Melville, I. A. and Gordon, I. A. (2005b) *Inspections and Reports on Dwellings: Reporting for Buyers.* London: Estates Gazette.

Melville, I. A. and Gordon, I. A. (2005c) *Inspections and Reports on Dwellings: Reporting for Sellers.* London: Estates Gazette.

Mercer, E. (1975) *English Vernacular Houses*, Royal Commission on Historical Monuments. London: HMSO.

Mietz, J. (ed.) *Electrochemical Rehabilitation Methods for Reinforced Concrete Structures – A State of the Art Report*, Number 24. London: European Federation of Corrosion Publications.

Miles, D. and Syagga, P. (1987) *Building Maintenance: A Management Manual.* London: Intermediate Technologies Ltd.

Mills, E. (ed.) (1994) *Building Maintenance and Preservation* (2nd edition). London: Architectural Press.

Mindham, C. N. (1999) *Roof Construction and Loft Conversion* (3rd edition). Oxford: Blackwell Science Ltd.

Mitchell, E. (1988) *Emergency Repairs for Historic Buildings*, English Heritage. London: Butterworth Architecture.

Mollison, B. and Slay, D. (1991) *Introduction to Permaculture*. London: R. M., Tagari.

Molnar, F. I. (2001) *Lofts: New Designs for Urban Living*. Massachusetts: Rockport Publications Inc.

Monodraught Ltd (2005) *Pitched Slate Roof, SunPipe Technical.* High Wycombe: Monodraught Ltd.

Moore, A. C. (2000) *The Powers of Preservation: New Life for Historic Urban Places.* New York: McGraw-Hill Professional.

Morgan, H. P. (1994) Basic Principles of Smoke Ventilation. In Stollard, P. and Johnston, L. (eds), *Design Against Fire (An Introduction to Fire Safety Engineering Design).* London: E & FN Spon.

Morgan, H. P. and Gardner, J. P. (1991) *Design Principles for Smoke Ventilation in Enclosed Shopping Centres.* BRE Report BR 186. London: Construction Research Communications Ltd.

Moss, G. et al. (2001) *HAPM Guide to Defect Avoidance*. London: Construction Audit Ltd, Spon Press.

Mostaedi, A. (2003) *Building Conversion and Renovation*. London: Links International.

Moy, S. S. J. (ed.) (2001) *FRP Composites: Life Extension and Strengthening of Metallic Structures.* London: Thomas Telford Publishing.

MSI Marketing Research for Industry Ltd (2002) *Retail Construction and Refurbishment* (Digital edn.). London: MarketResearch.com.

Murray, P. (2002) *Safety in Building Alteration and Demolition Schemes (A Student-Centred Learning Package).* Plymouth: SLICE.

Mynors, C. (1989) *Listed Buildings and Conservation Areas*. London: Longman.

Myton-Davies, P. (1988) *A Practical Guide to Repair and Maintenance of Houses*. London: International Thomson Business Publishing.

Nash, G. (2000) *Renovating Old Houses (For Pros by Pros).* New York: New Publishers.

Nash, G. (2003) *Renovating Old Houses: Bringing New Life to Vintage Homes*. Newtown CT: Taunton Press.

Nash, W. G. (1986) *Brickwork Repair and Restoration*. Eastborne: Attic Books.

Nastaras, M. (2003) *Studios and Lofts: One Room Living*. USA: Universe Publishing.

National Audit Office (1987) *Repair and Maintenance of School Buildings*. London: HMSO.

National Materials Advisory Board et al. (1982) *Conservation of Historic Stone Buildings and Monuments*. Washington, DC: National Academy Press.

National Park Service/Historic Preservation Education Foundation (1999) *Roofing Handbook for Historic Buildings*. Washington, DC: HPEF.

National Park Service/Heritage Preservation Services (1998) *Caring for Your Historic House.* Washington, DC: Harry N. Abrams, Inc.

National Research Council (1996) *Science and Judgment in Risk Assessment* (Student edition). London: Taylor & Francis Ltd.

National Trust for Historic Preservation (1980) *Old and New Architecture: Design Relationship*. Washington, DC: Preservation Press.

NBA (1980) *Annual Report 1979–1980*. London: The National Building Agency.

NBA (1985) *Maintenance Cycles and Life Expectancies of Building Components and Materials: A Guide to Data and Sources*. London: National Building Agency.

NCHA (1988) *Non-Traditional Housing in Northern England: Identification Booklet*. Newcastle: Northern Consortium of Housing Associations.

Needleman, L. (1969) The comparative economics of improvement and new building. *Urban Studies*, Vol. 6. Oxford: Blackwell Science.

Neiseward, N. (1998) *Converted Spaces*. London: Conran Octopus Ltd.

Newman, A. (2001) *Structural Renovation of Buildings: Methods, Details and Design Examples*. New York: McGraw-Hill Professional.

NHBC (2005) *Standards for Conversions and Renovations* (April 2005 edition). Amersham, Bucks: National House Building Council.

NHBC, BRE, EEO and BRECSU (1991) *Thermal Insulation and Ventilation: A Good Practice Guide*. London: National House-Building Council.

NHER (2005) *Home Energy Rating*. Milton Keynes: National Home Energy Rating.

NHS Estates (1991) *Refurbishment for Natural Ventilation, Health Facilities Note 26*. London: HMSO.

NHTPC (1992) *A Practical Guide to Living Over Shops*. London: National Housing and Town Planning Council.

NHTPC (1994) *No Losers – New Users: New Homes from Empty Properties*. London: National Housing and Town Planning Council.

NICEIC (2004) *part p fact sheet*. Dunstable: National Inspection Council for Electrical Installation Contracting.

Nicholson, J. (1994) Problems with replacement windows. *CEMicircular*, **2**(4), October 1994. Reading: The College of Estate Management.

Nicholson, P. (1823a) *New Practical Builder* Vol. 1 (1993 Facsimile edition). London: Beacon Books.

Nicholson, P. (1823b) *New Practical Builder* Vol. 2 (1993 Facsimile edition). London: Beacon Books.

Nolan, E. (2005) *Innovation in Concrete Frame Construction 1995–2015*. Garston: IHS Rapidoc (BRE Bookshop).

Noy, E. and Douglas, J. (2005) *Building Surveys and Reports* (3rd edition). Oxford: Blackwell Publishing.

NSSC (1999) *School Safety*. Westlake Village. CA, USA: National School Safety Center.

NTHP (1977) *Built to Last: A Handbook on Recycling Old Buildings*. Washington, DC: Preservation Press and National Trust for Historic Preservation.

Nutt, B. (c1993) The Adaptive Reuse of Buildings, Paper presented at a Conference on Building Refurbishment. London: Publisher Unknown.

Nutt, B., Walker, B., Holliday, S. and Sears, D. (1976) *Housing Obsolescence*. Hants: Saxon House.

ODPM (2003a) *Empty Homes: Temporary Management, Lasting Solutions*, A Consultation Paper, October 2003, Office of the Deputy Prime Minster. London: HMSO.

ODPM (2003b) *Preparing for Floods: Interim Guidance for Improving the Flood Resistance of Domestic and Small Business Properties*, October 2003, Office of the Deputy Prime Minister. London: HMSO.

ODPM (2003c) *Factsheet 7: The Home Information Pack (HIP)*, Office of the Deputy Prime Minister. London: HMSO.

ODPM (2004) *Delivering Housing Adaptations for Disabled People: A Good Practice Guide*, November 2004. London: Department for Education and Skills, ODPM and Department of Health.

ODPM (2005) *Factsheet 1: Decent Homes – The National Picture*, Office of the Deputy Prime Minister. London: HMSO.

O'Farrell, F. (1977) *Farm House Conversion: A Handbook for Renovating the Farm Home.* Dublin: Publisher Unknown.

Oliver, A. (1997) *Dampness in Buildings* (2nd edition revised by Douglas, J. and Stirling, J. S.). Oxford: Blackwell Science.

Oreszczyn, T. and Pretlove, S. E. C. (1999) *Condensation Targeter II: Modelling Surface Relative Humidity to Predict Bould Growth in Dwellings. Building Services Engineering Research and Technology*, **20**(3), 143–153. London: CIBSE.

Orton, A. (1992) *The Way We Build Now: Form, Scale, Techniques.* New York: Van Nostrand Reinhold.

Oseland, N. A. and Raw, G. (1993) 'Perceived air quality: Discussion on the new units'. Building Services Engineering Research and Technology, **14**(4), 137–141. London: CIBSE.

O'Sullivan, P. (ed.) (1988) Passive Solar Energy in Buildings, The Watt Committee Report No. 17. London: Elsevier Applied Science Publishers Ltd.

Oxley, R. (2003) *Survey and Repair of Traditional Buildings.* Shaftsbury: Donhead Publishing.

Oxley, T. A. and Gobert, E. G. (1994) *Dampness in Buildings (Diagnosis, Instruments, Treatment)* (2nd edition). Oxford: Butterworth-Heinemann.

Palfreyman, J. W. and Urquhart, D. (2002) The Environmental Control of Dry Rot. *Technical Advice Note 24.* Edinburgh: Historic Scotland.

Palfreyman, J. W., Smith, D. and Low, G. (2002) Studies of the Domestic Dry Rot Fungus *Serpula lacrymans* with Relevance to the Management of Decay in Buildings, Research Report. Edinburgh: Historic Scotland.

Palmer, A. and Rawlings, R. (eds) (2002) Building-Related Sickness: Causes, Effects and Ways to Avoid it. *Technical Note TN 2/2002.* London: BSRIA.

Park, S. (1993) *Mothballing Historic Buildings, Preservation Brief 31, National Parks Service (Preservation Assistance).* Washington, DC: US Department of the Interior.

Parnham, P. (1996) *Prevention of Premature Staining of New Buildings.* London: Chapman & Hall.

Pavey, N. (1995) Rules of Thumb. *Technical Note TN 17/95* (2nd edition). Reading: Building Services Research and Information Association.

Pearson, A. (1999b) Summertime blue. *Building*, **CCLXIV**(47), London.

Pearson, A. (2000) Noisy neighbour rules spell trouble for housebuilders. *Building*, **CCLXV**(43), London.

Pearson, D. (1989) *The Natural House Book.* London: Conran Octopus Ltd.

Pentagon Renovation Programme (2005) The History of the Pentagon and Pentagon Renovation Program. Available online at http://renovation.pentagon.mil/history.htm

Penton, J. H. (1999) *The Disability Discrimination Act: INCLUSION.* London: RIBA Publications.

Perry, J. G. (1994) *A Guide to the Management of Building Refurbishment*, Report 133. London: Construction Industry Research and Information Association.

Petrocelly, K. L. and Thumann, A. (1999) *Facilities Evaluation Handbook: Safety, Fire Protection and Environmental Compliance.* New York: Prentice Hall.

Pezzy, J. (1984) An Economic Assessment of Some Energy Conservation Measures in Housing and Other Buildings. BRE Report. Garston: Building Research Establishment.

Phillips, D. (1999) *Lighting Historic Buildings.* Oxford: Butterworth-Heinemann.

Pickard, R. D. (1996) *Conservation in the Built Environment.* Harlow: Addison Wesley Longman.

Platt, S., Martin, C., Hunt, S. and Lewis, C. (1989) Damp housing, mould growth and symptomatic health state. *British Medical Journal*, **298**, 1673–1678. London.

Pollard, R. (1998) Redundant government buildings. In Taylor, J. (ed.), *The Building Conservation Directory.* Cathedral Communications Ltd. England: Tisbury.

Potter, P. A. and Perry, G. A. (1999) *Fundamentals of Nursing* (5th edition). St Louis: Mosby.

Pout, C. H., MacKenzie, F. and Bettle, R. (2003) *Carbon dioxide Emissions from Non-domestic Buildings: 2000 and Beyond.* Garston: Building Research Establishment (BRE Energy Technology Centre).

Powell, K. (1999) *Architecture Reborn – The Conversion and Reconstruction of Old Buildings*. London: Lawrence King Publishing.

Powell, K. and de La Hey, C. (1987) *Churches: A Question of Conversion*. London: SAVE Britain's Heritage.

Powys, A. R. (1995) *Repair of Ancient Buildings* (3rd edition). London: SPAB.

Preiser, W. F. (1989) *Post Occupancy Evaluation*. New York: Van Nostrand Reinhold.

Preiser, W. F. and Vischer, J. (eds) (2004) *Assessing Building Performance*. New York: Butterworth-Heinemann.

Price, N. S., Talley, M. K. and Vaccaro, A. M. (eds) (1996) *Historical and Philosophical Issues in the Conservation of Cultural Heritage (Research in Conservation)*. New York: J Paul Getty Trust Publications.

Prior, J. J. (1999) *Sustainable Retail Premises: An Environmental Guide to Design, Refurbishment and Management of Retail Premises*. Garston: Building Research Establishment.

Procurement Group (1990) *Value for Money in Construction Procurement. Procurement Guidance No. 2*. London: HM Treasury.

Property Services Agency (1988) *The Conservation Handbook* (2nd edition). London: HMSO.

Proverbs, D. and Soetanto, R. (2004) *Flood Damaged Property*. Oxford: Blackwell Publishing.

Prudon, T. (1990) *Office Building Renovation Manual*. New York: Kluwer.

Pye, P. and Harrison, H. W. (2003) *Floors and Flooring: Performance, Diagnosis, Maintenance, Repair and the Avoidance of Defects, BRE Building Elements Series* (2nd edition). London: Construction Research Communications Ltd.

Quah, L. K. (1989) *Variability in Tender Bids for Construction Work, Occasional Paper 43*. Englemere: Chartered Institute of Building.

Rabun, J. S. (2002) *Structural Analysis of Historic Buildings: Restoration, Preservation and Adaptive Reuse Applications for Architects and Engineers*. New York: John Wiley & Sons.

Raftery, J. (1994) *Risk Analysis in Project Management*. London: E & FN Spon.

Ransom, W. (1987) *Building Failures: Diagnosis, Remedy and Prevention* (2nd edition). London: E & FN Spon.

Ranson, R. (1991) *Healthy Housing: A Practical Guide*. London: Chapman & Hall.

Ratay, R. (2000) *Forensic Structural Engineering Handbook*. New York: McGraw-Hill Professional.

Raw, G. J. and Hamilton, R. M. (1995) *Building Regulation and Health*, BRE Report 289. London: Construction Research Communications Ltd.

Raw, G. J., Aizlewood, C. E. and Hamilton, R. M. (2001) *Building Regulation, Health and Safety*, BRE Report 417. London: Construction Research Communications Ltd.

RCHM of England (1991) *Recording Historic Buildings – A Descriptive Specification* (2nd edition). London: HMSO.

Read, R. E. H. (ed.) (1999) *External Fire Spread: Building Separation and Boundary Distances*. London: BRE and Fire Research Station, CRC Ltd.

Read, R. E. H. and Morris, W. A. (1993) *Aspects of Fire Precautions in Buildings*, BRE Report BR 225. London: Construction Communications Ltd.

Reeves, B. R. and Martin, G. R. (1989) *The Structural Condition of Wimpey No-Fines Low-rise Dwellings*, BRE Report 153. Garston: Building Research Establishment.

Research, Analysis and Evaluation Division (2003) *English House Condition Survey 2001* (Main Report). London: Office of the Deputy Prime Minister.

Revell, K. and Leather, P. (2000) The State of Housing (2nd edition). In *A Factfile on Housing Conditions and Housing Renewal Policies in the UK*. London: The Policy Press.

Reyers, J. (2002) 'Toxic Mould', *The Journal*, RICS Building Surveying Faculty. Bristol: Origin Publishing Ltd.

Reyers, J. and Mansfield, J. (2000) Conservation Refurbishment Projects: A Comparative Assessment of Risk Management Approaches. *The Cutting Edge, The Real Estate Research Conference of the RICS Research Foundation*, 6–8 September 2000, London.

Richard Lees (c.1995) *Precast Concrete Lintels and Slabs, Technical Brochure.* Weston Underwood, Ashbourne, Derbyshire: Richard Lees Ltd.

Richards, I. (2002) *Manhattan Lofts.* USA: Academy Press.

Richardson, B. A. (2001) *Defects and Deterioration in Buildings* (2nd edition). London: Spon Press.

Richardson, B. A. (2004) *Remedial Treatment of Buildings* (2nd edition). Oxford: Butterworth-Heinemann.

Richardson, C. (1987) *AJ Guide to Structural Surveys.* London: Architectural Press.

Richardson, C. (1996) When the earth moves. *Architects Journal*, London.

Richardson, C. (2000) Moving structures. *Architects Journal*, London.

RICS (1973a) *Specifications for Rehabilitation Works for Residential Buildings, A Working Party Report Prepared by the Building Surveyors Division of the RICS, Surveyors Publications.* London: RICS Books.

RICS (1973b) *The Rehabilitation of Houses and Other Buildings, A Report Prepared for the Building Surveyors Committee of the RICS.* London: RICS Books.

RICS (1995a) *The Private Finance Initiative: The Essential Guide.* London: RICS Books.

RICS (1995b) *What is it Worth?* London: RICS Books.

RICS (1995c) *Flat Roof Covering Problems, RICS Guidance Note.* London: RICS Books.

RICS (1996) *Fire Damage Reinstatement, RICS Information Paper.* London: RICS Books.

RICS (1998) *Refurbishment in the Office Sector 1997/8, The Connaught Report.* London: The Royal Institution of Chartered Surveyors.

RICS (2000) *Building Maintenance: Strategy, Planning and Procurement, RICS Guidance Note.* London: RICS Books.

RICS (2002) *Building Surveys of Residential Property* (4th edition). Residential Property Faculty, RICS Guidance Note. London: RICS Books.

RICS and EEO (1992) *Energy Efficiency in Buildings (Energy Appraisal of Existing Buildings – A Handbook for Surveyors).* London: RICS Books.

Ridal, J., Reid, J. and Garvin, S. (2005) *Highly Glazed Buildings: Assessing and Managing the Risks.* Garston: IHS Rapidoc (BRE Bookshop).

Ridout, B. (1999) *Timber Decay in Buildings: The Conservation Approach to Treatment*, English Heritage and Historic Scotland. London: E & FN Spon.

Riha, J. (2000) *Attics: Your Guide to Planning and Remodelling.* New York: Meridith Books.

Riley, M. and Cotgrave, A. (2003) *Construction Technology 2: Commercial and Industrial Building.* London: Palgrave Macmillan.

Riley, M. and Cotgrave, A. (2004) *Construction Technology 3: The Technology of Refurbishment and Maintenance.* London: Palgrave Macmillan.

Riley, M. and Howard, C. (2002) *Construction Technology 1: House Construction.* London: Palgrave Macmillan.

Riyat, R. (1993) *Conservation and Preservation: A Definitive Statement.* Blackburn: R & J Services.

Roaf, S. (2001) *Eco House: A Design Guide.* Oxford: Butterworth-Heinemann.

Roaf, S. (2004) *Adapting Buildings and Cities for Climate Change.* Oxford: Butterworth-Heinemann.

Robson, P. (1991) *Structural Appraisal of Traditional Buildings.* Aldershot: Gower Technical.

Robson, P. (1999) *Structural Repair of Traditional Buildings.* Shaftsbury: Donhead Publishing.

Roger Trim and Partners (1991) Re-use of Upper Floor Town Centre Property, A report to Scottish Homes, April 1991. Edinburgh: Scottish Homes.

Rose, V. (1998) *Catalytic Conversion: REVIVE Historic Buildings to Regenerate Communities.* London: SAVE, AHF, IHBC, and UKABPT.

Ross, K. (2005) *Modern Methods of Construction: A Surveyor's Guide.* Garston: IHS Rapidoc (BRE Bookshop).

Rostron, J. (ed.) (1996) *Sick Building Syndrome.* London: E & FN Spon.

Roth, L. M. (1993) *Understanding Architecture: Its Elements, History and Meaning*. London: The Herbert Press.
Royal Society (1983) *Risk Assessment, A Study Group Report*. London: The Royal Society.
Royal Society (1992) *Risk: Analysis, Perception and Management, A Study Group Report*. London: The Royal Society.
Rural Development Commission (1996) *Old Buildings New Opportunities* (Booklet). London: Rural Development Commission.
Rural Development Commission (1997) *Redundant Building Grants* (Leaflet). London: Rural Development Commission.
Rus, M. and Warchol, P. (2002) *Loft*. USA: Monacelli.
Russell, T. (ed.) (1998) *Sustainable Business: Economic Development and Environmentally Sound Technologies*. London: The Regency Corporation Ltd.
Ryan, C. (1995) *Points for Successful and Unsuccessful Conversion to Residential Use, A Guidance Note*. Shropshire: Shropshire County Council.
Ryan, P. A., Wolstenholme, R. P. and Howell, D. M. (1994) *Durability of Cladding, A State of the Art Report*. London: Thomas Telford Services Ltd.
Sage, A. (2000) Housing plan makes a virtue of granny flat, *The Times*, Friday April 14. London: The Times.
Samuels, R. and Prasad, D. K. (eds) (1996) *Global Warming and the Built Environment*. London: E & FN Spon.
Sanderson, B. (2004) *Asbestos for Surveyors*. London: Estates Gazette.
Santamouris, M. and Asimakopoulos, D. (eds) (1996) *Passive Cooling of Buildings*. London: James & James (Science Publishers) Ltd.
Santini, C. (1999) Specifications – Metallic Luster, *Architecture*, June 1999, USA.
Saunders, G. K. (2004) *Reducing the Effects of Climate Change by Roof Design*. London: IHS Rapidoc and BRE.
SAVE (1995b) *Beacons of Learning: Breathing New Life into Old Schools*. London: SAVE Britain's Heritage.
Sayce, S., Walker, T. and McIntosh, A. (2004) *Building Sustainability in the Balance*. London: Estates Gazette.
Schittich, C. (ed.) (2003) *In Detail: Building in Existing Fabric (Refurbishment, Extensions, New Design)*. Berlin: Birkheuser.
Schmertz, M. F. and Architectural Record (2000) *Communication in the Design Process*. London: Spon Press.
Scottish Civic Trust (1981) *New Uses for Older Buildings in Scotland, Report for the Scottish Development Department*. Edinburgh: HMSO.
Scottish Development Department (1996) *Small Buildings Guide* (2nd edition). Edinburgh: The Stationery Office.
Scottish Executive (2000) *Fire Safety Scotland* (16 page tabloid Special Guide). Edinburgh: produced on behalf of the Scottish Executive by Tactica Solutions.
Scottish Executive (2003) *Modernising Scotland's Social Housing, Consultation Paper*, March 2003. Edinburgh: Scottish Executive.
Scottish Executive and Astron (2004) *Maintaining Houses – Preserving Homes, Consultation*. Edinburgh: Scottish Executive.
Scottish Executive Building Directorate (2001) *A Guide to Non-Traditional and Temporary Housing in Scotland 1923–1955* (2nd edition). Edinburgh: The Stationery Office.
Scottish Executive Central Research Unit (2001) *Evaluation of the Empty Homes Initiative*. Edinburgh: Scottish Executive.
Scottish Executive Development Department (2004) *Planning and Building Standards Advice on Flooding*, Planning Advice Note 69. Edinburgh: Scottish Executive.
Scottish Homes (1997) *Scottish House Condition Survey 1996*. Edinburgh: Scottish Homes.

Scottish Office (1989) *Housing Maintenance Kit*. Edinburgh: HMSO.

Scottish Office Building Directorate (1987) *Housing Conversion (To Meet Changing Needs in the Public Sector), A report prepared by Norman Raitt Architects*. Edinburgh: The Scottish Office.

Scottish Sports Council (1999) *Swimming Pools – Improvements and Alterations, Technical Digest*. London: The Scottish Sports Council.

SCPRC (1995) *Cathodic* Protection of Reinforced Concrete Status Report, Report No. SCPRC/001.95, 6 February 1995. London: Society for the Cathodic Protection of Reinforced Concrete.

Seeley, I. H. (1987) *Building Maintenance* (2nd edition). London: Palgrave Macmillan.

Select Committee on Environment, Transport and Regional Affairs (1999) 'Potential Risk of Fire Spread in Buildings via External Cladding Systems', *Special Report*. London: The Stationery Office.

Seward, D. (2002) *Understanding Structures: Analysis, materials, design* (2nd edition). London: Macmillan.

Sharpe, G. R. (2000) *Works to Historic Buildings: A Contractor's Manual*. London: Longman.

Shelter (1999) *Fire Safety Guide: For Multi-Occupied, Privately Rented Housing*. London: Shelter.

Shipway, J. S. (1987) *Repairs to Timber Frame Housing, Paper presented at Structural Faults and Repairs Conference*, 7–9 July 1987. London: City University.

Shopsin, W. C. (1986) *Restoring Old Buildings for Contemporary Use*. England: Watson Guptil Publications.

Sick, F. and Erge, T. (eds) (1996) *Photovoltaics in Buildings*. London: James & James (Scientific Publishers) Ltd.

Simmons, H. L. (1989) *The Architect's Remodelling, Renovation and Restoration Handbook*. New York: Van Nostrand Reinhold.

Simpson, J. (1992) Construction and the Generation of Style, *Aspect*, 2, Journal of the Alumni Association of the Institute of Advanced Architectural Studies, York.

Singh, J. (1993) Biological Contaminants in the built environment and their health implications. *Building Research and Information*, **21**(4). London: E & FN Spon.

Singh, J. (ed.) (1994) *Building Mycology: Management of Decay and Health in Buildings*. London: Chapman & Hall.

Singh, J. (1994) *The built environment and the development of fungi*. In Singh, J. (ed.), *Building Mycology*. London: E & FN Spon.

Singh, J. (2004) *Building Pathology: Timber*. Oxford: Blackwell Publishing.

Singh, J. and Walker, B. (eds) (1996) *Allergy Problems in Buildings*. Wilts: Quay Books.

Slesin, S., Cliff, S. and Rozensztroch (1986) *The International Book of Lofts*. New York: Clarkson N. Potter Inc.

Smeathie, P. H. and Smith, P. H. (1990) *New Construction for Older Buildings*. New York: John Wiley & Sons Inc.

Smith, B. J. and Warke, P. A. (2004) *Stone Decay Its Causes and Controls*. Shaftsbury: Donhead Publishing Limited.

Smith, D. J. (1994) *Underpinning and Repair of Subsidence Damage*. London: Camden Consultancy.

Smith, J. (1993) *Reliability, Maintainability and Risk* (4th edition). Oxford: Butterworth-Heinemann.

Smith, M., Whitelegg, J. and Williams, N. (1998) *Greening the Built Environment*. London: Earthscan Publications Ltd.

Smith, P. (ed.) (2004) *Rivington's Building Construction*, Vols 1–3. Shaftsbury: Donhead Publishing.

Smith, P. F. (2003) *Eco-Refurbishment: A Practical Guide to Creating an Energy Efficient Home*. Oxford: Butterworth-Heinemann.

Social Survey Division (1998) *Living in Britain (Results from the 1996 General Household Survey)*, Office of National Statistics. London: HMSO.

Son, T. and Yuen, G. (1996) *Building Maintenance Technology*. London: Macmillan.

SPAB (2004) *Church Extensions, Technical Statement*. London: Society for the Protection of Ancient Buildings.

Spain, B. (2000a) *Spon's Estimating Costs Guide to Minor Works, Refurbishment and Repairs*. London: Spon Press.

Spain, B. (2000b) *Spon's House Improvement Price Book: House Extensions, Alterations and Repairs, Loft Conversions, Insulation*. London: Spon Press.

SpeedDeck (2000) *The SpeedDeck Guide to Building Refurbishment*. Suffolk: SpeedDeck Building Systems Ltd.

Spengler, J. D., McCarthy, J. F. and Samet, J. M. (2000) *Indoor Air Quality Handbook*. New York: McGraw-Hill Professional.

Spittles, D. (2004) *Lofts of London*. London: Tectum.

Sports Council (1977) *Recreational Use of Church Buildings*. London: Sports Council.

Sports Council (1994) *Swimming Pools – Improvements and Alterations to Existing Pools, Guidance Note*. London: Sports Council.

Stahl, F. A. (1984) *A Guide to the Maintenance, Repair, and Alteration of Historic Buildings*. New York: Van Nostrand Reinhold.

Stansall, P. (1997) The quest for space. *Architects' Journal*, 18 September. London: Emap Business Communications.

Statham, R., Korczak, J. and Monaghan, P. (1986) *House Adaptations for People with Disabilities (A Guidance Manual for Practitioners)*. London: Department of the Environment, HMSO.

Stephen, G. (2002) *New Life for Old Houses* (Updated edition). New York: Dover Publications Inc.

Stephenson, J. (2004) *Building Regulations Explained* (7th edition). London: Spon.

Stevens Europe (2001) *TPO: Book of Knowledge*. Bellshill: Stevens Europe.

Stipe, R. E. (ed.) (2003) *A Richer Heritage: Historic Preservation in the Twenty-First Century*. Carolina: University of North Carolina Press.

Stitt, F. A. (1985) *Designing Buildings That Work*. New York: McGraw-Hill.

Stollard, P. and Abrahams, J. (1995) *Fire from First Principles: A Design Guide to Building Fire Safety* (2nd edition). London: E & FN Spon.

Stone, K. (2003) *Loft Design: Solutions for Creating a Liveable Space*. Massachusetts: Rockport Publishers Inc.

Strachan, D. (1993) Dampness, mould growth and respiratory disease in children. In Burridge, R. and Ormandy, D. (eds) *Unhealthy Housing: Research Remedies and Reforms*. London: E & FN Spon.

Strattan, M. (ed.) (1997) *Structure and Style (Conserving 20th Century Buildings)*. London: E & FN Spon.

Stratton, M. (1999) *Making Industrial Buildings Work: Initiatives in Conservation & Regeneration*. London: Spon Press.

Strong, S. J. (1999) Introduction to renewal energy technologies. In Kibert, C. J. (ed.) *Reshaping the Built Environment: Ecology, Ethics, and Economics*. Washington, DC: Island Press.

Sundell, J. (1994) On the association between building ventilation characteristics, some indoor environmental exposures, some allergic manifestations and subjective symptom reports, *Indoor Air, International Journal of Indoor Air Quality and Climate* (Suppl 2/94), May. Stockholm Institute of Environmental Medicine.

Sustainable Homes (2004) *Refurbishments – Good Practice Guide*. Middlesex: Sustainable Homes.

Swallow, P. (1999) Rehabilitation: getting to know the building. *Building Engineer*, July/August 1998. Northampton: Association of Building Engineers.

Swallow, P., Dallas, R., Jackson, S. and Watt, D. (2004) *Measurement and Recording of Buildings* (2nd edition). Shaftsbury: Donhead Publishing.

Swindells, D. J. and Hutchings, M. (1991) *A Checklist for the Structural Survey of Period Timber Framed Buildings*, Building Conservation Group. London: RICS Business Books.

Talbot, M. (1986) *Reviving Old Buildings and Communities*. Newton Abbot, England: David & Charles.

Tanqueray, R. (2002) *Smallspaces: Making the Most of the Space You Have*. USA: Ryland, Peters & Small Ltd.

Taylor, G. D. (1998) *Materials of Construction* (3rd edition). Essex: Longman Scientific and Technical.

Taylor, J. (ed.) (2004) *Historic Churches: The Conservation and Repair of Ecclesiastical Buildings, 2003*. London: Cathedral Publishing.

Teekaram, A. J. H. and Brown, R. G. (1999) *Retrofitting of Heating and Cooling Systems, Technical Note TN 15/99*. London: The Building Services Research and Information Association.

Teo, H. P. (1991) *Risk Perception of Contractors in Competitive Bidding for Refurbishment Work*. London: RICS Books.

Thomson, A. (1990) *The essence of facilities management, Facilities*, 8(8), August 1990. Bradford: MCB University Press.

Thompson, K. (2005) *Get Mould Solutions*. Available online at http://www. getmoldsolutions.com/hepa_vacuums_exposed.html

Thorburn, S. and Littlejohn, G. S. (eds) (1993) *Underpinning and Retention* (2nd edition). London: Blackie Academic and Professional.

Thorpe, S. (1998a) *Wheelchair Housing Design Guide*. London: CRC Ltd.

Thorpe, S. (1998b) *House Adaptations: A Working File for Occupational Therapists*. London: Centre for Accessible Environments.

Topliss, C. (1982) *Demolition*. London: Construction Press.

Torrington, J. (1996) *Care Homes for Older People*. London: E & FN Spon.

Torrington, J. (2004) *Upgrading Buildings for Older People*. London: RIBA Enterprises Ltd.

TRADA (1978) *Home Improvements and Conversions*. London: The Construction Press.

TRADA (1990) *Roof Space Conversions*. High Wycombe: Timber Research and Development Association.

TRADA (1993) *Rooms in the Roof Construction for New Houses*, Wood Information Sheet 12. High Wycombe, UK: Timber Research and Development Association.

TRADA (1996) *Loft Conversion Guide for Contractors and Designers*. High Wycombe: Timber Research and Development Association.

Trebilcock, P. J., Lawson, R. M. and Smart, C. (1994) *Adaptability in Steel*. Cleveland: British Steel.

Tricker, R. (2004) *Building Regulations in Brief* (2nd edition). Oxford: Butterworth-Heinemann.

Trotman, P., Sanders Harrison (2004) *Understanding Dampness – Effects, Causes, Diagnosis and Remedies*, BR 466. Garston: Building Research Establishment.

Trulove, J. G. (2002) *The Smart Loft*. USA: Harper Design International.

Trulove, J. G. and Kim, I. (2002) *Studio Apartments: Big Ideas for Small Spaces*. USA: Hearst Books.

Turner, A. (1997) *Building Procurement* (2nd edition). London: Macmillan.

Tweeds (2001) *Laxton's Trade Price Book: Small Works, Repairs and Maintenance*. Oxford: Butterworth-Heinemann.

Tyler, N. (2000a) *Historic Preservation: An Introduction to its History, Principles and Practice*. New York: W.W. Norton & Co.

Tyler, N. (2000b) *Historic Preservation: An Introduction to its History, Principles and Practice*. New York: W.W. Norton & Co.

UNEP (2001) *Climate Change 2001: Impacts, Adaptation and Vulnerability, UN Report on Global Climate Change*. New York: United Nations Environment Programme.

Universities of Sussex and Westminster (1996) *The Real Cost of Poor Homes*. London: The Royal Institution of Chartered Surveyors.

URA and PMB (1993) *Objectives, Principles and Standards for Preservation and Conservation*. Singapore: Urban Redevelopment Authority and Preservation of Monuments Board.

Urban Land Institute (1996) *Adaptive Reuse*. Washington, DC: Urban Land Institute.

Urban Task Force (1999) *Towards an Urban Renaissance, Report* (Chairman, Lord Rogers). London: The Department of the Environment, Transport and the Regions, HMSO.

URBED (1979) *Financing the Use of Old Buildings – A practical Guide to Sources of Finance, Materials and Labour for Projects Involving the Provision of Workspace.* London: The Department of the Environment (Urban Economic Development Ltd), HMSO.

URBED Ltd (1987) *Re-Using Redundant Buildings: Case Studies of Good Practice in Urban Regeneration.* London: The Department of the Environment (Urban Economic Development Ltd), HMSO.

US Department of Agriculture (2003) *Selecting and Renovating an Old House: A Complete Guide.* USA: Dover Publications.

US Department of Energy (2005) *Solar Energy Technologies Programme.* Available online at http://www.eere.energy.gov/solar/

Velux (1995) *Velux Roof Windows: The Complete System.* Bushy, Herts: The Velux Company Ltd.

Voss, K. (2000) Solar energy in building renovation – results and experience of international demonstration buildings. *Energy and Buildings*, **32**(2000), 291–302. London: Elsevier Press.

Wallace, J. and Whitehead, J. (1989) *Graffiti Removal and Control (Crime Prevention in Schools), Design Note 48.* London: Department of Education and Science (Architects and Buildings Group), CIRIA.

Walter, F. (1956) *House Conversion and Improvement.* London: Architectural Press.

Wanner, H. U. (1981) Indoor air quality and minimum ventilation. *Building Design for Minimum Air Infiltration, Proceedings of 2nd Air Infiltration Centre Conference.* Sweden: Royal Institute of Technology.

Warland, E. G. (1929) *Modern Practical Masonry.* London: Pitman.

Warran, J. (1998) *Conservation of Brick.* Oxford: Butterworth-Heinemann.

Warszawski, A. (1999) *Industrialized and Automated Building Systems* (2nd edition). London: E & FN Spon.

Watt, D. (1999) *Building Pathology: Principles and Practice.* Oxford: Blackwell Science.

Watt, D. (2004) *Architectural Conservation: An Introduction.* Oxford: Blackwell Publishing.

Watt, D. and Swallow, P. (1996) *Surveying Historic Buildings.* Shaftsbury: Donhead Publishing.

Watts and Partners (2005) *Watts Pocket Handbook 2005.* London: Watts & Partners.

Weaver, R. (1999) *Conservation Techniques.* New York: Van Nostrand Reinhold.

Weeks, K. D. (1986) *New Exterior Additions to Historic Buildings (Preservation Concerns)*, Preservation Brief 14. National Park Service. Available online at http://www2.cr.nps/tps/briefs/brief14.htm

Weeks, K. D. and Grimmer, A. E. (1995) *The Secretary of the Interior's Standards for the Treatment of Historic Properties with Guidelines for Preserving, Rehabilitating, Restoring and Reconstructing Historic Buildings.* Washington, DC: US Department of the Interior.

Weeks, K. D. and Maddex, D. (eds) (1982) *Respectful Rehabilitation: Answers to Your Questions on Historic Buildings.* National Park Service/National Trust for Historic Preservation. Somerset, NJ: John Wiley & Sons Inc.

Welsh Assembly (1999) *Welsh Housing Statistics.* Cardiff: National Assembly for Wales.

Wessex (2001) *Alterations and Refurbishment Building Price Book: 2002.* Dorset: Wessex.

WHO (1987) *Air Quality Guidelines for Europe.* Copenhagen: World Health Organisation.

Wilhide, E. (2002) *New Loft Living.* New York: Universe Publishing.

Williams, A. R. (1993) *A Guide to Small Extensions.* London: E & FN Spon.

Williams, A. R. (1995) *A Practical Guide to Alterations and Extensions.* London: E & FN Spon.

Williams, A. W. and Ward, G. C. (1991) *The Renovation of No-fines Housing*, BRE Report 191. Garston: Building Research Establishment.

Williams, B. (1978) *The Under-use of Upper Floors in Historic Town Centres*, Research Paper 15, September 1978. York: Institute of Advanced Architectural Studies, University of York.

Williamson, L. (2000a) *Home Extensions: Planning, Managing and Completing Your Home Extension*. New York: New Publishers.
Williamson, L. (2000b) *Loft Conversions: Planning, Managing and Completing Your Loft Conversion*. New York: New Publishers.
Willis, A. J. and Willis, C. J. (1983) *Specification Writing for Architects and Surveyors* (8th edition). Oxford: Blackwell Science.
Wood, B. (2003) *Building Care*. Oxford: Blackwell Publishing.
Wood, J. (ed.) (1994) *Buildings Archaeology: Applications in Practice*. Oxford: Oxbow Books.
Wood, M. (1965) *The English Medieval House* (1994 edition). London: Studio Editions Ltd.
Woolman, R. (1994) *Resealing of Buildings: A Guide to Good Practice*. Oxford: Butterworth-Heinemann.
Wordsworth, P. (2000) *Lee's Building Maintenance Management* (4th edition). Oxford: Blackwell Science.
World Commission on Environment and Development (1987) *Our Common Future*. London: OUP.
Wright, A. (1991) *Craft Techniques for Traditional Buildings*. London: BT Batsford Ltd.
Wulfinghoff, D. R. (2003) *Energy Efficiency Manual*. Wheaton, Maryland: Energy Institute Press.
XCO2 (2002) *Insulation for Sustainability – A Guide*. London: Bing.
Yeomans, D. (1997) *Construction Since 1900: Materials*. London: BT Batsford Ltd.
Young, M. (1986) *Architectural and Building Design: An Introduction*. London: Heinemann.
Zunde, J. M. (1982) *Design Procedures*. London: Longman.
Zunde, J. M. (1989) *Design Technology*. Sheffield: Pavic Publications.
Zurich Municipal (2002) *Solid Foundation: Building Guarantees Technical Manual* (Section 6 – Additional Guidance for Conversions). Zurich: Zurich Municipal.

BRE publications

Preamble

The Building Research Establishment (BRE) produces a wide range of very informative technical literature on many aspects of construction that relate to building adaptation, some of which are published on its behalf by Construction Research Communications Ltd (CRC Ltd) or IHS Rapidoc. The following *Digests, Good Building Guides, Good Repair Guides, Information Papers* and *Reports* are all relevant in some way to adaptation work.

Digests

Digest 45 (Part 1) (1971) *Design and appearance.*
Digest 46 (Part 2) (1971) *Design and appearance.*
Digest 54 (1971) *Damp-proofing solid floors.*
Digest 108 (1991) *Standards U-Values.*
Digest 110 (1972) *Condensation.*
Digest 125 (1988) *Colourless treatment of masonry.*
Digest 144 (1972) *Asphalt and built-up felt roofings: durability.*
Digest 145 (1972) *Heat losses through ground floors.*
Digest 152 (1973) *Repair and renovation of flood-damaged buildings.*
Digest 157 (1973) *Calcium silicate (sandlime, flintlime) brickwork.*
Digest 161 (1973) *Reinforced plastics cladding panels.*

Digest 163 (1974) *Drying out buildings.*
Digest 170 (1984) *Ventilation of internal bathrooms and WCs in dwellings.*
Digest 177 (1975) *Decay and conservation of stone masonry.*
Digest 180 (1986) *Condensation in roofs.*
Digest 190 (1976) *Heat losses from dwellings.*
Digest 192 (1976) *The acoustics of rooms for speech.*
Digest 196 (1976) *External rendered finishes.*
Digest 197 (1982) *Painting walls: Part 1 – choice of paint.*
Digest 198 (1977) *Painting walls: Part 2 – failures and remedies.*
Digest 200 (1977) *Repairing brickwork.*
Digest 208 (1985) *Increasing the fire resistance of existing floors.*
Digest 217 (1978) *Wall cladding defects and their diagnosis.*
Digest 223 (1983) *Wall cladding: designing to minimise defects due to inaccuracies and movements.*
Digest 232 (1983) *Energy conservation in artificial lighting.*
Digest 235 (1989) *Fixings for non-loadbearing precast concrete cladding panels.*
Digest 236 (1980) *Cavity insulation.*
Digest 238 (1992) *Reducing the risk of pest infestation: design recommendations and literature review.*
Digest 245 (1981) *Rising damp in walls – diagnosis and treatment.*
Digest 251 (1983) *Assessment of low-rise structures.*
Digest 254 (1983) *Reliability and performance of solar collector systems.*
Digest 268 (1988) *Common defects in low-rise buildings.*
Digest 270 (1983) *Condensation in insulated domestic roofs.*
Digest 280 (1983) *Cleaning external surfaces of buildings.*
Digest 282 (1983) *Structural appraisal of buildings with long spans.*
Digest 293 (1985) *Improving the sound insulation of separating walls and floors.*
Digest 295 (1985) *Stability under wind load of loose-laid external roof insulation boards.*
Digest 296 (1985) *Timbers – their natural durability and resistance to preservative treatments.*
Digest 297 (1985) *Surface condensation and mould growth in traditionally-built dwellings.*
Digest 299 (1985) *Dry rot: its recognition and control.*
Digest 300 (1985) *Toxic effects of fires.*
Digest 301 (1985) *Corrosion of metals by wood.*
Digest 304 (1985) *Preventing decay in external joinery.*
Digest 306 (1986) *Domestic draughtproofing: ventilation considerations.*
Digest 307 (1986) *Identifying damage by wood-boring insects.*
Digest 308 (1983) *Unvented domestic hot water systems.*
Digest 311 (1986) *Wind scour of gravel ballast on roofs.*
Digest 312 (1986) *Flat roof design: the technical options.*
Digest 313 (1986) *Mini piling for low-rise housing.*
Digest 320 (1987) *Fire doors.*
Digest 321 (1987) *Timber for joinery.*
Digest 324 (1987) *Flat roof design: insulation.*
Digest 327 (1992) *Insecticidal treatments against wood-boring insects.*
Digest 329 (2000) *Installing wall ties in existing construction.*
Digest 330 (1988) *Alkali aggregate reactions in concrete.*
Digest 331 (1983) *GRC.*
Digest 334 (1988) *Sound insulation of separating walls and floors, Part 1: walls.*
Digest 334 (1988) *Sound insulation of separating walls and floors, Part 2: floors.*
Digest 336 (1988) *Swimming pool roofs: minimizing the risks of condensation.*
Digest 337 (1983) *Sound insulation: basic principles.*

Digest 339 (1983) *Condensing boilers.*
Digest 340 (1989) *Choosing wood adhesives.*
Digest 345 (1989) *Wet rots: recognition and control.*
Digest 351 (1990) *Recovering old timber roofs.*
Digest 352 (1990) *Underpinning.*
Digest 354 (1990) *Painting of exterior wood.*
Digest 355 (1990) *Energy efficiency in dwellings.*
Digest 358 (1991) *CFCs in buildings.*
Digest 359 (1991) *Repairing brick and block masonry.*
Digest 361 (1983) *Why do buildings crack?*
Digest 362 (1991) *Building mortar.*
Digest 363 (1991) *Sulphate and acid resistance of concrete in the ground.*
Digest 364 (1991) *Design of timber floors to prevent decay.*
Digest 365 (1991) *Soakaway design.*
Digest 366 (1991) *Structural appraisal of existing buildings for change of use.*
Digest 369 (1992) *Interstitial condensation and fabric deterioration.*
Digest 370 (1992) *Control of lichens, moulds and similar growths.*
Digest 371 (1992) *Remedial wood preservatives: use them safely.*
Digest 372 (1992) *Flat roof design: waterproof membranes.*
Digest 373 (1992) *Wood chipboard.*
Digest 375 (1992) *Wood-based products: their contribution to the conservation of forest resources.*
Digest 378 (1993) *Wood preservatives: application methods.*
Digest 379 (1993) *Double glazing for heat and sound insulation.*
Digest 380 (1993) *Damp-proof courses.*
Digest 387 (1993) *Natural finishes for exterior wood.*
Digest 389 (1999) *Concrete – cracking and corrosion of reinforcement.*
Digest 392 (1994) *Assessment of existing high alumina cement construction in the UK.*
Digest 396 (1993) *Smoke control in buildings.*
Digest 398 (1993) *Continuous mechanical ventilation in dwellings: design, installation and operation.*
Digest 401 (1995) *Replacing wall ties.*
Digest 403 (1996) *Damage to structure from ground-borne vibration.*
Digest 404 (1996) *PVC-U windows.*
Digest 405 (1993) *Carbonation of concrete and its effects on durability.*
Digest 406 (1998) *Wind actions on buildings.*
Digest 410 (1994) *Cementitious renders for external walls.*
Digest 415 (1996) *Reducing the risk of pest infestations in buildings.*
Digest 418 (1996) *Bird, bee and plant damage to buildings.*
Digest 422 (1983) *Painting exterior wood.*
Digest 425 (1997) *List of excluded materials: a changing practice.*
Digest 428 (1983) *Protecting buildings against lightning.*
Digest 429 (1999) *Timbers: their natural durability and resistance to preservative treatment.*
Digest 430 (2000) *Plastics external glazing.*
Digest 432 (1999) *Corrosion of reinforcement in concrete: electrochemical monitoring.*
Digest 435 (1999) *Medium density fibreboard.*
Digest 437 (1999) *Industrialised and platform floors: mezzanine and raised storey.*
Digest 438 (1999) *Photovoltaics: integration into buildings.*
Digest 440 (1999) *Weathering and white PVC-U.*
Digest 443 Part 1 (1999) *Termites and UK buildings: biology, detection and diagnosis.*
Digest 443 Part 2 (1999) *Termites and UK buildings: control and management of subterranean termites.*

Digest 444 Part 1 (2000) *Corrosion of steel in concrete: investigation and assessment.*
Digest 444 Part 2 (2000) *Corrosion of steel in concrete: durability of reinforced concrete structures.*
Digest 444 Part 3 (2000) *Corrosion of steel in concrete: protection and remediation.*
Digest 447 (2000) *Waste minimisation on a construction sites.*
Digest 448 (2000) *Cleaning buildings: legislation and good practice.*
Digest 449 Part 1 (2000) *Cleaning external masonry: developing and implementing a strategy.*
Digest 449 Part 2 (2000) *Cleaning external masonry: methods and materials.*
Digest 452 (2000) *Whole life costing and life cycle assessment for sustainable building design.*
Digest 455 (2001) *Corrosion of steel in concrete: service life design and prediction.*
Digest 478 (2003) *Building performance feedback: getting started.*

Special digests

SD3 (2002) *HAC in the UK: assessment, durability management, maintenance, repair and refurbishment.*

Defect action sheets

This excellent series of concise but useful articles, which ran from May 1982 to March 1990, was superseded by the *Good Building/Repair Guide* series. Much of the material relevant to adaptation in the *Defect Action Sheets*, however, has been incorporated in the Carillion (2001) book referred to in the main part of this Bibliography as well as in many of the following:

Good building guides (GBG)

GBG 1 (1990) *Repairing or replacing lintels (Revision 1).*
GBG 2 (1990) *Surveying masonry chimneys for repair or rebuilding.*
GBG 3 (1990) *Damp proofing basements.*
GBG 4 (1990) *Repairing or rebuilding masonry chimneys.*
GBG 5 (1990) *Choosing between cavity, internal and external wall insulation.*
GBG 6 (1990) *Outline guide to assessment of traditional housing for rehabilitation.*
GBG 7 (1991) *Replacing failed plasterwork.*
GBG 8 (1991) *Bracing trussed rafter roofs.*
GBG 9 (1991) *Habitability guidelines for existing housing.*
GBG 10 (1992) *Temporary support: assessing loads above openings in external walls* (2nd edition).
GBG 11 (1993) *Supplementary guidance for assessment of timber framed houses, Part 1: Examination.*
GBG 12 (1993) *Supplementary guidance for assessment of timber framed houses, Part 2: Interpretation.*
GBG 13 (1992) *Surveying brick or block freestanding walls.*
GBG 14 (1992) *Building brick and block freestanding walls (Revision 1).*
GBG 15 (1992) *Providing temporary support during work on openings in external walls.*
GBG 17 (1993) *Freestanding brick walls – repairs to copings and cappings.*
GBG 18 (1994) *Choosing external rendering.*
GBG 20 (1999) *Removing internal loadbearing walls in older dwellings.*
GBG 21 (1996) *Joist hangers.*
GBG 22 (1995) *Maintaining exterior wood finishes.*

GBG 23 (1995) *Assessing external rendering for replacement or repair.*
GBG 24 (1995) *Repairing external rendering.*
GBG 25 (1995) *Buildings and radon.*
GBG 26 (1995) *Minimising noise from domestic fan systems.*
GBG 27 (1996) *Building brickwork and blockwork retaining walls.*
GBG 28 Part 1 (1997) *Domestic floors: construction, insulation and damp-proofing.*
GBG 28 Part 2 (1997) *Domestic floors: assessing them for replacement or repair – concrete floors, screeds and finishes.*
GBG 28 Part 3 (1997) *Domestic floors: assessing them for replacement or repair – timber floors.*
GBG 28 Part 4 (1997) *Domestic floors: repairing or replacing floors and flooring – magnesite, tiles, slabs and screeds.*
GBG 28 Part 5 (1997) *Domestic floors: repairing or replacing floors and flooring – Wood blocks and suspended timber.*
GBG 29 Part 1 (1999) *Connecting walls and floors – a practical guide.*
GBG 29 Part 2 (1999) *Connecting walls and floors – design and performance.*
GBG 30 (1999) *Carbon monoxide detectors.*
GBG 31 (1999) *Insulated external cladding systems.*
GBG 32 (1999) *Ventilating thatched roofs.*
GBG 33 (1999) *Building damp-free cavity walls.*
GBG 34 (1999) *Building in winter.*
GBG 35 (1999) *Building without cold spots.*
GBG 36 (1999) *Building a new felted flat roof.*
GBG 37 (2000) *Insulating roofs at rafter level.*
GBG 38 (2000) *Disposing of rainwater.*
GBG 39 Part 1 (2000) *Simple foundation design for low rise housing: site investigation.*
GBG 39 Part 2 (2000) *Simple foundation design for low rise housing: rule of thumb design.*
GBG 39 Part 3 (2000) *Simple foundation design for low rise housing: groundworks – getting it right.*
GBG 40 (2000) *Protecting pipes from freezing.*
GBG 41 (2000) *Installing wall ties.*
GBG 42 (2000) *Reed beds (two part set).*
GBG 43 (2000) *Insulated profiled metal roofs.*
GBG 44 Part 1 (2001) *Insulating masonry cavity walls: techniques and materials.*
GBG 44 Part 2 (2001) *Insulating masonry cavity walls: principal risks and guidance.*
GBG 45 (2001) *Insulating ground floors.*
GBG 46 (2001) *Domestic chimneys for solid fuel: flue design and installation.*
GBG 47 (2001) *Level external thresholds: reducing moisture penetration and thermal bridging.*
GBG 48 (2001) *Installing domestic automatic door controls.*
GBG 49 (2001) *Installing domestic automatic window controls.*
GBG 50 (2001) *Insulating solid masonry walls.*
GBG 51 (2001) *Ventilated and unventilated cold pitched roofs.*
GBG 52 Part 1 (2002) *Site cut pitched timber roofs: design.*
GBG 52 Part 2 (2002) *Site cut pitched timber roofs: construction.*
GBG 53 (2001) *Foundations for low-rise building extensions.*
GBG 54 Part 1 (2001) *Construction site communications: general.*
GBG 54 Part 2 (2001) *Construction site communications: masonry.*
GBG 55 (2002) *Quality mark scheme.*
GBG 56 (2002) *Off-site construction: an introduction.*
GBG 57 (2002) *Construction and demolition waste.*

Good repair guides (GRG)

GRG 1 (1996) *Cracks caused by foundation movement.*
GRG 2 (1996) *Damage to building caused by trees.*
GRG 3 (1996) *Repairing damage to brick and block walls.*
GRG 4 (1997) *Repairing masonry wall ties.*
GRG 5 (1997) *Diagnosing the causes of dampness.*
GRG 6 (1997) *Treating rising damp in houses.*
GRG 7 (1997) *Treating condensation in houses.*
GRG 8 (1997) *Treating rain penetration in houses.*
GRG 9 (1997) *Repairing and replacing rainwater goods.*
GRG 10 Part 1 (1997) *Repairing timber windows: investigating defects and dealing with water leakage.*
GRG 10 Part 2 (1997) *Repairing timber windows: dealing with draughty windows, condensation in sealed units, operating problems, and deterioration of frames.*
GRG11 Part 1 (1997) *Repairing flood damage: immediate action.*
GRG11 Part 2 (1997) *Repairing flood damage: ground floors and basements.*
GRG11 Part 3 (1997) *Repairing flood damage: foundations and walls.*
GRG11 Part 4 (1997) *Repairing flood damage: services, secondary elements, finishings and fittings.*
GRG12 (1997) Wood rot: assessing and treating decay.
GRG 13 Part 1 (1998) *Woodboring insect attack: identifying and assessing damage.*
GRG 13 Part 2 (1998) *Woodboring insect attack: treating damage.*
GRG 14 (1998) *Re-covering pitched roofs.*
GRG 15 (1998) *Repairing chimneys and parapets.*
GRG 16 Part 1 (1998) *Flat roofs: assessing bitumen felt and mastic asphalt roofs.*
GRG 16 Part 2 (1998) *Flat roofs: making repairs to bitumen felt and mastic asphalt roofs.*
GRG 17 (1998) *Repairing and replacing ground floors.*
GRG 18 (1998) *Replacing plasterwork.*
GRG 19 (1998) *Internal painting: hints and tips.*
GRG 20 Part 1 (1998) *Repairing frost damage: roofing.*
GRG 20 Part 2 (1998) *Repairing frost damage: walls.*
GRG 21 (1998) *Improving ventilation in housing.*
GRG 22 Part 1 (1999) *Improving sound insulation.*
GRG 22 Part 2 (1999) *Improving sound insulation.*
GRG 23 (1999) *Treating dampness in basements.*
GRG 24 (1999) *Repointing external brick and block walls.*
GRG 25 (1999) *Supporting temporary openings.*
GRG 26 Part 1 (2000) *Improving energy efficiency: thermal insulation.*
GRG 26 Part 2 (2000) *Improving energy efficiency: boilers and heating systems, draughtproofing.*
GRG 27 Part 1 (2000) *Cleaning external walls of buildings: cleaning methods.*
GRG 27 Part 2 (2000) *Cleaning external walls of building: removing dirt and stains.*
GRG 28 (2000) *Repairing brick and block freestanding walls.*
GRG 29 (2000) *Refixing ceramic tiles to internal walls.*
GRG 30 (2000) *Remedying condensation in domestic pitched tiled roofs.*
GRG 31 (2001) *Hot air repair of PVC-u window and door frames.*
GRG 32 (2001) *Dealing with noisy plumbing.*
GRG 33 Part 1 (2002) *Assessing moisture in building materials – sources of moisture.*
GRG 33 Part 2 (2002) *Assessing moisture in building materials – measuring moisture content.*
GRG 33 Part 3 (2002) *Assessing moisture in building materials – interpreting moisture data.*

Information papers (IP)

IP 4/81 (1981) *Performance of cavity wall ties.*
IP 11/81 (1981) *Maintenance of flat roofs.*
IP 13/81 (1981) *Conversions of older property to house single young people.*
IP 26/81 (1981) *Solar reflective paints.*
IP 10/84 (1984) *The structural condition of prefabricated reinforced concrete houses designed before 1960.*
IP 6/86 (1986) *Spacing of wall ties in cavity walls.*
IP 6/88 (1988) *Methods of improving sound insulation between converted flats.*
IP 9/88 (1988) *Methods of reducing impact sounds in buildings.*
IP 13/88 (1988) *Energy assessment for dwellings using BREDEM worksheets.*
IP 16/88 (1988) *Ties for cavity walls: new developments.*
IP 17/88 (1988) *Ties for masonry cladding.*
IP 2/89 (1989) *Thermal performance of lightweight inverted warm deck flat roofs.*
IP 20/89 (1989) *An energy-efficient refurbishment of electrically heated high-rise flats.*
IP 21/89 (1989) *Passive stack ventilation in dwellings.*
IP 7/90 (1990) *An introduction to infra-red thermography for building surveys.*
IP 8/90 (1990) *Retail warehouses: potential for increasing energy efficiency.*
IP 12/90 (1990) *Corrosion of steel wall ties: history, occurrence, background and treatment.*
IP 13/90 (1990) *Corrosion of steel wall ties: recognition and inspection.*
IP 15/90 (1990) *Defects in local authority housing: results of a building problems survey.*
IP 4/91 (1991) *Improving the energy-efficient performance of high-rise housing.*
IP 13/91 (1991) *Energy-efficient refurbishment of pre-1919 housing.*
IP 20/92 (1992) *Energy use in office buildings.*
IP 2/93 (1993) *Industrial building refurbishments: opportunities for energy efficiency.*
IP 10/93 (1993) *Avoiding latent mortar defects in masonry.*
IP 11/93 (1993) *Ecolabelling of building materials.*
IP 2/94 (1994) *Energy efficiency in schools.*
IP 12/94 (1994) *Assessing condensation risk and heat loss at thermal bridges round openings.*
IP 1/96 (1996) *Management of construction and demolition wastes.*
IP 12/97 (1997) *Plastics recycling in the construction industry.*
IP 9/99 (1999) *Cleaning exterior masonry: pretreatment assessment of a stone building.*
IP 7/00 (2000) *Reclamation and recycling of building materials: industry position report.*
IP 12/00 (2000) *Positive input ventilation in dwellings.*
IP 16/00 (2000) *Low-solvent primers: performance in construction steelwork.*
IP 17/00 (2000) *Advanced technologies for 21st century building services.*
IP 18/00 (2000) *Ammonia refrigerant in buildings: minimising the hazards.*
IP 4/01 (2001) *Reducing impact and structure-borne sound in buildings.*
IP 9/02 Part 1 (2002) *Refurbishment or redevelopment of office buildings. Sustainability comparisons.*
IP 9/02 Part 2 (2002) *Refurbishment or redevelopment of office buildings. Sustainability case histories.*
IP 3/03 (2003) *Dynamic insulation for energy saving and comfort.*
IP 13/03 Part 1 (2003) *Sustainable buildings: benefits for occupiers.*
IP 13/03 Part 2 (2003) *Sustainable buildings: benefits for designers.*
IP 13/03 Part 3 (2003) *Sustainable buildings: benefits for investors and developers.*

BRE leaflets

BR 156 (1989) *Cast rendered no-fines houses.*
BR 160 (1989) *No-fines houses.*

Reports (BR) (authored BRE reports are contained within the main texts listed above)

BR 113 (1987) *Steel-framed and steel-clad houses: inspection and assessment.*
BR 143 (1989) *Thermal insulation: avoiding risks.*
BR 148 (1989) *Atholl steel-framed, steel clad houses.*
BR 166 (1990) *Rehabilitation – a review of quality in traditional housing.*
BR 167 (1990) *Assessing traditional housing for rehabilitation.*
BR 168 (1990) *Surveyor's checklist for rehabilitation of traditional housing.*
BR 170 (1990) *Energy use in buildings and carbon dioxide emissions.*
BR 184 (1991) *Foundation movement and remedial underpinning in low-rise buildings.*
BR 185 (1991) *Overroofing: especially for large panel system dwellings.*
BR 238 (1993) *Upgrading floors – sound control for homes.*
BR 250 (1993) *Surveying dwellings with high indoor radon levels: a BRE guide to radon remedial measures in existing dwellings.*
BR 262 (2002) *Thermal insulation: avoiding risks.*
BR 268 (1988) *Common defects in low-rise traditional housing.*
BR 267 (1994) *Radon: major alterations and conversions.*
BR 299 (1996) *Indoor air quality in homes, Part 1.*
BR 349 (1999) *Impact of climate change on building.*
BR 364 (1999) *Solar shading of buildings.*
BR 366 (1999) *Sustainable retail premises: an environmental guide to design, refurbishment and management of retail premises.*
BR 389 (2000) *EcoHomes: the environmental rating for homes.*
BR 390 (2000) *The green guide to housing specification: an environmental profiling system for building materials and components.*
BR 402 (2000) *A risk assessment procedure for health and safety in buildings.*
BR 456 (2003) *Control of dust from construction and demolition activities.*

EST and carbon trust publications

Preamble

The Energy Saving Trust (EST) manages on behalf of the UK Government the Energy Efficiency Best Practice in Housing. It provides an extensive range of technical literature promoting energy efficiency in buildings as part of the former DETR's Energy Efficiency Best Practice Programme. These are excellent, practical publications, which provide impartial, authoritative information on energy efficiency on existing properties as well as new buildings.

The Carbon Trust, which is funded, by the UK Government, does the same for business and public sector organisations. It offers free, practical help and impartial advice on how to cut their energy costs. Until now, this help has been provided under the name of the **Action Energy** programme (formerly the Energy Efficiency Best Practice Programme) run by the Carbon Trust.

The following list comprises a selection of some of the main technical publications relevant to refurbishment and other forms of adaptation, many of which are available for downloading free of charge to relevant applicants on request from the EST or Carbon Trust via the following sources:

- Energy Saving Trust: Email: bestpractice@est.co.uk Web:www.est.org.uk/bestpractice
- The Carbon Trust: http://www.thecarbontrust.co.uk/energy/pages/publication_search.asp

General guides (CE)

CE 21 (2003) *Hard to treat homes and fuel poverty.*
CE 23 (2004) *Effective use of insulation in dwellings.*
CE 57 (2004) *Refurbishing cavity wall dwellings – A Summary of Best Practice.*
CE 58 (2004) *Refurbishing dwellings with solid walls – A Summary of Best Practice.*
CE 58 (2004) *Refurbishing timber framed dwellings – A Summary of Best Practice.*
CE 66 (2004) *Windows for new and existing housing – A Summary of Best Practice.*
CE 71 (2004) *Insulation materials chart – thermal properties and environmental ratings.*
CE 83 (2004) *Energy efficient refurbishment of existing housing.*
CE 84 (2004) *Scotland: assessing the U-values of existing housing.*
CE 97 (2005) *Advanced insulation in housing refurbishment.*
CE 104 (2005) *Energy efficient refurbishment of existing housing – case studies.*
CE 120 (2005) *Energy efficient loft conversions.*
CE 121 (2005) *Energy efficient garage extensions.*
CE 122 (2005) *Energy efficient domestic extensions.*
CE 126 (2005) *Frequently asked questions.*
CE 138 (2005) *Energy efficient historic homes – case studies.*

Good practice guides (GPG)

GPG 9 (1994) *Ground floor insulation in existing housing – a practical guide for specifiers.*
GPG 12 (1992) *Pitched roof insulation in existing housing – a practical guide for specifiers.*
GPG 26 (2003 edition). *Cavity wall insulation in existing housing.*
GPG 35 (1993) *Energy efficiency in offices. Energy efficient options for refurbished offices.*
GPG 40 (1993) *Saving energy in schools for school energy managers.*
GPG 41 (1993) *Opportunities for energy efficiency during refurbishment of schools.*
GPG 56 (1995) *Saving energy in school swimming pools: a guide to refurbishment and new pool design for headteachers, governors and local authorities.*
GPG 80 (1992) *Refurbishment of high rise dwellings – a strategic guide for local authority managers.*
GPG 81 (1992) *Refurbishment of existing dwellings – a strategic guide for private developers.*
GPG 82 (1992) *Energy efficiency in housing – guidance for local authorities.*
GPG 138 (1994) *Internal wall insulation in existing housing – a guide for specifiers and contractors.*
GPG 139 (1994) *Draught stripping of existing doors and windows. A guide for specifiers, installers and housing managers.*
GPG 150 (1995) *Energy efficient refurbishment of public houses – building fabric.*
GPG 151 (1995) *Energy efficient refurbishment of public houses – ventilation and air quality.*
GPG 152 (1995) *Energy efficient refurbishment of public houses – lighting.*
GPG 153 (1995) *Energy efficient refurbishment of public houses – cellar services.*
GPG 154 (1995) *Energy efficient refurbishment of public houses – heating and hot water.*
GPG 155 (2001 edition). *Energy efficient refurbishment of existing housing.*
GPG 156 (1995) *Energy efficient refurbishment of public houses – catering.*
GPG 157 (1995) *Energy efficient refurbishment of public houses – energy surveys.*
GPG 159 (1995) *Converting to compact fluorescent lighting – a refurbishment guide.*
GPG 173 (1997) *Energy efficient design of new school buildings and extensions – for schools and colleges.*

GPG 175 (1997) *Energy efficient refurbishment of low rise cavity wall housing (case studies with guidance for social housing landlords and contractors).*
GPG 179 (1997) *Energy efficient refurbishment of low rise solid wall housing (case studies with guidance for social housing landlords and contractors).*
GPG 182 (1997) *Heating system option appraisal – a managers guide.*
GPG 183 (1997) *Minimising thermal bridging when upgrading existing housing. A detailed guide for architects and building designers.*
GPG 184 (1997) *Energy efficient refurbishment of low rise non-traditional housing (case studies with guidance for social housing landlords and contractors).*
GPG 187 (1997) *Heating system option appraisal – an engineer's guide for existing buildings.*
GPG 190 (1997) *Energy efficient refurbishment of low rise housing.*
GPG 201 (1997) *Energy efficient refurbishment of retail buildings.*
GPG 205 (1997) *Energy efficient refurbishment of hotels and guesthouses – a guide for proprietors and managers.*
GPG 206 (1997) *Energy efficient refurbishment of hospitals.*
GPG 208 (1997) *Providing energy guide to householders – a guide for local authorities and housing associations.*
GPG 224 (1997) *Improving air tightness in existing homes.*
GPG 227 (1997) *Selecting energy-efficient windows.*
GPG 233 (1997) *Energy efficient refurbishment of schools.*
GPG 293 (1998) *External insulation systems for walls of dwellings.*
GPG 294 (1998) *Refurbishment guidance for solid wall houses – ground floors.*
GPG 295 (1998) *Refurbishment guidance for solid wall houses – windows and doors.*
GPG 296 (1998) *Refurbishment guidance for solid wall houses – roofs.*
GPG 297 (1998) *Refurbishment guidance for solid wall houses – walls.*

Good practice case studies (GPCS)

GPCS 1 (1993) *Energy efficiency in offices. Low cost major refurbishment – Policy Studies Institute, London.*
GPCS 3 (1993) *Retrofitted cavity wall insulation. Bournville Village Trust.*
GPCS 4 (1993) *Energy efficient rehabilitation of pre-1919 houses on Merseyside.*
GPCS 18 (1993) *Energy efficiency in 1980 high-rise through upgrading and energy management. Quadrant House, The Quadrant, Sutton, Surrey.*
GPCS 44 (1993) *Energy efficiency in public houses. Refurbishment of a large rural inn – The Marquis of Granby.*
GPCS 45 (1993) *Cavity wall insulation in existing dwellings – urea formaldehyde foam.*
GPCS 51 (1993) *Energy efficiency in public houses. Refurbishment of a small suburban public house – The Albion, Burton on Trent.*
GPCS 52 (1993) *Energy efficiency in public houses. Refurbishment of a large rural inn: River Wyre, Poulton le Fylde, Lancs.*
GPCS 53 (1993) *Energy efficiency in public houses. Refurbishment of a small suburban public house – The Tree Tops, Mapperley, Nottingham.*
GPCS 55 (1993) *Energy efficiency in public houses. Refurbishment of a small suburban public house – The Crown, Sedgley, Woverhampton.*
GPCS 56 (1993) *Energy efficiency in public houses. Refurbishment of a small suburban public house – The Engineer, Herpenden.*

GPCS 63 (1994) *Cavity wall insulation in existing dwellings – mineral wool insulation.*
GPCS 64 (1993) *Cavity wall insulation in existing dwellings – urea formaldehyde foam.*
GPCS 66 (1994) *Cavity wall insulation in existing dwellings – urea formaldehyde foam.*
GPCS 67 (1994) *Energy efficient refurbishment of high rise housing – York House, Bradford.*
GPCS 68 (1994) *Energy efficient refurbishment of high rise housing – Stannington Estate, Sheffield.*
GPCS 80 (1994) *Rejuvenation of community heating – pipework refurbishment in Manchester.*
GPCS 101 (1994) *Energy efficiency in schools – some simple energy conservation measures.*
GPCS 120 (1994) *Energy efficiency in refurbishment of industrial buildings – John Brown Engineering Ltd.*
GPCS 121 (1994) *Energy efficient refurbishment of high rise large panel system houses – Five case studies.*
GPCS 122 (1994) *Energy efficient refurbishment of high rise no-fines concrete housing – five case studies.*
GPCS 123 (1994) *Energy efficient refurbishment of high rise traditional construction housing.*
GPCS 150 (1994) *Energy efficient refurbishment of public house – building fabric.*
GPCS 155 (1994) *Benefits to the landlord of an energy efficient refurbishment of pre-1919 houses on Merseyside.*
GPCS 175 (1994) *Energy efficiency in refurbishment of industrial buildings – parts warehouse.*
GPCS 176 (1994) *Energy efficiency in existing housing – energy efficient refurbishment of low rise housing – Westburn Road, Glasgow.*
GPCS 180 (1994) *Energy efficient refurbishment of low rise housing – St. Andrews Gardens, Lincoln.*
GPCS 184 (1994) *Energy efficiency in schools – a head teacher speaks out.*
GPCS 188 (1994) *Energy efficient refurbishment of industrial buildings: GEC Alstholm Large Machines Ltd, Rugby.*
GPCS 191 (1995) *Energy efficiency in refurbishment of industrial buildings: Richards of Aberdeen.*
GPCS 192 (1994) *Energy efficiency in refurbishment of industrial buildings.*
GPCS 194 (1995) *Energy efficiency in refurbishment of industrial buildings: Alcan Plate Ltd. Birmingham.*
GPCS 204 (1995) *Energy efficient refurbishment of high rise housing.*
GPCS 205 (1995) *Energy efficient refurbishment of hotels, guesthouses – a guide for proprietors and managers.*
GPCS 244 (1995) *Energy efficient refurbishment of a medium sized hotel – Connaught Hotel, Bournemouth, Dorset.*
GPCS 258 (1995) *Energy efficient lighting in industrial buildings.*
GPCS 272 (1995) *Energy efficient refurbishment of industrial buildings: Dunlop Ltd, GRG Division, Cambridge Street, Manchester.*
GPCS 278 (1995) *British Telecomm maintains office refurbishment improvements as part of its energy management programme – Telephone House, Edinburgh.*
GPCS 283 (1995) *Energy efficient refurbishment of high rise housing.*
GPCS 294 (1995) *Energy efficient refurbishment of industrial buildings.*
GPCS 312 (1996) *Community heating in Nottingham – an overview of a rejuvenated system.*
GPCS 313 (1996) *Community heating in Nottingham – domestic refurbishment.*
GPCS 314 (1996) *Community heating in Nottingham – pipework refurbishment.*
GPCS 315 (1996) *Energy efficient refurbishment of solid walled houses.*
GPCS 316 (1996) *Energy efficient refurbishment of solid walled flats.*
GPCS 317 (1996) *Energy efficient refurbishment of cavity walled housing.*
GPCS 318 (1996) *Energy efficient refurbishment of cavity walled flats.*
GPCS 412 (2003) *Energy efficient refurbishment of existing housing – case studies.*

New practice final reports (NPFR)

NPFR 22 (1998) *Energy efficient refurbishment of high rise housing – Knowsley Heights, Liverpool.*
NPFR 84 (1998) *Energy efficient refurbishment of high rise housing – Lodge Lane, Croydon.*

General information reports (GIR)

GIR 16 (1993) *High efficiency condensing boilers (domestic applications) for installers in the UK.*
GIR 32 (1995) *Review and development of energy efficient refurbishment standards for housing associations.*
GIR 46 (1998) *Energy efficiency in Scottish housing association refurbishment projects.*
GIR 48 (1998) *Passive refurbishment at the open university – achieving staff comfort through improved ventilation.*
GIR 85 (2001) *New ways of cooling – information for building designers.*

General information leaflets (GIL)

GIL 1 (1992) *Condensing boilers in non-domestic buildings.*
GIL 4 (1992) *Energy efficiency in schools.*
GIL 12 (1993) *Energy efficiency in offices.*
GIL 9 (1993) *Domestic ventilation.*
GIL 16 (1993) *Using solar energy in schools.*
GIL 22 (1995) *Passive solar house design – Barratt Study.*
GIL 23 (1995) *Cavity wall insulation: unlocking the potential in existing buildings.*
GIL 31 (1998) *Building research establishment domestic energy model (BREDEM).*
GIL 63 (1998) *Saving energy in primary health buildings – an introduction for practice managers.*
GIL 70 (2002) *The effect of building regulations (Part L1 2002) on existing dwellings: information for installers and builders for extensions and alterations in England and Wales.*
GIL 72 (2002) *Energy efficiency standards – for new and existing dwellings.*
GIL 74 (2003) *Domestic condensing boilers – the benefits and the myths.*

Energy consumption guide (ECG)

ECG 6 (1992) *Energy efficiency in dwellings.*
ECG 28 (1992) *Saving energy in schools – a guide on lighting and IT equipment for headteachers, governors and school staff.*

Fuel efficiency booklet (FEB)

FEB 16 (1995) *Economic thickness of insulation for industrial buildings.*

THERMIE maxibrochures (1992)

- Energy efficient lighting in buildings.
- *Energy efficient lighting in schools.*
- *Energy efficient lighting in industrial buildings.*
- *Energy efficient lighting in offices.*

Energy efficiency in buildings – booklets (1993)

- Catering establishments.
- Courts, depots and emergency services buildings.
- Entertainment.
- Factories and warehouses.
- Health care buildings.
- Hotels.
- How to bring down energy costs in Schools.
- Libraries, museums, art galleries and churches.

National park service (NPS)

The Heritage Preservation Services of the NPS have produced a wide range of informative technical articles on the maintenance and adaptation of historic buildings. Although the focus is on North American construction, the principles of conservation (or preservation as it's known in the States) and restoration are applicable to other developed countries.

The publications listed below can be accessed from the Internet at the following URL: http://www2.cr.nps.gov/tps/tpscat.htm

Preservation briefs (PB)

PB 03 (1978) *Conserving energy in historic buildings* (BM Simth).

PB 11 (1982) *Rehabilitating historic storefronts* (HW Jandl).

PB 14 (1988) *New exterior additions to historic buildings – preservation concerns* (KD Weeks).

PB 17 (1988) *Architectural character – identifying the visual aspects of historic buildings as an aid to preserving their character-defining elements* (LH Nelson).

PB 18 (1988) *Rehabilitating interiors in historic buildings – identifying and preserving character-defining elements* (HW Jandl).

PB 31 (1993) *Mothballing historic buildings* (SC Park).

PB 32 (1993) *Making historic buildings accessible* (TC Jester and SC Park).

PB 35 (1994) *Understanding old buildings: the process of architectural investigation* (TC McDonald Jnr).

PB 38 (1996) *Holding the line: controlling unwanted moisture in historic buildings* (SC Park).

PB 43 (2005) *The preparation and use of historic structure reports* (D Slaton).

Historic preservation services guides and standards

The Secretary of the Interior's Standards for Rehabilitation (36 CFR Part 67) (1977).

The Secretary of the Interior's Standards for Rehabilitation and Illustrated Guidelines for Rehabilitating Historic Buildings (1977).

The Secretary of the Interior's Standards for the Treatment of Historic Properties (36 CFR 68)(1995).

The Secretary of the Interior's Standards for the Treatment of Historic Properties with Guidelines for Preserving, Rehabilitating, Restoring and Reconstructing Historic Buildings (2001 edition).

Historic Buildings – New Additions (2003).

Interpreting the standards (ITS) bulletins

ITS Number 1 (1999) *Interior plan: changes to shotgun interior plan.*
ITS Number 2 (1999) *Garage door openings: new infill for historic garage openings.*
ITS Number 3 (1999) *New additions: new additions to mid-size historic buildings 1.*
ITS Number 4 (1999) *Exterior doors: inappropriate replacement doors.*
ITS Number 5 (1999) *Exposing interior brick: removing interior plaster to expose brick.*
ITS Number 6 (1999) *Significant spaces: preserving historic church interiors.*
ITS Number 7 (1999) *Interior finishes: painting previously unpainted woodwork.*
ITS Number 8 (1999) *Interior alterations: interior alterations to detached residences to accommodate new functions.*
ITS Number 9 (1999) *Porches: inappropriate porch alterations.*
ITS Number 10 (1999) *Stair tower additions: exterior stair/elevator tower additions.*
ITS Number 11 (1999) *School buildings: interior alterations to school buildings to accommodate new uses.*
ITS Number 12 (1999) *School buildings: rehabilitation and adaptive reuse of schools.*
ITS Number 13 (2000) *Storefronts: repair/replacement of missing or altered storefronts.*
ITS Number 14 (2000) *Adding new openings: new openings in secondary elevations or introducing new windows in blank walls.*
ITS Number 15 (2000) *Industrial interiors: treatment of interiors in industrial buildings.*
ITS Number 16 (2000) *Loading door openings: new infill for historic door openings.*
ITS Number 17 (2000) *Interior parking: adding parking to the interior of historic buildings.*
ITS Number 18 (2001) *New additions: new additions to mid-size historic buildings 2.*
ITS Number 19 (2001) *Interior finishes: decorated plaster finishes.*
ITS Number 20 (2001) *School buildings: converting historic school buildings for residential use.*
ITS Number 21 (2001) *Adding new openings: adding new openings on secondary elevations.*
ITS Number 22 (2001) *Adding new openings: adding new entrances to historic buildings.*
ITS Number 23 (2001) *Windows: selecting new windows to replace non-historic windows.*
ITS Number 24 (2001) *Corridors: installing new systems in historic corridors.*
ITS Number 25 (2001) *Interior finishes: altering the character of historically finished interiors.*
ITS Number 26 (2001) *Entrances and doors: entrance treatments.*
ITS Number 27 (2001) *Awnings: adding awnings to historic storefronts and entrances.*
ITS Number 28 (2003) *Corridors in historic highrise apartment buildings and hotels.*
ITS Number 29 (2004) *Adding vehicular entrances and garage doors to historic buildings.*
ITS Number 30 (2004) *New entrances on mill buildings.*

Other relevant publications

A Checklist for Rehabilitating Historic Properties (1995).
Preservation Tech Notes (various topics over a number of years dating from the early 1990s).
Telling Historic Preservation Time: Using Illusion with Care to Reveal the Past (KD Weeks) (1995).
Four Approaches to Treatment – What They Are: Historic Preservation Treatment (Towards a Common Language) (KD Weeks) (1995).
The REHAB YES/NO Learning Program (1995).

British standards (BS)

BS 12 (1989) *Portland cements.*
CP 102 (1973) *Protection of buildings against water from the ground moisture.*

BS 144 (1990) *Wood preservation by means of pressure creosote.*
BS 402 (1990) *Plain tiles and fittings.*
BS 680-2 (1971) *Roofing slates.*
BS 743 (1970) *Materials for damp proof courses.*
BS 747 (1994) *Roofing felts.*
BS 847 (1973) *Methods of determining thermal insulation properties.*
BS 890 (1995) *Building limes.*
BS 1105 (1994) *Wood wool cement slabs up to 125 mm thick.*
BS 1142 (1989) *Fibre building boards.*
BS 1178 (1982) *Milled sheet lead for building purposes.*
BS 1191-1 (1994) *Gypsum building plasters.*
BS 1191-2 (1991) *Premixed lightweight plasters.*
BS 1196 (1989) *Clayware field drain pipes and junctions.*
BSs 1199 and 1200 (1984/86) *Building sands from natural sources.*
BS 1230 (1994) *Gypsum plasterboard.*
BS 1243 (1978) *Metal ties for cavity wall construction.*
BS 1521 (1972) *Specification for waterproof building papers.*
BS 2870 (1980) *Rolled copper and copper alloys.*
BS 3260 (1991) *Semi-flexible floor tiles.*
BS 3505 (1986) *Unplasticised PVC pipe for cold potable water.*
BS 3811 (1993) *Glossary of terms used in terotechnology.*
BS 3843 (1992) *Guide to terotechnology (the economic management of assets): Part 1 – Introduction to terotechnology.*
BS 3843-2 (1992) *Guide to terotechnology (the economic management of assets). Part 2 – Introduction to the techniques and applications.*
BS 3843-3 (1992) *Guide to terotechnology (the economic management of assets). Part 3 – Guide to the available techniques.*
BS 3921 (1995) *Clay bricks and blocks.*
BS 4016 (1972) *Specification for building papers (breather type).*
BS 4072 (1987) *Wood preservative by means of copper/chrome/arsenic compositions.*
BS 4514 (1983) *Upvc soil and ventilating pipes, fittings and accessories.*
BS 4551 (1980) *Tests on mortars, screeds and plasters.*
BS 4576-1 (1989) *Upvc rainwater goods and accessories.*
BS 4660 (1989) *Upvc pipes and plastic fittings for below ground drainage and sewers.*
BS 4756 (1991) *Ready-mixed aluminium priming paints for woodwork.*
BS 5051-1 (1988) *Bullet resistant glazing (specification for glazing for internal use).*
BS 5082 (1993) *Water-borne priming paints for wood.*
BS 5224 (1993) *Masonry cement.*
BS 5250 (1995) *Control of condensation in buildings.*
BS 5262 (1991) *External rendered finishes.*
BS 5268-5 (1989) *Preservative treatments for constructional timber.*
BS 5306-2 (1990) *Fire extinguishing installations and equipment on premises: specification for sprinkler systems.*
BS 5385 (1978) *External ceramic wall tiling and mosaics.*
BS 5446-1 (1997) *Specification for components of automatic alarm systems for residential premises.*
BS 5492 (1990) *Internal plastering.*
BS 5493 (1977) *Code of practice for protective coating of iron and steel structures against corrosion.*
BS 5534-1 (1990) *Slating and tiling: design.*
BS 5544 (1978) *Specification for anti-bandit glass (glazing resistant to manual attack).*

BS 5588 (1991) *Code of practice for fire precautions in the design, construction and use of buildings.*
BS 5589 (1989) *Preservation of timber.*
BS 5618 (1985) *Thermal insulation of cavity walls (with masonry or concrete inner and outer leaves) by filling with urea-formaldehyde (UF) foam systems.*
BS 5628 Part 3 (1985) *Structural use of masonry.*
BS 5669-1 (1989) *Particleboard.*
BS 5707-1,2,3 (1979/1990) *Wood preservatives in organic solvents.*
BS 5803-4 (1990) *Spread of fire.*
BS 5810 (1986) *Code of practice for access for the disabled to buildings.*
BS 5839-1 (1988) *Fire detection and alarm systems.*
BS 6073 (1978) *Precast concrete masonry units – specification for precast concrete masonry units.*
BS 6150 (1991) *Painting of buildings.*
BS 6187 (2000) *Code of practice for demolition.*
BS 6202 (1981) *Safety glazing.*
BS 6206 (1981) *Specification for impact performance requirements for flat safety glass and safety plastics for use in buildings.*
BS 6229 (1982) *Flats roofs with continuously supported coverings.*
BS 6232 (1982) *Thermal insulation of cavity walls by filling with blown man-made mineral fibre: Part 1 – 1982 specification for the performance of installation systems.*
BS 6297 (1993) *Design and installation of small sewage treatment works and cesspools.*
BS 6367 (1983) *Drainage of roofs and paved areas.*
BS 6398 (1983) *Bitumen damp-proof courses for masonry.*
BS 6399 (1990) *Dead and imposed loads.*
BS 6477 (1992) *Water repellents for masonry surfaces.*
BS 6515 (1984) *Polyethylene damp-proof courses for masonry.*
BS 6576 (1985) *Installation of chemical damp-proof courses.*
BS 6577 (1985) *Mastic asphalt for building (natural rock asphalt aggregate).*
BS 6676 (1982) *Thermal insulation of cavity walls by filling with blown man-made mineral fibre batts (slabs): Part 1 – 1982 code of practice for installation of batts (slabs) filling the cavity.*
BS 7036 (1996) *Code of safety for powered doors for pedestrian use.*
BS 7079 (1989) *Preparation of steel substrates before the application of paints and other related products.*
BS 7121 (1989) *Code of Practice for safe use of cranes.*
BS 7361-1 (1991) *Cathodic protection: Part 1 – code of practice for land and marine applications.*
BS 7543 (2003) *Guide to the durability of buildings and building elements, products and components.*
BS 7671 (2001) *16th Edition Wiring Regulations.*
BS 7913 (1998) *Guide to the principles of the conservation of historic buildings.*
BS 8000 (1990) *Workmanship of building sites (in 12 Parts).*
BS 8023 (1987) *Sheet and tile flooring.*
BS 8102 (1990) *Protection of structures against water from the ground.*
BS 8103-1 (1990) *Structural design of low-rise buildings: Code of practice for stability, site investigation, foundations and ground floor slabs for housing.*
BS 8203 (1987) *Code of practice for installation of sheet and tile flooring.*
BS 8204-1 (1987) *Concrete bases and screeds to receive* in-situ *floorings.*
BS 8208 (1985) *Guide to assessment of suitability of external cavity walls for filling with thermal insulants: Part 1 – existing traditional cavity construction.*
BS 8210 (1986) *Guide to building maintenance management.*
BS 8211-1 (1998) *Energy efficiency in housing: code of practice for energy efficient refurbishment of housing.*
BS 8213-1 (1990) *Windows, doors and rooflights: Part 1 – code of practice for safety in use and during cleaning of windows and doors (including guidance on cleaning materials and methods).*

BS 8213-4 (1990) *Windows, doors and rooflights: Part 4 – code of practice for the installation of replacement windows and doorsets in dwellings.*
BS 8215 (1991) *Design and installation of damp-proof courses.*
BS 8217 (1994) *Built up felt roofing.*
BS 8300 (2001) *Design of buildings and their approaches to meet the needs of disabled people.*
BS 8301 (1985) *Building drainage.*
BS EN ISO 13788 (2002) *Hygrothermal performance of building components and building elements – Internal surface temperature to avoid critical surface humidity and interstitial. Condensation – calculation methods.*

Other useful sources of information

Some Relevant Trade and Products Magazines

Several subscription-free magazines that deal with refurbishment and related works are currently available in the UK. They contain up-to-date information on materials, products, components as well as equipment and repair methods that are being used for adaptation works as well as in new build schemes.

- *ABC & D (Architect, Builder, Contractor and Developer)** – published by Ascent Publishing Ltd, 2 Sugar Brook Court, Ashton Road, Bromsgrove, Worcestershire B60 3EX.
- *Public Sector Building** – published by Ascent Publishing Ltd.
- *Refurb and Regeneration* – published by Waverley Communications, Waverley House, 11 Gate Lane, Sutton Coldfield, West Midlands B73 5TR.

Popular subscription-based magazines related to most types of residential adaptation work are:

- *Homebuilding & Renovating** – published by Ascent Publishing Ltd.
- *Move or Improve** – published by Ascent Publishing Ltd.

*see their websites below.

Relevant websites

The number of websites related to construction and refurbishment is extensive. The following list, therefore, is a sample of some of the most prominent sites available in this area.

www.abc-d.co.uk (product magazine website).
www.barbourexpert.com.uk (for access to a wide range of building products used in adaptation work and new build).
www.bbacerts.co.uk (website of the British Board of Agrément)
www.bcis.co.uk (for information on building costs).
www.bitc.org.uk/rth (website of the Business in the Community's Regeneration through heritage database).
www.bre.co.uk (website of the Building Research Establishment).
www.buildingconservation.com (website of the Building Conservation Directory).
www.buildingfutures.org.uk (website of a joint venture between CABE and RIBA).
www.buildingproducts.co.uk (website of the Building Products magazine).
www.cbppp.org.uk/cbpp (for information on the Construction Best Practice Programme).
www.cibworld.nl (website of the International Building Council – based in Holland).
www.cic.org.uk (construction industry council website).
www.ciria.org.uk (construction industry research and information association).
www.cisti.nrc.ca/irc (website of the Canadian Institute for Research in Construction, which contains an extensive list of useful articles on a whole range of building problems).

www.clasp.gov.uk (for full details of the various types of CLASP system buildings – including case studies of refurbished properties built using this method).
www.concreterepair.org.uk (website of the Concrete Repair Association).
www.constructsustainability.co.uk (for information about sustainability measures).
www.direct.gov.uk (for information on a wide range of sources relating to general and local government).
www.dti.gov.uk (for information relating to construction).
www.dryrotdogs.co.uk (Hutton + Rostron website for technical information on dry rot and other moisture-related problems).
www.energy-efficiency.gov.uk (for information and advice on energy efficiency best practice programme).
www.est.org.uk/bestpractice (for information on energy efficiency measures).
www.fugenex.co.uk (for information about dry rot sensors).
www.freewateruk.co.uk/ (for information on rainwater and greywater recycling technology).
www.getmoldsolutions.com (website of US mould eradication contractor).
www.homebuilding.co.uk (website of self-build and renovation magazine).
www.housedustmite.org.uk (for useful information on the risks and remedies associated with house dustmite infestation in dwellings).
www.info4study.com.uk (for information for students of the built environment).
www.leanconstruction.org (website of the Lean Construction Institute, USA).
www.m4i.org.uk (website of the movement for innovation in construction).
www.moveorimprove.co.uk (website of new quarterly magazine on extending, remodelling and updating dwellings).
www.newton-membranes.co.uk (website of John Newton Co Ltd, basement waterproofing specialists).
www.odpm.gov.uk (website of the Office of the Deputy Prime Minister, which contains information on the Building Regulations and other technical matters related to construction).
www.officescorer.info/ (website of the BRE's Sustainable Refurbishment/Redevelopment Decision Support Tool for office buildings).
www.planningportal.gov.uk (UK government's website on planning matters).
www.psb-info.co.uk (website of public sector building product magazine).
www.rethinkingconstruction.org.uk (website on modern and innovative construction practices).
www.scotland.gov.uk (for full access to the Scottish Building Regulations).
www.sustainableconstruction.co.uk (for information on sustainable adaptation projects).
www.thecarbontrust.co.uk (for information about how to reduce CO_2 emissions).
www.treasury.gov.uk (for information from their Procurement Group on value management, contracts, cost in use, etc.).
www2.cr.nps.gov/tps/briefs/presbhom.htm (website of the US National Park Service heritage preservation services containing an excellent range of articles and briefs on rehabilitation, restoration and conservation principles, problems and repair methods).

Index